Ocean-Atmosphere Interactions of Gases and Particles

Springer Earth System Sciences

For further volumes:
http://www.springer.com/series/10178

Peter S. Liss • Martin T. Johnson
Editors

Ocean-Atmosphere Interactions of Gases and Particles

Editors
Peter S. Liss
Martin T. Johnson
Centre for Ocean and Atmospheric Sciences
School of Environmental Sciences
University of East Anglia
Norwich
United Kingdom

This publication is supported by COST.

ESF provides the COST Office through an EC contract

COST is supported by the EU RTD Framework programme

ISBN 978-3-642-25642-4   ISBN 978-3-642-25643-1 (eBook)
DOI 10.1007/978-3-642-25643-1
Springer Heidelberg New York Dordrecht London

Library of Congress Control Number: 2013955001

© The Editor(s) (if applicable) and the Author(s) 2014. The book is published with open access at SpringerLink.com.

Open Access This book is distributed under the terms of the Creative Commons Attribution Noncommercial License which permits any noncommercial use, distribution, and reproduction in any medium, provided the original author(s) and source are credited.
All commercial rights are reserved by the Publisher, whether the whole or part of the material is concerned, specifically the rights of translation, reprinting, re-use of illustrations, recitation, broadcasting, reproduction on microfilms or in any other way, and storage in data banks. Duplication of this publication or parts thereof is permitted only under the provisions of the Copyright Law of the Publisher's location, in its current version, and permission for commercial use must always be obtained from Springer. Permissions for commercial use may be obtained through RightsLink at the Copyright Clearance Center. Violations are liable to prosecution under the respective Copyright Law.

Printed on acid-free paper

Springer is part of Springer Science+Business Media (www.springer.com)

Preface

This book is an outcome and an important part of the legacy of COST Action 735, whose overall aim was to develop tools for estimating air-sea fluxes of compounds important for climate, air quality and ocean productivity. The action was closely allied with the SOLAS (Surface Ocean – Lower Atmosphere Study) project of the International Geosphere-Biosphere Programme (IGBP), with both having very similar objectives. Because of this, they were mutually supportive in many ways.

The action ran for 5 years, starting in September 2006 and ending in September 2011. It involved more than 300 scientists mainly from Europe (77 % from the following 18 countries: Belgium, Cyprus, Denmark, Finland, France, Germany, Greece, Hungary, Ireland, Italy, Netherlands, Norway, Poland, Spain, Sweden, Switzerland, Turkey and the United Kingdom) but with a significant number from outside Europe, so that in total scientists from 30 countries participated. One third of the participants were female. The action had a reciprocal agreement with New Zealand. Additional information about the action can be found at http://www.cost-735.org/.

The action operated through two types of activity – working group meetings (21 being held in total) and short-term scientific missions, which allowed 19 younger scientists to work in other laboratories in the action. The work of the action was overseen and directed by a Management Committee that met eight times.

Many scientific publications have been produced from the work supported by the action; it has also enabled many young scientists working for their Ph.Ds. to broaden their vision and learn techniques not available in their home laboratories. In addition and central to the aims of the action, several important databases have been assembled and made available through public data centres. This has in general not involved making new measurements but the gathering together and collating in a coherent format of existing data, much of which was not readily available previously. Notable databases made with support from the action include IRONMAP (atmospheric aerosol iron measurements); HalOcAt (ocean and atmospheric organo-halocarbon data); MEMENTO (marine measurements of methane and nitrous oxide); SOCAT (global surface ocean carbon dioxide database), a comprehensive dimethyl sulphide database, which tripled the previous readily available number of surface ocean measurements; and finally an ongoing effort to assemble measurements of aerosol and rain composition (trace metals and nutrients) made from ships at sea. These various datasets will be an important legacy of the action, and we expect them to prove vital in the future development of the subject.

Given the large amount of research supported by the action, it was decided to produce this book to record the current state of knowledge in the area. Each of the five chapters has several Lead Authors and a larger number of Contributing Authors, using the IPCC authorship model. The Lead Authors together constituted the editorial board for the book. Lead authors are identified in the contacts list which follows and by the inclusion of their email addresses at the beginning of each chapter. Each chapter was reviewed by an external reviewer in addition to one member of the editorial board not associated with the chapter. We have tried to have as much consistency of nomenclature and units as possible but have not imposed this where non-standard usage is well established and accepted. Although we have tried to keep redundancy of material between chapters to a minimum, it has not been removed entirely. We consider this is acceptable and even desirable since each chapter can be downloaded independently and so needs to be complete within itself. Topics within each chapter are dealt with in considerable detail and at a research level; so we expect the book to be mainly of interest and constitute a fundamental text for research workers, including graduate students.

Many people should be thanked and congratulated for the success of the action and, consequently, for making this book possible. Firstly, thanks are due to the participants in the scientific meetings and members of the Management Committee, as well as the young researchers who took the opportunity to go on short-term scientific missions. The SOLAS office at UEA carried much of the administrative burden of the action, along with COST officers, rapporteurs and the administrative staff.

We are grateful to the reviewers of the chapters who made many very useful and perceptive suggestions. Rosie Cullington did a large amount of editorial work on the reference lists for several of the chapters. Kath Mortimer put in a huge and sustained effort at all stages of the production of the book; without her efforts it is unlikely that the project would have been completed. Her thoroughness, excellent planning, persuasiveness and stamina are truly remarkable. We are deeply indebted to all of these people.

UEA, Norwich, UK	Peter S. Liss
	COST 735 Chair and Book Editor
ICM-CSIC, Barcelona, Spain	Rafel Simó
	COST 735 Vice-Chair
UEA, Norwich, UK	Martin T. Johnson
	Book Editor

Acknowledgments

The authors would like to acknowledge the following organisations without whose funding this publication and the meetings which supported this work would not have been possible:
- COST Action ES0801: The ocean chemistry of bioactive trace elements and paleoclimate proxies; www.costaction.earth.ox.ac.uk/
- European Cooperation in Science and Technology (COST); www.cost.eu
- European Space Agency (ESA); www.esa.int
- International Geosphere-Biosphere Programme (IGBP); www.igbp.net
- Land-Ocean Interactions in the Coastal Zone (LOICZ); www.loicz.org
- Natural Environment Research Council (NERC); www.nerc.ac.uk
- Surface Ocean-Lower Atmosphere Study (SOLAS); www.solas-int.org
- University of East Anglia (UEA); www.uea.ac.uk

COST – European Cooperation in Science and Technology

COST – European Cooperation in Science and Technology – is an intergovernmental framework aimed at facilitating the collaboration and networking of scientists and researchers at European level. It was established in 1971 by 19 member countries and currently includes 35 member countries across Europe, and Israel as a cooperating state.

COST funds pan-European, bottom-up networks of scientists and researchers across all science and technology fields. These networks, called 'COST Actions', promote international coordination of nationally-funded research.

By fostering the networking of researchers at an international level, COST enables break-through scientific developments leading to new concepts and products, thereby contributing to strengthening Europe's research and innovation capacities.

COST's mission focuses in particular on:
- Building capacity by connecting high-quality scientific communities throughout Europe and worldwide;
- Providing networking opportunities for early career investigators;
- Increasing the impact of research on policy makers, regulatory bodies and national decision makers as well as the private sector.

Through its inclusiveness, COST supports the integration of research communities, leverages national research investments and addresses issues of global relevance.

Every year thousands of European scientists benefit from being involved in COST Actions, allowing the pooling of national research funding to achieve common goals.

As a precursor of advanced multidisciplinary research, COST anticipates and complements the activities of EU Framework Programmes, constituting a "bridge" towards the scientific communities of emerging countries. In particular, COST Actions are also open to participation by non-European scientists coming from neighbour countries (for example Albania, Algeria, Armenia, Azerbaijan, Belarus, Egypt, Georgia, Jordan, Lebanon, Libya, Moldova, Montenegro, Morocco, the Palestinian Authority, Russia, Syria, Tunisia and Ukraine) and from a number of international partner countries.

COST's budget for networking activities has traditionally been provided by successive EU RTD Framework Programmes. COST is currently executed by the European Science Foundation (ESF) through the COST Office on a mandate by the European Commission, and the framework is governed by a Committee of Senior Officials (CSO) representing all its 35 member countries.

More information about COST is available at www.cost.eu.

Contents

1 Short-Lived Trace Gases in the Surface Ocean and the Atmosphere ... 1
Peter S. Liss, Christa A. Marandino, Elizabeth E. Dahl, Detlev Helmig,
Eric J. Hintsa, Claire Hughes, Martin T. Johnson, Robert M. Moore,
John M.C. Plane, Birgit Quack, Hanwant B. Singh, Jacqueline Stefels,
Roland von Glasow, and Jonathan Williams

- 1.1 Introduction ... 1
- 1.2 Sulphur and Related Gases ... 2
    - 1.2.1 DMS(P) in the Surface Ocean ... 2
        - 1.2.1.1 Ecosystem Dynamics ... 2
        - 1.2.1.2 DMS Yield ... 3
        - 1.2.1.3 Predicted Impact of Climate Change ... 3
    - 1.2.2 Other Sulphur and Related Gases in the Surface Ocean ... 6
        - 1.2.2.1 Carbonyl Sulphide ... 6
        - 1.2.2.2 Carbon Disulphide ... 6
        - 1.2.2.3 Hydrogen Sulphide ... 6
        - 1.2.2.4 Methanethiol ... 7
        - 1.2.2.5 Dimethyl Selenide ... 7
    - 1.2.3 Atmospheric Sulphur and Related Gases ... 7
        - 1.2.3.1 Chemistry of Sulphur in the Marine Boundary Layer (MBL) ... 9
        - 1.2.3.2 CLAW Hypothesis ... 11
- 1.3 Halocarbon Gases ... 13
    - 1.3.1 Chlorinated Compounds ... 14
        - 1.3.1.1 Introduction ... 14
        - 1.3.1.2 Methyl Chloride ... 15
        - 1.3.1.3 Dichloromethane ... 15
        - 1.3.1.4 Tri- and Tetrachloroethylene ... 16
        - 1.3.1.5 Chloroform ... 16
    - 1.3.2 Brominated Compounds ... 16
        - 1.3.2.1 Methyl Bromide ... 16
        - 1.3.2.2 $CHBr_3$, CH_2Br_2 and Other Polybrominated Methanes ... 17
    - 1.3.3 Iodinated Compounds ... 19
        - 1.3.3.1 Iodomethane ... 19
        - 1.3.3.2 Other Mono-Iodinated Iodocarbons ... 20
        - 1.3.3.3 Di- and Tri-Halogenated Compounds ... 20
    - 1.3.4 Halogens in the Marine Atmospheric Boundary Layer ... 21

	1.4	Non-Methane Hydrocarbons (NMHCs)	26
		1.4.1 Oxygenated Volatile Organic Compounds (OVOCs)	26
		1.4.1.1 Atmospheric Importance of OVOCs	26
		1.4.1.2 Atmospheric Budget	27
		1.4.1.3 Surface Ocean Processes	29
		1.4.2 Alkanes and Alkenes	31
		1.4.3 Alkyl Nitrates	31
		1.4.4 Hydrogen Cyanide (HCN) and Methyl Cyanide (CH_3CN)	33
	1.5	Ozone	33
	1.6	Nitric Oxide	35
	1.7	Ammonia and Amines	35
		1.7.1 Ammonia	35
		1.7.2 Amines	37
	1.8	Hydrogen	37
	1.9	Carbon Monoxide	38
	1.10	Concluding Remarks	39
	References		40

2 Transfer Across the Air-Sea Interface … 55
Christoph S. Garbe, Anna Rutgersson, Jacqueline Boutin, Gerrit de Leeuw, Bruno Delille, Christopher W. Fairall, Nicolas Gruber, Jeffrey Hare, David T. Ho, Martin T. Johnson, Philip D. Nightingale, Heidi Pettersson, Jacek Piskozub, Erik Sahlée, Wu-ting Tsai, Brian Ward, David K. Woolf, and Christopher J. Zappa

	2.1	Introduction	55
	2.2	Processes	56
		2.2.1 Microscale Wave Breaking	56
		2.2.2 Small Scale Turbulence	60
		2.2.3 Bubbles, Sea Spray	61
		2.2.4 Wind-Generated Waves	65
		2.2.5 Large-Scale Turbulence	66
		2.2.6 Rain	67
		2.2.7 Surface Films	68
		2.2.8 Biological and Chemical Enhancement	69
		2.2.9 Atmospheric Processes	70
	2.3	Process Models	72
		2.3.1 Interfacial Models	72
		2.3.1.1 Thin (Stagnant) Film Model	72
		2.3.1.2 Surface Renewal Model	73
		2.3.1.3 Eddy Renewal Model	73
		2.3.1.4 Surface Penetration	73
		2.3.1.5 Air-Side Transfer	74
		2.3.2 Direct Numerical Simulations (DNS) and Large Eddy Simulations (LES)	74
	2.4	Exchanged Quantities	76
		2.4.1 Physical Quantities	76
		2.4.2 Gases	76

		2.4.3	Particles	78
			2.4.3.1 Dry Deposition	78
			2.4.3.2 Wet Deposition	79
	2.5	Measurement Techniques		80
		2.5.1	Small-Scale Measurements Techniques	80
			2.5.1.1 Particle-Based Techniques	80
			2.5.1.2 Thermographic Techniques	81
		2.5.2	Micrometeorological Techniques	81
		2.5.3	Mass Balance	83
			2.5.3.1 Techniques	83
			2.5.3.2 Scales (Spatial and Temporal)	84
			2.5.3.3 Accuracy and Limitations	84
			2.5.3.4 Current and Recent Field Studies	84
		2.5.4	Profiles of pCO_2 Near the Surface	85
		2.5.5	Method Evaluation	85
	2.6	Parameterization of Gas Exchange		87
		2.6.1	Wind Speed Relationships	87
		2.6.2	Surface Roughness, Slope	90
		2.6.3	NOAA-COARE	90
		2.6.4	Energy Dissipation	92
		2.6.5	Evaluating and Selecting Transfer Velocity Parameterisations	93
	2.7	Sea Ice		95
	2.8	Applications of Air-Sea Gas Transfer		96
		2.8.1	Models	97
		2.8.2	Remote Sensing	99
		2.8.3	Inventories, Climatologies Using In Situ Data	100
	2.9	Summary		101
	References			102
3	**Air-Sea Interactions of Natural Long-Lived Greenhouse Gases (CO_2, N_2O, CH_4) in a Changing Climate**			113

Dorothee C.E. Bakker, Hermann W. Bange, Nicolas Gruber,
Truls Johannessen, Rob C. Upstill-Goddard, Alberto V. Borges,
Bruno Delille, Carolin R. Löscher, S. Wajih A. Naqvi,
Abdirahman M. Omar, and J. Magdalena Santana-Casiano

	3.1	Introduction		113
		3.1.1	Atmospheric Greenhouse Gases from Ice Cores	116
	3.2	Surface Ocean Distribution and Air-Sea Exchange of CO_2		117
		3.2.1	Global Tropospheric CO_2 Budget	117
		3.2.2	Processes Controlling CO_2 Dynamics in the Upper Water Column	117
		3.2.3	Surface Ocean fCO_2 and Air-Sea CO_2 Fluxes in the Open Ocean	121
			3.2.3.1 Surface Ocean fCO_2 Distribution	121
			3.2.3.2 Multi-Year Changes and Trends	123
			3.2.3.3 Comparison of Air-Sea CO_2 Flux Estimates	124
			3.2.3.4 Sea Ice	125
			3.2.3.5 Coastal to Open Ocean Carbon Exchanges	126

	3.2.4	Air-Sea CO_2 Fluxes in Coastal Areas	126
		3.2.4.1 Continental Shelves	126
		3.2.4.2 Near-Shore Systems	129
		3.2.4.3 Multi-Year Changes and Trends	129
3.3	Marine Distribution and Air-Sea Exchange of N_2O		130
	3.3.1	Global Tropospheric N_2O Budget	130
	3.3.2	Nitrous Oxide Formation Processes	130
		3.3.2.1 Denitrification	130
		3.3.2.2 Nitrification	131
		3.3.2.3 N_2O Formation by Dissimilatory Nitrate Reduction to Ammonium	132
	3.3.3	Global Oceanic Distribution of Nitrous Oxide	132
	3.3.4	Coastal Distribution of Nitrous Oxide	134
	3.3.5	Marine Emissions of Nitrous Oxide	135
3.4	Marine Distribution and Air-Sea Exchange of CH_4		137
	3.4.1	Global Tropospheric CH_4 Budget	137
	3.4.2	Formation and Removal Processes for Methane	137
	3.4.3	Global Oceanic Distribution of Methane	139
	3.4.4	Coastal Distribution of Methane	139
		3.4.4.1 Coastal Sediments	139
		3.4.4.2 Coastal Waters	142
		3.4.4.3 Methane Hydrates	143
	3.4.5	Marine Emissions of Methane	146
3.5	Impact of Global Change		147
	3.5.1	Future Changes in the Physics of the Oceanic Surface Layer	147
		3.5.1.1 Carbon Dioxide in the Open Ocean	147
		3.5.1.2 Carbon Dioxide in Coastal Seas	149
		3.5.1.3 Nitrous Oxide and Methane	150
	3.5.2	Ocean Acidification	150
		3.5.2.1 Carbon Dioxide	150
		3.5.2.2 Nitrous Oxide and Methane	152
	3.5.3	Deoxygenation and Suboxia in the Open Ocean	152
	3.5.4	Coastal Euthrophication and Hypoxia	153
	3.5.5	Changes in Methane Hydrates	153
3.6	Key Uncertainties in the Air-Sea Transfer of CO_2, N_2O and CH_4		154
	3.6.1	Outgassing of Riverine Carbon Inputs	154
	3.6.2	Heterogeneity in Coastal Systems	155
	3.6.3	Sea Ice	155
	3.6.4	Parameterising Air-Sea Gas Transfer	155
	3.6.5	Data Collection, Data Quality and Data Synthesis	155
3.7	Conclusions and Outlook		156
	3.7.1	Carbon Dioxide	156
	3.7.2	Nitrous Oxide and Methane	156
References			157

4 Ocean–Atmosphere Interactions of Particles 171
Gerrit de Leeuw, Cécile Guieu, Almuth Arneth, Nicolas Bellouin,
Laurent Bopp, Philip W. Boyd, Hugo A.C. Denier van der Gon,
Karine V. Desboeufs, François Dulac, M. Cristina Facchini,
Brett Gantt, Baerbel Langmann, Natalie M. Mahowald,
Emilio Marañón, Colin O'Dowd, Nazli Olgun, Elvira Pulido-Villena,
Matteo Rinaldi, Euripides G. Stephanou, and Thibaut Wagener

- 4.1 Introduction 171
- 4.2 Aerosol Production and Transport in the Marine Atmosphere 174
  - 4.2.1 Sources of Aerosol in the Marine Atmosphere 174
    - 4.2.1.1 Sea Spray Aerosol Production 174
    - 4.2.1.2 Organic Enrichment of Particulate Organic Matter in Sea Spray Aerosol 176
      - Laboratory Studies 179
      - Global Distribution of Organic Enrichment 180
    - 4.2.1.3 Secondary Aerosol Formation in the Marine Atmospheric Boundary Layer 182
      - Secondary Inorganic Aerosol Formation 182
      - Secondary Organic Marine Aerosol 183
      - New Particle Formation in the Marine Boundary Layer? 184
  - 4.2.2 Non-Marine Sources 186
    - 4.2.2.1 Desert Dust 186
    - 4.2.2.2 Volcanic Gases, Aerosols and Ash 189
    - 4.2.2.3 Global Emissions of Biogenic Volatile Organis Compounds (BVOC's) from Terrestrial Ecosystems 191
    - 4.2.2.4 Anthropogenic Emissions 193
      - Anthropogenic Land-Based Emissions 193
      - Uncertainty in Global Anthropogenic Emissions ... 193
      - Global Biomass Burning Emissions 194
      - International Shipping Emissions 194
      - Comparison and Evaluation of Different Emission Datasets 195
  - 4.2.3 Ageing and Mixing of Aerosols During Transport 196
    - 4.2.3.1 Chemical Ageing of Organic Aerosols 196
    - 4.2.3.2 Internal Mixing 197
      - Dust/Inorganic Species 197
      - Dust/Organic Species 198
      - Sea Salt 200
      - Future Directions 200
  - 4.2.4 Dust-Mediated Transport of Living Organisms and Pollutants 201
- 4.3 Direct Radiative Effects (DRE) 202
- 4.4 Effects on Cloud Formation and Indirect Radiative Effects 204
- 4.5 Deposition of Aerosol Particles to the Ocean Surface and Impacts 206

		4.5.1	Deposition	206
			4.5.1.1 Iron	206
			4.5.1.2 Phosphorus	207
			4.5.1.3 Nitrogen	209
			4.5.1.4 Deposition of Other Species	209
		4.5.2	Elements of Biogeochemical Interest and Their Chemical Forms	209
		4.5.3	Dissolution-Scavenging Processes	211
		4.5.4	Atmospheric Impacts in HNLC and LNLC Areas	213
			4.5.4.1 Experimental: Large Scale Fertilisation Experiments (Fe, P)	213
			4.5.4.2 Experimental: Microcosms	215
			Main Results Obtained from the Microcosm Approach	215
			4.5.4.3 Experimental: In Situ Mesocosms	218
			4.5.4.4 Modelling	221
		4.5.5	Particulate Matter and Carbon Export	222
	4.6	Summary and Outlook		224
	References			227

5 Perspectives and Integration in SOLAS Science 247

Véronique C. Garçon, Thomas G. Bell, Douglas Wallace, Steve R. Arnold,
Alex Baker, Dorothee C.E. Bakker, Hermann W. Bange, Nicholas R. Bates,
Laurent Bopp, Jacqueline Boutin, Philip W. Boyd, Astrid Bracher,
John P. Burrows, Lucy J. Carpenter, Gerrit de Leeuw, Katja Fennel,
Jordi Font, Tobias Friedrich, Christoph S. Garbe, Nicolas Gruber,
Lyatt Jaeglé, Arancha Lana, James D. Lee, Peter S. Liss, Lisa A. Miller,
Nazli Olgun, Are Olsen, Benjamin Pfeil, Birgit Quack, Katie A. Read,
Nicolas Reul, Christian Rödenbeck, Shital S. Rohekar, Alfonso Saiz-Lopez,
Eric S. Saltzman, Oliver Schneising, Ute Schuster, Roland Seferian,
Tobias Steinhoff, Pierre-Yves Le Traon, and Franziska Ziska

	5.1	Perspectives: In Situ Observations, Remote Sensing, Modelling and Synthesis	248
		5.1.1 In Situ Observations	248
		5.1.1.1 ARGO (T, S, O_2)	248
		5.1.1.2 Ocean Observatories	250
		5.1.1.3 Atmospheric Observatories	250
		5.1.1.4 Monitoring Reactive Trace Species in the Marine Atmosphere: Highlights from the Cape Verde Observatory	251
		5.1.1.5 Conclusions	255
		5.1.2 Earth Observation Products	255
		5.1.2.1 Altimetry, SST, Winds, Sea State	256
		5.1.2.2 Sea Surface Salinity	260
		5.1.2.3 Marine Carbon Observations from Satellite Data: Ocean Color/PIC/POC	261
		5.1.2.4 Sea Ice	264
		5.1.2.5 Aerosols	266

			5.1.2.6	Satellite Measurements of Trace Gases Over the Oceans...............................	267
			5.1.2.7	Conclusions.................................	270
		5.1.3	Modelling...		270
			5.1.3.1	Global Perspective, Prognostic IPCC and Hindcast...........................	271
			5.1.3.2	Regional Perspectives from High-Resolution Modeling..................................	272
			5.1.3.3	Inverse Modelling.........................	274
			5.1.3.4	Conclusions.................................	275
		5.1.4	SOLAS/COST Data Synthesis Efforts.................		276
			5.1.4.1	MEMENTO (MarinE MethanE and NiTrous Oxide) Database....................	276
			5.1.4.2	HalOcAt (Halocarbons in the Ocean and Atmosphere)........................	276
			5.1.4.3	DMS-GO (DMS in the Global Ocean).........	278
			5.1.4.4	The Surface Ocean CO_2 ATlas (SOCAT)........	279
			5.1.4.5	Aerosol and Rainwater Chemistry Database.....	280
			5.1.4.6	A Data Compilation of Iron Addition Experiments.............................	281
			5.1.4.7	Conclusions.................................	283
	5.2	Examples of SOLAS Integrative Studies.....................			284
		5.2.1	DMS Ocean Climatology and DMS Marine Modelling....		284
			5.2.1.1	Global Climatologies Based on Observations.....	284
			5.2.1.2	Diagnostic Approaches: Based on Empirical Correlations...............................	284
			5.2.1.3	Prognostic Modelling: From 1D to 3D.........	284
			5.2.1.4	Examples of Applications...................	286
				Climate Change...........................	286
				Iron Fertilisation.........................	286
		5.2.2	North Pacific Volcanic Ash and Ecosystem Response.....		287
		5.2.3	CO_2 in the North Atlantic............................		289
		5.2.4	Global Distribution of Sea Salt Aerosols................		292
	5.3	Perspectives for the Future...............................			293
	References..				294
Index..					307

Contributors

Almut Arneth is Professor and Head of Division of Ecosystem-Atmosphere Interactions at the Karlsruhe Institute of Technology, Institute of Meteorology and Climate Research/Atmospheric Environmental Research. Her research interests are in the interactions of climate change, land use change, vegetation dynamics and terrestrial biogeochemical cycles, and the feedbacks existing in that system.

Steve Arnold is a Senior Lecturer in the School of Earth and Environment at the University of Leeds. His research interests are in the chemistry of the lower atmosphere and interactions between air quality, climate and the biosphere.

Alex Baker is a Reader in Marine and Atmospheric Chemistry in the School of Environmental Sciences at the University of East Anglia in Norwich, UK. He has studied the marine and industrial chemistry of humic material and the biogeochemistry of trace metals in seawater and, since 1997, has worked primarily on aerosol biogeochemistry in the marine boundary layer.

Dorothee Bakker is a Research Officer in the School of Environmental Sciences at the University of East Anglia in Norwich, UK. Her research is on processes controlling the carbon sink in shelf seas and the oceans, notably the roles of marine biota, ocean circulation, iron supply, sea ice and ocean acidification. Dorothee has a strong interest in creating better access to marine biogeochemical data, e.g. via the Surface Ocean CO_2 Atlas (SOCAT) (www.socat.info).

Hermann Bange is a Professor of Marine Chemistry in the Marine Biogeochemistry Research Division of GEOMAR in Kiel, Germany. His research interests are in the marine biogeochemistry of trace gases (N_2O, CH_4 and DMS) and the marine nitrogen cycle. He was Co-chair of Working Group 3 of COST Action 735. Currently, he is the Coordinator of the SOPRAN (Surface Ocean Processes in the Anthropocene; www.sopran.pangaea.de) project, which is a German contribution to SOLAS.

Contributors

Nick Bates is Senior Scientist at the Bermuda Institute of Ocean Sciences (BIOS), a US scientific research institution based in Bermuda. He is a Chemical Oceanographer whose research is focused on the ocean carbon cycle, marine biogeochemistry and ocean acidification. At present, he is Principal Investigator for two ocean time series near Bermuda, namely the Bermuda Atlantic Time-series Study (BATS) and Hydrostation S. Other projects include the investigations into the impact of ocean acidification on coral reef ecosystems and studies of the ocean carbon cycle in the North Atlantic Ocean, Arctic Ocean and surrounding shelf seas, and the Southern Ocean.

Tom Bell is a Senior Scientist at the Plymouth Marine Laboratory, Plymouth, UK. His research interests are in ocean–atmosphere interactions and surface ocean trace gas biogeochemistry. His recent work has been to make direct measurements of DMS air-sea flux by eddy covariance using Atmospheric Pressure-Chemical Ionisation Mass Spectrometry (API-CIMS).

Nicolas Bellouin is a Senior Climate Research Scientist at the Met Office Hadley Centre in Exeter, UK. He investigates the role of atmospheric aerosols within the Earth system by using large-scale numerical models and satellite observations.

Laurent Bopp is a Senior Scientist in the modelling department of the Laboratoire des Sciences du Climat et de l'Environnement (LSCE) at l'Institut Pierre-Simon Laplace (IPSL) in Paris, France. He graduated from Pierre-et-Marie-Curie University and was a Visiting Scientist at both the Max Planck Institute in Jena, Germany, and the University of East Anglia in Norwich, UK. Most of his research focuses on marine biogeochemical cycles and on how marine ecosystems and the ocean carbon cycle respond to natural and anthropogenic climate change.

Alberto Borges is FRS-FNRS Research Associate at the University of Liège where he heads the Chemical Oceanography Unit. His research interests are in carbon cycling across aquatic systems including freshwater ecosystems (lakes and rivers), coastal ecosystems (estuaries, seagrass beds, mangroves and continental margins), and open ocean.

Jacqueline Boutin is a Research Scientist at Laboratoire d'Océanographie et du Climat-Expérimentation et Approches Numériques in Paris, France. Her research interests are in air–sea exchange particularly involving upper ocean variability and carbon dioxide and freshwater fluxes.

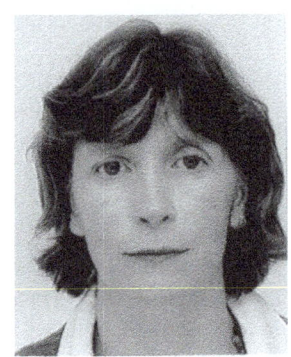

Philip Boyd is a Professor of Ocean Biogeochemistry based at the Institute for Marine and Antarctic Science, University of Tasmania. His research interests include trace metal biogeochemistry and exploring the links between dust, iron supply and altered ocean carbon biogeochemistry.

Astrid Bracher is leading the Helmholtz University Young Investigator's Group Phytooptics http://www.awi.de/en/go/phytooptics at the Climate Section of AWI http://www.awi.de/en/ and at the IUP at the University of Bremen http://www.iup.uni-bremen.de/eng/. She is an adjunct professor at the Department of Physics at the University of Bremen. Her studies are in the field of bio-optical in situ sampling techniques and ocean colour retrieval techniques. Her recent studies focus on retrieving new bio-optical information (phytoplankton functional types, vibrational Raman scattering, fluorescence) from hyperspectral satellite data.

John Burrows is Professor of Atmospheric and Oceanic Physics at the University of Bremen, Director of the Institute of Environmental Physics at the University of Bremen and a Fellow of the Natural Environmental Research Council's Centre for Ecology and Hydrology. His research is focused on Earth system and observation science, atmospheric physics and chemistry, kinetics and spectroscopy. He has pioneered the development of the remote sensing of tropospheric trace gases from space. He is also President of the International Commission on Atmospheric Chemistry and Global Pollution.

Lucy Carpenter received her BSc (hons) in Chemistry from the University of Bristol and studied for a PhD (awarded in 1996) in the subject of peroxy radicals in the lower atmosphere at the University of East Anglia. After postdoctoral research at the University of East Anglia and the University of Leeds, she moved to the Department of Chemistry, University of York, as a Lecturer in 2000 and was awarded a personal chair in 2009. Her research addresses the atmospheric impacts of marine-derived trace gases, particularly organohalogens. In 2006, she was awarded a Philip Leverhulme Prize in Earth, Ocean and Atmospheric Sciences.

Elizabeth Dahl is an Associate Professor of Chemistry at Loyola University, Maryland, USA. Her research focuses on the sources of alkyl nitrates in natural waters and the impact on tropospheric chemistry. She has a PhD in Earth System Science from the University of California, Irvine, and has mentored over a dozen undergraduate research students during her time at Loyola. When she is not working she enjoys spending time on kitchen chemistry with her future scientists.

Gerrit de Leeuw is a Research Professor at the Finnish Meteorological Institute and in the Department of Physics at the University of Helsinki, both in Helsinki, Finland, while being also affiliated to TNO, Utrecht, the Netherlands. His primary research interests are satellite remote sensing of aerosols, the production of sea spray aerosol and the application of satellite data to climate change and air quality as well as to ocean–atmosphere and land–atmosphere interactions.

Bruno Delille is FRS-FNRS Research Associate at the Chemical Oceanography Unit of the University of Liège. His research interests are in climate gas dynamics in polar ocean with an emphasis on sea ice.

Hugo Denier van der Gon is a Senior Researcher in the Department of Climate, Air and Sustainability of the Netherlands Organisation for Applied Scientific Research (TNO), Utrecht, the Netherlands. His current research interests are (i) improving emission inventories of non-CO_2 greenhouse gases and air pollutants for use in atmospheric chemistry and transport models, (ii) chemical speciation and source apportionment of particulate matter emissions and (iii) the use of satellite data to validate emission estimates.

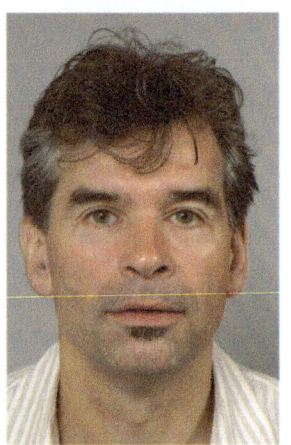

Karine Desboeufs is Assistant Professor in the LISA (Laboratoire Interuniversitaire des Systèmes Atmosphériques) at the University of Paris Diderot, France. Her research activities include the study of dust (and associated nutrients) cycling in the atmosphere and its biogeochemical impact after deposition to the ocean.

François Dulac is a Senior Research Scientist appointed by the French Commissariat à l'Energie Atomique et aux Energies Alternatives (CEA) in the Laboratoire des Sciences du Climat et de l'Environnement (LSCE) of CEA, CNRS, and the University of Versailles St-Quentin, at Gif-sur-Yvette, France. His research interests are in tropospheric chemistry with a focus on aerosols and their impacts. He is particularly experienced in desert dust and remote sensing and currently focuses on the Mediterranean region (http://charmex.lsce.ipsl.fr).

Cristina Facchini is Research Director at the Institute of Atmospheric Sciences and Climate (ISAC) of the National Research Council (CNR) in Bologna, Italy. Her research interest is in the effect of changes in atmospheric composition on global change and climate. Cristina's specific field of research in the last 15 years has been the organic composition of aerosol particles, their related physical properties and their interaction with atmospheric water. She is presently Lead Author of Intergovernmental Panel on Climate Change AR5-WG1.

Chris Fairall is a physicist at National Oceanic and Atmospheric Administration (NOAA)'s Earth System Research Laboratory in Boulder, Colorado, where he heads the Weather and Climate Physics Branch. He has spent decades developing/deploying air–sea interaction observing systems for ships and aircraft. His work is devoted to making direct measurements for verifying and improving the representation of air–sea interaction processes (surface evaporation, absorption of heat, generation of waves, uptake of carbon dioxide) in climate models used for climate change projections.

Katja Fennel holds a Canada Research Chair in Marine Prediction and is Associate Professor in the Department of Oceanography at Dalhousie University in Halifax, Canada. Her research interests are in biogeochemical modelling and data assimilation. She is a collaborator in major research projects, such as the Ocean Tracking Network and the U.S. Integrated Ocean Observing System (US IOOS) Modeling Testbed, and is a member of the IMBER/LOICZ Continental Margins Task Team and the CLIVAR Working Group on Ocean Model Development.

Jordi Font is a Research Professor in the Institute of Marine Sciences, Spanish Research Council, in Barcelona, Spain. He is a Physical Oceanographer now involved in ocean remote sensing, with the main focus of interest on the measurement of sea-surface salinity using microwave interferometric radiometry. At present, he is Co-lead Investigator for the Soil Moisture and Ocean Salinity (SMOS) satellite mission of the European Space Agency.

Tobias Friedrich is currently Postdoctoral Fellow at the International Pacific Research Center at the University of Hawaii. His research interests encompass natural and anthropogenic climate and carbon cycle variability, paleoceanography and Earth system modelling.

Brett Gantt is a Postdoctoral Researcher in the Department of Marine, Earth, and Atmospheric Sciences at North Carolina State University in Raleigh, North Carolina, USA. His research interests are in ocean–atmosphere interaction particularly involving sea spray aerosol and biogenic volatile organic compounds.

Christoph Garbe is the Head of an independent research group at the Interdisciplinary Center for Scientific Computing (IWR) at the University of Heidelberg. His research interests are in environmental transport processes and in digital image processing. Applications range from small-scale transport models at the air–water interface to satellite remote sensing. Christoph Garbe is member of the SOLAS scientific steering committee and principal investigator at the Intel Visual Computing Institute at Saarland University.

Véronique Garçon is a CNRS Senior Scientist at Laboratoire d'Etudes en Géophysique et Océanographie Spatiales (LEGOS) in Toulouse, France. Her research interests include marine biogeochemistry and ecosystem dynamics, large-scale circulation and tracers, global carbon and nitrogen cycles, physical–biological interactions, eastern boundary upwelling and Oxygen Minimum Zone (OMZ) systems, and biogeochemical climatic monitoring. She joined Centre national de la recherche scientifique (CNRS) in 1985 at the Institute of Physics of the Globe in Paris after a postdoctoral experience at MIT. She sits on many national and international Scientific Steering Committees (i.e. IFREMER, JGOFS, SOLAS, etc.)

Nicolas Gruber is Professor of Environmental Physics at ETH Zürich, Zürich. His research interests include the study of biogeochemical cycles on regional to global scales and on timescales from months to millennia, with a particular focus on the interaction of these cycles with the Earth's climate system. He has co-authored a textbook with Jorge Sarmiento, *Ocean Biogeochemical Dynamics*, and has received several awards, among which is the Rosenstiel Award from the University of Miami.

Cécile Guieu is a Senior Scientist at Centre National de la Recherche Scientifique and works at Laboratoire d'Océanographie in Villefranche sur Mer (LOV), France. She is a Marine Biogeochemist. Her research interests concern atmospheric inputs of nutrients and particles and how they impact marine biogeochemical cycles and carbon export. She has been a member of the SOLAS Scientific Steering Committee since 2009.

Jeff Hare is the Deputy Director for PIFSC Projects at the Joint Institute for Marine and Atmospheric Research (JIMAR) at the University of Hawaii at Manoa. His research interests are in the measurement and characterisation of air–sea mass and heat fluxes and in the development of transfer parameterisations. He previously served as Executive Officer for the Surface Ocean – Lower Atmosphere Study (SOLAS) International Project Office at the University of East Anglia in the UK (2005–2008).

Detlev Helmig is a Research Professor at the Institute of Arctic and Alpine Research at the University of Colorado, Boulder. His research entails studies of surface–atmosphere gas exchange and atmospheric chemistry and transport, with an emphasis on the ocean and cryosphere environment. He is Editor in Chief of the Atmospheric Science Domain of the non-profit, open access journal *Elementa*.

Eric Hintsa is a Research Scientist at the Cooperative Institute for Research in Environmental Sciences (CIRES) at the University of Colorado, Boulder, and the National Oceanic and Atmospheric Administration's Global Monitoring Division. His research interests include the sources, chemistry and transport of trace gases in the troposphere and stratosphere.

David Ho is a Professor in the Department of Oceanography at the University of Hawaii in Honolulu, Hawaii, USA. His research interests include air–sea gas exchange, tracer oceanography and coastal carbon cycling.

Claire Hughes is a Lecturer in Environmental Chemistry in the Environment Department at the University of York, UK. Her research interest is in understanding the mechanisms of biogenic trace gas production in the marine environment.

Lyatt Jaeglé is a Professor of the Atmospheric Sciences at the University of Washington. She is an atmospheric chemistry modeller, and her research interests are in long-range transport of pollution, biogeochemical cycling of mercury and surface emissions of chemical constituents.

Truls Johannessen is a Professor at the Geophysical Institute and the Bjerknes Centre for Climate Research, of Chemical Oceanography/Biogeochemistry/Earth Sciences at the University of Bergen and Uni Research. His special research interests are in marine climates, geochemistry, oceanography and earth systems. He is an author or co-author of about 70 scientific papers and a Steering Committee member in a number of international programmes under the auspices of JGOFS, SCOR and IOC. He was a member of the SOLAS Scientific Steering Committee from 2000 to 2008.

Martin Johnson is a Lecturer in Marine Science at the University of East Anglia in Norwich, UK. His main research interests lie in two related areas: the biogeochemical cycling of nitrogen compounds, particularly in aquatic environments, and their role in driving the biological carbon pump; and the physics, chemistry and biology of ocean–atmosphere trace gas exchange. He uses direct measurements and incubation experiments to quantify processes in the field and numerical and statistical modelling to constrain estimates of biogeochemical fluxes.

Arancha Lana is a postdoctoral researcher at the Mediterranean Institute for Advanced Studies (IMEDEA) in the Balearic Islands, Spain. She recently obtained her PhD, focused in marine aerosols, their precursors and their influence on clouds over the global ocean at Institut de Ciències del Mar (ICM-CSIC), in Barcelona. Her research interests are actually moving into Operational Oceanography, and the use of HF Radar to study ocean motion and surface currents.

Baerbel Langmann is a Senior Scientist at the Institute of Geophysics of the University of Hamburg, Germany. Her highly interdisciplinary research contributes to current topics in climate research. She investigates the importance of anthropogenic versus natural trace species emissions, atmospheric photochemical composition, aerosols, clouds and precipitation by regional-scale numerical modelling.

James Lee is an National Centre for Atmospheric Science (NCAS) Research Scientist in the Department of Chemistry, University of York. His research interest is in ozone photochemistry in remote and polluted environments, in particular the effect of long-term changes in NOx levels.

Pierre-Yves Le Traon is a Senior Oceanographer at IFREMER. His research interests include satellite and in situ observing systems, sea level and ocean circulation, and operational oceanography. He is the coordinator of the Euro-Argo Research Infrastructure and member of the International Argo Steering Team. He was Vice Chair of the Global Ocean Data Assimilation Experiment (GODAE) from 2002 to 2008. He received the Fridtjof Nansen Medal from the EGU in 2012.

Peter Liss is a Professorial Fellow in the School of Environmental Sciences at the University of East Anglia in Norwich, UK. His research interests are in ocean–atmosphere interaction particularly involving trace gases. He served as Chairman of the International Geosphere-Biosphere Programme (IGBP) for 5 years and was subsequently Chair of its Surface Ocean – Lower Atmosphere Study (SOLAS). He is a Fellow of the Royal Society.

Carolin Löscher works as a Postdoc in the Department of Microbiology at Christian-Albrecht University, Kiel, Germany. Her main research focus is set on the biological formation of nitrous oxide by microorganisms in tropical ocean areas. After studying biological oceanography at IFM-GEOMAR, Kiel, Germany, she worked on the sensitivity of the oceanic biological nitrogen cycle to changes in dissolved oxygen during her PhD from 2008 on and received her doctoral degree in December 2011.

Natalie Mahowald is an Associate Professor of Atmospheric Sciences in the Department of Earth and Atmospheric Sciences at Cornell University. Her research is focused on understanding global- and regional-scale atmospheric transport of biogeochemically important species such as desert dust. She is interested in how humans are perturbing the natural environment, especially through perturbations to aerosols. She uses a combination of three-dimensional global transport and climate models and analysis of satellite and in situ data.

Christa Marandino is a Junior Professor in the Chemical Oceanography Department at GEOMAR in Kiel, Germany. Her research focuses on performing direct open ocean flux measurements of dimethyl sulphide (DMS) and oxygenated volatile organic compounds (OVOCs, e.g. acetone and acetaldehyde) using the eddy covariance technique. She is also interested in surface ocean cycling of OVOCs.

Emilio Marañón is Associate Professor at the University of Vigo, Spain, since 1999. He studies the ecology and biogeochemical role of marine phytoplankton, with special interest in primary production, nitrogen fixation and calcification.

Lisa Miller is a Research Scientist at the Centre for Ocean Climate Chemistry at the Institute of Ocean Sciences in British Columbia. She studies the marine carbon cycle in relation to air–sea exchange, with particular focus on the roles of sea ice and the sea-surface microlayer.

Robert Moore is a Professor in the Department of Oceanography at Dalhousie University, Halifax, Canada. His research interest is in chemical oceanography with a focus on trace gases that are reactive in the atmosphere. A current focus of his research is the marine biogeochemistry of molecular hydrogen and the relationship between its oceanic distribution and nitrogen fixation.

Kath Mortimer is former Project Officer for the Surface Ocean Lower Atmosphere Study (SOLAS) where she focussed on projects such as the SOLAS Summer School, Open Science Conference, website and publications. As a part of this role she co-ordinated the activities of COST Action 735 including the publication of this book, Ocean-Atmosphere Interactions of Gases and Particles. She continues to work in research project management for the University of East Anglia, Norwich, UK.

Wajih Naqvi is the Director of the CSIR-National Institute of Oceanography, Dona Paula, Goa, India. His research interests are in the biogeochemistry of oxygen-deficient aquatic systems, especially transformations involving climatically important gases. He has actively participated in the planning and implementation of several IGBP projects (JGOFS, LOICZ and IMBER) at national and international levels. He is currently the Co-chair of Sustained Indian Ocean Biogeochemical and Ecological Research (SIBER).

Phil Nightingale is a scientist at the Plymouth Marine Laboratory (PML). His research interests include air-sea gas exchange, the cycling of trace gases in seawater and Lagrangian tracers. He obtained his BSc from the School of Chemical Sciences and PhD from the School of Environmental Sciences, both at the University of East Anglia. He has been working at PML since 1995.

Colin O'Dowd is Professor of Physics and Director of the Centre for Climate and Air Pollution Studies at the National University of Ireland, Galway, Ireland. His research interests include aerosol formation, transformation and climate effects with particular focus on marine aerosols. He has received the following honours: Smoluchowski Award, Fellow of the Institute of Physics and the Royal Meteorological Society, Doctorate of Science, Member of the Royal Irish Academy, and the Institute of Physics Appleton Medal.

Nazli Olgun finalised her PhD studies at the Helmholtz Centre for Ocean Research (GEOMAR) in Kiel, Germany. Her research interest is in the biogeochemical cycling of nutrients in the oceans involving the impacts of volcanic eruptions. She obtained her bachelor's degree in the Department of Geological Engineering, Middle East Technical University, Ankara, Turkey. She did her master's study at the Coastal Research Laboratory at the University of Kiel.

Are Olsen is Associate Professor at the Geophysical Institute/Bjerknes Center for Climate Research, University of Bergen. He spends his time doing research on marine biogeochemistry focusing in particular on ocean uptake of fossil fuel CO_2, air-sea CO_2 exchange and $\delta^{13}C$ dynamics. He also spends time on promoting ocean carbon data storage and access through SOCAT and GLODAPv2.

Abdirahman Omar is a Research Scientist and leader of the Biogeochemistry research group at Uni Climate, Uni Research Ltd. His research interests include the response of the oceanic carbon sink to increasing atmospheric carbon dioxide (CO_2) and the role of sea ice and brine formation in the oceanic uptake of atmospheric CO_2.

Heidi Pettersson is the Head of the Wave and Sea Level Group in the Marine Research Unit at the Finnish Meteorological Institute in Helsinki, Finland. Her research interests are in wind-generated waves and their role in the air–sea interactions, particularly in the exchange of carbon dioxide. She is an Adjunct Professor of Geophysics/Physical Oceanography (University of Helsinki).

Benjamin Pfeil is a Scientific Data Manager at the University of Bergen/Bjerknes Centre for Climate Research, Norway. His main interest is in data management in the field of chemical oceanography with an emphasis on data management of surface carbon dioxide measurements and producing data products with those data (e.g. SOCAT).

Jacek Piskozub is the Head of Physical Oceanography Department at the Institute of Oceanology, Polish Academy of Sciences, in Sopot, Poland. His research interests include ocean optics, air–sea interaction, marine aerosols as well as ocean-related processes influencing the climate.

John Plane is Professor of Atmospheric Chemistry at the University of Leeds, UK. His current research interests include halogens in the marine troposphere and the impact of cosmic dust on planetary atmospheres. He was a Research Fellow at Cambridge University from 1982 to 1985 and then Associate Professor at the University of Miami. In 1991, he moved to the University of East Anglia, becoming Professor of Environmental Science in 1999. He has been at Leeds since 2006.

Elvira Pulido-Villena is a Research Scientist at the Mediterranean Institute of Oceanography in Marseilles, France. Her research interests are in the effects of atmospheric deposition on bacterial activity and dissolved organic matter cycling in the surface ocean.

Birgit Quack is a Senior Scientist in the Chemical Oceanography Research Unit at the GEOMAR, Helmholtz Centre for Ocean Research, in Kiel, Germany. Her research interests are in the air–sea exchange of marine-chlorinated, -brominated and -iodinated halocarbons and in the investigation of their sources and the driving factors for their regional and temporal variable oceanic emission strengths.

Katie Read is a Postdoctoral Researcher in the National Centre of Atmospheric Science (NCAS) based at the University of York, UK. Her research interest is in the long-term monitoring of trace gases in the boundary layer.

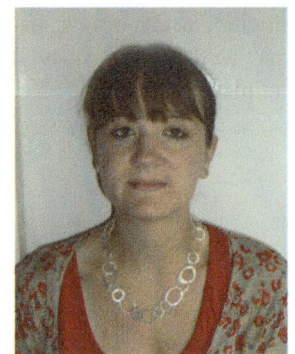

Nicolas Reul has been a Researcher in the Laboratory of Oceanography from Space at IFREMER, France, since 2001. His research interests are in ocean–atmosphere interactions and ocean remote sensing, particularly involving sea-surface salinity remote estimation from space.

Matteo Rinaldi is a Researcher at the Institute of Atmospheric Sciences and Climate (ISAC) of the Italian National Research Council (CNR) in Bologna, Italy. His research interests are in ocean–atmosphere interactions particularly involving the formation and atmospheric evolution of organic aerosols. He received his PhD in Environmental Sciences from the University of Urbino, Italy, in 2009, defending a thesis titled "Marine aerosol: distinguishing properties of primary and secondary components."

Christian Rödenbeck works as a Scientist at the Max Planck Institute for Biogeochemistry in Jena. His research interest is in the quantification of global biogeochemical fluxes from atmospheric, oceanic and remotely sensed data.

Shital Rohekar works as a SOLAS Project Integrator in the School of Environmental Sciences at the University of East Anglia, Norwich, UK. As a Project Integrator, she has contributed towards the development of an aerosol–rain chemistry database and the MEMENTO database. She plans to analyse the data products from these global databases and estimate the air–sea fluxes for the global oceans. Shital holds a PhD from the University of Cambridge and a master's degree from the University of Pune, India, specialising in Atmospheric Sciences.

Anna Rutgersson is a Professor of Meteorology in the Department of Earth Sciences at Uppsala University in Sweden. Her research interest is in air–sea interaction processes, particularly involving physical processes controlling the efficiency of gas exchange. She also works with the impact of moving surface waves on air–sea exchange and on turbulence in the atmosphere. Her research includes both measurements, mainly using micrometeorological methods, and modelling.

Erik Sahlée is a Senior Lecturer in Meteorology at Uppsala University. His research is related to boundary-layer processes at the air–water interface, including exchange processes of energy and matter and wave-atmosphere interactions.

Alfonso Saiz-Lopez received his PhD from the University of East Anglia in 2005. He was a Postdoctoral Scholar at the Jet Propulsion Laboratory, NASA, and at the Harvard-Smithsonian Center for Astrophysics. In 2009, he was appointed Senior Research Scientist and currently heads the Laboratory for Atmospheric and Climate Science at the Spanish Research Council (CSIC). Since 2009, he is Affiliate Scientist at NCAR. His research group focuses on atmospheric halogens and their impact on atmospheric chemistry and climate.

Eric Saltzman is Professor of Earth System Science at the University of California, Irvine. His research involves studying the ocean–atmosphere exchange and atmospheric cycling of climate-active trace gases and aerosols and reconstructing the history of the Earth's atmosphere from polar ice cores.

Magdalena Santana-Casiano is Professor of Chemical Oceanography in the Faculty of Marine Science at the Universidad de Las Palmas de Gran Canaria, Spain. Her research interests are in the chemical behaviour of iron in seawater and the changes in the carbon dioxide system with special focus on marine acidification.

Oliver Schneising is a Research Scientist at the Institute of Environmental Physics at the University of Bremen, Germany. His research interest is in satellite remote sensing particularly involving the atmospheric greenhouse gases carbon dioxide and methane.

Ute Schuster is a Senior Research Fellow in the College of Life and Environmental Sciences at the University of Exeter, Exeter, UK, having moved there in July 2013 from the School of Environmental Sciences at the University of East Anglia, Norwich, UK. Her research interests are in the marine carbon cycle, the biogeochemical drivers of its observed variability in both time and space, and its interaction with climate change.

Roland Séférian is a Researcher/Engineer at the Centre National de Recherche de Météo-France (CNRM-GAME) in the Laboratoire des Sciences du Climat et de l'Environnement (LSCE) in Gif-sur-Yvette, France. His research mainly focuses on climate marine carbon cycle feedbacks and contributions of the near-term or the long-term variability.

Hanwant Singh is a Senior Scientist in the Earth Science Division of NASA Ames Research Center. His primary research goal is to better understand the impact of human activities on the chemistry and climate of the Earth's atmosphere through direct observations and data analysis. He is also the Editor-in-Chief of the international journal *Atmospheric Environment*.

Jacqueline Stefels is a Senior Scientist and Lecturer at the University of Groningen, within the group Ecophysiology of Plants. She is a marine phytoplankton biologist. Her research interests are in the biological and environmental chemistry of dimethyl sulphide and related compounds in oceans and sea ice, the role of S-compounds in phytoplankton physiological adaptation to environmental stress and climate modelling. She is member of the SOLAS Scientific Steering Committee and Co-chair of SCOR-WG140 on Biogeochemical Exchange Processes at the Sea-Ice Interfaces.

Tobias Steinhoff is a Postdoc at the GEOMAR, Helmholtz Centre for Ocean Research, in Kiel, Germany. His research interests are in the ocean carbon and oxygen cycles and especially the air–sea gas exchange of CO_2. Furthermore, he is involved in the development of new observational techniques for the carbon and oxygen cycles.

Euripides Stephanou is a Professor of Environmental Chemistry at the University of Crete. His research interests are in the biogeochemical cycles of organic compounds and organic aerosol formation. He has been Coordinator and Principal Investigator in numerous European projects. He is a member of the ERC Panel Earth System Science and Editor of *Environmental Science and Pollution Research*. He is now the Rector of the University of Crete.

Wu-ting Tsai is a Professor in the Department of Engineering Science and Ocean Engineering at the National Taiwan University in Taipei, Taiwan. His research interests are in developing numerical models to simulate flows in close proximity to the air–sea interface which involve surface waves and turbulence and in applying these models to study the microscale air–sea interaction processes and their potential impact on a global scale.

Robert Upstill-Goddard is Professor of Marine Biogeochemistry in the School of Marine Science and Technology at Newcastle University. His research interests include the sea-surface microlayer, geophysical controls of air–sea gas exchange and marine emissions of methane and nitrous oxide. He has served on national and international committees, contributed to the European Nitrogen Assessment and the CARBOEUROPE report on GHG emissions from European waters and co-edited proceedings of an international symposium on GHG in mangrove coasts.

Roland von Glasow is a Professor of Atmospheric Sciences in the School of Environmental Sciences at the University of East Anglia in Norwich, UK. The work of his group focuses on tropospheric halogen chemistry (marine boundary layer, polar regions, volcanic plumes, salt lakes, free troposphere), the background chemistry of the marine boundary layer, and the microphysics of aerosols and clouds in the marine boundary layer.

Thibaut Wagener is an Assistant Professor at the Mediterranean Institute of Oceanography of the University of Aix-Marseille in Marseille, France. His research interest is in the deposition of atmospheric particles in the ocean with a special focus on trace metal fluxes.

Douglas Wallace is Canada Excellence Research Chair (CERC) in Ocean Science and Technology at Dalhousie University, Canada, and former Professor of Marine Chemistry at the Leibniz Institute of Marine Sciences, Germany. His research focuses on developing new observation instruments that are more sensitive in detecting biogeochemical oceanic change than current methods.

Brian Ward is a Lecturer in the School of Physics at National University of Ireland, Galway (NUIG) and is head of the AirSea Laboratory. His research interests are related to small-scale processes on both sides of the air–sea interface and their impact on ocean–atmosphere exchange. He is a member of the SOLAS Scientific Steering Committee and the WCRP Data Advisory Council.

Jonathan Williams is an Atmospheric Chemist. Presently, he is a research group leader at the Max Planck Institute with a focus on volatile organic compounds (VOC) from forests and oceans. He has participated in many international field campaigns on aircraft, ships and at ground stations.

David Woolf is a Reader in Marine Physics at Heriot-Watt University, based in Orkney, Scotland. His research interests include air–sea interaction, climate, Earth observation and renewable energy. He has been engaged in research on the influence of waves, wave breaking and bubbles on the air–sea exchange of gases and particles throughout his career. He is currently leading a European Space Agency project (www.oceanflux-ghg.org/) aimed at improving air–sea gas exchange climatologies.

Christopher Zappa is a Lamont Associate Research Professor at the Lamont-Doherty Earth Observatory of Columbia University in Palisades, NY, USA. He was an ONR Young Investigator from 2004 to 2007. He is dedicated to understanding the processes that affect ocean–atmosphere interaction and their boundary layers. His focus includes wave dynamics and wave breaking; the effect of near-surface turbulence on heat, gas and momentum transport; airborne infrared, multispectral and polarimetric remote sensing; upper ocean processes and polar ocean processes; and coastal and estuarine dynamics.

Franziska Ziska is a PhD student at the University of Kiel working at the GEOMAR, Helmholtz Centre for Ocean Research, in Kiel, Germany. Her research interest is in the global distribution of oceanic and atmospheric halocarbon concentrations including bromoform, dibromomethane and methyl iodide and their future trends.

Lead Authors

Dorothee C.E. Bakker Centre for Ocean and Atmospheric Sciences, School of Environmental Sciences, University of East Anglia, Norwich, NR4 7TJ, UK, d.bakker@uea.ac.uk

Hermann W. Bange GEOMAR Helmholtz Centre for Ocean Research Kiel, Düsternbrooker Weg 20, 24105 Kiel, Germany, hbange@geomar.de

Thomas G. Bell Plymouth Marine Laboratory, Prospect Place, The Hoe, Plymouth, PL1 3DH, UK. Previously at: Department of Earth System Science, University of California Irvine, CA 92697, USA; and Centre for Ocean and Atmospheric Sciences, School of Environmental Sciences, University of East Anglia, Norwich, NR4 7TJ, UK, tbe@pml.ac.uk

Gerrit de Leeuw Climate Change Unit, Finnish Meteorological Institute, Helsinki, Finland; Department of Physics, University of Helsinki, Helsinki, Finland; TNO, P.O. Box 800153508 TA Utrecht, The Netherlands, gerrit.leeuw@fmi.fi

Christoph S. Garbe IWR, University of Heidelberg, Speyerer Street 6, 69115 Heidelberg, Germany, christoph.garbe@iwr.uni-heidelberg.de

Véronique C. Garçon LEGOS, Centre National de la Recherche Scientifique, 31401 Toulouse, Cedex 9, France, veronique.garcon@legos.obs-mip.fr

Nicolas Gruber Institute of Biogeochemistry and Pollutant Dynamics, ETH Zürich, Universitätsstrasse 16, 8092 Zürich, Switzerland, nicolas.gruber@env.ethz.ch

Cécile Guieu Laboratoire d'Océanographie de Villefranche, CNRS, University Paris 6, Villefranche sur Mer, France, guieu@obs-vlfr.fr

Truls Johannessen Geophysical Institute & Bjerknes Centre for Climate Research, University of Bergen, Allégaten 70, NO-5007 Bergen, Norway
truls.johannessen@gfi.uib.no

Martin T. Johnson Centre for Ocean and Atmospheric Sciences, School of Environmental Sciences, University of East Anglia, Norwich, NR47TJ, UK; and Centre for environment, fisheries and aquaculture science, Pakefeld Road, Lowestoft, Suffolk, NR33 0HT, UK, martin.johnson@uea.ac.uk (Editorial board)

Peter S. Liss Centre for Ocean and Atmospheric Sciences, School of Environmental Sciences, University of East Anglia, Norwich, NR4 7TJ, UK; and Department of Oceanography, Texas A & M University, College Station, TX 77843, USA, p.liss@uea.ac.uk; liss@geos.tamu.edu

Christa A. Marandino GEOMAR Helmholtz Centre for Ocean Research Kiel, Düsternbrooker Weg 20, 24105 Kiel, Germany, cmarandino@geomar.de

Anna Rutgersson Department of Earth Sciences, Uppsala University, Villavägen 16, SE-752 36 Uppsala, Sweden, anna.rutgersson@met.uu.se

Rob C. Upstill-Goddard School of Marine Science and Technology, Newcastle University, Newcastle upon Tyne, NE1 7RU, UK, rob.goddard@ncl.ac.uk

Douglas Wallace Department of Oceanography, Dalhousie University, 1355 Oxford Street, PO BOX 15000, B3H 4R2 Halifax, NS, Canada douglas.wallace@dal.ca

Contributing Authors

Almuth Arneth Division of Ecosystem-Atmosphere Interactions, Karlsruhe Institute of Technology, Institute of Meteorology and Climate Research/Atmospheric Environmental Research, Kreuzeckbahn Str. 19, 82467 Garmisch-Partenkirchen, Germany, almut.arneth@kit.edu

Steve R. Arnold School of Earth & Environment, Institute for Climate and Atmospheric Science, University of Leeds, Leeds, LS2 9JT, UK s.arnold@leeds.ac.uk

Alex Baker Centre for Ocean and Atmospheric Sciences, School of Environmental Sciences, University of East Anglia, Norwich, NR4 7TJ, UK, alex.baker@uea.ac.uk

Dorothee C.E. Bakker Centre for Ocean and Atmospheric Sciences, School of Environmental Sciences, University of East Anglia, Norwich, NR4 7TJ, UK d.bakker@uea.ac.uk

Hermann W. Bange GEOMAR Helmholtz Centre for Ocean Research Kiel, Düsternbrooker Weg 20, 24105 Kiel, Germany, hbange@geomar.de

Nicholas R. Bates Bermuda Institute of Ocean Sciences, GE01 St George's, Bermuda, nick.bates@bios.edu

Nicolas Bellouin Met Office Hadley Centre, FitzRoy Road, Exeter, EX1 3PB, United Kingdom, nicolas.bellouin@metoffice.gov.uk

Laurent Bopp Centre National de la Recherche Scientifique (CNRS), Laboratoire des Sciences du Climat et l'Environnement (LSCE), Orme Des Merisiers, 91191 Gif sur Yvette, France, laurent.bopp@lsce.ipsl.fr

Alberto V. Borges Chemical Oceanography Unit, Institut de Physique (B5), University of Liège, B-4000 Liège, Belgium, alberto.borges@ulg.ac.be

Jacqueline Boutin LOCEAN/CNRS/UPMC/IRD, Université Pierre et Marie Curie, 75252 Paris Cedex 05, France, jb@locean-ipsl.upmc.fr

Philip W. Boyd Institute for Marine and Antarctic Science, University of Tasmania, 7005 Hobart, TAS, Australia, philip.boyd@utas.edu.au

Astrid Bracher Alfred-Wegener-Institute for Polar & Marine Research, Bremerhaven and Institute of Environmental Physics, University of Bremen, Bremen, Germany, astrid.bracher@awi.de

John P. Burrows Institute of Environmental Physics and Remote Sensing, University of Bremen, Postfach 330440, 28334 Bremen, Germany
burrows@iup.physik.uni-bremen.de

Lucy J. Carpenter Department of Chemistry, University of York, York, YO10 5DD, UK, lucy.carpenter@york.ac.uk

Elizabeth E. Dahl Loyola University Maryland, Baltimore, MD, USA, eedahl@loyola.edu

Gerrit de Leeuw Climate Change Unit, Finnish Meteorological Institute, Helsinki, Finland; Department of Physics, University of Helsinki, Helsinki, Finland; TNO, P.O. Box 800153508 TA Utrecht, The Netherlands, gerrit.leeuw@fmi.fi

Bruno Delille Chemical Oceanography Unit, University of Liège, Allée du 6 Août, 4000 Liège, Belgium, bruno.delille@ulg.ac.be

Karine V. Desboeufs LISA/IPSL, CNRS, Universités Paris Est Créteil and Paris Diderot, Créteil, France, desboeufs@lisa.u-pec.fr

François Dulac LSCE/IPSL, CNRS/CEA/UVSQ, Gif sur Yvette, France
francois.dulac@cea.fr

M. Cristina Facchini Institute of Atmospheric Sciences and Climate (ISAC), National Research Council (CNR), Via Gobetti 101, 40129 Bologna, Italy
mc.facchini@isac.cnr.it

Christopher W. Fairall Physical Sciences Division, NOAA Earth System Research Laboratory, Boulder, CO, USA, chris.fairall@noaa.gov

Katja Fennel Department of Oceanography, Dalhousie University, 1355 Oxford Street, PO BOX 15000, B3H 4R2 Halifax, NS, Canada, katja.fennel@dal.ca

Jordi Font Department of Physical Oceanography, Institut de Ciènces del Mar-CSIC, 08003 Barcelona, Spain, jfont@icm.csic.es

Tobias Friedrich School of Ocean and Earth Science and Technology, University of Hawai, 96822 Honolulu, HI, USA, tobiasf@hawaii.edu

Brett Gantt North Carolina State University, Raleigh, NC, USA
bdgantt@gmail.com

Christoph S. Garbe IWR, University of Heidelberg, Speyerer Street 6, 69115 Heidelberg, Germany, christoph.garbe@iwr.uni-heidelberg.de

Hugo A.C. Denier van der Gon TNO, P.O. Box 80015, 3508 TA Utrecht, The Netherlands, hugo.deniervandergon@tno.nl

Nicolas Gruber Institute of Biogeochemistry and Pollutant Dynamics, ETH Zürich, Universitätsstrasse 16, 8092 Zürich, Switzerland, nicolas.gruber@env.ethz.ch

Jeffrey Hare Joint Institute for Marine and Atmospheric Research (JIMAR), University of Hawaii at Manoa, 96822 Honolulu, HI, USA, jeff.hare@noaa.gov

Detlev Helmig Institute of Arctic and Alpine Research, Boulder, CO, USA
detlev.helmig@colorado.edu

Eric J. Hintsa University of Colorado and NOAA Global Monitoring Division, Boulder, USA, eric.j.hintsa@noaa.gov

David T. Ho Department of Oceanography, University of Hawaii, 1000 Pope Road, 96822 Honolulu, HI, USA, ho@hawaii.edu

Claire Hughes Environment Department, University of York, Heslington, York, YO10 5DD, UK, c.hughes@york.ac.uk

Lyatt Jaeglé Department of Atmospheric Sciences, University of Washington, BOX 351640, 98195 Seattle, WA, USA, jaegle@uw.edu

Martin T. Johnson Centre for Ocean and Atmospheric Sciences, School of Environmental Sciences, University of East Anglia, Norwich, NR4 7TJ, UK; Centre for environment, fisheries and aquaculture science, Pakefeld Road, Lowestoft, Suffolk, NR33 0HT, UK, martin.johnson@uea.ac.uk

Arancha Lana Department of Marine Technologies, Operational Oceanography and Sustainability, IMEDEA-CSIC, 07190 Balearic Islands, Spain alana@imedea.uib-csic.es

Baerbel Langmann Institute of Geophysics, University of Hamburg, KlimaCampus, Hamburg, Germany, baerbel.langmann@zmaw.de

James D. Lee Department of Chemistry, University of York, YO10 5DD York, UK, james.lee@york.ac.uk

Peter S. Liss Centre for Ocean and Atmospheric Sciences, School of Environmental Sciences, University of East Anglia, Norwich, NR4 7TJ, UK; Department of Oceanography, Texas A & M University, College Station, TX, 77843, USA p.liss@uea.ac.uk, liss@geos.tamu.edu

Carolin R. Löscher Institute of Microbiology, Christian-Albrechts University Kiel, Biologiezentrum, Am Botanischen Garten 1-9, D-24118 Kiel, Germany, cloescher@ifam.uni-kiel.de

Natalie M. Mahowald Department of Earth and Atmospheric Sciences, Cornell University, Snee 2140, 14853 Ithaca, NY, USA, mahowald@cornell.edu

Emilio Marañón Departamento de Ecología y Biología Animal, Facultad de Ciencias del Mar, University of Vigo, Vigo, Spain, em@uvigo.es

Lisa A. Miller Institute of Ocean Sciences, Fisheries and Oceans Canada, Sidney, BC V8L 4B2, Canada, lisa.miller@dfo-mpo.gc.ca

Robert M. Moore Department of Oceanography, Dalhousie University, 1355 Oxford Street, PO BOX 15000, B3H 4R2 Halifax, NS, Canada, robert.moore@dal.ca

S. Wajih A. Naqvi CSIR-National Institute of Oceanography, Dona Paula, Goa 403 004, India, naqvi@nio.org

Philip D. Nightingale Plymouth Marine Laboratory, Prospect Place, The Hoe, PL1 3DH Plymouth, UK, pdn@pml.ac.uk

Colin O'Dowd School of Physics, National University of Ireland, Galway, Ireland, colin.odowd@nuigalway.ie

Nazli Olgun   GEOMAR Helmholtz Centre for Ocean Research Kiel, Düsternbrooker Weg 20, 24105 Kiel, Germany, nolgun@ifm-geomar.de, nazliolgun.kiyak@itu.edu.tr

Are Olsen   Geophysical Institute and Bjerknes Centre for Climate Research, University of Bergen, Allégaten 70, NO-5007 Bergen, Norway
are.olsen@gfi.uib.no

Abdirahman M. Omar   Bjerknes Centre for Climate Research, Uni Research, Allégaten 55, 5007 Bergen, Norway, abdir.omar@uni.no

Heidi Pettersson   Marine Research Unit, Finnish Meteorological Institute, (Erik Palménin aukio 1, 00560 Helsinki), P.O.Box 503, FI-00101 Helsinki, Finland, heidi.pettersson@fmi.fi

Benjamin Pfeil   University of Bergen/Bjerknes Centre for Climate Research, Allégaten 55, 5007 Bergen, Norway, benjamin.pfeil@gfi.uib.no

Jacek Piskozub   Institute of Oceanology, Polish Academy of Sciences, Powstancow Warszawy 55, 81-712 Sopot, Poland, piskozub@iopan.gda.pl

John M.C. Plane   University of Leeds, Leeds, LS2 9JT, UK
j.m.c.plane@leeds.ac.uk

Elvira Pulido-Villena   Aix Marseille Université, CNRS/INSU, IRD, Mediterranean Institute of Oceanography (MIO), UM 110, 13288 Marseille (France)
elvira.pulido@univ-amu.fr

Birgit Quack   GEOMAR Helmholtz Centre for Ocean Research Kiel, Düsternbrooker Weg 20, 24105 Kiel, Germany, bquack@geomar.de

Katie A. Read   National Centre of Atmospheric Science (NCAS), University of York, York, YO10 5DD, UK, km519@york.ac.uk

Nicolas Reul   IFREMER, Laboratoire d'Océanographie Spatiale, Centre Méditerranée, Zone Portuaire de Brégaillon, CS20 330, 83507 La Seyne-sur-Mer Cedex, France, nicolas.reul@ifremer.fr

Matteo Rinaldi   Institute of Atmospheric Sciences and Climate (ISAC), National Research Council (CNR), Via Gobetti 101, 40129 Bologna, Italy
m.rinaldi@isac.cnr.it

Christian Rödenbeck   Max Planck Institute for Biogeochemistry, Jena, Germany, christian.roedenbeck@bgc-jena.mpg.de

Shital S. Rohekar   Centre for Ocean and Atmospheric Sciences, School of Environmental Sciences, University of East Anglia, Norwich, NR4 7TJ, UK, s.rohekar@uea.ac.uk

Erik Sahlée   Department of Earth Sciences, Uppsala University, Villavägen 16, SE-752 36 Uppsala, Sweden, erik.sahlee@met.uu.se

Alfonso Saiz-Lopez   Institute of Physical Chemistry Rocasolano - CSIC, 28006 Madrid, Spain, a.saiz@csic.es

Eric S. Saltzman   Department of Earth System Science, University of California, 92697 Irvine, CA, USA, eric.saltzman@uci.edu

J. Magdalena Santana-Casiano Facultad de Ciencias del Mar, Universidad de Las Palmas de Gran Canaria, 35017 Las Palmas de Gran Canaria, Spain, jmsantana@dqui.ulpgc.es

Oliver Schneising Institute of Environmental Physics and Remote Sensing, University of Bremen, Postfach 330440, 28334 Bremen, Germany
oliver.schneising@iup.physik.uni-bremen.de

Ute Schuster College of Life and Environmental Sciences, University of Exeter, Hatherly Laboratories, Prince of Wales Road, Exeter, EX4 4PS, UK
u.schuster@exeter.ac.uk

Roland Séférian CNRM-GAME/GMGEC/ASTER & IPSL/LSCE, Toulouse, France, roland.seferian@lsce.ipsl.fr, roland.seferian@meteo.fr

Hanwant B. Singh NASA Ames Research Centre, Mountain View, CA, USA, hanwant.b.singh@nasa.gov

Jacqueline Stefels Centre for Life Sciences, Ecophysiology of Plants, University of Groningen, Groningen, The Netherlands, j.stefels@rug.nl

Tobias Steinhoff GEOMAR Helmholtz Centre for Ocean Research Kiel, Düsternbrooker Weg 20, 24105 Kiel, Germany, tsteinhoff@geomar.de

Euripides G. Stephanou Department of Chemistry-University of Crete, Environmental Chemical Processes Laboratory (ECPL), GR-71003 Voutes-Heraklion, Greece, stephanou@chemistry.uoc.gr

Pierre-Yves Le Traon IFREMER and Mercator Ocean, Toulouse, France
pierre.yves.le.traon@ifremer.fr

Wu-ting Tsai Department of Engineering Science and Ocean Engineering, National Taiwan University, 10617 Taipei, Taiwan, wttsai@ntu.edu.tw

Roland von Glasow Centre for Ocean and Atmospheric Sciences, School of Environmental Sciences, University of East Anglia, Norwich, NR4 7TJ, UK,
r.von-glasow@uea.ac.uk

Thibaut Wagener Aix Marseille Université, CNRS/INSU, IRD, Mediterranean Institute of Oceanography (MIO), UM 110, 13288 Marseille, France
thibaut.wagener@univ-amu.fr

Brian Ward School of Physics, National University of Ireland, Galway, Ireland, bward@nuigalway.ie

Jonathan Williams Max Planck Institute for Chemistry, Department of Atmospheric Chemistry, Hahn-Meitner Weg 1, D-55128 Mainz, Germany
jonathan.williams@mpic.de

David K. Woolf International Centre for Island Technology, Heriot-Watt University, The Old Academy, Back Road, Stromness, Orkney, Scotland KW16 3AW
d.k.woolf@hw.ac.uk

Christopher J. Zappa Lamont-Doherty Earth Observatory, Columbia University, 61 Route 9W, 10964 Palisades, NY, USA, zappa@ldeo.columbia.edu

Franziska Ziska GEOMAR Helmholtz Centre for Ocean Research Kiel, Düsternbrooker Weg 20, 24105 Kiel, Germany, fziska@geomar.de

Short-Lived Trace Gases in the Surface Ocean and the Atmosphere

Peter S. Liss, Christa A. Marandino, Elizabeth E. Dahl, Detlev Helmig, Eric J. Hintsa, Claire Hughes, Martin T. Johnson, Robert M. Moore, John M.C. Plane, Birgit Quack, Hanwant B. Singh, Jacqueline Stefels, Roland von Glasow, and Jonathan Williams

Abstract

The two-way exchange of trace gases between the ocean and the atmosphere is important for both the chemistry and physics of the atmosphere and the biogeochemistry of the oceans, including the global cycling of elements. Here we review these exchanges and their importance for a range of gases whose lifetimes are generally short compared to the main greenhouse gases and which are, in most cases, more reactive than them. Gases considered include sulphur and related compounds, organohalogens, non-methane hydrocarbons, ozone, ammonia and related compounds, hydrogen and carbon monoxide. Finally, we stress the interactivity of the system, the importance of process understanding for modeling, the need for more extensive field measurements and their better seasonal coverage, the importance of inter-calibration exercises and finally the need to show the importance of air-sea exchanges for global cycling and how the field fits into the broader context of Earth System Science.

1.1 Introduction

Despite their seemingly low abundances, short-lived trace gases in the atmosphere critically influence global climate change, stratospheric ozone chemistry, and the oxidative capacity of the atmosphere. Because the ocean–atmosphere interface covers a large extent of the Earth's surface, the ocean is a major control on the atmospheric budget of many trace gases. Further, the chemical, biological, and physical processes that occur around this interface have a large impact on trace gas cycling between the oceanic and atmospheric reservoirs.

In this chapter we present current knowledge on surface ocean cycling processes, atmospheric reactivity and importance, and the influence of air-sea exchange for a suite of trace gases which generally have shorter atmospheric lifetimes than those discussed in Chap. 3 (i.e. CO_2, N_2O and CH_4). We focus not only on research from the past 10 years, but also on topics where much uncertainty remains. Unfortunately, not all of the important issues can be addressed here. Notably missing is detailed information about trace gas cycling in polar regions and particularly over-ice processes, including the role of frost flowers.

The chemical species discussed in the chapter are intimately related to the topics discussed in Chaps. 2, 4, and 5 of this book. More accurate parameterisations of gas exchange will allow for better calculations of

P.S. Liss (✉)
e-mail: p.liss@uea.ac.uk

C.A. Marandino (✉)
e-mail: cmarandino@geomar.de

oceanic emissions and uptake (Chap. 2). Many of the gases described, such as sulphur gases and non-methane hydrocarbons, are important for marine boundary layer particle formation and cloud coverage (Chap. 4). And, finally, predictive tools for the impacts of future environmental variability on these gases are required (Chap. 5).

In the chapter we have tried to rationalise the use of units wherever possible. However, different units are used by, for example, the atmospheric and ocean communities, the former often using volume units (e.g. ppt by volume), whereas the latter use mass or molar units (both of which are found widely). So, to avoid confusion, we have not tried to standardise on a common unit for all measurements reported here but have allowed the most common usage to remain. In the case of mass units we have tried to use molar units as much as possible but for instance have allowed mass units for global fluxes, since this is what is routinely done in that community.

1.2 Sulphur and Related Gases

Based largely on his own field measurements, Lovelock et al. (1972) suggested that dimethylsulphide (DMS) represents more than 50 % of the natural sulphur emission to the atmosphere, so closing the previous large gap in the global sulphur budget. This discovery stimulated much research as well as the so-called CLAW hypothesis (Charlson et al. 1987, see Sect. 1.2.3.2), which suggested an important role of DMS in climate regulation. Over the last decades, the biological sources of oceanic DMS and its precursor dimethylsulphoniopropionate (DMSP), the ocean–atmosphere flux of DMS, its atmospheric oxidation and climate relevance have been studied in detail. This section reviews the relevant processes of sulphur cycling, via DMS as well as other sulphur compounds, between the ocean and atmosphere and their potential climatic relevance.

1.2.1 DMS(P) in the Surface Ocean

1.2.1.1 Ecosystem Dynamics

DMS originates from dimethylsulphoniopropionate (DMSP), a compound solely produced by phytoplankton. DMSP is thought to regulate osmotic pressure in the cells of some plankton species and may also serve an anti-grazing function (Kirst et al. 1991). The conversion of DMSP to DMS is enzymatically mediated by algal or bacterial DMSP lyases. With respect to regulation of DMSP biosynthesis, currently two hypotheses exist: the anti-oxidant hypothesis (Sunda et al. 2002) and the overflow-metabolism hypothesis (Stefels 2000). Both hypotheses predict increased DMSP production under high-light conditions, but the difference is associated with UV-radiation (UVR), which would result in increased DMSP production in the view of the anti-oxidant hypothesis, but in a reduction according to the overflow hypothesis. The issue is currently unresolved, since the number of studies that differentiate between conditions with and without UVR is very limited.

In marine ecosystems, DMSP, DMS and its oxidation product dimethylsulphoxide (DMSO) are dynamically linked through a myriad of interactions among biological and chemical parameters (Stefels et al. 2007). As a result, algal biomass parameters often show different temporal transitions than the sulphur compounds. In fact, an overarching correlation was found between DMS surface-water concentration and the average radiation in the surface mixed layer, or solar radiation dose (SRD), and not with phytoplankton biomass (Vallina and Simo 2007). This mismatch between source and end product is called the DMS-summer-paradox and has been observed in large areas of the subtropical and temperate oceans (Dacey et al. 1998; Simo and Pedros-Alio 1999).

The simplicity of the correlation between SRD and DMS concentration makes it a useful parameter in large-scale models, but it cannot be used as a mechanistic explanation of underlying processes. Moreover, in other parts of the North Atlantic Ocean, no significant correlation was found between SRD and seasonal variations in DMS (Belviso and Caniaux 2009; Derevianko et al. 2009). On a global scale, the correlation only exists after removing the variance in the global DMS database through averaging into a single value per SRD range (Derevianko et al. 2009). Other factors may therefore play more important roles in determining the distribution of DMS, whereby interactions with light can take place.

Effects of changes in light condition on ecosystem processes are associated with shifts in species composition. The current view is that bacteria are more susceptible to high radiation than phytoplankton species (Herndl et al. 1993; Simo and Pedros-Alio 1999), but shifts also occur within the phytoplankton realm. The ability to produce DMSP may be of competitive advantage for algae, not only because of its potential protective physiological function under high-light conditions, but also because of adverse effects on grazers. Applying different scenarios of a dynamic model to the Sargasso Sea indicated that solar-radiation-induced DMS exudation by phytoplankton is the major contributor to the DMS summer paradox (Vallina et al. 2008). In a review, with the aim to assist sulphur modellers, Stefels et al. (2007) recommend distinguishing six functional phytoplankton groups, based on their size, DMSP production and conversion ability, and interactions with grazers. Shifts in dominance of these groups in association with shifts in bacterial community will have drastic consequences for the turnover of organic carbon and sulphur compounds.

1.2.1.2 DMS Yield

Given the many conversion pathways that originate from DMSP, it is difficult to assess what fraction of DMSP ultimately ends up as atmospheric DMS. As a general rule, the fraction of DMSP produced that is emitted to the atmosphere as DMS is thought to be between 1 % and 10 % (Bates et al. 1994). This far larger DMSP pool therefore leaves room for a several-fold change in the DMS flux, which could arise when shifts in any of the conversion pathways occur. Unfortunately, there is no database of DMSP concentrations comparable to the one for DMS, which makes it difficult to verify underlying ecosystem pathways (see Chap. 5 for discussion of such databases).

The potential for shifts in the DMS yield from DMSP can be illustrated with four scenarios (Fig. 1.1). The terms "Low" and "High DMSP" refer to variation in the species composition of a system, where one system is comprised of species that do not produce large amounts of DMSP – such as diatoms, prochlorophytes and cyanophytes – and the other of species that do produce DMSP – such as haptophytes, chrysophytes and dinoflagellates (Stefels et al. 2007). The term "High stress" refers to conditions that adversely affect algal or bacterial physiology, such as nutrient limitation and light inhibition.

Under low-stress conditions (Fig. 1.1a, b), DMS produced directly by algae is usually found to be low (compare exponentially growing algal cultures that have little DMS), and most of the DMSP is released from algal cells through grazing or viral lysis. DMSP has been found to be a major source of reduced sulphur for bacteria, but there is not always a good match between production and consumption. Although bacteria can also cleave DMSP into DMS, the more preferred metabolic pathway is demethylation, which does not yield DMS. About one-third of surface-ocean bacteria are capable of demethylating DMSP (Howard et al. 2006). It provides more energetic benefits than cleavage and is a relatively economic way to assimilate reduced sulphur. Kiene and co-workers proposed a model in which bacterioplankton will prefer the demethylation over the cleavage pathway at low dissolved DMSP concentrations (Kiene et al. 2000) (Fig. 1.1a). It is suggested that the total sulphur demand of bacteria can be derived in this way and, as such, this pathway can be directly linked to bacterial production. At higher dissolved DMSP concentrations (Fig. 1.1b), the DMSP fraction that is not assimilated would then be available to the cleavage pathway. In other words, the fraction that is converted to DMS depends on the biomass and growth of the bacterial community; if the bacterial sulphur demand were relatively small, a larger portion of DMSP would be converted to DMS, either by bacterial or algal enzymes.

Under high-stress conditions (Fig. 1.1c,d), such as nutrient limitation or light inhibition, the relative amount of DMS produced by algae directly is often higher. If DMS is released under conditions of high light, UV radiation will photo oxidise an important part of DMS into DMSO. High-stress conditions are also often detrimental to bacteria (Herndl et al. 1993), which would then result in a reduced S-demand and hence an increase in the relative contribution of the cleavage pathway that produces DMS. The net result of the different scenarios with respect to DMS emission is with our current knowledge difficult to quantify, but may vary by over an order of magnitude.

1.2.1.3 Predicted Impact of Climate Change

Accurate implementation of DMS cycling in global climate models requires the validation of such models with DMS field measurements. A first global DMS

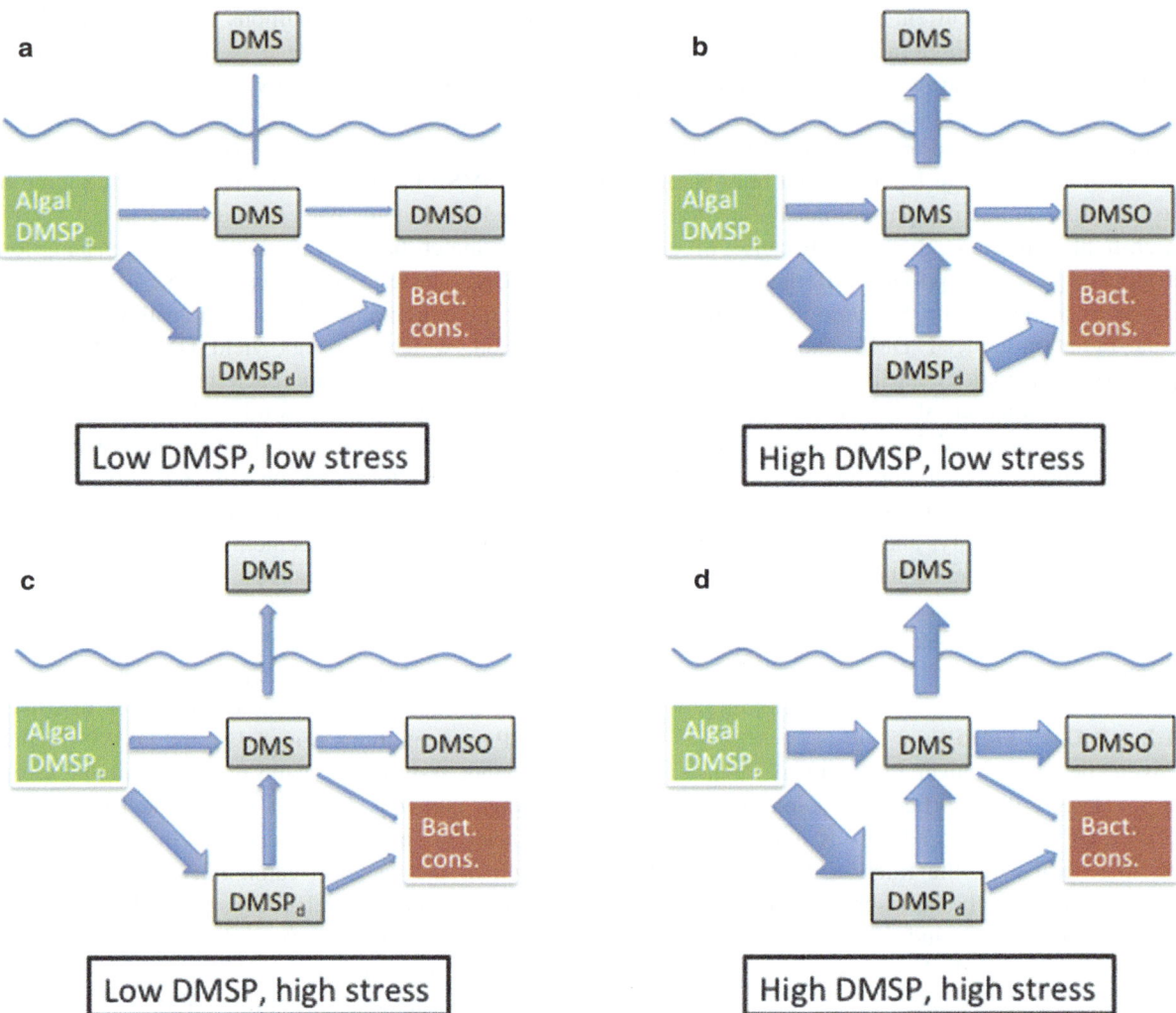

Fig. 1.1 Schematic representation of the major pathways within the marine sulphur cycle and the impact of four different regimes on the relative contribution of each pathway and ultimately on the fraction of DMSP that is emitted to the atmosphere as DMS. "Low DMSP" and "High DMSP" refer to variation in the species composition of a system under blooming or non-blooming conditions. The stress term refers to conditions that adversely affect algal or bacterial physiology, such as nutrient limitation and light inhibition. Thickness of the *arrows* represents an estimate of the magnitude of the process

climatology, derived exclusively from field data, was published more than a decade ago (Kettle et al. 1999; Kettle and Andreae 2000). The number of surface DMS data available at that time was approximately 17,000. They are deposited at the Global Surface Seawater (GSS) DMS database maintained by NOAA/PMEL, which is freely accessible at: http://saga.pmel.noaa.gov/dms/. Since the first data collection by Kettle and co-workers, the scientific community was encouraged to upload new data, which increased the number of DMS measurements threefold. Following this increase, a joint initiative of the SOLAS Project Integration, COST Action 735, and EUR-OCEANS was launched to produce an updated DMS climatology (Lana et al. 2011), which is illustrated by the global monthly climatology shown in Fig. 1.2. The DMS climatology is now available and posted for open access at the SOLAS Project Integration website (www.bodc.ac.uk/solas_integration/, see Chap. 5 for further detail).

The availability of a DMS climatology allowed verification of model results and hence the inclusion of DMS in global ocean models. A first evaluation of nine models has been published: four global 3D models and five local 1D models (Le Clainche et al. 2010).

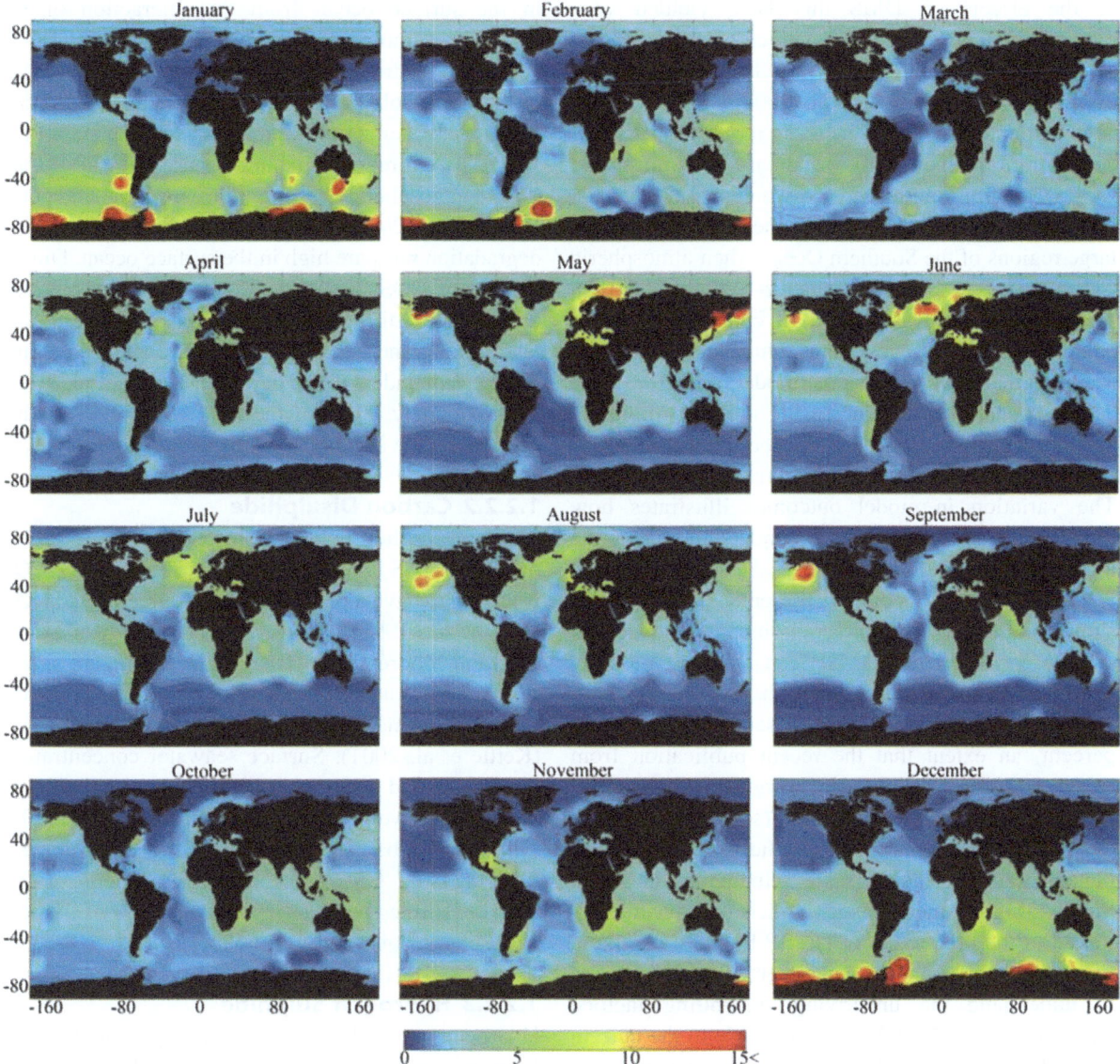

Fig. 1.2 Monthly climatology of DMS concentrations (nmol L^{-1}). Note that the scale is capped at 15 nmol L^{-1} to ensure readability of the plots, although only a few specific regions exceed 15 nmol L^{-1} DMS concentration (Reproduced from Lana et al. (2011) by permission of the American Geophysical Union)

The (in)ability of the models to predict the summer-DMS maximum that is often observed at low to mid latitudes, appeared to significantly affect estimates of global DMS emissions predicted by the models. A major conclusion of this intercomparison was that prognostic DMS models need to give more weight to the direct impact of environmental forcing (e.g. irradiance) on DMS dynamics to decouple them from ecological processes. Potential controlling effects of light and nutrients on DMSP biosynthesis by algae, resulting in shifts in C:S ratios or species composition, were also found to be key issues.

Global climate change is predicted to result in shallowing of the mixed layer depth at some places with increased storm events at others. This would have drastic impacts on the surface-ocean light environment and nutrient inputs and consequently on plankton development (Schmittner 2005). Predictions of the effects on DMS flux have yielded very different outcomes: Kloster and co-workers predict a 10 % reduction in global DMS flux by the end of the century (Kloster et al. 2007); Vallina et al. (2007) found only a 1 % increase in net global DMS concentrations upon a 50 % increase in atmospheric CO_2; Cameron-Smith and co-workers simulated a 50 % increase in DMS flux to the atmosphere over large regions of the Southern Ocean when atmospheric CO_2 increases to 970 ppm (Cameron-Smith et al. 2011). The latter result was related to concurrent sea-ice changes and ocean ecosystem composition shifts. In a comparison with previous model outcomes, the authors concluded that increasing model complexity appears to be associated with reduced DMS emission at the equator and increased emissions at high latitudes. The variation in model outcomes illustrates how important it is to improve our understanding of the underlying ecosystem processes.

The different scenarios presented in Fig. 1.1 show that under different environmental conditions the DMS yield from DMSP can shift by possibly an order of magnitude. This can potentially result in changes in DMS emission of several hundreds of percent, an extent that the recent publication from Quinn and Bates (2011), which argues for the retirement of the CLAW hypothesis (see Sect. 1.2.3.2), did not consider possible. Whether such shifts are plausible at the global scale remains to be seen. Since DMS in mechanistic models is dynamically linked to the production of DMSP by phytoplankton, validating this model parameter would help us to understand the underlying controlling factors. Building a DMSP database comparable to the current DMS database would therefore be of great value.

1.2.2 Other Sulphur and Related Gases in the Surface Ocean

1.2.2.1 Carbonyl Sulphide

Carbonyl sulphide (COS) is the most abundant sulphur gas in the atmosphere which has led to the study of air-sea exchange as a source. Measurements in surface seawaters indicate a net flux from the ocean to the atmosphere of approximately 0.1 Tg COS year^{-1} in the open ocean and 0.2 Tg COS year^{-1} in the coastal ocean (with considerable uncertainty, Watts 2000; Kettle et al. 2002). COS is produced photochemically in the surface ocean from the interaction of UV light with coloured dissolved organic matter (CDOM) (Uher and Andreae 1997) and its main loss mechanism in the water column is hydrolysis (Andreae and Ferek 1992). Seawater concentrations can vary between 1 and 100 pM (~2 orders of magnitude less than DMS) and typical atmospheric concentrations are about 500 ppt(v) (Kettle et al. 2001). It has been suggested that in situ degradation rates are high in the surface ocean. Diurnal cycling of COS in surface waters has been reported, during which the ocean acts as an atmospheric sink late at night and early in the morning and as a source for the remainder of the day (e.g. Kettle et al. 2001). Approximately 40 % of the total atmospheric source of COS can be accounted for by oceanic emissions.

1.2.2.2 Carbon Disulphide

There are limited published measurements of carbon disulphide (CS_2) in the surface ocean. CS_2, like COS, is known to have a photochemical source from CDOM (Xie et al. 1998), as well as a biological source (Xie and Moore 1999). There is no significant sink of CS_2 in the water column except air-sea exchange, although a small diurnal signal has been observed (Kettle et al. 2001). Surface seawater concentrations between 5 and 150 pM have been reported (highest values in upwelling areas, Kettle et al. 2001). The estimated global oceanic flux to the atmosphere is 0.13–0.24 Tg CS_2 year^{-1} (Xie and Moore 1999). This constitutes between 20 % and 35 % of the total atmospheric source of CS_2 (Watts 2000).

1.2.2.3 Hydrogen Sulphide

Hydrogen sulphide (H_2S), originally thought to be the dominant volatile sulphur compound in the oceans, makes only a minor contribution to the total marine flux of sulphur to the atmosphere (Watts 2000) and so to the global cycling of sulphur. The ocean source to the atmosphere has been estimated as 1.8 Tg year^{-1}, which is about 25 % of the total atmospheric source of H_2S (Watts 2000). In the ocean water column, H_2S is produced mainly as part of the hydrolysis of COS (Elliott 1989) and from degradation of particulate organic material. In addition, there is some evidence for a direct algal source (Andreae et al. 1991). H_2S is rapidly oxidised in surface waters (2–50 h), with an expected predawn maximum in its diurnal cycle (Andreae 1990). This fast oxidation is made manifest in particular environments, such as upwelled waters off Namibia, where microgranules of elemental sulphur arising from oxidation of sulphide can be detected by remote sensing (Weeks et al. 2002; Brüchert et al. 2009).

1.2.2.4 Methanethiol

Methanethiol (CH_3SH), also known as methylmercaptan, is produced from DMSP in a pathway that competes with DMS production and may even be the dominant product in some cases (Kiene et al. 2000). However, CH_3SH is thought to be rapidly removed from the water column due to assimilation into proteins by marine bacteria (Kiene 1996) and reaction with dissolved organic matter (Kiene et al. 2000). CH_3SH may also have a small photochemical sink in seawater (Flock and Andreae 1996). Seawater concentrations are assumed to be considerably lower than for DMS, although few measurements have been made (Kiene 1996). However, what data there are indicate that levels of CH_3SH in the Atlantic Ocean range from 150 to 1,500 pM and are typically about 10 % those of DMS (Kettle et al. 2001). Open ocean values were close to 300 pM, but increased dramatically in upwelling and coastal regions, such that the mean concentration of CH_3SH was as high as 20 % that of DMS. In some areas the ratio of CH_3SH to DMS was unity. This would make CH_3SH the second most dominant volatile sulphur compound in seawater and suggests that CH_3SH needs to receive greater consideration in future studies of DMS production or DMSP transformation and in estimates of the flux of biogenic sulphur from the oceans.

1.2.2.5 Dimethyl Selenide

There are considerable similarities between the biogeochemical cycles of the Group VIb elements sulphur and selenium in the marine environment. For example, both can be methylated by microorganisms to form volatile species which are required to balance their geochemical budgets. However, although DMS as the volatile S form has been extensively studied, there has been much less attention paid to volatile Se. One of the few studies is that of Amouroux et al. (2001), who measured volatile Se in surface waters of the North Atlantic in summer. They found that the dominant forms were DMSe (the direct analogue of DMS) and the mixed S/Se compound DMSeS, and calculate that emissions from this part of the oceans at the period of measurement (6.4 (range 1.4–17.9) nmol m^{-2} d^{-1}) were of the right order of magnitude to balance the Se budget, if they applied globally. As shown in Fig. 1.3, the positive relationships between seawater concentrations of DMSe and DMS and DMSe and coccolithophore carbon were interpreted to mean that, like DMS, DMSe is an algal product.

On emission to the atmosphere, the volatile Se compounds will be oxidised and incorporated into particulate phases in pathways which are probably analogous to those discussed in Sect. 1.2.3 for DMS, although the four orders of magnitude smaller flux of volatile Se compared to S means that it's significance for atmospheric properties will be negligible. However, some of the sea-to-air flux of Se will be dry and wet deposited to the land where it could have implications for human health (Rayman 2000). The pathway of Se from the oceans to land via the atmosphere is supported by measurements of the Se content of mosses in Norway which show decreased concentrations with distance from the sea, indicating a marine source (E. Steinnes, 2003, personal communication).

1.2.3 Atmospheric Sulphur and Related Gases

The recent reviews by Faloona (2009) and Vogt and Liss (2009) provide a detailed picture of the literature regarding sulphur cycling in the marine boundary layer. Here we focus on developments in the last 5–10 years after a few introductory remarks.

Sulphur is present in the atmosphere in both gaseous and particulate forms. It is emitted by natural processes such as from volcanoes, the ocean, vegetation and natural biomass burning as well as anthropogenic processing, mainly combustion of sulphur-containing fuels (oil, coal, anthropogenic biomass burning) and some smeltering processes. Sulphur is emitted in various oxidation states (e.g., -II: COS, H_2S, CS_2, DMS; +IV: SO_2; +VI: H_2SO_4) and is oxidised in the atmosphere, eventually resulting in sulphate (oxidation number + VI) unless deposited prior to complete oxidation.

The natural marine source is dominated by DMS. In some ocean regions, especially in the northern hemisphere, anthropogenic sulphur emissions from shipping (as SO_2 and particulate sulphate) are already of similar magnitude to natural marine sulphur emissions (Corbett et al. 1999) and shipping emissions

Fig. 1.3 Relationship between DMSe and coccolithophorid carbon concentrations (**a**) and DMSe and DMS (**b**) in surface waters of the North Atlantic in June 1998 around 60 °N, 20 °W. Error bars represent the relative standard deviation of the reproducibility of the analytical methods on duplicate samples (Reproduced from Amouroux et al. (2001) by permission of Elsevier)

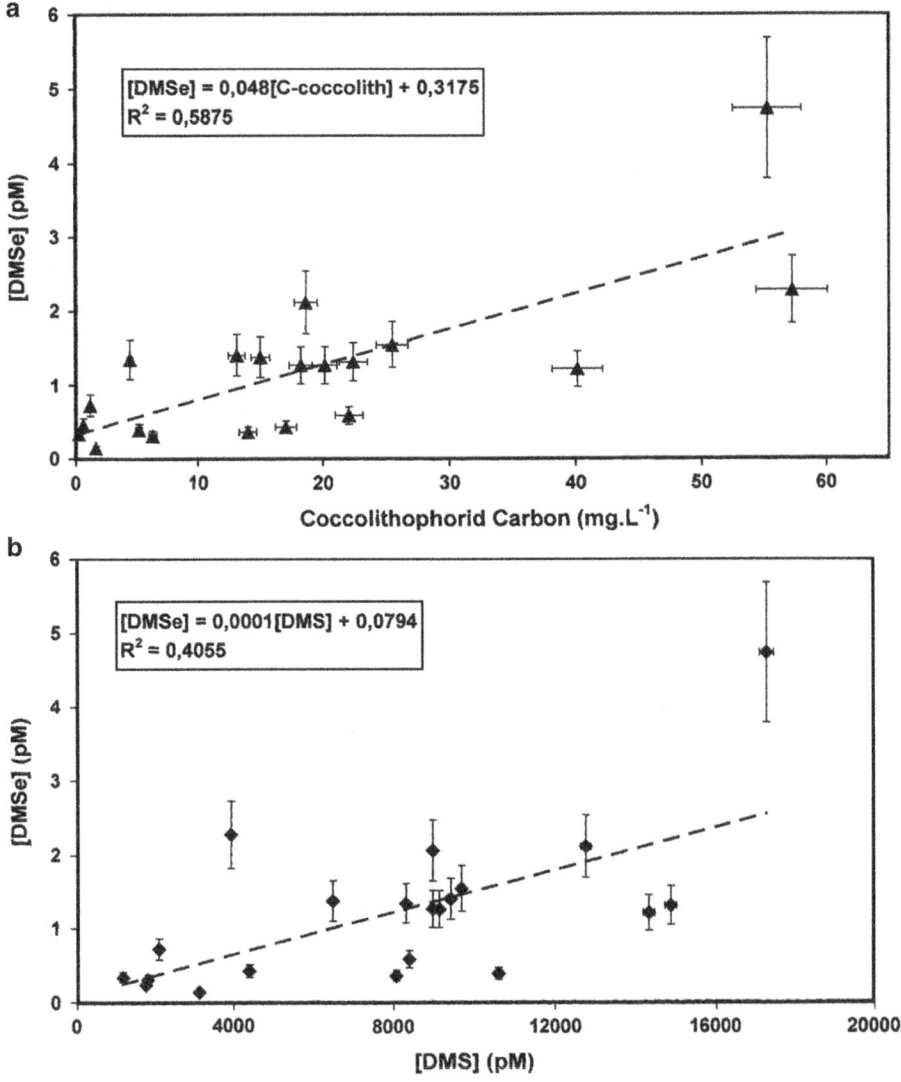

are expected to increase in the next decades (e.g. Eyring et al. 2010). The estimation of source strengths is not an easy task and comes with large uncertainties. For an oceanic gas, such as DMS, the "bottom-up" flux estimate requires reliable information about the oceanic concentrations (as indicated above) as well as the sea-air flux which depends on its temperature-dependent solubility, wind speed and other poorly quantified factors such as surfactants and sea state. The atmospheric concentration of DMS (typically a few nmol m^{-3}) is often ignored in estimating sea to air fluxes of DMS, since relative to seawater concentrations, it is very small and thereby an error of only a few percent is made. The sea-to-air flux of DMS and similar gases is a non-linear function of wind speed, so it is important to include wind speed variations and possibly even gusts in global estimates of its flux. Kettle and Andreae (2000) showed that depending on the choice of the sea-air-exchange parameterisation, the global flux can vary between 15 and 33 Tg(S) year^{-1} using the same database for oceanic DMS. Using an approximately threefold larger data set, Lana et al. (2011) estimated the global DMS flux to be 28.1 (17.6–34.4) Tg(S) year^{-1}, which is about 17 % higher than that calculated by Kettle and Andreae (2000).

The "top-down" approach to estimating fluxes involves the balancing of sources and sinks while reproducing atmospheric concentrations. This is usually done with global chemistry transport models. Faloona (2009) summarises the resulting estimates from a series of global models and highlights the uncertainties in the calculation of dry and wet deposition, showing that the largest individual source of variability in the sulphur cycle in global models is the wet deposition of sulphate aerosol, followed by deposition and heterogeneous oxidation of SO_2. Nevertheless, the models he cites have a range of DMS flux estimates, from 10.7 to 27.9 Tg(S) year^{-1}, which is roughly within the range of the bottom-up estimates given above. See also Chap. 5 for further information on oceanic DMS climatology and models.

1.2.3.1 Chemistry of Sulphur in the Marine Boundary Layer (MBL)

Once emitted, all reduced sulphur compounds are oxidised in the atmosphere until they are either deposited or reach the highest possible oxidation state for sulphur, +VI (H_2SO_4, sulphate). In the gas phase, the most important oxidant is the OH radical, which oxidises almost all sulphur compounds, albeit with different rates (see e.g. Atkinson et al. 2007, for a recent compilation of kinetic data). The oxidation of DMS follows either the abstraction pathway (OH at higher temperature, NO_3, Cl) or the addition pathway (OH at lower temperature, BrO). Only the abstraction pathway leads to the formation of H_2SO_4, which is the only sulphur compound with sufficiently low vapour pressure to allow the formation of new aerosol particles (e.g. Kreidenweis and Seinfeld 1988). Many oxidised sulphur compounds (e.g., dimethylsuphoxide (DMSO), methanesulphonic acid (MSA), sulphur dioxide (SO_2)) are soluble and can be oxidised further, and often more quickly, in aerosol particles and cloud droplets. The main aqueous oxidants are H_2O_2, O_3, O_2 (if metal catalysed) and hypohalous acids (HOCl, HOBr).

The aqueous oxidation of S (IV) by O_3 is very pH dependent and only efficient in particles with pH > 6; at high pH, it is orders of magnitude faster than oxidation by H_2O_2 (see, for example, Fig. 5 in Faloona 2009). This is most relevant for fresh sea-salt particles which initially have the pH of surface ocean water (pH 8.1) or slightly higher due to concentration effects immediately after emission. A particle with a pH 5.6 is in equilibrium with atmospheric CO_2 and strong acids, such as H_2SO_4 and HNO_3, often reduce the pH even further. Therefore, aqueous oxidation of S (IV) by O_3 is expected to be relevant only during the initial acidification of sea salt particles when the pH is still above the threshold for efficient oxidation by ozone. Depending on the availability of gas phase acidity, sea salt particles are usually acidified within minutes to tens of minutes for sub-micron particles but for super-micron particles this is extended to 1–3 h (von Glasow 2006) allowing for very efficient heterogeneous oxidation of S(IV), as earlier suggested by Sievering et al. (1999) and Chameides and Stelson (1992). Based on laboratory experiments, Laskin et al. (2003) suggested that the surface reaction of OH on sea salt might produce additional alkalinity in sea salt aerosol and hence increase the relevance of the O_3 oxidation pathway. However, their extrapolation to ambient conditions was criticised by Sander et al. (2004), Keene and Pszenny (2004), and von Glasow (2006) who all suggested that this pathway might only be minor under atmospheric conditions. Sievering et al. (2004) interpreted their measurements of calcium excesses in coarse aerosol at Baring Head, New Zealand, as evidence for increased buffering of sea salt which would strongly increase the relevance of the aqueous O_3 oxidation pathway. A larger scale view of the role of sea salt aerosol alkalinity was presented by Alexander et al. (2005) who measured sulphate aerosol isotopic composition from a cruise in the Indian Ocean that suggest a large role of the ozone pathway for sulphate formation. Using a global 3D model that includes the titration of alkalinity in sea salt aerosol, they managed to reproduce these data. They conclude that the formation of sulphate in sea salt aerosol is limited by the flux of alkalinity and its acidification mainly by HNO_3. Alexander et al. (2012) combined oxygen isotope measurements of non-sea salt sulphate from North Atlantic samples with simulations from a global chemical transport model and found that in-cloud oxidation of S (IV) by O_3 represents over one-third (36–37 %) of total in-cloud sulphate production on average. They also considered and found the importance of hypohalous acids to be negligible in winter conditions but contributing about 20 % to in-cloud S (VI) formation in summer.

Often it is assumed that OH is the only relevant oxidant for DMS, but this is not necessarily the case. In winter, night-time oxidation by NO_3 becomes important (Koga and Tanaka 1993, 1996; von Glasow and Crutzen 2004) and the confirmation of the presence of BrO in the MBL (e.g., Saiz-Lopez et al. 2007b; Read et al. 2008b; Martin et al. 2009) highlights the relevance of this oxidant (see below). The following list is meant to give an idea of DMS lifetimes (at T = 280 K, kinetic data from Atkinson et al. 2007) to reaction with the main gas phase oxidants: $\tau(OH = 10^6$ molecules $cm^{-3}) = 29$ h, $\tau(NO_3 = 5$ pmol $mol^{-1}) = 18$ h, $\tau(BrO = 1$ pmol $mol^{-1}) = 21$ h, $\tau(Cl = 10^4$ molecules $cm^{-3}) = 84$ h.

As the formation of SO_2 from the intermediates takes additional time, any local increases in DMS concentrations will affect SO_2 concentrations only days downwind, which usually equates to hundreds to thousands of kilometres in the MBL (e.g. Woodhouse et al. 2010). It is noteworthy that the oxidation of DMS by BrO is relevant on a global scale even when BrO is present only in mixing ratios of tenths of a pmol mol^{-1} (e.g. von Glasow et al. 2004; Breider et al. 2010).

The conversion efficiency of DMS to SO_2 is important for estimating any climate feedbacks (see Sect. 1.2.3.2). Reaction chamber experiments show the net SO_2 yields to be in the range of 0.7–0.8 at room temperature (Barnes et al. 2006 and references therein). However, the measurements were often performed at NO_x levels that are much higher than those typically found in the MBL. The laboratory study of NO_x oxidation by Patroescu et al. (1999) indicated an increase in the SO_2 yield at lower NO_x levels. Faloona (2009) compiled the conversion efficiency from many process and global models and found an average conversion efficiency of 71 % and a range of 31–98 %. Estimates of this conversion efficiency in an individual process model cover an even larger range (14–96 %, von Glasow and Crutzen 2004) due to remaining uncertainties in the kinetics, but also due to differences in meteorological, as well as chemical, conditions. The model study by von Glasow and Crutzen (2004) suggests higher conversion efficiencies for tropical regions and lower efficiencies for colder (winter) and cloudy conditions. The conversion efficiency of DMS to SO_2 from field studies is also large. Bandy et al. (1992) presented measurements from the northeast Pacific showing a low (but unquantified) DMS to SO_2 conversion efficiency. They concluded that SO_2 is not an important intermediate in the oxidation of DMS. Measurements on Christmas Island in July and August 1994 indicated a DMS to SO_2 conversion efficiency of about 62 % (Bandy et al. 1996), whereas Gray et al. (2010) calculated a DMS to SO_2 conversion efficiency of 73 % for more recent measurements at Christmas Island (see below).

The potential of the oxidation of DMS by BrO and Cl has received a lot of attention, initially mainly based on laboratory data and model calculations (e.g., Barnes et al. 1991; Toumi 1994; Ingham et al. 1999; von Glasow et al. 2002) and more recently increasingly based on field measurements. Saiz-Lopez et al. (2004) measured 2–6 pmol mol^{-1} of BrO at Mace Head, Ireland and stressed its importance for DMS oxidation. Saiz-Lopez et al. (2007b) presented a year-long time series of BrO at Halley, Antarctica showing its presence during most of the year from several pmol mol^{-1} up to 20 pmol mol^{-1}. Read et al. (2008b) calculated that these BrO levels have a substantial influence on DMS cycling, increasing the production of MSA ninefold. During 8 months of measurements at Cape Verde, Read et al. (2008a) detected on average 2.4 pmol mol^{-1} of BrO. Mahajan et al. (2010a) showed by comparing model results and observational data that under these conditions BrO would contribute about 50 % to DMS oxidation. Interestingly all published measurements of BrO in the MBL are from the Atlantic, which indicates the need for studies elsewhere. Lawler et al. (2009) presented measurements of Cl_2 and HOCl at Cape Verde and their modelling suggested that under these conditions BrO would contribute about 36 % and Cl about 9 % of the DMS oxidation rate, the rest being by OH. For the first time, Lawler et al. (2011) measured HOCl in addition to Cl_2 and their surprising results lead to even higher estimates of Cl atom concentrations. The global model study with a 3D aerosol-chemistry model by Breider et al. (2010) suggested a global contribution of BrO to DMS oxidation of about 16 %. Interestingly, they showed that an increase in DMS flux increased the abundance of BrO due to production of additional acidity and resulting enhanced release of bromine from sea salt aerosol. They suggest a possible feedback between DMS, acidity, sea salt and bromine that would act to reduce the sensitivity of the lifetime of DMS to increases in DMS emissions.

Sciare et al. (2000a) concluded, based on observations during a cruise in the Atlantic (October –

November 1996), that oxidants other than OH must have been involved in the oxidation of DMS. In contrast, data from Amsterdam Island (Sciare et al. 1998, 2000b, 2001) suggest that additional oxidants for DMS are not needed to close the budget for this location in the southern Indian Ocean. Nowak et al. (2001) discussed observations of DMS and DMSO during the PEM-Tropics B campaign (tropical Pacific, March-April 1999) and show that measured DMSO concentrations are inconsistent with DMS oxidation solely by OH. As the biggest mis-match occurs at night, they discount BrO as a potential reason. Wingenter et al. (2005) concluded, based on hydrocarbon ratios, that an additional oxidant must have been present during their measurements in the equatorial Pacific, equivalent to $6-8 \times 10^4$ atoms cm^{-3} of Cl or 1.3 pmol mol^{-1} of BrO. However, during the Pacific Atmospheric Sulfur Experiment (PASE, based on Christmas Island, August – September 2007), Conley et al. (2009) concluded that the DMS budget can be closed without invoking oxidants other than OH. Using a 1D, model Gray et al. (2010) also managed to close the DMS budget based on OH oxidation only, but they found that BrO up to levels of about 1 pmol mol^{-1} would still be consistent with the data. They estimate the conversion efficiency of DMS to SO_2 to be 73 %. The presence of 1 pmol mol^{-1} of BrO would require a 14 % higher ocean flux of DMS, which would also reduce the DMS to SO_2 conversion efficiency to 60 %. They found no evidence of relevant non-DMS sources of sulphur to the MBL. Furthermore, they calculated a dry deposition velocity of SO_2 that was only 50 % of that used in the global GEOS-Chem model hinting at further unknowns in the global budget of sulphur.

Yang et al. (2011) investigated the sulphur budget during the VOCALS campaign (October–November 2008) in the south-east Pacific (west of Chile). DMS was the predominant source of sulphur to the MBL outside the regions affected by continental outflow. The most important loss for SO_2 was in-cloud oxidation. They also found a distinct diurnal cycle in SO_4^{2-} in the MBL with a rapid rise after sunset and decay for the remainder of the day. The authors suggested that meteorology (night time recoupling of the MBL and precipitation scavenging) was the driving force for this variability

In summary, there appear to be large regional differences in oxidation pathways, which probably is not surprising given the large range of environmental conditions encountered in these different regional studies. In order to reach an improved understanding of the global relevance of the sulphur lifecycle (emission, transformation, deposition), more detailed studies in all seasons are required in the MBL of different oceans.

1.2.3.2 CLAW Hypothesis

Charlson et al. (1987) proposed a potential link between DMS emission from the oceans and production of new aerosol particles (see chapter 4 for a discussion of marine aerosols), resulting in changes in cloud albedo which could influence the temperature in the surface ocean and/or the amount of UV that reaches the ocean. This could, in turn, influence the production of DMS and hence its flux to the atmosphere. In their original paper, they mentioned that the sign of the impact of changes in cloud albedo was uncertain, but a decrease of oceanic DMS production following an increase in cloud albedo would constitute a negative feedback, i.e. a climate stabilising mechanism. This idea is usually referred to as the CLAW hypothesis (using the first letter of the authors' names, see Fig. 1.4) and has triggered much research since it was first published a quarter of a century ago (Charlson et al. 1987), especially in the field of ocean biogeochemistry and atmospheric particle formation. However, the fact that the hypothesis involves biological, chemical and physical interactions in the ocean and atmosphere over varying time and space scales makes it arguably untestable as a whole. So-called Earth System models that aim to include all components of the earth system (ocean and atmospheric physics and chemistry as well as ocean biogeochemistry) might, in principle, be able to test the CLAW hypothesis. However, many feedbacks are not well understood and are probably non-linear, hence their replication in a global model will be imperfect. Earth System models have shown inconclusive results, some predicting slight increases and others slight decreases in DMS fluxes for future climate simulations (e.g. Bopp et al. 2003; Gabric et al. 2001, 2004; Kloster et al. 2007; Vallina et al. 2007; Cameron-Smith et al. 2011). All studies found large regional differences, so a climate feedback – if present – may not be acting globally. Our limited knowledge of the strong interactions between physics, biogeochemistry and DMS make quantitative modeling of the whole system very challenging.

A crucial link for the CLAW hypothesis to act as a climate stabilising feedback is that an increase

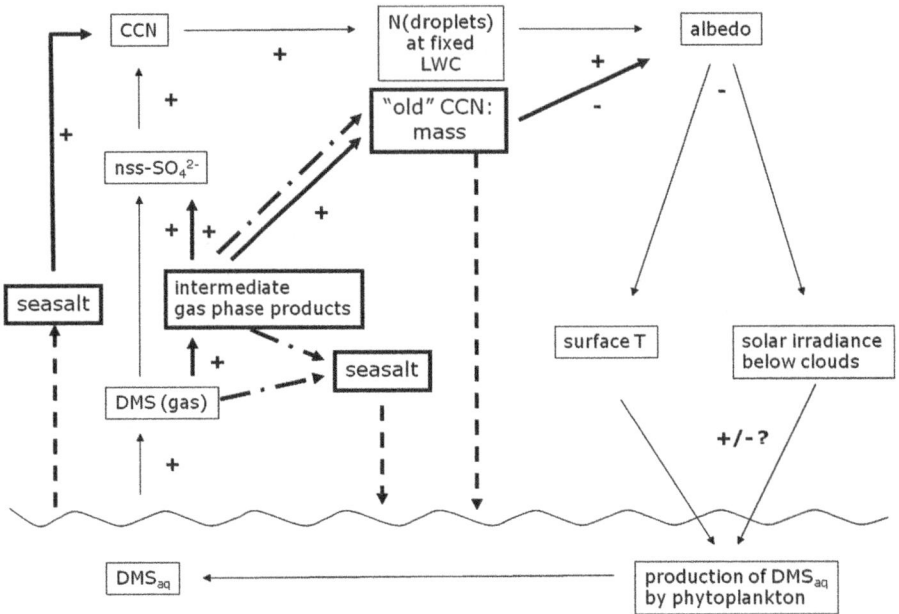

Fig. 1.4 Schematic of the CLAW hypothesis after Charlson et al. (1987) with additions (*in bold*) based on von Glasow and Crutzen (2004). Plus (minus) symbols denote positive (negative) feedbacks. *Solid arrows* indicate 'leads to', *dotted arrows* indicate fluxes (to/from the ocean), whereas *dash-dotted arrows* indicate uptake to aerosol particles and cloud droplets. Key additions to the CLAW hypothesis are the 'short-cut' in the atmospheric sulphur cycle by uptake of DMS and its gas-phase oxidation products on sea salt aerosol with a short lifetime and, therefore, rapid deposition to the ocean; the negative feedback of the growth of existing CCN on cloud albedo; and the importance of sea salt as CCN, thereby possibly reducing the relative role of sulphate aerosol as CCN (Reproduced with permission from von Glasow (2007), copyright CSIRO Publishing)

of solar radiation at the ocean surface leads to higher DMS concentrations in the ocean. As discussed earlier in this chapter, Vallina and Simó (2007) presented a study where they could in fact show that the "solar radiation dose" (SRD, a measure of the daily integral amount of UV radiation in the surface ocean mixed layer) correlates very well with surface ocean DMS concentrations all over the globe, irrespective of temperature, latitude or plankton biomass. Derevianko et al. (2009) re-examined this relationship and found that only 14 % of the variation in DMS could be explained by SRD, rather than 95 % as found by Vallina and Simó (2007). Derevianko et al. (2009) explain this striking difference by the fact that Vallina and Simó (2007) binned their data in latitude bands, whereas they used the individual data points in their analysis. Additionally, Derevianko et al. (2009) found that most of the correlation between SRD and DMS resulted from locations where the mixed layers deepens, resulting in a reduction of both SRD and DMS, highlighting the importance of ocean physics.

Many studies investigated potential links between DMS, MSA, aerosol and cloud properties especially in the mid and high southern latitudes, as this region is least influenced by anthropogenic activity (see references in Vogt and Liss 2009). These parameters were often found to co-vary but a causal link, as suggested by the CLAW hypothesis, could not be established definitively. Complexities of atmospheric chemistry were pointed out by von Glasow and Crutzen (2004) and von Glasow (2007), stressing that most sulphate derived from atmospheric DMS oxidation will likely be taken up on existing CCN rather than form new ones, which could lead to darker rather than brighter clouds. The role of sea salt as CCN might have also been underestimated in the past (e.g. Lewis and Schwartz 2004; Smith 2007) as well as that of primary organic marine aerosol (e.g. O'Dowd et al. 2004; Leck and Bigg 1999). These effects make detection of any CLAW-related feedbacks even more difficult.

Aerosol indirect effects on climate, by changing cloud reflectivity or lifetime, are potentially very important but are still associated with large uncertainties (e.g. IPCC 2007). Therefore, much research is aimed at elucidating such indirect aerosol effects. These studies have shown an amazing range of complex microphysical and dynamical feedbacks and have also led to the suggestion that on a global scale the aerosol–cloud–precipitation system is buffered (Stevens and Feingold 2009) and might be less sensitive to aerosol perturbations than previously thought. This further complicates the detection of any potential CLAW-like feedback, as the original proposal about a DMS–cloud link was rather simplistic in as much that an increase in DMS would increase cloud albedo. This is not always the case, as cloud dynamics (including heating, mixing with outside air) as well as aerosol size distribution and composition play major roles in the lifecycle of clouds and hence their climatic impacts (e.g. Stevens and Feingold 2009 and reference therein).

A recent study with a global aerosol microphysics model confirmed the importance of DMS in maintaining the aerosol burden (Korhonen et al. 2008). It also highlighted the non-linearity of the processes involved and that the effect of various sulphur sources on aerosol concentrations is not simply additive. They found that DMS emissions increase CCN number between November and April in the latitude band 30–40 °S but at higher latitudes only between December and February, where the effect was also less pronounced.

Wingenter et al. (2007) suggested a CLAW-based climate geo-engineering approach to cool climate by fertilising the ocean and hence increase DMS concentrations. Vogt et al. (2008) pointed out several shortcomings in the argument and calculations and questioned the effectivity of such an approach. Woodhouse et al. (2008) showed, using a global aerosol microphysics model, that an increase in CCN number would be only about 1.4 % for such a fertilised patch. Moreover, they showed how non-regional the impact of this geo-engineering approach would be on CCN number, which is easily explained by the timescales of DMS oxidation and CCN formation and growth, i.e. on the order of days or hundreds to thousands of kilometres. The same group (Woodhouse et al. 2010) used current and future sea surface DMS climatologies to estimate the potential effect of changes in DMS on CCN number. They found a very low sensitivity of CCN number to changes in DMS: a change of only 0.02 % in CCN number per 1 % change in ocean DMS concentrations in the Northern hemisphere, compared to 0.07 % for the Southern hemisphere. Given that expected future changes in ocean DMS concentrations are in the range of 1 %, their study suggests that any CLAW feedback on climate would be very small, similar to interannual variations. The uncertainty of current and future DMS flux estimates has to be noted (see earlier discussion in Sect. 1.2.1.3), but given the very low sensitivity of CCN number to the DMS flux the conclusion is unlikely to be affected.

In a recent review, Quinn and Bates (2011) questioned the impact of DMS on global climate. They discuss all three steps of the feedback loop: 1. DMS as a significant source of CCN to the marine boundary layer; 2. The impact of DMS-derived CCN on cloud albedo; and 3. The response of oceanic DMS production to changes in surface temperature and/or incident solar radiation. Quinn and Bates (2011) conclude that the role of DMS in each step is in fact very limited and that it is time to retire the CLAW hypothesis. They identify significantly more non-DMS sources of marine boundary layer CCN, such as sea salt and organic material. They cite studies indicating that there is little or no theoretical or observational basis for boundary layer nucleation due to DMS derived sulphur species. Finally, they determine that the initial hypothesis was too simple to capture the relationship between aerosols, clouds, and Earth's albedo. However, Quinn and Bates (2011) do acknowledge that it is possible that a direct link between biological production and climate effects can exist, but that it is more complicated than described in the CLAW hypothesis.

1.3 Halocarbon Gases

Halogenated hydrocarbons (halocarbons) affect the 'oxidizing capacity' of the atmosphere, primarily as a result of their influence on ozone concentration in both the troposphere and the stratosphere. Whereas a number of man-made compounds contribute to the chlorine supply, bromine and iodine are supplied to the atmosphere mainly as short-lived natural species. The natural sources exert a stronger influence on ozone now than in the past, since synergetic halogen cycles amplify ozone destruction when coupled with the increased levels of chlorine (Lary 1996; von Glasow 2008).

The temporal and spatial distribution of the highly variable oceanic emissions is uncertain and the oceanic sources of the organic halogen compounds are not well understood. Production and loss processes, both biological and chemical, will influence the oceanic distribution of halogenated compounds and are known to be closely coupled in the marine environment. Other processes affecting halogen distributions include physical transport within the ocean, air–sea exchange, phytoplankton abundance in the open ocean and macroalgae distributions in coastal waters.

Phytoplankton and macroalgae have been identified as sources for halocarbons in seawater. Sea surface temperature, which determines the solubility of the compounds, together with biological production and radiation, is a key factor in controlling seasonal and diurnal variability of concentrations in surface seawater. Anthropogenic sources from the atmosphere and water discharges also contribute to their environmental variability.

The supply of halocarbons to the atmosphere can be altered by changes within the surface ocean associated with climate change, and changes in surface ocean physics, chemistry and biology. Our ability to predict such effects is limited, in part because surface seawater sources and sinks are poorly known and quantified, and the relative importance of production and degradation processes, revealed in laboratory studies, are obscured in the natural environment. It is necessary to investigate the kinetics of photochemical, chemical and biological processes in the water, in organisms, on particles, and in the sea surface microlayer. While in this section we address mainly the oceanic emission of the compounds, it is the case that the background concentrations of many trace gases in ocean surface waters are dictated by the overlying atmospheric mixing ratios. The high atmospheric background concentrations of several gases lead to high surface ocean concentrations, which affects our ability to detect and quantify natural marine sources, particularly for some chlorinated compounds. To improve our understanding of future changes in oceanic emissions and thus their contribution to tropospheric and stratospheric chemistry we need: more measurements in oceanic hotspot regions; introduction of direct flux measurement techniques; the combining of existing data and intercalibrations within the global database (see Chap. 5 for information on the halocarbon database); and reliable predictive parameterisations. In this section, we outline some of the biogeochemical cycles of halocarbons in surface waters as a prerequisite for understanding future concentration and emission developments.

1.3.1 Chlorinated Compounds

1.3.1.1 Introduction

With the possible exception of methyl chloride (CH_3Cl), most of the compounds that contain Cl as the sole halogen do not have a primary ocean source. While CH_3Cl is mainly naturally occurring, trichloroethylene (TCE) and perchloroethylene (PCE) are predominantly anthropogenic, and dichloromethane (CH_2Cl_2) is about 70 % anthropogenic (Cox et al. 2003). The anthropogenic compounds have time varying concentrations in the atmosphere, typically increasing in the mid-twentieth century, and in some cases declining more recently, e.g. CH_2Cl_2 (at Mace Head, Simmonds et al. 2006), and chloroform $CHCl_3$ (from firn data, Trudinger et al. 2004). The predominance of non-marine sources in the global budget for these gases means that surface water concentrations will be largely controlled by concentrations in the atmosphere. A consequence is that detecting and quantifying natural marine sources of these chlorinated compounds is more challenging than for a gas such as CH_3I, which has a very low atmospheric background concentration. Chlorinated gases have widely differing atmospheric lifetimes, ranging from around 7 days for TCE (Simmonds et al. 2006), through 3–4 months for tetrachloroethylene (Olaguer 2002), ca. 5 months for CH_2Cl_2 (McCulloch and Midgley 1996), ca. 6 months for $CHCl_3$ (O'Doherty et al. 2001), 1–1.5 years for CH_3Cl, (Xiao et al. 2010) to ca. 35 years for carbon tetrachloride (CCl_4) (Prinn et al. 1999). As the lifetimes are dependent on atmospheric OH, concentrations vary with both season and latitude. Further influences on the latitudinal variation in concentration are the degree to which industrial emissions contribute to the global budget, and the importance of natural terrestrial sources and sinks. Lifetimes of the gases in ocean waters play an additional role in the distributions. Methyl chloride is relatively short-lived on account of microbial consumption, while PCE and TCE are apparently much longer lived in the ocean than in the atmosphere leading to concentrations that increase with depth.

1.3.1.2 Methyl Chloride

Our improving knowledge of sources and sinks of methyl chloride (CH_3Cl) has revealed a lesser role for ocean sources than previously thought. Chemical loss in the atmosphere of about 7×10^{10} mol year^{-1} is matched by a net ocean supply of about 5 % of that amount. It has been found that warm waters tend to be supersaturated with increasing temperature, while cool waters (below about 12 °C) are undersaturated (Moore et al. 1996a). It is clear from decreasing concentrations of CH_3Cl with depth in the ocean that CH_3Cl sinks exist within the water column. Measurements of isotopically labelled CH_3Cl have revealed biological uptake rate constants in the range 0–0.22 d^{-1} (mean 0.07 d^{-1}) in Southern Ocean surface waters (Tokarczyk et al. 2003a) and up to 0.3 d^{-1} in coastal waters (annual mean 0.07 d^{-1}, Tokarczyk et al. 2003b). Among the most firmly established marine sources of CH_3Cl are those from reactions of CH_3I and CH_3Br with Cl$^-$ in seawater, the rates of which are known to be strongly temperature dependent. There is evidence for more direct biological production of CH_3Cl from laboratory studies of algal cultures (e.g. Scarratt and Moore 1996, 1998), though these normally do not uniquely identify the source as the cultured alga rather than associated bacteria. Furthermore, the measured rates are very modest in comparison with estimates of CH_3Cl production in ocean waters. For example, taking the most prolific phytoplankton producer of CH_3Cl amongst those studied, *Phaeocystis*, was estimated to be able to account for only 0.5 % of oceanic production (Scarratt and Moore 1998) based on estimates of the abundance of the organism within oceanic blooms and their areal extent. Other limitations of these studies include the major differences that exist between culture conditions and the ocean. Even when production has been demonstrated, it is important to be able to relate it to the growth phase of the organism. Thus, while Ooki et al. (2010) report enhanced concentrations of CH_3Cl associated with elevated chlorophyll-*a* in the NW Pacific (SST 18 °C), the interpretation cannot be more specific than an indication of a biological source, with a wide range of algae, bacteria or other organisms potentially being responsible.

It has more recently been shown that photochemical breakdown of coloured dissolved organic matter in seawater is a source of CH_3Cl (Moore 2008). This source is likely to be more significant on a unit area basis in estuarine and coastal waters than in the open ocean, though Ooki et al. (2010) speculate that it could account for highly elevated concentrations in warm, (29 °C) low chlorophyll waters of the subtropical NW Pacific.

1.3.1.3 Dichloromethane

The oceanic distribution of dichloromethane (CH_2Cl_2) is remarkably different from that of methyl chloride. The short atmospheric lifetime (ca. 5 months) and anthropogenic sources together cause a large asymmetry in atmospheric concentrations between the two hemispheres and, in turn, different atmospherically-supported concentrations in the respective surface waters. Vertical profiles of CH_2Cl_2 can show increases in concentration with depth, as do the long-lived, anthropogenic Freons (Moore 2004). It appears that CH_2Cl_2, like Freons, enters the deep ocean at high latitude sites of vertical convection. The lifetime in seawater is apparently long enough to maintain a clear similarity in profile with the very long-lived Freon tracers. Seasonal variations of CH_2Cl_2 concentration in the atmosphere will tend to supply the compound to the ocean in winter and yield summertime supersaturations in surface waters that do not reflect in situ marine production but rather a recycling of the gas to the atmosphere. While the major atmospheric influences on CH_2Cl_2 concentration will make smaller natural sources difficult to discern and quantify, it may be predicted that some production occurs in seawater by reactions of chloride with precursors such as chloroiodomethane (CH_2ClI), which is both biogenic and formed by light catalysed reaction between di-iodomethane and chloride (Jones and Carpenter 2005). Additionally, the compound may be produced biologically by processes analogous to those involving haloperoxidases that yield dibromomethane. Ooki and Yokouchi (2011) provide evidence for in situ production by quantifying excess saturation anomalies which have had the influence of sea surface temperature rise on CH_2Cl_2

supersaturation removed, as well as the influence of seasonal change of CH_2Cl_2 in the atmosphere. Their results provide evidence for efflux to the atmosphere between 10 °S and 40 °S and also for the source being derived from the precursors CH_2ClI and CH_2Cl_2, but their claim for direct biogenic production is less well founded.

1.3.1.4 Tri- and Tetrachloroethylene

There are few published measurements of tri- and tetrachloroethylene (C_2HCl_3-TCE, C_2Cl_4-PCE, respectively) in seawater. These two compounds have primarily industrial sources, with production and release to the atmosphere displaying a declining trend in the case of PCE based on measurements at Mace Head (Simmonds et al. 2006). TCE was reported to have relatively constant emissions between 1988 and 1996, as estimated from production data. Surface waters will be influenced by the latitudinal and seasonal variations in atmospheric mixing ratio, tending towards higher and lower concentrations in winter and summer, respectively. Supersaturation has been reported for both compounds in the N. Atlantic during summer (Moore 2001) and it was argued that this is likely to reflect release of gases taken up during winter months. Deep, cooler waters had higher concentrations, increasing with depth, that are likely to result from high latitude ventilation and relative stability of both compounds in seawater. Moore (2001) reported concentration maxima of TCE at a depth of 50 m in the western Atlantic (ca. 21 °N) similar to those frequently seen for biogenic compounds such as isoprene, but physical processes (advection) could provide an alternative explanation. A report of TCE and PCE production by red algae (Abrahamsson et al. 1995) proved controversial with two later studies of the same algae showing no evidence for production (Marshall et al. 2000; Scarratt and Moore 1999). Marshall et al. (2000) suggested that TCE and PCE could potentially be produced in those experiments by dehydrohalogenation of 1,1,2,2-tetrachloroethylene and pentachloroethane, respectively; the significance of any such production is questionable in the absence of data on the precursors in ocean waters. Studies of these compounds in waters of the southern hemisphere, where the atmospherically-supported background concentration is lower, would be especially interesting and useful in identifying and quantifying any marine production.

1.3.1.5 Chloroform

The major sources of chloroform to the atmosphere are soils and anthropogenic emissions, which Trudinger et al. (2004) estimate from firn measurements account for more than half of current emissions. Nightingale et al. (1995) reported that early estimates of the flux of chloroform ($CHCl_3$) from the ocean, based on few measurements and with an assumed high transfer velocity (Khalil et al. 1983), are likely to be too high by about a factor of four. Little has been published on $CHCl_3$ distributions in the ocean, but there is evidence for higher concentrations in both intermediate and deep waters of the Atlantic than near the surface. In the case of TCE, PCE, and CH_2Cl_2, this has been attributed to influx of the gases in regions of deep convection, coupled with relatively long lifetimes in seawater compared with the atmosphere (Moore 2001, 2004); it is reasonable to interpret the $CHCl_3$ distribution in the same way. Seasonal and spatial fluctuations in both atmospheric concentrations and sea surface temperatures will lead to variations in magnitude, and perhaps direction, of $CHCl_3$ fluxes across the ocean surface.

Nightingale et al. (1995) reported the first direct evidence that some species of macrophytes release $CHCl_3$. It is possible that a chloroperoxidase is responsible, but if the mechanism involves HOCl release, the nature and concentration of halogenated products is evidence that the $CHCl_3$ must be have been produced intracellularly. A macrophyte source would tend to elevate concentrations in coastal waters. Two laboratory studies showed that the benthic microalga, *Porphyridium purpureum*, produced $CHCl_3$ (Scarratt and Moore 1999; Murphy et al. 2000) by an unidentified mechanism; these findings tell us little directly about producers in ocean waters since the organism studied occurs only in shallow waters.

1.3.2 Brominated Compounds

1.3.2.1 Methyl Bromide

The oceans play an important role in the geochemical cycle of methyl bromide (CH_3Br), contributing approximately half of the stratospheric bromine burden. Oceanic emissions contribute roughly 35 % of all known natural and anthropogenic sources to the atmosphere, while industrial production is decreasing due to the Montreal protocol (WMO 2011). The ocean is both a source and a sink of bromine for the atmosphere (0.5 and 0.6 Gmol Br year^{-1}, respectively) in the form of CH_3Br (King et al. 2002). Due to declining atmospheric concentrations (Yvon-Lewis et al. 2009), the ocean is now less undersaturated (i.e. closer to equilibrium between the ocean and the atmosphere) with CH_3Br than in the period 1996–1998, when atmospheric mixing ratios rose above 10 ppt. Preindustrial mixing ratios of $CHBr_3$ range from 5 to 5.5 ppt, (Saltzman et al. 2008), indicating substantial natural sources. Sea surface temperature (SST) can be used as proxy for the oceanic saturation state of CH_3Br (Groszko and Moore 1998; King et al. 2002). While polar and tropical regions are undersaturated the entire year, temperate waters show strong seasonal cycles (Baker et al. 1999), with supersaturations in summer months. Oceanic surface waters with temperatures between 12 °C and 20 °C are, in general, supersaturated. However, Lobert et al. (1995) concluded that coastal and upwelling regions are sources of methyl bromide (CH_3Br) to the atmosphere independent of SST, which can possibly be resolved if marine primary productivity is included in the modeling (Anbar et al. 1996). Hydrolysis and reaction with chloride, as well as biological degradation, are the major loss processes from the ocean, and the degradation rates (chemical: 0.04 d^{-1}, biological: 0–0.26 d^{-1}) can match the sea to air exchange rates (Zafiriou 1975; Elliott and Rowland 1995; Yvon-Lewis et al. 2002). Since the consumption reactions are extremely temperature-sensitive, all oceanic temperature variations have large effects on the concentration of CH_3Br in seawater and therefore the exchange between the atmosphere and the ocean.

The net flux of CH_3Br is also sensitive to variations in the rate of CH_3Br production. The natural oceanic production mechanism is unknown but likely phytoplankton in the surface layer are involved. Supersaturations of CH_3Br have been observed in coastal waters off Tasmania, especially in the presence of *Phaeocystis* (Sturrock et al. 2003), possibly due to nutrient limiting conditions. Laboratory studies have revealed CH_3Br production from a variety of phytoplankton species (Saemundsdottir and Matrai 1998; Scarratt and Moore 1998) and macroalgae (Gschwend et al. 1985; Manley and Dastoor 1987). However, all observed rates and inferred global estimates are insufficient to support the observed seawater concentrations and global fluxes. The coastal ocean can be a highly productive region for CH_3Br, due to enhanced biological processes, and including its emissions may increase the estimates of global oceanic emissions by 1–9 % (Hu et al. 2010). While abiotic production of CH_3Br can occur from the degradation of organic matter (Keppler et al. 2000), the relative importance of this process compared to biological generation, likely involving methyl transferases (Wuosma and Hager 1990), is not known.

Future work should include increasing the amount of data representing global coastal regimes and information on seasonal variations. Since the direction and magnitude of CH_3Br exchange between the atmosphere and ocean is very sensitive to temperature and marine productivity, future measurements of marine CH_3Br, temperature and primary production should be combined with models to determine the relationship between marine biological activity and CH_3Br production.

1.3.2.2 $CHBr_3$, CH_2Br_2 and Other Polybrominated Methanes

The oceans are also a source of reactive bromine to the atmosphere in the form of short-lived brominated methanes, including bromoform ($CHBr_3$) and dibromomethane (CH_2Br_2). These compounds represent the largest known natural contribution to atmospheric organic bromine and are recognised as an important source of reactive bromine to the troposphere and lower stratosphere, where they may contribute up to 40 % of O_3 depletion in mid latitudes (Salawitch et al. 2005; Salawitch 2006; Yang et al. 2005; WMO 2011).

The magnitude of the oceanic emissions is uncertain and ranges from 3 to 22 Gmol Br year^{-1}, with a recurring mean of around 10 Gmol Br year^{-1} for $CHBr_3$, and from 0.5 to 3.5 Gmol Br year^{-1} for CH_2Br_2 (Quack and Wallace 2003; WMO 2011). Elevated atmospheric concentrations have been observed over the tropical oceans (Class and Ballschmiter 1988a; Atlas et al. 1993; Schauffler et al. 1998) and were linked to oceanic supersaturations of the compounds and especially to productive upwelling areas. Equatorial surface waters of the tropical Atlantic were indeed identified as a significant $CHBr_3$ source to the atmosphere, with $CHBr_3$ production occurring in the deep chlorophyll maximum (Quack et al. 2004). Generally the oceanic distributions of the brominated compounds, bromoform, dibromomethane (CH_2Br_2), dibromochloromethane ($CHBr_2Cl$) and bromodichloromethane ($CHBrCl_2$), are associated with the abundance of phytoplankton in the open ocean and macroalgae in the coastal waters (Baker et al. 1999; Carpenter and Liss 2000; Nightingale et al. 1995; Arnold et al. 2010), with $CHBr_3$ showing an order of magnitude higher concentration compared with the other compounds in source regions. Pronounced seasonality of the brominated compounds has been observed in coastal regions, with elevated concentrations during summer months (Klick 1992; Hughes et al. 2009; Orlikowska and Schulz-Bull 2009). Diurnal variations are also observed with elevated concentrations around midday (Ekdahl et al. 1998; Karlsson et al. 2008).

Laboratory studies have shown $CHBr_3$ production from arctic diatom cultures (Moore et al. 1996b). Since then, high field concentrations of brominated trace gases have generally been found to be associated with diatom-rich open waters, especially upwelling systems (e.g. Baker et al. 2000). Indeed, significant correlations with $CHBr_3$ were found for low concentrations of diatom marker pigments in the Mauritanian upwelling. However, $CHBr_3$ concentrations do not continue to increase with high diatom abundances and chlorophyll *a* values (Quack et al. 2007). Biological sources of $CHBr_3$ also correlated with degradation pigments and the decay of organic matter. Cyanobacteria have been shown to be a bromocarbon source in the Baltic Sea (Karlsson et al. 2008). The parameterisation of $CHBr_3$ concentration with chlorophyll *a* and mixed layer depth has been tested (Palmer and Reason 2009) and found in some tropical areas to satisfactorily reproduce observed concentrations. In coastal regions macroalgae are thought to be the major marine sources (Manley et al. 1992; Nightingale et al. 1995; Carpenter and Liss 2000) of bromocarbons, while anthropogenic contamination by industrial or municipal effluents may overwhelm the natural sources in some areas (Quack and Wallace 2003). While advection of different water masses generally influences the oceanic distribution, oceanic supersaturations of the compounds can also be caused by the advection of coastal waters enriched in macroalgal releases (Carpenter and Liss 2000; Raimund et al. 2011) and emissions from floating macroalgae (Moore and Tokarczyk 1993).

The biological production of polybrominated compounds occurs during the enzymatic oxidation of bromine by bromoperoxidases and chloroperoxidases in the presence of hydrogen peroxide, resulting in the halogenation of organic compounds with activated terminal methyl groups (Theiler et al. 1978; Neidleman and Geigert 1986). The formation of $CHBr_3$ during the oxidation of organic matter is chemically favoured over CH_2Br_2, since intermediate products are stabilised by additional halogen atoms, and thus $CHBr_3$ is the major product in oceanic environments (Wade 1999). The other bromocarbons have been identified as minor products from phytoplankton and macro algae cultures (Theiler et al. 1978; Gschwend et al. 1985; Manley et al. 1992; Tokarczyk and Moore 1994; Nightingale et al. 1995). $CHBr_3$ has generally longer lifetimes compared with CH_2Br_2 in the oceanic environment under aerobic and anaerobic conditions, though in the case of the reductive hydrogenolysis, $CHBr_3$ reacts faster than CH_2Br_2 (Bartnicki and Castro 1994; Goodwin et al. 1997). Thus reductive hydrogenolysis of $CHBr_3$ could be a process that generates CH_2Br_2 in anoxic environments (Vogel et al. 1987; Tanhua et al. 1996). CH_2Br_2 concentrations have indeed been found to anticorrelate with $CHBr_3$ and increase in deeper waters, suggesting that this process may be possible (Quack et al. 2007). Alternatively, advection of subducted open ocean waters could possibly explain this pattern. Chemical conversion losses from hydrolysis, with half-lives of 183 years for CH_2Br_2 and 686 years for $CHBr_3$ (Mabey and Mill 1978), and halogen exchange, with a half-live of 5–74 years for $CHBr_3$ (Geen 1992), also serve as sources for other brominated compounds, especially for the mixed bromochloromethanes. The progressive abiotic substitution of bromine in $CHBr_3$ with chloride is suggested as the main oceanic source for dibromochloromethane ($CHBr_2Cl$) and bromodichloromethane ($CHBrCl_2$) in the ocean, and an in situ contributor of chloroform ($CHCl_3$) (Class and Ballschmiter 1988b).

The background concentration ratio of trace gases in the surface ocean is strongly influenced by air-sea exchange with the overlying atmosphere. Since CH_2Br_2 has a longer lifetime in the atmosphere (WMO 2011) than $CHBr_3$, the concentration ratio between CH_2Br_2 and $CHBr_3$ in both atmosphere and ocean increases away from coastal source regions towards the open ocean. Thus, remote open ocean surface waters contain relatively more CH_2Br_2 than $CHBr_3$, giving ratios up to >1, while ratios in source regions are around 0.1. Transport of halocarbons may occur through the atmosphere from coastal source regions to ocean surface waters (Carpenter and Liss 2000).

Future work in this area should include halocarbon measurements in water and air in concert with phytoplankton biomass and productivity, chemical and physical parameters. In this way their distributions and correlations can be evaluated and coupled with numerical modeling of source and sink processes (Hense and Quack 2009).

1.3.3 Iodinated Compounds

Transfer from the ocean reservoir to the atmosphere and then to land is an important pathway in the biogeochemical cycle of iodine, with important implications for human health. The iodocarbons, including volatile mono-, di-, and tri-halogenated organic compounds, are produced in seawater and are believed to play an important role in mediating this transfer.

1.3.3.1 Iodomethane

One volatile organic iodine compound that has generated much interest is iodomethane (CH_3I) with estimates of the sea-to-air transfer rate ranging from 0.9 to 9.2 Gmol year^{-1} (reviewed by Bell et al. 2002). These fluxes are far in excess of those reported for other sources of CH_3I including biomass burning (<0.1 Gmol year^{-1}, Andreae et al. 1996), rice paddies (0.1–0.5 Gmol year^{-1}, Muramatsu and Yoshida 1995), peatlands (< 0.1 Gmol year^{-1}, Dimmer et al. 2001) and wetlands (< 0.1 Gmol year^{-1}, Dimmer et al. 2001), making the oceans the dominant source of CH_3I to the atmosphere.

CH_3I production has been observed in experimental incubations of marine seaweeds (Nightingale et al. 1995; Carpenter et al. 2000), phytoplankton (Moore et al. 1996b; Manley and de la Cuesta 1997), cyanobacteria (Smythe-Wright et al. 2006; Brownell et al. 2010; Hughes et al. 2011), bacteria (Amachi et al. 2001; Fuse et al. 2003), detrital aggregates (Hughes et al. 2008b) and in irradiated sterile seawater, suggesting photochemical formation (Moore and Zafiriou 1994; Richter and Wallace 2004). The relative importance of each of these sources to the oceanic inventory of CH_3I is currently unknown. Estimates of global CH_3I production by seaweeds (0.4 × 10^{-2} Gmol year^{-1}, Manley and Dastoor 1988) and phytoplankton (0.8 × 10^{-2} Gmol year^{-1}, Manley and de la Cuesta 1997) suggest that these organisms do not contribute significantly. However, the calculations on which these estimates are based rely on the extrapolation of results obtained from surveys of a limited number of organisms under laboratory conditions which may not be a true reflection of the natural situation. This is an important consideration as algal CH_3I production has been found to vary depending on the species and environmental conditions, such as nutrient-limitation (Smythe-Wright et al. 2010), grazing (Smythe-Wright et al. 2010), light-levels (Moore et al. 1996b; Laturnus et al. 1998), mechanical damage (Nightingale et al. 1995) and desiccation (Nightingale et al. 1995). The only sources which some studies have suggested could contribute significantly to global or regional CH_3I production in the marine environment are photochemistry and cyanobacteria. Richter and Wallace (2004) suggest that 50 % of the average daily flux of CH_3I from the tropical Atlantic surface layer could be due to photochemistry.

The extrapolation of laboratory production rates suggests that the marine cyanobacterium *Prochlorococcus* could contribute significantly to CH_3I production in the eastern tropical Atlantic (Hughes et al. 2011) and calculations presented in Smythe-Wright et al. (2006) suggest that this organism is also an important global source producing 4.3 Gmol year^1. However, orders of magnitude variability in the CH_3I production rate have been observed in laboratory cultures of *Prochlorococcus* between the three studies on this topic published to date (Smythe-Wright et al. 2006; Brownell et al. 2010; Hughes et al. 2011). Hughes et al. (2011) suggest that CH_3I production by *Prochlorococcus* is strongly dependent on cell physiological state. Therefore, the different incubation conditions used in these three studies could explain this variability and should be considered when extrapolating laboratory production rates to the natural environment. Other sources such as bacteria (Amachi et al. 2001; Fuse et al. 2003) and detrital aggregates (Hughes et al. 2008b) are not understood sufficiently to allow accurate global flux estimates to be made.

Several mechanisms for CH_3I production in seawater have been proposed:
1. A photochemical CH_3I source has been suggested to involve a radical recombination pathway (Eq. 1.1, Moore and Zafiriou 1994) but more information on the mechanism is needed.

$$CH_3^\bullet + I^\bullet \longrightarrow CH_3I \quad (1.1)$$

2. Proposed pathways by which biological activity produces CH_3I include methyl group transfer to iodide via S-adenosyl-L-methionine (SAM)-dependent methyltransferases (Eq. 1.2, Amachi et al. 2001; Ohsawa et al. 2001), the production of methylating agents such as methylcobalamin or methyl vitamin B_{12} (Eq. 1.3, Manley 1994) and the breakdown of higher molecular weight organic compounds (Fenical 1982).

$$I^- + SAM \longrightarrow CH_3I + \text{s-adenosyl-L-homocysteine (SAH)} \quad (1.2)$$

$$CH_3\text{-}B_{12}(Co^{3+}) + I^- \longrightarrow CH_3I + B_{12s}(Co^{1+}) \quad (1.3)$$

Loss processes for CH_3I in seawater include gas exchange in surface waters (Liss and Slater 1974), nucleophilic (S_N2) substitution with Cl^- (Elliott and Rowland 1993), and possibly bacterial breakdown (Bell et al. 2002). Hydrolysis is generally an order of magnitude lower than nucleophilic substitution (Moelwyn-Hughes 1938) and the nature of the absorbance cross-section of CH_3I means that photolysis is not significant (Zika et al. 1984). The relative importance of gas exchange and the reaction with Cl^- varies depending on windspeed and seawater temperature, but under certain conditions loss rates due to these processes can be comparable (Bell et al. 2002). A wide range of methanotrophic and nitrifying bacteria have been found to be capable of breaking down CH_3I (McDonald et al. 2002). Methyl halides are molecular analogues of methane (CH_4) and ammonium (NH_4^+) and therefore act as competitive inhibitors of methane monooxygenase (MMO) and ammonium monooxygenase (AMO). To account for discrepancies between modelled and measured CH_3I concentrations in mid to high latitude ocean waters, Bell et al. (2002) propose that the bacterial breakdown rate is 0.24 d^{-1}. However, the mean non-chemical loss rate observed in incubations of North Atlantic seawater by Moore (2006) using $^{13}CH_3I$ as a tracer was 7 % d^{-1}, suggesting that this process is less significant.

1.3.3.2 Other Mono-Iodinated Iodocarbons

Other volatile mono-halogenated iodocarbons including iodoethane (C_2H_5I), 1- and 2-iodopropane (C_3H_7I) and the iodobutanes (C_4H_9I) have been detected in seawater (Klick and Abrahamsson 1992) but their generally low concentrations and their relatively higher solubility means that they make a much smaller contribution to sea-to-air iodine flux than CH_3I. Their production has been observed alongside CH_3I in incubations of marine seaweeds (Carpenter et al. 2000), phytoplankton (Moore et al. 1996b) and detrital aggregates (Hughes et al. 2008b).

1.3.3.3 Di- and Tri-Halogenated Compounds

The marine environment is also known to be a source of reactive di- and tri-halogenated iodocarbons including chloroiodomethane (CH_2ClI), di-iodomethane (CH_2I_2), and bromoiodomethane (CH_2BrI), which, due to their relatively short atmospheric lifetimes, play a more important role in the chemistry of the marine boundary layer (MBL) than the mono-iodinated organics. Martino et al. (2009) have measured iodoform in a laboratory study but, as far as we are aware, there are no published measurements in marine waters. Recent work has also examined the relative importance of organic versus inorganic iodine (i.e. I_2) flux to the marine boundary layer with varying results (e.g. Jones et al. 2010; Mahajan et al. 2010b). The production of the di- and tri-halogenated iodocarbons has been seen in laboratory experiments involving marine seaweeds (Carpenter et al. 2000), phytoplankton (Moore et al. 1996b), bacteria (Fuse et al. 2003) and ozone (O_3) deposition (Martino et al. 2009). Although they share common sources, the mechanisms of formation for the mono-, di- and tri-halogenated iodocarbons are different. The biological formation of reactive iodocarbons is known to be catalysed by haloperoxidase enzymes which have been detected in a wide range of marine organisms and catalyse the oxidation of halide ions by hydrogen peroxide (H_2O_2) and formation of HOI/I_2 (Butler and Walker 1993). The O_3 deposition pathway identified by Martino et al. (2009) involves the reaction between O_3 and I^- at the sea surface to form I_2/HOI (Eq. 1.4a, b). Another mechanism by which the reactive iodine precursor of di- and tri-halogenated iodocarbons could be formed in the marine environment is the reaction between reactive oxygen species (e.g. H_2O_2) and I^- released during oxidative stress, as was observed for the kelp *Laminaria* by Kupper et al. (2008). In all cases the I_2/HOI formed reacts with dissolved organic substrates (Truesdale and Luther 1995) to yield the di-/tri-halogenated organics.

$$O_3 + I^- + H_2O \longrightarrow HOI + O_2 + OH^- \quad (1.4a)$$

$$HOI + I^- \rightleftharpoons I_2 + OH^- \quad (1.4b)$$

Losses of the more reactive iodocarbons in seawater are dominated by gas exchange and photolysis. Photolysis is an important breakdown pathway in surface waters for compounds such as CH_2I_2 (Jones and Carpenter 2005; Martino et al. 2005). For example, Martino et al. (2005) report photolytic lifetimes of CH_2I_2 and CH_2ClI in surface waters (at 15 °C and a solar zenith angle of 0 °) are on the orders of minutes and hours, respectively. The photolytic breakdown of CH_2I_2 is also an iodocarbon source as this generates CH_2ClI with a yield of up to 30 % (Martino et al. 2005). As with all trace gases, the rate of exchange with the atmosphere can be quantified using known formulations (Liss and Slater 1974) and parameterisations (e.g. Nightingale et al. 2000). In general, published rates of sea-air transfer consider only the water-phase resistance to transfer (k_w). However, recent work (Archer et al. 2007; Johnson 2010) suggests that the air-phase resistance to transfer (k_a) is also an important control on the sea-to-air flux of the more soluble reactive iodocarbons, and that the total resistance to transfer (K_w) should be used in the calculation of emission rates. For example, the flux rate of CH_2I_2 is reduced by 40 % at 5 °C, and 20 % at 30 °C when K_w is used to calculate the flux instead of k_w (Johnson 2010). Further information on air-sea gas transfer rates can be found in Chap. 2.

1.3.4 Halogens in the Marine Atmospheric Boundary Layer

This section describes recent advances in understanding the impacts of halogen chemistry in the marine boundary layer. During the past decade there has been an explosion of interest in iodine chemistry, which is described here in some detail. For the chemistry of bromine and chlorine, on which there has been rather less work in the past 5 years and which is discussed in less detail here, the interested reader is referred to the review by von Glasow and Crutzen (2007). One point of note in chlorine chemistry is that the discovery during the last 3 years of the widespread abundance of nitryl chloride ($ClNO_2$) is rekindling interest in this halogen (Thornton et al. 2010).

The potential importance of iodine to the gas-phase chemistry of the troposphere was noted over 30 years ago (Chameides and Davis 1980), starting with its potential for changing the atmospheric oxidising capacity, both via ozone (see Sect. 1.5) destruction and changes to the chemistry of hydrogen oxide (OH and $HO_2 = HO_x$) and nitrogen oxide (NO and $NO_2 = NO_x$) radicals. However, one common conclusion of the early modeling studies on iodine chemistry in the atmosphere was an absence of kinetic data and sparseness of information concerning oceanic iodine sources, which made reliable assessments of the importance of iodine chemistry difficult. Furthermore, apart from measurements of iodine enrichment in sea-salt aerosol (Moyers and Duce 1972), no iodine species had actually been observed in the atmosphere.

During the past decade, the gas-phase iodine species atomic iodine (I), molecular iodine (I_2), iodine monoxide (IO) and iodine dioxide (OIO) have all been detected in the atmosphere for the first time, in locations ranging from Antarctica to the equatorial marine boundary layer. These observations have been made using the techniques of differential optical absorption spectroscopy (DOAS) (Alicke et al. 1999; Allan et al. 2000; Saiz-Lopez and Plane 2004), cavity ring-down spectroscopy (Wada et al. 2007), laser induced fluorescence (Whalley et al. 2007), resonance fluorescence (Bale et al. 2008; Gómez Martín et al. 2011), inductively coupled plasma mass spectrometry (Saiz-Lopez et al. 2006a), and atmospheric pressure chemical ionisation with tandem mass spectrometry (Finley and Saltzman 2008). Measurements of IO have very recently been extended to satellite-based DOAS (Saiz-Lopez et al. 2007a; Schönhardt et al. 2008). The resulting confirmation of active and widespread iodine chemistry has produced a huge increase of interest in this halogen.

These discoveries have been accompanied by improved knowledge regarding the fluxes and atmospheric concentrations of short-lived organic and inorganic iodine source compounds, as well as significant developments in laboratory and quantum theory studies of the reaction kinetics, photochemistry and heterogeneous chemistry of iodine species (Saiz-Lopez et al. 2012). In the past 5 years there has been active laboratory research on IO, OIO, $IONO_2$ and the higher iodine oxides (I_2O_x, x = 2–5) (Dillon et al. 2008; Gómez Martín et al. 2009; Gravestock et al. 2005; Joseph et al. 2007; Kaltsoyannis and Plane 2008; Plane et al. 2006).

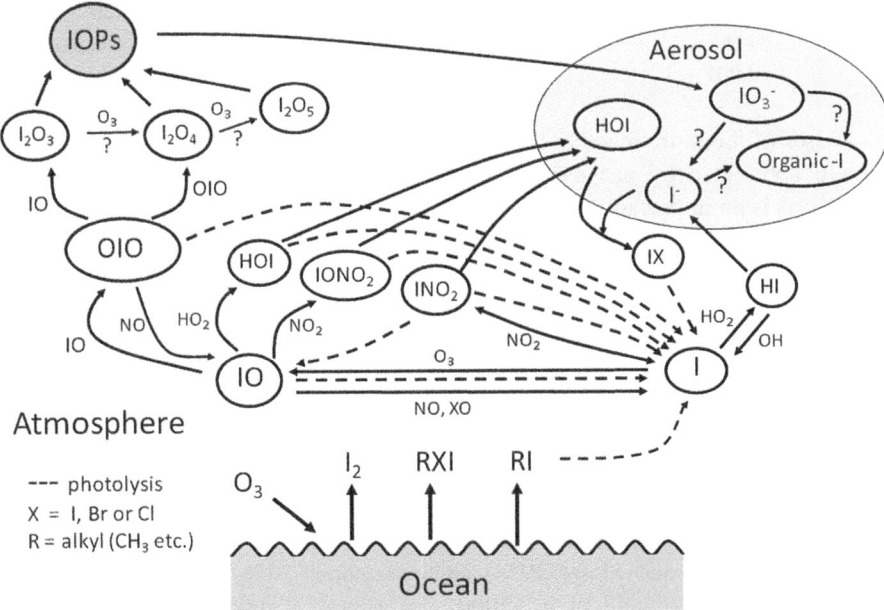

Fig. 1.5 Schematic diagram of iodine chemistry in the marine boundary layer. For the gas phase reactions shown rate coefficients and photolysis cross-sections have been measured in the laboratory. Reactions that are thermodynamically possible but for which rates and cross-sections have not been measured are indicated with a *question mark*. The night-time reaction between I_2 and NO_3 has been omitted for clarity (Figure adapted from Saiz-Lopez et al. (2012))

Figure 1.5 is a schematic diagram of our current understanding of tropospheric iodine chemistry. The elucidation of important reaction pathways has emerged from atmospheric models (containing the new laboratory reaction rate coefficients) that have been used to reproduce measured concentrations of I, IO, OIO, I_2 and observed new particle formation, as well as shifts in the observed ratios of OH/HO_2 and NO/NO_2. These modelling studies have been reviewed in detail recently (Saiz-Lopez et al. 2012). The major source of atmospheric iodine is through the evasion from the ocean of a range of compounds (see Sect. 1.3.3): monohalogenated organic compounds such as methyl iodide (CH_3I), ethyl iodide (C_2H_5I) and propyl iodide (1- and 2-C_3H_7I); reactive polyhalogenated compounds such as chloroiodomethane (CH_2ICl), bromoiodomethane (CH_2IBr) and diiodomethane (CH_2I_2); and I_2 (Saiz-Lopez et al. 2012). These compounds photodissociate in the atmosphere to generate iodine atoms. In the case of I_2, which photolyses most rapidly (< 10 s), I atoms are produced just above the ocean surface. The longer-lived compounds, such as CH_3I which has a lifetime of about 2 days, provide a source of iodine throughout the troposphere. It should be noted that, in contrast to iodine, the major source of bromine and chlorine in the marine boundary layer is sea-salt aerosol, from which the Br^- and Cl^- ions are removed to the gas phase by the uptake and subsequent reaction of species such as N_2O_5, HOI, HOBr and $IONO_2$ (von Glasow and Crutzen 2007).

As shown in Figure 1.5, I atoms produced by photolysis of these precursors mostly react with ozone to form the IO radical. A rich inorganic gas-phase chemistry then ensues, driven by rapid photochemistry and reactions with HO_x and NO_x species. The iodine-catalysed depletion of O_3 occurs via three catalytic cycles (identified so far). First, IO can deplete O_3 through its self reaction to form the OIO radical:

Cycle 1.

$$IO + IO \longrightarrow I + OIO \quad (1.5)$$

$$OIO + h\nu \longrightarrow I + O_2 \quad (1.6)$$

$$(I + O_3 \longrightarrow IO + O_2) \times 2$$
$$net : 2O_3 \longrightarrow 3O_2 \quad (1.7)$$

OIO has strong absorption bands in the visible from about 580 to 630 nm. Although the photodissociation of OIO has been a matter of recent controversy, it is now clear that absorption in these bands leads to photolysis to $I + O_2$ with a quantum yield close to 1 (Gómez Martín et al. 2009). It should be noted that

OIO also forms with a branching ratio of about 80 % from the reaction of IO and BrO. This cross-reaction therefore links bromine and iodine chemistry, and can significantly increase Br-catalysed ozone loss (Mahajan et al. 2010b; Read et al. 2008a). Cycle 1 is potentially very O_3-depleting, but its efficiency is second-order in IO (Eq. 1.5), so that it becomes most important when the IO (or BrO) mixing ratio is more than several parts per trillion (ppt). At lower halogen oxide concentrations, reaction with HO_2 radicals becomes important:

Cycle 2.
$$IO + HO_2 \longrightarrow HOI + O_2 \quad (1.8)$$
$$HOI + h\nu \longrightarrow I + OH \quad (1.9)$$
$$I + O_3 \longrightarrow IO + O_2$$
$$\text{net}: HO_2 + O_3 \longrightarrow OH + 2O_2 \quad (1.10)$$

In semi-polluted environments, $IONO_2$ formation can also lead to O_3 depletion:

Cycle 3.
$$IO + NO_2 \longrightarrow IONO_2 \quad (1.11)$$
$$IONO_2 + h\nu \longrightarrow I + NO_3 \quad (1.12)$$
$$I + O_3 \longrightarrow IO + O_2 \quad (1.13)$$
$$NO_3 + h\nu \longrightarrow NO + O_2 \quad (1.14)$$
$$NO + O_3 \longrightarrow NO_2 + O_2$$
$$\text{net}: 2O_3 \longrightarrow 3O_2 \quad (1.15)$$

However, although the quantum yield for photolysis of $IONO_2$ in the near-UV (Eq. 1.11) is close to 1 (Joseph et al. 2007), the major photolysis pathway (~80 %) of NO_3 produces $NO_2 + O$ (rather than the minor pathway, Eq. 1.12), leading to no overall O_3 depletion. Thus, this cycle can only operate at ~20 % efficiency.

The results in Figure 1.5 show that the HOI and $IONO_2$ formed in Eqs. 1.8 and 1.10, respectively, can also be removed from the gas phase by recycling through sea-salt aerosol. In the case of $IONO_2$ this provides an efficient route for converting NO_x to NO_3^- ions in the aerosol (Mahajan et al. 2009a; Stutz et al. 2007). Moreover, the uptake of HOI and $IONO_2$ enhances the release of chlorine and bromine from sea-salt particles into the gas phase, which can then cause further O_3 depletion (McFiggans et al. 2002). Two recent studies have revealed high levels of IO and OIO in polluted environments (Mahajan et al. 2009a; Stutz et al. 2007). Subsequent work using quantum chemistry calculations has shown that $IONO_2$ should rapidly recycle back to I_2 by reaction with I (Kaltsoyannis and Plane 2008):

$$IONO_2 + I \longrightarrow I_2 + NO_3 \quad (1.16)$$

The resulting I atoms, from I_2 photolysis, will react with $IONO_2$, rather than O_3, if the ratio $[IONO_2]/[O_3]$ is greater than about 0.01 (Kaltsoyannis and Plane 2008). This sequence represents an autocatalytic cycle that will limit the buildup of $IONO_2$, and explains why iodine chemistry is active even in a relatively high NO_x environment (Mahajan et al. 2009b).

The ratio of NO_2 to NO is controlled principally by the reactions:

$$NO_2 + h\nu \longrightarrow NO + O \quad (1.17)$$
$$NO + O_3 \longrightarrow NO_2 + O_2 \quad (1.18)$$

In the presence of significant IO (or BrO) concentrations, the balance will be shifted towards NO_2 (McFiggans et al. 2000, 2010; Saiz-Lopez et al. 2008):

$$IO + NO \longrightarrow I + NO_2 \quad (1.19)$$

where the halogen oxide plays a similar role to peroxy radicals such as HO_2 and CH_3O_2. The HO_2/OH ratio in the presence of IO will be reduced by the combination of Eqs. 1.8 and 1.9 (Bloss et al. 2005; McFiggans et al. 2000; Saiz-Lopez et al. 2008). Compared to the analogous bromine chemistry, this cycle is particularly important because the IO + HO_2 reaction is fast and HOI photolyses more readily than BrO (Atkinson et al. 2007).

The source of I atoms shown in Fig. 1.5 is the photolysis of oceanic precursors (see Sect. 1.3.3), which is therefore restricted to daytime. However, nocturnal IO has also been observed (Mahajan et al. 2009b; Saiz-Lopez and Plane 2004). It has been proposed that the source of IO is the reaction between the NO_3 radical and I_2, to give I atoms:

$$I_2 + NO_3 \longrightarrow I + IONO_2 \quad (1.20)$$

followed by reaction with O_3 (Eq. 1.7). The reaction between IO and NO_3 then produces OIO, which has also been observed at night (Mahajan et al. 2009b; Saiz-Lopez and Plane 2004; Saiz-Lopez et al. 2006b):

$$IO + NO_3 \longrightarrow OIO + NO_2 \quad (1.21)$$

Equation 1.21 has recently been studied in the laboratory (Dillon et al. 2008); the measured rate constant is close to the estimated value that is required to model the nocturnal concentration of OIO (Saiz-Lopez et al. 2006b).

The presence of fine particles (diameter < 100 nm) in the coastal atmosphere has been long established (Aitken 1895). Initially, it was postulated that the secondary production of these particles (i.e., growth through condensation of vapour, rather than primary emission into the atmosphere) involved sulphuric acid, likely derived from the oxidation of DMS (see Sect. 1.2.3.1). However, strong evidence for the important role of iodine compounds was provided by the correlation of daytime tidal cycles in iodocarbons (Carpenter et al. 1999), IO (Alicke et al. 1999; Allan et al. 2000), and ultrafine particle bursts (O'Dowd et al. 1999; O'Dowd et al. 1998). Probably the most comprehensive characterisation of coastal ultrafine particle properties was carried out during the Particle Formation in the Coastal Environment (PARFORCE) project at Mace Head, Ireland (O'Dowd et al. 2002a). The explosive appearance of particles between 3 and 10 nm diameter (referred to as particle bursts) was commonly observed during daytime low tide (Flanagan et al. 2005). The simultaneous reduction in measured gaseous H_2SO_4 concentration, and increase in particle surface area and hygroscopicity, indicated that the condensation of H_2SO_4 plays a role in the subsequent growth of the particles. This has since been confirmed in laboratory experiments (Saunders et al. 2010). Furthermore, particle analysis identified the presence of both iodine and sulphur in the majority of particles (Mäkelä et al. 2002). In the more recent North Atlantic Marine Boundary Layer Experiment (NAMBLEX), observations show that the dominant iodine precursor at Mace Head is actually I_2 (Saiz-Lopez and Plane 2004), released from exposed macroalgae (especially *Laminaria*) at low tide (McFiggans et al. 2004).

Since many coastal regions are semi-polluted, it has been important to establish whether iodine-mediated particles are formed at sites where NO_x is above 1 ppb. The Reactive Halogens in the Marine Boundary Layer (RHaMBLe) coastal experiment at Roscoff (France) showed that this was indeed the case: substantial particle bursts with concentrations up to 3×10^5 cm^{-3} were observed (McFiggans et al. 2010, Fig. 1.6). Importantly, the particles were then seen to grow to sizes at which they were active as cloud condensation nuclei (CCN). This experiment therefore demonstrated that the iodine chemistry involved in particle formation was not shut down in semi-polluted air by the NO_x reactions shown in Fig. 1.5 (Mahajan et al. 2009b).

The observation that iodine-mediated particles can grow to CCN size lends support to an earlier postulate that the iodine-enhanced marine particle number might be sufficient to affect global radiative forcing (O'Dowd et al. 2002b). To test this requires an understanding of the precise role that iodine plays in ultrafine particle formation. Almost uniquely in the atmosphere, iodine oxide vapours are able to form iodine oxide particles (IOPs) spontaneously (i.e. there is no free energy barrier). As shown in Fig. 1.5, laboratory studies have established that the higher iodine oxides I_2O_3 and I_2O_4 form from IO and OIO recombination (Saiz-Lopez et al. 2012). Although one laboratory study of IOP formation from the photolysis of CH_2I_2 in O_3 has indirectly inferred the tetra-oxide form (Jimenez et al. 2003), transmission electron microscope analysis of particles generated photochemically from molecular I_2 and O_3 in dry conditions showed that the IOPs were essentially I_2O_5 (Saunders and Plane 2005). Saunders and Plane speculated that I_2O_5 might form in the gas phase through the oxidation of I_2O_4 by O_3, and that this very stable iodine oxide then polymerised to produce IOPs. However, IOPs can also form without O_3 being present, almost certainly initiated by the spontaneous polymerisation of I_2O_3 and I_2O_4 (Saunders et al. 2010). These particles must then restructure in the solid phase to I_2O_5 and I_2. Under dry conditions, a kinetic study of IOP formation showed that the particles form fractal-like, aggregate structures

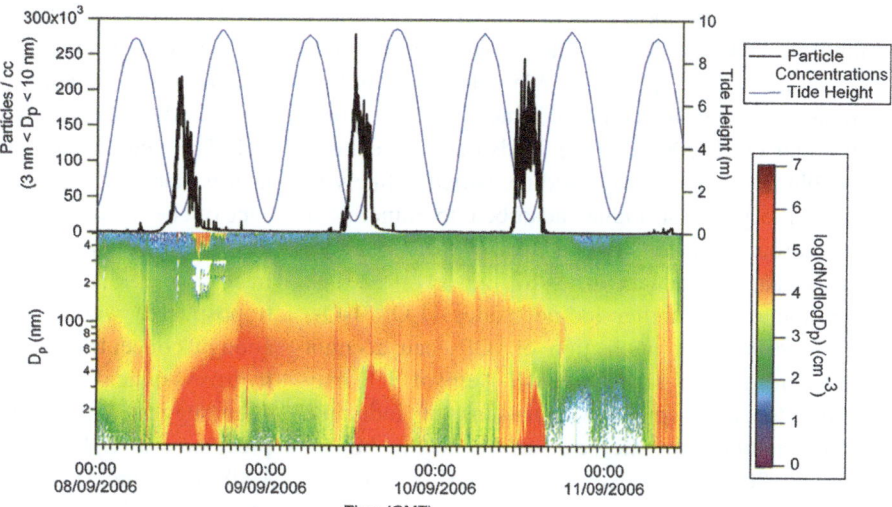

Fig. 1.6 Particle number concentrations and size distribution evolution during the RHaMBLe Roscoff field project (Reprinted with permission from Saiz-Lopez A, Plane JMC, Baker AR, Carpenter LJ, Glasow Rv, Martín JCG, McFiggans G, Saunders RW (2012) Atmospheric chemistry of iodine. Chem Rev 112:1773–1804. Copyright 2012 American Chemical Society)

(Saunders and Plane 2006), which collapse when humidified (Saunders et al. 2010).

One important question concerns the growth of IOPs in the presence of condensable vapours (water, ammonia and both mineral and organic acids) in the marine atmosphere. This is because the supply of iodine oxides is limited, so that once IOPs form they will only grow by condensation of these other vapour species to sizes (> 50 nm diameter) where the particles can have a significant impact on climate either directly (scattering and absorption of solar radiation) or indirectly (enhancement of CCN) (McFiggans et al. 2006). A recent laboratory study has shown that the accommodation of H_2SO_4 vapour on IOPs is very efficient, particularly at high relative humidities (Saunders et al. 2010). The growth of IOPs in the remote MBL to sizes at which their role as CCN may become important is therefore likely to be governed by the uptake of H_2SO_4, accompanied by H_2O and NH_3 to maintain the pH close to neutral (Kulmala and Kerminen 2008). A recent modelling study (Mahajan et al. 2010a) examined the rate of production of iodine-mediated CCN in the MBL, as a function of the mean daytime IO concentration. There is a highly non-linear relationship, as expected from the chemistry of IOP formation (Fig. 1.5). Another critical factor is the background aerosol surface area, because the loss to background aerosols through uptake is usually faster than growth by coagulation and condensation (except at high IO concentrations). The result is that there is an extremely small probability of forming new potential CCN particles when [IO] < 5 ppt, which seems to be the case in the remote open-ocean MBL (Allan et al. 2000; Mahajan et al. 2010a; Read et al. 2008a). However, IOPs may well produce CCN in environments with higher IO concentrations, such as mid-latitude coastal areas (Mahajan et al. 2009b; Saiz-Lopez and Plane 2004; Whalley et al. 2007) and Antarctica (Saiz-Lopez et al. 2007b).

Iodine occurs at the low ng m^{-3} level in atmospheric aerosols from a wide variety of environments; these concentrations are considerably enriched over the seawater composition, as indicated by I/Cl and I/Na ratios (Baker et al. 2000; Duce et al. 1967). This enrichment most likely arises from the uptake of gas-phase inorganic iodine species and small IOPs (Fig. 1.5). A range of inorganic iodine species are potentially present in the aerosol. These include I^-, HOI, I_2, ICl, IBr and IO_3^-. However, because HOI reacts rapidly with halide ions and the resulting IX (X = Cl, Br, I) species are insoluble (and photochemically active), only the ionic species I^- and IO_3^- are expected to accumulate appreciably. Of these, I^- participates in halogen activation reactions to yield I_2, and so model studies concluded that IO_3^- should be the only stable iodine species in aerosols (McFiggans et al. 2000; Vogt et al. 1999). However, measurements of I^-/IO_3^- speciation in marine aerosol have shown that the ratio between these two species is highly variable, for reasons that are poorly understood (Saiz-Lopez et al. 2012).

Soluble organic iodine (SOI) in marine aerosol is ubiquitous (Baker 2005). The proportion of SOI relative to IO_3^- is quite variable, although SOI is frequently found to constitute the major fraction (Gilfedder et al. 2008). There is also an appreciable insoluble fraction of aerosol iodine, which is likely to be either organic or iodine adsorbed to mineral or black carbon surfaces (Baker et al. 2000; Gilfedder et al. 2010; Tsukada et al. 1987; Xu et al. 2010). Hydration of the higher iodine oxides (I_2O_4, I_2O_5) associated with the nucleation of IOPs can account for the presence of aerosol iodate. Aerosol SOI may be introduced by primary emissions of iodinated organic matter from the sea surface during bubble bursting (Seto and Duce 1972). The sea surface microlayer is enriched in organic matter and reactions of O_3 and I^- at the sea surface are known to produce iodinated organic matter (Martino et al. 2009). SOI could also form via the reaction of aerosol organic matter with HOI (Baker 2005). The interactions of iodine and organic matter, and the degree to which aerosol iodine can be recycled to the gas phase, are the most significant unknowns in aerosol iodine chemistry at present.

1.4 Non-Methane Hydrocarbons (NMHCs)

1.4.1 Oxygenated Volatile Organic Compounds (OVOCs)

Oxygenated volatile organic compounds (OVOCs) are a subgroup of non-methane hydrocarbons consisting of alcohols, aldehydes, ketones, and carboxylic acids (e.g. methanol, acetaldehyde, acetone, and formic acid). In this section only short chain OVOCs (<5 C atoms) will be considered. Many of these gases are ubiquitous in the atmosphere (Singh et al. 1995) and can also play a significant role in atmospheric trace gas chemistry. For example, Lewis et al. (2005) found that acetone, acetaldehyde and methanol constituted up to 85 % of total NMHCs and 80 % of the OH radical sink at Mace Head Ireland under maritime conditions. In the following section, we discuss the atmospheric importance and budgets of OVOCs, as well as the role of the ocean as a control on their atmospheric concentrations.

1.4.1.1 Atmospheric Importance of OVOCs

The build-up of biogenic and anthropogenic chemical compounds in the atmosphere is in many cases largely controlled by their reaction with the hydroxyl radical, OH. The abundance of OH depends on a variety of factors, such as light levels, temperature, and the presence of chemical precursors (Spivakovsky et al. 1990), most importantly water vapour (H_2O) and ozone (O_3). In certain regions of the atmosphere, such as the upper troposphere, water vapour is not present at high enough concentrations to account for the amount of hydroxyl radical observed there. In these regions it has been suggested that OVOCs are important chemical precursors of OH (Singh et al. 1995; Wennberg et al. 1998; Lary and Shallcross 2000, Fig. 1.7). In the case of acetone, Elias et al. (2011) show that there is enhancement of this compound, although variable, in the upper troposphere in convective regions during summer, indicating transport of acetone from surface sources. However, the acetone source of upper tropospheric OH is still poorly quantified (Sprung and Zahn 2010).

OVOCs can also be significant sources of RO_2 (Eq. 1.23, Monks 2005; Mueller and Brasseur 1999; Singh et al. 1994) and are therefore an important control on tropospheric ozone (See Sect. 1.5), a harmful pollutant when present in large quantities (Prather et al. 2003 and references therein, see Sect. 1.5). The principal reactions involved in stratospheric ozone formation are the Chapman reactions (Lelieveld and Dentener 2000). However, these reactions cannot account for the high concentrations of ozone in some regions of the troposphere, such as urban areas. Initially, the source of tropospheric ozone was believed to be stratospheric-tropospheric exchange. Later it was found that in highly polluted regions ozone is principally formed by chemical reactions involving hydrocarbons and nitrogen oxide chemical species ($NO_x = NO + NO_2$),

$$RH + OH \xrightarrow{O_2} RO_2 + H_2O \quad (1.22)$$

$$RO_2 + NO \xrightarrow{O_2} NO_2 + HO_2 + R(C=O) \quad (1.23)$$

$$HO_2 + NO \longrightarrow NO_2 + OH \quad (1.24)$$

$$2NO_2 \xrightarrow{h\nu, O_2} 2NO + 2O_3 \quad (1.25)$$

1.4 Non-Methane Hydrocarbons (NMHCs)

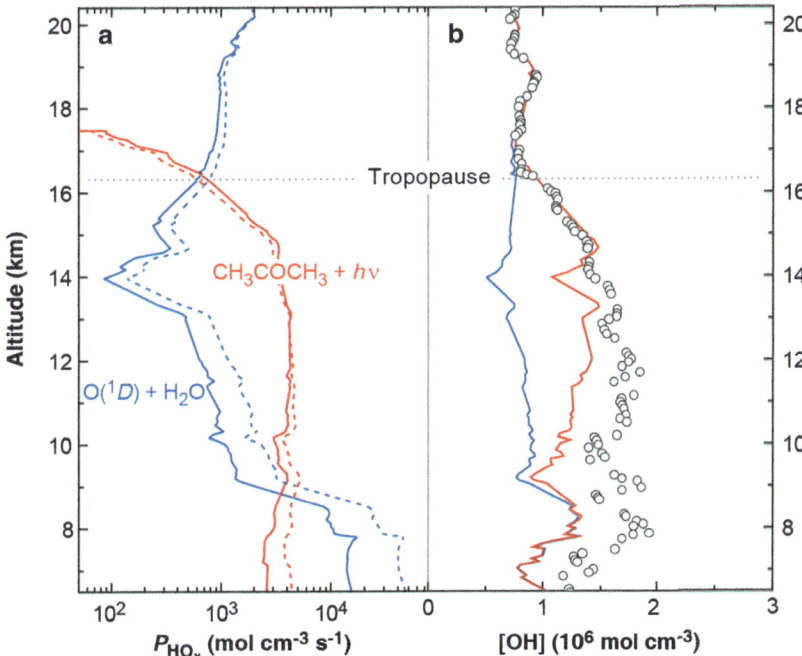

Fig. 1.7 (a) Comparison of HO_x (OH + HO_2) production rates (*dashed lines*) with model outputs (*solid lines*). (b) Measured OH concentrations (*open symbols*) in a vertical atmospheric profile over the Pacific Ocean. *Solid lines* are model output. In both panels the *red line* includes acetone photolysis and the *blue* does not (From Wennberg PO, et al. (1998) Hydrogen radicals, nitrogen radicals, and the production of O_3 in the upper troposphere. Science 279:49–53. Reprinted with permission from AAAS)

Additionally, reactions of OVOCs have been shown to form peroxyaceylnitrate (PAN), a reservoir for reactive nitrogen compounds, and to form secondary organic aerosols and cloud condensation nuclei (see Chap. 4), which influence the radiative budget of the atmosphere (Singh et al. 1995; Blando and Turpin 2000).

1.4.1.2 Atmospheric Budget

The atmospheric sources and sinks of OVOCs have only recently begun to be investigated. There are more detailed budgets for certain compounds, such as acetone and methanol (Heikes et al. 2002; Jacob et al. 2002, 2005; Millet et al. 2008, Schade and Goldstein 2006), but not for others, such as C2–C5 aldehydes and carboxylic acids, although a few exist (Razavi et al. 2011-formic acid, Millet et al. 2010 – acetaldehyde, and Naik et al. 2010 – ethanol). Several studies use models and atmospheric measurements from aircraft to determine budgets (e.g. Jacob et al. 2002), while more recently measurements from satellites have become available (e.g. Elias et al. 2011 – methanol, Dufour et al. 2007 – methanol, Razavi et al. 2011 – methanol and formic acid, Rinsland et al. 2006, 2007 – methanol and formic acid; Harrison et al. 2011a, b – acetone). Tables 1.1 and 1.2 present current estimates of acetone and methanol atmospheric budgets, respectively, and it is evident that there is considerable discrepancy in the literature regarding these compounds. In general, the main sources of OVOCs are primary anthropogenic and biogenic emissions, biomass burning, atmospheric oxidation of precursors (such as methane, isoalkanes, monoterpenes and methylbutanol), and plant decay. The main sinks are wet and dry deposition, reaction with OH, and photochemical oxidation. It is possible to produce OVOCs from atmospheric transformations of other OVOCS. For example, acetaldehyde can be produced from methyl ethyl ketone (Nadasdi et al. 2010; Lewis et al. 2005) and ethanol (Millet et al. 2010; Lewis et al. 2005), as well as formaldehyde from methanol (Heikes et al. 2002), and ethanol from propanal (Naik et al. 2010).

The large extent and productivity of the global oceans dictate their potential to be a significant source or sink for these compounds. Remote marine areas are also of interest since they are the main convective regions that cycle chemical species between the upper and lower troposphere. In these regions oceanic sources of OVOCs can be especially important for upper tropospheric O_3, OH, and NO_x chemistry. Nonetheless, the role of the oceans in the atmospheric budgets of OVOCs is largely unknown and intensely debated.

Table 1.1 Acetone atmospheric source and sink estimates from several publications (as % of total)

	Jacob et al. (2002)	Marandino et al. (2005) –Sinks only	Shim et al. (2007) – sources only
Sources			
Atmospheric oxidation	30		–
Biogenic	35		20–40
Biomass burning	5		19–55
Industry/Urban	–		8–30
Ocean	29		2–4
Sinks			
Photolysis	48	20	
OH oxidation	28	16	
Dry deposition to land	24	8	
Ocean		62	

Table 1.2 Methanol atmospheric source and sink estimates from several publications (in Tg year^{-1})

	Jacob et al. (2005)	Heikes et al. (2002)	Millet et al. (2008)
Sources	170–330	90–490	242
Ocean	–	0–80	85
Terrestrial plant growth	100–160	50–280	80
Atmospheric oxidation	50–100	18–30	37
Plant decay	5–40	10–40	23
Biomass/biofuel buring	10–20	2–32	12
Urban emissions	1–10	5–11	5
Sinks	206	160–570	242
Ocean	10	60–150	101
OH oxidation (gas phase)	129	25–150	88
Dry deposition to land	55	35–210	40
Wet deposition	12	4–36	13
In cloud OH oxidation (aqueous phase)	<1	5–20	<1
Inventory (Tg)	4	3.5–6.9	3.1
Lifetime (days)	7	9	4.7

In the case of acetone, Singh et al. (2001, 2003, 2004) infer that the ocean is both a source and a sink from tropospheric gradient measurements. Sinha et al. (2007) show that the ocean can be both a source and a sink depending on regional biological productivity and light levels. Marandino et al. (2005) directly measured the flux of acetone using the eddy correlation technique (see Chap. 2) over the north and equatorial Pacific Ocean. They found that the flux (F, Fig. 1.8) was always into the ocean (negative) and, when normalised to wind speed (U), is directly proportional to the atmospheric concentration of acetone (C_a). The results of Mao et al. (2006) are in agreement with the findings of Marandino et al. (2005).

For methanol, the ocean appears to be both a large source and a large sink. Carpenter et al. (2004) use DMS (solely oceanic in origin because it is biogenically produced in the surface ocean, see Sect. 1.2) and wind speed to calculate a net ocean sink. Williams et al. (2004) also calculate an ocean sink for methanol of 66 ± 266 μmol m^{-2} d^{-1} in the N. Atlantic, while Millet et al. (2008) agree that the ocean is a net sink for methanol but can be a weak source when background concentrations are low and the sea surface temperature is warm.

For acetaldehyde, Lewis et al. (2005) state that there is no statistically significant relationship between DMS and acetaldehyde, indicating the ocean is probably not a source for acetaldehyde. Millet et al. (2010) find large regional ocean sources for acetaldehyde. More studies, including direct flux measurements, are needed to constrain the role of the ocean in the atmospheric budgets of a suite of OVOCs. Because flux is directly proportional to the concentration

1.4 Non-Methane Hydrocarbons (NMHCs)

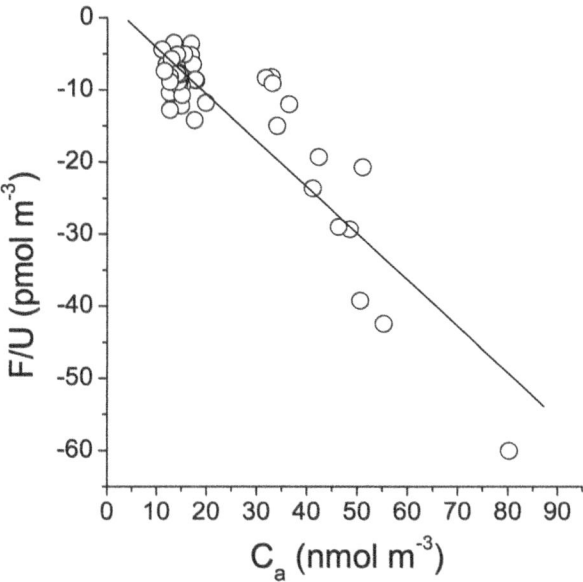

Fig. 1.8 Empirical relationship between the acetone air/sea flux (F), horizontal wind speed (U), and the atmospheric concentration of acetone (C_a). Measurements were made over the northern and equatorial Pacific Ocean during the summer season (Reproduced from Marandino et al. (2005) by permission of the American Geophysical Union)

gradient, more information is needed regarding surface ocean processes and OVOC cycling.

1.4.1.3 Surface Ocean Processes

There have been limited investigations of the surface ocean concentrations, sources, and sinks of OVOCs. This is mainly due to the fact that it is difficult to reliably measure these compounds in seawater. Measurement problems include loss from seawater sample to the headspace during extraction, contamination of samples from surrounding air (especially in laboratories that extensively use these compounds as solvents), large Milli-Q blanks from UV photolysis, and high solubility of these compounds in humid sample lines (Hudson et al. 2011).

Low molecular weight carbonyl compounds have been detected in seawater as early as 1955 (Mopper and Stahovec 1986 and references therein). Since then the concentration of a variety of <5 C OVOCs have been measured in the Delaware and Biscayne Bays, the Caribbean, Sargasso, and Aegean Seas, and in the Atlantic and Pacific Oceans. These compounds include methanol, ethanol, 1- and 2-propanol, tert-butyl alcohol, formaldehyde, acetaldehyde, propional, butyraldehyde, acetone, and pyruvate (Beale et al.

2010; de Bruyn et al. 2011; Dixon et al. 2011a, b; Kieber and Mopper 1990; Marandino et al. 2005; Mezcua et al. 2003; Mopper and Stahovec 1986; Obernosterer et al. 1999; Williams et al. 2004; Zhou and Mopper 1997). An example of measured acetone atmospheric mixing ratios and surface ocean concentrations from the Pacific Ocean (Marandino et al. 2005) is shown in Fig. 1.9. These findings illustrate that the atmospheric acetone levels increase with latitude (going north) but that the seawater levels remain nearly constant. The atmospheric sources and sinks of acetone are relatively well understood and the measured change in atmospheric levels is probably due to anthropogenic emissions. However, the explanation for the lack of trend in ocean concentrations is unknown and points to the lack of understanding of surface ocean processes that influence acetone concentrations. In the case of methanol, Carpenter et al. (2004) reports that the biggest uncertainty in the methanol atmospheric budget is related to its concentration in seawater, which cannot be accounted for by their calculated air-to-sea flux.

Most oceanic source process studies have focused on the production of OVOCs from the photochemical/photosensitized oxidation of dissolved organic matter (DOM) (de Bruyn et al. 2011; Ehrhardt and Weber 1991; Kieber et al. 1990; Mopper and Stahovec 1986; Mopper et al. 1991; Moran and Zepp 1997; Obernosterer et al. 1999; Sinha et al. 2007; Zhou and Mopper 1997). This process has been highlighted as the most likely source of OVOCs in the surface ocean (see also Sect. 1.5). Other measured sources of OVOCs include biological production, air/sea exchange, and oxidation of DOM via radical chemistry (Dixon et al. 2011a, b; Singh et al. 2003b, 2004; Nemecek-Marshall et al. 1995; Mopper and Stahovec 1986). It is most likely that the production of OVOCs in the surface ocean is from a combination of the above processes. de Bruyn et al. (2011) found a combined effect between the photoproduction of acetaldehyde and acetone and the addition of NO_3 radicals to seawater and hypothesised that radical production enhances OVOCs production. When they increased oxygen levels in the seawater samples they measured increased apparent quantum yields (AQY) for formaldehyde and acetone. In addition, while de Bruyn et al. (2011) saw an increase in photoproduction of formaldehyde, acetone, and acetaldehyde with increasing coloured DOM (CDOM), they found a non-linear

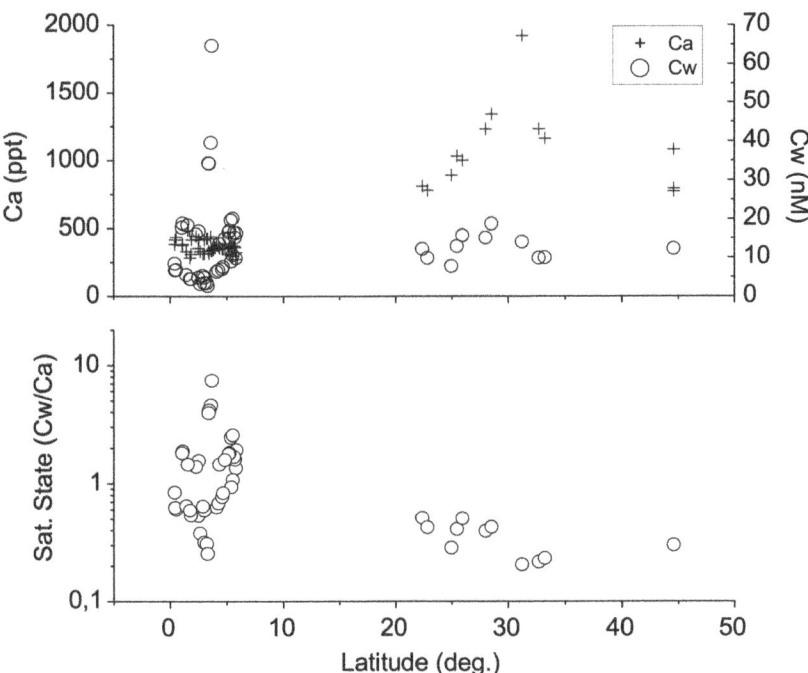

Fig. 1.9 Shipboard measurements of acetone over the equatorial and northern Pacific Ocean during summer, 2004, *top*) Atmospheric mixing ratios (C_a) and surface ocean (C_w) acetone concentrations, *bottom*) Computed saturation state from measured acetone values (Figure adapted from Marandino et al. (2005))

relationship at low CDOM levels (i.e. those more representative of the open ocean). The AQY increased with decreasing CDOM levels indicating that there is enhanced production efficiency with CDOM photobleaching/humic substance breakdown. This is in contrast to Kieber et al. (1990) who report a linear relationship. The Kieber et al. (1990) study included UV radiation while de Bruyn et al. (2011) did not. Beale et al. (2010), although they did not perform direct process studies, found that the ocean may be an important source of ethanol (concentration 2–33 nmol L^{-1}), and 1- and 2-propanol (concentrations 2–22 and 1–19 nmol L^{-1}, respectively) to the atmosphere. They found enhanced concentrations of the propanols in more biologically productive regions (e.g. upwelling areas). Ethanol concentrations followed different trends in different regions. In one area they saw pronounced diurnal cycles with peaks for ethanol at predawn, possibly indicating bacterial production. They propose that when biological activity is low there is a productivity switch to more photochemical control on ethanol production. In the case of methanol (discussed in more detail below), Dixon et al. (2011b) measured concentrations of 151–296 nmol L^{-1}. This is the highest ever reported value for methanol in oligotrophic waters, with the peak in concentration at predawn. They calculated the production rate of methanol needed to sustain these levels and the uptake rates, as discussed below, and concluded that there must be a large in situ source, unrelated to air-sea exchange.

Surface ocean OVOCs sink processes include biological uptake, air/sea exchange, and oxidation (both radical and photochemical) (Obernosterer et al. 1999; Singh et al. 2003b, 2004; Sinha et al. 2007; Sluis and Ensign 1997). Methanol uptake by bacteria has recently been reported by Dixon et al. (2011a, b). They have found that methanol can be used as both an energy and carbon source. Reported uptake rates are 2–146 nmol L^{-1} d^{-1} in surface waters, with enhanced uptake in more biologically productive waters. They hypothesise that methanol is an important carbon source for bacteria in oligotrophic waters (up to 54 % and concentrations are a power function of chlorophyll *a* concentrations when they are lower than 0.2 µg L^{-1}), where there is not enough DOM produced by phytoplanktion to sustain bacterial populations. Beale et al. (2010) also found evidence of removal of alcohol by marine organisms, this time ethanol. They found an anti-correlation between biologically productive regions and ethanol concentrations in the Atlantic Ocean.

Further process studies looking into the role of biology, photochemistry, and air-sea exchange simultaneously are needed for a range of OVOCs to better surprisingly low spatial and seasonal variability of ozone fluxes to the oceans, due to compensating effects in the main drivers, i.e. atmospheric turbulence, and oceanic chemical and physical properties (Ganzeveld et al. 2009). The most fundamental finding of this research is that chemical reactants in the surface ocean are the primary drivers of ozone uptake. This is a well-known mechanism, as ozone is often used as an oxidising agent in domestic, municipal and industrial water disinfection. Those applications have shown that the ozone reaction with organic constituents results in production of soluble oxygenated organics, such as aldehydes, ketones, and acids (von Gunten 2003). Similar processes are expected to arise from ozone deposition to the ocean, suggesting that ozone reactions at or near the ocean surface oxidise organic material and constitute a source of oxygenated organic compounds (see Sect. 1.4.1). At present it is uncertain if and how reactions with ozone affect the oceanic microlayer, dissolved organic material, and ocean biogeochemistry. The increase in the ozone burden seen in the MBL implies that there has been an increase in the ozone ocean uptake rate since pre-industrial times. Following the arguments presented above nurtures speculation that this may have also driven changes in ocean surface chemical processing and possibly aquatic life, though experimental evidence for this effect is lacking thus far.

1.6 Nitric Oxide

Nitric oxide (NO) is important in the atmosphere due to its role in the cycling of ozone (see Sect. 1.5). However, little is known about its transfer across the air-sea interface. There is evidence from measurements in the equatorial Pacific that NO can be formed photochemically in the surface oceans (Zafiriou et al. 1980). The formation of NO is by photolysis of nitrite ions, with the major destruction pathway being reaction with O_2^-, which can also mop up any NO from the atmosphere entering the surface microlayer (Blough and Zafiriou 1985). Extrapolation of the limited field data to wider areas is problematic since both light of the correct wavelength (295–410 nm) and the presence of nitrite, whose concentration is very variable, are required. The tentative conclusion is that the oceans are likely to be a source of NO to the atmosphere, although as far as we are aware the size of the flux has not been estimated.

1.7 Ammonia and Amines

Ammonia (NH_3) is a polar, strongly soluble compound which is ubiquitous in the environment and is important in the atmosphere as the dominant gas phase base (e.g. Quinn et al. 1992). It has been shown to neutralise between 50 % and 100 % of acidity in aerosol in marine air (e.g. Savoie et al. 1993, Johnson and Bell 2008). The amines, derivatives of ammonia (mostly organic, general formula R_xNH_y) are also basic, soluble and contribute to neutralisation of atmospheric acidity. In the aqueous phase, ammonia and the amines (NH_x) are partitioned between their (basic) molecular forms and their protonated (weakly acidic) ionic forms, e.g. ammonium, NH_4^+ and dimethylammonium, $(CH_2)_2NH_2^+$. The base dissociation constants (pK_b) for ammonia and the aliphatic amines tends to be similar and of the order of 5 (with the amines generally somewhat more basic than ammonia); pK_b values for aromatic amines can vary substantially, but they tend to be weaker bases than ammonia.

1.7.1 Ammonia

Ammonia is produced as a byproduct of biological activity, e.g. from the breakdown of nitrogen-containing organic compounds. There are natural terrestrial and marine sources of ammonia to the atmosphere (Bouwman et al. 1997; Dentener and Crutzen 1994) and strong anthropogenic sources, including agricultural, domestic and industrial wastes (e.g. Sutton et al. 2008; Dentener et al. 2006). In the atmosphere it readily reacts with acidic gases and particles to enter the particulate phase as NH_4^+ (either in aqueous or dry particles, see chap. 4 for a discussion of particles), from which it is deposited to land or water surfaces. This is the major sink for NH_3, while oxidation by OH radical is relatively slow (Dentener and Crutzen 1994). The half-life of NH_3 in the troposphere is dependent on the net acidity of the atmosphere but is thought to be of the order of 1–2 days, whereas the lifetime of NH_4^+ in aerosol particles and cloud water is of the order of 5 days or more (Dentener and Crutzen 1994).

It is thought that in the past the ocean would have been a net source of NH_x ($NH_3 + NH_4^+$) to the terrestrial environment, but since the industrial revolution this net flux has reversed (Holland et al. 1999; Galloway 2004). Thus, NH_x is transported from the terrestrial to the marine atmosphere and deposited on the ocean surface. However, given the relatively short lifetimes, concentrations in the atmosphere decrease rapidly away from terrestrial source regions, so the large anthropogenic flux of NH_x probably affects only the coastal and shelf seas significantly. Typically, NH_3 concentrations in the marine atmosphere are of the order of 0.1–10 nmol m^{-3} (mid-to-high ppt concentrations), but they can be orders of magnitude higher in polluted terrestrial environments (Johnson et al. 2008; Dentener et al. 2006).

In open ocean seawater, NH_x exists predominantly as ammonium with concentrations in the nmol L^{-1} range; concentrations of 1 µmol L^{-1} or higher may occur frequently in some coastal and shelf seas but are uncommon in the open ocean (Johnson et al. 2008). Ammonium is both a nutrient and a waste product. As a nutrient its uptake by both phytoplankton and bacteria is energetically favourable relative to nitrate and it can be released or produced at various stages in the marine microbial nitrogen cycle, e.g. from 'leaky' algal cells to heterotrophic breakdown of nitrogen-containing dissolved organic matter (e.g. Capone 2000). As such, during productive periods, it tends to be cycled extremely rapidly (turnover of the entire standing stock on the order of hours to days) and temporary decoupling of uptake from remineralisation can lead to substantial transient increases in NH_4^+ concentration over wide areas (Johnson et al. 2007). There is some evidence of photochemical production of ammonium from dissolved organics, which could also contribute to the turnover rate (Kitidis et al. 2006, but see also Grzybowski 2003). The amount of NH_3 available to transfer across the air-sea interface depends strongly on pH and temperature (Bell et al. 2007), as does the Henry's law solubility of ammonia in water (Johnson et al. 2008). The partial pressure of NH_3 over the open ocean calculated from typical concentrations, pH and temperatures is in the range 0.1–5 nmol m^{-3}, which is of the same order as atmospheric concentrations, demonstrating that ammonia tends to be close to equilibrium and that the magnitude *and direction* of the flux can be very sensitive to factors controlling the acid dissociation and Henry's law equilibrium positions. Johnson et al. (2008) find that the major factor controlling the flux of NH_3 at a global scale is temperature, with higher seawater concentrations but downward fluxes at high latitudes, and lower seawater concentrations but ocean emissions at low latitudes. However, transient biologically-driven peaks in NH_3 concentration in the surface ocean can drive emission events in regions where the flux might typically be into the ocean (Johnson et al. 2007). The finding that the high latitudes appear to be predominantly a sink for ammonia is somewhat at odds with strong circumstantial evidence for a North Atlantic marine ammonia source presented by Jickells et al. (2003). This may be due to advection from lower latitudes or sampling during transient emission periods. However, further simultaneous measurements of concentrations in the ocean and atmosphere are required to resolve this apparent discrepancy, and to improve estimates of the net global emission of ammonia from the ocean, previous estimates of which lie between 5 (Galloway 2004) and 13 (Schlesinger and Hartley 1992) Tg NH_3-N year^{-1}. In the light of evidence of strong downward fluxes at high latitudes (Johnson et al. 2008), future estimates are likely to be lower as low latitude emissions will to some extent be balanced by high latitude uptake.

An alternative mechanism controlling NH_3 flux has been proposed by Johnson and Bell (2008) to explain the near-constant ratio of ammonium to non-sea-salt sulphate in aerosol particles in the remote marine environment (see Sect. 1.2.1). They invoke titration of acidic DMS oxidation products with gas phase ammonia, the rate of which in inversely proportional to the degree of neutralisation in the particle phase (i.e. to the NH_4^+: $nssSO_4^{2-}$ ratio). Thus, the DMS emission (and subsequent oxidation) can control the gas phase ammonia concentration and thus the oceanic emission, given that the atmosphere and surface ocean are generally close to equilibrium with respect to ammonia.

It is important to note that, as a soluble gas, the air-sea exchange of ammonia is controlled by transfer on the air-side of the interface, particularly as its water phase transport is likely to be significantly chemically enhanced by protonation (Johnson et al. 2011) (as further discussed in Chap. 2). The same applies for the majority of the amines, discussed below. This means that calculating the rate of transfer requires application of a k_a (air-side) transfer velocity term rather than the typically applied k_w (water side) term used for less soluble gases such as CO_2, N_2O, DMS etc. (e.g. Liss and Slater 1974; Johnson 2010).

1.7.2 Amines

A recent review has identified >150 amine species which have been measured in the atmosphere or are known to be produced (Ge et al. 2011). Many of these have only industrial sources, but a large number are also known to be biogenically produced and are therefore likely to be formed by biological activity in the surface ocean in a similar manner to ammonium. Others have specific natural production pathways, such as hydrazine (N_2H_4) which is an intermediary species in the anaerobic bacterial oxidation of ammonia with nitrite (Anammox, e.g. Ward 2003).

In both ocean and atmosphere the most abundant volatile amines are the low molecular weight aliphatic amines of carbon number <6 (Ge et al. 2011), typically occurring at between 1 and 3 orders of magnitude lower concentration than ammonia (Gibb et al. 1999) in the ocean and between 2 and 5 orders of magnitude lower in the atmosphere. There is an increasing body of atmospheric concentration data for the low molecular weight alkyl amines (e.g. Müller et al. 2009).

The significance of both ammonia and the amines in particle formation is currently poorly understood (see Chap. 4). For ammonia the literature is contradictory, with some experimental and modelling studies suggesting that ammonia significantly enhances ternary new particle formation from water and sulphuric acid (Coffman and Hegg 1995; Korhonen et al. 1999); and other evidence suggesting little effect (e.g. Yu 2006). However, recent works suggest that ammonia can play an important role in particle formation and growth (Kirkby et al. 2011; Benson et al. 2011). The amines have been postulated as even stronger agents than ammonia in ternary nucleation (Kurten et al. 2008), due to their higher basicity and higher molecular weight; as well as being significant components of the secondary organic aerosol fraction (e.g. Facchini 2008; Wang et al. 2010). In order to better understand the role of the oceans as sources/sinks of ammonia and the amines and their importance in atmospheric chemistry, substantial further observational data is required.

1.8 Hydrogen

There is growing interest in hydrogen (H_2) in view of its role in affecting the oxidising capacity of the atmosphere and also because of the potential for increased emissions in a future 'hydrogen economy'. The production and flux to the atmosphere of H_2 formed in the oceans constitutes about 5 % of the total current global production of the gas (IPCC 2007). The majority of the production is from the photochemical oxidation of methane in the atmosphere (see Chap. 3: Sect. 3.4), with deposition to soils being by far the major sink (Rhee et al. 2005).

The near surface waters of the ocean are frequently at, or well above, saturation with the atmosphere, indicating net production. Below the upper tens of meters of the water column, concentrations decline and, in the case of 8 profiles to 1,000 m in the central equatorial Pacific, become strongly undersaturated, indicating net consumption, probably bacterial (Moore et al. 2009). Processes having the potential to account for supersaturations have long been recognised as anaerobic bacterial activity in detrital particles, nitrogen fixation by cyanobacteria, and photo-oxidation of dissolved organic material (Conrad and Seiler 1986). However, there is no definitive answer as to the relative importance of these three processes (or others) for the formation of H_2 in surface seawater. In the study by Moore et al. (2009) it was apparent that the highest concentrations observed in the water column coincided with high levels of nitrogen fixation. A more recent study (Moore, 2012, personal communication) including about 3,000 surface measurements along a meridional transect in the Atlantic (cruise AMT 20) showed marked elevations in hydrogen concentrations in surface waters between about 30 °N and 30 °S. A laboratory study by Wilson et al. (2010) showed that diazotrophic bacteria varied greatly in their net release of H_2 during nitrogen fixation, with *Trichodesmium erythraeum* yielding an order of magnitude more H_2 than the unicellular cyanobacterium *Cyanothece*, and about two orders more than *Crocosphaera watsonii*. With regard to photoproduction in ocean surface waters, Punshon and Moore (2008) showed from a study of lake and coastal waters that hydrogen is produced during irradiation with sunlight, with the rate being dependent on the concentration of coloured dissolved organic matter (CDOM). They concluded that while it is unlikely that photochemistry can account fully for the widespread supersaturation of H_2 in low latitude oligotrophic ocean waters, it may be a contributor.

Whatever the operative production processes, the measured supersaturations lead to a net flux of H_2 from the oceans to the atmosphere, estimates of which range from 3 to 6 Tg year^{-1}, with an uncertainty on each estimate of 50 % or more (Schmidt 1974; Conrad and Seiler 1986; Novelli et al. 1999; Rhee et al. 2005).

1.9 Carbon Monoxide

Since the pioneering work of Swinnerton et al. (1970), the ocean is known to be a source of carbon monoxide (CO) to the atmosphere. This compound is important for atmospheric chemistry as its oxidation by the hydroxyl radical, OH (its main sink) leads, in presence of nitrogen oxide, to the formation of ozone (see Sect. 1.5), a pollutant and greenhouse gas. Although the evaluation of its global oceanic source led to estimates ranging over two orders of magnitude (from 4 to 600 Tg CO year^{-1}), the more recent estimates lie in the lower part of this range (Bates et al. 1995; Stubbins et al. 2006b). Therefore, the oceanic source of CO is likely to play a minor role in its global budget (surface emissions > 1,000 Tg CO year^{-1}, Duncan et al. (2007)), which is dominated by fossil fuel, biomass burning and secondary emissions. Nevertheless, in remote regions, oceanic emissions from CO may affect the regional oxidising capacity of the troposphere and therefore deserve to be better characterised and quantified.

More than three decades ago, the main source and sink of oceanic CO were established as photo-oxidation of dissolved organic matter under UV light and microbial consumption, respectively (Conrad and Seiler 1980; Wilson et al. 1970). Further work on the main CO source and sink in the ocean has since been performed, allowing a better understanding and quantification of these processes, which are dependent on several biogeochemical or physical parameters, such as UV irradiance at the seawater surface, quantum efficiency of the conversion of DOC to CO, the attenuation coefficient of UV radiation in the water layers in relation to their DOC content, and microbial consumption (Day and Faloona 2009; Johnson and Bates 1996; Stubbins et al. 2006a, 2008; Valentine and Zepp 1993; Xie et al. 2005; Zafiriou et al. 2003; Zhang et al. 2006; Ziolkowski and Miller 2007; Zuo and Jones 1995). A biotic source of CO originating from macro-algae was also investigated by different groups (Loewus and Delwiche 1963; Troxler and Dokos 1973), but King (2001) concluded that this source is not a significant fraction of the global oceanic source. Recently, laboratory studies of trace gas emissions by various phytoplankton species have shown that CO is also emitted directly by phytoplankton under photosynthetically available radiation (Gros et al. 2009). Cyanobacteria and diatoms were the largest CO emitters and this source could account for up to 20 % of the CO production in seawater. Finally, another source of oceanic CO is the so-called "dark production" which has been reported by several groups (Day and Faloona 2009 and references therein) and which is needed to explain non-zero CO values below the euphotic zone (Kettle 2005).

The first detailed measurements of CO in the mixed layer of the oceanic water column in the Pacific Ocean showed that it varies seasonally and regionally with daily means between 0.1 and 4.7 nmol L^{-1} (Bates et al. 1995). CO showed pronounced diurnal cycles with maxima in the afternoon, in agreement with the photochemical nature of the main source in water. Since then, measurements of CO have been performed only over a few additional areas in the open ocean. Stubbins et al. (2006b) determined values between 0.2 and 2.6 nmol L^{-1} during an Atlantic Meridional cruise (from 35 °S to 54 °N). Gros et al. (unpublished data) measured values around 1 nmol L^{-1} in the Southern Atlantic Ocean, with levels rising to 20 nmol L^{-1} in a phytoplankton bloom near the Argentinian coast. Xie et al. (2009) conducted measurements in the open waters of the South Eastern Beaufort Sea at two different seasons with mean concentrations of 0.45 and 4.7 nmol L^{-1} in spring and autumn, respectively.

In shelf waters, CO concentrations were also found to be variable but with values in the upper range of those measured in the open ocean. Day and Faloona (2009) found values between 2.7 and 17 nmol L^{-1} for a Californian upwelling system; Xie et al. (2005) measured values around 1 nmol L^{-1} in the Beaufort Sea and Gulf Stream and 2–18 nmol L^{-1} in the coastal NW Atlantic. Yang et al. (2010, 2011) have recently investigated the East China Sea and the Yellow Sea and observed values between 0.13 and 2.3 nmol L^{-1} (mean of 0.68 nmol L^{-1}) in November and values between 0.12 and 6.99 nmol L^{-1} (mean of 2.24 nmol L^{-1}) in April.

Sea-to-air CO fluxes have been shown to vary over two orders of magnitude, depending on the source (which is variable geographically and seasonally) and sink strengths, but also strongly dependent on wind speed. Based on concomitant measurements of CO in seawater and the atmosphere over the Pacific Ocean, Bates et al. (1995) calculated zonal average fluxes ranging from 0.25 to 13 µmol CO m^{-2} day^{-1} and Stubbins et al. (2006b) have estimated fluxes from 0.4 to 12.1 µmol CO m^{-2} day^{-1}.

New approaches are becoming available to estimate photoproduction of CO in water on a global scale, as shown recently by the study of Fichot and Miller (2011) who derived CO production fluxes from remotely sensed ocean colour and modelled solar irradiance.

Despite the large temporal and spatial variability of the oceanic source of CO, it is usually considered a constant yearly mean in global model studies. Further work combining field measurements (in less documented areas) and new approaches to derive global fluxes will help to refine estimates of the marine CO source.

1.10 Concluding Remarks

In this chapter we have attempted to summarise in some detail present knowledge of the main trace gases whose transfer across the air-sea interface is environmentally important from local to global scales. This has been done with sections for each gas or group of gases. But we are aware that the real system does not follow such arbitrary chemical sectioning. For example, it is clear that in the atmosphere there is much interaction between various gases and also between the gas and particulate phases (both secondary particle formation and reactions on existing particles). We have tried to deal with this by cross-referencing to relevant sections both within the chapter and across to other chapters, but a more integrated approach would have been desirable.

An example of how focusing on just one gas can lead to misunderstanding is DMS and the CLAW hypothesis. Although the idea that evasion of a biologically-derived gas, DMS, might affect the formation of CCN was on its promulgation a quarter of a century ago a hugely powerful one, at that time the complexity of particle formation in the atmosphere was not well appreciated. Now we are aware that several gases and possibly pre-existing particles are likely to be involved in CCN production and so the original concept needs reframing in the light of the complexity of our present understanding (see, for example, Sect. 1.3.4). This is still a very active research area, as is clear from several earlier sections, and we have consciously not tried to come to an overall conclusion to the ongoing discussion.

We have tried to show how air-sea gas fluxes are driven by processes in both the atmosphere and oceans and thereby affect and are affected by the properties of both. In the future, research effort will need to be put into understanding processes at work in both the surface oceans and marine troposphere because the sophisticated modelling discussed in almost every part of the chapter is highly dependent on such understanding.

Another thread that comes out very strongly from our synthesis is the importance of field measurements. The chemical nature and low concentrations of many of the trace gases makes them difficult to measure accurately. Further, in terms of remoteness and difficulty of sampling, the sea surface and overlying atmosphere are some of the most difficult places from which to obtain reliable measurements. This is particularly acute in terms of remote locations, such as the Arctic and Southern Oceans, and the lack of data covering all seasons is often due to the difficulty of fieldwork in winter and under other adverse conditions. Maybe in the future, satellite and other autonomous measurement platforms will lessen the dependence on ships and aircraft and so, increase both the temporal and spatial coverage of our measurements. This requires investment in miniaturising, automating, and development of new sensitive and stable sensors that can be deployed unattended. If for no other reason, environmental measurements are vital for calibrating and validating modelling studies. Other aspects that need attention are the need for international inter-calibration of measurement techniques and the importance of developing the reliability of techniques for direct flux measurements and their application to a greater range of gases than is possible at present.

In the chapter authors have, where possible, indicated how air-sea trace gas fluxes may change under expected future global change, although this is necessarily quite speculative. One area that has come into prominence recently is ocean acidification and there has been some study of how lowered future ocean pH may alter trace gas concentrations in seawater. For example, mesocosm experiments have shown significant downward changes in DMS and DMSP concentrations under doubled CO_2 (Avgoustidi et al. 2012; Hopkins et al. 2010), and for DMS only (Archer et al. 2012). However, this is not always the case and increases in concentrations with lowered pH have been reported by Kim et al. (2010) for both DMS and DMSP and by Archer et al. (2012) for DMSP only. Hopkins et al. 2010 found substantial downward changes in concentrations of organo-halogen gases in a mesocosm experiment with seawater under elevated CO_2.

Another aspect of global change that may affect air-sea trace gas fluxes is the role of iron and other nutrients depositing from the atmosphere into the ocean (discussed in detail in Chap. 4) or coming into surface waters from below. The 13 in situ iron fertilisation experiments conducted over the last 20 years give some indications of how changing iron amounts from dust or other sources might affect ocean biogeochemistry and so lead to changed air-sea fluxes of trace gases (see Williamson et al. 2012 for a general overview of these experiments and Law 2008 for a review of the effect on trace gases). The best studied gas in this context is DMS where in many of the experiments concentrations of the gas (and its precursor DMSP) have been found to increase substantially (Turner et al. 1996, 2004). But this is not always the case since in the sub-Arctic Pacific fertilisation with iron was found to decrease or show no change in DMS (Levasseur et al. 2006; Nagao et al. 2009). For a variety of other trace gases including organo-halogens, isoprene and NMHCs the situation is similarly mixed (Wingenter et al. 2004; Liss et al. 2005; Hashimoto et al. 2009; Kato et al. 2009; Nagao et al. 2009). This means that predicting the impact of changes in dust inputs on production of these other gases and hence on, for example, atmospheric oxidising capacity and climate is extremely difficult to predict given present understanding of the complex interactions involved. As noted in Liss et al. 2005, this complexity would need to be taken into account in any proposal to ameliorate global climate by fertilising the oceans with iron.

Throughout the chapter we have endeavoured to show how gas transfer between ocean and atmosphere fits both conceptually and quantitatively into the global cycling of elements. In the future, further attention will need to be paid to how air-sea exchange contributes to the concept of Earth System Science, particularly for forecasting how environmental variables (e.g. wind, pH, dust deposition) are likely to alter in the future under the substantial global changes predicted for this century.

Open Access This chapter is distributed under the terms of the Creative Commons Attribution Noncommercial License, which permits any noncommercial use, distribution, and reproduction in any medium, provided the original author(s) and source are credited.

References

Abrahamsson KA, Ekdahl J, Collen M, Pedersen M (1995) Marine algae: a source of trichloroethylene and perchloroethylene. Limnol Oceanogr 40:1321–1326

Aitken J (1895) On the number of dust particles in the atmosphere of certain places in Great Britain and on the continent, with remarks on the relation between the amount of dust and meteorological phenomena. Trans Roy Soc Edinb 37:17–49

Alexander B, Park RJ, Jacob DJ, Li QB, Yantosca RM, Savarino J, Lee CCW, Thiemens MH (2005) Sulfate formation in sea-salt aerosols: constraints from oxygen isotopes. J Geophys Res 110, D10307. doi:10.1029/2004JD005659

Alexander B, Allman DJ, Amos HM, Fairlie TD, Dachs J, Hegg DA, Sletten RS (2012) Isotopic constraints on the formation pathways of sulfate aerosol in the marine boundary layer of the subtropical northeast Atlantic Ocean. J Geophys Res 117. doi:10,1029/2011JD016773

Alicke B, Hebestreit K, Stutz J, Platt U (1999) Iodine oxide in the marine boundary layer. Nature 397:572–573

Allan BJ, McFiggans G, Plane JMC, Coe H (2000) Observations of iodine monoxide in the remote marine boundary layer. J Geophys Res 105:14363–14369

Allen G et al (2011) South East Pacific atmospheric composition and variability sampled along 20 degrees S during VOCALS-Rex. Atmos Chem Phys 11:5237–5262

Amachi S, Kamagata Y, Kanagawa T, Muramatsu Y (2001) Bacteria mediate methylation of iodine in marine and terrestrial environments. Appl Environ Microbiol 67:2718–2722

Amouroux D, Liss PS, Tessier E, Hamren-Larsson M, Donard OFX (2001) Role of oceans as biogenic sources of selenium. Earth Planet Sci Lett 5878:1–7

References

Anbar AD, Yung YL, Chavez FP (1996) Methyl bromide: ocean sources, ocean sinks, and climate sensitivity. Glob Biogeochem Cycle 10:175–190

Andreae MO (1990) Ocean–atmosphere interaction in the global biogeochemical sulfur cycle. Mar Chem 30:1–29

Andreae MO, Ferek RJ (1992) Photochemical production of carbonyl sulfide in seawater and its emission to the atmosphere. Global Biogeochem Cycle 6:173–175

Andreae TW, Cutter GA, Husain RN, Radford-Knoery J, Andreae MO (1991) Hydrogen-sulfide and radon in an over the western North-Atlantic Ocean. J Geophys Res 96:18753–18760

Andreae MO, Atlas E, Harris GW, Helas G, deKock A, Koppmann R, Maenhaut W, Mano S, Pollock WH, Rudolph J, Scharle D, Schebeske G, Welling M (1996) Methyl halide emissions from savanna fires in Southern Africa. J Geophys Res 101:23603–23613

Archer SD, Goldson LE, Liddicoat MI, Cummings DG, Nightingale PF (2007) Marked seasonality in the concentrations and sea-to-air flux of volatile iodocarbon compounds in the western English Channel. J Geophys Res 112, C08009. doi:10.1029/2006JC003963

Archer SD, Kimmance SA, Stephens JA, Hopkins FE, Bellerby RGJ, Schulz KG, Piontek J, Engel A (2012) Contrasting responses of DMS and DMSP to ocean acidification in Arctic waters. Biogeosciences 9:12803–12843

Arnold SR, Spracklen DV, Williams J, Yassaa N, Sciare J, Bonsang B, Gros V, Peeken I, Lewis AC, Alvain S, Moulin C (2009) Evaluation of the global oceanic isoprene source and its impacts on marine carbon aerosol. Atmos Chem Phys 9:1253–1262

Arnold SR, Spracklen DV, Gebhardt S, Custer T, Williams J, Peeken I, Alvain S (2010) Relationships between atmospheric organic compounds and air-mass exposure to marine biology. Environ Chem 7:232–241

Atkinson R, Carter WPL, Winer AM (1982) Kinetics of the gas-phase reactions of OH radicals with alkyl nitrates at $299 \pm 2K$. Int J Chem Kinet 14:919–926

Atkinson R, Baulch DL, Cox RA, Crowley JN, Hampson RF, Hynes RG, Jenkin ME, Rossi MJ, Troe J (2007) Evaluated kinetic and photochemical data for atmospheric chemistry: vol. III gas phase reactions of inorganic halogens. Atmos Chem Phys 7:981–1191

Atlas EL, Ridley BA, Hubler G, Walega JG, Carroll MA, Montzka DD, Huebert BJ, Norton RB, Grahek FE, Schauffler S (1992) Partitioning and budget of NOy species during the Mauna Loa observatory photochemistry experiment. J Geophys Res 97:10449–10462

Atlas E, Pollock W, Greenberg J, Heidt L, Thompson AM (1993) Alkyl nitrates, nonmethane hydrocarbons and halocarbon gases over the equatorial Pacific-Ocean during Saga-3. J Geophys Res Atmos 98:16933–16947

Avgoustidi V, Nightingale PD, Joint I, Steinke M, Turner SM, Hopkins FE, Liss PS (2012) Decreased marine dimethyl sulfide production under elevated CO_2 levels in mesocosm and in vitro studies. Environ Chem 9:399–404

Baker AR (2005) Marine aerosol iodine chemistry: the importance of soluble organic iodine. Environ Chem 2:295–298

Baker JM, Reeves CE, Nightingale PD, Penkett SA, Gibb SW, Hatton AD (1999) Biological production of methyl bromide in the coastal waters of the North Sea and open ocean of the northeast Atlantic. Mar Chem 64:267–285

Baker AR, Thompson D, Campos MLAM, Perry SJ, Jickells TD (2000) Iodine concentration and availability in atmospheric aerosol. Atmos Environ 34:4331–4336

Bale C, Ingham T, Commane R, Heard D, Bloss W (2008) Novel measurements of atmospheric iodine species by resonance fluorescence. J Atmos Chem 60:51–70

Ballschmiter K (2002) A marine source for alkyl nitrates. Science 197:1127–1128

Bandy AR, Scott DL, Blomquist BW, Chen SM, Thornton DC (1992) Low yields of SO_2 from dimethyl sulfide oxidation in the marine boundary layer. Geophys Res Lett 19:1125–1127

Bandy AR, Thornton DC, Blomquist BW, Chen S, Wade TP, Ianni JC, Mitchell GM, Nadler W (1996) Chemistry of dimethyl sulfide in the equatorial Pacific atmosphere. Geophys Res Lett 23:741–744

Bange HW, Williams J (2000) New directions: acetonitrile in atmospheric and biogeochemical cycles. Atmos Environ 34:4959–4960

Bariteau L, Helmig D, Fairall CW, Hare JE, Hueber J, Lang EK (2010) Determination of oceanic ozone deposition by shipborne eddy covariance flux measurements. Atmos Meas Tech 3:441–455

Barnes I, Bastian V, Becker KH, Overath RD (1991) Kinetic studies of the reactions of IO, BrO and ClO with DMS. Int J Chem Kinet 23:579–591

Barnes I, Hjorth J, Mihalopoulos N (2006) Dimethyl sulfide and dimethyl sulfoxide and their oxidation in the atmosphere. Chem Rev 106:940

Bartnicki EW, Castro CE (1994) Biodehalogenation – rapid xxidative-metabolism of monohalomethanes and polyhalomethanes by methylosinus-trichosporium Ob-3b. Environ Toxicol Chem 13:241–245

Bates TS, Kiene RP, Wolfe GV, Matrai PA, Chavez FP, Buck KR, Blomquist BW, Cuhel RL (1994) The cycling of sulfur in surface seawater of the Northeast Pacific. J Geophys Res Ocean 99:7835–7843

Bates TS, Kelly KC, Johnson JE, Gammon RH (1995) Regional and seasonal variations in the flux of oceanic carbon monoxide to the atmosphere. J Geophys Res 100:23093–23101

Beale R, Liss PS, Nightingale PD (2010) First oceanic measurements of ethanol and propanol. Geophys Res Lett 37, L24607. doi:10.1029/2010GL045534

Bell N, Hsu L, Jacob DJ, Schultz MG, Blake DR, Butler JH, King DB, Lobert JM, Maier-Reimer E (2002) Methyl iodide: atmospheric budget and use as a tracer of marine convection in global models. J Geophys Res 107(D17):4340. doi:10.1029/2001JD001151

Bell TG, Johnson MT, Jickells TD, Liss PS (2007) Ammonia/ammonium dissociation coefficient in seawater: a significant numerical correction. Environ Chem 4:183–186. doi:10.1071/EN07032

Belviso S, Caniaux G (2009) A new assessment in North Atlantic waters of the relationship between DMS concentration and the upper mixed layer solar radiation dose. Glob Biogeochem Cycle 23. doi:Gb1014 10.1029/2008gb003382

Benson DR, Yu JH, Markovich A, Lee SH (2011) Ternary homogeneous nucleation of H_2SO_4, NH_3, and H_2O under conditions relevant to the lower troposphere. Atmos Chem Phys 11:4755–4766

Beyersdorf AJ, Blake DR, Swanson A, Meinardi S, Rowland FS, Davis S (2010) Abundances and variability of tropsopheric volatile organic compounds at the South Pole and other Antarctic locations. Atmos Environ 44:4565–4574

Blake NJ et al (1999) Aircraft measurements of the latitudinal, vertical, and seasonal variations of NMHCs, methyl nitrate, methyl halides, and DMS during the First Aerosol Characterization Experiment (ACE 1). J Geophys Res 104:21803–21817

Blake NJ, Blake DR, Swanson AL, Atlas E, Flocke F, Rowland FS (2003) Latitudinal, vertical, and seasonal variations of C1–C4 alkyl nitrates in the troposphere over the Pacific Ocean during PEM-Tropics A and B: Oceanic and continental sources. J Geophys Res 108:8242. doi:10.1029/2001J D001444

Blando JD, Turpin BJ (2000) Secondary organic aerosol formation in cloud and fog droplets: a literature evaluation of plausibility. Atmos Environ 34:1623–1632

Bloss WJ, Lee JD, Johnson GP, Sommariva R, Heard DE, Saiz-Lopez A, Plane JMC, McFiggans G, Coe H, Flynn M, Williams P, Rickard AR, Fleming ZL (2005) Impact of halogen monoxide chemistry upon boundary layer OH and HO_2 concentrations at a coastal site. Geophys Res Lett 32, L06814

Blough NV (1997) Photochemistry in the sea-surface microlayer. In: Duce R, Liss PS (eds) The sea-surface and global change. Cambridge University Press, Cambridge, UK

Blough NV, Zafiriou OC (1985) Reactions of superoxide with nitric oxide to form peroxonitrate in alkaline aqueous solution. Inorg Chem 24:3502–3504

Bonsang B, Kanakidou M, Lambert G, Monfray P (1988) The marine source of C2–C6 aliphatic hydrocarbons. J Atmos Chem 6:3–20

Bonsang B, Polle C, Lambert G (1992) Evidence for marine production of isoprene. Geophys Res Lett 19:1129–1132

Bonsang B, Al Aarbaour A, Sciare J (2008) Diurnal variation of nonmethane hydrocarbons in the subantarctic atmosphere. Environ Chem 5:16–23. doi:10.1071/EN07018

Bonsang B, Gros V, Peeken I, Yassaa N, Bluhm K, Zoellner E, Sarda-Esteve R, Williams J (2010) Isoprene from phytoplankton monocultures: the relationship with chlorophyll-a, cell volume and carbon content. Environ Chem 7:554–563. doi:10.1071/EN09156

Bopp L, Aumont O, Belviso S, Monfray P (2003) Potential impact of climate change on marine dimethyl sulfide emissions. Tellus 55B:11–22

Bottenheim JW, Netcheva S, Morin S, Nghiem SV (2009) Ozone in the boundary layer air over the Arctic Ocean: measurements during the TARA transpolar drift 2006–2008. Atmos Chem Phys 9:4545–4557

Bouwman A, Lee D, Asman W, Dentener F, Van Der Hoek K, Olivier J (1997) A global high-resolution emission inventory for ammonia. Glob Biogeochem Cycle 11:561–587

Breider TJ, Chipperfield MP, Richards NAD, Carslaw KS, Mann GW, Spracklen DV (2010) Impact of BrO on dimethylsulfide in the remote marine boundary layer. Geophys Res Lett 37, L02807. doi:10.1029/2009GL040868

Broadgate W, Liss PS, Penkett SA (1997) Seasonal emissions of isoprene and other reactive hydrocarbon gases from the oceans. Geophys Res Lett 24:2675–2878

Brownell DK, Moore RW, Cullen JJ (2010) Production of methyl halides by *Prochlorococcus* and *Synechococcus*. Glob Biogeochem Cycle 24, GB2002. doi:10.1029/2009 GB003671

Brüchert V, Currie B, Peard K (2009) Hydrogen sulphide and methane emissions on the central Namibian shelf. Prog Oceanogr 83:169–179

Butler A, Walker JV (1993) Marine haloperoxidases. Chem Rev 93:1937–1944

Cameron-Smith P, Elliott S, Maltrud M, Erickson D, Wingenter O (2011) Changes in dimethyl sulfide oceanic distribution due to climate change. Geophys Res Lett 38. L07704. doi:10.1029/2011gl047069

Capone DG (2000) The marine microbial nitrogen cycle. In: Kirchman DL (ed) Microbial ecology of the oceans. Wiley, New York

Carpenter LJ, Liss PS (2000) On temperate sources of bromoform and other reactive organic bromine gases. J Geophys Res Atmos 105:20539–20547

Carpenter LJ, Sturges WT, Penkett SA, Liss PS, Alicke B, Hebestreit K, Platt U (1999) Short-lived alkyl iodides and bromides at Mace Head, Ireland: links to biogenic sources and halogen oxide production. J Geophys Res Atmos 104:1679–1689

Carpenter LJ, Malin G, Liss PS, Kupper FC (2000) Novel biogenic iodine-containing trihalomethanes and other short-lived halocarbons in the coastal East Atlantic. Glob Biogeochem Cycle 14:1191–1204

Carpenter LJ, Lewis AC, Hopkins JR, Read KA, Longley ID, Gallagher MW (2004) Uptake of methanol to the North Atlantic Ocean surface. Global Biogeochem Cycle 18, GB4027. doi:10.1029/2004GB002294

Chameides WL, Davis D (1980) Iodine: its possible role in tropospheric photochemistry. J Geophys Res 85:7383–7398

Chameides WL, Stelson AW (1992) Aqueous-phase chemical processes in deliquescent sea-salt aerosols: a mechanism that couples the atmospheric cycles of S and sea salt. J Geophys Res 97:20565–20580

Chang W, Heikes BG, Lee M (2004) Ozone deposition to the sea surface: chemical enhancement and wind speed dependence. Atmos Environ 38:1053–1059

Charlson RJ, Lovelock JE, Andreae MO, Warren SG (1987) Oceanic phytoplankton, atmospheric sulphur, cloud albedo and climate. Nature 326:655–661

Chuck AL, Turner SM, Liss PS (2002) Direct evidence for a marine source of alkyl nitrates. Science 297:1151–1154

Class TH, Ballschmiter K (1988) Chemistry of organic traces in air: sources and distribution of bromo- and bromochloromethanes in marine air and surface water of the Atlantic Ocean. J Atmos Chem 6:35–46

Coffman DJ, Hegg DA (1995) A preliminary study of the effect of ammonia on particle nucleation in the marine boundary layer. J Geophys Res Atmos 100:7147. doi:10.1029/94JD03253

Conley SA, Faloona I, Miller GH, Lenschow D, Blomquist B, Bandy A (2009) Closing the dimethyl sulfide budget in the tropical marine boundary layer during the Pacific atmospheric sulfur experiment. Atmos Chem Phys 9:8745–8756

Conrad R, Seiler W (1980) Photo-oxidative production and microbial consumption of carbon monoxide in seawater. FEMS Microbiol Lett 9:61–64

Conrad R, Seiler W (1986) Exchange of CO and H_2 between ocean and atmosphere. In: Buat-Menard P (ed) The role of air-sea exchange in geochemical cycling. Reidel, Dordrecht

Corbett JJ, Fischbeck PS, Pandis SN (1999) Global nitrogen and sulfur inventories for oceangoing ships. J Geophys Res 104:3457–3470

Cox ML, Sturrock GA, Fraser PJ, Siems S, Krummel PB, O'Doherty S (2003) Regional sources of methyl chloride, chloroform and dichloromethane identified from AGAGE observations at Cape Grim, Tasmania, 1998–2000. J Atmos Chem 45:79–99

Dacey JWH, Howse FA, Michaels AF, Wakeham SG (1998) Temporal variability of dimethylsulfide and dimethylsulfoniopropionate in the Sargasso Sea. Deep Sea Res Part I 45:2085–2104

Dahl EE (2005) Photochemical production of oceanic alkyl nitrates. Dissertation, University of California, Irvine, 178pp

Dahl EE, Saltzman ES (2008) Alkyl nitrate photochemical production rates in North Pacific seawater. Mar Chem 112:137–141

Dahl EE, Saltzman ES, de Bruyn WJ (2003) The aqueous phase yield of alkyl nitrates from ROO + NO: implications for photochemical production in seawater. Geophys Res Lett 30:1271

Dahl EE, Yvon-Lewis SA, Saltzman ES (2005) Saturation anomalies of alkyl nitrates in the tropical Pacific Ocean. Geophys Res Lett 32, L20817

Dahl EE, Yvon-Lewis SA, Saltzman ES (2007) Alkyl nitrate (C1–C3) depth profiles in the tropical Pacific Ocean. J Geophys Res 112, C01012

Dahl EE, Heiss EM, Murawski K (2012a) The effects of dissolved organic matter on alkyl nitrate production during GOMECC and laboratory studies. Mar Chem 142–144:11–17. doi:10.1016/j.marchem.2012.08.001

Dahl EE, Kellogg D, Escobar C (2012b) Are diatoms a source of oceanic alkyl nitrates?. SOLAS Open Science Conference, Cle Elum Washington, May 7–10

Day DA, Faloona I (2009) Carbon monoxide and chromophoric dissolved organic matter cycles in the shelf waters of the Northern California upwelling system. J Geophys Res Ocean 114, CO1006. doi:10.1029/2007JC004590

de Bruyn WJ, Clark PL, Takehara C (2011) Photochemical production of formaldehyde, acetaldehyde, and acetone from chromophoric dissolved organic matter in coastal waters. J Photoch Photobio A 226:16–22

de Gouw JA, Warneke C, Parrish DD, Holloway JS, Trainer M, Fehsenfeld FC (2003) Emission sources and ocean uptake of acetonitrile (CH_3CN) in the atmosphere. J Geophys Res 108 (D11):4329. doi:10.1029/2002JD002897

Dentener FJ, Crutzen PJ (1994) A three-dimensional model of the global ammonia cycle. J Atmos Chem 19:331–369. doi:10.1007/BF00694492

Dentener F et al (2006) Nitrogen and sulfur deposition on regional and global scales: a multimodel evaluation. Global Biogeochem Cycle 20, GB4003. doi:10.1029/2005GB002672

Derevianko GJ, Deutsch C, Hall A (2009) On the relationship between ocean DMS and solar radiation. Geophys Res Lett 36, L17606. doi:10.1029/2009GL039412

Dillon TJ, Tucceri ME, Sander R, Crowley JN (2008) LIF studies of iodine oxide chemistry Part 3. Reactions IO + NO_3 -> OIO + NO_2, I + NO_3 -> IO + NO_2, and $CH_2I + O_2$ -> (products): implications for the chemistry of the marine atmosphere at night. Phys Chem Chem Phys 10:1540–1554

Dimmer CH, Simmonds PG, Nickless G, Bassford MR (2001) Biogenic halomethanes from Irish peatland ecosystems. Atmos Environ 35:321–330

Dixon JL, Beale R, Nightingale PD (2011a) Microbial methanol uptake in the northeast Atlantic waters. ISME J 5:704–716

Dixon JL, Beale R, Nightingale PD (2011b) Rapid biological oxidation of methanol in the tropical Atlantic: significance as a microbial carbon source. Biogeosciences 8:2707–2716

Donahue NM, Prinn RG (1993) Non-methane hydrocarbon chemistry in the remote marine boundary layer. J Geophys Res 95:18387–18411

Duce RA, Woodhouse AH, Moyers JL (1967) Variation of ion ratios with size among particles in tropical oceanic air. Tellus 19:367–379

Dufour G, Szopa S, Hauglustaine DA, Boone CD, Rinsland CP, Bernath PF (2007) The influence of biogenic emissions on uppert-tropospheric methanol as revealed from space. Atmos Chem Phys 7:6119–6129

Duncan BN, Logan JA, Bey I, Megretskaia IA, Yantosca RM, Novelli PC, Jones NB, Rinsland CP (2007) Global budget of CO, 1988–1997: source estimates and validation with a global model. J Geophys Res Atmos 112, D22301. doi:10.1029/2007JD008459

Ehrhardt M, Weber RR (1991) Formation of low molecular weight carbonyl compounds by sensitized photochemical decomposition of aliphatic hydrocarbons in seawater. Fresenius J Anal Chem 339:772–776

Ekdahl A, Pedersen M, Abrahamsson K (1998) A study of the diurnal variation of biogenic volatile halocarbons. Mar Chem 63:1–8

Elias T, Szopa S, Zahn A, Schuck T, Brenninkmeijer C, Sprung D, Slemr F (2011) Acetone variability in the upper troposphere: analysis of CARIBIC observations and LMDz-INCA chemistry-climate model simulations. Atmos Chem Phys 11:8053–8074

Elliott S (1989) The effect of hydrogen peroxide on the alkaline hydrolysis of carbon disulfide. Environ Sci Technol 24:264–267

Elliott S, Rowland FS (1993) Nucleophilic substitution rates and solubilities for methyl halides in seawater. Geophys Res Lett 20:1043–1046

Elliott S, Rowland FS (1995) Methyl halide hydrolysis rates in natural-waters. J Atmos Chem 20:229–236

Eyring V, Isaksen ISA, Berntsen T, Collins WJ, Corbett JJ, Endresen O, Grainger RG, Moldanova J, Schlager H, Stevenson DS (2010) Transport impacts on atmosphere and climate: shipping. Atmos Environ 44:4735–4771

Facchini MC (2008) Important source of marine secondary organic aerosol from biogenic amines. Environ Sci Tech 42:9116–9121. doi:10.1021/es8018385

Fairall CW, Helmig D, Ganzefeld L, Hare J (2007) Water-side turbulence enhancement of ozone deposition to the ocean. Atmos Chem Phys 7:443–451

Faloona I (2009) Sulfur processing in the marine atmospheric boundary layer: a review and critical assessment of modeling uncertainties. Atmos Environ 43:2841–2854

Fenical W (1982) Natural products chemistry in the marine environment. Science 215:923–928

Fichot CG, Miller WL (2011) An approach to quantify depth-resolved marine photochemical fluxes using remote sensing:

application to carbon monoxide (CO) photoproduction. Rem Sens Environ 114:1363–1377

Finley BD, Saltzman ES (2008) Observations of Cl_2, Br_2, and I_2 in coastal marine air. J Geophys Res 113, D21301

Fischer R, Weller R, Jacobi HW, Ballschmiter K (2002) Levels and pattern of volatile organic nitrates and halocarbons in the air at Neumayer Station (70°S) Antarctica. Chemosphere 48:981–992

Flanagan RJ, Geever M, O'Dowd CD (2005) Direct measurements of new-particle fluxes in the coastal environment. Environ Chem 2:256–259

Flock OR, Andreae MO (1996) Photochemical and non-photochemical formation and destruction of carbonyl sulfide and methyl mercaptan in ocean waters. Mar Chem 54:11–26

Fuse H, Inoue H, Murakami K, Takimura O, Yamaoko Y (2003) Production of free and organic iodine by *Roseovarius* spp. FEMS Microbiol Lett 229:189–194

Gabric A, Gregg W, Najjar R, Erickson D, Matrai P (2001) Modeling the biogeochemical cycle of dimethylsulfide in the upper ocean: a review. Chemosphere 3:377–392

Gabric AJ, Simo R, Cropp RA, Hirst AC, Dachs J (2004) Modeling estimates of the global emission of dimethylsulfide under enhanced greenhouse conditions. Global Biogeochem Cycle 18, GB2014. doi:10.1029/2003GB002183

Galloway JN (2004) Nitrogen cycles: past, present and future. Biogeochemistry 70:153–226. doi:10.1007/s10533-004-0370-0

Gantt B, Meskhidze N, Kamykowski D (2009) A physically based quantification of marine isoprene and primary organic aerosol emissions. Atmos Chem Phys 9:4915–4927

Gantt B, Meskhidze N, Carlton AG (2010) The contribution of marine organics to the air quality of the western United States. Atmos Chem Phys 10:7415–7423

Ganzeveld L, Helmig D, Fairall CW, Hare J, Pozzer A (2009) Atmosphere–ocean ozone exchange: a global modeling study of biogeochemical, atmospheric, and waterside turbulence dependencies. Glob Biogeochem Cycle 23, GB4021. doi:10.1029/2008GB003301

Ge X, Wexler AS, Clegg SL (2011) Atmospheric amines – part I. A review. Atmos Environ 45:524–546. doi:16/j.atmosenv.2010.10.012

Geen CE (1992) Selected marine sources and sinks of bromoform and other low molecular weight organobromines. Dalhousie University, Halifax, 109pp

Gibb SW, Mantoura RFC, Liss PS (1999) Ocean–atmosphere exchange and atmospheric speciation of ammonia and methylamines in the region of the NW Arabian Sea. Glob Biogeochem Cycle 13:161–178

Gilfedder BS, Lai S, Petri M, Biester H, Hoffmann T (2008) Iodine speciation in rain, snow and aerosols. Atmos Chem Phys 8:6069–6084

Gilfedder BS, Chance R, Dettmann U, Lai SC, Baker AR (2010) Determination of total and non-water soluble iodine in atmospheric aerosols by thermal extraction and spectrometric detection (TESI). Anal Bioanal Chem 398:519–526

Gómez Martín JC, Ashworth SH, Mahajan AS, Plane JMC (2009) Photochemistry of OIO: laboratory study and atmospheric implications. Geophys Res Lett 36, L09802. doi:09810.01029/02009GL037642

Gómez Martín JC, Blahins J, Gross U, Ingham T, Goddard A, Mahajan AS, Ubelis A, Saiz-López A (2011) In situ detection of atomic and molecular iodine using Resonance and Off-Resonance Fluorescence by Lamp Excitation: ROFLEX. Atmos Meas Tech 4:29–45

Goodwin KD, Lidstrom ME, Oremland RS (1997) Marine bacterial degradation of brominated methanes. Environ Sci Tech 31:3188–3192

Gravestock T, Blitz MA, Heard DE (2005) Kinetics study of the reaction of iodine monoxide radicals with dimethyl sulfide. Phys Chem Chem Phys 7:2173–2181

Gray BA, Wang Y, Gu D, Bandy A, Mauldin L, Clarke A, Alexander B, Davis DD (2010) Sources, transport, and sinks of SO_2 over the equatorial Pacific during the Pacific atmospheric sulfur experiment. J Atmos Chem. doi:10.1007/s10874-010-9177-7

Gros V, Peeken I, Bluhm K, Zollner E, Sarda-Esteve R, Bonsang B (2009) Carbon monoxide emissions by phytoplankton: evidence from laboratory experiments. Environ Chem 6:369–379

Groszko W, Moore RM (1998) Ocean–atmosphere exchange of methyl bromide: NW Atlantic and Pacific Ocean studies. J Geophys Res Atmos 103:16737–16741

Grzybowski W (2003) Are data on light-induced ammonium release from dissolved organic matter consistent? Chemosphere 52:933–936

Gschwend PM, Macfarlane JK, Newman KA (1985) Volatile halogenated organic-compounds released to seawater from temperate marine macroalgae. Science 227:1033–1035

Guenther A, Hewitt N, Erickson D, Fall R, Geron C, Graedel T, Harley P, Klinger L, Lerdau M, Mckay WA, Pierce T, Scholes B, Steinbrecher R, Tallamraju R, Taylor J, Zimmerman P (1995) A global model of natural volatile organic compound emissions. J Geophys Res 100:8873–8892

Harrison JJ, Allen NDC, Bernath PF (2011a) Infrared absorption cross sections for acetone (propanone) in the 3 μm region. J Quant Spectrosc Rad 112:53–58

Harrison JJ, Humpage N, Allen NDC, Waterfall AM, Bernath PF, Remedios JJ (2011b) Mid-Infrared absorption cross sections for acetone (propanone). J Quant Spectrosc Rad 112:457–464

Hashimoto S et al (2009) Production and air-sea flux of halomethanes in the western subarctic Pacific in relation to phytoplankton pigment concentrations during the iron fertilization experiment (SEEDS II). Deep Sea Res II 56:2928–2935

Heikes BG et al (2002) Atmospheric methanol budget and ocean implication. Glob Biogeochem Cycle 16:1133. doi:10.1029/2002GB001895

Helmig D, Lang EK, Bariteau L, Boylan P, Fairall CW, Ganzeveld L, Hare JE, Hueber J, Pallandt M (2012) Atmosphere–ocean ozone fluxes during the TexAQS 2006, STRATUS 2006, GOMECC 2007, GasEX 2008, and AMMA 2008 cruises. J Geophys Res 117, D04305. doi:10.1029/2011JD015955

Hense I, Quack B (2009) Modelling the vertical distribution of bromoform in the upper water column of the tropical Atlantic Ocean. Biogeoscience 6:535–544

Herndl GJ, Mulleriklas G, Frick J (1993) Major role of ultraviolet-B in controlling bacterioplankton growth in the surface-layer of the ocean. Nature 361:717–719

Holland EA, Dentener FJ, Braswell BH, Sulzman JM (1999) Contemporary and pre-industrial global reactive nitrogen budgets. Biogeochemistry 46:7–43. doi:10.1007/BF01007572

Hopkins FE, Turner SM, Nightingale PD, Steinke M, Bakker D, Liss PS (2010) Ocean acidification and marine trace gas emissions. Proc Natl Acad Sci USA 107:760–765

Howard EC, Henriksen JR, Buchan A, Reisch CR, Buergmann H, Welsh R, Ye WY, Gonzalez JM, Mace K, Joye SB, Kiene RP, Whitman WB, Moran MA (2006) Bacterial taxa that limit sulfur flux from the ocean. Science 314:649–652

Hu L, Yvon-Lewis SA, Liu Y, Salisbury JE, O'Hern JE (2010) Coastal emissions of methyl bromide and methyl chloride along the eastern Gulf of Mexico and the east coast of the United States. Glob Biogeochem Cycle 24, GB1007. doi:10.1029/2009GB003514

Hudson ED, Ariya PA, Gelinas Y (2011) A method for the simultaneous quantification of 23 C_1–C_9 trace aldehydes and ketones in seawater. Environ Chem 8:441–449

Hughes C, Chuck AL, Turner SM, Liss PS (2008a) Methyl and ethyl nitrate saturation anomalies in the Southern Ocean (36–65°S, 30–70°W). Environ Chem 5:11–15

Hughes C, Chuck AL, Rossetti H, Mann PJ, Turner SM, Clarke A, Chance R, Liss PS (2009) Seasonal cycle of seawater bromoform and dibromomethane concentrations in a coastal bay on the western Antarctic Peninsula. Glob Biogeochem Cycle 23:2024. doi:10.1029/2008GB003268

Hughes C, Malin G, Turley CM, Keely BJ, Nightingale PD, Liss PS (2008b) The production of volatile iodocarbons by biogenic marine aggregates. Limnol Oceanogr 53:867–872

Hughes C, Kettle AJ, Unazi GA, Weston K, Jones MR, Johnson MT (2010) Seasonal variations in the concentrations of methyl and ethyl nitrate in a shallow freshwater lake. Limnol Oceanogr 55:305–314

Hughes C, Franklin D, Malin G (2011) Iodomethane production by two important marine cyanobacteria; *Prochlorococcus marinus* (CCMP 2389) and *Synechococcus* sp. (2370). Mar Chem 125:19–25

Ingham T, Bauer D, Sander R, Crutzen PJ, Crowley JN (1999) Kinetics and products of the reactions BrO + DMS and Br + DMS at 298 k. J Phys Chem A 103:7199–7209

IPCC, Climate Change (2007) The physical science basis. Contribution of working group I to the fourth assessment, report of the intergovernmental panel on climate change, Cambridge University Press, Cambridge,UK/New York

Jacob DJ, Field BD, Jin EM, Bey I, Li QB, Logan JA, Yantosca RM, Singh HB (2002) Atmospheric budget of acetone. J Geophys Res 107(D10). doi:10.1029/2001JD000694

Jacob DJ, Field BD, Li QB, Blake DR, de Gouw J, Warneke C, Hansel A, Wisthaler A, Singh HB, Genther A (2005) Global budget of methanol: constraints from atmospheric observations. J Geophys Res 110, D08303. doi:10.1029/2004JD005172

Jickells TD, Kelly SD, Baker AR, Biswas K, Dennis PF, Spokes LJ, Witt M, Yeatman SG (2003) Isotopic evidence for a marine ammonia source. Geophys Res Lett 30. doi:10.1029/2002GL016728

Jimenez JL, Bahreini R, Cocker DR III, Zhuang H, Varutbangkul V, Flagan RC, Seinfeld JH, O'Dowd CD, Hoffmann T (2003) New particle formation from photooxidation of diiodomethane (CH2I2). J Geophys Res 108:4733. doi:10.1029/2003JD004249

Johnson MT (2010) A numerical scheme to calculate temperature and salinity dependent air-water transfer velocities for any gas. Ocean Sci 6:913–932. doi:10.5194/os-6-913-2010

Johnson JE, Bates TS (1996) Sources and sinks of carbon monoxide in the mixed layer of the tropical South Pacific Ocean. Glob Biochem Cycle 10:347–359

Johnson MT, Bell TG (2008) Coupling between dimethylsupfide emissions and the ocean–atmosphere exchange of ammonia. Environ Chem 5:259–267. doi:10.1071/EN08030

Johnson MT, Sanders R, Avgoustidi V, Lucas MI, Brown L, Hansell DA, Moore CM, Gibb SW, Liss PS, Jickells TD (2007) Ammonium accumulation during a silicate-limited diatom bloom indicates the potential for ammonia emission events. Mar Chem 106:63–75. doi:10.1016/j.marchem.2006.09.006

Johnson MT, Liss PS, Bell TG, Lesworth TJ, Baker AR, Hind AJ, Jickells TD, Biswas KF, Woodward EMS, Gibb SW (2008) Field observations of the ocean–atmosphere exchange of ammonia: fundamental importance of temperature as revealed by a comparison of high and low latitudes. Glob Biogeochem 22, GB1019. doi:10.1029/2007GB003039

Johnson MT, Hughes C, Bell TG, Liss PS (2011) A Rumsfeldian analysis of uncertainty in air-sea gas exchange. In: Komori S, McGillis W, Kurose R (eds) Gas transfer at water surfaces 2010. Kyoto University Press, Kyoto

Jones CE, Carpenter LJ (2005) Solar photolysis of CH_2I_2, CH_2ClI and CH_2BrI in water, saltwater and seawater. Environ Sci Technol 39:6130–6138

Jones AE, Weller R, Minikin A, Wolff EW, Sturges WT, Mcintyre HP, Leonard SR, Schrems O, Bauguitte S (1999) Oxidized nitrogen chemistry and speciation in the Antarctic troposphere. J Geophys Res 104:21355–21366

Jones CE, Hornsby KE, Sommariva R, Dunk RM, von Glasow R, McFiggans G, Carpenter LJ (2010) Quantifying the contribution of marine organic gases to atmospheric iodine. Geophys Res Lett 37, L18804. doi:10.1029/2010GL043990

Joseph DM, Ashworth SH, Plane JMC (2007) On the photochemistry of $IONO_2$: absorption cross section (240–370 nm) and photolysis product yields at 248 nm. Phys Chem Chem Phys 9:5599–5607

Kaltsoyannis N, Plane JMC (2008) Quantum chemical calculations on a selection of iodine-containing species (IO, OIO, INO_3, $(IO)_2$, I_2O_3, I_2O_4 and I_2O_5) of importance in the atmosphere. Phys Chem Chem Phys 10:1723–1733

Karlsson A, Auer N, Schulz-Bull D, Abrahamsson K (2008) Cyanobacterial blooms in the Baltic – a source of halocarbons. Mar Chem 110:129–139

Kato S, Watiri M, Nagao I, Uematsu M, Kajii Y (2009) Atmospheric trace gas measurements during SEEDS-II over the northwesternPacific. Deep Sea Res Part II: Topical Stud Oceanography 56:2918–2927. doi:http://dx.doi.org/10.1016/j.dsr.2009.07.002

Keene WC, Pszenny AAP (2004) Comment on: Laskin, et al., reactions at interfaces as a source of sulfate formation in seasalt particles. Science 303:628a–628b

Keppler F, Eiden R, Niedan V, Pracht J, Scholer HF (2000) Halocarbons produced by natural oxidation processes during degradation of organic matter. Nature 403:298–301

Kettle AJ (2005) Diurnal cycling of carbon monoxide (CO) in the upper ocean near Bermuda. Ocean Model 8:337–367

Kettle AJ, Andreae MO (2000) Flux of dimethylsulfide from the oceans: a comparison of updated data sets and flux models. J Geophys Res 105:26793–26808

Kettle AJ, Andreae MO, Amouroux D, Andreae TW, Bates TS, Berresheim H, Bingemer H, Boniforti R, Curran MAJ, DiTullio GR, Helas G, Jones GB, Keller MD, Kiene RP, Leck C, Levasseur M, Malin G, Maspero M, Matrai P, McTaggart AR, Mihalopoulos N, Nguyen BC, Novo A, Putaud JP, Rapsomanikis S, Roberts G, Schebeske G, Sharma S, Simo R, Staubes R, Turner S, Uher G (1999) A global database of sea surface dimethylsulfide (DMS) measurements and a procedure to predict sea surface DMS as a function of latitude, longitude, and month. Glob Biogeochem Cycle 13:399–444

Kettle AJ, Rhee TS, von Hobe M, Poulton A, Aiken J, Andreae MO (2001) Assessing the flux of different volatile sulfur gases from the ocean to the atmosphere. J Geophys Res 106:12193–12209

Kettle AJ, Kuhn U, von Hobe M, Kesselmeier J, Liss PS, Andreae MO (2002) Comparing forward and inverse models to estimate the seasonal variation of hemisphere-integrated fluxes of carbonyl sulfide. Atmos Chem Phys 2:343–361

Khalil A, Rasmussen RA, Hoyt SD (1983) Atmospheric chloroform ($CHCl_3$): ocean-air exchange and global mass balance. Tellus 35B:226–274

Kieber RJ, Mopper K (1990) Determination of picomolar concentrations of carbonyl compounds in natural waters, including seawater, by liquid chromatography. Environ Sci Technol 24:1477–1481

Kieber RJ, Zhou X, Mopper K (1990) Formation of carbonyl compounds from UV induced photodegradation of humic substances in nature waters: fate of riverine carbon in the sea. Limnol Oceanogr 35:1503–1515

Kiene RP (1996) Production of methanethiol from dimethylsulfoniopropionate in marine surface waters. Mar Chem 54:69–83

Kiene RP, Linn LJ, Bruton JA (2000) New and important roles for DMSP in marine microbial communities. J Sea Res 43:209–224

Kim D, Yamaguchi K, Oda T (2006) Nitric oxide synthase-like enzyme mediated nitric oxide generation in harmful red tide phytoplankton *Chattonella marina*. J Plankton Res 28:613–620

Kim J-M, Lee K, Yang EJ, Shin K, Noh JH, Park K-T, Hyun B, Jeong H-J, Kim J-K, Kim KY, Kim M, Kim H-C, Jang P-G, Jang M-C (2010) Enhanced production of oceanic dimethylsulfide resulting from CO_2-induced grazing activity in a high CO_2 world. Environ Sci Technol 44:8140. doi:10.1021/ES102028K

King GM (2001) Aspects of carbon monoxide production and oxidation by marine macroalgae. Mar Ecol Prog Ser 224:69–75

King DB, Butler JH, Yvon-Lewis SA, Cotton SA (2002) Predicting oceanic methyl bromide saturation from SST. Geophys Res Lett 29:2199. doi:10.1029/2002GL016091

Kirkby J et al (2011) Role of sulphuric acid, ammonia and galactic cosmic rays in atmospheric aerosol nucleation. Nature 476:429–433. doi:10.1038/nature10343

Kirst GO, Thiel C, Wolff H, Nothnagel J, Wanzek M, Ulmke R (1991) Dimethylsulfoniopropionate (DMSP) in ice-algae and its possible biological role. Mar Chem 35:381–388

Kitidis V, Uher G, Upstill-Goddard RC, Mantoura RFC, Spyres G, Woodward EMS (2006) Photochemical production of ammonium in the oligotrophic Cyprus Gyre (Eastern Mediterranean). Biogeosciences 3:439–449. doi:10.5194/bg-3-439-2006

Klick S (1992) Seasonal-variations of biogenic and anthropogenic halocarbons in seawater from a coastal site. Limnol Oceanogr 37:1579–1585

Klick S, Abrahamsson K (1992) Biogenic volatile iodated hydrocarbons in the ocean. J Geophys Res 97 (C8):12683–12687. doi:10.1029/92JC00948

Kloster S, Six KD, Feichter J, Maier-Reimer E, Roeckner E, Wetzel P, Stier P, Esch M (2007) Response of dimethylsulfide (DMS) in the ocean and atmosphere to global warming. J Geophys Res Biogeosci 112 (G3), G03005. doi:10.1029/2006jg000224

Knepp TN, Bottenheim J, Carlsen M, Donohoue D, Friederich G, Matrai PA, Netcheva S, Perovich DK, Santini R, Shepson PB, Simpson W, Stehle T, Valentic T, Williams C, Wyss PJ (2010) Development on an autonomous sea ice tethered buoy for the study of ocean–atmosphere-sea ice-snow pack interaction: the O-buoy. Atmos Meas Tech 3:249–261

Koga S, Tanaka H (1993) Numerical study of the oxidation process of dimethylsulfide in the marine atmosphere. J Atmos Chem 17:201–228

Koga S, Tanaka H (1996) Simulation of seasonal variations of sulphur compounds in the remote marine atmosphere. J Atmos Chem 23:163–192

Korhonen P, Kulmala M, Laaksonen A, Viisanen Y, McGraw R, Seinfeld JH (1999) Ternary nucleation of H_2SO_4, NH_3, and H_2O in the atmosphere. J Geophys Res 104(D21):26349. doi:10.1029/1999JD900784

Korhonen H, Carslaw KS, Spracklen DV, Mann GW, Woodhouse MT (2008) Influence of oceanic dimethyl sulfide emissions on cloud condensation nuclei concentrations and seasonality over the remote Southern Hemisphere oceans: a global model study. J Geophys Res 113, D15204. doi:10.1029/2007JD009718

Kreidenweis SM, Seinfeld JH (1988) Nucleation of sulfuric acid-water solution particles: implications for the atmospheric chemistry of organosulphur species. Atmos Environ 22:283–296

Kulmala M, Kerminen V-M (2008) On the formation and growth of atmospheric nanoparticles. Atmos Res 90:132–150

Kurten T, Loukonen V, Vehkkamaki H, Kulmala M (2008) Amines are likely to enhance neutral and ion-induced sulfuric acid-water nucleation in the atmosphere more effectively than ammonia. Atmos Chem Phys 8:4095–4103

Küpper FC, Carpenter LJ, McFiggans GB, Palmer CJ, Waite TJ, Boneberg EM, Woitsch S, Weiller M, Abela R, Grolimund D, Potin P, Butler A, Luther GW III, Kurtén T, Loukonen V, Vehkamäki H, Kulmala M (2008) Amines are likely to enhance neutral and ion-induced sulfuric acid-water nucleation in the atmosphere more effectively than ammonia. Atmos Chem Phys 8:4095–4103. doi:10.5194/acp-8-4095-2008

Lamarque JF, Hess P, Emmons L, Buja L, Washington W, Granier C (2005) Tropospheric ozone evolution between 1890 and 1990. J Geophys Res Atmos 110, D08304. doi:10.1029/2004JD005537

Lana A, Bell TG, Simo R, Vallina SM, Ballabrera-Poy J, Kettle AJ, Dachs J, Bopp L, Saltzman ES, Stefels J, Johnson JE, Liss PS (2011) An updated climatology of surface dimethylsulfide concentrations and emission fluxes in the global ocean. Glob Biogeochem Cycle 25, GB1004. doi:10.1029/2010GB003850

Lary DJ (1996) Gas phase atmospheric bromine photochemistry. J Geophys Res 101:1505–1516

Lary DJ, Shallcross DE (2000) Centrol role of carbonyl compounds in atmospheric chemistry. J Geophys Res 105:19771–19778

Laskin A, Gaspar DJ, Wang W, Hunt SW, Cowin JP, Colson SD, Finlayson-Pitts BJ (2003) Reactions at interfaces as a source of sulfate formation in sea-salt particles. Science 301:340–344

Laturnus F, Wiencke C, Adams FC (1998) Influence of light conditions on the release of volatile halocarbons by Antarctic macroalgae. Mar Environ Res 45:285–294

Law CS (2008) Predicting and monitoring the effects of large-scale ocean iron fertilization on marine trace gas emissions. Mar Ecol Progr Ser 364:283–288

Lawler MJ, Finley BD, Keene WC, Pszenny AAP, Read KA, von Glasow R, Saltzman ES (2009) Pollution-enhanced reactive chlorine chemistry in the eastern tropical Atlantic boundary layer. Geophys Res Lett 36, L08810. doi:10.1029/2008GL036666

Lawler MJ, Sander R, Carpenter LJ, Lee JD, von Glasow R, Sommariva R, Saltzman ES (2011) HOCl and Cl_2 observations in marine air. Atmos Chem Phys 11:7617–7628

Le Clainche Y, Vezina A, Levasseur M, Cropp RA, Gunson JR, Vallina SM, Vogt M, Lancelot C, Allen JI, Archer SD, Bopp L, Deal C, Elliott S, Jin M, Malin G, Schoemann V, Simo R, Six KD, Stefels J (2010) A first appraisal of prognostic ocean DMS models and prospects for their use in climate models. Glob Biogeo Cycles 24, Gb3021. doi:10.1029/2009gb 003721

Leck C, Bigg EK (1999) Aerosol production over remote marine areas – a new route. Geophys Res Lett 26:3577–3581

Lelieveld J, Dentener FJ (2000) What controls tropospheric ozone? J Geophys Res 105:3531–3551

Lelieveld J, Van Aardenne J, Fisher H, De Reus M, Williams J, Winkler P (2004) Increasing ozone over the Atlantic Ocean. Science 304:1483–1487

Lenschow DH, Pearson R, Stankov BB (1982) Measurements of ozone vertical flux to ocean and forest. J Geophys Res Ocean Atmos 87(NC11):8833–8837

Levasseur M, Scarratt MG, Michaud S, Merzouk A, Wong CS, Arychuk M, Richardson W, Rivkin RB, Hale M, Wong E, Marchetti A, Kiyosawa H (2006) DMSP and DMS dynamics during a mesoscale iron fertilization experiment in the Northeast Pacific – Part 1: temporal and vertical distributions. Deep Sea Res Part II 53:2353–2369

Lewis ER, Schwartz SE (2004) Sea salt aerosol production, vol 152, Geophysical monograph. American Geophysical Union, Washington, DC

Lewis AC, Hopkins JR, Carpenter LJ, Stanton J, Read KA, Pilling MJ (2005) Sourcs and sinks of acetone methanol, and acetaldehyde in North Atlantic marine air. Atmos Chem Phys 5:1963–1974

Li Q, Jacob DJ, Bey I, Yantosca RM, Zhao Y, Kondo Y, Notholt J (2000) Atmospheric hydrogen cyanide (HCN): biomass burning source, oceanic sink? Geophys Res Lett 27:357–360

Li QB et al (2002) Transatlantic transport of pollution and its effects on surface ozone in Europe and North America. J Geophys Res Atmos 107:4166. doi:10.1029/2001JD001422

Liss PS, Slater PG (1974) Flux of gases across the air-sea interface. Nature 247:181–184. doi:10.1038/247181a0

Liss PS, Chuck A, Bakker D, Turner S (2005) Ocean fertilization with iron: effects on climate and air quality. Tellus 57B:269–271

Lobert JM, Butler JH, Montzka SA, Geller LS, Myers RC, Elkins JW (1995) A net sink for atmospheric CH_3Br in the East Pacific-Ocean. Science 267(5200):1002–1005

Loewus MW, Delwiche CC (1963) Carbon monoxide production by algae. Plant Physiol 38:371–374

Lovelock JE, Maggs RJ, Rasmussen RA (1972) Atmospheric dimethyl sulphide and the natural sulphur cycle. Nature 237:452–453

Luo G, Yu F (2010) A numerical evaluation of global oceanic emissions of pinene and isoprene. Atmos Chem Phys 10:2007–2015

Lupu A, Kaminski JW, Neary L, McConnell JC, Toyota K, Rinsland CP, Bernath PF, Walker KA, Boone CD, Nagaham Y, Suzuki K (2009) Hydrogen cyanide in the upper troposphere: GEM-AQ simulation and comparison with ACE-FTS observations. Atmos Chem Phys 9:4301–4313

Mabey W, Mill T (1978) Critical-review of hydrolysis of organic-compounds in water under environmental-conditions. J Phys Chem Ref Data 7:383–415

Mahajan AS, Oetjen H, Saiz-Lopez A, Lee JD, McFiggans GB, Plane JMC (2009) Reactive iodine species in a semi-polluted environment. Geophys Res Lett 36, L16803. doi:16810.11029/12009GL038018

Mahajan AS, Plane JMC, Oetjen H, Mendes L, Saunders RW, Saiz-Lopez A, Jones CE, Carpenter LJ, McFiggans GB (2010a) Measurement and modelling of tropospheric reactive halogen species over the tropical Atlantic Ocean. Atmos Chem Phys 10:4611–4624

Mahajan AS, Shaw M, Oetjen H, Hornsby KE, Carpenter LJ, Kalescheke L, Tian-Kunze X, Lee JD, Moller SJ, Edwards P, Commane R, Ingham T, Heard DE, Plane JMC (2010b) Evidence of reactive iodine chemistry in the Arctic boundary layer. J Geophys Res 115, D20303. doi:10.1029/2009JD013665

Mäkelä JM, Hoffmann T, Holzke C, Väkevä M, Suni T, Mattila T, Aalto PP, Tapper U, Kauppinen EI, O'Dowd CD (2002) Biogenic iodine emissions and identification of end-products in coastal ultrafine particles during nucleation bursts. J Geophys Res 107:8110

Manley SL (1994) The possible involvement of methylcobalamin in the production of methyl iodide. Mar Chem 46:361–369

Manley SL, Dastoor MN (1987) Methyl halide (CH_3X) production from the giant-kelp, Macrocystis, and estimates of global CH_3X production by kelp. Limnol Oceanogr 32:709–715

Manley SL, Dastoor MN (1988) Methyl iodide production by kelp and associated microbes. Mar Biol 88:447–482

Manley SL, de la Cuesta J (1997) Methyl iodide production from marine phytoplankton cultures. Limnol Oceanogr 42:142–147

Manley SL, Goodwin K, North WJ (1992) Laboratory production of bromoform, methylene bromide, and methyl-iodide by macroalgae and distribution in nearshore Southern California waters. Limnol Oceanogr 37:1652–1659

Mao H, Talbot R, Nielsen C, Sive B (2006) Controls on methanol and acetone in the marine and contintental atmospheres. Geophys Res Lett 33, L02803. doi:10.1029/2005GL024810

Marandino CA, de Bruyn WJ, Miller SD, Prather MJ, Saltzman ES (2005) Oceanic uptake and the global atmospheric acetone budget. Geophys Res Lett 32, L15806. doi:10.1029/2005GL023285

Marshall RA, Hamilton RTJ, Dring MJ, Harper DB (2000) The red alga Asparagopsis taxiformis/Falkenbergia hillebradiii – a possible source of trichloroethylene and perchloroethylene? Limnol Oceanogr 45:516–519

Martin M, Pohler D, Seitz K, Sinreich R, Platt U (2009) BrO measurements over the eastern North-Atlantic. Atmos Chem Phys 9:9545–9554

Martino M, Liss PS, Plane JMC (2005) The photolysis of dihalomethanes in surface seawater. Env Sci Technol 39:7097–7101. doi:10.1021/es048718s

Martino M, Mills GP, Woeltjen J, Liss PS (2009) A new source of volatile organoiodine compounds in surface seawater. Geophys Res Lett 36, L01609. doi:10.1029/2008GL036334

Martino M, Lézé B, Baker AR, Liss PS (2012) Chemical controls on ozone deposition to water. Geophys Res Lett 39, L05809. doi:10.1029/2011GL050282

McCulloch A, Midgley PM (1996) The production and global distribution of emissions of trichloroethene, tetrachloroethene and dichloromethane over the period 1988–1992. Atmos Environ 30:601–608

McDonald IR, Warner KL, Mcanulla C, Woodall CA, Oremland RS, Murrell JC (2002) A review of bacterial methyl halide degradation: biochemistry, genetics and molecular ecology. Environ Microbiol 4:193–203

McFiggans G, Plane JMC, Allan BJ, Carpenter LJ, Coe H, O'Dowd C (2000) A modeling study of iodine chemistry in the marine boundary layer. J Geophys Res Atmos 105:14371–14385

McFiggans G, Cox RA, Mössinger JC, Allan BJ, Plane JMC (2002) Active chlorine release from marine aerosols: roles for reactive iodine and nitrogen species. J Geophys Res Atmos 107:4271. doi:10.1029/2001JD000383

McFiggans G, Coe H, Burgess R, Allan J, Cubison M, Alfarra MR, Saunders R, Saiz-Lopez A, Plane JMC, Wevill DJ, Carpenter LJ, Rickard AR, Monks PS (2004) Direct evidence for coastal iodine particles from Laminaria macroalgae – linkage to emissions of molecular iodine. Atmos Chem Phys 4:701–713

McFiggans G, Artaxo P, Baltensperger U, Coe H, Facchini MC, Feingold G, Fuzzi S, Gysel M, Laaksonen A, Lohmann U, Mentel TF, Murphy DM, O'Dowd CD, Snider JR, Weingartner E (2006) The effect of physical and chemical aerosol properties on warm cloud droplet activation. Atmos Chem Phys 6:2593–2649

McFiggans G, Bale CSE, Ball SM, Beames JM, Bloss WJ, Carpenter LJ, Dorsey J, Dunk R, Flynn MJ, Furneaux KL, Gallagher MW, Heard DE, Hollingsworth AM, Hornsby K, Ingham T, Jones CE, Jones RL, Kramer LJ, Langridge JM, Leblanc C, LeCrane JP, Lee JD, Leigh RJ, Longley I, Mahajan AS, Monks PS, Oetjen H, Orr-Ewing AJ, Plane JMC, Potin P, Shillings AJL, Thomas F, von Glasow R, Wada R, Whalley LK, Whitehead JD (2010) Iodine-mediated coastal particle formation: an overview of the Reactive Halogens in the Marine Boundary Layer (RHaMBLe) Roscoff coastal study. Atmos Chem Phys 10:2975–2999

Meskhidze N, Nenes A (2006) Phytoplankton and cloudiness in the Southern Ocean. Science 314:1419–1423

Mezcua M, Aguera A, Hernando MD, Piedra L, Fernandez-Alba AR (2003) Determination of methyl tert.-butyl ether and ter.-butyl alcohol in seawater samples using purge-and-trap enrichment coupled to gas chromatography with atomic emission and mass spectrometric detection. J Chromatogr A 999:81–90

Millet DB, Jacob DJ, Custer TG, de Gouw JA, Goldstein AH, Karl T, Singh HB, Sive BC, Talbot RW, Warneke C, Williams J (2008) New constraints on terrestrial and oceanic sources of atmospheric methanol. Atmos Chem Phys 8:6887–6905

Millet DB, Guenther A, Siegel DA, Nelson NB, Singh HB, de Gouw JA, Warneke C, Williams J, Eerdekens G, Sinha V, Karl T, Flocke F, Apel E, Riemer DD, Palmer PI, Barkley M (2010) Global atmospheric budget of acetaldehyde: 3-D model analysis and constraints from in-situ and satellite observations. Atmos Chem Phys 10:3405–3425

Moelwyn-Hughes EA (1938) The hydrolysis of the methyl halides. Proc R Soc A164:295–306

Monks PS (2005) Gas-phase radical chemistry in the troposphere. Chem Soc Rev 34:376–395

Moore RM (2001) Trichloroethylene and tetrachloroethylene in Atlantic waters. J Geophys Res 106:135–227

Moore RM (2004) Dichloromethane in North Atlantic waters. J Geophys Res 109, C09004. doi:10.1029/2004JC002397

Moore RM (2006) Methyl halide production and loss rates in seawater from field incubation experiments. Mar Chem 101:213–219

Moore RM (2008) A photochemical source of methyl chloride in saline waters. Environ Sci Technol 42(6):1933–1937. doi:10.1021/es0719201

Moore RM, Blough NV (2002) A marine source of methyl nitrate. Geophys Res Lett 29:1–4

Moore RM, Tokarczyk R (1993) Volatile biogenic halocarbons in the Northwest Atlantic. Glob Biogeochem Cycle 7:195–210

Moore RM, Zafiriou O (1994) Photochemical production of methyl iodide in seawater. J Geophys Res 99:16415–16420

Moore RM, Groszko W, Niven S (1996a) Ocean–atmosphere exchange of methyl chloride: results from N.W. Atlantic and Pacific Ocean studies. J Geophys Res 101:28529–28538

Moore RM, Webb M, Tokarczyk R, Wever R (1996b) Bromoperoxidase and iodoperoxidase enzymes and production of halogenated methanes in marine diatom cultures. J Geophys Res 101:20899–20908

Moore RM, Punshon S, Mahaffey C, Karl D (2009) The relationship between dissolved hydrogen and nitrogen fixation in ocean waters. Deep Sea Res I 56:1449–1458

Mopper K, Stahovec WL (1986) Sources and sinks of low molecular weight organic carbonyl compounds in seawater. Mar Chem 19:305–321

Mopper K, Zhou X, Kieber RJ, Kieber DJ, Sikorski RJ, Jones RD (1991) Photochemical degradation of dissolved organic carbon and its impact on the oceanic carbon cycle. Nature 353:60–62

Moran MA, Zepp RG (1997) Role of photoreactions in the formation of biologically labile compounds from dissolved organic matter. Limnol Oceanogr 42:1307–1316

Moyers JL, Duce RA (1972) Gaseous and particulate iodine in the marine atmosphere. J Geophys Res 77:5229–5238

Mueller J-F, Brasseur G (1999) Sources of upper tropospheric HO_x: a three-dimensional study. J Geophys Res 104 (D1):1705–1715

Müller C, Iinuma Y, Karstensen J, Van Pinxteren D, Lehmann S, Gnauk T, Herrmann H (2009) Seasonal variation of aliphatic amines in marine sub-micrometer particles at the Cape Verde islands. Atmos Chem Phys 9:9587–9597

Muramatsu Y, Yoshida S (1995) Volatilization of methyl iodide from the soil plant system. Atmos Environ 29:21–25

Murphy CD, Moore RM, White RL (2000) An isotopic labeling method for determining production of volatile organohalogens by marine microalgae. Limnol Oceanogr 45:1868–1871

Nadasdi R, Zuegner GL, Farkas M, Dobe S, Maeda S, Morokuma K (2010) Photochemistry of methyld ethyl ketone: quantum yields and S_1/S_0-diradical mechanism of photodissociation. Chem Phys Chem 11:3883–3895

Nagao I, Hashimoto S, Suzuki K, Toda S, Narita Y, Tsuda A, Saito H, Kudo I (2009) Responses of DMS in the seawater and atmosphere to iron enrichment in the subarctic western North Pacific (SEEDS-II). Deep Sea Res Part II 56:2899–2917

Naik V, Fiore AM, Horowitz LW, Singh HB, Wiedmeyer C, Guenther A, de Gouw JA, Millet DB, Goldan PD, Kuster WC, Goldstein A (2010) Obsevational constraints on the global atmospheric budget of ethanol. Atmos Chem Phys 10:925–945

Neidleman SL, Geigert J (1986) Biohalogenation: principles basic roles and applications. Ellis Horwood, Chichester

Nemecek-Marshall M, Wojciechowski C, Kuzma J, Silver GM, Fall R (1995) Marine *Vibrio* species produce the volatile organic compound acetone. Appl Environ Microbiol 61:44–47

Neu JL, Lawler MJ, Prather MJ, Saltzman ES (2008) Oceanic alkyl nitrates as a natural source of tropospheric ozone. Geophys Res Lett 35(L13814)

Nightingale PD, Malin G, Liss PS (1995) Production of chloroform and other low-molecular weight halocarbons by some species of marine algae. Limnol Oceanogr 40:680–689

Nightingale PD, Malin G, Law CS, Watson AJ, Liss PS, Liddicoat MI, Boutin J, Upstill-Goddard RC (2000) In situ evaluation of air-sea gas exchange parameterizations using novel conservative and volatile tracers. Glob Biogeochem Cycle 14:373–387

Novelli PC, Lang PM, Masarie KA, Hurst DM, Myers R, Elkins JW (1999) Molecular hydrogen in the troposphere: global distribution and budget. J Geophys Res 104:30427–30444

Nowak JB, Davis DD, Chen G, Eisele FL, Mauldin RL, Tanner DJ, Cantrell C, Kosciuch E, Bandy A, Thornton D, Clarke A (2001) Airborne observations of DMSO, DMS and OH at marine tropical latitudes. Geophys Res Lett 28:2201–2204

O'Doherty S et al (2001) In situ chloroform measurements at AGAGE atmospheric research stations from 1994–1998. J Geophys Res 106:20429–20444

O'Dowd CD, Geever M, Hill MK, Smith MH, Jennings SG (1998) New particle formation: nucleation rates and spatial scales in the clean marine coastal environment. Geophys Res Lett 25:1661–1664

O'Dowd C, McFiggans G, Creasey DJ, Pirjola L, Hoell C, Smith MH, Allan BJ, Plane JMC, Heard DE, Lee JD, Pilling MJ, Kulmala M (1999) On the photochemical production of new particles in the coastal boundary layer. Geophys Res Lett 26:1707–1710

O'Dowd CD, Hämeri K, Mäkelä JM, Pirjola L, Kulmala M, Jennings SG, Berresheim H, Hansson H-C, de Leeuw G, Kunz GJ, Allen AG, Hewitt CN, Jackson A, Viisanen Y, Hoffmann T (2002a) A dedicated study of New Particle Formation and Fate in the Coastal Environment (PARFORCE): overview of objectives and achievements. J Geophys Res 107:8108. doi:10.1029/JD000555

O'Dowd CD, Jimenez JL, Bahreini R, Flagan RC, Seinfeld JH, Hameri K, Pirjola L, Kulmala M, Jennings SG, Hoffmann T (2002b) Marine aerosol formation from biogenic iodine emissions. Nature 417:632–636

O'Dowd CD, Facchini MC, Cavalli F, Ceburnis D, Mircea M, Decesari S, Fuzzi S, Yoon YJ, Putaud J-P (2004) Biogenically driven organic contribution to marine aerosol. Nature 431:676–680

Obernosterer I, Kraay G, de Ranitz E, Herndl GJ (1999) Concentrations of low molecular weight carboxylic acids and carbonyl compounds in the Aegean Sea (Eastern Mediterranean) and the turnover of pyruvate. Aquat Microb Ecol 20:147–156

Oh IB, Byun DW, Kim HC, Kim S, Cameron B (2008) Modeling the effect of iodide distribution on ozone deposition to seawater surface. Atmos Environ 42:4453–4466

Ohsawa N, Tsujita M, Morikawa S, Itoh N (2001) Purification and characterisation of a monohalomethane-producing enzyme S.adenosyl-L-methionine: halide ion methyltransferase from a marine microalga, Pavlova pinguis. Biosci Biotechnol Biochem 65:2397–2404

Olaguer EP (2002) The distribution of the chlorinated solvents dichloromethane, perchloroethylene, and trichloroethylene in the global atmosphere. Environ Sci Pollut Res 9:175–182

Oltmans SJ et al (2006) Long-term changes in tropospheric ozone. Atmos Environ 40:3156–3173

Ooki A, Yokouchi Y (2011) Dichloromethane in the Indian Ocean: evidence for in-situ production in seawater. Mar Chem 124:119–124

Ooki A et al (2010) Methyl halides in surface seawater and marine boundary layer of the northwest Pacific Source. J Geophys Res Ocean 115, C10013. doi:10.1029/2009JC005703

Orlikowska A, Schulz-Bull DE (2009) Seasonal variations of volatile organic compounds in the coastal Baltic Sea. Environ Chem 6:495–507

Padmaja S, Huie RE (1993) The reaction of nitric oxide with organic peroxyl radicals. Biochem Bioph Res Co 195:539–544

Palmer CJ, Reason CJ (2009) Relationship of surface bromoform concentrations with mixed layer depth and salinity in the tropical oceans. Glob Biogeochem Cycle 23, doi:10.1029/2008GB003338

Patroescu IV, Barnes I, Becker KH (1999) FT-IR product study of the OH-initiated oxidation of DMS in the presence of NO_x. Atmos Environ 33:25–35

Plane JMC, Vondrak T, Broadley S, Cosic B, Ermoline A, Fontijn A (2006) Kinetic study of the reaction $Ca^+ + N_2O$ from 188 to 1207 K. J Phys Chem A 110:7874–7881

Plass-Dülmer C, Koppmann R, Ratte M, Rudolph J (1995) Light nonmethane hydrocarbons in seawater. Global Biogeochem Cycle 9:79–100

Prather M et al (2003) Fresh air in the 21st century? Geophys Res Lett 30:1100. doi:10.1029/2002GL016285

Prinn RG et al (1999) Long-lived ozone-related compounds in scientific assessment of ozone depletion: 1998. World Meteorological Organization, Geneva

Pumphrey HC, Santee MJ, Livesey NJ, Schwartz MJ, Read WG (2011) Microwave Limb Sounder observations of biomass-burning products from the Australian bush fires of February 2009. Atmos Chem Phys 11:6285–6296

Punshon S, Moore RM (2008) Photochemical production of molecular hydrogen in lake water and coastal seawater. Mar Chem 108:215–220

Quack B, Wallace DWR (2003) Air-sea flux of bromoform: controls, rates, and implications. Global Biogeochem Cycle 17:1023. doi:10.1029/2002GB001890

Quack B, Atlas E, Petrick G, Stroud V, Schauffler S, Wallace DWR (2004) Oceanic bromoform sources for the tropical atmosphere. Geophys Res Lett 31, L23S05. doi:10.1029/2004GL020597

Quack B, Peeken I, Petrick G, Nachtigall K (2007) Oceanic distribution and sources of bromoform and dibromomethane in the Mauritanian upwelling. J Geophys Res Ocean 112, C10006. doi:10.1029/2006JC003803

Quinn PK, Bates TS (2011) The case against climate regulation via oceanic phytoplankton sulphur emissions. Nature 480:51–56

Quinn PK, Asher WE, Charlson RJ (1992) Equilibria of the marine multiphase ammonia system. J Atmos Chem 14:11–30. doi:10.1007/BF00115219

Raimund S, Quack B, Bozec Y, Vernet M, Rossi V, Garcon V, Morel Y, Morin P (2011) Sources of short-lived bromocarbons in the Iberian upwelling system. Biogeosciences 8:1551–1564

Ratte M, Plass-Dülmer C, Koppmann R, Rudolph J, Denga J (1993) Production mechanism of C2–C4 hydrocarbons in seawater: field measurements and experiments. J Glob Biogeochem Cycle 7:369–378

Ratte M, Plass-Dülmer C, Koppmann R, Rudolph J (1995) Horizontal and vertical profiles of light hydrocarbons in sea water related to biological, chemical and physical profiles. Tellus B 47:607–623

Ratte M, Bujok O, Spitzy A, Rudolph J (1998) Photochemical alkene formation in seawater from dissolved organic carbon: results from laboratory experiments. J Geophys Res 103:5707–5717

Rayman MP (2000) The importance of selenium to human health. Lancet 356:233–241

Razavi A, Karagulian F, Clarisse L, Hurtmans D, Coheur PF, Clerbaux C, Mueller JF, Stavrakou T (2011) Global distributions of methanol and formic acid retrieved for the first time from the IASI/MetOp thermal infrared sounder. Atmos Chem Phys 11:857–872

Read KA, Mahajan AS, Carpenter LJ, Evans MJ, Faria BVE, Heard DE, Hopkins JR, Lee JD, Moller SJ, Lewis AC, Mendes L, McQuaid JB, Oetjen H, Saiz-Lopez A, Pilling MJ, Plane JMC (2008a) Extensive halogen-mediated ozone destruction over the tropical Atlantic Ocean. Nature 453:1232–1235

Read KA, Lewis AC, Bauguitte S, Rankin AM, Salmon RA, Wolff EW, Saiz-Lopez A, Bloss WJ, Heard DE, Lee JD, Plane JMC (2008b) DMS and MSA measurements in the Antarctic boundary layer: impact of BrO on MSA production. Atmos Chem Phys 8:2985–2997

Rhee TS, Brenninkmeijer CAM, Rockmann T (2005) The overwhelming role of soils in the global atmospheric hydrogen cycle. Atmos Chem Phys Discuss 5:11215–11248

Richter U, Wallace DWR (2004) Production of methyl iodide in the tropical Atlantic Ocean. Geophys Res Lett 31, L23S03. doi:10.1029/2004GL020779

Riemer DD, Milne P, Zika RG, Pos WH (2000) Photoproduction of nonmethane hydrocarbons (NMHC) in seawater. Mar Chem 71:177–198

Rinsland C, Boone C, Bernath P, Mahieu E, Zander R, Dufour G, Clerbaux C, Turquety S, Chiou L, Mc-Connel J, Neary L, Kaminski JW (2006) First space-based observations of formic acid (HCOOH): atmospheric chemistry experiment austral spring 2004 and 2005 Southern Hemisphere tropical-midlatitude upper tropospheric measurements. Geophys Res Lett 33, L23804. doi:10.1029/2006GL027128

Rinsland C, Dufour G, Boone C, Bernath P, Chiou L, Coheur P, Turquety S, Clerbaux C (2007) Satellite boreal measurements over Alaska and Canada during June-July 2004: simulataneous measurements of upper tropospheric CO, C_2H_6, HCN, CH_3Cl, CH_4, C_2H_2, CH_3OH, HCOOH, OCS, and SF_6 mixing ratios. Glob Biogeochem Cycle 21, GB3008. doi:10.1029/2006GB002795

Russo RS, Zhou Y, Haase KB, Wingenter OW, Frinak EK, Mao H, Talbot RW, Sive BC (2010) Temporal variability, sources, and sinks of C1–C5 alkyl nitrates in coastal New England. Atmos Chem Phys 10:1865–1883

Saemundsdottir S, Matrai PA (1998) Biological production of methyl bromide by cultures of marine phytoplankton. Limnol Oceanogr 43:81–87

Saiz-Lopez A, Plane JMC (2004) Novel iodine chemistry in the marine boundary layer. Geophys Res Lett 31, L04112

Saiz-Lopez A, Plane JMC, Shillito JA (2004) Bromine oxide in the mid-latitude marine boundary layer. Geophys Res Lett 31. doi: 10.1029/2003GL018956

Saiz-Lopez A, Plane JMC, McFiggans G, Williams PI, Ball SM, Bitter M, Jones RL, Hongwei C, Hoffmann T (2006a) Modelling molecular iodine emissions in a coastal marine environment: the link to new particle formation. Atmos Chem Phys 6:883–895

Saiz-Lopez A, Shillito JA, Coe H, Plane JMC (2006b) Measurements and modelling of I_2, IO, OIO, BrO and NO_3 in the mid-latitude marine boundary layer. Atmos Chem Phys 6:1513–1528

Saiz-Lopez A, Chance K, Liu X, Kurosu TP, Sander SP (2007a) First observations of iodine oxide from space. Geophys Res Lett 34, L12812

Saiz-Lopez A, Mahajan AS, Salmon RA, Bauguitte SJ-B, Jones AE, Roscoe HK, Plane JMC (2007b) Boundary layer halogens in coastal Antarctica. Science 317:348–351

Saiz-Lopez A, Plane JMC, Mahajan AS, Anderson PS, Bauguitte SJ-B, Jones AE, Roscoe HK, Salmon RA, Bloss WJ, Lee JD, Heard DE (2008) On the vertical distribution of boundary layer halogens over coastal

Antarctica: implications for O_3, HO_x, NO_x and the Hg lifetime. Atmos Chem Phys 8:887–900

Saiz-Lopez A, Plane JMC, Baker AR, Carpenter LJ, Glasow Rv, Martín JCG, McFiggans G, Saunders RW (2012) Atmospheric chemistry of iodine. Chem Rev 112:1773–1804. doi: http://dx.doi.org/10.1021/cr200029u

Salawitch RJ (2006) Atmospheric chemistry – biogenic bromine. Nature 439:275–277

Salawitch RJ, Weisenstein DK, Kovalenko LJ, Sioris CE, Wennberg PO, Chance K, Ko MKW, McLinden CA (2005) Sensitivity of ozone to bromine in the lower stratosphere. Geophys Res Lett 32, L05811. doi:10.1029/2004GL021504

Saltzman ES, Aydin M, Tatum C, Williams MB (2008) 2,000-year record of atmospheric methyl bromide from a South Pole ice core. J Geophys Res Atmos 113, D05304. doi:10.1029/2007JD008919

Sander R, von Glasow R, Crutzen PJ (2004) Comment on: Laskin, et al. Reactions at interfaces as a source of sulfate formation in sea-salt particles. Science 303:628c

Saunders RW, Plane JMC (2005) Formation pathways and composition of iodine oxide ultra-fine particles. Environ Chem 2:299–303

Saunders RW, Plane JMC (2006) Fractal growth modelling of I_2O_5 nanoparticles. J Aerosol Sci 37:1737

Saunders RW, Mahajan AS, Gómez Martín JC, Kumar R, Plane JMC (2010) Studies of the formation and growth of aerosol from molecular iodine precursor. Z Phys Chem (Munich) 224:1095–1117

Savoie DL, Prospero JM, Larsen RJ, Huang F, Izaguirre MA, Huang T, Snowdon TH, Custals L, Sanderson CG (1993) Nitrogen and sulfur species in Antarctic aerosols at Mawson, Palmer Station, and Marsh (King George Island). J Atmos Chem 17:95. doi:10.1007/BF00702821

Scarratt MG, Moore RM (1996) Production of methyl bromide and chloride in laboratory cultures of marine phytoplankton. Mar Chem 54:263–272

Scarratt MG, Moore RM (1998) Production of methyl bromide and chloride in laboratory cultures of marine phytoplankton II. Mar Chem 59:311–320

Scarratt MG, Moore RM (1999) Production of chlorinated hydrocarbons by the red microalga, Porphyridium purpureum. Limnol Oceanogr 44:703–707

Schade GW, Goldstein AH (2006) Seasonal measurements of acetone and methanol: abundances and implications for atmospheric budgets. Global Biogeochem Cycle 20, GB1011. doi:10.1029/2005GB002566

Schauffler SM, Atlas EL, Flocke F, Lueb RA, Stroud V, Travnicek W (1998) Measurements of bromine containing organic compounds at the tropical tropopause. Geophys Res Lett 25:317–320

Schlesinger WH, Hartley AE (1992) A global budget for atmospheric NH_3. Biogeochemistry 15:191–211. doi:10.1007/BF00002936

Schmidt U (1974) Molecular hydrogen in the atmosphere. Tellus 26:78–90

Schmittner A (2005) Decline of the marine ecosystem caused by a reduction in the Atlantic overturning circulation. Nature 434:628–633

Schönhardt A, Richter A, Wittrock F, Kirk H, Oetjen H, Roscoe HK, Burrows JP (2008) Observations of iodine monoxide columns from satellite. Atmos Chem Phys 8:637–653

Sciare J, Baboukas E, Hancy R, Mihalopoulos N, Nguyen BC (1998) Seasonal variation of dimethylsulfoxide in rainwater at Amsterdam Island in the Southern Indian Ocean; Implications on the biogenic sulfur cycle. J Atmos Chem 30:229–240

Sciare J, Baboukas E, Kanakidou M, Krischke U, Belviso S, Bardouki H, Mihalopoulos N (2000a) Spatial and temporal variability of atmospheric sulfur-containing gases and particles during the Albatross campaign. J Geophys Res 105:14433–14448

Sciare J, Kanakidou M, Mihalopoulos N (2000b) Diurnal and seasonal variation of atmospheric dimethylsulfoxide at Amsterdam Island in the southern Indian Ocean. J Geophys Res 105:17257–17265

Sciare J, Baboukas E, Mihalopoulos N (2001) Short-term variability of atmospheric DMS and its oxidation products at Amsterdam Island during summer time. J Atmos Chem 39:281–302

Seto FYB, Duce RA (1972) A laboratory study of iodine enrichment on atmospheric sea-salt particles produced by bubbles. J Geophys Res 77:5339–5349

Shaw SL, Chisholm SW, Prinn RG (2003) Isoprene production by *Prochlorococcus*, a marine cyanobacterium, and other phytoplankton. Mar Chem 80:227–245

Shaw SL, Gantt B, Meskhidze N (2010) Production and emissions of marine isoprene and monoterpenes: a review. Adv Meteor 2010:4048696. doi:10.1155/2010/408696

Shim C, Wang Y, Singh HB, Blake DR, Guenther AB (2007) Source characteristics of oxygenated volatile organic compounds and hydrogen cyanide. J Geophys Res 112, D10305. doi:10.1029/2006JD007543

Sievering H, Lerner B, Slavich J, Anderson J, Posfai M, Cainey J (1999) O3 oxidation of SO2 in sea-salt aerosol water: size distribution of non-sea-salt sulfate during the first aerosol characterization experiment (ACE 1). J Geophys Res 104:21707–21717

Sievering H, Cainey J, Harvey M, McGregor J, Nichol S, Quinn P (2004) Aerosol non-sea-salt sulfate in the remote marine boundary layer under clear-sky and normal cloudiness conditions: ocean-derived biogenic alkalinity enhances sea-salt sulfate production by ozone oxidation. J Geophys Res 109, D19317. doi:10.1029/2003JD004315

Simmonds PG et al (2006) Global trends, seasonal cycles, and European emissions of dichloromethane, trichloroethene, and tetrachloroethene from the AGAGE observations at Mace Head, Ireland, and Cape Grim, Tasmania. J Geophys Res 111, D18304. doi:10.1029/2006JD007082

Simo R, Pedros-Alio C (1999) Role of vertical mixing in controlling the oceanic production of dimethyl sulphide. Nature 402:396–399

Singh HB, Ohara D, Herlth D, Sachse W, Blake DR, Bradshaw JD, Kanakidou M, Critzen PJ (1994) Acetone in the atmosphere: distribution, sources, and sinks. J Geophys Res 99:1805–1819

Singh HB, Kanakidou M, Crutzen PJ, Jacob DJ (1995) High concentrations and photochemical fate of oxygenated hydrocarbons in the global troposphere. Nature 378:50–54

Singh H, Chen Y, Staudt A, Jacob D, Blake D, Heikes B, Snow J (2001) Evidence from the Pacific troposphere for large global sources of oxygenated organic compounds. Nature 410:1078–1081

Singh HB et al (2003a) In situ measurements of HCN and CH_3CN over the Pacific Ocean: sources, sinks and budgets. J Geophys Res 108:8795. doi:10.1029/2002JD003006

Singh HB, Tabazadeh A, Evans MJ, Field BD, Jacob DJ, Sachse G, Crawford JH, Sette R, Brune WH (2003b) Oxygenated volatile organic chemicals in the oceans: inferences and implications based on atmospheric observations and air-sea exchange models. Geophys Res Lett 30:1862–1866

Singh HB et al (2004) Analysis of the atmospheric distribution, sources, and sinks of oxygenated volatile organic chemicals based on measurements over the Pacific during TRACE-P. J Geophys Res 109, D15D07. doi:10.1029/2003JD003883

Singh HB et al (2010) Pollution influences on atmospheric composition and chemistry at high northern latitudes: Boreal and California forest fire emissions. Atmos Environ 44:4553–4564

Sinha V, Williams J, Meyerhoefer M, Riebesell U, Paulino AI, Larsen A (2007) Air-sea fluxes of methanol, acetone, acetaldehyde, isoprene, and DMS from a Norwegian fjord following a phytoplankton bloom in a mesocosm experiment. Atmos Chem Phys 7:739–755

Sluis MK, Ensign SA (1997) Purification and characterization of acetone carboxylase from *Xanthobacter* strain Py2. Proc Natl Acad Sci USA 94:8456–8461

Smith MH (2007) Sea-salt particles and the CLAW hypothesis. Environ Chem 4:391–395

Smythe-Wright D, Boswell M, Breithaupt P, Davidson RD, Dimmer CH, Eiras-Diaz LB (2006) Methyl iodide production in the ocean: implications for climate change. Glob Biogeochem Cycle 20, GB3003. doi:10.1029/2005GB 002642

Smythe-Wright D, Peckett C, Boswell S, Harrison R (2010) Controls on the production of organohalogens by phytoplankton: effect of nitrate concentration and grazing. J Geophys Res 115, G03020. doi:10.1029/2009JG001036

Spivakovsky CM, Yevich R, Logan JA, Wofsy SC, McElroy MB (1990) Tropospheric OH in a three-dimensional chemical tracer model: an assessment based on observations of CH_3CCl_3. J Geophys Res 95:18441–18471

Sprung D, Zahn A (2010) Acetone in the upper troposphere/lowermost stratosphere measured by the CARIBIC passenger aircraft: distribution, seasonal cycle, and variability. J Geophys Res 115, D16301. doi:10.1029/2009JD012099

Stefels J (2000) Physiological aspects of the production and conversion of DMSP in marine algae and higher plants. J Sea Res 43:183–197

Stefels J, Steinke M, Turner S, Malin G, Belviso S (2007) Environmental constraints on the production and removal of the climatically active gas dimethylsulphide (DMS) and implications for ecosystem modeling. Biogeochemistry 83:245–275

Stevens B, Feingold G (2009) Untangling aerosol effects on clouds and precipitation in a buffered system. Nature 461:607–613

Stubbins A, Uher G, Law CS, Mopper K, Robinson C, Upstill-Goddard RC (2006a) Open-ocean carbon monoxide photoproduction. Deep Sea Res II 53:1695–1705

Stubbins A, Uhera G, Kitidis V, Law CS, Upstill-Goddard RC, Woodward EMS (2006b) The open-ocean source of atmospheric carbon monoxide. Deep Sea Res II 53:1685–1694

Stubbins A, Hubbard V, Uher G, Law CS, Upstill-Goddard RC, Aiken GR, Mopper K (2008) Relating carbon monoxide photoproduction to dissolved organic matter functionality. Environ Sci Technol 42:3271–3276

Sturrock GA, Reeves CE, Mills GP, Penkett SA, Parr CR, McMinn A, Corno G, Tindale NW, Fraser PJ (2003) Saturation levels of methyl bromide in the coastal waters off Tasmania. Global Biogeochem Cycle 17:1107. doi:10.1029/2002GB002024

Stutz J, Pikelnaya O, Hurlock SC, Trick S, Pechtl S, von Glasow R (2007) Daytime OIO in the Gulf of Maine. Geophys Res Lett 34, L22816

Sunda W, Kieber DJ, Kiene RP, Huntsman S (2002) An antioxidant function for DMSP and DMS in marine algae. Nature 418:317–320

Sutton MA, Erisman JW, Dentener F, Möller D (2008) Ammonia in the environment: from ancient times to the present. Environ Poll 156:583–604. doi:10.1016/j.envpol.2008.03.013

Swanson AL, Davis DD, Arimooto R, Robert P, Atlas EL, Flocke F, Meinardi S, Rowland FS, Blake DR (2004) Organic trace gases of oceanic origin observed at South Pole during ISCAT 2000. Atmos Environ 38:5462–5472

Swinnerton J, Linnenbom V, Lamontagne R (1970) Ocean: a natural source of carbon monoxide. Science 167:984–986

Tanhua T, Fogelqvist E, Basturk O (1996) Reduction of volatile halocarbons in anoxic seawater, results from a study in the Black Sea. Mar Chem 54:159–170

Theiler R, Cook JC, Hager LP (1978) Halohydrocarbon synthesis by bromoperoxidase. Science 202:1094–1096

Thornton JA, Kercher JP, Riedel TP, Wagner NL, Cozic J, Holloway JS, Dube WP, Wolfe GM, Quinn PK, Middlebrook AM, Alexander B, Brown SS (2010) A large atomic chlorine source inferred from mid-continental reactive nitrogen chemistry. Nature 464:271–274

Tiefenau HK (1973) The specific ozone destruction rate of the ocean surface and its dependence on horizontal wind velocity. Pure Appl Geophys 106–108:1116–1123

Tokarczyk R, Moore RM (1994) Production of volatile organohalogens by phytoplankton cultures. Geophys Res Lett 21:285–288

Tokarczyk R, Goodwin KD, Saltzman ES (2003a) Methyl chloride and methyl bromide degradation in the Southern Ocean. Geophys Res Lett 30:1808. doi:10.1029/2003GL017459

Tokarczyk R, Saltzman ES, Moore RM, Yvon-Lewis SA (2003b) Biological degradation of methyl chloride in coastal seawater. Global Biogeochem Cycle 17:1057. doi:10.1029/2002GB001949

Toumi R (1994) BrO as a sink for dimethylsulphide in the marine atmosphere. Geophys Res Lett 21:117–120

Troxler RF, Dokos JM (1973) Formation of carbon-monoxide and bile pigment in red and blue-green-algae. Plant Phys 51:72–75

Trudinger CM, Etheridge DM, Sturrock GA, Fraser PJ, Krummel PB, McCulloch A (2004) Atmospheric histories of halocarbons from analysis of Antarctic firn air: methyl bromide, methyl chloride, chloroform, and dichloromethane. J Geophys Res 109, D22310. doi:10.1029/2004JD004932

Truesdale VW, Luther GW III (1995) Molecular iodine reduction by natural and model organic substances in seawater. Aquat Geochem 1:89–104

Tsukada H, Hara H, Iwashima K, Yamagata N (1987) The iodine content of atmospheric aerosols as determined by the use of a Fluoropore filter for collection. Bull Chem Soc Japan 60:3195–3198

Turner SM, Nightingale PD, Spokes LJ, Liddicoat MI, Liss PS (1996) Increased dimethyl sulphide concentrations in sea water from in situ iron enrichment. Nature 383:513–517

Turner SM, Harvey MJ, Law CS, Nightingale PD, Liss PS (2004) Iron-induced changes in oceanic sulfur biogeochemistry. Geophys Res Lett 31:doi:10.1029/2004GL020296

Uher G, Andreae MO (1997) Photochemical production of carbonyl sulfide in North Sea water: a process study. Limnol Oceanogr 42:432–442

Valentine RL, Zepp RG (1993) Formation of carbon-monoxide from the photodegradation of terrestrial dissolved organic-carbon in natural-waters. Environ Sci Technol 27:409–412

Vallina SM, Simo R (2007) Strong relationship between DMS and the solar radiation dose over the global surface ocean. Science 315:506–508

Vallina SM, Simo R, Manizza M (2007) Weak response of oceanic dimethylsulfide to upper mixing shoaling induced by global warming. Proc Natl Acad Sci USA 104:16004–16009

Vallina SM, Simo R, Anderson TR, Gabric A, Cropp R, Pacheco JM (2008) A dynamic model of oceanic sulfur (DMOS) applied to the Sargasso Sea: Simulating the dimethylsulfide (DMS) summer paradox. J Geophys Res-Biogeosciences 113 (G1). doi:10.1029/2007JG000415

Vardi A, Formiggini F, Casotti R, De Martino A, Ribalet F, Miralto A, Bowler C (2006) A stress surveillance system based on calcium and nitric oxide in marine diatoms. PLoS Biol 4:411–419

Vogel TM, Criddle CS, McCarty PL (1987) Transformations of halogenated aliphatic-compounds. Environ Sci Tech 21:722–736

Vogt M, Liss PS (2009) Dimethylsulfide and climate. In: Le Quéré C, Saltzman ES (eds) Surface ocean-lower atmosphere processes. American Geophysical Union, Washington, DC, pp 197–232

Vogt R, Sander R, Glasow RV, Crutzen PJ (1999) Iodine chemistry and its role in halogen activation and ozone loss in the marine boundary layer: a model study. J Atmos Chem 32:375–395

Vogt M, Vallina S, von Glasow S (2008) New directions: Correspondence on "Enhancing the natural cycle to slow global warming". Atmos Environ 42:4803–4805

von Glasow R (2006) Importance of the surface reaction OH + Cl- on sea salt aerosol for the chemistry of the marine boundary layer – a model study. Atmos Chem Phys 6:3571–3581

von Glasow R (2007) A look at the CLAW hypothesis from an atmospheric chemistry point of view. Environ Chem 4:379–381. http://www.publish.csiro.au/nid/188/paper/EN07064.htm, http://www.publish.csiro.au/nid/188/paper/EN07064.htm

Von Glasow R (2008) Sun, sea and ozone destruction. Nature 453:1195–1196

von Glasow R, Crutzen PJ (2004) Model study of multiphase DMS oxidation with a focus on halogens. Atmos Chem Phys 4:589–608

von Glasow R, Crutzen PJ (2007) Tropospheric halogen chemistry. In: Heinrich DH, Karl KT (eds) Treatise on geochemistry. Pergamon, Oxford

von Glasow R, Sander R, Bott A, Crutzen PJ (2002) Modeling halogen chemistry in the marine boundary layer. 1. Cloud-free MBL. J Geophys Res 107:4341

von Glasow R, von Kuhlmann R, Lawrence MG, Platt U, Crutzen PJ (2004) Impact of reactive bromine chemistry in the troposphere. Atmos Chem Phys 4:2481–2497

von Gunten U (2003) Ozonation of drinking water: Part I. Oxidation kinetics and product formation. Water Res 37:1443–1467

Wada R, Beames J, Orr-Ewing A (2007) Measurement of IO radical concentrations in the marine boundary layer using a cavity ring-down spectrometer. J Atmos Chem 58:69–87

Wade LG (1999) Organic chemistry, 4th edn. Prentice-Hall, Upper Saddle River

Wang L, Lal V, Khalizov AF, Zhang R (2010) Heterogeneous chemistry of alkylamines with sulfuric acid: implications for atmospheric formation of alkylaminium sulfates. Environ Sci Tech 44:2461–2465. doi:10.1021/es9036868

Ward BB (2003) Significance of anaerobic ammonium oxidation in the ocean. Trends Microbiol 11:408–410. doi:10.1016/S0966-842X(03)00181-1

Warneck P, Williams J (2011) The atmospheric chemist's companion, 1st edn. Springer, Dordrecht. ISBN 10: 9400722745

Watts SF (2000) The mass budgets of carbonyl sulfide, dimethyl sulfide, carbon disulfide and hydrogen sulfide. Atmos Environ 34:761–779

Weeks SJ, Currie B, Bakun A (2002) Massive emissions of toxic gas in the Atlantic. Nature 415:493–494

Wennberg PO et al (1998) Hydrogen radicals, nitrogen radicals, and the production of O_3 in the upper troposphere. Science 279:49–53

Wesely ML, Hicks BB (2000) A review of the current status of knowledge on dry deposition. Atmos Environ 34:2261–2282

Whalley L, Furneaux K, Gravestock T, Atkinson H, Bale C, Ingham T, Bloss W, Heard D (2007) Detection of iodine monoxide radicals in the marine boundary layer using laser induced fluorescence spectroscopy. J Atmos Chem 58:19–39

Williams J, Holzinger R, Gros V, Xu X, Atlas E, Wallace DWR (2004) Measurements of organic species in air and seawater from the tropical Atlantic. Geophys Res Lett 31, L23S06. doi:10.1029/2004GL020012

Williams J, Custer T, Riede H, Sander R, Jöckel P, Hoor P, Pozzer A, Wong-Zehnpfennig S, Hosaynali-Beygi Z, Fischer H, Gros V, Colomb A, Bonsang B, Yassaa N, Peeken I, Atlöas EL, Waluda CM, van Aardenne JA, Lelieveld J (2010) Assessing the effect of marine isoprene and ship emissions on ozone, using modeling and measurements from the South Atlantic Ocean. Environ Chem 7:171–182. doi:10.1071/EN09154

Williamson P, Wallace DWR, Law CS, Boyd BW, Collos Y, Croot P, Denman K, Riebesell U, Takeda S, Vivian C (2012) Ocean fertilization for geoengineering: a review of effectiveness, environmental impacts and emerging governance. Process Saf Environ Prot. doi:10.1016/j.psep.2012.10.007

Wilson DF, Swinnerton J, Lamontagne R (1970) Production of carbon monoxide and gaseous hydrocarbons in seawater – relation to dissolved organic carbon. Science 168:1576–1577

Wilson ST, Foster RA, Zehr JP, Karl DM (2010) Hydrogen productioin b *Trichodesmium erhthraeum Cyanothece* sp. and *Crocosphaera watsonii*. Aquat Microb Ecol 59:197–206

Wingenter OW, Haase KB, Strutton P, Friederich G, Meinardi S, Blake DR, Rowland FS (2004) Changing concentrations of

CO, CH$_4$, C$_5$H$_8$, CH$_3$Br, CH$_3$I and dimethyl sulfide during the southern ocean iron enrichment experiments. Proc Natl Acad Sci USA 101:8537–8541

Wingenter OW, Sive BC, Blake NJ, Blake DR, Rowland FS (2005) Atomic chlorine concentrations derived from ethane and hydroxyl measurements over the Equatorial Pacific Ocean: implication for dimethyl sulfide and bromine monoxide. J Geophys Res 110, D20308. doi:10.1029/2005JD005875

Wingenter OW, Elliot SM, Blake DR (2007) New directions: enhancing the natural sulfur cycle to slow global warming. Atmos Environ 41:7373–7375

WMO (2011) Scientific assessment of ozone depletion: 2010, Global ozone research and monitoring project-report no. 52 Rep. World Meteorological Organization, Geneva

Woodhouse MT, Mann GW, Carslaw KS, Boucher O (2008) New directions: the impact of oceanic iron fertilization on cloud condensation nuclei. Atmos Environ 42:5728–5730

Woodhouse MT, Carslaw KS, Mann GW, Vallina SM, Vogt M, Halloran PR, Boucher O (2010) Low sensitivity of cloud condensation nuclei to changes in the sea-air flux of dimethyl-sulphide. Atmos Chem Phys 10:7545–7559

Wuosma AM, Hager PL (1990) Methylchloride transerfrase. A carbocation route for biosynthesis of halometabolites. Science 249:160–162

Xiao X et al (2010) Optimal estimation of the surface fluxes of methyl chloride using a 3-D global chemical transport model. Atmos Chem Phys 10:5515–5533

Xie HX, Moore RM, Miller WL (1998) Photochemical production of carbon disulphide in seawater. J Geophys Res 103:5635–5644

Xie HX, Moore RM (1999) Carbon disulfide in the North Atlantic and Pacific Ocean. J Geophys Res 104:5393–5402

Xie HX, Zafiriou OC, Umile TP, Kieber DJ (2005) Biological consumption of carbon monoxide in Delaware Bay, NW Atlantic and Beaufort Sea. Mar Ecol Prog Ser 290:1–14

Xie HX, Belanger S, Demers S, Vincent WF, Papakyriakou TN (2009) Photobiogeochemical cycling of carbon monoxide in the southeastern Beaufort Sea in spring and autumn. Limnol Oceanogr 54:234–249

Xu S, Xie Z-Q, Li B, Sun L, Kang H, Yang H, Zhang P (2010) Iodine speciation in marine aerosols along a 15000-km round-trip cruise path from Shanghai, China, to the Arctic Ocean. Environ Chem 7:406–412

Yang M, Huebert BJ, Blomquist BW, Howell SG, Shank LM, McNaughton CS, Clarke AD, Hawkins LN, Russell LM, Covert DS, Coffman DJ, Bates TS, Quinn PK, Zagorac N, Bandy AR, de Szoeke SP, Zuidema PD, Tucker SC, Brewer WA, Yang X, Cox RA, Warwick NJ, Pyle JA, Carver GD, O'Connor FM, Savage NH (2005) Tropospheric bromine chemistry and its impacts on ozone: a model study. J Geophys Res Atmos 110, D23311. doi:10.1029/2005JD006244

Yang GP, Wang WL, Lu XL, Ren CY (2010) Distribution, flux and biological consumption of carbon monoxide in the Southern Yellow Sea and the East China Sea. Mar Chem 122:74–82

Yang GP, Ren CY, Lu XL, Liu CY, Ding HB (2011) Distribution, flux, and photoproduction of carbon monoxide in the East China Sea and Yellow Sea in spring. J Geophys Res Ocean 116, CO2001. doi:10.1029/2010JC006300

Yassaa N, Peeken I, Zöllner E, Bluhm K, Arnold S, Spracklen D, Williams J (2008) Evidence for marine production of monoterpenes. Environ Chem 5:391–401. doi:10.1071/EN08047

Yu F (2006) Effect of ammonia on new particle formation: a kinetic H$_2$SO$_4$-H$_2$O-NH$_3$ nucleation model constrained by laboratory measurements. J Geophys Res D111, D01204. doi:10.1029/2005JD005968

Yvon-Lewis SA, Butler JH, Saltzman EH, Matrai PA, King DB, Tokarczyk R, Moore RM, Zhang JZ (2002) Methyl bromide cycling in a warm-core eddy of the North Atlantic Ocean. Glob Biogeochem Cycle 16:1141. doi:10.1029/2002GB001898

Yvon-Lewis SA, Saltzman ES, Montzka SA (2009) Recent trends in atmospheric methyl bromide: analysis of post-montreal protocol variability. Atmos Chem Phys 9:5963–5974

Zafiriou OC (1975) Reaction of methyl halides with seawater and marine aerosols. J Mar Res 33:75–81

Zafiriou OC, MacFarland M (1981) Nitric oxide formation from nitrite photolysis in the central equatorial Pacific. J Geophys Res 86:3173–3182

Zafiriou OC, McFarland M, Bromund RH (1980) Nitric oxide in seawater. Science 207:637–639

Zafiriou OC, Andrews SS, Wang W (2003) Concordant estimates of oceanic carbon monoxide source and sink processes in the Pacific yield a balanced global "bluewater" CO budget. Glob Biogeochem Cycle 17. doi:10.1029/2001GB001638

Zhang Z, Liu C, Wu Z, Xing L, Li P (2006) Detection of nitric oxide in culture media and studies on nitric oxide formation by marine microalgae. Med Sci Monit 12: BR75–BR85

Zhou X, Mopper K (1997) Photochemical production of low-molecular-weight carbonyl compounds in seawater and surface microlayer and their air-sea exchange. Mar Chem 56:201–213

Zika RG, Gidel LT, Davis DD (1984) A comparison of photolysis and substitution decomposition rates of methyl iodide in the ocean. Geophys Res Lett 11:353–356

Ziolkowski L, Miller W (2007) Variability of the apparent quantum efficiency of CO photoproduction in the Gulf of Maine and Northwest Atlantic. Mar Chem 105:258–270

Zuo Y, Jones RD (1995) Formation of carbon monoxide by photolysis of dissolved marine organic material and its significance in the carbon cycling of the oceans. Naturwissenschaften 82:472–474

Transfer Across the Air-Sea Interface

Christoph S. Garbe, Anna Rutgersson, Jacqueline Boutin,
Gerrit de Leeuw, Bruno Delille, Christopher W. Fairall,
Nicolas Gruber, Jeffrey Hare, David T. Ho, Martin T. Johnson,
Philip D. Nightingale, Heidi Pettersson, Jacek Piskozub, Erik Sahlée,
Wu-ting Tsai, Brian Ward, David K. Woolf, and Christopher J. Zappa

Abstract

The efficiency of transfer of gases and particles across the air-sea interface is controlled by several physical, biological and chemical processes in the atmosphere and water which are described here (including waves, large- and small-scale turbulence, bubbles, sea spray, rain and surface films). For a deeper understanding of relevant transport mechanisms, several models have been developed, ranging from conceptual models to numerical models. Most frequently the transfer is described by various functional dependencies of the wind speed, but more detailed descriptions need additional information. The study of gas transfer mechanisms uses a variety of experimental methods ranging from laboratory studies to carbon budgets, mass balance methods, micrometeorological techniques and thermographic techniques. Different methods resolve the transfer at different scales of time and space; this is important to take into account when comparing different results. Air-sea transfer is relevant in a wide range of applications, for example, local and regional fluxes, global models, remote sensing and computations of global inventories. The sensitivity of global models to the description of transfer velocity is limited; it is however likely that the formulations are more important when the resolution increases and other processes in models are improved. For global flux estimates using inventories or remote sensing products the accuracy of the transfer formulation as well as the accuracy of the wind field is crucial.

2.1 Introduction

The transfer of gases and particles across the air-sea interface depends not only on the concentration difference between the water and the air, but also on the efficiency of the transfer process. The efficiency of the transfer is controlled by complex interaction of a variety of processes in the air and in the water near the interface. Here we treat both gases and particles since the transfer, to some extent, is governed by similar mechanisms.

Studies of transfer across the air-sea interface include a variety of methods and techniques ranging from laboratory studies, modeling and large-scale field studies. Various methods reach somewhat different conclusions, due to representation of different

C.S. Garbe (✉)
e-mail: christoph.garbe@iwr.uni-heidelberg.de

A. Rutgersson (✉)
e-mail: anna.rutgersson@met.uu.se

mechanisms, but also because of uncertainties in the different methodologies.

Much of the interest in air-sea gas and particle transfer relates to global or regional phenomena, by using models or remote sensing products to determine oceanic sinks and sources of gases and particles. Upscaling of transfer involves relating the transfer to the environmental factors that influence the exchange. Figure 2.1 summarises factors most likely to be of great importance. Most commonly the transfer is described by a wind speed dependent function. There is, however, increasing understanding that a variety of processes influence gases of varying properties (like solubility) and particles differently.

The transport of a quantity is characterised by the *transfer velocity*, k. The transfer velocity is given by the flux F of the transported quantity divided by the "concentration difference" of the quantity between surface (C_0) and the bulk (C_{bulk}):

$$k = \frac{F}{C_0 - C_{bulk}}. \quad (2.1)$$

For the transport of CO_2, the water-sided transfer velocity k_w is often related to the partial pressure difference ΔpCO_2 between air and water by $k_w = F/(\alpha \Delta pCO_2)$. The transfer velocity k will then depend both on the Schmidt number (Sc) and the dimensionless solubility α. It is related to the friction velocity u_* and the Schmidt number $Sc = \nu/D$ given by

$$k = u_* \beta(s) Sc^{n(s)}, \quad (2.2)$$

where both $\beta(s)$ and $n(s)$ depend on parameters describing the surface conditions and the related transport processes. If heat is the transported quantity of interest, then Sc is substituted by the Prandtl number $Pr = \nu/\varkappa$ in this equation. Here ν is the kinematic viscosity, D is the molecular mass diffusivity and $\varkappa$ is the thermal diffusivity.

There exist a number of previous books and reviews concerning air-sea gas and particle transfer (e.g. Fairall et al. 2000; Wanninkhof et al. 2009; Le Quéré and Saltzman 2009; de Leeuw et al. 2011). Here we summarise mechanisms of importance, measurement techniques frequently used to investigate the transfer, present state-of-the-art parameterisations and applications of the air-sea transfer description. The accuracy of the various methods and the importance of measurement uncertainty for global flux estimates are further discussed.

2.2 Processes

The major processes influencing the air-sea transfer velocity are described here. This includes processes important for a relatively insoluble gas like CO_2: micro scale wave breaking, small and large scale turbulence in the water, waves, bubbles, sea spray, rain and surface films (see Fig. 2.1). In addition biological and chemical enhancement is described having been recognised as important for a small number of gases (including O_3 and SO_2). Atmospheric processes are also briefly described due to the importance for turbulence generation at the surface as well as their direct importance for the transfer velocity of air-side controlled (soluble) gases.

2.2.1 Microscale Wave Breaking

Microscale wave breaking, or microbreaking (Banner and Phillips 1974), is the breaking of very short wind waves without air entrainment and begins at wind speeds well below the level at which whitecaps appear (Melville 1996). Laboratory and field observations show microbreaking waves are ~ 0.1–1 m in length and ~ 0.01–0.1 m in amplitude and have a bore-like crest with parasitic capillary waves riding on the forward face (Zappa et al. 2001).

Other defining characteristics exist between microbreaking and whitecapping. Oh et al. (2008) have shown that a single strong coherent structure develops beneath a large-scale whitecapping breaking crest. The coherent structure rotates in the same sense as the wave orbital motion. In contrast, a series of coherent structures whose rotation sense is not fixed are generated beneath microscale breakers. However, the overall characteristics of the spatio-temporal evolution of the coherent vortical structures are qualitatively the same between the two types of wave breaking. It is also significant that unlike energetic whitecapping breakers, no jet is formed at the crest of a microscale breaking wave, suggesting that surface

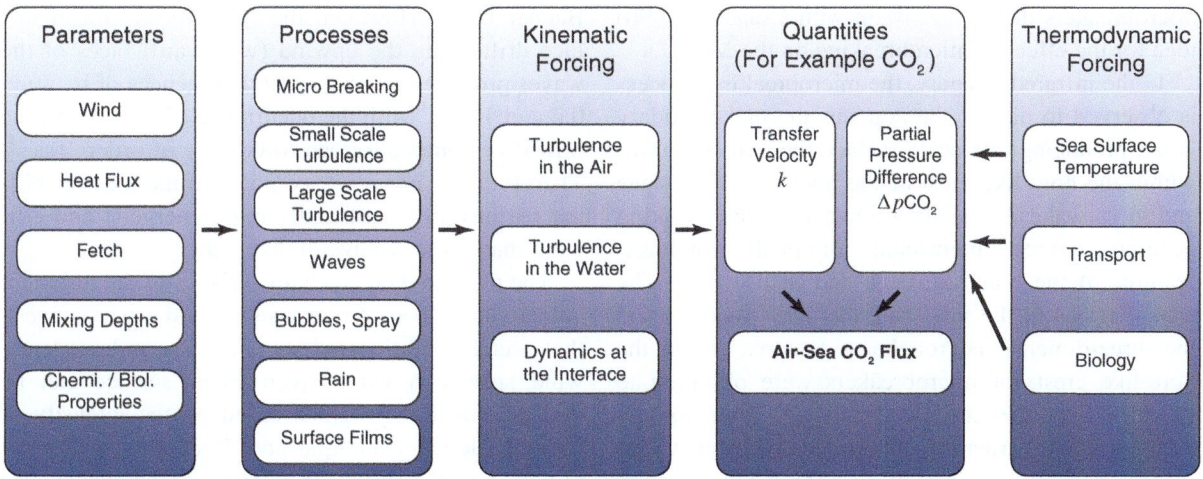

Fig. 2.1 Simplified schematic of factors influencing air-sea CO_2 fluxes. On the *right* are factors that affect the air-sea pCO_2 difference (thermodynamic forcing). On the *left* are environmental forcing factors that control the efficiency of the transfer (kinematic forcing)

tension has a significant impact on the breaking of waves shorter than 1 m (Tulin and Landrini 2001). Instead, the bore-like crest of the microbreaker can propagate for a considerable distance without significant change of shape. Microbreakers inherently lose a significant portion of their height and dissipate their energy during the breaking.

Laboratory experiments in wind-wave tunnels provide a growing body of evidence that turbulence generated by microscale wave breaking is the dominant mechanism for air-water gas transfer at low to moderate wind speeds. Laboratory measurements indicate that a wave-related mechanism regulates gas transfer because the transfer velocity correlates with the total mean-square wave slope, $\langle S^2 \rangle$ (Jähne et al. 1987). Wave slope characterises the stability of water waves (e.g. Longuet-Higgins et al. 1994) and certain limiting values of slope are typically used to detect and define wave breaking (e.g. Banner 1990). Therefore, Jähne et al. (1987) have argued that wave slope is representative of the near-surface turbulence produced by microscale wave breaking.

Microbreaking is widespread over the oceans, and Csanady (1990) has proposed that the specific manner by which it affects k_w is the thinning of the aqueous mass boundary layer (MBL) by the intense surface divergence generated during the breaking process. Infrared (IR) techniques have been successfully implemented in the laboratory to detect this visually ambiguous process and have been used to quantify regions of surface renewal of the MBL caused by microbreakers (Jessup et al. 1997a). Analogous to the MBL, Wells et al. (2009) suggest that an appreciable change in skin temperature of the thermal boundary layer requires a large strain rate, such as might be generated by high vorticity wavelets (Okuda 1982) or microbreaking waves. Skin layer straining may be important for large Péclet (Pe) number flows. The Péclet number is defined to be the ratio of the rate of advection to the rate of diffusion of a physical quantity. In the context of heat transport, the Péclet number is given by the product of the Reynolds number and the Prandtl number. In the context of mass diffusion, the Péclet number is the product of the Reynolds number and the Schmidt number. For example, Peirson and Banner (2003) and Turney et al. (2005) measured surface divergence of approximately 1–10 s^{-1} for microbreaking surface waves, yielding Pe ~ 10–100 for a 0.1 cm skin layer. In addition, flow separation can produce high vorticity wavelets with Pe ~ 200 (Csanady 1990). On the basis of the experimental data, the variation of the skin temperature for steady large Péclet number flows is reasonably well described by $\frac{Q}{\rho C_p}\sqrt{\frac{\pi}{2\kappa E}}$, where Q is the net heat flux, ρ is the density, C_p is the specific heat capacity at constant pressure, κ is the thermal diffusivity and E is the strain rate (Wells et al. 2009). The IR signature provides qualitative evidence of the turbulent wakes of microbreakers that is consistent with Csanady (1990) idea for the effect of microbreaking on the MBL.

In the infrared imagery, the microbreaking process is observed to disrupt the aqueous thermal boundary layer, producing fine-scale surface thermal structures within the bore-like crest of the microbreaking wave and in its wake. In the slope imagery, microbreaking generates three-dimensional dimpled roughness features of the bore-like crest and in its wake. The spatial scales of the fine-scale thermal structures and the three-dimensional roughness features within the bore-like crests of microbreakers were observed to be $\sim 10^{-2}$ m, the same as the length scale for the vortical eddy structures measured beneath wind-forced waves by Siddiqui et al. (2004). Furthermore, the dimpled roughness features of the bore-like crest correspond directly to the warmest fine-scale features of the skin-layer disrupted by microbreaking. The implication is that the fine-scale thermal structures and the dimpled roughness features are directly related to the near-surface turbulence generated at the bore-like crests. Because near-surface turbulence increases k_w, microbreakers are likely to be the mechanism that enhances both heat and mass transfer.

Understanding the nature and key features of the interfacial flows associated with small-scale waves is fundamental to explaining the prominent enhancement of air-water exchange that occurs in their presence. Within a few hundred microns of the surface of laboratory microscale breaking wind waves, the mean surface drift directly induced by the wind on the upwind faces and crests is $(0.23 \pm 0.02)u_{*a}$ in the trough increasing to $(0.33 \pm 0.07)u_{*a}$ at the crest, where u_{*a} is the wind friction velocity (Peirson and Banner 2003). Substantial variability in the instantaneous surface velocity exists up to approximately $\pm 0.17u_{*a}$ in the trough and $\pm 0.37u_{*a}$ at the crest. Peirson and Banner (2003) attribute this variability primarily to the modulation of the wave field, although additional contributions arise due to the influence of transient microscale breaking or from near-surface turbulence generated by shear in the drift layer or from fluctuations in wind forcing.

Microscale wind-wave breaking plays an important role in the direct transport of fluid from the surface to the turbulent domain below. The toe of a microscale breaker spilling region exhibits an intense and highly localized convergence of surface fluid, with convergence rates generally exceeding 100 s^{-1} (Peirson and Banner 2003). However, the variability in mean surface drift along the upwind (windward) faces of the waves produces mean surface divergences of between 0.2 and 1.3 s^{-1} with the occurrence of locally intense flow divergence observed to be only of order 10 s^{-1}. Therefore, the convergence zones at the toes of spilling regions are significantly more energetic and efficient than these windward divergence zones.

The phase-averaged characteristics of the turbulent velocity fields beneath steep short wind waves suggest, that under conditions of short fetches and moderate wind speeds, a wind-driven water surface can be divided into three regions based on the intensity of the turbulence (Siddiqui and Loewen 2010). The regions are the crests of microbreaking waves, the crests of non-breaking waves and the troughs of all waves. The turbulence is most intense beneath the crests of microbreaking waves. In the crest region of microbreaking waves, coherent structures are observed that are stronger and occur more frequently than beneath the crests of non-breaking waves. Beneath the crests of non-breaking waves the turbulence is a factor of 2–3 times weaker and beneath the wave troughs it is a factor of 6 weaker.

The mean velocity profiles in the wind drift layer formed beneath short wind waves have been shown to be logarithmic and the flow is hydrodynamically smooth at short fetch up to moderate wind speeds (Siddiqui and Loewen 2007). The turbulent kinetic energy dissipation rate for these conditions is significantly greater in magnitude than would occur in a comparable wall-layer. At a depth of 1 mm, the dissipation rate is 1.7–3.2 times greater beneath microbreaking waves compared to non-breaking waves. In the crest–trough region, 40–50 % of the dissipation is associated with microbreakers. These results demonstrate that the enhanced near-surface turbulence in the wind drift layer is the result of microscale wave breaking.

Terray et al. (1996) proposed that the turbulent kinetic energy dissipation rate due to wave breaking is a function of depth, friction velocity, wave height and phase speed. Vertical profiles of the rate of dissipation showed that beneath microscale breaking waves there were two distinct layers (Siddiqui and Loewen 2007). Immediately beneath the surface, the dissipation decayed as $\zeta^{-0.7}$ and below this in the second layer it decayed as ζ^{-2} where ζ is the vertical coordinate in the wave-following Eulerian system.

The enhanced turbulence associated with microscale wave breaking was found to extend to a depth of approximately one significant wave height. The only similarity between the flows in these wind drift layers and wall-layers is that in both cases the mean velocity profiles are logarithmic.

A comprehensive study was performed to show that microbreaking is the dominant mechanism governing k_w at low to moderate wind speeds (i.e. nominally up to 10 ms^{-1}) (Zappa et al. 2001, 2004). Simultaneous and collocated IR and wave slope imagery demonstrated that the IR signature of the disruption of the aqueous thermal boundary layer corresponds directly to the wakes of microbreakers (Zappa et al. 2001). Concurrently, simultaneous particle image velocimetry and IR imagery showed that vortices are generated behind the leading edge of microbreakers and that the vorticity correlated with k_w (Siddiqui et al. 2001). Furthermore, k_w was shown to be linearly correlated with the fraction of water surface covered by the wakes of microbreaking (A_B) and this correlation was invariant with the presence of surfactants (Zappa et al. 2001). This correlation is evidence for a causal link between microbreaking and gas transfer. Zappa et al. (2004) showed that the surface renewal in the wakes of microbreakers enhanced transfer by a factor of 3.4 over that in the background. Furthermore, microbreaking directly contributed up to 75 % of the transfer across the air-water interface under moderate wind speed conditions. Zappa et al. (2004) argued that these results show conclusively that microbreaking is an underlying mechanism that explains the observation of enhanced gas transfer in the presence of waves and may govern air-sea gas transfer at low to moderate wind speeds.

Microbreaking events directly enhance heat and gas transfer and produce surface roughness elements that contribute directly to $\langle S^2 \rangle$. Also, A_B was correlated with $\langle S^2 \rangle$. Since many capillary and non-breaking gravity waves may dominate $\langle S^2 \rangle$, these results suggest there is an indirect link between microbreaking and the correlation of transfer velocity with $\langle S^2 \rangle$. Furthermore, the modulation of capillary waves as microscale breaking waves evolve and the transient roughness features associated with microbreaking have shown promise in identifying individual microbreaking events. A comprehensive examination of the IR and wave slope imagery data should clarify the relationship between microbreaking and $\langle S^2 \rangle$.

The correlation of k_w with wave slope has shown results similar to the correlation of k_w with A_B. Jähne et al. (1987) observed a correlation of k_w with $\langle S^2 \rangle$ for both fetch-limited and unlimited fetch cases, and Frew (1997) and Bock et al. (1999) showed the correlation was independent of surfactant for the unlimited fetch case. Zappa et al. (2004) suggest that the reason k_w correlates with wave slope is that microbreaking is the wave-related mechanism controlling gas transfer and contributes to $\langle S^2 \rangle$.

The greatest contribution of microbreaking to $\langle S^2 \rangle$ occurs in areas not directly affected by active microbreaking. The density of capillary waves present during the microscale wave breaking process is less than for non-breaking waves. Specifically, the capillary waves on the forward face of the bore-like crest become extremely short during the most intense initiation of microscale breaking when steep slopes occupy the dimpled bore-like crest. The wave field slope characteristics are constantly evolving throughout the growth process of the wave packet from the moment capillary waves are formed to the moment that microscale wave breaking occurs. The wave evolution up to microbreaking incorporates contributions of slopes from waves of all scales. Microscale wave breaking is of short duration and merely one component of this wind-wave cycle. Dense packets of capillary-gravity waves ubiquitous in regions not affected by microbreaking dominate the contribution to the total mean-square wave slope.

Since the bore-like crest produces the signature of breaking detected as A_B in the infrared imagery, only the steep slopes associated with the front of the actively breaking crest and the wake of a microscale breaker influence $\langle S^2 \rangle$. Capillary waves have been shown to contribute significantly to $\langle S^2 \rangle$ (Bock et al. 1999) and are observed to be transient during the microscale breaking process. The fact that capillary waves are damped by surfactants, coupled with the fact that the smallest scale waves have been shown to correlate with the gas transfer velocity, suggests that capillary waves characterise the importance of the wave field to gas transfer. The potential for capillary waves as a direct mechanism for gas transfer has been demonstrated both experimentally (Saylor and Handler 1997) and theoretically (Coantic 1986; Szeri 1997; Witting 1971). However, microbreaking clearly dominates over these capillary wave processes. The contribution of capillary waves to $\langle S^2 \rangle$ is symptomatic of a simultaneous increase in overall microbreaking and the densely structured capillary-gravity wave system that increases with wind forcing. The link between k_w and $\langle S^2 \rangle$ is more complicated than simply stating that microbreaking is the direct link. It is more likely that microbreaking controls k_w, the capillary-gravity wave system facilitates microbreaking, and the non-breaking capillary and gravity waves contribute significantly to $\langle S^2 \rangle$.

Breaking of small-scale waves can be affected by longer waves in a number of ways that provide insights into the hydrodynamics of formation, propagation and evolution of microbreaking waves. One of them is modulation of the train of the short waves riding the underlying large-scale waves (Babanin et al. 2009). The large-scale waves compress the short-wave lengths at their front face and extend those at the rear face. As a result, the front-face small waves become steeper and frequently break (Donelan 2001). Small-scale breaking is also affected by the breaking of large waves (Banner et al. 1989; Manasseh et al. 2006; Young and Babanin 2006). The growth of short waves is affected by the existence of energetic breaking waves, which not only modulate the microbreakers, but virtually eliminate them as the microbreakers are overcome by the energetic breakers. In principle, short waves can be dissipated without breaking, because of interaction with the turbulent wake of the large breaker (Banner et al. 1989). Regardless, microbreakers begin growing again from very short waves and their eventual length is determined by the effective fetch between energetic breakers.

2.2.2 Small Scale Turbulence

The kinematic viscosity and thermal diffusivity of water differ by an order of magnitude. Hence, the thermal sublayer is embedded within the viscous sublayer, its momentum analogue.

This points to the significance of small-scale processes, such as near-surface turbulence, for the transport of heat and gas though the interface (McKenna and McGillis 2004). Eddies close to the surface bring "fresh" bulk water to the surface from which gases can diffuse in or out of the water. Overall this leads to enhanced gas exchange.

Wind generates turbulence at the air-water interface directly through shear or indirectly through the wave field. Turbulence generated by wind-induced waves will be covered in Sect. 2.2.4. In this section, we are focusing on shear induced turbulence.

The turbulence generated at the air-water interface can be isotropic or, more importantly, coherent. Numerical simulations performed by Tsai et al. (2005) and experiments conducted by Rashidi and Banerjee (1990) have shown that shear stress alone without waves is sufficient for coherent turbulent structures to evolve. The structure of coherent shear induced turbulence has implications for turbulent kinetic energy dissipation (c.f. Sect. 2.6.4). The shear induced turbulence is of a similar scale to the shear layer thickness (Hunt et al. 2011). Infrared thermography will be introduced in Sect. 2.5.1.2. In infrared images of the water surface, the near surface turbulence becomes visible because areas of surface divergence consist of water parcels that have been in contact with the water surface for a longer time. In the presence of a net heat flux away from the interface, such water parcels cool down and evolve into cold streaks. The opposite holds true for upwelling regions. The temperature distribution of such coherent structures visualized with thermography can be seen in Fig. 2.2. Such patterns were termed fish-scales by Tsai et al. (2005) and Handler et al. (2001) and consist of narrow cold bands of high speed fluid.

The narrow dark streaks form the characteristic intersections when they encounter slow warm plumes that "burst" into the surface. Schnieders et al. (2013) performed a statistical analysis of the streak spacing. These spacings were extracted accurately using an image processing approach based on supervised learning.

The streaky pattern observed at the water surface is very similar to the pattern of low speed streaks near no-slip walls, as described by Nakagawa and Nezu (1981) and Smith and Paxson (1983). They found the spacing between these streaks to be lognormally distributed with a mean dimensionless streak spacing of $l^+ = lu_{*w}/\nu = 100$, where the mean streak spacing is given by l, the friction velocity by u_{*w} and the kinematic viscosity by ν. The factor u_{*w}/ν which when divided by l is dimensionless is estimated to be the inverse of the thickness of the thermal boundary layer (Grassl 1976).

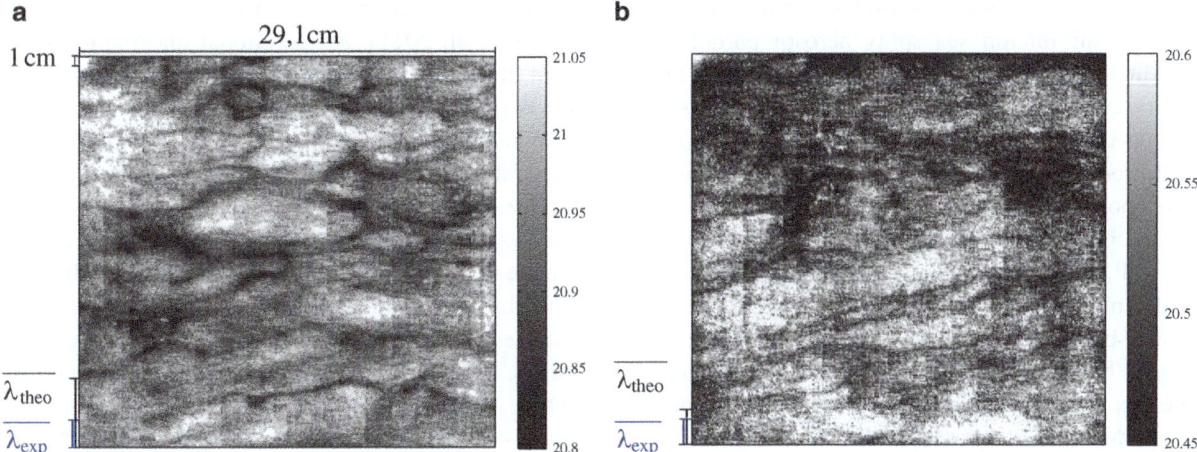

Fig. 2.2 Thermographic images of the water surface at two different wind conditions. $\overline{\lambda_{\text{theo}}} = 100 u_{*w}/\nu$ describes the expected value of the mean dimensionless streak spacing in the case of low speed streaks near a no-slip wall, with the friction velocity given by u_{*w}, and the kinematic viscosity by ν. $\overline{\lambda_{\text{exp}}} = \overline{l_{\text{exp}}} u_{*w}/\nu$ describes the mean dimensionless streak spacing, where l is the determined mean streak spacing for each case (Images reproduced from Schnieders et al. (2013))

The generally agreed scaling of $l^+ = l u_{*w}/\nu = 100$ leads to the scaling of the streak spacing with the friction velocity in water u_{*w}. With increasing u_{*w} the spacing of the streaks is therefore supposed to decrease linearly in the regime of low wind speeds (Scott et al. 2008). Schnieders et al. (2013) performed a statistical analysis of the streak spacing in different laboratory faciliets in different conditions. They found the spacing to plateau for higher wind speeds.

Although the fish-scale pattern seems to be universal and has been described by several authors, the process that causes these streaks has not been conclusively identified. It has been suggested (Tsai et al. 2005) that low speed streaks near no-slip walls and high speed streak near the water surface are caused by a similar process. Tsai et al. (2005) point out that one possible mechanism causing these streaks is that of horseshoe vortices that form in the turbulent shear layer by turning and stretching the spanwise vortices (see Kline et al. (1967)). These coherent turbulent structures move upward until their upper ends burst into the surface and subsequently lead to upwelling of warmer and slower water at the surface.

Chernyshenko and Baig (2005) suggest another mechanism for the formation of streaks. In their scheme, the lift-up of the mean profile in combination with shear stress and viscous diffusion lead to the formation of the streaky pattern. Currently, no conclusive measurements of these processes have been conducted.

2.2.3 Bubbles, Sea Spray

Transfer of matter across the air-sea interface may occur not only by molecular diffusion of volatiles across the sea surface but also through the mediation of bubbles and sea spray. Non-volatile material can only be ejected from the sea into the lower atmosphere within sea spray. Sea spray can be produced by "tearing" of the sea surface by the action of the wind ("spume drops"); which is most efficient near the crests of breaking waves and sharply crested waves. This tearing process is mostly effective at wind speeds greater than 10 ms^{-1} (Monahan et al. 1983), though it can be observed at the crest of breaking waves at moderate wind speeds. The spume droplets are generally large, ranging from tens of micrometres to a few millimetres. These large droplets will generally only be airborne for a short time (seconds to minutes) but are significant for their role in the transfer of moisture and latent heat (Andreas 1992). Smaller droplets that are more likely to evolve over hours or days in the marine atmosphere are almost entirely associated with the bursting of bubbles at the sea surface. A comprehensive review of sea salt aerosol production was presented by Lewis and Schwartz (2004), the production of sub-micron sea spray aerosol particles was recently reviewed by de Leeuw et al. (2011). An overview of both production of sea spray aerosol and effects of aerosols produced over land on biogeochemical processes in the ocean is presented in Chap. 4 of this book (de Leeuw et al. 2013).

Bubble generation at the sea surface is associated with all types of precipitation falling on the sea surface and can result from supersaturation of air in the upper ocean, but is primarily associated with air entrainment within breaking waves. When these bubbles burst at the sea surface they produce two types of drops, jet drops and film drops (Blanchard 1963; Woolf et al. 1987). Jet drops are pinched off from the "Worthington jet" that projects from the open cavity of a bursting bubble and their initial radii are typically a tenth of the radius of the parent bubble. Jet drops are inferred to be the main source of the important subset of marine aerosol particles with radii between 1 and 25 μm radius (de Leeuw et al. 2011). Film drops were originally assumed to be formed from the shattering of the film cap of large bubbles, but a more complicated picture emerged from later investigations (Spiel 1998). Nevertheless the numerous small sea spray particles in the lower atmosphere (approximately 10 nm–1 μm) are associated with film drop production.

The study of small sea spray particles underpins our understanding of direct and indirect radiative effects of this aerosol (the latter through their role in cloud microphysics) and chemical reactions and pathways in the lower atmosphere involving both the sea spray particles and natural and anthropogenic volatiles. Models and quantitative estimates of these processes requires a "sea spray source function" (SSSF) and estimates of this function have recently been reviewed (de Leeuw et al. 2011). Several methods have been proposed for estimating SSSF including those based on a balance between production and deposition (wet or dry), micrometeorological methods and the whitecap method. Micrometeorological methods are showing considerable promise, but most estimates are based on the whitecap method. The whitecap method requires an assessment of aerosol production by a "standard whitecap", usually by measuring the aerosol production by a simulated whitecap in the laboratory, which is then scaled to the ocean using predictions of whitecapping. A serious problem remains since "order-of-magnitude variation remains in estimates of the size-dependent production flux per white area" (de Leeuw et al. 2011). An additional uncertainty arises from the challenge of parameterising whitecap coverage W (see below).

When bubbles are formed they will accumulate material on their surface that was previously on the sea surface. As they are mixed through the upper ocean, the bubbles may scavenge further material (primarily surface-active) from the water column. Some bubbles will dissolve leaving fragments (a "microbubble" enclosed in a shell of organics, or particles), but most surface and burst. Material carried to the surface on the bubble, or skimmed from the sea surface in the bursting process may be ejected on the sea spray. The cycling of organic material described above represents another role of bubbles in geochemical transport and air-sea transfer. It is recognised that the sea surface microlayer is highly dynamic when waves are breaking and constantly renewed by the action of bubbles (Liss et al. 1997). It follows that bubbles and spray have an indirect effect on interfacial transfer through their effect on the composition of the sea surface (see Sect. 2.2.7). The sea spray aerosol will be enriched in organic material and much of the current interest and progress in studying that aerosol focuses on the organic content and its effects (de Leeuw et al. 2011).

Air-sea transfer of volatiles is also dependent on bubbles and spray. In principle, sea spray should have some effect on the deposition or exchange of soluble gases across the atmospheric surface layer, but this pathway is usually not explicitly identified in parameterisations of air-side transfer velocity (Johnson 2010). Far more attention has been given to the role of bubbles in the air-sea exchange of poorly soluble gases. As already mentioned, bubbles may influence air-sea gas transfer through their influence on the composition of the sea surface. Both breaking waves and surfacing bubbles generate turbulence and this may enhance transfer at the sea surface. The greatest influence on poorly soluble gases appears to be through "bubble-mediated transfer", defined as the net transfer of gas across the surface of bubbles while they are submerged (Woolf and Thorpe 1991). Both laboratory experiments and numerical calculations suggest that bubble-mediated transfer is a highly effective mechanism for poorly soluble gases. The effect of bubbles on gas transfer can be seen in Fig. 2.3 where the transfer velocity k_w is shown for both CO_2 and dimethylsulfide (DMS). DMS has a much higher solubility coefficient

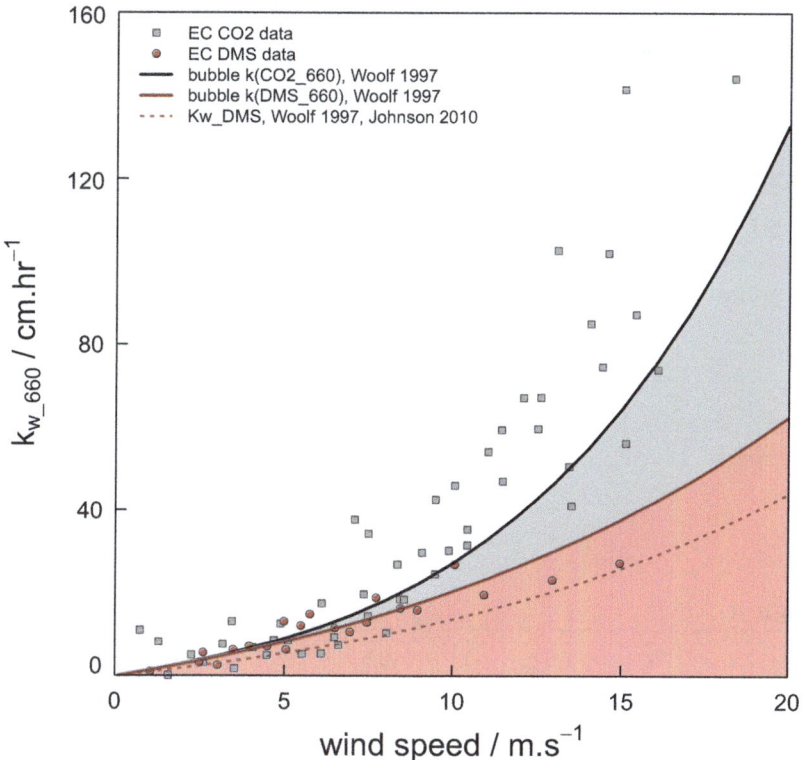

Fig. 2.3 The magnitude of the bubble effect on transfer velocity for the relatively soluble gas dimethylsulpide (*DMS*) compared to less soluble CO_2. *Pink dots* summarise eddy covariance estimates of DMS transfer velocity to date and *grey squares* those for CO_2 (detailed in Fig. 2.10). These are plotted over predicted k_{660}-normalized transfer velocities from the bubble model of Woolf (1997) for DMS (*dark pink solid line*) and CO_2 (*solid black line*), using solubility at 15 °C from Johnson (2010). For DMS, the bubble-mediated transfer is very small, so the *pink* shaded area effectively represents the diffusive transfer, which is approximately equal for both CO_2 and DMS (at the same Schmidt number), with the *grey* shaded area showing the additional bubble-mediated transfer for CO_2. Total transfer velocity, K_W, for DMS is also calculated using the scheme of Johnson (2010), applying the Woolf (1997) bubble parameterization for k_w and the Johnson (2010) k_a term (Figure by M.T. Johnson, shared under creative commons license at http://dx.doi.org/10.6084/m9.figshare.92419)

than CO_2. When the effect is scaled to the ocean by variants of the whitecap method, bubble-mediated gas transfer is calculated to contribute substantially to the transfer of poorly soluble gases (Keeling 1993; Woolf 1993, 1997; Asher et al. 1996). The inclusion of bubble-mediated transfer is crucial to the successful application of physically-based models such as NOAA-COARE (see Sect. 2.6.3) to the air-sea transfer of carbon dioxide and other poorly soluble gases (Fairall et al. 2011).

Bubble-mediated gas transfer is distinct in several respects from direct transfer of gas across the sea surface. Firstly, since bubbles may dissolve and are always subject to additional pressure, bubble-mediated transfer is asymmetric with a bias to invasion of gas (Woolf and Thorpe 1991). Secondly, through the change in composition of a bubble while it is submerged, the contribution of bubble-mediated transfer to the water-side transfer velocity of a gas is dependent on solubility and has a complicated dependence on Schmidt number. This feature complicates the interpretation of dual tracer experiments (Sect. 2.5.3; Asher and Wanninkhof 1998) and affects the applicability of gas transfer parameterisations across gases and the range of water temperature (see Sect. 2.6.5). Some estimates of the sensitivity of transfer velocity to whitecap coverage are depicted in Fig. 2.4, emphasising particularly the theoretical sensitivity to solubility and structural uncertainty relating to the specific model.

The precise dependences of bubble-mediated transfer on the environment and the molecular properties of

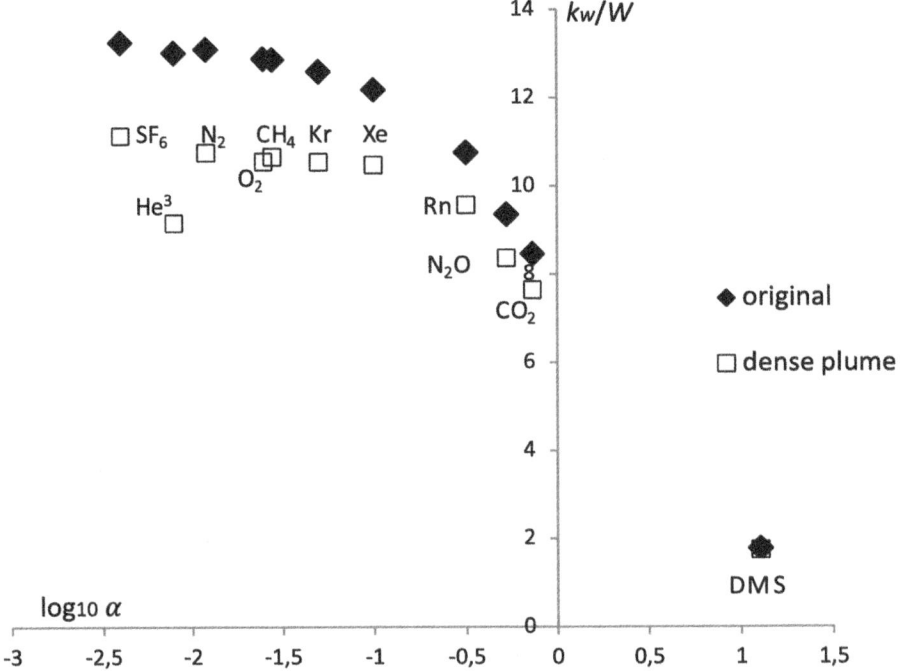

Fig. 2.4 Sensitivity of gas transfer velocity to whitecap coverage (k_w/W in cm(h %)$^{-1}$) plotted against logarithm of Ostwald solubility ($\log_{10} \alpha$). Based on the original "independent bubble model" (*filled diamonds*) and "dense plume model" (*open squares*) of Woolf et al. (2007). Dense plume model calculated for a plume void fraction of 20 %. All transfer velocities calculated for seawater at 20 °C using gas constants suggested by Wanninkhof et al. (2009) and normalised to a Schmidt number of 660

the dissolved gas require further investigation (Woolf et al. 2007). A third important feature of bubble-mediated gas transfer is that since it is dependent on wave breaking, whitecapping and the dispersion by mixing processes in the upper ocean (see Sect. 2.2.5), its environmental dependence (on wind speed, sea state, water temperature ...) is a function of the environmental dependence of these processes. In particular, in common with the sea spray source function, the environmental dependence of whitecap coverage is critical for quantification of bubble-mediated gas transfer.

Whitecap coverage W is a practical but enigmatic measure of the whitening of the sea surface associated with wave breaking, air entrainment, bubble plumes and surface foam. Its indefinite nature and practical obstacles to its systematic measurement make applying field measurements and parameterisations of whitecapping within the whitecap method difficult (de Leeuw et al. 2011). Nevertheless, studies of historical data sets (Bortkovskii and Novak 1993; Zhao and Toba 2001) have elucidated the environmental dependence of whitecapping. There have also been significant advances in its systematic measurement (Callaghan and White 2009) that enable more detailed analysis of environmental dependence (Goddijn-Murphy et al. 2011). Wind speed is the main driver of whitecapping and wind-speed-only parameterisations are useful (Monahan and O'Muircheartaigh 1980; Goddijn-Murphy et al. 2011), but their accuracy for instantaneous whitecap coverage is poor. Recent measurements all indicate that the whitecap fraction is lower than that predicted by Monahan and O'Muircheartaigh (1980) which is most frequently used in models (de Leeuw et al. 2011). It may be inferred that the sea spray source function and gas transfer velocities are also likely to vary greatly at a given wind speed. Water temperature and sea state appear to be major additional factors in determining whitecap coverage, and by implication aerosol production and gas transfer. Long et al. (2011) suggests scaling particle production by wave-breaking energy.

2.2.4 Wind-Generated Waves

Wind-generated waves affect processes at the air-sea interface in several ways. In the form of variable roughness elements and group speeds, they interact with and modify the wind field above the sea surface (c.f. Sects. 2.2.9 and 2.4.1). The Stoke's drift caused by the waves is involved in the generation of the Langmuir circulation which, in turn, is one of the mixing processes of the surface layer of the ocean (Sect. 2.2.5) and affects the distribution of surfactants (Sect. 2.2.7). Breaking waves inject bubbles (Sect. 2.2.3) and cause turbulence in the surface waters contributing to the surface processes and to the mixing of the surface layer. Another source of turbulence is the microscale breaking of short waves (Sect. 2.2.1).

The properties of the irregular wave field cannot be parameterised by wind speed alone. For example, wave breaking depends on the wave steepness and not on the wind speed. In addition to the wind speed, the development of waves depends also on wind duration, fetch, the shape of the basin, water depth and atmospheric stratification. Typically, the wave field consists of one or several wave systems, e.g. locally generated waves and a swell originating from a distant storm. Their influence on the interaction with the atmosphere varies with state of development, the relative dominance of the wave systems and the difference of their directions (e.g. Donelan et al. 1997; Drennan et al. 1999; Veron et al. 2008). The properties of the wave field can be quite different in the oceans to that in marginal seas. While in the oceans the waves more often reach full development, in areas closer to the shore waves are typically strongly forced due to the limited fetch and are thus steeper. Depending on the fetch geometry, the direction of the waves can differ from the wind direction (Pettersson et al. 2010). If the sea is shallow enough for the waves to 'feel' the bottom, there will be further changes in the steepness of the waves and changes in wave directions leading to areas of wave energy convergence and divergence. In high latitudes the seasonal ice cover forms a changing fetch and fetch geometry affecting the properties of the wave field adjacent to the ice edge.

For transfer across the air-sea interface the turbulence in the subsurface layer is one of the key factors. When present, the breaking waves contribute strongly to the subsurface turbulence. Babanin (2006) has suggested that the orbital motion of the waves causes turbulence also in the absence of the breaking waves.

In view of the importance of the water-side turbulence for the processes at the interface and the surface layer of the ocean, studies have been undertaken on the dissipation of the turbulent kinetic energy in the upper layer. The experimental data needed in these studies are not easy to obtain (see Terray et al. (1996) and Soloviev et al. (2007) for a review), but the limited data shows that in the presence of breaking waves the dissipation values are higher than given by the shear-driven wall layer approach (e.g. Agrawal et al. 1992; Osborne et al. 1992; Anis and Moum 1995; Drennan et al. 1996; Terray et al. 1996; Phillips et al. 2001; Zappa et al. 2007). Anis and Moum (1995) measured high dissipation rates at depths greater than the height of the breaking waves and suggested that the orbital motions of swell could be one possible mechanism that transports the wave induced turbulence to deeper depths. On the other hand, Terray et al. (1996) found that the dissipation has a constant value to a certain depth below which the wall-layer behaviour is found. Their premise was that the dissipation is balanced with the wind input and the depth of the constant dissipation layer is of the order of the significant wave height. Based on his similarity analysis, Kitaigorodskii (1984, 2011) proposed the existence of a constant dissipation rate under breaking waves. He concluded that the dissipation is dependent on the water-side friction velocity and the turbulent viscosity constant in the layer of constant dissipation. In both approaches (Terray et al. 1996; Kitaigorodskii 1984, 2011), the dissipation is dependent on the state of development of the waves. Sutherland et al. (2013) compared measured dissipation rates in the surface ocean boundary layer to various scalings (e.g. Terray et al. (1996) and Huang and Qiao (2010)). They found that the depth dependence was consistent with that expected for a purely shear-driven wall layer. Many dissipation profiles scaled with a Stokes drift-generated shear, suggesting there may be occasions where the shear in the mixed layer are dominated by wave-induced currents.

So far there have been only a few experimental studies on the dependence of gas exchange on the dissipation caused by the breaking waves. Zappa et al. (2007) demonstrated the dependence of gas transfer velocity on the subsurface dissipation rate.

Presently there are on-going field campaigns aiming to study this question (e.g. Pettersson et al. (2011) and using the ASIP profiler see Sect. 2.6.4). Kitaigorodskii (2011) proposed a wave-age-dependent transfer velocity based on his considerations of the dissipation rate and previously published data. Soloviev et al. (2007) have proposed a gas exchange parameterisation based on present knowledge of the different processes in the exchange. Their model includes three sources of turbulence, the convective, shear and wave-induced turbulence. The model predicted that at wind speeds of up to $10\,\mathrm{ms^{-1}}$ the exchange is not dependent on the wave age due to factors that cancel each other. At higher wind speeds, the bubble-mediated transfer is dominating. Due to the lack of coincident observations, this behaviour could not be confirmed and also hindered a detailed analysis of the relative importance of different sources of the turbulence.

2.2.5 Large-Scale Turbulence

Upper ocean dynamics is dominated by shear-generated eddies (small-scale turbulence) that coexist with buoyant plumes and wave-generated coherent structures (large-scale turbulence). The buoyant large eddies are generated by cooling at the surface resulting in convection extending to the bottom of the mixed layer, at which depth stability suppresses turbulence (Csanady 1997).

The surface cooling (net heat flux and evaporation) increases density of the surface water, thereby enhancing buoyancy. The buoyancy flux is defined according to e.g. Jeffery et al. (2007) as:

$$B = \frac{gaQ_{net}}{c_{pw}\rho_w} + \frac{g\beta_{sal}Q_{lat}}{\lambda\rho_w} \qquad (2.3)$$

where a is the thermal expansion coefficient, g acceleration due to gravity, Q_{net} is the net surface heat flux (i.e. sensible and latent heat flux plus net long-wave radiation), c_{pw} is the specific heat of water, ρ_w is the density of water, β_{sal} is the saline expansion coefficient, Q_{lat} is the latent heat flux, and λ is the latent heat of vaporisation. Following convective scaling in the atmosphere (see Eq. 2.11), the characteristic velocity scale of the turbulence generated by convection (the water-side convective velocity scale) is defined as follows (MacIntyre et al. 2002):

$$w_* = (Bz_{ml})^{1/3} \qquad (2.4)$$

where z_{ml} is the depth of the mixed layer. According to Eq. 2.4, a stronger buoyancy flux (a larger value of B) and a deeper mixed layer (a larger value of z_{ml}) produce enhanced convective mixing in the water. The convective velocity scale exhibits diurnal as well as seasonal cycles depending on variations in surface heating and variations of mixed layer depth. Rutgersson et al. (2011) suggested relating the convective velocity scale to the friction velocity of the water (u_{*w}) as u_{*w}/w_*. This parallels the description of atmospheric flow by combined convective and shear-generated turbulence, and characterises the comparative energetic roles of surface shear and buoyancy forces (Zilitinkevich 1994). This scaling can be used to express the additional parallel resistance to transfer initiated by the water-side convection. When wind is in the low to intermediate speed regime, convection is important for mixing, and it has in several studies been shown to enhance surface gas transfer (MacIntyre et al. 2002; Eugster et al. 2003; McGillis et al. 2004b; Rutgersson and Smedman 2010; Rutgersson et al. 2011). Water-side convection is expected to dominate in situations with cooling at the water surface (during the night, or during advection of cold air masses) and deepening of the mixed layer.

The interaction between the mean particle drift of surface waves (Stokes drift) and wind-driven surface shear current generates Langmuir circulation consisting of counter-rotating vortices roughly parallel to the wind direction (Langmuir 1938). The Langmuir circulation can be seen at the surface by the collection of surface foam in meandering lines in the along-wind direction or by sub-surface observations, following bubbles trapped in downwelling regions between vortices and current profiles (Smith 1998; Thorpe 2004; Gargett and Wells 2007). McWilliams et al. (1997) defined a turbulent Langmuir number describing the relative influence of directly wind-driven shear and Stokes drift:

$$La = \left(\frac{u_{*a}}{u_S}\right)^{1/2} \qquad (2.5)$$

where u_{*a} is the atmospheric friction velocity and u_s is the surface Stokes velocity. Sullivan and McWilliams (2010) summarise several studies focusing on the Langmuir mechanism and it is clear that Langmuir circulation (or Langmuir turbulence) greatly enhances turbulent vertical fluxes of momentum and heat at the surface.

A regime diagram for classifying turbulent large eddies in the upper ocean was suggested by Li et al. (2005) using the Hoennecker number (Li and Garrett 1995) as a dimensionless number comparing the unstable buoyancy force driving thermal convection with the wave forcing driving the Langmuir circulation. According to Li et al. (2005), the wind driven upper ocean is dominated by Langmuir turbulence under typical sea state conditions. Transition from Langmuir to convective turbulence occurs with strong thermal convection, relatively low winds and small surface velocity. An alternative description of the relative role of Langmuir and convectively generated turbulence is given by Belcher et al. (2012) as z_{ml}/L_L, where L_L is a convective-Langmuir number stability length as an analogue to Obukhov length for convective-shear turbulence.

Large scale turbulence disrupts the molecular sublayer and initiates a more efficient gas transfer at the surface. The large scale turbulence is dominated by Langmuir circulation or buoyancy depending on environmental conditions.

2.2.6 Rain

Because rain events in nature are episodic and rain rates are variable, only a few field studies have been conducted. These experiments, using natural rain and geochemical mass balances of O_2 invasion and SF_6 evasion in small plastic pools (Belanger and Korzun 1990, 1991; Ho et al. 1997), demonstrated that rain could significantly enhance $k_w(660)$, although the exact relationship between rain and $k_w(660)$ was not established in these studies.

Most systematic studies on the effect of rain on air-water gas exchange have been conducted in the laboratory using a rain simulator and a receiving tank. These studies began in the 1960s, where initial laboratory studies using O_2 invasion employed a rain simulators that had only 8–12 nozzles so only produced a few raindrops at a time, and only investigated a range of rain rates up to 17 mm h^{-1} (Department of Scientific and Industrial Research 1964; Banks and Herrera 1977; Banks et al. 1984). In the last two decades, systematic laboratory studies examining the full range of rain rates encountered in nature have quantified the effect of rain on gas exchange (Ho et al. 1997; Takagaki and Komori 2007), the interaction between rain and wind (Ho et al. 2007), the mechanism behind the enhancement (Ho et al. 2000), the effect in saltwater Ho et al. (2004); Zappa et al. (2009). Furthermore, some modeling studies have been conducted to examine the potential effect of rain on air-sea CO_2 exchange (Komori et al. 2007; Turk et al. 2010).

In laboratory experiments using evasion of He, N_2O, and SF_6 or invasion of CO_2, rain has been shown to enhance the rate of gas exchange significantly, and the relationship between $k_w(660)$ and rain can easily be related to either the kinetic energy flux or momentum flux of rain, both of which encompass variability in rain rate and drop size (Ho et al. 1997; Takagaki and Komori 2007). Enhancement in $k_w(660)$ is due mostly to increased near surface turbulence, whereas bubbles play a minor role (Ho et al. 2000). Experiments in saltwater demonstrate that the relationship between rain and $k_w(660)$ is the same as in freshwater, but density stratification could inhibit vertical mixing and decrease the overall gas flux (Ho et al. 2004; Zappa et al. 2009).

Some SF_6 evasion experiments have been conducted to examine the combined effects of rain and wind on gas exchange. Initial results indicated that the effect of rain and wind might be linearly additive (Ho et al. 2007), but further experiments have shown that the enhancement effects of rain fades with increasing wind speeds (Harrison et al. 2012), and wind speeds in the initial experiments were not high enough to exhibit that effect. While the exact mechanistic interaction between wind and rain has not been determined, it has been shown that when Kinetic Energy Flux (KEF) from wind (calculated from u_*) is greater than KEF from rain, the enhancement effects of rain is diminished (Harrison et al. 2012).

Modeling studies using relationships derived in the laboratory shows that the effect of rain on air-sea CO_2 exchange is insignificant on a global scale, but could be important on regional scales (Komori et al. 2007; Turk et al. 2010). However, those studies made simple assumptions that remain to be tested, including those about the dynamics of rain falling on the ocean, and about how rain and wind interact.

Future experiments should examine the effect of temperature change caused by rainfall on gas exchange, the interaction of rain with surfactants, and detail the mechanism behind the interaction of rain and wind and how they affect gas exchange. Field experiments should also be conducted in areas likely to be impacted by rain.

2.2.7 Surface Films

Surfactants can influence air-sea gas transfer rates via several mechanisms. Firstly, the presence of concentrated insoluble surfactant films (slicks) can act as a barrier to gas exchange, either by forming a condensed monolayer on the sea surface (Springer and Pigford 1970) or by providing an additional liquid phase that provides a resistance to mass transfer (Liss and Martinelli 1978). However, in the field, this effect is believed to be important only at low wind speeds, as slicks are easily dispersed by wind and waves (Liss 1983). The main effect of surface-active material is believed to be due to the occurrence of soluble surfactants that alter the hydrodynamic properties of the sea surface and hence turbulent energy transfer. Their presence reduces the roughness of the sea surface and hence the rate of micro-scale wave breaking (see Sect. 2.2.1), lowers sub-surface turbulence (see Sect. 2.2.2) and impacts on rates of surface renewal (see Sect. 2.3.1).

Early experiments in wind/wave tanks showed that artificial surfactants could cause reductions in k_w of up to 60 % for a given wind speed (Broecker et al. 1978). Indeed, contamination of the air-water interface by surface films was a common problem, particularly in circular tanks where there was no "beach" at the end of the tunnel for the film to collect (Jähne et al. 1987). Later experiments found that the presence of surfactants influenced the point at which small scale waves were observed at the water surface, co-incident with a rapid increase in k_w (Frew 1997). Even more interesting were observations that k_w could vary with biological activity (Goldman et al. 1988) due to the exudation of soluble surface-active material (including carbohydrates, lipids and proteins) by phytoplankton (Frew et al. 1990).

Supporting evidence came from laboratory experiments using seawater collected on a transect from the United States of America to Bermuda (Frew 1997) indicating that the decrease in k_w correlated inversely with bulk-water chlorophyll, dissolved organic carbon (DOC) and coloured dissolved organic matter (CDOM), see Fig. 2.5. The relationship of k_w with CDOM is important because this parameter can be determined remotely by satellite.

Although the importance of surfactants in determining air-sea gas transfer rates has long been recognised, it has proven exceedingly difficult to demonstrate that they exert a measurable effect in the field. Perhaps the first direct evidence came from a study of air-sea gas transfer rates using the heat flux technique (see Sect. 2.5.1) off the coast of New England in fairly light winds (Frew et al. 2004). The transfer of heat was measured inside and outside of a naturally occurring CDOM-rich slick and was found to be substantially inhibited by the presence of the slick (from 4.1 to 1.3 cm hr^{-1}). Due to the patchy nature of the slick and changeable wind speed the exact magnitude of the inhibition due to the slick is unclear.

It is commonly thought that surfactants may only be important in retarding gas transfer at very low winds. However, a big unknown is whether exchange rates might be reduced at higher winds when, despite wave breaking, surfactants are predicted to be brought back to the sea surface via bubble scavenging (Liss 1975). Support for this hypothesis came from Asher et al. (1996) who demonstrated, in a series of laboratory experiments, that soluble surfactants inhibited gas transfer even under wave-breaking conditions. Very recently, Salter et al. (2011) reported on a deliberate large scale release of an artificial surfactant (oleyl alcohol), in the north-east Atlantic Ocean. Gas transfer rates were measured by both the dual tracer technique (see Sect. 2.5.3) and direct covariance measurements of the DMS flux (see Sect. 2.5.2). Air-sea gas transfer rates were reduced by about 50 % at 7 ms^{-1} and were still impacted by the presence of the surfactant at high wind speeds; k_w was lowered by 25 % at 11 ms^{-1}.

Whilst the Salter et al. (2011) study has proven that surfactants can significantly impinge on gas transfer in the world's oceans, it is not yet clear that they do so at ambient levels. A recent study (Wurl et al. 2011) found many of the world's oceans (subtropical, temperate and polar) are covered to a significant extent by surfactants. Enrichments in the sea surface microlayer compared to bulk seawater were observed to persist at wind speeds of up to 10 ms^{-1}

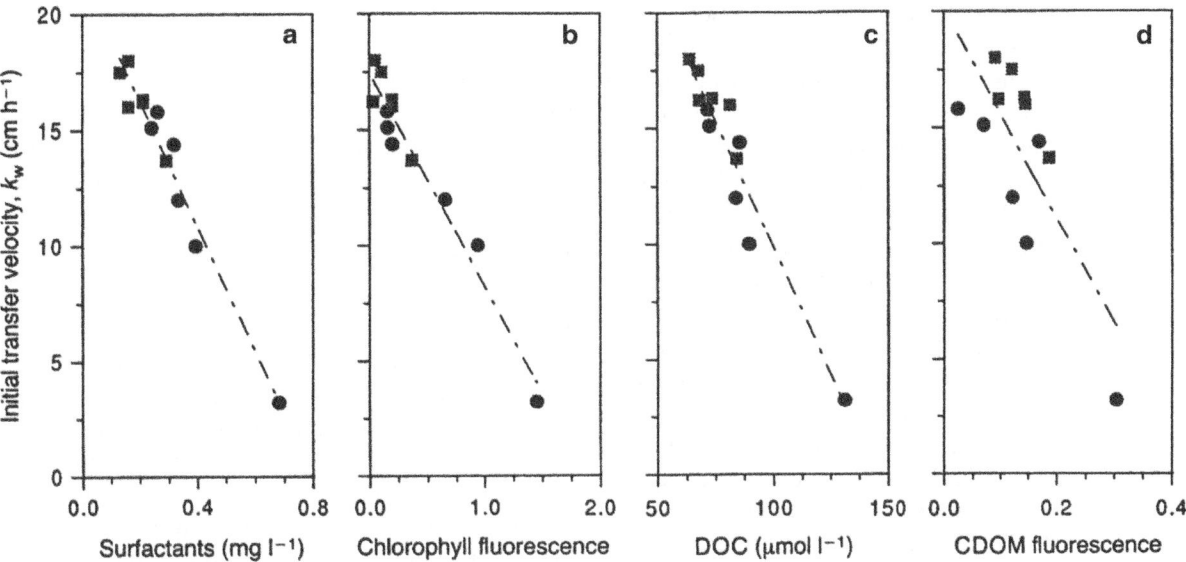

Fig. 2.5 Correlations of k_w with (**a**) surfactant concentration, (**b**) in situ fluorescence, (**c**) dissolved organic carbon (DOC) and (**d**) coloured dissolved organic matter (CDOM) fluorescence at 450 nm (Figure reproduced from Frew (1997))

consistent with the observations of the Salter et al. (2011) study. Clearly, this is an area worthy of further effort.

How important might surfactants be on a global basis? Asher (1997) predicted a global decrease of 20 % in the net sea to air flux of carbon dioxide based on the reasonable assumption that surfactant concentrations would scale with primary productivity. Tsai and Liu (2003) estimated that the net uptake of CO_2 could be reduced by between 20 % and 50 %; the range reflecting the uncertainty in measurements of surfactants and their impact on gas transfer. However, a simple relationship between chlorophyll and a reduction in k_w may be unrealistic as sea surface surfactant enrichments have been found to be greatest in oligotrophic (i.e. low productivity) waters rather than, as might have been expected, in highly productivity waters (Wurl et al. 2011); presumably due to the greater occurrence of bacterial degradation in the latter waters. Nightingale et al. (2000a) found no decline of gas transfer rates during the development of a large algal bloom in the equatorial Pacific. The implication of these studies of surfactants and air-sea gas transfer is that wind speed may not be the best parameter with which to parameterise k_w in the oceans, particularly in biologically productive regions.

A reasonable correlation with the total mean square wave slope for both filmed and film-free surfaces, particularly of shorter wind waves, suggested that this parameter, although difficult to measure at sea, might be a more useful predictor of k_w (Jähne et al. (1987) – see Sect. 2.6.2). This has been shown experimentally in the field by Frew et al. (2004) who found that k_w was better correlated to the mean square wave slope than to wind speed, specifically when winds were below 6 ms^{-1} and when CDOM levels were enhanced in the microlayer.

2.2.8 Biological and Chemical Enhancement

The enhancement of gas transfer by chemical reaction within the mass boundary layer(s) has been recognised as important for a small number of gases including O_3 (Fairall et al. 2007) and SO_2 (Liss 1971), and potentially important for CO_2 (Hoover and Berkshire 1969; Wanninkhof and Knox 1996; Boutin and Etcheto 1995). Suitably rapid reactions serve to 'steepen' the concentration gradient of the gas and thus reduce the effective depth of the mass boundary layer, thus leading to an enhancement in the transfer velocity (Johnson et al. 2011).

Note that whilst reversible reactions such as the hydration of CO_2 will act to buffer a reaction in either direction and thus lead to enhancement of fluxes into and out of the ocean (albeit assymetrically where forward and reverse reactions occur at different rates), irreversible reactions have the capacity to inhibit fluxes by a similar mechanism (i.e. to decrease the steepness of the concentration gradient where either a production reaction acts against a flux into the ocean or a breakdown reaction acts against a flux out) (Johnson et al. 2011). Here we will refer only to enhancement, but the reader should be aware that enhancement factors can be negative in some situations.

The enhancement of CO_2 exchange by hydration to carbonic acid and subsequent acid dissociation has been estimated to account for between 0 % and 20 % of the global CO_2 flux estimated from the ^{14}C technique (Keller 1994). Unlike the near-instantaneous hydration of SO_2 (Liss 1971), the hydration of CO_2 is relatively slow and its enhancement is thought to be important only when turbulent forcing is weak (and thus the timescale of transport across the mass boundary layer is relatively large). The effect on global CO_2 fluxes is rather complex. Boutin and Etcheto (1995) estimate that the net global atmosphere-to-ocean CO_2 flux is reduced by approximately 5 % due to the bias introduced by outgassing areas being generally associated with low average winds. On a local scale, CO_2 hydration must enhance the flux at low or zero wind speeds. The magnitude of this effect is represented in the hybrid parameterisation of Wanninkhof et al. (2009) as a constant component of k_w of 2.3 cm h^{-1}. Interactions between different compounds and reactions might lead to more complex behaviour than simple first-order enhancement. For instance, air-water mass transfer of CO_2 can be inhibited or enhanced by the mass transfer of NH_3 due to the reversible and pH dependent formation of ammonium carbamate (Budzianowski and Koziol 2005), although this phenomena has not been studied in the natural environment.

Just as physico-chemical processes may enhance mass transfer by modifying the concentration gradient in the mass boundary layers, biological activity might achieve the same, at least on the water side of the interface. Microbial communities at the ocean surface tend to be different from bulk water communities, often with considerably enhanced population densities (Cunliffe et al. 2009), which would lead to potentially rapid processing of bio-active compounds. There is some circumstantial evidence for the asymmetrical biologically-mediated transfer of methane (Upstill-Goddard et al. 2003), and O_2/CO_2 (Garabetian 1991; Matthews 1999), but these processes are not well studied.

2.2.9 Atmospheric Processes

Within a few millimetres of the water surface there is a thin sublayer dominated by molecular diffusion. Above the molecular diffusion layer is the atmospheric surface layer where the vertical transport is dominated by turbulent eddies; it extends upwards from the molecular layer to a rather poorly defined distance (ranging from approximately 10–100 m). Further away from the surface (in the Ekman layer) the Coriolis effect gradually changes the flow. These layers make up the Atmospheric Boundary Layer (ABL), in which the presence of the surface has a profound effect on the flow. Traditionally the atmospheric surface layer is described by the Monin-Obukhov Similarity Theory (MOST), it assumes stationary and homogeneous conditions and a solid surface (Panofsky and Dutton 1984). The fluxes are approximated to be constant with height (within 10 %) and it is thus enough to describe the flux at just one height. Using MOST the turbulent surface fluxes are often expressed using the bulk aerodynamic formula. Stress (τ), heat and scalar fluxes can be written as:

$$\tau = \rho_a u_{*a}^2 = \rho_a C_D (U_z - U_0)^2 \quad (2.6)$$

$$H = \rho_a c_p \overline{w'\theta'} = \rho_a c_p C_H U_z (T_z - T_0) \quad (2.7)$$

$$C = \rho_a c_p \overline{w'c'} = D_C U_z (C_z - C_0) \quad (2.8)$$

where ρ_a is the air density, u_{*a} the friction velocity on the air-side, C_D, C_H and D_C are the transfer coefficients for momentum heat and scalars at the specific height z. Transfer coefficients for scalars (Dalton number) can be related to transfer velocity by $D_C = k/U_z$. Wind speed, temperature and scalar values at height z are U_z, T_z and C_z, corresponding parameters at the surface are U_0, T_0 and C_0.

In the atmosphere, gradients of wind, temperature, and scalars are dependent on the atmospheric stratification, which also influences the transfer coefficients. Neutral stratification (giving logarithmic profiles, assuming MOST) is used as the reference state and

2.2 Processes

flux coefficients are then normalised using the actual stratification. Stratification can be expressed in terms of the Monin-Obukhov length ($L = -\frac{u_{*a}^3 T}{\kappa g \overline{w'\theta'_v}}$, where $\overline{w'\theta'_v}$ is the surface virtual potential temperature flux). For unstable atmospheric stratification $L < 0$ and for stable atmspheric stratification $L > 0$. Turbulence in the atmosphere (and thus the vertical gradients) is stability dependent and the non-dimensional gradients of wind, temperature and scalars are expressed as:

$$\phi_\alpha = \frac{d\chi}{dz} \frac{\kappa z}{\chi_*} \qquad (2.9)$$

where $\alpha =$ m, h, c and $\chi =$ U, θ and C. The ϕ-functions are expressed by empirical expressions for wind (ϕ_m) and temperature (ϕ_h); see Högström (1996) for a review of expressions for momentum and heat, and in Edson et al. (2004) functions for humidity are given. McGillis et al. (2004a) suggest the same expressions for CO_2 as for humidity. For stable stratification, expressions from Holtslag and De Bruin (1988) are frequently used.

Over the sea in the presence of surface gravity waves the wave boundary layer (WBL) is the atmospheric layer that is directly influenced by surface waves. For a growing sea the WBL is of the order of 1 m (Janssen 2004) and for swell waves it is significantly larger, it can even extend throughout the ABL (Smedman et al. 1994).

The neutral transfer coefficients are related to the roughness length defined as the intersect of the logarithmic profiles with the surface value. Roughness length for momentum (z_0) is crudely related to the roughness of the surface. Charnock (1955) expressed z_0 over the ocean as:

$$z_0 = \alpha \frac{u_{*a}^2}{g} \qquad (2.10)$$

where the Charnock coefficient, α, is a constant or described as a function of the state of the waves, where younger waves are expected to give a rougher surface (Fairall et al. 2003; Drennan et al. 2003; Carlsson et al. 2009). For temperature and scalars it is more complicated (Garratt 1992). The scalar roughness lengths can basically be expressed by the velocity roughness length, friction velocity and Schmidt number (see Fairall et al. (2000) for a discussion of different approaches). When the flow is aerodynamically smooth, a thin viscous sublayer exists adjacent to the surface.

Stability is a dominating parameter in the atmosphere since it determines the scale of the turbulence and thus the efficiency of eddy transport. For stable stratification, turbulence is suppressed, being dominated by intermittent turbulent events and atmospheric gravity waves and with a low boundary layer height. For unstable stratification, the convection at the surface enhances turbulence initiating convective eddies. The mean wind can be close to zero, but there is a non-zero wind component due to the gustiness. Godfrey and Beljaars (1991) suggested adding a gustiness wind component proportional to the convective scaling velocity (Deardorff 1970):

$$w_{*a} = \left(\frac{g}{T} z_i \overline{w'\theta'_v}\right)^{1/3} \qquad (2.11)$$

where z_i is the height of the ABL. For specific conditions (during free convective conditions, swell or low boundary layer height) the wind gradients are altered (Beljaars 1995; Fairall et al. 2003; Guo et al. 2004; Högström et al. 2008). For gradients of temperature and scalars, free convection is important, but less is known about swell and boundary layer height.

Over land, stratification is mainly determined by the diurnal cycle due to effective radiative heating and cooling of land surfaces. Over sea, the diurnal cycle is not as dominating due to the larger heat capacity of water. Then stronger atmospheric stratification occurs during advection of air masses with a different temperature to the water surface. Strongly stratified conditions thus occur in areas close to coasts or with great horizontal temperature gradients.

For momentum, heat, and gases with high solubility (like water vapour) the atmosphere induces the major resistance to transfer. For gases with low solubility, processes in the water contribute the main resistance to transfer. Taking atmospheric stability into account when determining transfer coefficients makes a significant difference when calculating fluxes of momentum, heat and humidity. For CO_2 the effect is relatively minor, up to about 20 % for low wind speeds (Rutgersson and Smedman 2010).

2.3 Process Models

To gain a deeper understanding of relevant transport mechanisms, several models have been developed. These range from conceptual models to numerical models based on first principles. Conceptual models are major simplifications of the actual processes and frequently address one dominant process taking place. Nevertheless, they are appealing as they can be used for deriving certain properties of the transport, such as gradients, fluxes or Schmidt number exponents. Such models will be discussed in Sect. 2.3.1. These simplistic models cannot describe interfacial properties, such as temperature distributions or the wave field. Numerical simulations based on first principles may be used for addressing such problems. The current state-of-the-art of such models is presented in Sect. 2.3.2.

2.3.1 Interfacial Models

The ocean–atmosphere exchange of insoluble gases and sparingly soluble chemical species such as CO_2, as well as other properties such as heat and momentum, is controlled by the mass boundary layer occupying the upper 10–100 µm of the ocean surface. Within this boundary layer, molecular diffusive transport tends to dominate over turbulent transport, with increased turbulent forcing leading to an increase in the rate of exchange by a reduction of the thickness of the diffusion-dominated domain. Various models have been proposed to represent such diffusion-mediated transport across the air-sea interface, and these lead to different dependency of exchange kinetics (i.e. the transfer velocity) on the diffusivity of the tracer (due to differences in the balance between the diffusive and turbulent processes controlling exchange). These models are described below, along with brief consideration of analogous models for the transport on the air-side of the interface for soluble gases.

2.3.1.1 Thin (Stagnant) Film Model

The simple thin film model (Whitman 1923) applied to the air-sea interface by Liss and Slater (1974), represents the sea-surface as a flat, solid boundary, with stagnant mass boundary layers on either side, through which diffusion is the sole transport processes. The diffusive flux through the water side stagnant film can be written as

$$F = -D\frac{dC}{dz} \quad (2.12)$$

where D is the diffusivity of the gas in the medium, C is concentration of the gas of interest and z is the vertical depth of the mass boundary layer. The thickness of the stagnant film must also be a function of the properties of the medium, notably the viscosity. We can rewrite the equation in terms of a transfer velocity of the water side, k_w as follows:

$$F = -k_w \Delta C \quad (2.13)$$

where

$$k_w = \frac{D}{\delta_c} \quad (2.14)$$

It can be seen from this approach that the transfer velocity is a function of both the gas properties (D) and the thickness of the thin film (δ_c), and has units of velocity ($m^2 s^{-1}\, m^{-1} = m s^{-1}$). This transfer velocity (also known as piston velocity) represents the rate of vertical equilibration between water column and atmosphere. Equation 2.14 is directly proportional to the diffusivity of the gas. The effect of (e.g. wind-driven) turbulence is to reduce the effective depth of the mass boundary layer and thus reduce the resistance to transfer.

Where the stagnant film models describe discrete layers where turbulent and diffusive mixing dominate transfer, the rigid boundary or solid wall models (e.g. Deacon 1977; Hasse and Liss 1980) apply a velocity profile at either side of the interface to describe a smooth transition between molecular and turbulent transport regimes. Such models predict that the transfer velocity is proportional to $D^{2/3}$, demonstrating that as turbulence plays a (modest) role in determining the rate of transfer in such a model, the diffusivity of the tracer becomes somewhat less important in determining the rate of exchange than for the stagnant film model. This diffusivity dependency has been demonstrated rather conclusively for a smooth (laminar flow) water surface by tank and wind tunnel heat and gas exchange experiments (see e.g. Liss and Merlivat 1986, for a summary).

2.3.1.2 Surface Renewal Model

A second class of models has historically been applied to interfacial exchange problems, where the diffusion-limited surface water layer is episodically and instantaneously replaced by bulk water from below (e.g. Higbie 1935; Danckwerts 1951). This episodic replacement leads to an increased role for turbulence in bringing water to the surface to exchange with the atmosphere. The details of the various surface renewal models differ; for example Higbie (1935) assumes a single turbulence-dependent renewal rate, whereas Dankwerts describes a statistical distribution of possible renewal timescales which is modulated by turbulent forcing. However, they all demonstrate the same dependence of the transfer velocity on $D^{1/2}$ which results from the description of an average flux resulting from episodic replacement of the surface with water of maximum disequilibrium and subsequent reduction in flux as this disequilibrium is reduced prior to the next renewal event. The general form of the surface renewal model transfer velocity term is

$$k_w \propto \left(\frac{D}{\tau}\right)^{\frac{1}{2}} \quad (2.15)$$

where τ is the renewal timescale. Surface renewal models have been widely used in studies of heat fluxes from water surfaces, including in the ocean, with generally good agreement at moderate to high winds (e.g. Garbe et al. 2004). This is strong validation of the dependency of the transfer velocity on the square root of the diffusivity for non-smooth water surfaces.

2.3.1.3 Eddy Renewal Model

The surface renewal model describes periodic 'disturbances' of the surface by discrete events, which are not physically described. Fortescue and Pearson (1967) and Lamont and Scott (1970) developed the surface renewal approach by explicitly modelling the physical processes at the interface assuming eddy turbulence to be the dominating process in transporting bulk water to the interface. This treatment models the eddy turbulence as a series of stationary cells of rotating fluid in long 'rollers' in the along-wind direction, alternatively converging and diverging, leading to regions of upwelling and downwelling to and from the surface. The transfer velocity in this model is related to ε, the energy dissipation rate, (along with molecular diffusivity) and can be generalised as

$$k_w = D/\delta_c \quad \text{with}$$
$$\delta_c = (D/\nu)^{1/2}\delta_\nu = (D/\nu)^{1/2}(\nu^3/\varepsilon)^{1/4} \quad (2.16)$$

$$\Rightarrow k_w = bS_c^{-1/2}(\varepsilon\nu)^{1/4} \quad (2.17)$$

where $f(\varepsilon\nu)$ is some function of ε and the kinematic viscosity of water, ν. As with instantaneous surface renewal models, the predicted transfer velocity is found to vary with $D^{\frac{1}{2}}$. Recently, eddy renewal models have been used with considerable success in predicting not only observed heat fluxes from environmental and laboratory water surfaces but also patterns of heat distribution – parallel streaks of warmer and colder water on the surface (e.g. Hara et al. 2007; Veron et al. 2011). The parallel development of eddy renewal models and surface infrared imaging has the potential to directly relate surface turbulence measurements, mean surface renewal timescales and transfer velocities leading to alternative parameterisations of transfer velocities.

2.3.1.4 Surface Penetration

Whilst good agreement has been found between surface and eddy renewal models and heat fluxes and distributions at water surfaces, it has been found that these do not scale well to gas transfer velocities via the diffusivity dependency (e.g. Atmane et al. 2004). This led to the application of the surface penetration model of Harriott (1962) to detailed heat and mass flux data by Asher et al. (2004). The surface penetration model differs from renewal models as it considers incomplete replacement of the surface by eddy transport from the bulk. This means that as well as eddy lifetime, the transfer velocity is a function of the 'approach distance'. The implication of this is that the diffusivity dependence of transfer is not constant with turbulent forcing. This is demonstrated by Asher et al. (2004) by using a special case of the model of Harriott (1962) for constant eddy approach distance and lifetime:

$$\frac{1}{k_w} = \frac{h}{D} + \frac{1}{1.13}\left(\frac{t}{D}\right)^{\frac{1}{2}} \quad (2.18)$$

where h is the eddy approach distance and t is the renewal timescale. The application of the surface penetration model is found by Asher et al. (2004) to explain the apparent discrepancy between transfer velocities of heat and mass through the surface water layer from surface/eddy renewal models.

The diffusivity dependence of the transfer velocity as predicted by surface penetration theory is not constant with turbulent forcing. This implies that as well as step changes in the diffusivity transfer velocity relationship (from $k_w \propto D^{2/3}$ to $k_w \propto D^{1/2}$ at the transition between smooth and rough surface regimes) there is a continuous change in the exponent with changing turbulent forcing.

2.3.1.5 Air-Side Transfer

A similar array of physical models of the transfer velocity on the air-side of the interface exist (e.g. Fairall et al. 2003; Jeffery et al. 2010), which are principally concerned with the flux of water vapour from water surfaces. Such models are also applicable to other soluble gases or those whose transfer is significantly chemically enhanced in the water phase (and thus under gas-phase control) and also to the dry deposition of particles to the water surface. Generally these models show that, as in the water phase, the diffusivity dependence on transfer is between $D^{0.5}$ and $D^{0.7}$ (e.g. Fairall et al. 2003; Johnson 2010).

2.3.2 Direct Numerical Simulations (DNS) and Large Eddy Simulations (LES)

High-resolution numerical simulation of the turbulent flow and heat/gas transport in the vicinity of an air-water interface is an alternative approach for studying interfacial transfer processes. The computational approach, which is also called direct numerical simulation (DNS), solves the posed initial-boundary-value problem and resolves the flow fields down to that of dissipation and diffusion scales, and examines the detailed dynamics and transfer processes. The boundary-value problem consists of conservation equations of mass and momentum, and advection–diffusion equations of temperature and gas in the bulk fluids, subject to stress/flux balance and kinematic constraint at the air-water interface. The interfacial boundary conditions also govern the dynamics of surface waves ranging from capillary wavelets to gravity waves. Imposing the nonlinear conditions accurately on the interface, which is also an unknown to be solved, is the most challenging part in solving such a "fully-nonlinear" free-surface problem. Owing to the limit in previous computing capacity, DNS of three-dimensional turbulent flow underneath a free-moving water surface was not feasible until recently. The numerical model of Fulgosi et al. (2003) (also Lakehal et al. 2003) is among the first attempts to consider the dynamical effect of a deformable interface in numerical simulation of interfacial turbulence and transfer processes. The simulation, however, is confined to very small surface deformation and the maximum wave steepness ak_{wave} (a is the amplitude and k_{wave} the wavenumber) never exceeds 0.01.

A more recent development by Tsai and Hung (2007) successfully simulates the turbulent boundary layer bounded by wind-generated surface waves and the accompanying transfer processes up to immediate surface steepness (maximum $ak_{\text{wave}} \approx 0.25$). The simulation resolves all modes of the turbulent flow, including the coherent vortical structures (the renewal eddies), the laminar sublayer immediately next to the interface, and the Kolmogoroff scale turbulence, as well as the interfacial motions, including the gravity-dominant waves and the parasitic capillary ripples. A representative three-dimensional surface profile from the simulation is shown in Fig. 2.6 with the corresponding distributions of temperature and surface gas-flux density, streamwise vorticity is shown on a vertical plane. The result clearly reveals high correlation between the surface temperature and the gas-flux density. Two distinct signatures appear in these distributions: elongated streaks and random localised spots, indicative of different "surface renewal" processes in the underlying aqueous flow. The computed distribution of surface temperature resembles the measured thermal imagery of a wind-driven water surface (e.g. Garbe et al. 2004; Smith et al. 2007). Parasitic capillary wavelets form on the forward face of the carrier gravity wave, as commonly observed in wind-generated waves (e.g. Jähne and Riemer 1990). The impact of these capillary wavelets on cross-interface transfer, however, is not as significant as expected previously (e.g. Jähne et al. 1979). Note that the initial condition in posing the numerical simulation consists of a pure gravity wave and random velocity fluctuations without any prescribed vortical structures and surface capillaries. The appearance of these coherent eddies and parasitic ripples all arise from the nonlinear dynamics of the surface waves and the underlying turbulent boundary layer.

2.3 Process Models

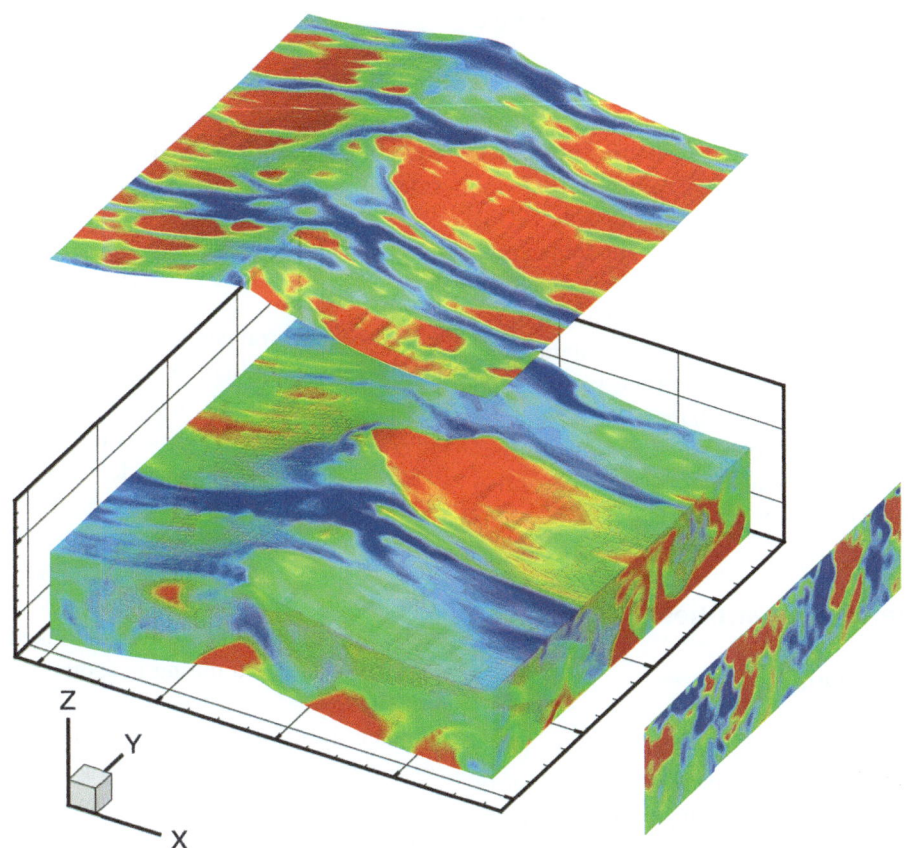

Fig. 2.6 Representative surface profile of a simulated wind-driven surface flow at the time instance of ten linear wave periods from the start of the computation. The wave propagates in the x direction. The wavelength of the carrier gravity wave is 7.5 cm. The contours superimposed on the surface, the front and side planes depict the temperature distributions. The corresponding distribution of gas-flux density is shown on an additional surface above the air-water interface. The contours of streamwise vorticity in the flow is also depicted on a cross-stream vertical plane

Despite the success of the high-resolution numerical model in simulating realistic transfer processes of wind-generated waves, DNS of interfacial turbulent boundary-layer flow is still limited to low Reynolds number (Re of the order of 10^3). The wavelength of the simulated gravity wave is also restricted to centimetres scale, and the wind speeds are limited to low to immediate regime. This is due to the fact that in DNS of the turbulent boundary layer the computation grid must be fine enough to resolve all spatial scales of the velocity, temperature and dissolved gas fields. For isotropic turbulence, the number of spatial modes increases as Reynolds number to the power 3/4, meaning the number of computation grids increases as $Re^{9/4}$. The minimum spatial scale of dissolved gas field is even smaller than that of velocity fluctuation (e.g., the diffusivities of dissolved CO_2 and O_2 are $\sim 10^{-3}$ of the viscosity of water). If all of these modes must be computed accurately, the required computer capacity becomes overwhelming. As such, the current DNS model is mainly applied to qualitative study of transfer process (e.g. Tsai et al. 2013) or testing of hypotheses in parameterization development (e.g. Hung et al. 2011).

The new advances in computer hardware and the use of parallel computing can certainly increase the size of the physical domain and speed up the computation. An alternative approach will be to adopt the method of large-eddy simulation (LES) in combination with new large-wave simulation (LWS). The methodology of LES, which was first introduced in atmospheric boundary layer research in the 1970s, is increasingly used in process studies of the ocean mixed later. However, unlike the LES model for ocean mixed layer, in which both the subgrid turbulence and the surface wave dynamics are parameterized, the proposed LES-LWS model for simulating air-water interfacial turbulent flow resolves not only the energetic turbulent eddies but also the gravity-dominant surface waves. In contrast to the well developed LES for "wall" turbulent boundary layer, subgrid-scale (SGS) parameterization for turbulent flow underneath a deformable wavy surface is yet to be developed. In addition, parameterization models of wave dynamics of sub-surface-grid scales, including the interactions among the SGS wavelets, the interactions between the SGS wavelets and the resolved "large waves", and the interaction between the SGS wavelets and the turbulence underneath (e.g. Tsai and Hung 2010), also require further development.

2.4 Exchanged Quantities

2.4.1 Physical Quantities

Transfer across the air-sea interface involves of a number of properties, the most prominent is the exchange of momentum. Basically the exchange of momentum originates from friction at the surface extracting energy from the mean atmospheric flow. This is the major source of energy for turbulence in the atmosphere and the ocean, as well as for surface gravity waves. The transfer of momentum over the sea is different to that over land since the surface changes form as a response to the atmospheric forcing. The transfer of momentum can be expressed in terms of kinematic stress (τ) and can be partitioned into turbulent, wave induced and viscous components (Phillips 1977):

$$\tau = \tau_t + \tau_w + \tau_\nu \quad (2.19)$$

the turbulent part (τ_t) depends on the wind shear, wave induced part (τ_w) generated by the waves; the viscous part (τ_ν) is neglected from some millimetres above the surface. Since the contribution from the different components varies with height, it is possible that the vertical flux is not constant with height within the wave boundary layer (the layer directly influenced by the waves). This also means that the surface stress is highly wave dependent.

The kinematic stress τ is directly related to the friction velocity u_* by

$$u_* = \sqrt{\frac{\tau}{\rho}}, \quad (2.20)$$

where ρ is the density of the fluid. The friction velocity can be seen from the air-side (u_{*a}) or from the water-side (u_{*w}) and are related according to

$$u_{*a} = \sqrt{\frac{\rho_w}{\rho_a}} u_{*w}. \quad (2.21)$$

For near neutral conditions one can consider heat to act as a passive scalar, however, when the heat flux is larger it is dynamically active and influences the turbulence in the atmosphere and the ocean. An upward buoyancy flux influences the density of the lower atmosphere and acts to make the atmospheric stratification unstable, in contrast a buoyancy flux directed to the surface makes the atmospheric stratification stable. The turbulent heat flux has a sensible component (direct) and a latent component (indirect, linked to the energy being used/released during evaporation/condensation). For the water surface the buoyancy flux consists of turbulent heat fluxes (sensible, latent heat) as well as long wave radiation lowering the temperature of the surface water. Together with evaporation acting to make the surface more saline, this can act to make the water surface buoyantly stable or unstable.

The transfer of momentum close to the surface is caused by viscosity and pressure perturbations, but the transfer of heat is ultimately caused by diffusivity. Thus the transfer of momentum is more efficient and exhibits a stronger wind speed dependence than the transfer of heat. The latent heat flux is generally greater during warmer conditions (in the tropics) as the specific humidity at saturation (controlling the surface value) is strongly temperature dependent. Skin-temperature as well as sea spray have been shown to influence the sensible and latent heat (e.g. Fairall et al. 1996, 2003).

2.4.2 Gases

Similarly to heat, mass is a scalar quantity. However, different gases will be transported at significantly different rates. Review articles exists on the issue (Jähne and Haußecker 1998; Wanninkhof et al. 2009), hence only a brief overview is given here.

The transfer velocity k_w (defined in Sect. 2.1) will depend both on the Schmidt number (Sc) and the dimensionless solubility α. Both these quantities are

2.4 Exchanged Quantities

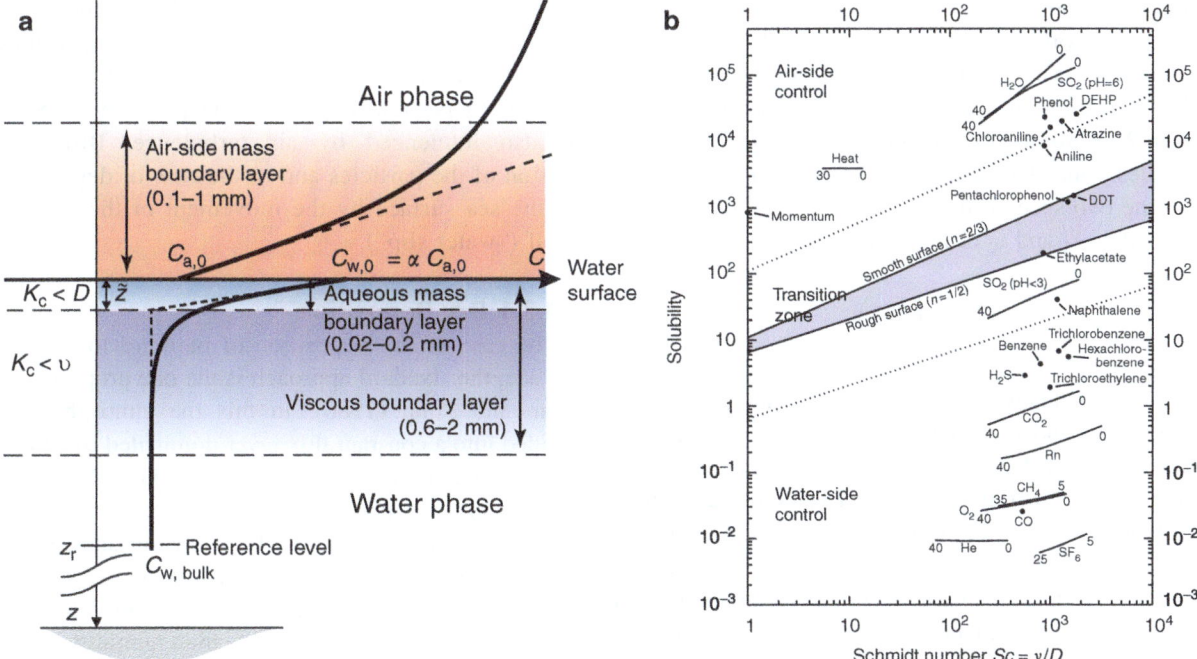

Fig. 2.7 (a) Schematic graph of the mass boundary layers at a gas-liquid interface for a tracer with a solubility $\alpha = 3$. (b) Schmidt number/solubility diagram including various volatile tracers, momentum, and heat for a temperature range (1 °C) as indicated. *Filled circles* refer to a temperature of 20 °C. The regions for air-side, mixed, and waterside control to the transfer process between gas and liquid phase are marked. At the *solid lines*, the transfer resistance is equal in both phases (Adapted from Jähne and Haußecker (1998))

material properties and depend on temperature and salinity (Johnson 2010). A Schmidt number/solubility diagram can be seen in Fig. 2.7. The solubility α is also termed a partition coefficient. It generally (i.e not when $\alpha = 1$) causes a concentration jump at the surface (c.f. Fig. 2.7). This is due to the thermodynamic solubility equilibrium that is established between the tracer concentration C_a in the gas phase and C_w in the liquid phase. Directly at the interface, this results in

$$C_{w,0} = \alpha \cdot C_{a,0}. \qquad (2.22)$$

$C_{x,0}$ indicates the concentration in the phase x directly at the surface.

A mass boundary layer exists on both sides of the interface. This leads to a transfer resistance R, the inverse of the transfer velocity k. The total transfer resistance R_t is the sum of the transfer resistance R_w in the water and R_a in the air-phase. Due to the solubility, this transfer resistance will be different when viewed from the air-side or the water-side of the interface:

$$\text{air-side}: \quad R_{ta} = R_a + R_w/\alpha,$$
$$\frac{1}{k_{ta}} = \frac{1}{k_a} + \frac{1}{\alpha \cdot k_w} \qquad (2.23)$$

$$\text{water-side}: \quad R_{tw} = \alpha \cdot R_a + R_w,$$
$$\frac{1}{k_{tw}} = \frac{\alpha}{k_a} + \frac{1}{k_w} \qquad (2.24)$$

The ratio $\alpha k_w/k_a$ indicates which boundary layer controls the transfer process, i.e. on which side of the interface is the dominant transfer resistance. Figure 2.7 indicates which side of the interface dominates the transfer resistance for a range of relevant gases.

The transfer velocity is related to the friction velocity u_{*w} and the Schmidt number Sc by

$$k_w = u_{*w}\beta(s)Sc^{n(s)} \qquad (2.25)$$

where both $\beta(s)$ and $n(s)$ depend on parameters describing the surface conditions. The Schmidt number exponent $n(s)$ transitions from $n(s) = -2/3$ for a smooth interface to $n(s) = -1/2$ for wavy conditions. The exact functional form and details of the transition are still subject to research. Equation 2.25 allows the scaling between different gas species. Since the surface conditions will be the same for the two species, the quotient between the transfer velocities k_1 and k_2 of the species results in

$$\frac{k_1}{k_2} = \left(\frac{Sc_1}{Sc_2}\right)^{n(s)} \quad (2.26)$$

From the knowledge of the Schmidt numbers of two species Sc_1 and Sc_2 and the transfer velocity of one (k_1), the transfer velocity of the other (k_2) can easily be calculated. Generally, the transfer velocities are scaled to that of CO_2 at 20 °C in seawater with a Schmidt number of $Sc = 660$, while $Sc = 600$ for CO_2 in freshwater at 20 °C. In the remainder of this chapter, we will refer to $Sc = 660$. In most studies, the Schmidt number exponent is assumed to be $n(s) = -1/2$.

2.4.3 Particles

Particle fluxes are governed by a set of processes of different temporal and spatial scales. They are produced at the sea surface (primary aerosol), as condensates from gas molecules in the atmosphere (secondary aerosol), are advected from continents (c.f. de Leeuw et al. 2013), are removed by direct deposition at the sea surface as explained below and wet depostion when they are entrained in precipitation rain droplets. The particles are also changed due to cloud processing. Here we follow the historical approach dividing the intertwined processes into sources and sinks. The sources are covered in Chap. 4 while the sinks are the focus of this section.

The process of particle removal from the atmosphere by their collection on a solid or fluid surface is called deposition. The particles can be scavenged by precipitation (wet deposition) or be deposited directly on a surface (dry deposition). A measure of the efficiency of the deposition process is the deposition velocity v_d, defined as the ratio of particle flux F to the particle concentration. It has be noted that although deposition velocity has the unit of ms^{-1}, in general case it is not the velocity of any physical object, not even necessarily the average vertical speed of the particles. Especially, it should not be confused with the gravitational settling velocity v_g. The two parameters would have identical values only if gravitational settling was the only process leading to particle deposition. Deposition velocity is also influenced by air turbulence, Brownian motion of the particles and in the case of deposition on the sea surface by the movement of the surface itself ("water slip").

2.4.3.1 Dry Deposition

In the case of dry deposition of particles to the sea surface, the standard approach is the one proposed by Slinn and Slinn (1980). In this the atmosphere is divided into a constant flux layer dominated by turbulence and a thin "deposition layer" just above the water interface where wind speed is close to zero and air movement is dominated by diffusion and microphysical processes such as diffusiophoresis. The particles in the deposition layer are assumed to have their radius increased (a_w) due to high humidity, compared to dry particle radius a_d. The particle concentration at the sea surface is assumed to be zero (no resuspension). In such a case, assuming that both the turbulent and deposition layers have their respective transfer velocities (analogous to gas transfer), k'_C and k'_D which can be calculated separately, the deposition velocity v_d may be derived from

$$\frac{1}{v_d} = \frac{1}{k_C} + \frac{1}{k_D} - \frac{v_g(a_d)}{k_C k_D} \quad (2.27)$$

where $k_C = k'_C + v_g(a_d)$ and $k_D = k'_D + v_g(a_w)$.

The formula is analogous to the one for parallel resistance of electric current where the "overall transfer resistance", v_d^{-1}, is the sum of the transfer resistances for both layers reduced by gravitational settling. This approach, similar to one used for gas transfer velocity, allows for adding additional transfer resistances for other processes, as well as to adjust the two transfer velocity parameters. This explains its popularity; it has been cited almost 300 times according to the ISI Web of Science database.

However in the 32 years since Eq. 2.27 was created it has also been the target of criticism from the theoretical point of view. The electrical analogy was shown by Venkatram and Pleim (1999) to be incorrect if transport cannot be represented in terms of a concentration gradient, as in the case of particle settling. Equation 2.27 is not consistent with mass conservation. A rigorous derivation based on mass conservation (which implies fluxes being constant with altitude) gives the following formula

$$v_d = \frac{v_g}{1 - \exp(r\, v_g)} \quad (2.28)$$

where r is total resistance to transport.

A special case of the general Eq. 2.28 occurs assuming the viscous deposition layer not being rate-determining has been independently derived by Carruthers and Choularton (1986) for conditions of neutral stability. It has been introduced to the marine aerosol research field by Smith et al. (1991). Its use in the gradient method of flux measurements by Petelski and Piskozub (2006) has been recently criticised by de Leeuw et al. (2011) on the ground that it does not include processes such as impaction, molecular diffusion, or growth of particles due to increased relative humidity (RH) near the sea surface, yielding dry deposition velocities that are considerably greater than those from most other formulations for particles with radii smaller than several micrometers.

Another point where the Slinn and Slinn (1980) description seems an oversimplification for most marine conditions is the assumption of zero particle concentration at the sea surface. Because the surface is not only a sink but also source of particles (see Chap. 4 and de de Leeuw et al. (2011) for a review), this condition is rarely true and particle fluxes are the net result of two processes: aerosol production at the sea surface and particle deposition (Nilsson et al. 2001). Hoppel et al. (2002) suggested that when the two processes are in equilibrium, deposition velocity at any fixed height equals the gravitational settling velocity, allowing turbulent mixing processes to be neglected. However, the time needed to achieve equilibrium for particles with radii smaller than about 5–10 µm is so long that other processes (such as wet deposition) will remove them before the two fluxes are in equilibrium. In a follow-up study, Hoppel et al. (2005) proposed that in the case of a surface source, both the source function and deposition velocity should be considered above the interfacial (deposition) layer. The logic behind this proposition is that measurements of the surface source function are made above the interface layer while the microscopic processes occurring in the interface layer are poorly characterised.

In situ direct measurements of deposition velocity from ships and other platforms are inherently difficult. Any such study would need to measure both particle concentrations and deposition fluxes. Direct measurements of fluxes (eddy correlation and similar techniques) measure net fluxes that are the difference between source and deposition fluxes. This method also needs correction for platform motion, a non-trivial problem on research ships, but which could be avoided by using spectral techniques (Sorensen and Larsen 2010). This lack of experimental verification of deposition velocity parameterisation (Andreas et al. 2010) is worrying, particularly considering that its theoretical values are used to estimate source fluxes using equilibrium methods (de Leeuw et al. 2011). These assume that the source and sink terms are in balance (an assumption unreliable for particles with radii smaller than 10 µm (Hoppel et al. 2002)).

2.4.3.2 Wet Deposition

A separate process of aerosol particle removal from the atmosphere is wet deposition, scavenging by water droplets both within clouds and by precipitation below them (see Zhang and Vet (2006) for a review). In-cloud scavenging consists of nucleation scavenging (where particles become condensation nuclei for cloud droplets) and impaction scavenging (particles are collected by cloud droplets, raindrops or snow crystals). The form of mathematical description of wet deposition is usually based on the concept of a scavenging ratio formulation (Iversen 1989), where the dimensionless parameter is the ratio of solution increase of the particles in the precipitation element over the effective scavenging height to its air concentration. This parameter is expected to have higher values inside the cloud than below it.

The details of the scavenging processes are still poorly constrained due to scarcity of in situ measurements from aircraft platforms. A climatology of wet deposition scavenging ratios for the continental USA (Hicks 2005) showed that the ratios for individual rain events are distributed over two orders of magnitude, with standard deviations for weekly averages being twice the mean value (for daily predictions, the factor increases to about five). Similarly, comparison of aerosol concentrations measured over the oceans with the results of aerosol transport models (Witek et al. 2007) shows that model-measurements discrepancies increase with the amount of precipitation, showing that wet deposition is still the weakest link in aerosol transport models. However, at the same time this result suggests that presently used parametrisations of the aerosol production function and deposition velocity, with all the still existing uncertainties, are able to provide reasonably good agreement with shipboard measurements.

It has been recently proposed (Piskozub and Petelski 2009) that marine aerosol composed of water droplets is able to scavenge small particles in a similar manner to precipitation droplets. Because this (still unconstrained) process would be dependent on the surface (aerosol production function rather than precipitation) it should be included in deposition velocity formulations as it is expected to affect the flux by changing vertical concentration gradients.

2.5 Measurement Techniques

There exists a variety of measurement techniques to estimate the sea spray aerosol production (de Leeuw et al. 2011) and the gas transfer velocity. Here we mainly focus on the three most frequently used types of methods for gas transfer velocity, small-scale techniques, micrometeorological methods and mass-balance techniques. The focus is on the methodology, accuracy, advantages and disadvantages. In addition a section on pCO_2 near-surface profiles is included due to the importance when estimating the transfer velocity or calculating air-sea CO_2 fluxes. In the case of large near-surface gradients, measurement depth is an issue.

2.5.1 Small-Scale Measurements Techniques

2.5.1.1 Particle-Based Techniques

In particle-based flow-measurement techniques, the fluid is seeded with small (µm) particles for visualizing the flow. In standard Particle Imaging Velocimetry (PIV) techniques, the particles are illuminated with a 2D laser light sheet (Raffel et al. 2007). Existing particle-based flow-measurement techniques can be grouped as proposed by Adrian (1991) in three main categories:

Laser Speckle Velocimetry (LSV) which assesses flow-information by analysing random interference patterns caused by a high seeding density of tracer particles,

Particle Tracking Velocimetry (PTV) which uses lower seeding densities to enable tracking of single particles through a long exposure time or multiple images in a sequence, and

Particle Image Velocimetry (PIV) that extracts the flow information from a fluid with a medium seeding density where the displacement of small groups of particles within an interrogation window are analysed statistically.

Most proposed approaches in the third category use correlation based methods to find similar particle constellations within an interrogation window in consecutive image pairs. This mode is often referred to as 'standard PIV'. These PIV techniques have been widely used in laboratory measurements at the air-water interface, both on the water-side (Siddiqui and Loewen 2010; Siddiqui et al. 2004; Turney et al. 2005; Banner and Peirson 1998) or, more challengingly, on the air-side (Reul et al. 2008; Shaikh and Siddiqui 2010; Troitskaya et al. 2011). Particularly for anisotropic turbulent flow, 2D measurement of velocities is problematic as out of plane velocities are not captured (Adrian 2005). Therefore, a strong effort has been put into extending the measurement techniques to volumetric three-dimensional three-component (v3D3C). These approaches can also be distinguished according to the technique used to obtain 3D-velocity information and the methods used to measure volumetric datasets. The most popular method to access three-components (3C) of the velocity field is to measure the out-of-plane velocity using Stereoscopic methods (Prasad 2000), extended by a technique called multi-plane PIV by using multiple laser-sheets (Kähler and Kompenhans 2000; Müller et al. 2001; Liberzon et al. 2004; Cenedese and Paglialunga 1989; Brücker 1996) or intensity graded light sheets (Dinkelacker et al. 1992) for the reconstruction of the third velocity component. In the recent past many methods were proposed to access volumetric information from 3D flow fields. This was achieved using holographic measurements (Hinsch 2002; Sheng et al. 2008), by combining PIV with Doppler global velocimetry (PIV/DGV) (Wernet 2004), using tomography

2.5 Measurement Techniques

(Elsinga et al. 2006; Schröder et al. 2008), defocussing-based approaches (Pereira et al. 2000, 2007; Willert and Gharib 1992), scanning-light-sheet methods (Burgmann et al. 2008; Hoyer et al. 2005; Brücker 1995) and absorption based methods (Jehle and Jähne 2008; Berthe et al. 2010; Voss et al. 2012).

For gaining a better understanding of transport processes at the air-water interface, flow structures relative to the location of the interface are of interest. Most techniques measure in a Eulerian reference frame. This makes an accurate extraction of the interface location necessary. Absorption based methods do not need this additional step, which may introduce further sources of errors. They encode the depth below the interface by light absorption in the dyed water body (Jehle and Jähne 2008; Berthe et al. 2010; Voss et al. 2012). Recently, an approach was presented for using one single plenoptic camera for assessing v3D3C flows (Garbe et al. 2012). A micro–lens array in front of the camera's sensor allows the extraction of depth and all-in-focus from computational photography.

While Steinbuck et al. (2010) have employed PIV techniques in an autonomous profiler at depth 10–60 m in the ocean, such measurements directly at the air-water interface are not practical.

2.5.1.2 Thermographic Techniques

Infrared measurements of the sea surface have been used to detect breaking waves (Jessup et al. 1997a), microscale breaking waves (Jessup et al. 1997b; Zappa et al. 2001), internal wave structures (Zappa and Jessup 2005), interfacial turbulence (Smith et al. 2007; Scott et al. 2008; Handler and Smith 2011; Schnieders et al. 2013), the momentum flux (Garbe et al. 2007; Garbe and Heinlein 2011) and also used to infer gas flux (Jähne and Haußecker 1998; Garbe et al. 2003, 2004; Schimpf et al. 2004; Asher et al. 2004).

In thermographic technique, an infrared camera is used for visualizing thermal patterns directly at the interface. Generally, midrange (sensitive spectral range: $\lambda = 3$–5 µm) or longrange (sensitive spectral range: $\lambda = 8$–10 µm) thermal imagers are used. In these spectral ranges, the penetration depth of the radiation is ≈ 20 µm. Thermographic techniques are distinguished as active or passive techniques. In passive techniques, visualized temperature fluctuations occur from a natural net heat flux at the interface. For active techniques, a heat source such as a laser is used for imposing an external heat flux.

One such active thermographic technique is the active controlled heat flux method proposed by Jähne (1989). Haußecker and Jähne (1995) used an IR-camera to track a small patch at the water surface heated up by a short pulse from a CO_2 laser. The temporal temperature decay of the patch was fitted based on solving the diffusive transport equation and utilising the surface renewal model (Higbie 1935; Danckwerts 1951) as a first-order process. The time constant of the decay was identified with the surface renewal time scale and the heat transfer rate calculated. A further method for the analysis of the decay curves was proposed by Atmane et al. (2004), in which the diffusive transport is combined with a Monte Carlo simulation of the renewal process based on the surface penetration model (Harriott 1962). Zappa et al. (2004) and Asher et al. (2004) measured a scaling factor of roughly 2.5 between the gas and heat transfer velocity when they applied the active controlled flux technique. Following Asher et al. (2004), the surface penetration theory provides a more accurate conceptual model for air-sea gas exchange. This is supported by the work of Jessup et al. (2009) who found evidence for complete and partial surface renewal at an air-water interface.

Passive techniques were used successfully to estimate the net heat flux, the temperature difference across the thermal skin layer and parameters of the surface renewal model by Garbe et al. (2003, 2004). Recently, Schnieders et al. (2013) extracted characteristics of interfacial turbulence from passive thermography.

2.5.2 Micrometeorological Techniques

Turbulence is an efficient transport mechanism for both physical quantities such as heat and momentum as well as mass transport i.e. gases and particles. If a gradient of some air constituent exists and a turbulent element, an eddy, moves in this gradient, the eddy mixing of the air will result in transport, a flux, of this air constituent from high concentration to low. Micrometeorological techniques are commonly used to measure the turbulent flux. In this section we will briefly present two of these techniques. The most direct technique is called the eddy-covariance (EC) method (also called eddy-correlation method). The turbulent flux can be directly obtained from the turbulent fluctuations of the wind speed components and the air constituent in question. The turbulent fluctuations are extracted from the measurement signal by the use of Reynolds decomposition:

$$x = \bar{x} + x' \quad (2.29)$$

where x is a measurement signal, overbar represents a mean value and the prime deviation from the mean, a perturbation. The mean value represents the mean flow and the perturbation the turbulent fluctuations. Applying Reynolds averaging (see e.g. Stull 1988) to the product between the vertical wind, w, and the air constituent, c, yields the covariance term $\overline{w'c'}$ which represents the vertical turbulent flux.

In order to capture all relevant fluctuations contributing to the flux, the sampling needs to be fast, usually > 10 Hz is required. In addition the measurement period needs to be long enough to include all relevant scales but not too long since this risks including slower variations not related to the local turbulence, e.g. diurnal cycles and mesoscale variations. Typically periods between 10 and 60 min are used.

Alternatively, if a fast response analyzer is unavailable the EC method can be modified by using disjunct eddy sampling, originally presented by Haugen (1978). The air constituent is sampled discontinuously, e.g. every 10 s, allowing the instrumentation a longer time to analyse the sample. As long as the sampling period is sufficiently short, the covariance between the vertical wind and the air constituent will still represent the flux. This version of the EC method is called Disjunct Eddy Covariance (DEC) and has been verified experimentally with the EC method (e.g. Rinne et al. 2008).

The flux measured using eddy-covariance techniques is representative for some surface area upstream of the sensor, called the footprint. The spatial and intensity distribution of the footprint is described by the footprint function and is dependent on several variables such as sensor height, atmospheric stability, surface roughness length, wind speed and wind direction. Different approaches have been used to calculate the footprint: analytical, Lagrangian and backward trajectory models, see Vesala et al. (2008) for an overview. The footprint is important in regions where one would expect systematic variations of the signal source at the surface. Typically most important for coastal sites but also in experiments investigating effects of e.g. surfactants on fluxes, researchers must be aware of where the flux footprint is positioned.

Although the EC-method is straightforward in principle, some issues need to be dealt with in the processing of the signals such as: motion correction (if measuring on a moving platform), dilution effects (the so called Webb or WPL-correction (Webb et al. 1980)), salt contamination, correction for sensor separation, corrections for flow distortion (if measurements are made from a bulky platform) and possibly corrections related to flux losses in closed path systems (see below). Some of these issues will be presented here. The motion correction is needed to remove effect of platform motions from the measured wind speed. This can be done by measuring the platform motion, see e.g. Edson et al. (1998) or Anctil et al. (1994) and is a necessary requirement when using the EC-method on ships and buoys.

The dilution correction is applied when calculating fluxes from density measurements of some air constituent like CO_2. Density fluctuations may arise not only by concentration variation of the air constituent but also through variations in water vapour, temperature and pressure (Webb et al. 1980; Fairall et al. 2000). This correction can be applied in the post processing or directly on the measured time series (Sahlée et al. 2008).

Measuring over the open ocean will cause salt contamination on the sensors, and if not dealt with properly will cause poor quality measurements. The instrumentation must be washed regularly with fresh water to remove salt contamination. Salt build up on open path gas analysers may cause cross-talk between the water vapour measurements and the CO_2 measurements. A method to correct for this was suggested by Peter K. Taylor and is presented in Prytherch et al. (2010).

When two sensors are used to measure fluxes, problems arise since the instruments are separated by some distance and thus don't measure in the same air volume. This will cause an underestimation of the flux since decorrelation arises between the measured velocity and the measured scalar. This flux attenuation is a function of separation distance, atmospheric stability and measurement height (determining the scale of the turbulent eddies), see e.g. Nilsson et al. (2010) or Horst and Lenschow (2009) for more details on this issue.

If measurements are planned over areas where small fluxes are expected, i.e. in areas with a relatively small air-sea gradient, a gas analyser should be chosen accordingly. The resolution should be sufficient to resolve the expected fluctuations. Otherwise the measurements will be dominated by instrumental noise (Rowe et al. 2011). Data with a low signal-to-noise ratio should be excluded from the analysis.

For measurements of gas fluxes, two types of gas analysers are available, closed or open path sensors. Open path sensor measure directly in the ambient air, whereas closed path sensors measure in air that has been drawn through some distance of tubing before reaching the analyser. Both types have their merits and complications. The open path is a more direct measurement, but the sensor is exposed and thus has problems with salt contamination and may be unable to function during precipitation. The closed path sensor is protected and may function also during precipitation events. Also, if the tube is long enough heat loss in the tubing makes the WPL-correction independent of temperature (Rannik et al. 1997; Sahlée and Drennan 2009). However, flux loss related to signal damping in the tube might be significant and need to be corrected. The closed path system requires pumps which are energy consuming and thus these type of instrument might be unsuitable on buoys where power is limited.

When measuring gas fluxes for which no fast response sensor exists one can utilise the eddy accumulation method (or the DEC method mentioned above). The eddy accumulation method is based on conditional sampling where the upward and downward moving gases are collected in two separate containers and analysed later. The total flux is then the sum of the two covariances calculated separately for the two containers (Desjardins 1977). Difficulties arise since the two separate terms in the total covariance may be much larger than the total term i.e. they need to be measured with great accuracy. Also, the air flow to the containers needs to be proportional to the magnitude of the vertical wind, which makes this a difficult method to implement in practice. The relaxed eddy accumulation method, REA, (Businger and Oncley 1990) is based on the same principle but combines it with the flux-variance similarity, thus the flux can be written as:

$$\overline{w'c'} = b\sigma_w(c^+ - c^-) \qquad (2.30)$$

where σ_w is the standard deviation of the vertical velocity, c is the mean concentration of the sampled gas and the sign denotes the direction of the vertical wind, b is a coefficient ≈ 0.6, with little stability dependence. Using Eq. 2.30 the air flow to the two containers can be set at a constant rate making the sampling easier and only dependent on the actual sign of the vertical wind.

Commercially-available gas analysers used with EC systems have typically been developed for use on fixed platforms on land. Some instruments have been found to introduce artificial signals when mounted on moving platforms. Such inaccuracy can be dealt with methodologically (e.g. Yelland et al. 2009) but there is also room for improvement by creating sturdier sensors. Recent development in methodology also allows for correcting water-vapour cross-talk on open-path sensors due to salt contamination Prytherch et al. (2010). It is very likely that this methodology will continue to develop. Future versions of current sensors will likely be more sensitive allowing the instrumentation to resolve smaller variations. This will reduce the flux uncertainty (particular for the smallest fluxes). Current efforts are ongoing to develop new gas analysers. For example, one such approach is based on photoacoustic spectroscopy with a target sensitivity of 20 ppb for CO_2.

2.5.3 Mass Balance

2.5.3.1 Techniques

On a global scale, inventories of radiocarbon (^{14}C) in the ocean (Broecker et al. 1985; Naegler et al. 2006; Peacock 2004) produced by spallation in the atmosphere due to cosmic rays (natural) or aboveground thermonuclear (bomb) testing have been used to determine the global mean gas transfer velocity for CO_2. Inventories are combined with wind data to predict how this transfer might vary with wind speed (Wanninkhof 1992). There has been some debate about the accuracy of the ^{14}C inventories and the relationships of CO_2 transfer with wind speed have since been re-evaluated (Sweeney et al. 2007).

On timescales of process studies, two main geochemical mass balance technique utilising natural and deliberately-injected tracers have been used in the ocean to determine gas transfer velocities.

The first technique, called the radon deficit method, assumes that ^{222}Rn and ^{226}Ra are in secular equilibrium in the mixed layer and below, and the observed deficit of radon in the surface ocean is due to evasion of ^{222}Rn out of the ocean (e.g. Bender et al. 2011; Peng et al. 1979; Roether and Kromer 1984).

The second technique involves the deliberate simultaneous injection of two volatile tracers with very different diffusion coefficients (^{3}He and SF_6) in the mixed layer of the ocean or into coastal seas, and the subsequent monitoring of their concentrations with time (e.g. Wanninkhof et al. 1993; Watson et al. 1991). Dilution will not alter the ^{3}He/SF_6 ratio, whereas the gas exchange of ^{3}He is faster than SF_6, and so the ^{3}He/SF_6 ratio will decrease with time, and allow the gas transfer velocity to be determined.

2.5.3.2 Scales (Spatial and Temporal)

The ^{222}Rn-deficit method has a time scale of about a week, as this is dictated by the ^{222}Rn half-life of 3.8 days, yielding a mean life of ca. 5.5 days, and has a spatial resolution 100s of km. The ^{3}He/SF_6 technique has a temporal resolution of at best 12–24 h, since there needs to be a detectable change in the ^{3}He/SF_6 ratio. The method typically has a spatial resolution of 10s of km.

2.5.3.3 Accuracy and Limitations

Besides the limitations imposed by the sampling methods and ability of analytical instruments to measure ^{222}Rn, ^{226}Ra, ^{3}He, and SF_6 accurately and precisely, both methods have other complications.

The dual tracer technique was originally developed for use in non-stratified coastal seas where the tracers would be well mixed to the seafloor (Watson et al. 1991). However, when used in the open ocean, both mass balance techniques depend on knowledge of the mixed layer depths, see Ho et al. (2011a) for a detailed discussion of mixed layer depth estimates derived using different methods. Because a profile of ^{222}Rn reflects processes taking place over days to weeks, a one-time estimate of the mixed layer depth may not reflect its variability over time. This has been overcome by repeated sampling over a period of time (Emerson et al. 1991). Also, with both methods, there is the possibility of entrainment of water from the pycnocline underlying the mixed layer either through convection or internal waves, which could alter the ^{222}Rn deficit or ^{3}He/SF_6 ratios.

In addition, the radon deficit method is confounded by many complications, as documented by Roether and Kromer (1984) and others. Among them:
1. The method relies on the assumption that ^{226}Ra concentrations throughout and below the mixed layer is constant, which is not always the case;
2. The relatively long half life of ^{222}Rn (3.8 days) compared to gas exchange means that both the wind history and the ^{222}Rn deficit history have to be known;
3. Because the ^{222}Rn deficit is not horizontally homogeneous, advection can perturb the "steady state" assumption.

A complication of the dual tracer technique is that it actually determines the difference in the transfer velocities of ^{3}He and SF_6. A knowledge of the Schmidt number dependence is required to obtain an estimate of the transfer velocity of either gas. Although lake experiments (Watson et al. 1991) and multiple tracer experiments in the North Sea. (Nightingale et al. 2000b) have shown that the value for n is close to 0.5 for the ^{3}He/SF_6 pair, this may not hold true under high wind conditions when significant populations of bubbles are expected to enhance air-sea gas transfer and might cause a deviation from $n = 0.5$ for ^{3}He/SF_6 (see Sect. 2.2.3 and Asher (2009)).

2.5.3.4 Current and Recent Field Studies

Most of the field studies with ^{222}Rn were conducted in the 1970s and 1980s (e.g. Kromer and Roether 1983; Peng et al. 1979; Smethie et al. 1985). The method has fallen into disuse because of the complications confounding its use as described above, and the difficulty in determining a relationship between wind speed and gas exchange. This is primarily due to the multi-day half life of ^{222}Rn, averaging significantly different sea-surface and wind conditions. This makes establishing a relationship with wind speed particularly difficult.

The ^{3}He/SF_6 dual tracer technique has emerged as the mass balance technique of choice in the ocean and in coastal seas. Experiments have been conducted in a variety of locations, the North Sea (Watson et al. 1991; Nightingale et al. 2000b), the Georges Bank (Asher and Wanninkhof 1998), the Florida Shelf (Wanninkhof et al. 1997), the Atlantic Ocean (McGillis et al. 2001; Salter et al. 2011), the equatorial Pacific (Nightingale et al. 2000a) and the Southern Ocean (Wanninkhof et al. 2004; Ho et al. 2006, 2011b).

2.5.4 Profiles of pCO₂ Near the Surface

To quantify the global exchange of CO_2 between the atmosphere and ocean with great accuracy requires the inclusion of factors which influence the air–sea flux, such as diurnal warming. This occurs at the sea surface when incoming shortwave radiation can cause stratification in the absence of wind-induced mixing. Temperature differences of up to 3 K can occur across this warm layer, which is ~ 10 m in depth (Ward 2006). This warm layer influences the air-sea flux of CO_2 because the partial pressure of CO_2 in seawater has an experimentally determined temperature dependence under isochemical conditions (Takahashi et al. 1993):

$$\Delta pCO_2^{WL} = pCO_2 \left[1 + e^{0.0423 \Delta T^{WL}}\right] \quad (2.31)$$

The diurnal warm layer enhances the flux of CO_2 in regions where the ocean acts as a CO_2 source and therefore reduces the oceanic carbon uptake (Olsen et al. 2004).

An experiment was conducted off Martha's Vineyard in 2002 (McNeil et al. 2006) in which in situ profiles of temperature/salinity were acquired with a seabird microcat, and pCO_2 with water pumped from depth to an equilibrator, where a sample of the headspace was analysed with a LI-COR LI-7000 IRGA. The profile data and subsequent pCO_2/temperature relationship is shown in Fig. 2.8. There is a relatively strong diurnal warm layer, with a ΔT of about 0.6 °C over the upper 4.5 m. There is a corresponding pCO_2 gradient of about 12 µatm and the best fit provides $(\partial pCO_2/\partial T)/pCO_2 = 0.062$ °C^{-1}, compared to the value from Takahashi et al. (1993) of 0.0423 °C^{-1}.

For a global net flux of CO_2 from the atmosphere to ocean, atmospheric pCO_2 levels are about 2 % (approximately 6 µatm) greater than the oceanic partial pressure for CO_2, making measurements susceptible to small biases in the determination of the correct partial pressure difference (McGillis and Wanninkhof 2006). Therefore the diurnal warm layer can mask the true air-sea fluxes when calculated with measurements of ΔpCO_2 when the water concentration is measured several meters below the sea surface.

Further work is required to understand the dependency of the ocean carbon system on temperature so that the most appropriate value for the change in seawater pCO_2 with temperature can be determined (McGillis and Wanninkhof 2006).

2.5.5 Method Evaluation

There are a large number of techniques to estimate gas transfer velocities from measurements, here we mainly discuss three different types, small-scale technique, micrometeorological technique (mainly Eddy-Covariance) and mass-balance technique (mainly Dual Tracer). The spatial and temporal scales resolved by each method are shown in Fig. 2.9.

Small scale techniques as outlined in Sect. 2.5.1 have significant advantages in laboratory conditions, making spatio-temporally resolved water-side concentration measurements at the frame rate of the camera used in the visualisation system feasible (up to 2,000 frames per second). Concentration gradients, flow dynamics and turbulence in the air and in the water can be resolved. This leads to a better understanding of processes, their relevance to the transport and their variability. To extract parameters from these techniques, inverse problems have to be solved, which introduces uncertainties. For field measurements, most small-scale techniques are not applicable, either because they are too delicate for the field, because the set up and the lighting conditions cannot be controlled, or because the water body cannot be seeded with the tracer. Thermographic techniques have been applied successfully in field campaigns. However, besides having to deal with artefacts in challenging conditions, they mainly measure the transport of heat with the transfer of gases being determined indirectly. Not all transfer mechanisms can be captured from thermography. The most dominant of these is bubble-mediated gas transfer, making thermographic techniques mainly applicable in low to medium wind speed ranges. However, when thermographic measurement techniques are applicable in the field, they make measuring natural variability or spatial homogeneity of fluxes possible. Not capturing bubble-mediated transfer can also be an advantage since its contribution to gas exchange can be assessed through comparison with complementary measurement techniques. Small-scale measurement techniques may therefore be very beneficial in combination with other techniques such as eddy-correlation, as the

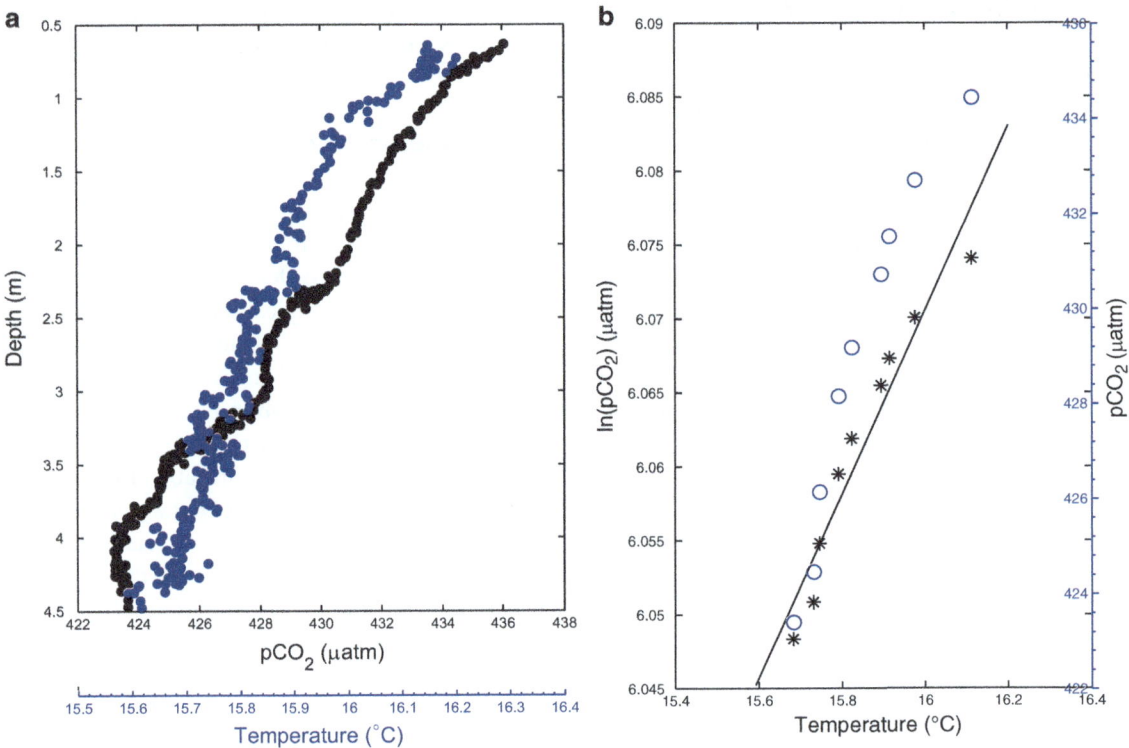

Fig. 2.8 *Left*: In situ profiles of temperature and pCO$_2$. *Right*: Relationship between temperature and pCO$_2$ for this profile. Although this relationship is non-linear, the best fit provides $(\partial pCO_2/\partial T)/pCO_2 = 0.062\ °C^{-1}$

assumptions underlying these techniques may be verified under the measurement conditions.

For dual tracer techniques, 2–3 stations are typically used and gas exchange on a time scale of 12–24 h can be resolved, in addition to integrated measurements over days to weeks. Eddy-Covariance measurements resolve fluxes on a 30 min–1 h time-scale. However, even during perfect conditions and no instrumental problems, a single flux estimate for a 30-min period using the covariance method would still be uncertain. This is due to the fact that turbulence is not sampled properly i.e. we cannot measure the ensemble mean, instead we are limited to time averages at a fixed point (i.e. tower and ship measurements) or spatial averages (aircraft measurements) which are bound to be limited. This means that we do not sample all possible realisations that a turbulent field can attain given the ambient atmospheric conditions. During conditions with large variability of surface conditions within the flux foot-print the variability of EC data is significantly larger.

All methods have methodological problems as well as instrumental uncertainties. For the dual tracer methods relevant instrumental accuracy is 0.2–0.5 % for ^{3}He and 2 % for SF$_6$, methodological difficulties relate to determination of mixed layer depth which typically adds 10 % uncertainty to the calculation of k_w.

Discussions of uncertainty in eddy-covariance flux measurements can be found in the literature, see e.g. Fairall et al. (2000) and Vickers et al. (2010) and references therein. Uncertainties arise from both instrumental problems and methodological issues. For CO$_2$ the relative instrumental uncertainty is estimated to be 17–20 % for tower-based fluxes (Rutgersson et al. 2008; Vickers et al. 2010). The uncertainty is higher for moving platforms (ship-based) in high-salinity environments. Better accuracy is, however, acquired with gases that can be measured with a larger signal-to-noise-ratio over sea, like DMS (Huebert et al. 2004).

Eddy-Covariance is the most direct method since the flux of interest is directly measured, it does not rely

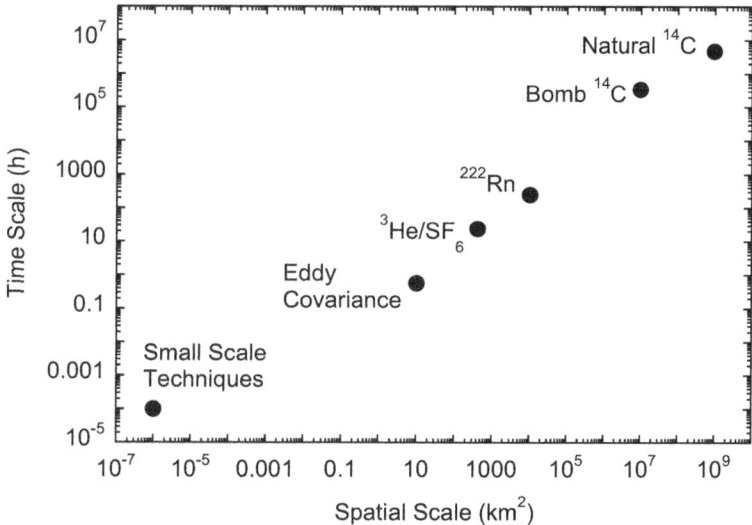

Fig. 2.9 Spatial and temporal scales of different measurement techniques frequently applied to gas exchange at the air-water interface

on assumptions about gas properties or approximations concerning the turbulent structure of the atmospheric boundary layer (Wanninkhof et al. 2009); the problem is presently the relatively high instrumental uncertainty in marine conditions, in particular for CO_2. Dual Tracer is an indirect method with the advantage of an integrated measurement of the gas transfer velocity with relatively high accuracy.

Wind stress at the ocean surface is responsible for kinetic energy input into the water, which generates turbulence. Several studies have shown a relationship with the transfer velocity and the dissipation of turbulence kinetic energy ε (Asher and Pankow 1986; Zappa et al. 2007). One can thus infer k_w by measuring dissipation in the water (see Sect. 2.6.4), although operational measurements of ε (e.g. remote sensing for assessing inventories), will be challenging.

2.6 Parameterization of Gas Exchange

Measuring transfer rates or transfer velocities of gas exchange directly is challenging. Therefore it is appealing to relate them to readily accessible parameters that dominate the transfer. These parameters might be the wind speed at a certain reference height or surface roughness. Such parameters can also be assessed globally from satellite remote sensing or are available in simulations. In this section, different parameterisations are examined. The most obvious ones are wind speed dependent $k_w - U$ relationships or sea state dependent ones through the surface roughness. The NOAA-COARE algorithm is a mixture of basic interfacial physics along with empirical additions such as bubbles and tuneable parameters. It requires more parameters than simple $k_w - U$ relationships but also probably results in more accurate estimates. Parameterising gas transfer with energy dissipation is appealing as it relates the transport directly to turbulence but this parameter is challenging to measure in itself.

2.6.1 Wind Speed Relationships

The processes mediating diffusive gas exchange across the air-sea interface are complex, with multiple physical controls (e.g. capillary waves, bubbles, small scale eddy turbulence etc., cf Sect. 2.2). In the ocean environment these processes are directly or indirectly related to wind via turbulent forcing and wave breaking and as such wind speed is found to correlate well with transfer velocity in observational data (Ho et al. 2011b). Wind is routinely measured and can be derived from remotely sensed data and ground-truthed by high precision instruments on ships (Wanninkhof et al. 2009; Bender et al. 2011; Ho et al. 2011b); hence it is likely to remain the key variable used in empirical parameterisations of transfer velocities for the foreseeable future.

The first parameterisation of environmental air-water transfer velocity from wind speed (U) were theoretical or based on simple laboratory experiments (e.g. Liss and Slater (1974) and references therein, Deacon (1977)). For the water-side transfer velocity (k_w), these have since been validated to first order by wind tunnel experiments (e.g. Liss and Merlivat 1986) and field studies as outlined previously in this chapter.

All of these methods yield a net flux over some scale of area and time and thus enable calculation of transfer velocity from flux (F) and the magnitude of the concentration gradient between air and sea (ΔC), following Eqs. 2.12 and 2.13. There is a considerable range in the estimates of the $k_w - U$ relationship from these methods (c.f. Fig. 2.10) – approximately a factor of 2 at a typical average wind speed of 7 ms^{-1} (e.g. Yang et al. 2011; Ho et al. 2011b; Johnson et al. 2011). However, this might be considered rather good agreement considering the differing scales of space and time over which the various methods operate e.g. eddy covariance (10^2–10^3 m^2 and minutes) and global ^{14}C estimates (10^{15} m^2 and decades).

Generally, at least for mass balance techniques, the field is tending to converge around a smaller range of uncertainty in the $k_w - U$ relationship (for CO$_2$), bringing estimates from ^{14}C inventory (e.g. Sweeney et al. 2007), dual/multiple tracer experiments (e.g. Ho et al. 2011b) and ^{222}Rn distributions (e.g. Bender et al. 2011) into reasonable agreement, all broadly in line with the commonly used parametrisation of Nightingale et al. (2000b), at least to within the experimental uncertainty in mass balance techniques (Asher 2009; Johnson et al. 2011). The Nightingale et al. (2000b) parametrisation is largely based on the results of a multiple tracer study conducted in the coastal North Sea and as such is potentially subject to surfactant and limited fetch effects which might inhibit k_w. However, their parametrisation compares well with that presented for the equatorial Pacific (Nightingale et al. 2000a) and more recent measurements in the Southern Ocean (Ho et al. 2006, 2011b); suggesting that the $k_w - U$ relationship as quantified by mass balance techniques is remarkably robust to local and regional conditions.

One key uncertainty in the parameterisation of k_w with U is the functional form of the relationship i.e. linear, quadratic, cubic, polynomial. A simple quadratic function ($k_w \approx U^2$) has been observed in wind-wave tanks (e.g. Wanninkhof and Bliven 1991). Such relationships have been assumed in scaling global ^{14}C inventories (and inferred fluxes) to global average wind speed (Wanninkhof 1992; Sweeney et al. 2007). Simple and polynomial quadratic functions have been found to fit dual tracer and ^{222}Rn deficiency data well (e.g. Wanninkhof 1992; Nightingale et al. 2000b; Ho et al. 2011b).

The considerably greater transfer velocities observed at high wind speeds in direct flux measurements of CO$_2$ via the eddy covariance technique (see Sect. 2.5.2) have led to suggestions of a cubic relationship to wind speed as a best fit to the data (e.g. McGillis et al. 2001; Prytherch et al. 2010), although given the large uncertainties in the measurements, a quadratic could also be fitted in most cases. This has been explained by the proposed cubic relationship between windspeed and whitecapping (e.g. Monahan and Spillane 1984) and the importance of whitecaps on bubble-mediated gas exchange (e.g. Woolf 1997) at high winds plus a reduction in gas transfer at low winds due to the supposed presence of surfactants (see Sect. 2.2.7). The cubic relationship is not supported by water-side eddy covariance measurements of O$_2$ and N$_2$ exchange by D'Asaro and McNeil (2007) using autonomous floats deployed in hurricane-force winds. They show a relationship that demonstrates much lower k_w at hurricane-force winds than predicted by the cubic parameterisations of Wanninkhof and McGillis (1999) and McGillis et al. (2001), although with k_w scaling to approximately the 6th power of windspeed (McNeil and D'Asaro 2007). McNeil and D'Asaro (2007) suggest that this is due to the much greater bubble-mediated transfer for O$_2$ and N$_2$, which are considerably more insoluble than CO$_2$. Other studies of oxygen (Kihm and Körtzinger 2010) and oxygen and nitrogen (Vagle, McNeil and Steiner, 2010) suggest considerably stronger bubble effects.

Evidence that CO$_2$ exchange may be considerably enhanced by bubbles comes from comparison with the eddy covariance measurements of DMS fluxes at high winds (see Fig. 2.3). This is summarised in Yang et al. (2011), where a near-linear relationship with wind speed is seen, and is quantitatively close to the linear relationship predicted for the non-bubble mediated component of the transfer velocity in the

2.6 Parameterization of Gas Exchange

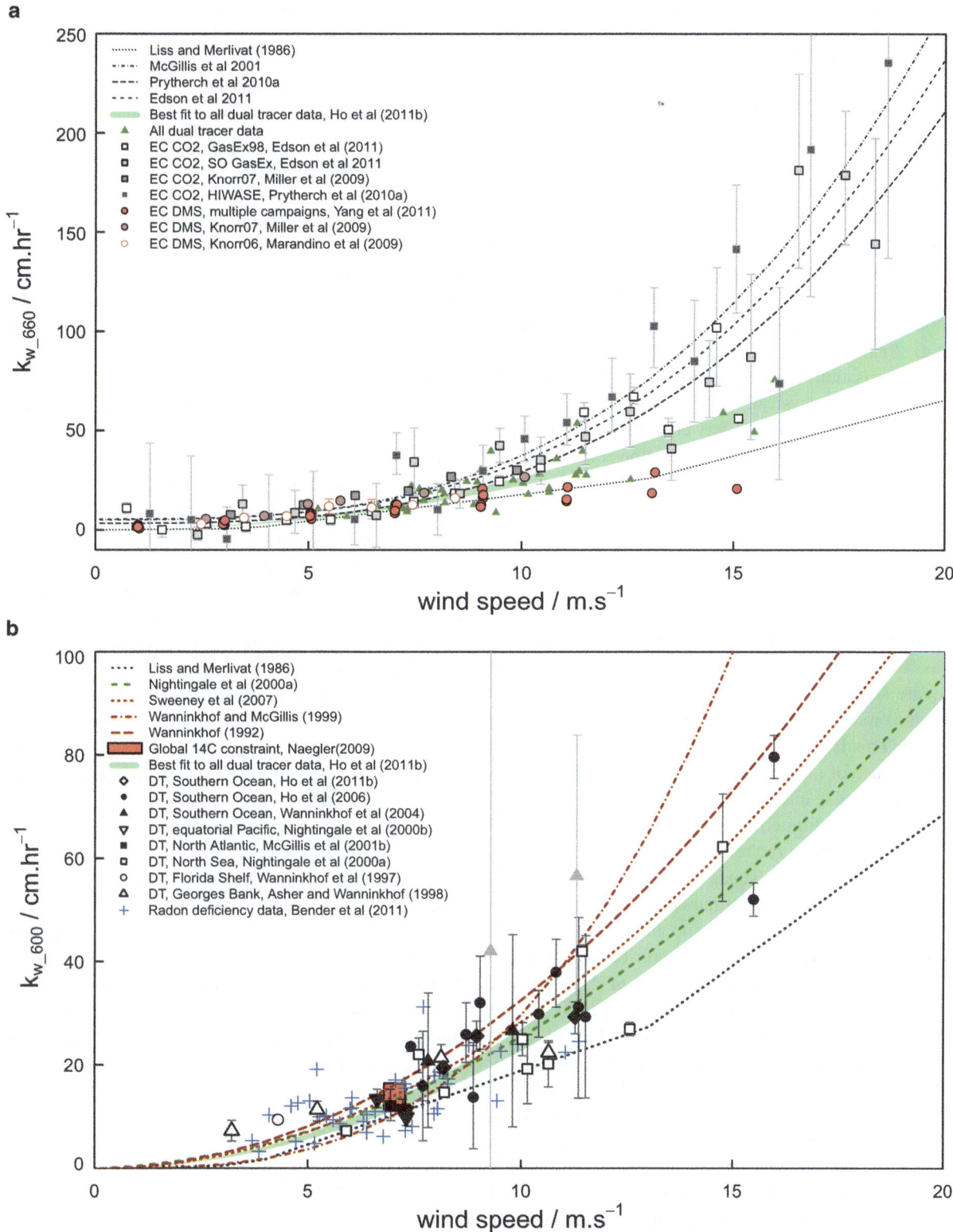

Fig. 2.10 A comparison of different windspeed relationships of the water side transfer velocity, k_w. Measurements from eddy covariance techniques are presented in (**a**), those from mass balance techniques are shown in (**b**) (Figures by M.T. Johnson, shared under creative commons license at http://dx.doi.org/10.6084/m9.figshare.92419)

physically-based bubble model of Woolf (1997), suggesting that DMS is sufficiently soluble to be unaffected by the bubble effect (Sect. 2.2.3), whilst CO_2 is not. This also suggests that, for gases of intermediate solubility such as DMS, non-linear parameterisations such as Nightingale et al. (2000b) may lead to substantial overestimates of fluxes at higher wind speeds.

For gases more soluble than DMS, where the air-side transfer velocity becomes limiting, the relationship of interest is that between wind speed and the air-side transfer velocity (k_a). This is less well constrained than k_w, with very little field validation of the empirical models, particularly for gases other than water vapour (e.g. Johnson 2010). Typically, however, $k_a - U$ parameterisations are close to linear (e.g. Johnson 2010; Sect. 2.6.3) and physically-based model predictions (e.g. Duce and Tindale 1991; Jeffery et al. 2010) tend to agree reasonably with wind-tunnel observations for trace gases by Mackay and Yeun (1983).

2.6.2 Surface Roughness, Slope

Frew et al. (2004) have shown from field measurements in coastal and offshore waters that the gas transfer velocity better correlates with mean square slope computed for the wave number range of 40–800 rad m^{-1} than with the classically used wind speed. This is consistent with laboratory observations and the use of mean square slope (mss) was shown to be of particular interest in case of surface films. This study was performed from low to moderate wind speed and the observed relationship between gas transfer velocity and mean square slope was linear.

Dual frequency measurements of altimeter like TOPEX and JASON in Ku-band and C-band, provide a mean of retrieving the mss contribution of small scale waves between 40 and 100 rad m^{-1}, as derived by a geometric optics model. Frew et al. (2007), after calibrating satellite mss with field mss, derived k-mss relationships and global k_w maps. The main difficulty is to calibrate satellite mss relationships and k-mss relationships, due to few existing field measurements of k_w and mss and to even fewer collocations between altimeter mss with these field measurements. Other difficulties arise from the fact that at high wind speed the effect of bubbles is not implicitly included in such algorithms so that these algorithms underestimate k_w at high wind speed (Wanninkhof et al. 2009). k_w fields derived from altimeters are very undersampled compared to k_w-fields deduced from satellite wind speeds due to the poor spatial coverage of altimeter measurements compared to scatterometer and microwave radiometer wind speeds.

2.6.3 NOAA-COARE

Details of the original development of the COARE algorithm for meteorological fluxes (sensible heat, latent heat, momentum) can be found in Fairall et al. (1996, 2000), and the subsequent development of the algorithm to include gas transfer is shown in Hare et al. (2004) and Fairall et al. (2011).

Advancements in sensor technologies led to the application of the micrometeorological direct covariance method to estimate air-sea fluxes at hourly time scales on the atmospheric side of the interface. This method was first successfully applied in the GasEx field programmes beginning in 1998 (McGillis et al. 2001; Fairall et al. 2000).

The short time scale of the covariance estimates enables observational investigations of the relationship of k to physical/chemical forcing beyond wind speed – examples include wind stress, buoyancy flux, surfactants, or surface gravity wave properties. Physically-based parameterisations (Hare et al. 2004; Soloviev 2007; Vlahos and Monahan 2009) are now available that incorporate these additional forcing factors and may lead to quite different transfer properties for different gases. For example, the dependence of bubble-mediated exchange on gas solubility implies the Wanninkhof (1992) formula may not be appropriate for the fairly soluble biogenic gas dimethylsulfide (DMS) (Woolf 1993; Blomquist et al. 2006). Thus, application of a single wind-speed formula for all gases is inconsistent with current understanding of gas transfer physics (Fig. 2.11).

In the COARE gas transfer algorithm (COAREG), the flux of a trace gas on the atmospheric side of the interface is estimated as

$$F_c = \alpha k_a (C_w/\alpha - C_a) = \alpha k_a \Delta C$$
$$= \frac{\alpha k_a}{u_{*a}} u_{*a} \Delta C \qquad (2.32)$$

where k_a is the transfer velocity on the air-side, α is dimensionless solubility, C_w and C_a the mean concentration of the gas in the water and air at reference depth

2.6 Parameterization of Gas Exchange

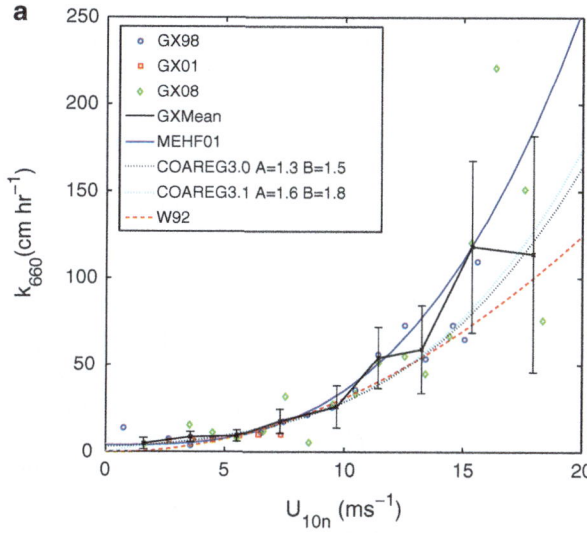

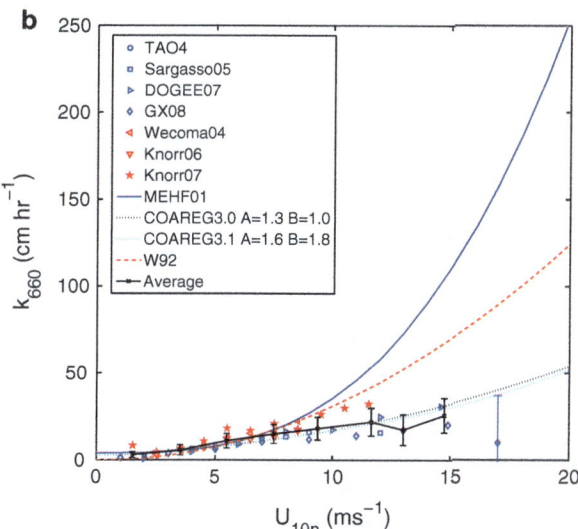

Fig. 2.11 (a) Gas transfer velocity for CO_2 as a function of 10-m neutral wind speed U_{10n} from direct surface-based observations. The *black line* is the mean of the data sets; the *error bars* are statistical estimates of the uncertainty in the mean computed. Symbols are: ○ – GASEX98, * – GASEX01, ◇ – SOGASEX08. The parameterisations shown are: *blue solid line* – McGillis et al. (2001); *black dotted line* – COAREG3.0 CO_2; *cyan dotted line* – COAREG3.1 CO_2 using tangential u_*; *red dashed line* – W92 (Wanninkhof 1992). (b) DMS gas transfer coefficient as a function of 10-m neutral wind speed U_{10n} from direct surface-based observations. The *black line* is the mean of the data sets; the *error bars* are statistical estimates of the uncertainty in the mean computed. Symbols are: ○ – TAO04 (Equatorial Pacific); □ – Sargasso Sea 05; ▷ – DOGEE07, ◇ – SOGASEX08, ◁ – Wecoma04, ▽ – Knorr06, * – Knorr07. The parameterisations shown are: *blue solid line* – McGillis et al. (2001), *black dotted line* – COAREG3.0 DMS; *cyan dotted line* – COAREG3.1 DMS using tangential u_*; *red dashed line* – W92

and height. The COAREG algorithm (current public version is COAREG3.0) gives a simple form for the transfer velocity (see Hare et al. 2004),

$$k_a = \frac{u_{*a}}{r_w + \alpha r_a} \quad \text{where} \quad r_w = \left(r_{wt}^{-1} + k_b/u_{*a}\right)^{-1}. \quad (2.33)$$

Here u_{*a}/r_{wt} represents molecular-turbulent transfer which has a basis in air-sea physics and k_b represents bubble transfer based on empirical formulation. On the atmospheric side we neglect spray-mediated gas transfer, so there is only molecular-turbulent transfer $r_a = r_{at}$.

In the COAREG model the individual terms are represented as

$$r_{wt} = \sqrt{\frac{\rho_w}{\rho_a}} \left(h_w Sc_w^{1/2} + \kappa^{-1} \ln\left(\frac{z_w}{\delta_w}\right) \right), \quad (2.34)$$

where z_w is the water depth of the reference measurements, δ_w the molecular sublayer thickness, Sc_w the Schmidt number of water, ρ_w the density of water, and $h_w = 13.3/(A\varphi)$.

In this expression, A is an empirical constant and φ accounts for surface buoyancy flux enhancement of the transfer. A similar expression is used for r_{at}, but the molecular sublayer thickness is explicitly approximated by incorporating the velocity drag coefficient, C_d, which is a function of the atmospheric measurement height, z_a,

$$r_{at} = \left(h_a Sc_a^{1/2} + C_d^{-1/2}(z_a) - 5 + \ln(Sc_a)/(2\kappa) \right) \quad (2.35)$$

where $h_a = 13.3$. The bubble-driven part of the transfer is taken from Woolf (1997) as

$$k_b = BV_0 f_{wh} \alpha^{-1} \left[1 + \left(e\alpha Sc_w^{-1/2} \right)^{-1/n} \right]^{-n} \quad (2.36)$$

where B is a second adjustable constant, $V_o = 2{,}450$ cm h^{-1}, f_{wh} is the whitecap fraction, $e = 14$, and

$n = 1.2$ for CO_2. The parameters A and B have been adjusted to fit observations with values between 1 and 2. Note there is no capability at present to describe the presence of surface films.

Recently COAREG was extended to include the case of an atmospheric gas (such as ozone) that reacts strongly in the ocean (Fairall et al. 2007). Using the notation from Fairall et al. (2000), the budget equation for the oceanic concentration of a chemical, C_w, is

$$\frac{\partial C_w}{\partial t} + U \nabla C_w = \frac{\partial[-(D_c + K)\partial C_w/\partial z]}{\partial z} - aC_w \quad (2.37)$$

where z is the vertical coordinate (depth for the ocean), D_c is the molecular diffusivity of C_w in water, K the turbulent eddy diffusivity, and the last term is the loss rate of C_w due to reactions with chemical Y. Thus, $a = R_{cy}Y$, where Y is the concentration of the reacting chemical, and R_{cy} the reaction rate constant. Assuming that the concentration of Y is much larger than C_w so that is remains effectively constant and that a is sufficiently large that C_w is completely removed within the molecular sublayer, Garland et al. (1980) showed that the water-side resistance is

$$\frac{u_{*a}}{r_w} = \sqrt{aD_c} \quad (2.38)$$

Fairall et al. (2007) relaxed the requirement that the reaction was confined to the molecular sublayer and obtained a solution that allowed the ozone deposition velocity to depend on the oceanic turbulence

$$\frac{u_{*w}}{r_w} = \sqrt{aD_c} \frac{K_1(\xi_0)}{K_0(\xi_0)} \quad (2.39)$$

Here K_0 and K_1 are modified Bessel functions of order 0 and 1 and

$$\xi_0 = \frac{2}{k_w u_{*w}} \sqrt{aD_c}. \quad (2.40)$$

COARE3.0 is a good fit to mean observed momentum and heat transfer coefficients over a considerable range of wind speeds (more observations are above 18 ms^{-1}). However, at any specific wind speed bin there is considerable scatter – more than can be explained by atmospheric sampling variability alone.

It is also clear that over the open ocean there is, on average, a systematic change in the influence of waves on air-sea turbulent exchange with increasing wind speed. Attempts to find simple scaling expressions that explain variations in surface roughness length (or drag coefficient) due to variations in surface wave properties have been of limited success (Drennan et al. 2005).

The most recent results from application of the algorithm to CO_2 and DMS transfer for SO GasEx can be found in Edson et al. (2011), while an analysis of the implementation of COAREG for ozone is outlined in Bariteau et al. (2010) and Helmig et al. (2011). Current application of COAREG includes input of satellite fields of sea surface temperature, near surface humidity and temperature, and wind speed to obtain global estimates of air-sea gas transfer (Jackson et al. 2011).

One major technical problem is the poor performance of fast CO_2 sensors at sea, which limit field work to regions with very large air-sea differences in CO_2 concentration. Some clarity in scientific issues could also come from high-quality eddy covariance measurements of gases more soluble than DMS and less soluble than CO_2. There is considerable room for progress on bubble-mediated transfer through field, laboratory, and numerical modeling studies.

COAREG is publicly available from the authors of (Fairall et al. 2011). The latest version of the code (COAREG3.1) includes a scheme to partition the wind stress into tangential and wave-induced components (see Fairall et al. (2011) for more details).

2.6.4 Energy Dissipation

The flux of slightly soluble, non-reactive gases (i.e. water side-controlled) between the ocean and atmosphere is determined from knowledge of the concentration difference across the diffusive layer as well as the gas transfer velocity k_w (see Eq. 2.14).

The ease of availability of wind speed U (m s^{-1}) measurements has resulted in many attempts to develop parameterisations between k_w and U (see Sect. 2.6.1). However, the primary mechanism for controlling k_w is shear-produced turbulence at the air-sea interface (Salter et al. 2011; Asher 1997; Wanninkhof et al. 2009), whose generation mechanism is wind stress acting on the ocean surface (Asher 1997; Kitaigorodskii and Donelan 1984).

There has been much effort to describe k_w in terms of molecular diffusivity D (m² s⁻¹), kinematic viscosity ν (m² s⁻¹), and a function based on a velocity Q (m s⁻¹) and length L (m) (Danckwerts 1951; Fortescue and Pearson 1967; Lamont and Scott 1970; Deacon 1977; Ledwell 1984; Jähne et al. 1987; Brumley and Jirka 1988; Soloviev and Schlüssel 1994; Fairall et al. 2000):

$$k_w = \beta Sc^n f(Q, L) \quad (2.41)$$

where β is an experimentally determined constant and $Sc = \nu/D$ is the Schmidt number. The Schmidt number exponent has been shown to vary between $n = -2/3$ for conditions of low winds (e.g. Deacon 1977; Davies 1972) or where surfactants are present at the water surface (Asher and Pankow 1986), to $n = -1/2$ at winds greater than 5 ms⁻¹ (Jähne et al. 1987). Lamont and Scott (1970) used an eddy cell model to describe $f(Q, L)$ which resulted in a scaling relation:

$$k_w = 0.4 Sc^{-1/2}(\varepsilon \nu)^{1/4} \quad (2.42)$$

where ε (m³ s⁻²) is the dissipation rate of kinetic energy, a parameter that can be measured in the field.

Zappa et al. (2007) provided the first field investigations that supported the mechanistic model in Eq. 2.42 based on surface water turbulence that predicts gas exchange for a range of aquatic and marine processes. Their findings indicated that the gas transfer rate varies linearly with the turbulent dissipation rate to the ¼ power in a range of systems with different types of forcing – in the coastal ocean, in a macro-tidal river estuary, in a large tidal freshwater river, and in a model ocean.

An underlying assumption in Eq. 2.42 is that ε is measured directly at the water surface. Since the profile of turbulence near the air-water interface may be complicated by the interplay between wind, waves, current shear and other processes, measurements at depth will not be representative of ε at the surface because the profile changes nonlinearly with environmental forcing. Zappa et al. (2009) implement the Craig-Banner turbulence model, modified for rain instead of breaking-wave turbulence, to successfully predict the near-surface dissipation profile that varies by two orders of magnitude over the top 50 cm at the onset of the rain event before stratification plays a dominant role. This result is important for predictive modeling of k_w as it allows inferring the surface value of ε fundamental to gas transfer.

Lorke and Peeters (2006) took the approach of using law of the wall scaling to describe $\varepsilon(z)$ as a function of depth z (m):

$$\varepsilon = \frac{u_{*w}^3}{\kappa z} \quad (2.43)$$

where $\kappa \sim 0.41$ is the von Kármán constant and $u_{*w} = \sqrt{\tau/\rho_w}$ is the friction velocity, τ is the wind stress acting on the ocean surface which is parameterised with knowledge of the drag coefficient C_D and ρ_w is the water density. Although this provides a mechanism to derive the parameters to parameterise k_w according to Eq. 2.42, it is unlikely that law of the wall scaling will adequately describe ε at the ocean surface (e.g. Agrawal et al. 1992; Terray et al. 1996; Babanin and Haus 2009).

Compared to wind speed, dissipation in the ocean is much more challenging to measure as it requires sophisticated instumentation and an ability to provide data close to the air-sea interface for the application of air-sea transfer parameterisation. Future progress requires a new generation of experiments which combine turbulence data with estimates of gas transfer velocities (Lorke and Peeters 2006).

2.6.5 Evaluating and Selecting Transfer Velocity Parameterisations

The above sections demonstrate that various different turbulent forcing terms can be used to parameterise k_w. Wind speed, which is the most commonly used predictor for k_w is only indirectly related to the key forcing terms and hence a great deal of variability is expected (and observed) in $k_w - U$ relationships. The choice of most appropriate parameterisation depends on both the forcing data available and the application to which k_w is to be used (e.g. spatial and temporal scale, gas of interest).

For example, when considering fluxes of CO_2 at the global scale there is a constraint on the transfer velocity given the bomb ^{14}C inventory of the ocean (e.g. Sweeney et al. 2007). Thus, notwithstanding uncertainties in both the inventory (Wanninkhof 1992; Naegler et al. 2006) and the global average wind speed, which may be as much as 1.3 m s^{-1} between different datasets (Naegler et al. 2006), there is a 'known point' on the $k_w - U$ curve for global CO_2 fluxes. Note that whilst parameterisations derived from this value have tended to apply a quadratic relationship, as described in Sect. 2.6.1, there is no evidence from this global approach to support any particular form for the $k_w - U$ relationship. Furthermore, whilst this is a valuable constraint on long timescale global fluxes, it is not necessarily a constraint for regional or local fluxes or over shorter timescales, due to variable and non-wind speed dependent forcings (e.g. fetch/wave slope, thermal stability, etc.) or the effects of other processes such as chemical or biological enhancement, rainfall etc.

Where wind speed is the only available forcing term for a particular study on smaller-than-global scale (as is commonly the case), an empirical $k_w - U$ parameterisation must be employed to quantify the flux of a gas given wind speeds averaged over an appropriate timescale. Data on k_w over a range of wind speeds in the natural environment come primarily from the dual tracer and eddy covariance approaches. As noted above, significant variability due to processes not directly related to wind speed is to be expected in observations of such relationships, but even accounting for this there are apparent inconsistencies which must be addressed. First and foremost is the apparently greater transfer velocities observed by CO_2 eddy covariance measurements than by the dual tracer technique, particularly at higher wind speeds (Sect. 2.6.1). It is possible that these two methods are measuring empirically different properties – it has been suggested by Ho et al. (2011b) that, whilst on the timescale of eddy covariance measurements (minutes-hours) the apparent transfer velocity varies with the cube of the wind speed, over the timescale of a tracer study (days) the vertical mixing of the bulk water becomes limiting to the total tracer mass balance and thus the higher-order dependence is not apparent. Alternatively, Asher (2009) suggests that the increasing importance of bubbles and thus increasingly turbulence dominated transfer at higher winds means that the scaling of k_w to $Sc^{-0.5}$ might break down, which would lead to errors in the relationship calculated by the dual tracer method.

The dual tracer method also fails to account for the effect of the solubility dependence of bubble-mediated transfer. Bubble fluxes are predicted to be lower for more soluble gases, and this is supported by laboratory experiment (e.g. Rhee et al. 2007; Woolf et al. 2007) and circumstantially by the lower and more linear transfer velocities of DMS (compared to CO_2) observed in the field by various studies (Sects. 2.6.1 and 2.6.3, Figs. 2.3 and 2.10). Such effects, however, should mean that the dual tracer method (based on the exchange of ^{3}He and SF_6, which are considerably less soluble than CO_2) should predict greater fluxes than those directly observed for CO_2, which is the opposite of what is observed. It has been suggested that such inconsistencies may be reconciled by the application of an improved representation of the bubble flux such as that outlined by Woolf et al. (2007), which accounts for the high void fraction in dense bubble plumes. This may lead to 'suffocation' of the bubble flux (i.e. the flux of gas into the bubbles starts to be limited by the decreasing concentration in the water around the bubble plumes). This effect will be largest for the most diffusive gases and as such there is likely to be a differential in the effect between the highly diffusive ^{3}He and rather less diffusive SF_6. The result of this effect when calculating k_w from dual tracer data would be an underestimation of k_w (Woolf et al. 2007).

Such explanations may be able to reconcile all the available data and may also satisfy the ^{14}C and ^{222}Rn derived values for k_w where appropriate, given that they are subject to significant uncertainty themselves. Given the current uncertainties associated with eddy covariance and dual tracer methods and the magnitude of the effect of bubbles on CO_2 fluxes in the ocean environment, it is impossible to determine which of the above explanations for observed discrepancies between methods is correct. It is possible that under different regimes of thermal stability, wave field, fetch and other forcings, the observed range of $k_w - U$ relationships may all be valid in particular situations. With forthcoming advances in CO_2 sensors for eddy covariance measurements, progress may be made towards better understanding and resolution of the above discrepancies.

With sufficient environmental forcing data, probably the best estimates of transfer velocity can be achieved by using physically-based models of boundary layer interactions such as NOAA-COARE. The latest incarnation, COARE G 3.0 (Fairall et al. (2011), including a generalised scheme to enable application to any gas following Johnson (2010)) might be considered the 'state of the art' in quantifying k_w. Nonetheless it still contains some physical parameterisations which are not state of the art. For instance it applies the 'classic' bubble model of Woolf (1997) rather than the dense bubble plume model mentioned above, and relies on rather uncertain empirical parameterisations such as the empirical windspeed-whitecapping relationship of Monahan and O'Muircheartaigh (1980) which is subject to significant uncertainty (Johnson et al. 2011) although recent developments are leading to considerable constraint of said uncertainty (Goddijn-Murphy et al. 2011). Recent measurements of whitecap fraction were discussed in de Leeuw et al. (2011) as outlined in Sect. 2.2.3.

Selection of an appropriate transfer velocity must be based on the requirements of the particular study. It is important to recognise that parameterisations are not necessarily universally applicable to gases of differing solubilites, or to differing environmental situations. However, it must also be acknowledged that in the context of the wide range of other uncertainties associated with quantifying air-sea gas fluxes (summarised by Johnson et al. 2011), not least the uncertainty in concentrations and selection of appropriate wind speed averaging, the uncertainty in k_w parameterisations is often relatively small.

2.7 Sea Ice

Sea ice affects air-sea exchanges of greenhouse gases in ice covered waters in several ways. The first effect is that sea ice prevents air-sea exchange of gases. First studies of the vertical distribution of anthropogenic CO_2 in the Weddell Sea showed limited air-sea exchange of CO_2 in the winter surface water when it is subducted and mixed with other water masses to form Weddell Bottom Water (Weiss 1987; Poisson and Chen 1987). This suggested that sea ice was acting as an inert and impermeable barrier for gas exchange so that most carbon cycle research has not considered the possibility of either direct air-sea CO_2 exchange in the presence of sea ice or indirect air-ice-sea CO_2 exchange related to sea ice melting (Bates and Mathis 2009; Tison et al. 2002). However, it is well established that sea ice is a permeable medium, depending of its brine volume fraction (Golden 2003; Golden et al. 1998). Golden et al. (1998) showed that sea ice is permeable to brine transport when brine volume fraction is above 5 %, corresponding to a temperature of -5 °C and sea ice salinity of 5 – the so-called "law of fives"-. Gas transport within sea ice is therefore possible as observed by (Gosink et al. 1976) who reported CO_2 and SF_6 diffusion in sea ice at -7 °C. More recently, Loose et al. (2011b) measured accurately diffusion of SF_6 and O_2 through columnar sea-ice and reported diffusion coefficient values within permeable artificial columnar ice of $1.3 \cdot 10^{-4}$ cm^2 s^{-1} and $3.9 \cdot 10^{-5}$ cm^2 s^{-1} for SF_6 and O_2, respectively. These values are rather low compared to air-sea transfers in open water. However, it must be borne in mind that sea ice covers about 7 % of the Earth surface at its maximal seasonal extent and represents one of the largest biomes on Earth. Even though fluxes through sea ice are modest, integrated over the global sea ice cover can potentially be significant (Delille et al. 2006).

Several studies reported measurements of direct air-ice CO_2 exchanges. Most studies reported air-ice CO_2 fluxes in spring and summer (Delille et al. 2006; Semiletov et al. 2004; Zemmelink et al. 2006; Nomura et al. 2010a, b; Papakyriakou and Miller 2011). Strikingly, some studies reported air-ice CO_2 exchange during the winter season when the air-ice interface is supposed to be impermeable to gas exchanges (Heinesch et al. 2009; Miller et al. 2011). Indeed, despite the fact that sea ice can be seen as an ideal environment for micro-meteorological measurement of CO_2 fluxes (Loose et al. 2011a) since the surface is relatively smooth and levelled, air-ice CO_2 fluxes measurements are still in their infancy. Different methods are used, without inter-comparison, and temporal and spatial coverage is scarce. In addition, the early studies of Semiletov et al. (2004) and Zemmelink et al. (2006) should be considered cautiously as they do not take into account bias of open path sensor CO_2 analysers in cold environments, as pointed out by Burba et al. (2008). For the time being, sea ice heterogeneity, gaps in understanding of sea ice biogeochemistry, poor understanding of the role of snow in controlling air-ice gas fluxes and poor temporal and spatial coverage prevent robust integration of air-ice gas fluxes at large scale. Further studies are required to provide such assessments.

Whatever the magnitude of air-ice gases fluxes, the role of sea ice in gas exchange is not limited to the ability to transfer gases from the ocean to the atmosphere. O_2, CO_2 and DMS observations within sea ice show clearly that gas dynamics within sea ice are different to that in the underlying layer. Several peculiar processes e.g. temperature change, brine concentration/dilution, brine transport, primary production and respiration by sympagic[1] microbial communities and others biogeochemical processes affect gases dynamics within sea ice (Delille et al. 2006, 2007; Tison et al. 2010; Trevena and Jones 2006; Trevena et al. 2000, 2003; Glud et al. 2002; Rysgaard and Glud 2004; Rysgaard et al. 2008), so that sea ice cannot be seen as a simple open/closed pathway between ocean and the atmosphere. Peculiar gas dynamics within sea ice determine gas contents of sea ice and subsequently air-ice partial pressure gradients of each gas. Therefore, the way and potential magnitude of air-ice gas fluxes are controlled by gas dynamics within sea ice rather than the gas content of the underlying water.

Rysgaard et al. (2011) recently reviewed the significance of sea ice related processes in terms of air-sea CO_2 exchange and pointed out the role of rejection of CO_2-rich brines in the underlying layer during sea ice growth. Sea ice expels about 80 % of solutes, including gases, out of the ice matrix, mainly into the underlying water. In the Southern Ocean, this brine rejection is one of the main drivers of Antarctic Deep Water formation. Gases transported together with brines can therefor potentially sink towards deep layers, providing an efficient path-way for CO_2 sequestration (Rysgaard et al. 2011). In addition, Loose et al. (2009) observed in artificial ice experiments that rejection of gases (SF_6 and O_2) is enhanced compared to salts.

$CaCO_3$ precipitation within sea ice as ikaite (Papadimitriou et al. 2004; Dieckmann et al. 2008, 2010) has also been suggested as an efficient pathway for atmospheric CO_2 sequestration (Rysgaard et al. 2007) depending on the conditions where precipitation takes place (Delille et al. 2006). In certain sea ice permeability conditions, assuming that $CaCO_3$ precipitates in the early sea ice growth phase (Assur 1958; Marion et al. 2009), CO_2 produced by $CaCO_3$ precipitation is rejected with brine to the underlying layer, while $CaCO_3$ crystals remain trapped within the ice matrix. In summer, $CaCO_3$ dissolution and related CO_2 consumption within the ice or release of alkalinity depleted meltwater into the underlying water promote atmospheric CO_2 uptake by the ice and underlying water, respectively (Delille et al. 2006; Rysgaard et al. 2007). In parallel to CO_2-rich brine rejection, this precipitation acts as an effective pathway for atmospheric CO_2 sequestration (Rysgaard et al. 2007, 2011; Delille et al. 2006).

The impact of sea ice transfer estimates can be very important regionally. This is less the case at the global scale, assuming impermeability of ice to be appropriate for many applications. Rysgaard et al. (2011) recently provided a first tentative budget of the global significance of sea ice related processes for atmospheric CO_2 uptake. They assessed that sea-ice related atmospheric CO_2 uptake is about 45 TgC y^{-1} and 71 TgC y^{-1} in the Arctic and Southern Ocean, respectively. Oceanic CO_2 up-take during the seasonal cycle of sea ice growth and decay in ice-covered oceanic regions therefore equals almost half of the net atmospheric CO_2 uptake in ice-free polar seas. Finally, remobilisation of CH_4 held in East Siberian Arctic Shelf sediments has been assessed to be a major contributor to the oceanic CH_4 flux to the atmosphere (Shakhova et al. 2010a, b). Shakhova et al. (2010b) reported large bubbles of methane entrapped in fast sea ice probably originating from ebullition from sediments. Either the ice acts as a simple inert transient buffer for the ebullition CH_4 flux or as more complex pathway has still to be determined.

2.8 Applications of Air-Sea Gas Transfer

There exist a wide range of applications in which air-sea transport are important. In this section we focus on applications in global models, large-scale measurements from satellite remote sensing and computations of global inventories. These applications are not exclusive but are the dominant topics from air-sea gas exchange studies to global environmental research.

---

[1] A sympagic environment is one where water exists mostly as solid ice.

2.8 Applications of Air-Sea Gas Transfer

2.8.1 Models

Nearly all numerical models employed to date to simulate the cycling of carbon in the ocean are using the parameterisation of Wanninkhof (1992) to estimate the gas-transfer velocity, k_w, i.e. a parameterisation that suggests a quadratic dependence of k_w on the windspeed (see Sect. 2.6.1). Different coefficients have been used with the quadratic dependence, but we are not aware of any ocean model that uses a linear (Krakauer et al. 2006), a cubic (Wanninkhof and McGillis 1999), or piecewise linear function (Liss and Merlivat 1986) for k_w. The very small diversity of parameterisations used today is a legacy of the Ocean Carbon-cycle Model Intercomparison Project (OCMIP-2), where it was decided to use the same parameterisations for all models (Najjar and Orr 1998). In this particular study, k_w was proposed to be parameterised as follows:

$$k_w = (1 - f_{ice})[a \cdot (U^2 + v)](Sc/660)^{-1/2} \quad (2.44)$$

Where f_{ice} is the sea-ice fraction, U^2 the square of the windspeed at 10 m, v is the variance of the windspeed, added to reflect the non-linear nature of the relationship, and Sc the Schmidt number. For OCMIP, the instantaneous windspeed was first averaged to form a monthly climatology of U, and then squared. v was similarly computed from the instantaneous windspeed. The coefficient a was then chosen such that the global mean k_w matched the global bomb radiocarbon-derived estimate of Broecker et al. (1986). This gave a coefficient a of 0.336 cm h^{-1} s^2 m^{-2}, which is slightly larger than the value of 0.31 cm h^{-1} s^2 m^{-2} proposed by Wanninkhof (1992) due to the use of different winds.

This parameterisation has been carried forward into nearly all ocean carbon cycle models since then, with the only major development being the suggestion that the coefficient a needs to be downscaled substantially to about 0.24 cm h^{-1} s^2 m^{-2} in order to take into consideration that Broecker et al. (1986) bomb radiocarbon inventory estimate was too large (Peacock 2004; Naegler et al. 2006; Sweeney et al. 2007; Müller et al. 2008; Nägler 2009). This gives a global mean transfer velocity k_w for CO$_2$ of about 15 cm h^{-1} instead of Broecker's mean value of 19 cm h^{-1}.

Very few sensitivity studies have been performed to investigate the influence of different gas exchange parameterisations on the exchange of trace gases. This is likely a consequence of the influential paper by Sarmiento et al. (1992) where the authors demonstrated little sensitivity of the uptake of anthropogenic CO$_2$ with respect to different formulations of the gas transfer velocity. This is because the rate limiting step for the uptake of this transient tracer is its transport from the near surface ocean into the ocean's interior. But this lack of sensitivity does not apply to the exchange of natural CO$_2$, where a higher transfer velocity generally leads to stronger fluxes (Fig. 2.12). However, the modelled flux changes are less than proportional to the change in the gas transfer velocity, since the surface ocean concentration of CO$_2$ adjusts to the change in the "resistance" across the air-sea interface by generally decreasing the air-sea gradient when k_w is increased (Fig. 2.12c). Due to lateral transport and mixing, the altered CO$_2$ concentration also affects regions away from regions of strong air-sea gradients, even leading to a change in the sign of the gradient.

An even larger sensitivity to the choice of k_w exists for the air-sea transfer of the isotopes of CO$_2$, i.e., ^{14}CO$_2$ and ^{13}CO$_2$, since the characteristic exchange time-scale for these isotopes is about ten times longer than that for CO$_2$ itself (Broecker and Peng 1974). This is a consequence of the fact that isotopic equilibrium across the air-sea interface needs to be established by equilibrating all carbonate species in the surface ocean, while for CO$_2$, the shortcut reaction involving HCO$_3^-$ and CO$_3^{-2}$ establish a faster equilibrium. Few numerical simulations have been performed to investigate this sensitivity explicitly, but Broecker and Maier-Reimer (1992) demonstrated that a doubling of the gas transfer velocity resulted in a substantial change in the surface ocean distribution of the ^{13}C/^{12}C ratio with a near doubling of the range.

For DMS and N$_2$O, virtually all models have used the gas transfer formulation of Wanninkhof (1992) to represent the air-sea flux of these gases (Le Clainche et al. 2010; Jin and Gruber 2003; Suntharalingam et al. 2012). In contrast, for less soluble gases, such as argon and neon, a broader set of parameterisations have been employed, particularly to reflect the importance of bubble mediated transfer. For the latter, most modellers follow the suggestion of Asher and

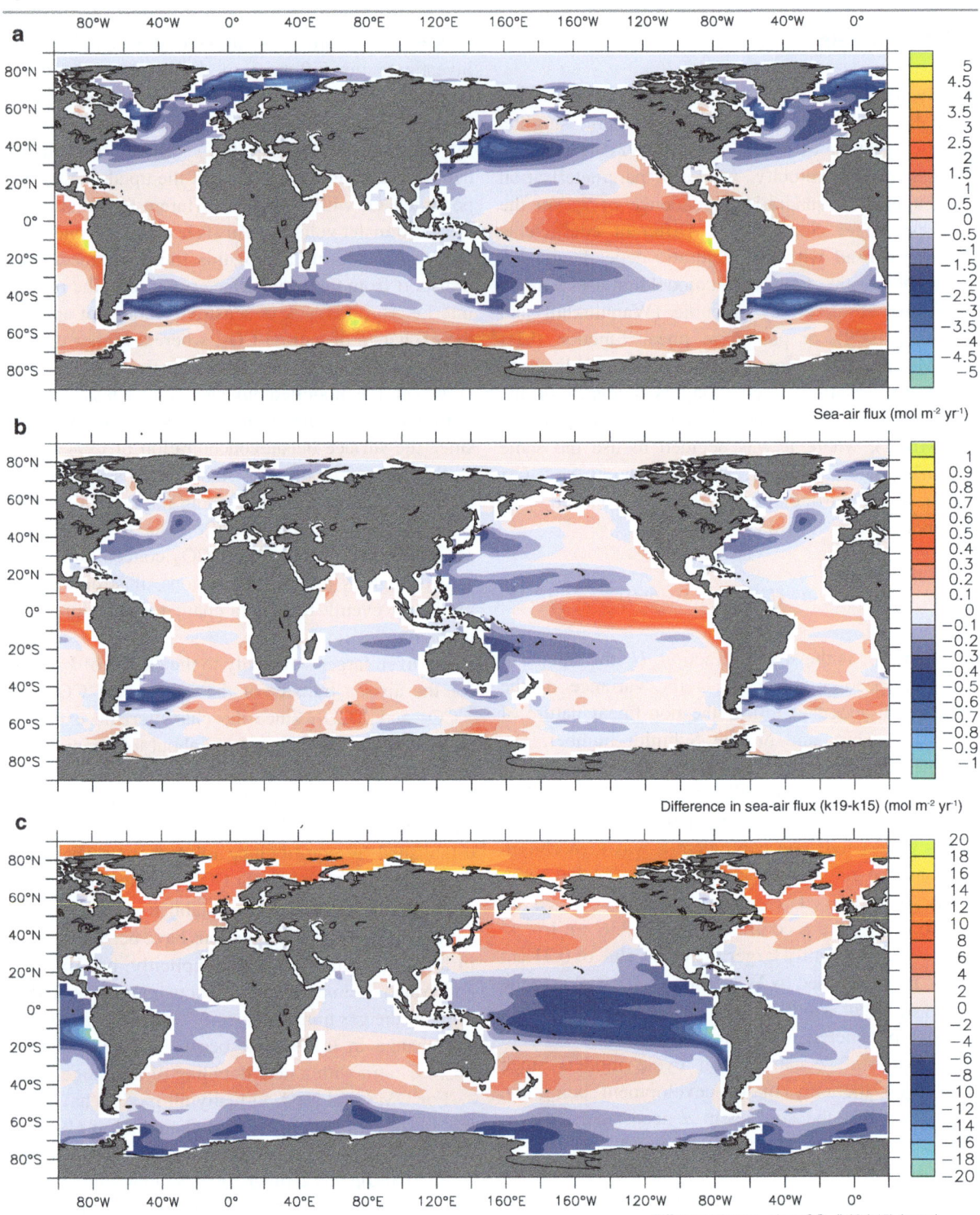

Fig. 2.12 Impact of the gas transfer velocity parameterisation on the model simulated air-sea flux of CO_2. (**a**) Annual mean sea-to-air flux of CO_2 in pre-industrial times simulated using a global mean gas transfer velocity of 15 cm h^{-1}. (**b**) Difference in the annual mean sea-to-air flux of CO_2 in pre-industrial times between a simulation using a global mean gas transfer velocity of 19 cm h^{-1} and the standard one with 15 cm h^{-1}. (**c**) Difference in the annual mean sea-air partial pressure difference in preindustrial times between the 19 cm h^{-1} and 15 cm h^{-1} cases. Based on model simulations executed with the global NCAR CCSM model (Graven et al. 2012)

Wanninkhof (1998) to scale the bubble-mediated flux with the cubic power of the wind speed (Hamme and Severinghaus 2007; Ito et al. 2011), but also other powers have been used (Spitzer and Jenkins 1989).

In the last decade, global and regional modellers have paid relatively little attention to the parameterisations used for the modelling the transfer of gases across the air-sea interface. This is likely to change in the coming years. On the one hand, the increasing computational power permits modellers to increase the spatial and temporal resolutions of their models, pushing them to the level where some of the assumptions underlying the classical parameterisations may no longer be fulfilled. For example, when employing the Wanninkhof (1992) parameterisation, most (global) models use the parameters for long-term average winds, while models are now being run with winds that change every 6 h, requiring the parameters for short-term winds. This change has not been undertaken in most models, yet. But the need for a reconsideration of the appropriateness of the chosen parameterisations needs to go further. Processes such as diurnal heating and cooling, eddies and fronts, and organic slicks that were considered of secondary importance at basin to global scales, may be important at local to regional scales, thus requiring renewed attention.

Furthermore, models are now increasingly being used to assimilate ocean observations, including pCO_2. Given the substantial sensitivity of surface ocean pCO_2 to the value of k_w (Fig. 2.12), it will be necessary to better handle gas transfer in order to exploit also the usefulness of pCO_2 to determine the rate of net community production, for example. With the possible advent of large-scale observations of dissolved oxygen by floats (Johnson et al. 2009), this need will increase even further, as the interpretation of the biologically generated O_2 signal requires accurate estimates of the amount of O_2 that is exchanged with the atmosphere.

2.8.2 Remote Sensing

Gas transfer velocity has been retrieved at global scale from satellite measurements using mainly two approaches:

1. The use of satellite wind speed and $k_w - U$ relationships (see Boutin et al. (2009) for a review).
2. Or using mean square slope (mss) derived from dual frequency altimeters (Glover et al. 2007).

Whatever the approach used, the big advantage of the satellite measurements is that they provide a better description of the spatio-temporal variability than meteorological models or at-sea measurements. In particular, a correct determination of k_w following point (1) requires a good knowledge of the statistical moments of the wind speed (at least its average and variance (see Wanninkhof et al. (2009))).

The main advantages of point (1) are related to the excellent spatial and temporal coverage of the wind speed owing to the large swaths of scatterometers and microwave radiometers, and to the mature algorithms for retrieving wind speed from satellite measurements (the rms difference of satellite wind speed with respect to buoy neutral wind speeds being on the order of 1 m s^{-1} (Bourassa et al. 2010)), while field measurements almost always provide k_w and U simultaneously, but not always surface stress. Although satellite wind speeds are expressed as neutral wind speeds to facilitate the comparison with in situ measurements, a scatterometer measures a backscatter coefficient that is related to resonant Bragg scattering by centimetric waves. Hence it is expected that scatterometer measurements would better correlate with k_w than wind speed at 10 m height. But this is almost impossible to demonstrate with field measurements due to the spatio-temporal variability of U (scatterometer measurements are integrated over typically 25 km while k_w is obtained from point measurements) and temporal undersampling of satellite measurements (at best two measurements per day while measurements of k_w by dual tracer techniques are integrated over several days). It is remarkable that the use of $k_w - U$ relationship derived from dual tracer experiments (Ho et al. 2006; Nightingale et al. 2000b) and scatterometer wind speeds give global k_w average within the new ^{14}C constraints without a-posteriori adjustments (Boutin et al. 2009) as shown in Fig. 2.13. Nevertheless, given the 30 % uncertainty in the global k_w average deduced from the ^{14}C inventory (Sweeney et al. 2007), the remaining uncertainty on the functional form of the $k_w - U$ relationship (quadratic, cubic, or hybrid, see Wanninkhof et al. (2009) for a review) and the remaining uncertainty

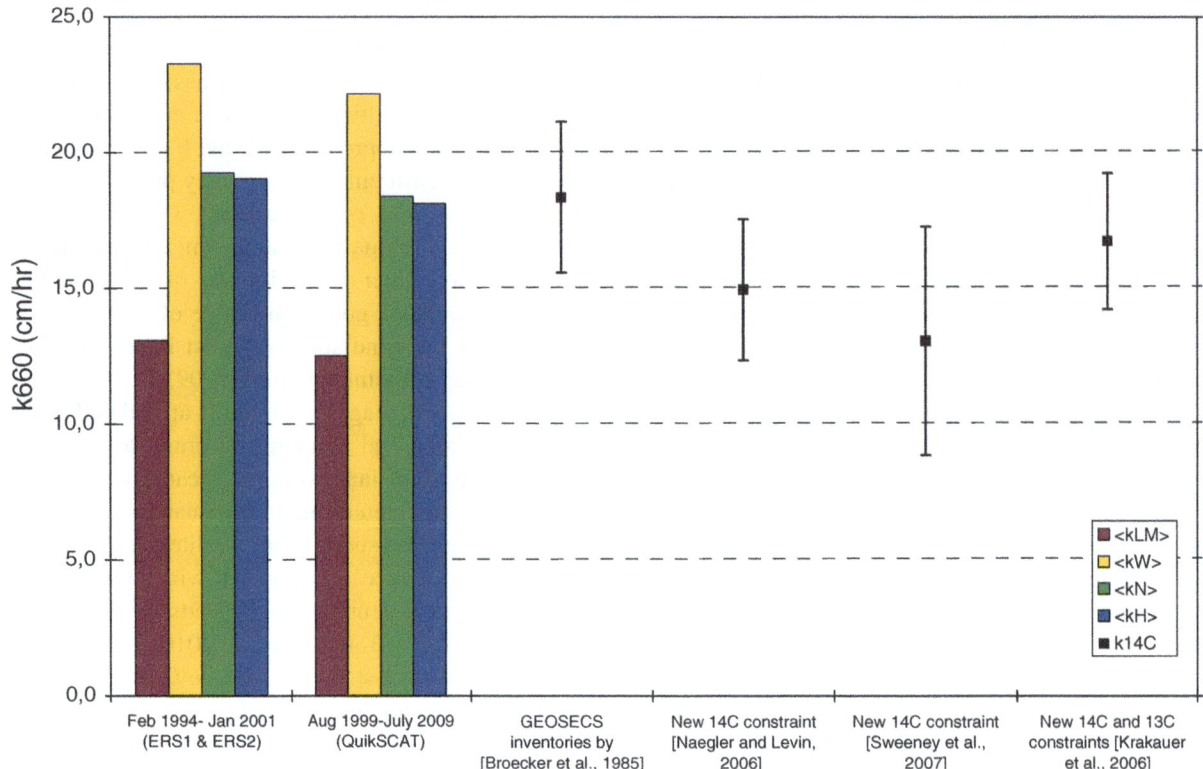

Fig. 2.13 Global averages of k_w (in cm h^{-1}) deduced from long time series of satellite wind speeds and $k_w - U$ relationships (*bar charts*) (*maroon bars* indicate k_{LM} (Liss and Merlivat 1986), *yellow bars* indicate k_W (Wanninkhof 1992), *green bars* indicate k_N (Nightingale et al. 2000a), and *blue bars* indicate k_H (Ho et al. 2006)) and deduced from ^{14}C global inventories (*black squares*). Methodology to derive k_w from satellite wind speed is described in Boutin et al. (2009); K_W fields are available at http://cersat.ifremer.fr/Data/Discovery/By-product-type/Gridded-products/

on the wind speed, the new ^{14}C estimates are not sufficient to fully constrain a $k_w - U$ relationship.

Method (2) has been developed because of experimental evidences that k_w is better correlated to small scale mss than to wind speed or wind stress (as reviewed in Frew et al. (2007)). Nevertheless, it has some limitations described in Sect. 2.6.2. Future large swath altimeters will improve the spatial coverage but in situ measurements of both mss and k_w in various regions of the globe are critical to definitely assess the utility of this method with respect to the $k_w - U$ method.

2.8.3 Inventories, Climatologies Using In Situ Data

Until now, long time series of k_w or of CO_2 exchange coefficients, K_W, (K_W is usually preferred when building maps as it minimises the temperature dependence due to the CO_2 Schmidt number and to the CO_2 solubility in the air-sea CO_2 flux computation) have mostly been built from a single satellite instrument. K_W derived from ERS1, ERS2 and QuickSCAT wind speeds between 1994 and 2009 are available at http://cersat.ifremer.fr/Data/Discovery/By-product-type/Gridded-products/K-CO2-QuikSCAT. Global averages of k_w derived from these scatterometer wind speeds are presented on Fig. 2.13.

Using k_w-mss relationships and after a thorough cross-validation of Jason-1 and TOPEX, Glover et al. (2007) have produced global k_w fields from 1993 to 2005. Because of the quadratic or cubic dependency of k_w with U, the accuracy of the wind speed fields used to build k_w fields is a big issue and the merging of several satellite wind speeds requires a thorough cross-validation (e.g. Fangohr et al. 2008). Not only the average wind speed must be well known but also at least the second moment of the wind speed. On the other hand, biases of a few tenths of meter per second remain between the various types of satellite wind speeds because of the difficulty of validating satellite wind speeds integrated over typically 25 km with point measurements of wind speeds in the field and because the various satellite instruments are not sensitive to the same surface state. For instance, a microwave radiometer or an altimeter will better sense high wind speeds than a scatterometer; a Ku-band scatterometer will be closer to friction velocity than a C-band scatterometer but will be more affected by rain.

Recently, the numerous satellite wind speeds available since 1987, and the existence of biases between various satellite wind products have motivated the development of cross-calibrated multiplatform surface wind products (CCMP) (e.g. Atlas et al. 2011). Such merged products are necessary to ensure long term continuity of wind speed records (in particular scatterometer measurements are not continuous), and are very useful for monitoring K_W using existing $k_w - U$ relationships and for computing the air-sea CO_2 flux from CO_2 partial pressure. Mean k_w deduced with a quadratic $k_w - U$ relationship and CCMP wind speeds between 2000 and 2009 and between 50 °S and 50 °N agrees to within 2.5 % with mean k_w deduced from QuickSCAT products available at Ifremer.[2]

2.9 Summary

Transfer across the air-sea interface is traditionally described by a wind-speed dependent function, where different studies indicate different functional forms and scaling dependencies. Measurements from mass-balance techniques seem to converge at a quadratic wind speed dependence (e.g. Nightingale et al. 2000b), agreeing with estimates from ^{14}C inventories. Eddy covariance techniques tend to give higher values, in particularly at higher winds, partly explained by the enhancement due to bubble mediated transfer (not fully captured by mass-balance techniques). It is generally agreed that wind is not the only forcing factor of importance. For better understanding additional processes need to be taken into account (micro breaking waves, small- and large scale turbulence, waves, bubbles and sea spray, surface films, chemical and biological enhancement, rain and atmospheric processes etc.). This is particularly important when more detailed information is required in time and space. Further understanding of the transfer can be obtained by measurements, where it is necessary to consider the strengths/weaknesses of the various methods and the scale being resolved by each method. High-resolution numerical simulation is an alternative approach for studying transfer mechanisms with the potential of giving detailed information on the action/interaction of various mechanisms.

Global transfer velocities are often derived from remote sensing or various inventories and use wind speed estimates from satellites or surface wind speed products and $k_w - U$ based relationships, although mean squares slopes from altimeter data are sometimes used. Here the accuracy of the satellite products and resolution in time and space are important factors to consider when used for flux calculations. Air-sea transfer velocities are frequently used in models. Fluxes derived from models or modelled marine biogeochemical processes have shown relatively small sensitivity to the formulation of transfer velocity, partly due to feed-back mechanisms in the models (the air-sea gradient is slightly adjusted) but the modelling community have not focused much on the issue. Global gas flux estimates calculated using air-sea gradients estimated from global inventories (from in situ data or remote sensing) are more sensitive to the formulation of transfer velocity than modelled estimates.

To be able to derive more accurate regional and global methods of determining the transfer velocity it is necessary to increase our understanding of the dependence of various processes and their relative importance. The variation in time and space can then be better explained and described. To achieve this, it is suggested to combine different measurement techniques and use gases of different solubility.

Open Access This chapter is distributed under the terms of the Creative Commons Attribution Noncommercial License, which permits any noncommercial use, distribution, and reproduction in any medium, provided the original author(s) and source are credited.

---

[2] http://cersat.ifremer.fr/data/discovery/by_product_type/gridded_products/k_co2_quikscat

References

Adrian R (1991) Particle-imaging techniques for experimental fluid mechanics. Annu Rev Fluid Mech 23:261–304

Adrian R (2005) Twenty years of particle image velocimetry. Exp Fluids 39(2):159–169. doi:10.1007/s00348-005-0991-7

Agrawal Y, Terray E, Donelan M, Hwang P, Williams AJ III, Drennan W, Kahma K, Kitaigorodskii S (1992) High dissipation beneath surface waves due to breaking. Nature 359:219–220. doi:10.1038/359219a0

Anctil F, Donelan M, Drennan W, Graber H (1994) Eddy-correlation measurements of air-sea fluxes from a discus buoy. J Atmos Ocean Technol 11:1144–1150

Andreas E (1992) Sea spray and the turbulent air–sea heat fluxes. J Geophys Res 97:11429–11441

Andreas EL, Jones KF, Fairall CW (2010) Production velocity of sea spray droplets. J Geophys Res 115(C12):C12065. doi:10.1029/2010JC006458

Anis A, Moum J (1995) Surface wave-turbulence interactions: scaling $\varepsilon(z)$ near the sea surface. J Phys Oceanogr 25:2025–2045

Asher W (1997) The sea-surface microlayer and its effect on global air-sea gas transfer. In: Liss PS, Duce RA (eds) The sea surface and global change. Cambridge University Press, Cambridge, pp 251–285

Asher WE (2009) The effects of experimental uncertainty in parameterizing air-sea gas exchange using tracer experiment data. Atmos Chem Phys 9(1):131–139

Asher WE, Pankow JF (1986) The interaction of mechanically generated turbulence and interfacial films with a liquid phase controlled gas/liquid transport process. Tellus 38B:305–318

Asher W, Wanninkhof R (1998) The effect of bubble-mediated gas transfer on purposeful dual-gaseous tracer experiments. J Geophys Res 103. doi:10.1029/98JC00245

Asher W, Karle L, Higgins B, Farley P (1996) The influence of bubble plumes on air-seawater gas transfer velocities. J Geophys Res 101:12027–12041

Asher WE, Jessup AT, Atmane MA (2004) On the use of the active controlled flux technique for in situ measurement oh the air–sea transfer velocity of heat and gas. J Geophys Res 109:C08S14. doi:10.1029/2003JC001805

Assur A (1958) Composition of sea ice and its tensile strength. In: Arctic sea ice, vol 598, Publication. National Academy of Sciences-National Research Council, Washington, DC, pp 106–138

Atlas R, Hoffman R, Ardizzone J, Leidner S, Jusem J, Smith D, Gombos D (2011) A cross-calibrated, multiplatform ocean surface wind velocity product for meteorological and oceanographic applications. Bull Am Meteorol Soc 92:157–174. doi:10.1175/2010BAMS2946.1

Atmane MA, Asher W, Jessup AT (2004) On the use of the active infrared technique to infer heat and gas transfer velocities at the air-water interface. J Geophys Res 109: C08S14

Babanin A (2006) On a wave-induced turbulence and a wave-mixed upper ocean layer. Geophys Res Lett 33:L20605. doi:10.1029/2006GL027308

Babanin AV, Haus BK (2009) On the existence of water turbulence induced by nonbreaking surface waves. J Phys Oceanogr 39(10):2675–2679

Babanin AV, Ganopolski A, Phillips WR (2009) Wave-induced upper-ocean mixing in a climate model of intermediate complexity. Ocean Model 29(3):189–197. doi:10.1016/j.ocemod.2009.04.003

Banks RB, Herrera FF (1977) Effect of wind and rain on surface reaeration. J Environ Eng 103:489–504

Banks RB, Wickramanayake GB, Lohani BN (1984) Effect of rain on surface reaeration. J Environ Eng 110:1–14

Banner ML (1990) The influence of wave breaking on the surface pressure distribution in wind-wave interactions. J Fluid Mech 211:463–495

Banner M, Peirson W (1998) Tangential stress beneath wind-driven air-water interfaces. J Fluid Mech 364:115–145

Banner ML, Phillips OM (1974) On the incipient breaking of small-scale waves. J Fluid Mech 65:647–656

Banner ML, Jones ISF, Trinder JC (1989) Wavenumber spectra of short gravity waves. J Fluid Mech 198:321–344

Bariteau L, Hueber J, Lang K, Helmig D, Fairall CW, Hare JE (2010) Ozone deposition velocity by ship-based eddy correlation flux measurements. Atmos Meas Tech 3:441–455. doi:10.5194/amt-3-441-2010

Bates N, Mathis J (2009) The arctic ocean marine carbon cycle: evaluation of air-sea co2 exchanges, ocean acidification impacts and potential feedbacks. Biogeosciences 6:2433–2459

Belanger TV, Korzun EA (1990) Rainfall-reaeration effects. J Irrig Drain 116:582–587. doi:10.1061/(asce)0733-9437(1990)116:4(582)

Belanger TV, Korzun EA (1991) Rainfall-reaeration effects. In: Wilhelms SC, Gulliver JS (eds) Air-water mass transfer. ASCE, New York, pp 388–399

Belcher SE, Grant ALM, Hanley KE, Fox-Kemper B, Van Roekel L, Sullivan PP, Large WG, Brown A, Hines A, Calvert D, Rutgersson A, Pettersson H, Bidlot J-R, Janssen PAEM, Polton JA (2012) A global perspective on Langmuir turbulence in the ocean surface boundary layer. Geophys Res Lett 39(18):L18605. doi:10.1029/2012GL052932

Beljaars ACM (1995) The parameterization of surface fluxes in large-scale models under free convection. Q J R Meteorl Soc 121:255–270

Bender ML, Kinter S, Cassar N, Wanninkhof R (2011) Evaluating gas transfer velocity parameterizations using upper ocean radon distributions. J Geophys Res 116(C2): C02010. doi:10.1029/2009JC005805

Berthe A, Kondermann D, Christensen C, Goubergrits L, Garbe C, Affeld K, Kertzscher U (2010) Three-dimensional, three-component wall-PIV. Exp Fluids 48(6):983–997. doi:10.1007/s00348-009-0777-4

Blanchard D (1963) The electrification of the atmosphere by particles from bubbles in the sea. Prog Oceanogr 1:71–202

Blomquist BW, Fairall CW, Huebert B, Kleiber DJ (2006) Dms sea-air transfer velocity: direct measurements by eddy covariance and parameterization based on the NOAA/COARE gas transfer model. Geophys Res Lett 33(7): L07601. doi:10.1029/2006GL025735

Bock EJ, Hara T, Frew NM, McGillis WR (1999) Relationship between air-sea gas transfer and short wind waves. J Geophys Res Ocean 104(C11):25821–25831, j NOV 15

Bortkovskii R, Novak V (1993) Statistical dependencies of sea state characteristics on water temperature and wind-wave age. J Mar Syst 4:161–169

Bourassa M, Stoffelen A, Bonekamp H, Chang P, Chelton D, Courtney J, Edson R, Figa J, He Y, Hersbach H, Hilburn K, Jelenak Z, Kelly K, Knabb R, Lee T, Lindstrom E, Liu W, Long D, Perrie W, Portabella M, Powell M, Rodriguez E, Smith D, Swail V, Wentz F (2010) Remotely sensed winds and wind stresses for marine forecasting and ocean modeling. In: Hall J, Harrison D, Stammer D (eds) Proceedings of OceanObs'09: sustained ocean observations and information for society, vol 2. ESA Publication WPP-306, Venice, 21–25 Sept 2009. doi:10.5270/OceanObs09.cwp.08

Boutin J, Etcheto J (1995) Estimating the chemical enhancement effect on the air-sea CO_2 exchange using the ERS1 scatterometer wind speeds. In: Monahan EC (ed) Air-water gas transfer. AEON Verlag & Studio, Hanau, pp 827–841

Boutin J, Quilfen Y, Merlivat L, Piolle J (2009) Global average of air-sea CO_2 transfer velocity from QuikSCAT scatterometer wind speeds. J Geophys Res 114:C04007. doi:10.1029/2007JC004168

Broecker W, Maier-Reimer E (1992) The influence of air and sea exchange on the carbon isotope distribution in the sea. Global Biogeochem Cycles 6:315–320. doi:10.1029/92GB01672

Broecker WS, Peng TH (1974) Gas exchange rates between air and sea. Tellus 26:21–35. doi:10.1111/j.2153-3490.1974.tb01948.x

Broecker HC, Petermann J, Siems W (1978) The influence of wind on co_2 exchange in a wind wave tunnel, including the effects of monolayers. J Mar Res 36:595–610

Broecker WS, Peng T-H, Ostlund G, Stuiver M (1985) The distribution of bomb radiocarbon in the ocean. J Geophys Res 90:6953–6970. doi:10.1029/JC090iC04p06953

Broecker WS, Ledwell JR, Takahashi T, Weiss R, Merlivat L, Memery L, Jähne B, Münnich KO (1986) Isotopic versus micrometeorologic ocean CO_2 fluxes: a serious conflict. J Geophys Res 91(C9):10517–10528. doi:10.1029/JC091iC09p10517

Brücker C (1995) Digital-particle-image-velocimetry (dpiv) in a scanning light-sheet: 3D starting flow around a short cylinder. Exp Fluids 19:255–263

Brücker C (1996) 3D PIV via spatial correlation in a color-coded light-sheet. Exp Fluids 21:312–314

Brumley BH, Jirka GH (1988) Air-water transfer of slightly soluble gases: turbulence, interfacial processes and conceptual models. Physicochem Hydrodyn 10:295–319

Budzianowski W, Koziol A (2005) Stripping of ammonia from aqueous solutions in the presence of carbon dioxide: effect of negative enhancement of mass transfer. Trans ICHemE Part A Chem Eng Res Des 83:196–204. doi:10.1205/cherd.03289

Burba G, McDermitt D, Grelle A, Anderson D, Xu L (2008) Addressing the influence of instrument surface heat exchange on the measurements of co2 flux from open-path gas analyzers. Glob Chang Biol 14:1854–1876

Burgmann S, Dannemann J, Schröder W (2008) Time-resolved and volumetric piv measurements of a transitional separation bubble on an sd7003 airfoil. Exp Fluids 44:609–622. doi:10.1007/s00348-007-0421-0

Businger J, Oncley S (1990) Flux measurement with conditional sampling. J Atmos Ocean Technol 7:349–352

Callaghan A, White M (2009) Automated processing of sea surface images for the determination of whitecap coverage. J Atmos Ocean Technol 26:384–394

Carlsson B, Rutgersson A, Smedman A (2009) Impact of swell on simulations using a regional atmospheric climate model. Tellus 61A:527–538

Carruthers DJ, Choularton TW (1986) The microstructure of hill cap clouds. Q J R Meteorol Soc 112:113–129

Cenedese A, Paglialunga A (1989) A new technique for the determination of the third velocity component with piv. Exp Fluids 8:228–230. doi:10.1007/BF00195799

Charnock H (1955) Wind stress on a water surface. Q J R Meteorol Soc 81:639–640

Chernyshenko SI, Baig MF (2005) The mechanism of streak formation in near-wall turbulence. J Fluid Mech 544:99–131. doi:10.1017/S0022112005006506

Coantic M (1986) A model of gas transfer across air–water interfaces with capillary waves. J Geophys Res 91:3925–3943. doi:10.1029/JC091iC03p03925

Csanady GT (1990) The role of breaking wavelets in air-sea gas transfer. J Geophys Res 95(C1):749–759

Csanady G (1997) The "slip law" of the free surface. J Oceanogr 53:67–80

Cunliffe M, Whitely A, Schafer H, Newbold L (2009) Comparison of bacterioneuston and bacterioplankton dynamics during a phytoplankton bloom in the a fjord microcosm. Appl Environ Microbiol 75:7173–7181

D'Asaro E, McNeil C (2007) Air–sea gas exchange at extreme wind speeds measured by autonomous oceanographic floats. J Mar Syst 66(1–4):92–109. doi:10.1016/j.jmarsys.2006.06.007

Danckwerts PV (1951) Significance of a liquid-film coefficients in gas absorption. Ind Eng Chem 43:1460–1467. doi:10.1021/ie50498a055

Davies JT (1972) Turbulence phenomena. An introduction to the eddy transfer of momentum, mass, and heat, particularly at interfaces. Academic, New York

de Leeuw G, Andreas EL, Anguelova MD, Fairall CW, Lewis ER, O'Dowd C, Schulz M, Schwartz SE (2011) Production flux of sea spray aerosol. Rev Geophys 49:RG2001. doi:10.1029/2010RG000349

de Leeuw G, Guieu C, Arneth A, Bellouin N, Bopp L, Boyd P, Denier van de Gon HAC, Desboeufs K, Dulac F, Facchini C, Gantt B, Langmann B, Mahowald N, Maranon E, O'Dowd C, Olgun N, Pulido-Villena E, Rinaldi M, Stephanou E, Wagener T (2013) Ocean–atmosphere interactions of particles. In: Liss P, Johnson M (eds) Ocean–atmosphere interactions of gases and particles. Springer, this volume

Deacon EL (1977) Gas transfer to and across an air-water interface. Tellus 29:363–374. doi:10.1111/j.2153-3490.1977.tb00746.x

Deardorff JW (1970) Convective velocity and temperature scales for the unstable planetary boundary layer and for rayleigh convection. J Atmos Sci 27(8):1211–1213. doi:10.1175/1520-0469(1970)027<1211:CVATSF>2.0.CO;2

Delille B, Tilbrook B, Lannuzel D, Schoemann V, Borges A, Lancelot C, Chou L, Dieckmann G, Tison J (2006) Air – sea ice exchange of carbon dioxide: the end of a long-lived paradigm? Science Submitted

Delille B, Jourdain B, Borges A, Tison J, Delille D (2007) Biogas (CO_2, O_2, dimethylsulfide) dynamics in spring Antarctic fast ice. Limnol Oceanogr 52:1367–1379

Department of Scientific and Industrial Research (1964) Effects of polluting discharges on the thames estuary: the reports of the Thames Survey Committee and of the Water Pollution, Research Laboratory. Her Majesty's Stationery Office, London, p 609

Desjardins R (1977) Description and evaluation of a sensible heat flux detector. Bound-Lay Meteorol 11:147–154

Dieckmann G, Nehrke G, Papadimitriou S, Göttlicher J, Steininger R, Kennedy H, Wolf-Gladrow D, Thomas D (2008) Calcium carbonate as ikaite crystals in Antarctic sea ice. Geophys Res Lett 35:L08501

Dieckmann G, Nehrke G, Uhlig C, Göttlicher J, Gerland S, Granskog M, Thomas D (2010) Ikaite ($CaCO_3*6H_2O$) discovered in Arctic sea ice. Cryosphere Discuss 4:153–161

Dinkelacker F, Schäfer M, Ketterle W, Wolfrum J, Stolz W, Köhler J (1992) Determination of the third velocity component with PTA using an intensity graded light sheet. Exp Fluids 13:357–359. doi:10.1007/BF00209511

Donelan MA (2001) A nonlinear dissipation function due to wave breaking. In: ECMWF workshop on ocean wave forecasting, The European Centre for Medium-Range Weather Forecasts, pp 87–94

Donelan M, Drennan W, Katsaros K (1997) The air-sea momentum flux in conditions of wind sea and swell. J Phys Oceanogr 27:2087–2099

Drennan W, Donelan M, Terray E, Katsaros K (1996) Oceanic turbulence dissipation measurements in SWADE. J Phys Oceanogr 26:808–815. doi:10.1175/1520-0485(1996)026<0808:OTDMIS>2.0.CO;2

Drennan W, Kahma K, Donelan M (1999) On momentum flux and velocity spectra over waves. Bound Lay Meteorol 92:489–515

Drennan W, Graber H, Hauser D, Quentin C (2003) On the wave age dependence of wind stress over pure wind seas. J Geophys Res 108. doi:10.1029/2000JC000715

Drennan W, Taylor P, Yelland M (2005) Parameterizing the sea surface roughness. J Phys Oceanogr 35:835–848

Duce RA, Tindale NW (1991) Atmospheric transport of iron and its deposition in the ocean. Limnol Oceanogr 36:1715–1726

Edson J, Hinton A, Prada K, Hare J, Fairall C (1998) Direct covari-ance measurements from mobile platforms at sea. J Atmos Ocean Technol 15:547–562

Edson J, Zappa C, Ware J, McGillis W, Hare J (2004) Scalar flux profile relationships over the open ocean. J Geophys Res 109:C08S09. doi:10.1029/2003JC001960

Edson J, Fairall C, Bariteau L, Helmig D, Zappa C, Cifuentes-Lorenzen A, McGillis W, Pezoa S, Hare J (2011) Direct covariance measurement of CO_2 gas transfer velocity during the 2008 Southern Ocean Gas Exchange Experiment: wind speed dependency. J Geophys Res. doi:10.1029/2011JC007022 (in press)

Elsinga GE, Scarano F, Wieneke B (2006) Tomographic particle image velocimetry. Exp Fluids 41:933–947

Emerson S, Quay P, Stump C, Wilbur D, Knox M (1991) O_2, ar, n_2, and ^{222}rn in surface waters of the subarctic Ocean: net biological o_2 production. Global Biogeochem Cycles 5(1):49–69. doi:10.1029/90GB02656

Eugster W, Kling G, Jonas T, McFadden J, Wuest A, MacIntyre S, Chapin F (2003) CO_2 exchange between air and water in an Arctic Alaskan and midlatitude Swiss lake: importance of convective mixing. J Geophys Res 108(D12):4362. doi:10.1029/2002JD002653

Fairall C, Bradley E, Rogers D, Edson J, Young G (1996) Bulk parameterization of air-sea fluxes for TOGA COARE. J Geophys Res 101:3747–3764

Fairall C, Hare J, Edson J, McGillis W (2000) Parameterization and micrometeorological measurements of air-sea gas transfer. Bound-Lay Meteorol 96:63–105. doi:10.1023/A:1002662826020

Fairall C, Bradley E, Hare J, Grachev A, Edson J (2003) Bulk parameterization of air-sea fluxes: updates and verification for the coare algorithm. J Climate 16:571–591

Fairall CW, Hare JE, Helmig D, Ganzveld L (2007) Water-side turbulence enhancement of ozone deposition to the ocean. Atmos Chem Phys 7:443–451. doi:10.5194/acp-7-443-2007

Fairall C, Yang M, Bariteau L, Edson J, Helmig D, McGillis W, Pezoa S, Hare J, Huebert B, Blomquist B (2011) Implementation of the coupled ocean–atmosphere response experiment flux algorithm with CO_2, dimethyl sulfide, and O_3. J Geophys Res 116:C00F09. doi:10.1029/2010JC006884

Fangohr S, Woolf D, Jeffery C, Robinson I (2008) Calculating long-term global air-sea flux of carbon dioxide using scatterometer, passive microwave, and model reanalysis wind data. J Geophys Res 113:C09032. doi:10.1029/2005JC003376

Fortescue GE, Pearson JRA (1967) On gas absorption into a turbulent liquid. Chem Eng Sci 22:1163–1176

Frew N (1997) The role of organic films in air-sea gas exchange. In: Liss PS, Duce RA (eds) The sea surface and global change. Cambridge University Press, Cambridge, pp 121–172

Frew N, Goldman J, Dennett M, Johnson A (1990) Impact of phytoplankton-generated surfactants on airsea gas-exchange. J Geophys Res 95:3337–3352

Frew N, Bock E, Schimpf U, Hara T, Haussecker H, Edson J, McGillis W, Nelson R, McKenna S, Uz B, Jähne B (2004) Air-sea gas transfer: its dependence on wind stress, small-scale roughness, and surface films. J Geophys Res Ocean C08S17:17. doi:10.1029/2003JC002131

Frew N, Glover D, Bock E, McCue S (2007) A new approach to estimation of global air-sea gas transfer velocity fields using dual-frequency altimeter backscatter. J Geophys Res 112:C11003. doi:10.1029/2006JC003819

Fulgosi M, Lakehal D, Banerjee S, Angelis VD (2003) Direct numerical simulation of turbulence in a sheared air–water flow with a deformable interface. J Fluid Mech 482:319–345

Garabetian F (1991) 14 c-glucose uptake and 14 C-CO_2 production in surface microlayer and surface water samples: influence of uv and visible radiation. Mar Ecol Prog Ser 77:21–26

Garbe C, Heinlein A (2011) Friction velocity from active thermography and shape analysis. In: Komori S, McGillis W, Kurose R (eds) Gas transfer at water surfaces 2010. Kyoto University Press, Kyoto, pp 535–543

Garbe C, Spies H, Jähne B (2003) Estimation of surface flow and net heat flux from infrared image sequences. J Math Imaging Vis 19(3):159–174. doi:10.1023/A:1026233919766

Garbe CS, Schimpf U, Jähne B (2004) A surface renewal model to analyze infrared image sequences of the ocean surface for the study of air-sea heat and gas exchange. J Geophys Res 109:C08S15. doi:10.1029/2003JC001802

Garbe C, Degreif K, Jähne B (2007) Estimating the viscous shear stress at the water surface from active thermography. In: Garbe C, Handler RA, Jähne B (eds) Transport at the air sea interface – measurements, models and parametrizations. Springer, Berlin, pp 223–239. doi:10.1007/978-3-540-36906-6_16

Garbe C, Voss B, Stapf J (2012) Plenoptic particle streak velocimetry (ppsv): 3d3c fluid flow measurement from light fields with a single plenoptic camera. In: 16th international symposium on applications of laser techniques to fluid mechanics, Instituto Superior Técnico, Lisbon, pp 1–12

Gargett AE, Wells JR (2007) Langmuir turbulence in shallow water. Part 1. Observations. J Fluid Mech 576:27–61

Garland JA, Etzerman AW, Penkett SA (1980) The mechanism for dry deposition of ozone to seawater surfaces. J Geophys Res 85:7488–7492

Garratt JR (1992) The atmospheric boundary layer. Cambridge University Press, Cambridge

Glover D, Frew N, McCue S (2007) Air-sea gas transfer velocity estimates from the Jason-1 and TOPEX altimeters: prospects for a long-term global time series. J Mar Syst 66:173–181

Glud R, Rysgaard S, Kuhl M (2002) A laboratory study on o-2 dynamics and photosynthesis in ice algal communities: quantification by microsensors, o-2 exchange rates, c-14 incubations and a pam fluorometer. Aquat Microb Ecol 27:301–311

Goddijn-Murphy L, Woolf D, Callaghan A (2011) Parameterizations and algorithms for oceanic whitecap coverage. J Phys Oceanogr 41:742–756. doi:10.1175/2010JPO4533.1

Godfrey J, Beljaars A (1991) On the turbulent fluxes of buoyancy, heat and moisture at the air-sea interface at low wind speeds. J Geophys Res 96:22043–22048

Golden K (2003) Critical behavior of transport in sea ice. Phys B Condens Matter 338:274–283

Golden K, Ackley S, Lytle V (1998) The percolation phase transition in sea ice. Science 282:2238–2241

Goldman J, Dennett M, Frew N (1988) Surfactant effects on air sea gas-exchange under turbulent conditions. Deep-Sea Res 35:1953–1970

Gosink T, Pearson J, Kelley J (1976) Gas movement through sea ice. Nature 263:41–42

Grassl H (1976) The dependence of the measured cool skin of the ocean on wind stress and total heat flux. Bound-Lay Meteorol 10:465–474

Graven H, Gruber N, Key RF, Khatiwala S (2012) Changing controls on oceanic radiocarbon: new insights on shallow-to-deep ocean exchange and anthropogenic CO_2 uptake. J Geophys Res 117:C10005

Guo LX, Smedman A-S, Högström U (2004) Air-sea exchange of sensible heat over the baltic sea. Q J R Meteorol Soc 130:519–539

Hamme RC, Severinghaus JP (2007) Trace gas disequilibria during deep-water formation. Deep Sea Res Part I 54:939–950. doi:10.1016/j.dsr.2007.03.008

Handler RA, Smith GB (2011) Statistics of the temperature and its derivatives at the surface of a wind-driven air-water interface. J Geophys Res 116(C6):C06021. doi:10.1029/2010JC006496

Handler RA, Smith GB, Leighton RI (2001) The thermal structure of an air–water interface at low wind speeds. Tellus 53(A):233–244

Hara T, VanInwegen E, Wendelbo J, Garbe CS, Schimpf U, Jähne B, Frew N (2007) Estimation of air-sea gas and heat fluxes from infrared imagery based on near surface turbulence models. In: Garbe CS, Handler RA, Jähne B (eds) Transport at the air sea interface – measurements, models and parameterizations. Springer, Berlin. doi:10.1007/978-3-540-36906-6_17

Hare J, Fairall C, McGillis W, Edson J, Ward B, Wanninkhof R (2004) Evaluation of the NOAA/COARE air-sea gas transfer parameterization using GasEx data. J Geophys Res 109:C08S02. doi:10.1029/2003/C002256

Harriott P (1962) A random eddy modification of the penetration theory. Chem Eng Sci 17:149–154

Harrison EL, Veron F, Ho DT, Reid MC, Eggleston SS, Orton P, McGillis WR (2012) Nonlinear interaction between rain and wind induced air-water gas exchange. J Geophys Res 117:C03034

Hasse L, Liss P (1980) Gas exchange across the air-sea interface. Tellus 32:470–481. doi:10.1111/j.2153-3490.1980.tb00974.x

Haugen D (1978) Effects of sampling rates and averaging periods on me-teorlogical measurements. In: Fourth Symp Meteorol Observ Instr, Am Meteorol Soc, pp 15–18

Haußecker H, Jähne B (1995) In situ measurements of the air-sea gas transfer rate during the MBL/CoOP west coast experiment. In: Jähne B, Monahan EC (eds) Air-water gas transfer – selected papers from the third international symposium on air-water gas transfer. AEON Verlag & Studio, Hanau/Heidelberg, pp 775–784

Heinesch B, Yernaux M, Aubinet M, Geilfus N, Papakyriakou T, Carnat G, Eicken H, Tison J, Delille B (2009) Measuring air-ice CO_2 fluxes in the Arctic. FluxLetter Newsl FLUXNET 2:9–10

Helmig D, Lang E, Bariteau L, Ganzeveld L, Fairall C, Hare J, Boylan P, Hueber J (2011) Atmosphere-ocean ozone fluxes during the TexAQS 2006, STRATUS 2006, GOMECC 2007, GasEX 2008, and AMMA 2008 cruises. J Geophys Res 117:D04305

Hicks B (2005) A climatology of wet deposition scavenging ratios for the united states. Atmos Environ 39(9):1585–1596. doi:10.1016/j.atmosenv.2004.10.039

Higbie R (1935) The rate of absorption of a pure gas into a still liquid during short periods of exposure. Trans Am Inst Chem Eng 31:365–389

Hinsch KD (2002) Holographic particle image velocimetry. Meas Sci Technol 13:R61–R72

Ho DT, Bliven LF, Wanninkhof R, Schlosser P (1997) The effect of rain on air-water gas exchange. Tellus 49:149–158

Ho DT, Asher WE, Bliven LF, Schlosser P, Gordan EL (2000) On mechanisms of rain-induced air-water gas exchange. J Geophys Res 105:24045–24057

Ho DT, Zappa CJ, McGillis WR, Bliven LF, Ward B, Dacey JWH, Schlosser P, Hendricks MB (2004) Influence of rain on air-sea gas exchange: lessons from a model ocean. J Geophys Res 109:C08S18. doi:10.1029/2003JC001806

Ho DT, Law CS, Smith MJ, Schlosser P, Harville M, Hill P (2006) Measurements of air-sea gas exchange at high wind speeds in the southern ocean: implications for global parameterizations. Geophys Res Lett 33:16611–16616. doi:10.1029/2006GL026817

Ho DT, Veron F, Harrison E, Bliven LF, Scott N, McGillis WR (2007) The combined effect of rain and wind on air-water gas exchange: a feasibility study. J Mar Syst 66:150–160. doi:10.1016/j.jmarsys.2006.02.012

Ho DT, Sabine CL, Hebert D, Ullman DS, Wanninkhof R, Hamme RC, Strutton PG, Hales B, Edson JB, Hargreaves BR (2011a) Southern ocean gas exchange experiment: setting the stage. J Geophys Res 116:C00F08. doi:10.1029/2010JC006852

Ho DT, Wanninkhof R, Schlosser P, Ullman DS, Hebert D, Sullivan KF (2011b) Toward a universal relationship between wind speed and gas exchange: gas transfer velocities measured with ^{3}He/SF$_6$ during the Southern Ocean Gas Exchange Experiment. J Geophys Res 116:C00F04. doi:10.1029/2010JC006854

Högström U (1996) Review of some basic characteristics of the atmospheric surface layer. Bound-Lay Meteorol 78:215–246

Högström U, Sahlée E, Drennan WM, Kahma KK, Smedman A-S, Johansson C, Pettersson H, Rutgersson A, Tuomi L, Zhang F, Johansson M (2008) Momentum fluxes and wind gradients in the marine boundary layer – a multi platform study. Boreal Environ Res 13:475–502

Holtslag A, De Bruin H (1988) Applied modeling of the nighttime surface energy balance over land. J Appl Meteorol 27:689–704

Hoover TE, Berkshire DC (1969) Effects of hydration on carbon dioxide exchange across an air-water interface. J Geophys Res 74(2):456–464

Hoppel WA, Frick GM, Fitzgerald JW (2002) Surface source function for sea-salt aerosol and aerosol dry deposition to the ocean surface. J Geophys Res 107:4382. doi:10.1029/2001JD002014

Hoppel WA, Caffrey PF, Frick GM (2005) Particle deposition on water: surface source versus upwind source. J Geophys Res 110:D10206. doi:10.1029/2004JD005148

Horst T, Lenschow D (2009) Attenuation of scalar fluxes measured with spatially displaced sensors. Bound-Lay Meteorol 130:275–300. doi:10.1007/s10546-008-9348-0

Hoyer K, Holzner M, Lüthi B, Guala M, Liberzon A, Kinzelbach W (2005) 3D scanning particle tracking velocimetry. Exp Fluids 39:923–934. doi:10.1007/s00348-005-0031-7

Huebert B, Blomquist B, Hare JE, Fairall CW, Johnson J, Bates T (2004) Measurements of the sea-air DMS flux and transfer velocity using eddy correlation. J Geophys Res Lett 31:L23113. doi:10.1029/2004GL021567

Hung L-P, Garbe CS, Tsai W-T (2011) Validation of eddy-renewal model by numerical simulation. In: Komori S, McGillis W, Krose R (eds) Gas transfer at water surfaces 2010. Kyoto University Press, Kyoto, pp 165–176

Hunt J, Belcher S, Stretch D, Sajjadi S, Clegg J (2011) Turbulence and wave dynamics across gas-liquid interfaces. In: Komori S, McGillis W, Kurose R (eds) Gas transfer at water surfaces 2010. Kyoto University Press, Kyoto

Ito T, Hamme RC, Emerson S (2011) Temporal and spatial variability of noble gas tracers in the north pacific. J Geophys Res 116:C08039. doi:10.1029/2010JC006828

Iversen T (1989) Numerical modelling of the long range atmospheric transport of sulphur dioxide and particulate sulphate to the arctic. Atmos Environ 23(11):2571–2595. doi:10.1016/0004-6981(89)90267-9

Jackson D, Wick G, Hare J (2011) A comparison of satellite-derived carbon dioxide transfer velocities from a physically-based model with GasEx cruise observations. J Geophys Res. doi:10.1029/2011JC007329 (in press)

Jähne B (1989) Energy balance in small-scale waves: an experimental approach using optical slope measuring technique and image processing. In: Komen GJ, Oost WA (eds) Radar scattering from modulated wind waves. Kluwer, Dordrecht, pp 105–120

Jähne B, Haußecker H (1998) Air-water gas exchange. Annu Rev Fluid Mech 30:443–468. doi:10.1146/annurev.fluid.30.1.443

Jähne B, Riemer KS (1990) Two-dimensional wave number spectra of small-scale water surface waves. J Geophys Res 95:11531–11546

Jähne B, Münnich KO, Siegenthaler U (1979) Measurements of gas exchange and momentum transfer in a circular wind-water tunnel. Tellus 31:321–329

Jähne B, Munnich K, Bosinger R, Dutzi A, Huber W, Libner P (1987) On the parameters influencing air-water gas exchange. J Geophys Res 92:1937–1949

Janssen PAEM (2004) The interaction of ocean waves and wind. Cambridge University Press, Cambridge

Jeffery CD, Woolf DK, Robinson IS, Donlon CJ (2007) One-dimensional modelling of convective CO_2 exchange in the Tropical Atlantic. Ocean Model 19:161–182. doi:10.1016/j.ocemod.2007.07.003

Jeffery C, Robinson I, Woolf D (2010) Tuning a physically-based model of the air-sea gas transfer velocity. Ocean Model 31:28–35. doi:10.1016/j.ocemod.2009.09.001

Jehle M, Jähne B (2008) A novel method for three-dimensional three-component analysis of flow close to free water surfaces. Exp Fluids 44:469–480. doi:10.1007/s00348-007-0453-5

Jessup AT, Zappa C, Loewen MR, Hesany V (1997a) Infrared remote sensing of breaking waves. Nature 385(6611):52–55. doi:10.1038/385052a0

Jessup AT, Zappa CJ, Yeh HH (1997b) Defining and quantifying microscale wave breaking with infrared imagery. J Geophys Res 102(C10):23145–23153

Jessup AT, Asher WE, Atmane M, Phadnis K, Zappa CJ, Loewen MR (2009) Evidence for complete and partial surface renewal at an air-water interface. Geophys Res Lett 36:1–5. doi:10.1029/2009GL038986

Jin X, Gruber N (2003) Offsetting the radiative benefit of ocean iron fertilization by enhancing N_2O emissions. Geophys Res Lett 30:2249. doi:10.1029/2003GL018458

Johnson M (2010) A numerical scheme to calculate temperature and salinity dependent air-water transfer velocities for any gas. Ocean Sci 6:913–932. doi:10.5194/os-6-913-2010

Johnson KS, Berelson WM, Boss ES, Chase Z, Claustre H, Emerson SR, Gruber N, Körtzinger A, Perry MJ, Riser SC (2009) Observing biogeochemical cycles at global scales with profiling floats and gliders: prospects for a global array. Oceanography 22:216–225

Johnson M, Hughes C, Bell T, Liss P (2011) A rumsfeldian analysis of uncertainty in air-sea gas exchange. In: Gas transfer at water surface 2010. Kyoto University Press, Kyoto, pp 464–484

Kähler C, Kompenhans J (2000) Fundamentals of multiple plane stereo particle image velocimetry. Exp Fluids 29:70–77

Keeling R (1993) On the role of large bubbles in air-sea gas exchange and supersaturation in the ocean. J Mar Res 51:237–271

Keller K (1994) Chemical enhancement of carbon dioxide transfer across the air-sea interface. Ph.D. thesis, Massachusetts Institute of Technology. http://dspace.mit.edu/bitstream/handle/1721.1/35997/32162323.%pdf?sequence=1

Kihm C, Körtzinger A (2010) Air-sea gas transfer velocity for oxygen derived from float data. J Geophys Res 115:C12003. doi:10.1029/2009JC006077

Kitaigorodskii S (1984) On the fluid dynamical theory of turbulent gas transfer across an air-sea interface in the presence of breaking wind waves. J Phys Oceanogr 14:960–972. doi:10.1175/1520-0485(1984)014<0960:OTFDTO>2.0.CO;2

Kitaigorodskii S (2011) The calculation of the gas transfer between the ocean and atmosphere. In: Komori S, McGillis W, Kurose R (eds) Gas transfer at water surfaces 2011. Kyoto University Press, Kyoto, pp 13–28

Kitaigorodskii S, Donelan MA (1984) Wind–wave effects on gas transfer. In: Brutsært W, Jirka GH (eds) Gas transfer at water surfaces. Reidel, Dordrecht, pp 147–170

Kline SJ, Reynolds WC, Schraub FA, Runstadler PW (1967) The structure of turbulent boundary layers. J Fluid Mech 30(04):741–773. doi:10.1017/S0022112067001740

Komori S, Takagaki N, Saiki R, Suzuki N, Tanno K (2007) The effect of raindrops on interfacial turbulence and air-water gas transfer. In: Handler RA, Garbe C, Jähne B (eds) Transport at the air-sea interface. Springer, Berlin/Heidelberg, pp 169–179

Krakauer NY, Randerson JT, Primeau FW, Gruber N, Menemenlis D (2006) Carbon isotope evidence for the latitudinal distribution and wind speed dependence of the air-sea gas transfer velocity. Tellus B 58(5):390–417

Kromer B, Roether W (1983) Field measurements of air-sea gas exchange by the radon deficit method during jasin 1978 and fgge 1979. In: Meteor Forschungsergebnisse, Reihe A/B Allgemeines, Physik und Chemie des Meeres, Gebrüder Borntrãger, vol A/B24, Deutsche Forschungsgemeinschaft, Berlin/Stuttgart, pp 55–76

Lakehal D, Fulgosi M, Yadigaroglu G, Banerjee S (2003) Direct numerical simulation of turbulent heat transfer across a mobile, sheared gas-liquid interface. J Heat Transf 15:1129–1139

Lamont JC, Scott DS (1970) An eddy cell model of mass transfer into the surface of a turbulent liquid. AIChE J 16:512–519. doi:10.1002/aic.690160403

Langmuir I (1938) Surface motion of water induced by wind. Science 87(2250):119–123. doi:10.1126/science.87.2250.119

Le Clainche Y, Vézina A, Levasseur M, Cropp RA, Gunson JR, Vallina SM, Vogt M, Lancelot C, Allen JI, Archer SD, Bopp L, Deal C, Elliott S, Jin M, Malin G, Schoemann V, Simo R, Six KD, Stefels J (2010) A first appraisal of prognostic ocean DMS models and prospects for their use in climate models. Global Biogeochem Cycles 24(3):GB3021. doi:10.1029/2009GB003721

Le Quéré C, Saltzman ES (eds) (2009) Surface ocean-lower atmosphere processes, vol 187, Geophysical monograph series. AGU, Washington, DC

Ledwell J (1984) The variation of the gas transfer coefficient with molecular diffusivity. In: Brutsært W, Jirka GH (eds) Gas transfer at water surfaces. Reidel, Dordrecht, pp 293–303

Lewis ER, Schwartz SE (2004) Sea salt aerosol production: mechanisms, methods, measurements, and models. American Geophysical Union, Washington, DC

Li M, Garrett C (1995) Is langmuir circulation driven by surface waves or surface cooling? J Phys Oceanogr 25(1):64–76. doi:10.1175/1520-0485(1995)025<0064:ILCDBS>2.0.CO;2

Li M, Garrett C, Skyllingstad E (2005) A regime diagram for classifying turbulent large eddies in the upper ocean. Deep-Sea Res 52(2):259–278. doi:10.1016/j.dsr.2004.09.004

Liberzon A, Gurka R, Hetsroni G (2004) XPIV-multi-plane stereoscopic particle image velocimetry. Exp Fluids 36:355–362

Liss PS (1971) Exchange of SO_2 between the atmosphere and natural waters. Nature 233(5318):327–329. doi:10.1038/233327a0

Liss P (1975) Chemistry of the sea surface microlayer. In: Riley J, Skirrow G (eds) Chemical oceanography, vol 2. Academic, London, pp 192–244

Liss P (1983) Gas transfer: experiments and geochemical implications. In: Liss P, Slinn W (eds) Air-sea exchange of gases and particles. Springer, Dordrecht, pp 241–298

Liss P, Martinelli F (1978) The effect of oil films on the transfer of oxygen and water vapour across an air-water interface. Thalass Jugosl 14:215–220

Liss PS, Merlivat L (1986) Air-sea gas exchange rates: introduction and synthesis. In: Buat-Menard P (ed) The role of air-sea exchange in geochemical cycling. Reidel, Boston, pp 113–129

Liss PS, Slater PG (1974) Flux of gases across the air-sea interface. Nature 247:181–184

Liss P, Watson A, Bock E, Jähne B, Asher W, Frew N, Hasse L, Korenowski G, Merlivat L, Phillips L, Schlüssel P, Woolf D (1997) Report group 1 – physical processes in the microlayer and the air-sea exchange of trace gases. In: Liss P, Duce R (eds) The sea surface and global change. Cambridge University Press, Cambridge, pp 1–33

Long MS, Keene WC, Kieber DJ, Erickson DJ, Maring H (2011) A sea-state based source function for size- and composition-resolved marine aerosol production. Atmos Chem Phys 11:1203–1216. doi:10.5194/acp-11-1203-2011

Longuet-Higgins MS, Cleaver RP, Fox MJH (1994) Crest instabilities of gravity waves. Part 2. Matching and asymptotic analysis. J Fluid Mech 259:333–344

Loose B, McGillis W, Schlosser P, Perovich D, Takahashi T (2009) Effects of freezing, growth, and ice cover on gas transport processes in laboratory seawater experiments. Geophys Res Lett 36. doi:10.1029/2008gl036318

Loose B, Miller L, Elliott S, Papakyriakou T (2011a) Sea ice biogeochemistry and material transport across the frozen interface. Oceanography 24:202–218. doi:10.5670/oceanog.2011.72

Loose B, Schlosser P, Perovich D, Ringelberg D, Ho D, Takahashi T, Reynolds C, McGillis W, Tison J (2011b) Gas diffusion through columnar laboratory sea ice: implications for mixed-layer ventilation of CO_2 in the seasonal ice zone. Tellus B 63. doi:10.1111/j.1600-0889.2010.00506.x

Lorke A, Peeters F (2006) Toward a unified scaling relation for interfacial fluxes. J Phys Oceanogr 36(5):955–961

MacIntyre S, Eugster W, Kling GW (2002) The critical importance of buoyancy flux for gas flux across the air-water interface. In: Donelan MA, Drennan WM, Saltzman ES, Wanninkhof R (eds) Gas transfer at water surfaces, vol 127, Geophysical monograph. American Geophysical Union, Washington, DC, pp 13–28

Mackay D, Yeun ATK (1983) Mass transfer coefficient correlations for volatilization of organic solutes from water. Environ Sci Technol 17:211–217

Manasseh R, Babanin AV, Forbes C, Richards K, Bobevski I, Ooi A (2006) Passive acoustic determination of wave-breaking events and their severity across the spectrum. J Atmos Ocean Technol 23:599–618

Marion G, Millero F, Feistel R (2009) Precipitation of solid phase calcium carbonates and their effect on application of seawater SA-T -P models. Ocean Sci 5:285–291

Matthews B (1999) The rate of air-sea CO_2 exchange: chemical enhancement and catalysis by marine microalgae. Ph.D. thesis, University of East Anglia, Norwich

McGillis WR, Wanninkhof R (2006) Aqueous CO_2 gradients for air-sea flux estimates. Mar Chem 98:100–108

McGillis W, Edson J, Hare J, Fairall C (2001) Direct covariance air-sea CO_2 fluxes. J Geophys Res 106:16729–16745

McGillis W, Edson J, Zappa C, Ware J, McKenna S, Terray E, Hare J, Fairall C, Drennan W, Donelan M, DeGrandpre M, Wanninkhof R, Feely R (2004a) Air-sea CO_2 exchange in the equatorial Pacific. J Geophys Res 109:C08S02. doi:10.1029/2003JC002256

McGillis WR, Asher WE, Wanninkhof R, Jessup AT, Feely RA (2004b) Air-sea CO_2 exchange in the equatorial pacific. J Geophys Res 109:C08S01

McKenna SP, McGillis WR (2004) The role of free-surface turbulence and surfactants in air-water gas transfer. Int J Heat Mass Transf 47:539–553. doi:10.1016/j.ijheatmasstransfer.2003.06.001

McNeil C, D'Asaro E (2007) Parameterization of air sea gas fluxes at extreme wind speeds. J Mar Syst 66:110–121. doi:10.1016/j.jmarsys.2006.05.013

McNeil CL, Ward B, McGillis WR, DeGrandpre MD, Marcinowski L (2006) Fluxes of N_2, O_2, and CO_2 in nearshore waters off Martha's Vineyard. Cont Shelf Res 26:1281–1294

McWilliams J, Sullivan P, Moeng C (1997) Langmuir turbulence in the ocean. J Fluid Mech 334:1–30

Melville WK (1996) The role of surface-wave breaking in air-sea interaction. Annu Rev Fluid Mech 28:279

Miller L, Papakyriakou T, Collins R, Deming J, Ehn J, Macdonald R, Mucci A, Owens O, Raudsepp M, Sutherland N (2011) Carbon dynamics in sea ice: a winter flux time series. J Geophys Res 116:C02028. doi:10.1029/2009jc006058

Monahan E, O'Muircheartaigh I (1980) Optimal power-law description of oceanic whitecap coverage dependence on wind speed. J Phys Oceanogr 10:2094–2099

Monahan EC, Spillane MC (1984) The role of whitecaps in air-sea gas exchange. In: Brutsaert W, Jirka GH (eds) Gas transfer at water surfaces. Reidel, Hingham, pp 495–503

Monahan E, Fairall C, Davidson K, Jones-Boyle P (1983) Observed interrelations between 10 m winds, ocean whitecaps and marine aerosols. Q J R Meteorol Soc 109:379–392

Müller D, Müller B, Renz U (2001) Three-dimensional particle-streak tracking (PST) velocity measurements of a heat exchanger inlet flow. Exp Fluids 30(6):645–656. doi:10.1007/s003480000242

Müller SA, Joos F, Plattner GK, Edwards NR, Stocker TF (2008) Modeled natural and excess radiocarbon: sensitivities to the gas exchange formulation and ocean transport strength. Global Biogeochem Cycles 22(3):GB3011. doi:10.1029/2007GB003065

Naegler T, Ciais P, Rodgers K, Levin I (2006) Excess radiocarbon constraints on air-sea gas exchange and the uptake of CO_2 by the oceans. Geophys Res Lett 33(11):L11802. doi:10.1029/2005GL025408

Nägler T (2009) Reconciliation of excess ^{14}C-constrained global CO_2 piston velocity estimates. Tellus B 61(2):372–384. doi:10.1111/j.1600-0889.2008.00408.x

Najjar R, Orr J (1998) Design of OCMIP-2 simulations of chlorofluorocarbons, the solubility pump and common biogeochemistry, http://www.cgd.ucar.edu/oce/klindsay/OCMIP/design.pdf

Nakagawa H, Nezu I (1981) Structure of space-time correlation of bursting phenomena in an open-channel flow. J Fluid Mech 104:1–43. doi:10.1017/S0022112081002796

Nightingale PD, Liss PS, Schlosser P (2000a) Measurements of air-sea gas transfer during an open ocean algal bloom. Geophys Res Lett 27(14):2117–2120. doi:10.1029/2000GL011541

Nightingale PD, Malin G, Law CS, Watson AJ, Liss PS, Liddicoat MI, Boutin J, Upstill-Goddard RC (2000b) In situ evaluation of air-sea gas exchange parameterizations using novel conservative and volatile tracers. Global Biogeochem Cycles 14(1):373–387. doi:10.1029/1999GB900091

Nilsson ED, Rannik U, Swietlicki E, Leck C, Aalto PP, Zhou J, Norman M (2001) Turbulent aerosol fluxes over the Arctic Ocean. 2. wind driven sources from the sea. J Geophys Rev 106:32129–32154. doi:10.1029/2000JD900747

Nilsson E, Rutgersson A, Sullivan P (2010) Flux attenuation due to sensor separation over sea. J Atmos Ocean Technol 27:856–868. doi:10.1175/2010JTECHA1388.1

Nomura D, Eicken H, Gradinger R, Shirasawa K (2010a) Rapid physically driven invesrion of the air-sea ice CO_2 flux in the seasonal landfast ice off Barrow, Alaska after onset of surface melt. Cont Shelf Res 30:1998–2004

Nomura D, Yoshikawa-Inoue H, Toyota T, Shirasawa K (2010b) Effects of snow, snowmelting and refreezing processes on air-sea-ice CO_2 flux. J Glaciol 56:262–270

Oh S-H, Mizutani N, Suh K-D (2008) Laboratory observation of coherent structures beneath microscale and large-scale breaking waves under wind action. Exp Therm Fluid Sci 32:1232–1247

Okuda K (1982) The internal structure of short wind waves. Part I: on the internal vorticity structure. J Oceanogr Soc Jpn 38:28–42

Olsen A, Omar AM, Stuart-Menteth AC, Triñanes JA (2004) Diurnal variations of surface ocean pCO_2 and sea–air CO_2 flux evaluated using remotely sensed data. Geophys Res Lett 31:L20304. doi:10.1029/2004GL020583

Osborne T, Farmer D, Vagle S, Thorpe S, Cure M (1992) Measurements of bubble plumes and turbulence from a submarine. Atmos Ocean 30:419–440. doi:10.1080/07055900.1992.9649447

Panofsky H, Dutton J (1984) Atmospheric turbulence, models and methods for engineering applications. Wiley, New York

Papadimitriou S, Kennedy H, Kattner G, Dieckmann G, Thomas D (2004) Experimental evidence for carbonate precipitation and CO_2 degassing during sea ice formation. Geochim Cosmochim Acta 68:1749–1761. doi:10.1016/j.gca.2003.07.004

Papakyriakou T, Miller L (2011) Springtime CO_2 exchange over seasonal sea ice in the Canadian Arctic Archipelago. Ann Glaciol 52:215–224

Peacock S (2004) Debate over the ocean bomb radiocarbon sink: closing the gap. Global Biogeochem Cycles 18(2):GB2022. doi:10.1029/2003GB002211

Peirson WL, Banner ML (2003) Aqueous surface layer flows induced by microscale breaking wind waves. J Fluid Mech 479:1–38. doi:10.1017/S0022112002003336

Peng TH, Broecker WS, Mathieu GG, Li Y-H, Bainbridge A (1979) Radon evasion rates in the Atlantic and Pacific oceans as determined during the geosecs program. J Geophys Res 84(C5):2471–2487

Pereira F, Gharib M, Dabiri D, Modarress M (2000) Defocusing PIV: a three component 3D DPIV measurement technique. Application to bubbly flows. Exp Fluids 29:S78–S84

Pereira F, Lu J, Castaño GE, Gharib M (2007) Microscale 3D flow mapping with µDDPIV. Exp Fluids 42:589–599. doi:10.1007/s00348-007-0267-5

Petelski T, Piskozub J (2006) Vertical coarse aerosol fluxes in the atmospheric surface layer over the north polar waters of the atlantic. J Geophys Res 111:C06039. doi:10.1029/2005JC003295

Pettersson H, Kahma K, Tuomi L (2010) Wave directions in a narrow bay. J Phys Oceanogr 40:155–169. doi:10.1175/2009JPO4220.1

Pettersson H, Kahma K, Rutgersson A, Perttilä M (2011) Air-sea carbon dioxide exchange during upwelling. In: Komori S, McGillis W, Kurose R (eds) Gas transfer at water surfaces 2011. Kyoto University Press, Kyoto, pp 420–429

Phillips O (1977) The dynamics of the upper ocean, 2nd edn. Cambridge University Press, Cambridge

Phillips O, Posner F, Hansen J (2001) High range resolution radar measurements of the speed distribution of breaking events in wind-generated ocean waves: surface impulse and wave energy dissipation rates. J Phys Oceanogr 31:450–460. doi:10.1175/1520-0485(2001)031<0450:HRRRMO>2.0.CO;2

Piskozub J, Petelski T (2009) Scavenging by marine aerosol. In: SOLAS open science conference, Barcelona

Poisson A, Chen C (1987) Why is there little anthropogenic CO_2 in the Antarctic bottom water? Deep-Sea Res Part A 34:1255–1275

Prasad AK (2000) Stereoscopic particle image velocimetry. Exp Fluids 29:103–116

Prytherch J, Yelland M, Pascal R, Moat B, Skjelvan I, Neill C (2010) Direct measurements of the co2 flux over the ocean: development of a novel method. Geophys Res Lett 37:L03607. doi:10.1029/2009GL041482

Raffel M, Willert CE, Wereley ST, Kompenhans J (2007) Particle image velocimetry: a practical guide. Springer, Heidelberg

Rannik U, Vesala T, Keskinen R (1997) On the damping of temperature fluctuations in a circular tube relevant to the eddy covariance technique. J Geophys Res 102:12789–12794. doi:10.1029/97JD00362

Rashidi M, Banerjee S (1990) The effect of boundary conditions and shear rate on streak formation and breakdown in turbulent channel flows. Phys Fluids 2:1827–1838. doi:10.1063/1.857656

Reul N, Branger H, Giovanangeli J-P (2008) Air flow structure over short-gravity breaking water waves. Bound-Lay Meteorol 126(3):477–505

Rhee T, Nightingale P, Woolf D, Caulliez G, Bowyer P, Andreae M (2007) Influence of energetic wind and waves on gas transfer in a large wind-wave tunnel facility. J Geophys Res (Ocean) 112:5027. doi:10.1029/2005JC003358

Rinne J, Douffet T, Prigent Y, Durand P (2008) Field comparison of disjunct and conventional eddy covariance techniques for trace gas flux measurements. Environ Pollut 152:630–635. doi:10.1016/j.envpol.2007.06.063

Roether W, Kromer B (1984) Optimum application of the radon deficit method to obtain air–sea gas exchange rates. In: Brutsaert W, Jirka GH (eds) Gas transfer at water surfaces. Reidel, Hingham, pp 447–457

Rowe M, Fairall C, Perlinger J (2011) Chemical sensor resolution requirements for near-surface measurements of turbulent fluxes. Atmos Chem Phys 11:5263–5275. doi:10.5194/acp-11-5263-2011

Rutgersson A, Smedman A (2010) Enhanced air–sea CO_2 transfer due to water-side convection. J Mar Syst 80(1–2):125–134. doi:10.1016/j.jmarsys.2009.11.004

Rutgersson A, Norman M, Schneider B, Pettersson H, Sahlée E (2008) The annual cycle of carbon-dioxide and parameters influencing the air-sea carbon exchange in the Baltic Proper. J Mar Syst 74:381–394. doi:10.1016/j.jmarsys.2008.02.005

Rutgersson A, Smedman A, Sahlée E (2011) Oceanic convective mixing and the impact on air-sea gas transfer velocity. Geophys Res Lett 38:L02602. doi:10.1029/2010GL045581

Rysgaard S, Glud R (2004) Anaerobic n-2 production in arctic sea ice. Limnol Oceanogr 49:86–94

Rysgaard S, Glud R, Sejr M, Bendtsen J, Christensen P (2007) Inorganic carbon transport during sea ice growth and decay: a carbon pump in polar seas. J Geophys Res 112:C03016. doi:10.1029/2006jc003572

Rysgaard S, Glud R, Sejr M, Blicher M, Stahl H (2008) Denitrification activity and oxygen dynamics in arctic sea ice. Polar Biol 31:527–537. doi:10.1007/s00300-007-0384-x

Rysgaard S, Bendtsen J, Delille B, Dieckmann G, Glud R, Kennedy H, Mortensen J, Papadimitriou S, Thomas D, Tison J-L (2011) Sea ice contribution to the air–sea CO_2 exchange in the Arctic and Southern Oceans. Tellus B 63(5). doi:10.1111/j.1600-0889.2011.00571.x

Sahlée E, Drennan W (2009) Measurements of damping of temperature fluctuations in a tube. Bound-Lay Meteorol 132:339–348. doi:10.1007/s10546-009-9396-0

Sahlée E, Smedman A, Rutgersson A, Högström U (2008) Spectra of CO_2 and water vapour in the marine atmospheric surface layer. Bound-Lay Meteorol 126:279–295. doi:10.1007/s10546-007-9230-5

Salter M, Upstill-Goddard R, Nightingale P, Archer S, Blomquist B, Ho D, Huebert B, Schlosser P, Yang M (2011) Impact of an artificial surfactant release on air-sea gas fluxes during Deep Ocean Gas Exchange Experiment II. J Geophys Res 116:C11016. doi:10.1029/2011JC007023

Sarmiento J, Orr J, Siegenthaler U (1992) A perturbation simulation of CO_2 uptake in an ocean general circulation model. J Geophys Res 97(C3):3621–3645. doi:10.1029/91JC02849

Saylor JR, Handler RA (1997) Gas transport across an air/water interface populated with capillary waves. Phys Fluids 9:2529–2541

Schimpf U, Garbe C, Jähne B (2004) Investigation of transport processes across the sea surface microlayer by infrared imagery. J Geophys Res-Ocean 109(C8):C08S13. doi:10.1029/2003JC001803

Schnieders J, Garbe C, Peirson W, Smith G, Zappa C (2013) Analyzing the footprints of near surface aqueous turbulence – an image processing based approach. J Geophys Res 118(3):1272–1286. doi:10.1002/jgrc.20102

Schröder A, Geisler R, Elsinga G, Scarano F, Dierksheide U (2008) Investigation of a turbulent spot and a tripped turbulent boundary layer flow using time-resolved tomographic piv. Exp Fluids 44:305–316. doi:10.1007/s00348-007-0403-2

Scott NV, Handler RA, Smith GB (2008) Wavelet analysis of the surface temperature field at an air–water interface subject to moderate wind stress. Int J Heat Fluid Flow 29(4):1103–1112. doi:10.1016/j.ijheatfluidflow.2007.11.002

Semiletov I, Makshtas A, Akasofu S, Andreas E (2004) Atmospheric co2 balance: the role of arctic sea ice. Geophys Res Lett 31:L05121. doi:10.1029/02003GL017996

Shaikh N, Siddiqui K (2010) An experimental investigation of the near surface flow over air-water and air-solid interfaces. Phys Fluids 22(2):025103

Shakhova N, Semiletov I, Leifer I, Salyuk A, Rekant P, Kosmach D (2010a) Geochemical and geophysical evidence of methane release over the east siberian arctic shelf. J Geophys Res C08007. doi:10.1029/2009jc005602

Shakhova N, Semiletov I, Salyuk A, Yusupov V, Kosmach D, Gustafsson O (2010b) Extensive methane venting to the atmosphere from sediments of the east siberian arctic shelf. Science 327. doi:10.1126/science.1182221

Sheng J, Malkiel E, Katz J (2008) Using digital holographic microscopy for simultaneous measurements of 3d near wall velocity and wall shear stress in a turbulent boundary layer. Exp Fluids 45:1023–1035. doi:10.1007/s00348-008-0524-2

Siddiqui MHK, Loewen MR (2007) Characteristics of the wind drift layer and microscale breaking waves. J Fluid Mech 573:417–456. doi:10.1017/S0022112006003892

Siddiqui K, Loewen M (2010) Phase-Averaged flow properties beneath microscale breaking waves. Bound-Lay Meteorol 134(3):499–523. doi:10.1007/s10546-009-9447-6

Siddiqui M, Loewen M, Jessup A, Asher W (2001) Infrared remote sensing of microscale breaking waves andnear-surface flow fields. In: Geoscience and Remote Sensing Symposium, 2001. IGARSS 01, vol 2. IEEE 2001 International, pp 969–971. doi:10.1109/IGARSS.2001.976697

Siddiqui M, Loewen MR, Asher WE, Jessup AT (2004) Coherent structures beneath wind waves and their influence on air-water gas transfer. J Geophys Res 109:C03024

Slinn SA, Slinn VGN (1980) Predictions for particle deposition on natural waters. Atmos Environ 14:1013–1016

Smedman A-S, Tjernström M, Högström U (1994) The near-neutral marine atmospheric boundary layer with no surface shearing stress: a case study. J Atmos Sci 51:3399–3411

Smethie WM Jr, Takahashi T, Chipman DW, Ledwell JR (1985) Gas exchange and co2 flux in the tropical atlantic ocean determined from ^{222}rn and pco2 measurements. J Geophys Res 90(C4):7005–7022. doi:10.1029/JC090iC04p07005

Smith J (1998) Evolution of Langmuir circulation during a storm. J Geophys Res-Ocean 103:12649–12668. doi:10.1029/97JC03611

Smith CR, Paxson RD (1983) A technique for evaluation of three-dimensional behavior in turbulent boundary layers using computer augmented hydrogen bubble-wire flow visualization. Exp Fluids 1:43–49

Smith MH, Park PM, Consterdine IE (1991) North-atlantic aerosol remote concentrations measured at a hebridean coastal site. Atmos Environ 25A:547–555

Smith GB, Handler RA, Scott N (2007) Observations of the structure of the surface temperature field at an air-water interface for stable and unstable cases. In: Garbe CS, Handler RA, Jähne BH (eds) Transport at the air sea interface. Springer, Berlin, pp 205–222

Soloviev A (2007) Coupled renewal model of ocean viscous sublayer, thermal skin effect and interfacial gas transfer velocity. J Mar Syst 66:19–27. doi:10.1016/j.jmarsys.2006.03.024

Soloviev AV, Schlüssel P (1994) Parameterization of the cool skin of the ocean and of the air–ocean gas transfer on the basis of modelling surface renewal. J Phys Oceanogr 24:1339–1346

Soloviev A, Donelan M, Graber H, Haus B, Schlüssel P (2007) An approach to estimation of near-surface turbulence and co2 transfer velocity from remote sensing data. J Mar Syst 66:182–194. doi:10.1016/j.jmarsys.2006.03.023

Sorensen LL, Larsen SE (2010) Atmosphere-surface fluxes of CO_2 using spectral techniques. Bound-Lay Meteorol 136:59–81

Spiel D (1998) On the births of film drops from bubbles bursting on seawater surfaces. J Geophys Res 103:24907–24918

Spitzer WS, Jenkins WJ (1989) Rates of vertical mixing, gas exchange and new production: estimates from seasonal gas cycles in the upper ocean near bermuda. J Mar Res 47(1):169–196. doi:10.1357/002224089785076370

Springer T, Pigford R (1970) Influence of surface turbulence and surfactants on gas transport through liquid interfaces. Ind Eng Chem Fundam 9:458–465

Steinbuck JV, Roberts PLD, Troy CD, Horner-Devine AR, Simonet F, Uhlman AH, Jaffe JS, Monismith SG, Franks PJS (2010) An autonomous open-ocean stereoscopic piv profiler. J Atmos Oceanic Tech 27(8):1362–1380. doi:10.1175/2010JTECHO694.1

Stull R (1988) An introduction to boundary layer meteorology. Kluwer, Dordrecht

Sullivan PP, McWilliams JC (2010) Dynamics of winds and currents coupled to surface waves. Annu Rev Fluid Mech 42:19–42. doi:10.1146/annurev-fluid-121108-145541

Sutherland G, Christensen KH, Ward B (2013) Wave-turbulence scaling in the ocean mixed layer. Ocean Sci 9:597–608. doi:10.5194/os-9-597-2013

Suntharalingam P, Buitenhuis ET, Quere CL, Dentener F, Nevison CD, Butler JH, Bange H, Forster GL (2012)

Quantifying the impact of anthropogenic nitrogen deposition on oceanic nitrous oxide. Geophys Res Lett. doi:10.1029/2011GL050778 (in press)

Sweeney C, Gloor E, Jacobson A, Key R, McKinley G, Sarmiento J, Wanninkhof R (2007) Constraining global air-sea gas exchange for CO_2 with recent bomb ^{14}C measurements. Global Biogeochem Cycles 21:GB2015. doi:10.1029/2006GB002784

Szeri AJ (1997) Capillary waves and air-sea transfer. J Fluid Mech 332:341–358

Takagaki N, Komori S (2007) Effects of rainfall on mass transfer across the air-water interface. J Geophys Res 112: C06006. doi:10.1029/2006jc003752

Takahashi T, Olafsson J, Goddard J, Chipman D, Sutherland S (1993) Seasonal variations of CO_2 and nutrients in the high–latitude surface oceans: a comparative study. Global Biogeochem Cycles 7:843–878

Terray E, Donelan M, Agrawal Y, Drennan W, Kahma K, Hwang P, Kitaigorodskii S (1996) Estimates of kinetic energy dissipation under breaking waves. J Phys Oceanogr 26:792–807. doi:10.1175/1520-0485(1996)026<0792: EOKEDU>2.0.CO;2

Thorpe S (2004) Langmuir circulation. Annu Rev Fluid Mech 36:55–79. doi:10.1146/annurev.fluid.36.052203.071431

Tison J, Haas C, Gowing M, Sleewaegen S, Bernard A (2002) Tank study of physico-chemical controls on gas content and composition during growth of young sea ice. J Glaciol 48:177–191

Tison J-L, Brabant F, Dumont I, Stefels J (2010) High resolution DMS and DMSP time series profiles in decaying summer first-year sea ice at ISPOL (Western Weddell Sea, Antarctica). J Geophys Res Submitted

Trevena A, Jones G (2006) Dimethylsulphide and dimethylsulphoniopropionate in Antarctic sea ice and their release during sea ice melting. Mar Chem 98:210–222

Trevena A, Jones G, Wright SW, van den Enden R (2000) Profiles of DMSP, algal pigments, nutrients and salinity in pack ice from eastern Antarctica. J Sea Res 43:265–273

Trevena A, Jones G, Wright S, van den Enden R (2003) Profiles of dimethylsulphoniopropionate (DMSP), algal pigments, nutrients, and salinity in the fast ice of Prydz Bay, Antarctica. J Geophys Res 108:3145

Troitskaya Y, Sergeev D, Ermakova O, Balandina G (2011) Statistical parameters of the air turbulent boundary layer over steep water waves measured by the PIV technique. J Phys Oceanogr 41:1421–1454. doi:10.1175/2011JPO4392.1

Tsai W-T, Hung L-P (2007) Three-dimensional modeling of small-scale processes in the upper boundary layer bounded by a dynamic ocean surface. J Geophys Res 112. doi:10.1029/2006JC003686

Tsai W-T, Hung L-P (2010) Enhanced energy dissipation by parasitic capillaries on short gravity–capillary waves. J Phys Oceanogr 40:2435–2450

Tsai W, Liu K (2003) An assessment of the effect of sea-surface surfactant onglobal atmosphere-ocean CO_2 flux. J Geophys Res 108:3127. doi:10.1029/2000JC000740

Tsai W-T, Chen S-M, Moeng C-H (2005) A numerical study on the evolution and structure of a stress-driven, free-surface turbulent shear flow. J Fluid Mech 545:163–192

Tsai W-T, Chen S-M, Lu G-H, Garbe C (2013) Characteristics of interfacial signatures on a wind-driven gravity-capillary wave. J Geophys Res 118. doi:10.1002/jgrc.20145

Tulin MP, Landrini M (2001) Breaking waves in the ocean and around ships. In: Proceedings of the 23rd symposium of naval hydrodynamics, The National Academies Press, pp 713–745

Turk D, Zappa CJ, Meinen CS, Christian JR, Ho DT, Dickson AG, McGillis WR (2010) Rain impacts on co2 exchange in the western equatorial pacific ocean. Geophys Res Lett 37: L23610. doi:10.1029/2010gl045520

Turney D, Smith W, Banerjee S (2005) A measure of near-surface fluid motions that predicts air-water gas transfer in a wide range of conditions. Geophys Res Lett 32(4): L04607

Upstill-Goddard R, Frost T, Henry G, Franklin M, Murrell J, Owens N (2003) Bacterioneuston control of air-water methane exchange determined with a laboratory gas exchange tank. Global Biogeochem Cycles 17(4):1108. doi:10.1029/2003GB002043

Vagle S, McNeil C, Steiner N (2010) Upper ocean bubble measurements from the NE Pacific and estimates of their role in air-sea gas transfer of the weakly soluble gases nitrogen and oxygen. J Geophys Res 115:C12054. doi:10.1029/2009JC005990

Venkatram A, Pleim J (1999) The electrical analogy does not apply to modeling dry deposition of particles. Atmos Environ 33:3075–3076

Veron F, Melville W, Lenain L (2008) Wave-coherent air-sea heat flux. J Phys Oceanogr 38:788–802. doi:10.1175/2007JPO3682.1

Veron F, Melville WK, Lenain L (2011) The effects of small-scale turbulence on air-sea heat flux. J Phys Oceanogr 41(1):205–220. doi:10.1175/2010JPO4491.1

Vesala T, Kljun N, Rannik U, Rinne J, Sogachev A, Markkanen T, Sabel-feld K, Foken T, Leclerc M (2008) Flux and concentration footprint model-ling: state of the art. Environ Pollut 152:653–666. doi:10.1016/j.envpol.2007.06.070

Vickers D, Göckede M, Law B (2010) Uncertainty estimates for 1-h averaged turbulence fluxes of carbon diocide, latent heat and sensible heat. Tellus 62B:87–99. doi:10.1111/j.1600-0889.2009.00449.x

Vlahos P, Monahan EC (2009) A generalized model for the air-sea transfer of dimethylsulfide at high wind speeds. Geophys Res Lett 36:L21605. doi:10.1029/2009GL040695

Voss B, Stapf J, Berthe A, Garbe C (2012) Bichromatic particle streak velocimetry bpsv – interfacial, v3c3d velocimetry using a single camera. Exp Fluids 53:1405–1420. doi:10.1007/s00348-012-1355-8

Wanninkhof R (1992) Relationship between gas exchange and wind speed over the ocean. J Geophys Res 97 (C5):7373–7382. doi:10.1029/92JC00188

Wanninkhof RH, Bliven LF (1991) Relationship between gas exchange, wind speed and radar backscatter in a large wind wave tank. J Geophys Res 96(C2):2785–2796

Wanninkhof R, Knox M (1996) Chemical enhancement of CO_2 exchange in natural waters. Limnol Oceanogr 41(4):689–697

Wanninkhof R, McGillis WR (1999) A cubic relationship between gas transfer and wind speed. Geophys Res Lett 26:1889–1892

Wanninkhof R, Asher WE, Wepperning R, Hua C, Schlosser P, Langdon C, Sambrotto R (1993) Gas transfer experiment on georges bank using two volatile deliberate tracers. J Geophys Res 98(C11):20237–20248

Wanninkhof R, Hitchcock G, Wiseman WJ et al (1997) Exchange, dispersion, and biological productivity on the west florida shelf: results from a lagrangian tracer study. Geophys Res Lett 24(14):1767–1770

Wanninkhof R, Sullivan KF, Top Z (2004) Air-sea gas transfer in the southern ocean. J Geophys Res 109:C08S19. doi:10.1029/2003JC001767

Wanninkhof R, Asher WE, Ho DT, Sweeney C, McGillis WR (2009) Advances in quantifying air-sea gas exchange and environmental forcing. Annu Rev Mar Sci 1:213–244. doi:10.1146/annurev.marine.010908.163742

Ward B (2006) Near–surface ocean temperature. J Geophys Res 111:C02005. doi:10.1029/2004JC002689

Watson AJ, Upstill-Goddard RC, Liss PS (1991) Air-sea exchange in rough and stormy seas measured by a dual tracer technique. Nature 349(6305):145–147

Webb E, Pearman G, Leuning R (1980) Correction of the flux measure-ments for density effects due to hear and water vapour transfer. Q J R Meteorol Soc 106:85–100

Weiss R (1987) Winter weddell sea project 1986: trace gas studies during legs ant v/2 and ant v/3 of polarstern. Antarct J US 22:99–100

Wells AJ, Cenedese C, Farrar JT, Zappa CJ (2009) Variations in ocean surface temperature due to near surface flow: straining the cool skin layer. J Phys Oceanogr 39:2685–2710

Wernet MP (2004) Planar particle imaging doppler velocimetry: a hybrid piv/dgv technique for three-component velocity measurements. Meas Sci Technol 15(10):2011

Whitman WG (1923) The two-film theory of absorption. Chem Met Eng 29:147

Willert CE, Gharib M (1992) Three-dimensional particle imaging with a single camera. Exp Fluids 12:353–358

Witek ML, Flatau PJ, Quinn PK, Westphal DL (2007) Global sea-salt modeling: results and validation against multicampaign shipboard measurements. J Geophys Res 112:D08215. doi:10.1029/2006JD007779

Witting J (1971) Effects of plane progressive irrotational waves on thermal boundary layers. J Fluid Mech 50:321–334

Woolf D (1993) Bubbles and the air-sea transfer velocity of gases. Atmos-Ocean 31:517–540

Woolf D (1997) Bubbles and their role in air-sea gas exchange. In: Liss P, Duce R (eds) The sea surface and global change. Cambridge University Press, Cambridge, pp 173–205

Woolf D, Thorpe S (1991) Bubbles and the air-sea exchange of gases in near-saturation conditions. J Mar Res 49:435–466

Woolf D, Bowyer P, Monahan E (1987) Discriminating between the film drops and jet drops produced by a simulated whitecap. J Geophys Res 92:5142–5150

Woolf D, Leifer I, Nightingale P, Rhee T, Bowyer P, Caulliez G, de Leeuw G, Larsen S, Liddicoat M, Baker J, Andreae MO (2007) Modelling of bubble-mediated gas transfer; fundamental principles and a laboratory test. J Mar Syst 66:71–91

Wurl O, Wurl E, Miller L, Johnson L, Vagle S (2011) Formation and global distribution of sea-surface microlayers. Biogeosciences 8:121–135

Yang M, Blomquist B, Fairall C, Archer S, Huebert B (2011) Effects of sea surface temperature and gas solubility on air-sea exchange of Dimethylsulfide (DMS). J Geophys Res 116:C00F05. doi:10.1029/2010JC006526

Yelland M, Pascal R, Taylor P, Moat B (2009) Autoflux: an autonomous system for the direct measurement of the air-sea fluxes of co_2, heat and momentum. J Op Oceanogr 2:15–23

Young IR, Babanin AV (2006) Spectral distribution of energy dissipation of Wind-Generated waves due to dominant wave breaking. J Phys Oceanogr 36(3):376–394. doi:10.1175/JPO2859.1

Zappa C, Jessup AT (2005) High-resolution airborne infrared measurements of ocean skin temperature. IEEE Geosci Remote Sens Lett 2(2):146–150. doi:10.1109/LGRS.2004.841629

Zappa CJ, Asher WE, Jessup AT (2001) Microscale wave breaking and air-water gas transfer. J Geophys Res-Ocean 106(C5):9385–9391

Zappa CJ, Asher WE, Jessup AT, Klinke J, Long SR (2004) Microbreaking and the enhancement of air-water transfer velocity. J Geophys Res 109:C08S16. doi:10.1029/2003JC001897

Zappa CJ, McGillis WR, Raymond PA, Edson JB, Hintsa EJ, Zemmelink HJ, Dacey JWH, Ho DT (2007) Environmental turbulent mixing controls on the air-water gas exchange in marine and aquatic systems. Geophys Res Lett 34:L10601. doi:10.1029/2006GL028790

Zappa CJ, Ho DT, McGillis WR, Banner ML, Dacey JWH, Bliven LF, Ma B, Nystuen J (2009) Rain-induced turbulence and air-sea gas transfer. J Geophys Res 114:C07009. doi:10.1029/2008JC005008

Zemmelink H, Delille B, Tison J, Hintsa E, Houghton L, Dacey J (2006) CO_2 deposition over the multi-year ice of the western weddell sea. Geophys Res Lett 33:L13606. doi:10.1029/2006gl026320

Zhang L, Vet R (2006) A review of current knowledge concerning size-dependent aerosol removal. China Particuology 4:272–282

Zhao D, Toba Y (2001) Dependence of whitecap coverage on wind and wave properties. J Oceanogr 57:603–616

Zilitinkevich S (1994) A generalized scaling for convective shear flows. Bound-Lay Meteorol 70:51–78. doi:10.1007/BF00712523

Air-Sea Interactions of Natural Long-Lived Greenhouse Gases (CO_2, N_2O, CH_4) in a Changing Climate

Dorothee C.E. Bakker, Hermann W. Bange, Nicolas Gruber, Truls Johannessen, Rob C. Upstill-Goddard, Alberto V. Borges, Bruno Delille, Carolin R. Löscher, S. Wajih A. Naqvi, Abdirahman M. Omar, and J. Magdalena Santana-Casiano

Abstract

Understanding and quantifying ocean–atmosphere exchanges of the long-lived greenhouse gases carbon dioxide (CO_2), nitrous oxide (N_2O) and methane (CH_4) are important for understanding the global biogeochemical cycles of carbon and nitrogen in the context of ongoing global climate change. In this chapter we summarise our current state of knowledge regarding the oceanic distributions, formation and consumption pathways, and oceanic uptake and emissions of CO_2, N_2O and CH_4, with a particular emphasis on the upper ocean. We specifically consider the role of the ocean in regulating the tropospheric content of these important radiative gases in a world in which their tropospheric content is rapidly increasing and estimate the impact of global change on their present and future oceanic uptake and/or emission. Finally, we evaluate the various uncertainties associated with the most commonly used methods for estimating uptake and emission and identify future research needs.

3.1 Introduction

Carbon dioxide (CO_2), nitrous oxide (N_2O) and methane (CH_4) are long-lived atmospheric greenhouse gases, whose global budgets are substantially determined by the marine system. Understanding and accurately predicting the evolution of the marine CO_2 sink and the marine emissions of N_2O and CH_4 is of great importance for future climate change scenarios as used in studies for the Intergovernmental Panel on Climate Change (Denman et al. 2007).

The tropospheric dry mole fractions of these three greenhouse gases (Box 3.1) have been increasing since the industrial revolution, principally reflecting anthropogenic inputs, but also comparatively small fluctuations in the balance of natural sources and sinks. The tropospheric abundance of CO_2 has been regularly monitored since the late 1950s, and those of N_2O and CH_4 since the 1970s (http://agage.eas.gatech.edu/index.htm; http://www.esrl.noaa.gov/gmd) (Fig. 3.1). Table 3.1 summarises the tropospheric abundances, lifetimes, and radiative forcings of CO_2, N_2O and CH_4 for 2005

D.C.E. Bakker (✉)
e-mail: d.bakker@uea.ac.uk

H.W. Bange (✉)
e-mail: hbange@geomar.de

N. Gruber (✉)
e-mail: nicolas.gruber@env.ethz.ch

T. Johannessen (✉)
e-mail: truls.johannessen@gfi.uib.no

R.C. Upstill-Goddard (✉)
e-mail: rob.goddard@ncl.ac.uk

> **Box 3.1**
> Atmospheric gases are quantified by their dry mole fraction, given in units of ppm or μmol mol^{-1} for CO_2 and in units of ppb or nmol mol^{-1} for N_2O and CH_4. In this chapter we report annual fluxes in Pg C year^{-1} for CO_2, in Tg C year^{-1} for CH_4 and in Tg N year^{-1} for N_2O. One Pg (Petagram) is equivalent to 10^{15} g and one Tg (Teragram) is equivalent to 10^{12} g. The troposphere is the lower part of the atmosphere and extends from the Earth's surface to the tropopause at 10–15 km height. In this chapter we are mainly concerned with the troposphere, unless specified otherwise.

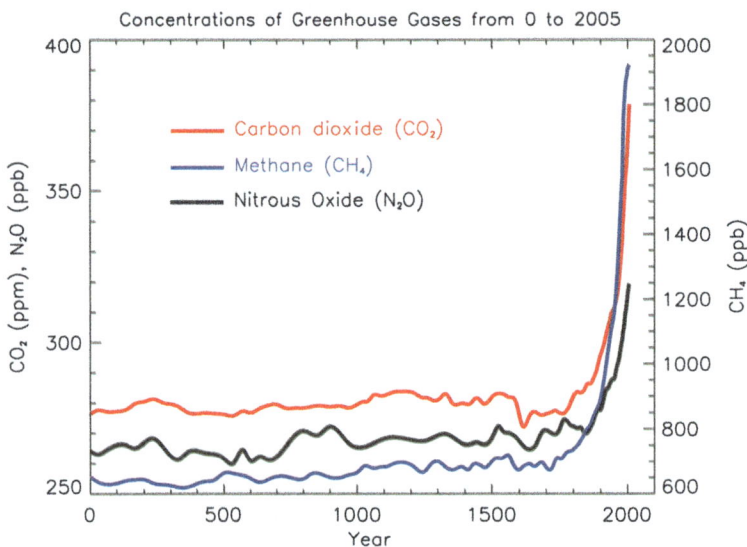

Fig. 3.1 Atmospheric concentrations of carbon dioxide, methane and nitrous oxide over the last 2,000 years (Reproduced from Forster et al. (2007) by permission of the IPCC)

Table 3.1 The tropospheric dry mole fractions, radiative forcings (RF) and lifetimes (adjustment time) of CO_2, N_2O and CH_4 (After Forster et al. 2007; WDCGG 2012)

Gas	Abundance in 2010	Abundance in 2005	RF in 2005 (W m^{-2})	Lifetime (years)
CO_2	389.0 ppm	379 ± 0.65 ppm	1.66	See below[a]
N_2O	323.2 ppb	319 ± 0.12 ppb	0.16	114
CH_4	1,808 ppb	1,774 ± 1.8 ppb	0.48	12

[a] No single adjustment time exists for CO_2 (Joos et al. 2001), as the rate of removal of CO_2 from the troposphere is determined by the removal rate of carbon from the surface ocean (Annexe I in IPCC 2007). An approximate value of 100 years may be given, while decay constants of 172.9 to 18.51 and 1.186 years have been used in models (Forster et al. 2007).

(Forster et al. 2007). An increase in the dry mole fractions (Box 3.1) of these long-lived greenhouse gases leads to tropospheric warming and stratospheric cooling, which may impact on chemical reaction rates and atmospheric dynamics (Wayne 2000). Other effects of changes in the dry mole fractions of these gases are listed in Table 3.2.

Global CO_2 emissions are currently increasing exponentially, primarily reflecting the accelerating development of large emerging economies such as China and India (Friedlingstein et al. 2010). If sustained, this recently rapid growth in tropospheric CO_2 may precipitate critical climate and other global environmental changes, possibly faster than previously identified (e.g. IPCC 2007).

Given its current tropospheric growth and the ongoing decline in chlorofluorocarbon (CFC) emissions, N_2O may soon replace CFCs as the fourth most important greenhouse gas after water vapour (H_2O), CO_2 and CH_4 (Forster et al. 2007). The major sink of N_2O is

3.1 Introduction

Table 3.2 Air-sea exchange and impact of the long-lived greenhouse gases CO_2, N_2O and CH_4 in a changing climate, as discussed in this chapter. See Table 3.1 for the radiative forcings and atmospheric lifetimes

Gas	Role in atmospheric chemistry	Oceanic contribution to contemporary atmospheric budget	Impact of environmental change on air-sea gas exchange in the twenty-first century			
			Global warming	Ocean acidification	Open ocean deoxygenation	Coastal eutrophication and hypoxia
CO_2	Inert	Net ocean sink for about 30 % of CO_2 emissions from human activity partly mitigates climate change; the ocean ultimately controls atmospheric CO_2 content	The net ocean sink likely to decrease by 2100	Small effect on ocean CO_2 sink	Small effect	Coastal CO_2 sink likely to increase
N_2O	Largely inert in the troposphere; depletion of stratospheric O_3	Open ocean natural N_2O source, coastal regions (incl. rivers) anthropogenic N_2O source, equivalent to 20 % and 10 % of the global N_2O emissions	Increase of subsurface production	Unknown	Negligible or small increase in the open ocean N_2O source	Large increase in the coastal N_2O source
CH_4	Regulates tropospheric O_3 and OH radical; affects stratospheric O_3 chemistry; CO_2 source	Natural open ocean and coastal CH_4 sources equivalent to 10 % of the global, natural CH_4 sources; continental CH_4 seeps poorly known	Increased CH_4 release from CH_4 hydrates and increased CO_2 release upon oxidation of hydrate-derived CH_4	Unknown	Negligible or small increase in the open ocean CH_4 source	Negligible or small increase in coastal CH_4 source

Table 3.3 Sources and sinks of tropospheric CH_4. Ranges are derived from estimates for the period 1983–2004, as compiled by Denman et al. (2007)

	CH_4 sources and sinks (Tg C year^{-1})
Natural sources:	
Wetlands	75–173
Termites	15–22
Oceans, incl coastal regions	3–11
Hydrates	3–4
Geological sources	3–11
Wild animals	11
Wildfires	2–4
Total natural sources	**109–195**
Anthropogenic sources:	
Energy	56–58
Coal mining	23–36
Gas, oil, industry	36–68
Landfills and waste	26–52
Ruminants	57–142
Rice agriculture	23–84
Biomass burning	11–66
Vegetation	27
Total anthropogenic sources	**198–321**
Total sources:	**377–458**
Sinks:	
Tropospheric OH radical	321–383
Soils	20–26
Stratospheric loss	22–34
Total sinks	**386–436**

the stratosphere where it photochemically decomposes via reaction with $O(^1D)$ (oxygen singlet D) to form nitric oxide (NO) radicals. The latter represent a major removal pathway for stratospheric ozone (O_3) (Crutzen 1970; Ravishankara et al. 2009). Indeed, N_2O is expected to become the dominant O_3 depleting compound during the twenty-first century (Ravishankara et al. 2009).

CH_4 is the most abundant organic species in the troposphere, where it influences oxidising capacity and regulates levels of O_3 and OH (hydroxyl free radical). Oxidation of CH_4 by OH to CO_2 and CO (carbon monoxide) is its major tropospheric sink (Table 3.3). In the stratosphere photo-oxidation of CH_4 is a major source of stratospheric H_2O, which influences both tropospheric warming and stratospheric cooling (Michelsen et al. 2000), and a small source of stratospheric CO_2. CH_4 plays a complex role in stratospheric O_3 chemistry (Wayne 2000). Additional stratospheric CO_2 arises from rapid CO oxidation. However, these two CO_2 sources are minor and as there are no recognised stratospheric sinks for CO_2 (Hall and Prather 1993); any variation in stratospheric CO_2 principally reflects the inflow of tropospheric air masses.

3.1.1 Atmospheric Greenhouse Gases from Ice Cores

Analysis of the composition of fossil air trapped in ice cores has extended the tropospheric histories of all three gases, to ~800,000 years before present (YBP) for CO_2 (Petit et al. 1999; EPICA community members 2004; Lüthi et al. 2008) and to ~650,000 YBP for N_2O and CH_4 (Spahni et al. 2005). These data show that during the last 650,000 years CO_2 has varied from ~170 ppm during glacials to ~280 ppm during interglacials, while during the preceeding 100,000 years the range was somewhat smaller. For comparison, tropospheric CO_2 increased from 280 ppm pre-industrially to 389 ppm in 2010 (Forster et al. 2007; WDCGG 2012). Changes in ocean circulation and biology and the feedbacks between them have been invoked to explain the glacial/interglacial fluctuations of tropospheric CO_2 but understanding the precise mechanistic details remains a substantial challenge (Jansen et al. 2007). Over the last 420,000 years, tropospheric CO_2 has tracked reconstructed changes in Antarctic temperature with a time lag of several hundred to a thousand years (Mudelsee 2001), implying that changes in the physical climate system such as temperature and the extent of glaciers have initiated changes in the global carbon cycle and tropospheric CO_2. The carbon cycle then has responded by amplifying these initial perturbations through positive carbon-climate feedbacks. Today the situation is fundamentally different in that the increasing greenhouse gas content drives changes in climate and environment.

Variation in stratospheric N_2O between 200 and 280 ppb during the past 650,000 years (Spahni et al. 2005) can be attributed to concurrent natural changes in both the terrestrial and the oceanic sources (Sowers et al. 2003; Flückiger et al. 2004). Since the pre-industrial era the mean tropospheric N_2O dry mole fraction has increased from 270 ± 7 to ~323 ppb. The current tropospheric N_2O growth rate of about 0.7 ppb year^{-1} can primarily be attributed to the continued increased use of nitrogen fertilisers (Forster et al. 2007; Montzka et al. 2011).

The tropospheric dry mole fraction of CH_4 has varied from ~400 ppb during glacials to ~700 ppb during interglacials. The current average tropospheric CH_4 dry mole fraction is ~1,808 ppb, reflecting large and growing anthropogenic CH_4 fluxes since the pre-industrial era (Table 3.3). Even so, tropospheric CH_4 growth is temporally quite variable. High annual growth rates of ~20 ppb year^{-1} during the 1970s were followed by growth rates of ~9–13 ppb year^{-1} through the 1980s, 0–13 ppb year^{-1} through most of the 1990s, almost zero growth during the late 1990s to early 2000s (Dlugokencky et al. 2003) and renewed growth rates of ~10 ppb year^{-1} during the late 2000s (Rigby et al. 2008). This complex behaviour reflects short-term source variability that has been variously ascribed to decreased fossil fuel output following the economic collapse of the former Soviet Union, volcanic activity, wetland and rice paddy emissions, biomass burning, changes in the global distributions of temperature and precipitation, and reduced microbial sources in the Northern Hemisphere (Denman et al. 2007; Dlugokencky et al. 2009; Aydin et al. 2011; Kai et al. 2011).

A consequence of this ocean CO_2 uptake is a decrease in ocean pH, known as ocean acidification (Sect. 3.5.2) (Feely et al. 2004; Raven et al. 2005). If anthropogenic CO_2 emissions were to cease now, the oceans would eventually absorb 70–80 % of the anthropogenic CO_2 so far added to the troposphere, but this would take several hundred years (Archer et al. 1997; Watson and Orr 2003). Dissolution of calcium carbonate ($CaCO_3$) in deep ocean sediments and on land would further reduce tropospheric CO_2 to within 8 % of its pre-industrial level over thousands of years (Archer et al. 1997). Given the importance of the oceans in moderating human-induced climate change, quantifying net oceanic CO_2 uptake and estimating its long-term evolution are of critical importance. Although much progress has been made in quantifying CO_2 air-sea fluxes over the past decade, considerable uncertainties remain, in particular relating to inter-annual variability and long-term trends. The current state of knowledge is discussed here for the open ocean (Sect. 3.2.3) and for coastal seas (Sect. 3.2.4), with emphasis on the principal uncertainties (Sect. 3.6).

3.2 Surface Ocean Distribution and Air-Sea Exchange of CO_2

3.2.1 Global Tropospheric CO_2 Budget

In 2010 alone the tropospheric CO_2 increase was equivalent to 5.0 ± 0.2 Pg C (Box 3.1), principally due to the release of 9.1 ± 0.5 Pg C from fossil fuel burning and cement manufacture and 0.9 ± 0.7 Pg C from land use change (Fig. 3.2) (Global Carbon Project 2011; Peters et al. 2012). The ocean absorbs a substantial fraction of CO_2 emissions to the troposphere. From pre-industrial times to 1994 the oceans are estimated to have taken up 118 ± 19 Pg C from the troposphere, corresponding to roughly 50 % of fossil fuel CO_2 or about 30 % of the total anthropogenic emissions that include CO_2 emissions from land use change (Fig. 3.2; Table 3.2) (Sabine et al. 2004). Scientists are debating whether regional and global ocean CO_2 uptake has increased, remained constant or decreased in recent decades (Le Quéré et al. 2007, 2010; Schuster and Watson 2007; McKinley et al. 2011; Ballantyre et al. 2012).

3.2.2 Processes Controlling CO_2 Dynamics in the Upper Water Column

The air-sea exchange fluxes of CO_2 show high spatial and temporal variability, reflecting a complex interplay between the biological and physical processes affecting surface water fCO_2 (Box 3.2) (Takahashi et al. 2002; Sarmiento and Gruber 2006). In addition, observations show surface water fCO_2 to rarely be in equilibrium with tropospheric fCO_2 (see below and Box 3.3). Key to understanding the behaviour of CO_2 with regard to equilibration is CO_2 chemistry, which we briefly review next. We also discuss the key processes controlling the CO_2 dynamics of the upper ocean.

Dissolved CO_2 in seawater chemically equilibrates with carbonic acid (H_2CO_3) and the bicarbonate (HCO_3^-) and carbonate (CO_3^{2-}) ions:

$$CO_2 + H_2O \leftrightarrow H_2CO_3 \qquad (3.1)$$

$$H_2CO_3 \leftrightarrow H^+ + HCO_3^- \qquad (3.2)$$

$$HCO_3^- \leftrightarrow H^+ + CO_3^{2-} \qquad (3.3)$$

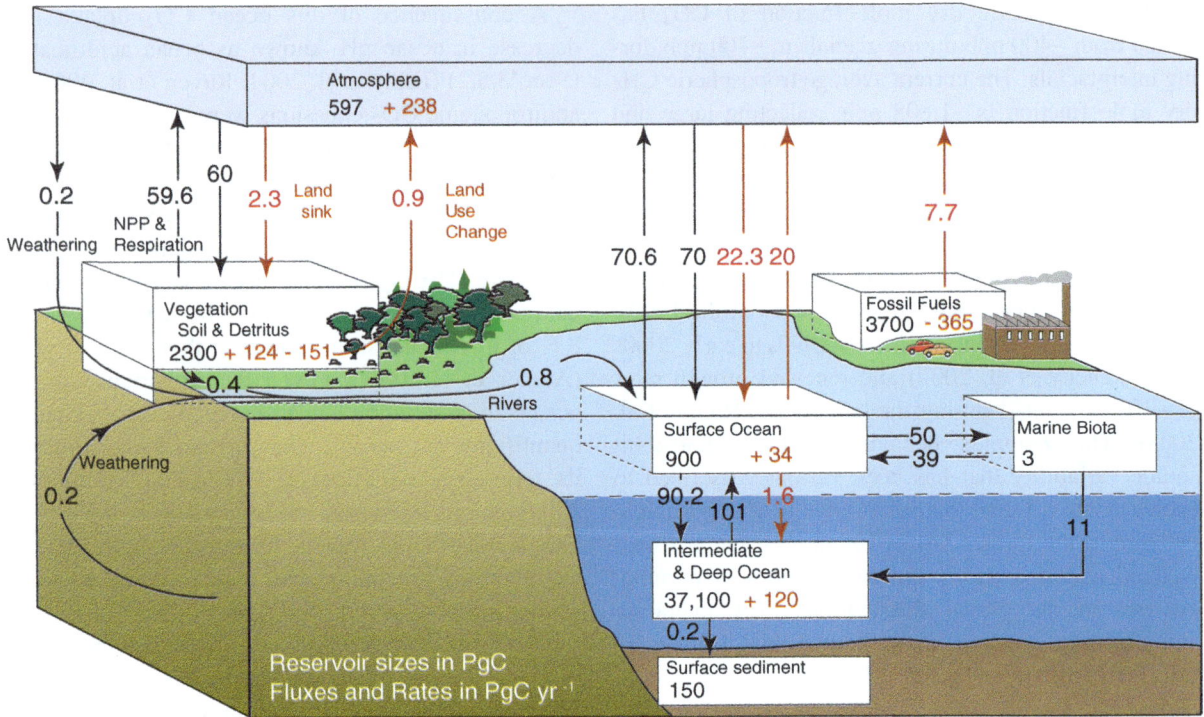

Fig. 3.2 The global carbon cycle with annual fluxes (in Pg C year^{-1}) for the years 2000–2009. Pre-industrial, natural fluxes are in *black* and anthropogenic fluxes are in *red*. Integrated fluxes and standing stocks are from 1850 to 2011. NPP is annual net terrestrial primary production. Cumulative changes are for end 2011 (The figure updates those in Sarmiento and Gruber (2002) and Denman et al. (2007). Figure courtesy of N Gruber)

Box 3.2
Whereas the amount of CO_2 dissolved in seawater is generally reported in terms of its partial pressure pCO_2 (unit: µatm or 0.101325 Pa) or fugacity fCO_2 (unit: µatm), N_2O and CH_4 are more commonly presented in concentration units (nM or nmol kg^{-1} seawater) or as percent (%) saturation. The latter is calculated from the ratio of the measured concentration to the theoretical equilibrium concentration, as determined by ambient water temperature, salinity and air pressure, and the atmospheric dry mole fraction corresponding to the time of last atmospheric contact. A surface saturation of 100 % indicates a water mass in equilibrium with overlying air, values below 100 % indicate undersaturation and values above 100 % indicate supersaturation. For all three gases, deviations from the air-sea equilibrium value are expressed as a negative or positive partial pressure difference (i.e. ΔpCO_2, ΔpN_2O, ΔpCH_4) or for N_2O and CH_4, as a negative or positive "concentration anomaly". In this chapter, positive values denote a partial pressure that is higher in the water than in the overlying air.

The fugacity of a gas is its partial pressure after correcting for any non-ideal behaviour by applying a fugacity coefficient γ (Weiss 1974). The equation for CO_2 is:

$$fCO_2 = \gamma\, pCO_2 \qquad (3.4)$$

In practice the fugacity and partial pressure of CO_2 differ by only about 0.4 %. In this chaper we refer to fCO_2 throughout.

Box 3.3

Air-sea fluxes (F) can be quantified as the product of a gas transfer velocity (k), the gas solubility (K_0) and the difference in the gas fugacity (e.g. fCO_2) across the air-sea interface. For N_2O and CH_4 the difference in the partial pressure is usually applied.

$$F_{CO2} = k_{CO2} K_{0,CO2} (fCO_{2water} - fCO_{2air}) \quad (3.5)$$

$$F_X = k_X K_{0,X} (pX_{water} - pX_{air}) \text{ (with } X = N_2O \text{ or } CH_4) \quad (3.6)$$

By convention, positive flux values indicate emission from the ocean and negative flux values indicate uptake by the ocean. For CO_2 these simplified equations neglect its possible chemical enhancement, although this effect is thought to be small (Wanninkhof 1992; Matthews 1999). They further assume that away from the air-sea interface both the lower troposphere and upper ocean are well mixed, so that bulk measurements within them can be used to define a gas concentration gradient at the interface. A final assumption is that the water temperature at the interface (the skin temperature) is the same as that of the well-mixed upper ocean (Robertson and Watson 1992; Van Scoy et al. 1995).

Atmospheric CO_2 needs to equilibrate with the large pool of dissolved inorganic carbon in seawater, resulting in an equilibration time-scale of the surface ocean for gas exchange of nearly 1 year (Broecker and Peng 1982), i.e. much longer than the time-scale typically associated with upper ocean perturbations (such as by the seasonal cycle). The equilibration time-scale for CH_4 and N_2O is about 10 times shorter than for CO_2 so that the deviations of these gases from equilibrium are generally smaller unless strong sources are present.

The gas transfer velocity (k) is a function of turbulence at the sea surface and is often parameterised as a function of wind speed, as discussed in detail in Chap. 2. Several parameterisations of k as a function of wind speed have been proposed (e.g. Liss and Merlivat 1986; Wanninkhof 1992; Wanninkhof and McGillis 1999; Nightingale et al. 2000; Ho et al. 2006; Sweeney et al. 2007; Prytherch et al. 2010). The uncertainty in k, which has been estimated at 30 % (Sweeney et al. 2007), adds further uncertainty to estimates of net gas uptake and/or emission determined from surface water measurements (Sect. 3.6).

On average surface seawater dissolved inorganic carbon (DIC) (alternatively referred to as total CO_2, ΣCO_2 and C_T) comprises about 90 % HCO_3^-, 9 % CO_3^{2-}, 1 % dissolved CO_2 and 0.001 % H_2CO_3. Thus, in order to equilibrate across the air-sea interface, CO_2 needs to equilibrate not only with the dissolved CO_2 pool, but with all chemical species making up DIC, explaining the long equilibration time scale. The dominant presence of HCO_3^- and CO_3^{2-} are also key to explaining the large uptake capacity of the ocean with regard to the anthropogenic perturbation of tropospheric CO_2, as it is the reaction of CO_3^{2-} with the dissolved CO_2 taken up to form two HCO_3^- ions that gives seawater its large capacity to take up CO_2 and that will enable the ocean to eventually take up nearly 80 % of total anthropogenic emissions. An important metric for this reaction is the oceanic buffer (or Revelle) factor, which is a measure of the degree to which this titration reaction occurs. The larger the concentration of the CO_3^{2-} ion, the higher this factor is, and thus the larger is the oceanic uptake capacity.

However, the current net rate of oceanic CO_2 uptake is overall set by its transport from the surface to the deep oceans (Fig. 3.2), leading to the observation that the current uptake fraction (about 30 %) is considerably smaller than the long-term potential (about 80 %).

The anthropogenic perturbation occurs on top of an intense but largely internal cycling of "natural" carbon, which is the primary driver of the high spatio-temporal variability of CO_2 in the surface ocean. This natural internal cycling is often conceptualised as a number of "pumps", namely the solubility pump, the soft tissue or organic carbon pump and the carbonate or hard tissue pump (Volk and Hoffert 1985; Heinze et al. 1991). The reason for the pump analogy is that the associated processes act as gradient makers in that they tend to reduce the surface concentration of DIC and enhance its concentration at depth, thereby acting against the tendency for these gradients to be eliminated by transport and mixing. The net effect of these pumps on the air-sea exchange of CO_2 is controlled by the interaction and relative importance of the downward pump component relative to upward mixing and transport (Gruber and Sarmiento 2002). Regions where the downward component dominates over upward transport are sinks for tropospheric CO_2, while regions where upward transport dominates are CO_2 sources. Given the need to consider both the downward and upward components, the concept of biogeochemical loops has been proposed (Gruber and Sarmiento 2002).

The solubility pump is maximal at high-latitudes during winter when cold surface water rich in DIC (due to higher CO_2 solubility at lower temperatures) sinks to depth, resulting in a net downward transport of DIC (Fig. 3.2) (Volk and Hoffert 1985; Heinze et al 1991). In contrast, the solubility pump acts in quasi reverse order, when colder waters rich in DIC are brought to the surface and warm, giving rise to reduced CO_2 solubility.

The soft tissue pump is initiated by the photosynthetic incorporation of CO_2 as phytoplankton cellular organic carbon. As this organic carbon travels up the food chain, a fraction of it is "lost" at each trophic step by respiration, excretion and the death of organisms (Kaiser et al. 2011). Bacteria and other microorganisms are critical to the recycling of carbon in the upper ocean (Fig. 3.2). Nevertheless, a significant fraction of the photosynthetically fixed carbon leaves the upper ocean as "export production" in the form of sinking organic particles, by vertical migration of zooplankton or as dissolved organic carbon (DOC) in sinking water (Volk and Hoffert 1985; Heinze et al. 1991; Sarmiento and Gruber 2006).

The carbonate or hard tissue pump involves the biological formation of calcium carbonate ($CaCO_3$) in near surface waters, its downward export primarily by sinking and its subsequent dissolution in deep water. The initial step is incorporation of the CO_3^{2-} ion into the shells of calcifying organisms:

$$Ca^{2+} + CO_3^{2-} \leftrightarrow CaCO_3 \quad (3.7)$$

Equation 3.7 describes $CaCO_3$ precipitation at a physiological level; however, the uptake of CO_3^{2-} leads to a chemical re-adjustment of DIC species with the overall equation:

$$Ca^{2+} + 2HCO_3^- \leftrightarrow CaCO_3 + CO_2 + H_2O \quad (3.8)$$

The precipitation of $CaCO_3$ leads to a shift from the HCO_3^- pool to the CO_2 pool and a release of CO_2 to the surrounding water. The "released" CO_2 subsequently equilibrates with HCO_3^-, so that for each mole of $CaCO_3$ precipitated, less than one mole of CO_2 is "released". The fraction for average surface sea water is 0.6 (Frankignoulle et al. 1994). Consequently, where the ratio between net organic carbon production or net community production (NCP) and calcification is below 0.6, the waters are a CO_2 source and where this ratio exceeds 0.6, they are a CO_2 sink (Suzuki and Kawahata 2004).

In coral reefs NCP is close to zero (Gattuso et al. 1998). Hence the CO_2 "released" by $CaCO_3$ precipitation generally exceeds the CO_2 drawdown by NCP and coral reefs tend to act as CO_2 sources to the troposphere (Gattuso et al. 1993, 1997; Frankignoulle et al. 1996; Ohde and van Woesik 1999; Bates et al. 2001). In the pelagic realm, where the vast majority of calcification is carried out by the coccolithophore component of the phytoplankton (Buitenhuis et al. 1996; Harlay et al. 2010, 2011; Suykens et al. 2010), the average ratio of NCP to net $CaCO_3$ precipitation is between 11 and 16 (Sarmiento et al. 2002; Jin et al. 2006). The net consequence of biological production and the export of organic carbon and $CaCO_3$ from the pelagic realm is a tendency towards CO_2 uptake from the troposphere.

Mineral $CaCO_3$ in seawater occurs in two forms: calcite and the more soluble aragonite (Mucci 1983). For both, solubility increases with increased pressure (depth) and decreased temperature. The saturation state Ω describes whether sea-water is supersaturated ($\Omega > 1$) or undersaturated ($\Omega < 1$) with respect to the solubility product, K_{sp}, of either of these two $CaCO_3$ forms:

$$\Omega = [Ca^{2+}] [CO_3^{2-}]/K_{sp} \quad (3.9)$$

$$K_{sp} = [Ca^{2+}]_{sat} [CO_3^{2-}]_{sat} \qquad (3.10)$$

At present nearly the entire upper ocean is supersaturated with regard to both calcite and aragonite, while most of the deep ocean is undersaturated. Organisms that form $CaCO_3$ shells and structures therefore do so largely in waters that are supersaturated, while the exported $CaCO_3$ eventually sinks into regions of undersaturation and dissolves.

A reduction in ocean pH due to anthropogenic activities (Feely et al. 2004; Orr et al. 2005; Raven et al. 2005) is one consequence of increased tropospheric CO_2 and its transfer to the ocean. While the term "ocean acidification" (OA) describes a decrease in ocean pH, this is not expected to fall below 7 (Kleypas et al. 2006). The uptake of CO_2 since pre-industrial times has led to a reduction in surface seawater pH of 0.1 units relative to the pre-industrial value of about 8.2 (Orr et al. 2005). This is equivalent to a 30 % increase in the hydrogen ion (H^+) concentration.

In situ pH measurements at the European Station for Time-series in the Ocean (ESTOC, 29°N 15°W) show a progressive reduction of pH and other changes in the carbonate chemistry of surface waters since 1995 (González-Dávila et al. 2010; Santana-Casiano and González-Dávila 2011). Figure 3.3 highlights a decrease in surface water pH_T (the pH corrected to a constant temperature of 25°C) of 0.0019 pH units year^{-1} from 1995 to 2010, accompanied by increases in salinity normalised DIC (NC_T) and fCO_2. Similar trends in pH, DIC and fCO_2 have been observed at the Bermuda Atlantic Time-series Study, BATS (Gruber et al. 2002), and the Hawaii Ocean Time-Series site, HOT (Brix et al. 2004; Denman et al. 2007).

An important consequence of the net oceanic uptake of anthropogenic CO_2 from the troposphere is a decrease in the saturation states with regard to calcite and aragonite. This is due to the aforementioned titration of the CO_3^{2-} ion by the CO_2 taken up, which leads to a fall in the CO_3^{2-} concentration. These chemical changes are accompanied by an increase in the concentration of H^+ and CO_2 (Feely et al. 2004; Raven et al. 2005). The ESTOC time series demonstrates how the saturation states for calcite and aragonite have decreased at rates of 0.018 ± 0.006 units year^{-1} and 0.012 ± 0.004 units year^{-1}, respectively, from 1995 to 2004 (Santana-Casiano and González-Dávila 2011).

Ocean acidification is suspected to lead to a reduction in calcification by calcifying organisms, such as coral reefs, coccolithophores, foraminifera, pteropods and shell fish (Sect. 3.5.2) (Raven et al. 2005). In addition, diminishing calcification would reduce net $CaCO_3$ transfer to the deep ocean (Feely et al. 2004; Denman et al. 2007).

3.2.3 Surface Ocean fCO_2 and Air-Sea CO_2 Fluxes in the Open Ocean

3.2.3.1 Surface Ocean fCO_2 Distribution

The seasonal cycle in surface water fCO_2 is relatively weak in tropical regions (14°S–14°N), which are strong CO_2 sources throughout the year (Takahashi et al. 2009) (Box 3.3). Surface water fCO_2 in temperate ocean regions (14–50°N and 14–50°S) has a strong seasonal cycle with high values in summer and low values in winter, as the seasonal effects of warming and cooling outweigh biological effects (Fig. 3.3) (Bates et al. 1996a; Dore et al. 2003; González-Dávila et al. 2003; Takahashi et al. 2009). The temperate Indian Ocean north of 14°N also has high fCO_2 in summer, but here seasonal upwelling in the southwest monsoon is the main driver (Takahashi et al. 2009). High latitude northern hemisphere waters have strong fCO_2 undersaturation in spring and summer as a result of biological CO_2 drawdown in the upper ocean (Takahashi et al. 2009). Biological activity equally creates a CO_2 sink in Southern Ocean waters from 50°S to 60°S during austral spring and summer (Takahashi et al. 2009). Seasonally ice covered waters south of ~60°S rapidly change from strong CO_2 supersaturation below sea ice to strong undersaturation upon ice melt, most likely driven by biological carbon uptake (Bakker et al. 2008).

Surface water fCO_2 data coverage has improved greatly over the past decade (Takahashi et al. 2009; Watson et al. 2009; Pfeil et al. 2013; Sabine et al. 2013). For example, a basin-wide network of fCO_2 measurements on Voluntary Observing Ships (VOS) and buoys has been operational in the North Atlantic Ocean since 2004, which allows the creation of basin-wide monthly fCO_2 maps, annual flux estimates and trend analyses (Schuster et al. 2009; Telszewski et al. 2009; Watson et al. 2009). Data coverage is similarly

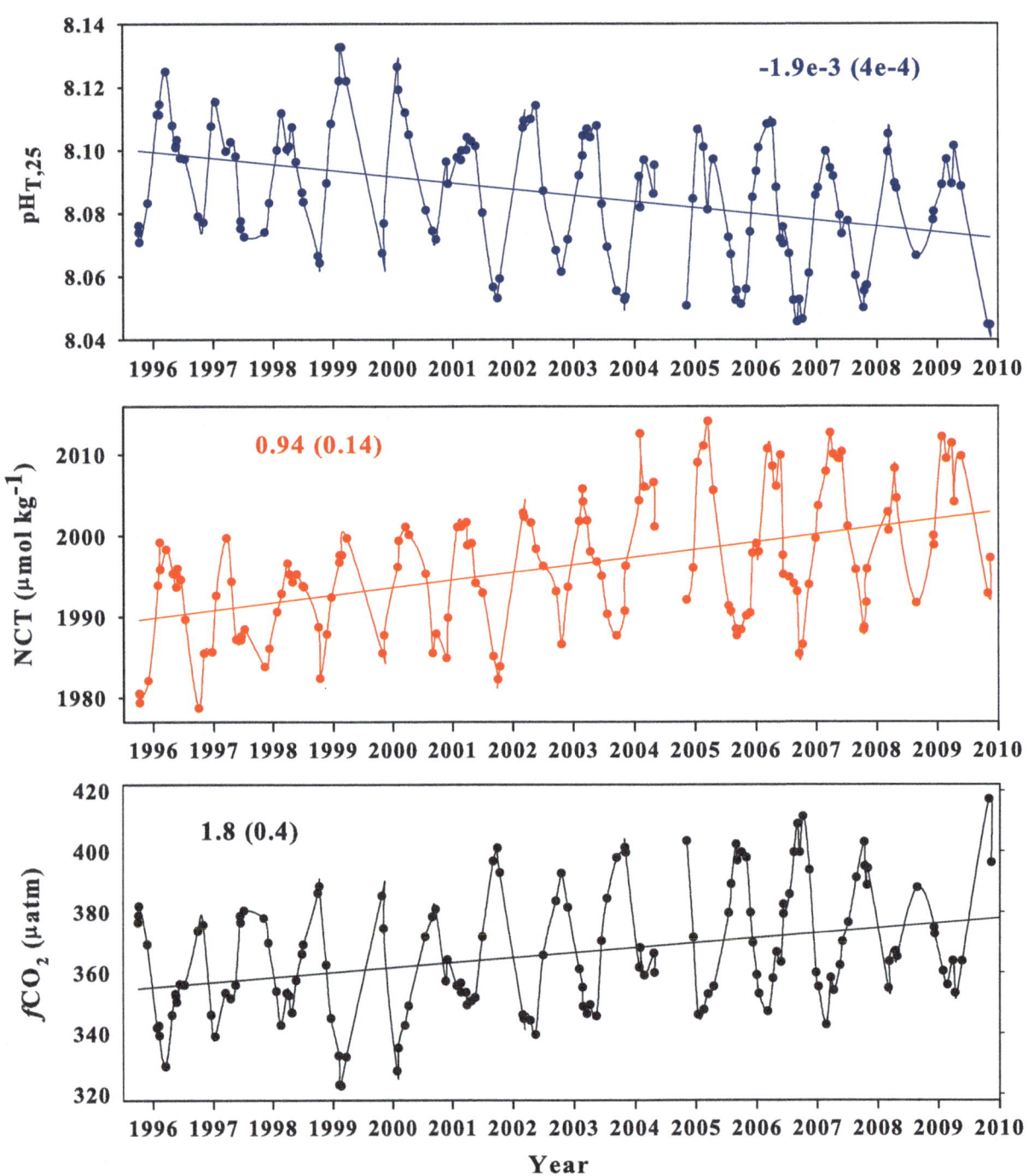

Fig. 3.3 Changes in total pH at 25 °C, salinity normalised dissolved inorganic carbon (NC_T) and fCO_2 from 1995 to 2010 at the European Station for Time-series in the Ocean (ESTOC, 29°N 15°W) for the full set of surface data (*upper* 10 m). The regression lines have slopes of -0.0019 ± 0.0004 pH units year^{-1} for $pH_{T,25}$, of 0.94 ± 0.14 µmol kg^{-1} year^{-1} for NC_T and of 1.8 ± 0.4 µatm year^{-1} for fCO_2. (Figure courtesy of M González-Dávila and JM Santana-Casiano)

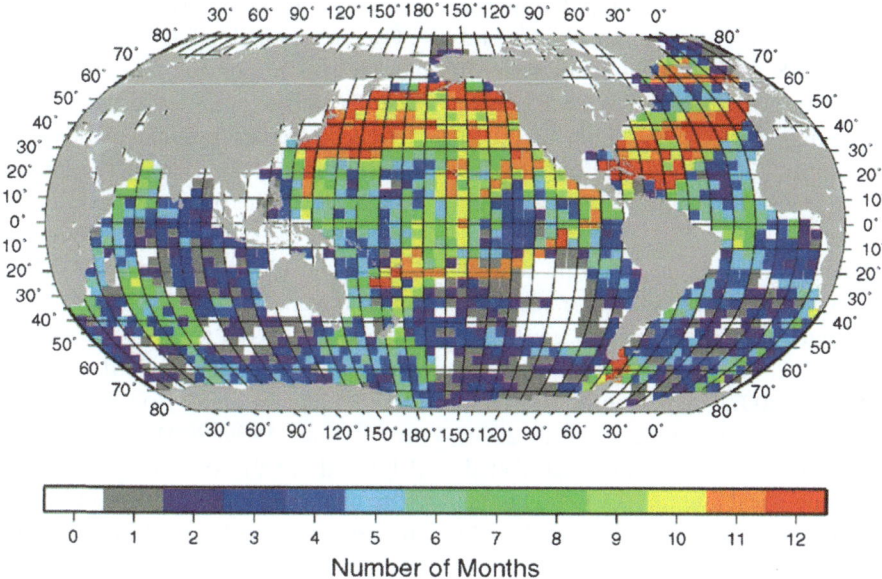

Fig. 3.4 Number of months in each 4° latitude by 5° longitude box with at least one surface water fCO_2 measurement between 1970 and 2007 (Reproduced from Takahashi et al. (2009) by permission of Elsevier)

high in the North Pacific (Feely et al. 2006; Ishii et al. 2009). Elsewhere data coverage has increased, but many regions remain data sparse, e.g. the Indian Ocean, the South Pacific Ocean, the South Atlantic Ocean and the Southern Ocean, notably in autumn and winter (Fig. 3.4) (Takahashi et al. 2009, 2011; Bakker et al. 2012; Pfeil et al. 2013; Sabine et al. 2013).

A variety of techniques have been applied to interpolate between surface ocean fCO_2 data, including a diffusion–advection based interpolation scheme (Takahashi et al. 1997, 2009), (multiple) linear regression (Boutin et al. 1999; Rangama et al. 2005; Olsen et al. 2008) and a neural network approach (Lefèvre et al. 2005; Telszewski et al. 2009). The principle of many of these methods is to correlate sparse fCO_2 data with more widely available parameters such as satellite-derived chlorophyll a concentrations, sea surface temperatures and mixed layer depths and then to use these correlations to predict fCO_2 where measurements are lacking.

The 'true' spatial distributions of surface water fCO_2 and air-sea CO_2 fluxes are unknown and the above methods only deliver approximations of them. Interestingly, however, Watson et al. (2009) derived similar air-sea CO_2 fluxes for the North Atlantic Ocean (10–65°N) using multiple linear regression and a neural network. For both the standard deviation of the annual mean fCO_2 was ~10 %. It was concluded that if the flux uncertainty arising from uncertainty in k is ignored, the overall air-sea CO_2 flux in this region is well constrained by fCO_2 observations and is thus relatively insensitive to the mapping technique used. Further development and testing of interpolation methods should be a priority.

3.2.3.2 Multi-Year Changes and Trends

Analysis of the decadal evolution of fCO_2 provides information on the evolution of the oceanic CO_2 sink. If the rate of increase of surface ocean fCO_2 matches the increase in tropospheric CO_2 the oceanic CO_2 sink is at steady state, but if it is higher, then the oceanic CO_2 sink is decreasing (Schuster et al. 2009). For example, Fig. 3.3 shows fCO_2 from 1995 to 2010 for the upper 10 m at ESTOC (29°N 15°W). Regression of the data reveals an increase in fCO_2 of 1.8 ± 0.4 µatm year^{-1}.

Globally, surface water fCO_2 increased at a mean rate of 1.5 µatm year^{-1} from 1970 to 2007, similar to the pace of the tropospheric CO_2 increase of 1.5 µatm year^{-1} from 1972 to 2005 (Takahashi et al. 2009). Relatively low rates of increase were found in the Equatorial Pacific Ocean (1.26 ± 0.55 µatm year^{-1}) and the North Pacific Ocean (1.28 ± 0.46 µatm year^{-1}), while fCO_2 increased more rapidly in the North Atlantic Ocean (1.80 ± 0.37 μatm year^{-1}) and between 50°S and 60°S (2.13 ± 0.64 μatm year^{-1}) (Takahashi et al. 2009). Similarly, surface water fCO_2 in the Southern Indian Ocean (south of 20°S) increased more rapidly (2.11 ± 0.07 μatm year^{-1}) than did tropospheric CO_2 (1.72 μatm year^{-1}) between 1991 and 2007 (Metzl 2009).

Regional and temporal differences in the rate of increase of surface water fCO_2 are not well understood but have been attributed to changes in seawater buffer capacity (Thomas et al. 2007), mixing and stratification (Schuster and Watson 2007), temperature (Corbière et al. 2007), biological activity (Lefèvre et al. 2004) and lateral and vertical water transport (Takahashi et al. 2009). The expanding database for fCO_2 highlights considerable year-to-year and multi-year variations in ocean carbon cycling.

Theory and biogeochemical models predict an increase in air-sea fCO_2 disequilibrium over time in high latitude regions. Here water from the interior ocean reaches the surface. This water has a relatively low DIC content, as it equilibrated with an atmospheric CO_2 mixing ratio below the present one, when the water last was at the surface. One might expect that the increase in surface water fCO_2 of these waters lags the increase in tropospheric CO_2 (Takahashi et al. 1997, 2002), given the long equilibration time for CO_2 of almost a year. Such an increase in the air-sea fCO_2 disequilibrium would be accompanied by an increase in the net oceanic CO_2 sink. However, the observation that surface water fCO_2 in some regions of the Southern Ocean is currently increasing more rapidly than tropospheric CO_2 (Metzl 2009; Takahashi et al. 2009) runs counter to these predictions. Air-sea CO_2 flux estimates derived from the inversion of tropospheric CO_2 data suggest that this may be a more wide-spread phenomenon in the Southern Ocean, extending to the entire region south of 45°S (Le Quéré et al. 2007). This hypothesis of a weakening relative sink strength in the Southern Ocean is supported by several ocean modelling studies (Wetzel et al. 2005; Le Quéré et al. 2007; Lovenduski et al. 2007) and is attributed to a trend of increasing Southern Ocean wind speeds, which enhance the upwelling of deeper waters with high concentrations of "natural" DIC (Lovenduski et al. 2008). The changing wind regime may be related to a strengthening of the Southern Annular Mode in response to increasing greenhouse gases and the depletion of stratospheric ozone (Lenton et al. 2009). These trends in Southern Ocean fCO_2, the strength of the oceanic CO_2 sink and the mechanisms responsible are currently topics of much scientific debate.

Recent studies provide evidence of multi-annual variation in surface water fCO_2 growth rates and CO_2 air-sea fluxes in other regions, notably the Pacific Ocean and the North Atlantic Ocean (Corbière et al. 2007; Schuster and Watson 2007; Ishii et al. 2009; Schuster et al. 2009; Watson et al. 2009). For example, the growth rates of surface water fCO_2 in the western Equatorial Pacific were different from 1985–1990 (0.3 ± 1.3 μatm year^{-1}) to 1990–1999 (2.2 ± 0.7 μatm year^{-1}) and 1999–2004 (−0.2 ± 1.0 μatm year^{-1}) (Ishii et al. 2009). Annual CO_2 uptake along a shipping route between the United Kingdom and the Caribbean strongly decreased from the early 1990s to 2002–2005 (Schuster and Watson 2007; Schuster et al. 2009). Annual air-sea CO_2 fluxes varied by more than a factor two for the period 2002–2007, with values rising and falling over several years (Fig. 3.5) (Watson et al. 2009). These gradual changes suggest multi-year or possibly decadal variation that might be linked to the North Atlantic Oscillation (Thomas et al. 2008).

3.2.3.3 Comparison of Air-Sea CO_2 Flux Estimates

Independent estimates of the global oceanic uptake of anthropogenic CO_2 for the 1990s and early 2000s range from 1.8 to 2.4 Pg C year^{-1} with model-based values often exceeding observation-based estimates (Gruber et al. 2009). An uptake of 1.8 ± 1.0 Pg C year^{-1} has been obtained by inversion of tropospheric CO_2 (Gurney et al. 2004; adjusted by Gruber et al. 2009), while a net ocean sink of 1.9 ± 0.7 Pg C year^{-1} has been estimated from a surface water CO_2 climatology (Takahashi et al. 2009; adjusted by Gruber et al. 2009). Ocean inversion of DIC has given an oceanic CO_2 sink of 2.2 ± 0.3 Pg C year^{-1}, (Gruber et al. 2009) and 2.4 ± 0.5 Pg C year^{-1} has been estimated using ocean biogeochemical models (Watson and Orr 2003). Other methods give a similar range of estimates (Joos et al. 1999; Gruber and Keeling 2001; Bender et al. 2005; Manning and Keeling 2006; Jacobson et al. 2007; Gruber et al. 2009).

Measurement and modelling techniques vary in whether they quantify anthropogenic CO_2 fluxes or net contemporary CO_2 fluxes and a correction needs to be made for the outgassing of carbon from rivers and for

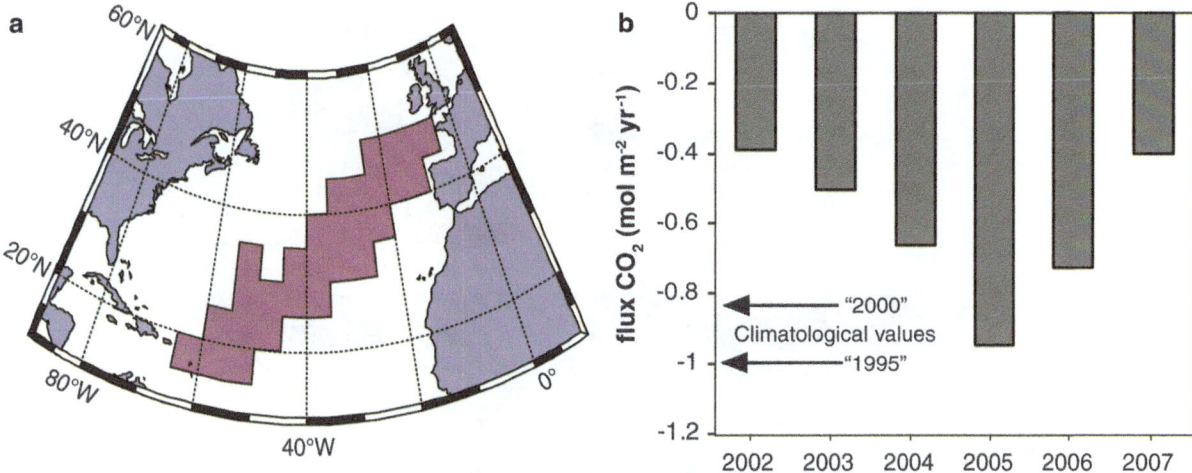

Fig. 3.5 Annual CO$_2$ uptake for the *pink shaded* ocean area between the UK and the Caribbean for 2002–2007. The size of climatological fluxes for the years 1995 and 2000 (Takahashi et al. 2002, 2009) is indicated on the *left axis* of the *right figure*. (Reproduced from Watson et al. (2009) by permission of Science)

other natural CO$_2$ fluxes when comparing such flux estimates (Gruber et al. 2009; Takahashi et al. 2009). The open ocean source of natural CO$_2$ arising from river inputs has been estimated as 0.5 ± 0.2 Pg C year^{-1} (Gruber et al. 2009, after Sarmiento and Sundquist 1992). However, this value could be too high by ~0.2 Pg C year^{-1} due to the substantial outgassing of river inputs during estuarine mixing (Sect. 3.2.4).

Figures 3.6 and 3.7 show the spatial distribution of net contemporary CO$_2$ fluxes as determined from a pCO$_2$-based climatology (Takahashi et al. 2009), ocean inversion (Gruber et al. 2009), atmospheric inversion (Baker et al. 2006) and ocean biogeochemistry models (Watson and Orr 2003). The fluxes from the four methods are in reasonable agreement for most ocean regions. The notable exception is the Southern Ocean (here south of 44°S), where marine biogeochemistry models predict a much larger CO$_2$ sink than the other methods, mainly as a result of a weak outgassing of natural CO$_2$ (Mikaloff Fletcher et al. 2007). A comparison of the ocean inverse results with the pCO$_2$ climatology shows that while both methods indicate a similar net contemporary CO$_2$ sink of 0.3 Pg C year^{-1} south of 44°S, the estimates disagree in the spatial distribution of the flux (Gruber et al. 2009). The climatology-derived flux estimates indicate a Southern Ocean sink between 44°S and 58°S and a small source south of 58°S (Takahashi et al. 2009), while the ocean inversion suggests a more uniform CO$_2$ sink south of 44°S (Gruber et al. 2009). It is worth noting that the recent addition of further surface water fCO$_2$ data in the Southern Ocean, and in particular in seasonally ice covered waters, has led to a revision of air-sea CO$_2$ flux estimates for 50–62°S (from −0.34 to −0.06 Pg C year^{-1}) and south of 62°S (from −0.04 to +0.01 Pg C year^{-1}) in successive climatologies (Takahashi et al. 2002, 2009).

The separation of contemporary air-sea CO$_2$ fluxes into natural CO$_2$ fluxes (here excluding river-induced fluxes), river borne fluxes and anthropogenic CO$_2$ fluxes, using an inversion of interior ocean inorganic carbon data, is shown in Fig. 3.7 (Gruber et al. 2009). Natural CO$_2$ fluxes in this study vary from CO$_2$ sources in the tropics and the Southern Ocean to CO$_2$ sinks in global temperate regions and the high latitude northern hemisphere (Gruber et al. 2009). On a global scale these natural fluxes (excluding river borne fluxes) cancel out. Anthropogenic CO$_2$ is taken up by all ocean regions, with the largest sinks in the tropics and the Southern Ocean.

3.2.3.4 Sea Ice

Sea ice influences marine DIC cycling and the air-sea exchange of CO$_2$ through physical processes such as brine rejection (e.g. Anderson et al. 2004; Omar et al. 2005; Rysgaard et al. 2011) and air-ice-sea exchange (e.g. Miller et al. 2011) (Chap. 2), in addition to

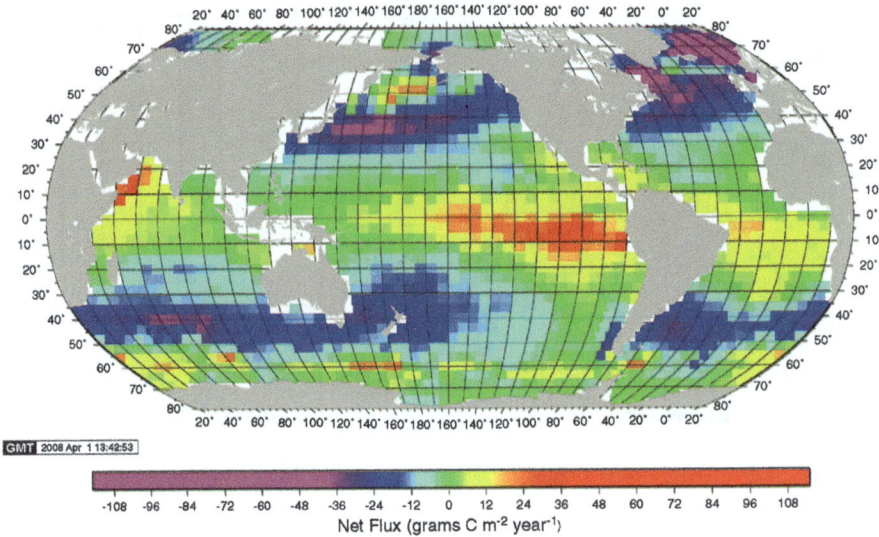

Fig. 3.6 Contemporary annual air-sea CO_2 fluxes for the year 2000 from a pCO_2 climatology (Reproduced from Takahashi et al. (2009) by permission of Elsevier)

biological and chemical processes (Delille et al. 2007; Bakker et al. 2008; Geibert et al. 2010). Although our understanding of the underlying processes is limited and quantitative estimates are scarce, the physical processes are thought to result in a net sink for tropospheric CO_2 during sea ice formation in the polar oceans. Recently, Rysgaard et al. (2011) estimated the net influx of CO_2 into the polar oceans at 33 Tg C year^{-1}, a flux resulting from the rejection of carbon from the ice crystal matrix during winter and subsequent formation of a surface layer of melt-water, undersaturated in CO_2 during summer. The sink would be much stronger (83 Tg C year^{-1}), if $CaCO_3$ crystals form in the sea ice. Omar et al. (2005) suggested a wintertime CO_2 sink of 5.2 g C m^{-2} associated with the formation of seasonal sea ice and brine rejection in the Arctic. With a seasonal sea ice extent of 14 × 10^6 km^2 (in 2005) this translated into a wintertime sink of 36 Tg C year^{-1}, which is on the higher end of estimates of 14 Tg C year^{-1} (no $CaCO_3$ precipitation) and 31 Tg C year^{-1} (with $CaCO_3$ precipitation) for the Arctic Ocean by Rysgaard et al. (2011). Sea ice related tropospheric CO_2 uptake was estimated as 19 Tg C year^{-1} for the Southern Ocean, which would increase to 52 Tg C year^{-1}, if $CaCO_3$ crystals form in the ice (Rysgaard et al. 2011). Oceanic CO_2 uptake during the seasonal cycle of sea ice growth and decay is thus equivalent to 17–42 % of net tropospheric CO_2 uptake in ice-free polar seas (Rysgaard et al. 2011).

3.2.3.5 Coastal to Open Ocean Carbon Exchanges

Exchanges of organic and inorganic carbon between coastal shelves and the deep ocean remain poorly quantified (Biscaye et al. 1988; Monaco et al. 1990; Biscaye and Anderson 1994; Wollast and Chou 2001), even though such exchange may be an important conduit for transferring tropospheric carbon to the interior ocean (Tsunogai et al. 1999; Thomas et al. 2004). For example an efficient 'continental shelf carbon pump', as proposed for the East China Sea and the North Sea, critically depends on the off-shelf transport of carbon-rich subsurface water (Tsunogai et al. 1999; Thomas et al. 2004) to below the permanent pycnocline of the deep ocean (Holt et al. 2008; Huthnance et al. 2009; Wakelin et al. 2012), but these carbon transports have not been verified in situ.

3.2.4 Air-Sea CO_2 Fluxes in Coastal Areas

3.2.4.1 Continental Shelves

Contemporary air-sea exchange fluxes of CO_2 in the coastal environment have been estimated by adjusting local flux data to the global scale using procedures of varying complexity (Box 3.4; Table 3.4). These range from extrapolating a single flux estimate from a continental shelf to the global scale, such as in the East China Sea (Tsunogai et al. 1999) or the North Sea

3.2 Surface Ocean Distribution and Air-Sea Exchange of CO_2

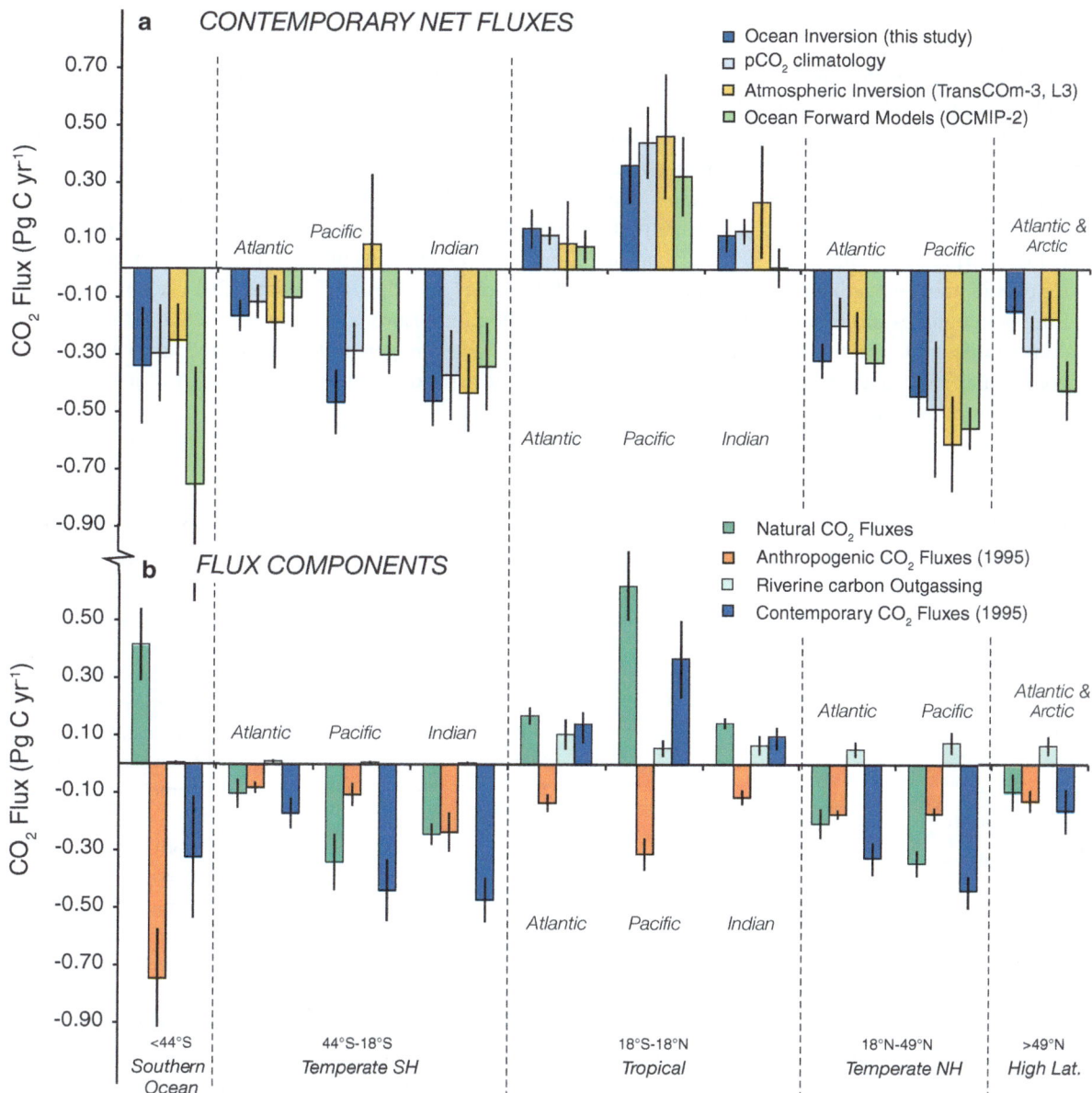

Fig. 3.7 Air-sea CO_2 fluxes for 10 regions per latitude range and per basin with positive fluxes for CO_2 leaving the ocean. (**a**) Net contemporary air-sea CO_2 fluxes from ocean inversion estimates (Gruber et al. 2009), a surface ocean pCO_2 climatology (Takahashi et al. 2009), mean estimates from 13 ocean biogeochemistry models (Watson and Orr 2003) and mean estimates from atmospheric inversion of CO_2 (Baker et al. 2006). Error bars for the ocean biogeochemistry model estimates are the unweighted standard deviation of the model outputs. The uncertainties in the atmospheric inversion estimates are based on the square of the errors within and between models. (**b**) Natural, anthropogenic, river-induced and contemporary air-sea CO_2 fluxes by ocean inversion. *Error bars* indicate the cross-model weighted standard deviation of the mean. Anthropogenic and contemporary fluxes are for the nominal year 1995. (Reproduced from Gruber et al. (2009) by permission of the American Geophysical Union)

(Thomas et al. 2004), to approaches that compiled flux values for several continental shelf systems with scaling by surface area. Areas have been grouped by latitudinal bands (Borges 2005; Borges et al. 2005), oceanic provinces (Cai et al. 2006) and surface areas derived from bathymetry (Chen and Borges 2009; Laruelle et al. 2010). The estimate of Laruelle et al. (2010) was based on a typological approach, whereby

Box 3.4
For the purpose of this work the coastal zone is defined to include near-shore systems, such as estuaries, and the continental shelf as far offshore as the 200 m depth contour (Walsh 1988; Gattuso et al. 1998; Liu et al. 2010). The continental margin consists of the coastal zone and the continental slope (from 200 to 2,000 m depth) (Liu et al. 2010). Inner estuaries are characterised by large salinity gradients, complex mixing and varying anthropogenic inputs. Laruelle et al. (2010) give a global inner estuarine area of $\sim 1 \times 10^6$ km^2. Outer estuaries (estuarine plumes) have restricted salinity ranges and salinities typically below 34 (Frankignoulle et al. 1998). Estuarine plumes may extend tens of km offshore and may account for considerable estuarine mixing (Naudin et al. 1997). Barnes and Upstill-Goddard (2011) estimated that European outer estuaries account for more than 75 % of the total European estuarine area, but there is no comparable global estimate. Nonetheless, estuarine plumes are generally considered as part of the continental shelf and there is a concern that the characteristic distributions of CO_2, CH_4 and N_2O in such plumes may not be sufficiently sampled.

Table 3.4 CO_2 fluxes scaled globally for continental shelves and estuaries: n is the number of data points used in the up-scaling

CO_2 flux (Pg C year^{-1})	CO_2 flux (mol m^{-2} year^{-1})	Surface area (10^6 km^2)	n	Reference
Continental shelves globally				
−0.95	−2.90	27.0	1	Tsunogai et al. (1999)
−0.40	−1.33	25.2	1	Thomas et al. (2004)
−0.37	−1.17	25.8	15	Borges (2005)
−0.45	−1.44	25.8	17	Borges et al. (2005)
−0.22	−0.71	25.8	29	Cai et al. (2006)
−0.34	−0.92	30.0	58	Chen and Borges (2009)
−0.22	−0.71	24.7	37	Laruelle et al. (2010)
Estuaries globally				
+0.60	+36.5	1.40	13	Abril and Borges (2004)
+0.43	+38.2	0.94	16	Borges (2005)
+0.32	+28.6	0.94	16	Borges et al. (2005)
+0.36	+32.1	0.94	32	Chen and Borges (2009)
+0.27	+21.0	1.10	62	Laruelle et al. (2010)

continental shelves were defined as one of three types: enclosed, upwelling, and open. In these studies the Arctic Ocean was included in estimates for the coastal ocean, but other deep marginal seas were excluded.

The first global estimate of the continental shelf sink for CO_2, based on East China Sea data (Tsunogai et al. 1999) was 1.0 Pg C year^{-1}, whereas most recent estimates converge to a value $\sim$0.3 Pg C year^{-1} (Chen and Borges 2009; Laruelle et al. 2010; Cai 2011). While continental shelves cover less than 10 % of the total ocean surface area, their air-sea CO_2 flux density is about twice as large (Laruelle et al. 2010) as the global average for the open oceans based on the most recent CO_2 climatology (Takahashi et al. 2009). This is consistent with higher biogeochemical reaction rates on continental shelves; rates of net primary production and export production are twice as high as in the open ocean, for example (Wollast 1998). Even so, the zonal variability in air-sea CO_2 fluxes over continental shelves (Borges 2005; Borges et al. 2005; Laruelle et al. 2010) follows the patterns of the open ocean (Takahashi et al. 2009), with low latitude continental shelves being CO_2 sources and temperate and high latitude shelves being sinks for tropospheric CO_2. This suggests that the direction of air-sea CO_2 fluxes on continental shelves is to some extent dictated by a "background" signal of "incoming" open ocean waters, and that the intensity of the flux is further modulated (enhanced) by biogeochemical processes on the continental shelf.

Lee et al. (2011) recently evaluated the anthropogenic carbon inventory in four marginal seas (Arctic Ocean, Mediterranean Sea, Sea of Okhotsk, and East/Japan Sea). These authors conclude that each of these marginal seas stores proportionally more anthropogenic CO_2 than the global open ocean and they attribute this to a dynamic over-turning circulation in these marginal seas.

3.2.4.2 Near-Shore Systems

Near-shore systems such as estuaries are known to significantly modify the fluxes of organic carbon from land to sea (e.g. Smith and Hollibaugh 1993; Gattuso et al. 1998; Battin et al. 2008) and to also emit large quantities of N_2O and CH_4 (Sects. 3.3.5 and 3.4.5) (Barnes et al. 2006; Denman et al. 2007; Upstill-Goddard 2011). Estuaries are also characterised by a net annual emission of CO_2 to the troposphere with intense flux densities (Frankignoulle et al. 1998). Various estimates of the global emission of CO_2 to the troposphere from inner estuaries are based on scaling exercises (Table 3.4) (Abril and Borges 2004; Borges 2005; Borges et al. 2005; Chen and Borges 2009; Laruelle et al. 2010). All are based on the global surface area estimate of Woodwell et al. (1973) with the exception of Laruelle et al. (2010), which is based on the typology of estuaries from Dürr et al. (2011). The first estimate by Abril and Borges (2004) of the emission of CO_2 to the troposphere was 0.6 Pg C year^{-1} and the most recent estimates converge to ~0.3 Pg C year^{-1} (Laruelle et al. 2010; Cai 2011). The estimate of Laruelle et al. (2010) relies on an estuarine typology with four types (small deltas and small estuaries, tidal systems and embayments, lagoons, fjords and fjärds (sea inlets, which have been subject to glacial scouring, in a rocky area of low topography)). This is an important innovation relative to previous scaling attempts, since estuarine morphology and physical structure strongly modulate the exchange of CO_2 with the troposhere (Borges 2005; Koné et al. 2009; Borges and Abril 2011). Fjords and fjärds constitute the most abundant estuarine type (~43 %), although CO_2 flux data have been reported for only one system. This highlights the limitation of using scaling approaches that are too complex with regards to the available data, and the need to obtain further data in near-shore systems to improve estimates of air-sea exchange of CO_2.

In most macro-tidal estuaries the river input of DIC can only sustain a small fraction of the observed CO_2 emission (Borges et al. 2006), implying that the bulk of estuarine CO_2 emission is sustained by the degradation of allochthonous organic matter, in agreement with the net heterotrophic nature of these systems established from measurements of community metabolic rates (Odum and Hoskin 1958; Odum and Wilson 1962; Heip et al. 1995; Kemp et al. 1997; Gattuso et al. 1998; Gazeau et al. 2004; Hopkinson and Smith 2005). This implies that near-shore coastal environments are effective sites (or 'bypasses') for returning to the troposphere as CO_2, a fraction of the carbon passing from continents (through rivers) to the ocean. The removal of river borne organic carbon during estuarine transit can be roughly evaluated at ~60 % based on the above, given a global CO_2 emission of ~0.3 Pg C year^{-1} from near shore waters (Laruelle et al. 2010; Cai 2011) and known global organic carbon river inputs of ~0.4 Pg C year^{-1} (Schlünz and Schneider 2000). This is in general agreement with the analysis of organic carbon in estuaries (e.g. Abril et al. 2002). Such a bypass of carbon has important consequences for understanding and quantifying the global carbon cycle. For instance, the pre-industrial ocean is assumed to have been a CO_2 source driven by degradation of river borne organic carbon (Smith and Mackenzie 1987; Sarmiento and Sundquist 1992). In budget studies the contemporary ocean air-sea CO_2 flux is typically corrected for the pre-industrial air-sea CO_2 flux of 0.5 ± 0.2 Pg C year^{-1}, so as to derive the anthropogenic CO_2 flux (Sarmiento and Sundquist 1992; Gruber et al. 2009; Takahashi et al. 2009). However, if most of the degradation of river borne organic carbon occurs in near-shore coastal environments rather than in the open ocean, this correction may be overestimated by ~0.2 Pg C year^{-1}, corresponding to much of the estuarine CO_2 emissions of ~0.3 Pg C year^{-1} (Laruelle et al. 2010; Cai 2011).

3.2.4.3 Multi-Year Changes and Trends

Based on the decadal analysis of surface water fCO_2 in a very limited number of coastal regions, the coastal CO_2 sink could be increasing in some regions (Wong et al. 2010), while decreasing elsewhere (Thomas et al. 2007). Gypens et al. (2009) used a model reconstruction of the biogeochemistry of the Southern North Sea during the last 50 years to evaluate how the change of river nutrient loads has affected the annual exchange of CO_2 with the troposphere. These authors concluded that carbon sequestration in the southern North Sea increased from the 1950s to the mid 1980s due to an increase in primary production fuelled by eutrophication with an N to P (nitrogen to phosphorus) ratio close to Redfield of 16 to 1 (Redfield et al. 1963). In consequence, the system shifted from a source to a sink of tropospheric CO_2. During this period pH and calcite saturation increased, rather than decreased as one would have expected from ocean acidification alone (Borges and Gypens 2010). During a period of eutrophication reversal from the mid 1980s onwards, in which river borne nitrogen inputs continued to increase but phosphorus inputs were reduced, primary production in the southern North Sea decreased due to phosphorus limitation and the system shifted back to being a source of tropospheric CO_2. During this period, the carbonate chemistry changed faster than that expected from ocean acidification alone, i.e. ocean acidification was enhanced.

3.3 Marine Distribution and Air-Sea Exchange of N_2O

3.3.1 Global Tropospheric N_2O Budget

N_2O emissions from oceanic and coastal waters play a major role in the tropospheric N_2O budget (Table 3.5). According to the IPCC 4th Assessment Report (Denman et al. 2007) the oceans are a natural N_2O source of 3.8 Tg N year^{-1} (range 1.8–5.8 Tg year^{-1}), while coastal waters, estuaries, rivers and streams together are an anthropogenic N_2O source of 1.7 Tg N year^{-1} (range 0.5–2.9 Tg year^{-1}). These sources thus contribute 20 % and 10 % respectively, of total global N_2O emissions (Tables 3.2 and 3.5). Considerable uncertainties arise over these emission estimates for reasons that are discussed in Sects. 3.3.5 and 3.6. The quantification of oceanic N_2O emissions and the identification of the marine pathways of N_2O formation and consumption have received increased attention in recent decades (Bange 2008, 2010b).

3.3.2 Nitrous Oxide Formation Processes

Oceanic N_2O is formed exclusively by prokaryotes (bacteria and archaea) via two major processes: nitrification (i.e. oxidation of ammonium, NH_4^+, to nitrate, NO_3^-) and denitrification (i.e. reduction of NO_3^- to N_2) (Fig. 3.8). Nitrification is the dominant N_2O formation process whereas denitrification contributes about 7–35 % to the overall N_2O budget of the oceans

Table 3.5 Anthropogenic and natural sources of N_2O to the troposphere with the range of estimates between brackets (Denman et al. 2007)

	N_2O source (Tg N year^{-1})
Anthropogenic sources	
Fossil fuel combustion and industrial processes	0.7 (0.2–1.8)
Agriculture	2.8 (1.7–4.8)
Biomass burning	0.7 (0.2–1.0)
Human excreta	0.2 (0.1–0.3)
Rivers, estuaries, coastal zones	1.7 (0.5–2.9)
Atmospheric deposition	0.6 (0.3–0.9)
Total anthropogenic sources	6.7
Natural sources	
Soils under natural vegetation	6.6 (3.3–9.0)
Oceans	3.8 (1.8–5.8)
Atmospheric chemistry	0.6 (0.3–1.2)
Total natural sources	11.0
Total sources	17.7

(Bange and Andreae 1999; Freing et al. 2012). The contributions to oceanic N_2O production from other microbial processes such as dissimilatory nitrate reduction to ammonia (DNRA) are largely unknown. In general, biological N_2O production strongly depends on the availability of dissolved oxygen (O_2). Under oxic conditions, as found in the majority of oceanic waters, N_2O formation occurs via nitrification. Suboxic to anoxic conditions, which occur in about 0.1–0.2 % of the ocean volume, favour the net formation of N_2O via denitrification (Box 3.5) (Codispoti 2010).

3.3.2.1 Denitrification

During denitrification N_2O occurs as an intermediate which can be both produced and consumed. The denitrification pathway consists of the four step reduction of NO_3^- to N_2 (Fig. 3.8), thus it constitutes a net loss of bioavailable (or "fixed") nitrogen (N). Denitrification is catalysed by four independent metallo-enzymes (Zumft 1997). Both bacterial and archaeal denitrifiers (Philippot 2002; Cabello et al. 2004) are able to respire NO_3^- when O_2 becomes limiting. Denitrification may therefore be considered the ancestor of aerobic respiration (Cabello et al. 2004). The O_2 sensitivity of the enzymes involved in denitrification increases step by step along the reduction chain. The enzymes are induced sequentially and a complete denitrification process can only take place at O_2 concentrations below 2–10 µM (Fig. 3.9) (Codispoti et al. 2005). With the observed

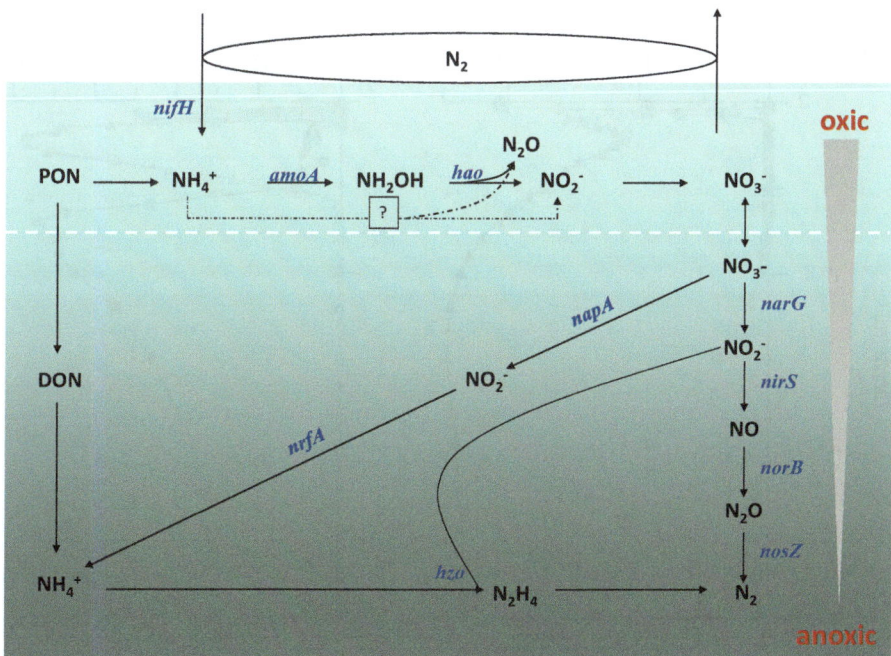

Fig. 3.8 The nitrogen cycle in the oceanic water column along a vertical oxygen gradient. Key functional genes are shown in *blue italic letters* for the transformations, the oxycline is indicated by a *horizontal, white dashed line*, and archaeal ammonia-oxidation is indicated by a *horizontal, thin, black dashed-dotted line*. (Modified from Francis et al. (2007))

Box 3.5
Dissolved oxygen concentrations play an important role in N_2O and CH_4 cycling. Here we define hypoxic or low oxygen conditions as an O_2 concentration below ~60 µM, suboxic conditions as an O_2 concentration below 5 µM (Deutsch et al. 2011) and anoxic conditions where O_2 is undetectable.

---

expansion of Oxygen Minimum Zones (OMZs) in the open ocean (Stramma et al. 2010) and the ongoing deoxygenation of highly productive eastern boundary upwelling areas (Codispoti 2010), net N_2O formation by denitrification may increase in the future (Sect. 3.5.3).

3.3.2.2 Nitrification

Under the oxic conditions present in more than 90 % of the ocean, N_2O is formed as a metabolic by-product during nitrification, the stepwise oxidation of NH_4^+ to nitrite (NO_2^-) by both ammonia-oxidising bacteria (AOB) and archaea (AOA). Bacteria form N_2O during the oxidation of NH_4^+ via hydroxylamine (NH_2OH) to NO_2^- (Fig. 3.8). Alternatively, N_2O can be formed during the reduction of NO_2^- via nitric oxide (NO) to N_2O, the so-called nitrifier-denitrification pathway (Cantera and Stein 2007). However, the enzymes involved in the nitrifier-denitrification pathway are different from those involved in classical denitrification (Sect. 3.3.2.1). The production of N_2O during nitrification increases with decreasing O_2 concentrations (Goreau et al. 1980; Codispoti et al. 1992). This implies that a significant in situ N_2O production in the upper mixed layer is unlikely, as this layer tends to be well oxygenated.

Until recently the formation of N_2O by nitrification was regarded as an exclusive property of AOB. This view has subsequently been revised in the light of recent work showing that AOA are the key organisms for oceanic nitrification (Wuchter et al. 2006, 2007) and that AOA are able to produce N_2O in large amounts (Santoro et al. 2011; Löscher et al. 2012). Experiments using AOA enriched cultures and pure cultures of *Nitrosopumilus maritimus*, as well as onboard incubation

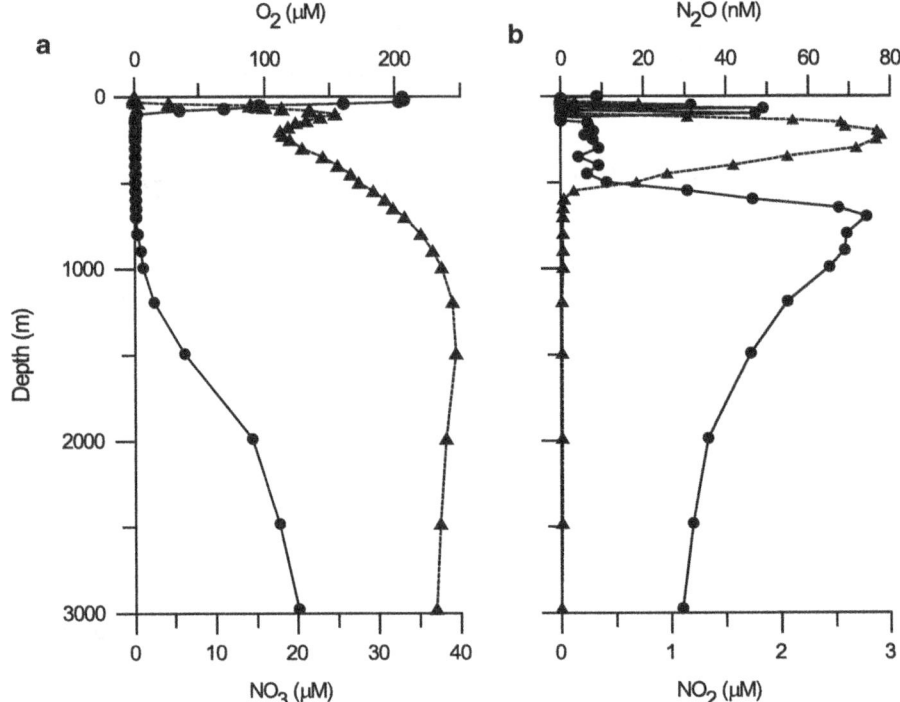

Fig. 3.9 Vertical profiles of (**a**) dissolved oxygen (*circles*) and nitrate (*triangles*), and (**b**) nitrite (*triangles*) and nitrous oxide (*circles*) at 19°N 67°E in the Arabian Sea. Note the pronounced minimum in nitrous oxide within the denitrifying zone, characterised by a minimum in nitrate and a maximum in nitrite. Maxima in nitrous oxide are found at the peripheries of this zone. (Reproduced from Naqvi (2008) by permission of Elsevier)

experiments with the archaea inhibitor GC7 (Jansson et al. 2000), demonstrated that AOA are the key organisms for N_2O production (Löscher et al. 2012). Nevertheless, the precise metabolic pathway remains unknown. The high affinity of archaea for NH_4^+ indicates their potential to outcompete AOB even under the nutrient depleted (oligotrophic) conditions (Martens-Habbena et al. 2009) which characterise large areas of the open (surface) ocean. It can thus be hypothesised that archaeal NH_4^+ oxidation is the major source of oceanic N_2O formation.

3.3.2.3 N_2O Formation by Dissimilatory Nitrate Reduction to Ammonium

Dissimilatory nitrate reduction to ammonium (DNRA) via NO_2^- is a known source of N_2O (Cole 1988), but was previously considered unimportant in the oceanic water column. However, it was recently found to significantly impact nitrogen cycling in OMZs (Lam et al. 2009) and so may be a more important source of N_2O than previously thought. A variety of bacteria (*Bacillus* sp., *Clostridium* sp., *Enterobacter* sp.) are able to carry out DNRA (Fazzolari et al. 1990a, b) and are widespread in the ocean and in other environments. N_2O is produced during the second stage of DNRA, the reduction of NO_2^- to NH_4^+ catalysed by NO_2^- reductase (Jackson et al. 1991). Nevertheless, information on the biochemical regulation of DNRA in oceanic environments remains sparse (Baggs and Philippot 2010).

3.3.3 Global Oceanic Distribution of Nitrous Oxide

Global maps of N_2O in the surface ocean have been computed by Nevison et al. (1995) (N95) and by Suntharalingam and Sarmiento (2000) (SS00) (Fig. 3.10). The N95 map is based on more than 60,000 measurements mainly made by the Scripps Institution of Oceanography between 1977 and 1993. Oceanic regions with no measurements were filled with a simple statistical routine. The SS00 map was derived from the same N_2O data set, but employed a multivariate adaptive regression spline method using mixed

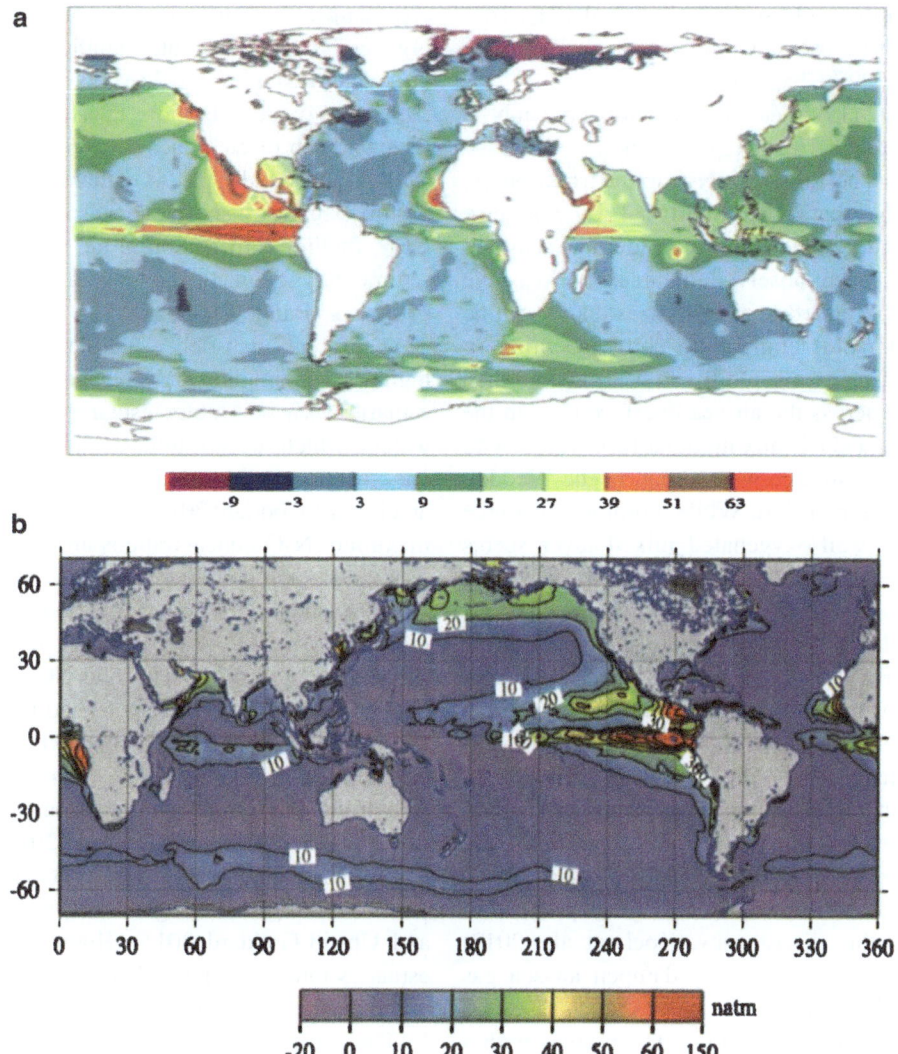

Fig. 3.10 Maps of ΔpN_2O (in natm) in the surface layer of the world's oceans: (**a**) map by Nevison et al. (1995) and (**b**) map by Suntharalingam and Sarmiento (2000). Note that the colour coding is non-linear and different for both maps. (Reproduced from Nevison et al. (1995) and Suntharalingam and Sarmiento (2000) by permission of the American Geophysical Union)

layer depth, O_2, sea surface temperature and upwelling rate as predictor variables. Differences in the two maps result mainly from the different computation methods but they share important common features: (i) enhanced N_2O anomalies (i.e. supersaturation of N_2O) in the equatorial upwelling regions of the eastern Pacific and Atlantic Oceans and in coastal upwelling regions such as along the west coasts of North and Central America, off Peru, off Northwest Africa and in the northwestern Indian Ocean (Arabian Sea); (ii) N_2O anomalies close to zero (i.e. near equilibrium) in the North and South Atlantic Ocean, the South Indian Ocean and the central gyres of the North and South Pacific Oceans. Both maps are biased by insufficient data coverage in some ocean regions (for example in the Indian and western Pacific Oceans). Since the studies of N95 and SS00 the number of available N_2O measurements has been steadily increasing. With this in mind the MEMENTO (MarinE MethanE and NiTrous Oxide) initiative was launched with the aim of collecting and archiving N_2O (and CH_4)

data sets and to provide surface N_2O (and CH_4) concentration fields for use in deriving emissions estimates (Bange et al. 2009) (Chap. 5).

Nevison et al. (1995) calculated a global mean N_2O surface saturation of 103.5 %, which indicates that the global ocean is a net source of N_2O to the troposphere. Apart from the spatial variability of N_2O surface concentrations described above, considerable seasonal variability has been observed in areas such as the Greenland and Weddell Seas. This seasonality can be caused by (i) rapid temperature shifts resulting in pronounced changes in solubility at a faster rate than N_2O exchange across the air-sea interface (e.g. in the Greenland Sea) and (ii) mixing of surface waters with N_2O enriched subsurface waters (e.g. in the Weddell Sea) (Nevison et al. 1995). While a biological source of N_2O in the well-oxygenated mixed layer seems unlikely, some studies suggested in situ mixed layer production based on a mismatch between the N_2O air-sea flux and the diapycnal flux into the mixed layer (Dore and Karl 1996; Morell et al. 2001). In a recent study of N_2O air-sea and diapycnal fluxes in the eastern tropical North Atlantic (including the upwelling off Mauritania, NW Africa) the mean air-sea flux, calculated using a common gas exchange approach, was about three to four times larger than the mean diapycnal flux into the mixed layer (Kock et al. 2012). Neither vertical advection nor biological production could explain this discrepancy. Kock et al. (2012) speculated that surfactants may dampen air-sea gas exchange of N_2O and other gases such as CO_2 (see for example Tsai and Liu 2003) in areas with a high biological productivity.

3.3.4 Coastal Distribution of Nitrous Oxide

During the last two decades coastal areas such as estuaries, upwelling regions, and mangrove ecosystems have received increased attention as sites of intense N_2O formation and release to the troposphere. Studies of the N_2O pathways in coastal regions have mostly been undertaken in European and North American coastal regions but the number of studies from other coastal regions (e.g. from Asia and South America) has recently been increasing. In general, strongly positive N_2O anomalies are found in nitrogen-rich estuaries (Zhang et al. 2010; Barnes and Upstill-Goddard 2011) and in coastal upwelling systems (Nevison et al. 2004). Coastal N_2O emissions contribute significantly to the overall oceanic emission (Table 3.6).

Nitrous oxide saturations in estuaries are highly variable and can reach values up to 6,500 % (Zhang et al. 2010; Barnes and Upstill-Goddard 2011). N_2O formation in estuaries heavily depends on the availability of NH_4^+ fuelling nitrification in the water column and/or sedimentary denitrification as major N_2O formation pathways (Bange 2006b; Barnes and Upstill-Goddard 2011). In nitrogen-rich estuarine systems, extremely high N_2O anomalies are usually only found in inner estuaries, whereas outer estuaries and adjacent shelf waters, which are not influenced by the river plumes, are close to equilibrium with the troposphere (Barnes and Upstill-Goddard 2011). In some European estuaries maximum N_2O concentrations are associated with the turbidity maximum zone at low salinities (Barnes and Upstill-Goddard 2011). The traditional view of a simple relationship between river inputs of dissolved inorganic nitrogen (the sum of NH_4^+ and NO_3^-) and estuarine N_2O formation has been challenged by recent findings that resuspended NH_4^+ and/or NH_4^+ derived from ammonification of particulate organic nitrogen in the turbidity maximum zone might dominate N_2O production (Barnes and Upstill-Goddard 2011). This implies that N_2O formation may not be related to river inputs of dissolved inorganic nitrogen in any simple way (Barnes and Upstill-Goddard 2011). High N_2O saturations in estuaries (and rivers) are also found at sites of sewage and industrial effluents.

The narrow bands of coastal upwelling systems such as those found in the northwestern Indian Ocean (Arabian Sea) and in the southeastern Pacific Ocean (off central Chile) have been identified as 'hot spots' of extremely high N_2O concentrations with N_2O saturations of up to 8,250 % and 2,426 %, respectively (Naqvi et al. 2005; Cornejo et al. 2007). The high N_2O saturations in coastal upwelling regions appear to be caused by the upwelling of N_2O enriched subsurface waters (Naqvi et al. 2005; Cornejo et al. 2007).

Some coastal upwelling areas show a rapid seasonal transition from oxic via suboxic to anoxic conditions and vice versa. In these systems, significant amounts of N_2O (up to several hundred nM) (Fig. 3.11) can accumulate temporarily during the short transition time, when the system is changing its oxygen regime. This phenomenon has been observed at different coastal time-series sites associated with coastal upwelling, such as off central

Table 3.6 N$_2$O emissions from marine waters

	Area (10^6 km^2)	N$_2$O emission (Tg N year^{-1})	Reference
Open ocean	–	1.8–5.8	Denman et al. (2007)
	313	0.6–1.1	Rhee et al. (2009)
Total open ocean	**348**[a]	**0.6–5.8**	
Coastal upwellings	1.75	0.05–0.2	Nevison et al. (2004)
	0.4[b]	0.0015–0.0035	Rhee et al. (2009)
Estuaries	1.4[c]	0.25	Kroeze et al. (2005)
Mangroves	0.2	0.1	Barnes et al. (2006)
Other coastal regions	10[d]	0.08 ($\pm$ >25 %)	Nevison et al. (2004)
	48[b]	0.4–0.9	Rhee et al. (2009)
Total coastal regions (incl. upwelling, estuaries and mangroves)	**13.35**[e]	**0.4–1.45**	
Total ocean (open ocean and coastal regions)	**361**	**1.0–7.25**[f]	

[a]Estimated as the difference between the total ocean area (361 × 10^6 km^2) and the sum of the areas (in 10^6 km^2) of coastal upwelling (1.75), continental shelves with a depth <200 m (10), estuaries (1.4) and mangroves (0.2)
[b]Source not given
[c]As cited in Nevison et al. (2004)
[d]Continental shelves with a depth <200 m
[e]Estimated as the sum of the areas (in 10^6 km^2) of coastal upwelling (1.75), continental shelves with a depth <200 m (10), estuaries (1.4) and mangroves (0.2)
[f]Estimated as the sum of the regional minimum and maximum values, respectively

Chile and off West India, and in the western Baltic Sea (Naqvi et al. 2010). During the transition stages, the accumulation of N$_2$O does not occur in the anoxic zones, but at the oxic to anoxic boundaries. The exact cause of this extreme accumulation of N$_2$O is not well understood, although inhibition of the activity of N$_2$O reductase through frequent incursion of O$_2$ into the O$_2$-deficient layer has been proposed as one possible explanation. In anoxic zones, N$_2$O is usually found at very low or even undetectable concentrations.

Another intriguing feature is the much smaller N$_2$O accumulation at the upper boundary of the suboxic zone of enclosed anoxic basins (Black Sea, Baltic Proper) and some anthropogenically-formed anoxic zones (Tokyo Bay and Chesapeake Bay). In the hypoxic bottom waters of the East China Sea and the Gulf of Mexico, the observed N$_2$O build-up is modest. Overall, these results do not show comparable N$_2$O build-up in the anthropogenically-formed coastal hypoxic zones to those in naturally-formed, upwelling-related coastal suboxic zones (Naqvi et al. 2010). However, a large number of anthropogenically-formed anoxic zones remain to be investigated.

Mangrove ecosystems cover ~75 % of tropical coasts and are among the world's most productive ecosystems. Their open waters cover ~0.2 × 10^6 km^2, equivalent to ~20 % of the global estuarine area (Borges et al. 2003). The ecosystems of mangrove forests have a high potential of N$_2$O formation and release to the troposphere (Barnes et al. 2006). There seems to be no dominant formation process: N$_2$O in mangrove sediments from Puerto Rico was mainly produced by nitrification (Bauzá et al. 2002), whereas incubation experiments with mangrove soils from the east coast of Australia revealed denitrification to be the main N$_2$O formation pathway (Kreuzwieser et al. 2003). In a seasonal study of N$_2$O emissions from a pristine mangrove creek on South Andaman Island (Gulf of Bengal), Barnes et al. (2006) found that N$_2$O emissions were negatively correlated with tidal height, indicating that N$_2$O (and CH$_4$) is released from sediment pore waters during "tidal pumping", i.e. during cyclic decrease and increase of the hydrostatic pressure between low and high water (Fig. 3.12; Sect. 3.4.4).

3.3.5 Marine Emissions of Nitrous Oxide

The N$_2$O emission estimates in Tables 3.5 and 3.6 imply that coastal areas contribute significantly to total marine N$_2$O emissions, which is in line with previous emission estimates (Bange 2006a). The

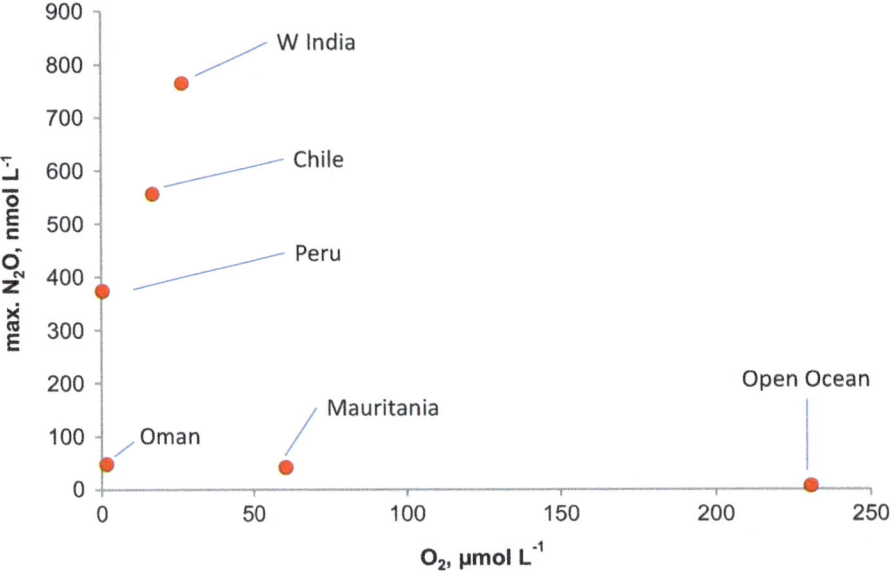

Fig. 3.11 Maximum N_2O concentrations and associated O_2 concentrations in coastal upwelling regions. A typical N_2O surface concentration in the tropical open ocean is also shown. (Data sources: W. India and Oman – S.W.A. Naqvi, pers. comm.; Mauritania – A. Kock and H.W. Bange, unpublished; Chile – Cornejo et al. (2006); Peru – C.R. Löscher and H.W. Bange, unpublished)

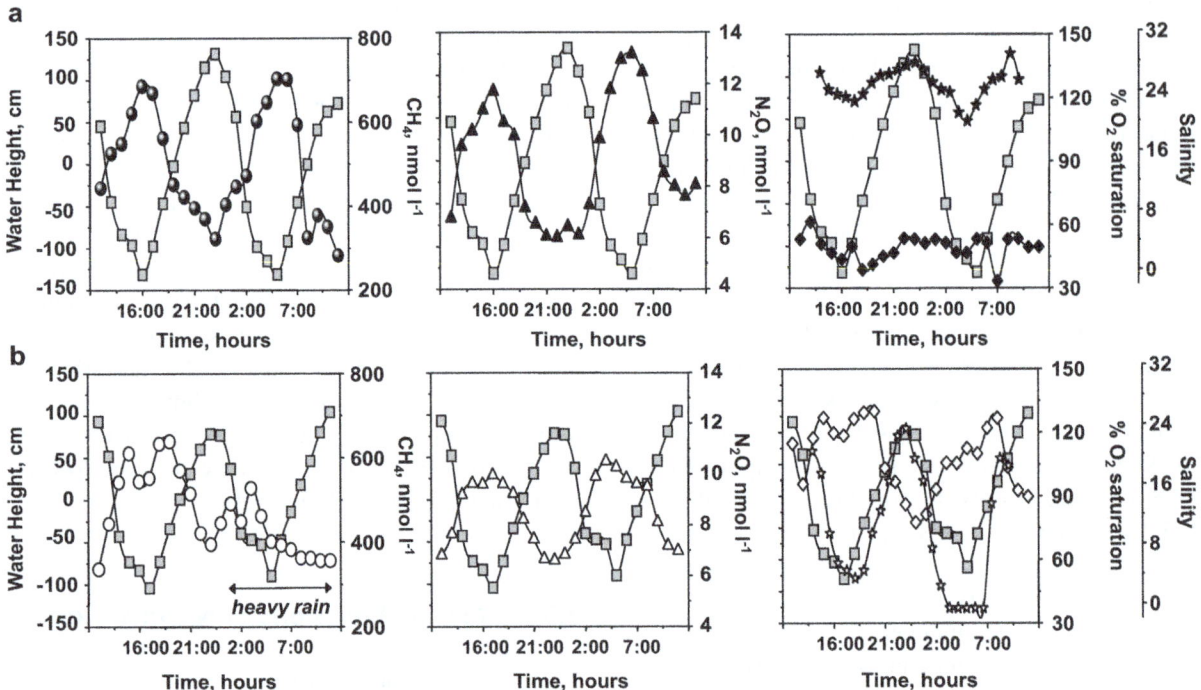

Fig. 3.12 Variation of CH_4 (*circles*), N_2O (*triangles*), tidal height (*squares*), O_2 saturation (*stars*) and salinity (*diamonds*) in a tropical mangrove creek (Wright Myo, Andaman Island) during (**a**) the dry season (January 2004) and (**b**) the wet season (July 2004) (Reproduced from Barnes et al. (2006) by permission of the American Geophysical Union)

emission estimates for coastal areas (upwelling regions, shelves, estuaries and mangroves) have a large uncertainty because of the small number of available measurements. In particular, natural coastal suboxic zones are strong N_2O sources to the troposphere. The total N_2O emission from these areas could be as high as 0.56 Tg N year^{-1} (Naqvi et al. 2010), comparable to global emissions from estuaries (0.25 Tg N year^{-1}) and continental shelves (0.4–1.45 Tg N year^{-1}) (Table 3.6) (Seitzinger and Kroeze 1998). As is the case for open ocean emissions (see below), flux estimates from coastal areas need to consider seasonal variability (Wittke et al. 2010; Zhang et al. 2010; Barnes and Upstill-Goddard 2011).

The recent open ocean estimate by Rhee et al. (2009) is considerably lower than that of the widely used IPCC 4th Assessment Report (Table 3.6) (Denman et al. 2007). However, because the estimate by Rhee et al. (2009) is based on a single meridional transect in the Atlantic Ocean, it almost certainly includes an unquantified seasonal and regional bias.

Unfortunately, the seasonality of surface water N_2O concentrations over large regions of the ocean remains unknown because ship campaigns are limited in space and time and N_2O sensors are not yet available on gliders, floats or moorings. Neglecting the seasonality of surface N_2O concentrations introduces severe bias into the N_2O flux estimates. Freing (2009) demonstrated how both N_2O concentrations in surface water and N_2O fluxes in the North Atlantic Ocean (19–42°N) follow a seasonal cycle similar to that of fCO_2 (Fig. 3.13; Sect. 3.2.3). This seasonal cycle can be described by a harmonic function and is mainly controlled by temperature. The presence of such seasonal variation renders a mean flux, if calculated from a seasonally-biased dataset, a potentially poor estimate of the true annual net flux (Freing 2009). Integrating the harmonic function over a full annual cycle gives a better estimate of the net annual flux.

In ocean regions where the upper boundary of the OMZ is shallow, minor changes in the hydrographic or meteorological conditions can lead to entrainment of N_2O from the OMZ into the surface layer, thereby enhancing N_2O sea-to-air fluxes (Naik et al. 2008). As a result of high N_2O concentrations close to the sea surface, N_2O emissions in open-ocean regions with substantial N_2O accumulation at mid-depth (associated with O_2 depletion) (e.g. in the Eastern Tropical South Pacific and the Arabian Sea), are quite high. The N_2O emissions from these regions (0.8–1.35 Tg N year^{-1}) (Naqvi et al. 2010) make up a significant fraction of the overall N_2O emission from the oceans (Table 3.6).

3.4 Marine Distribution and Air-Sea Exchange of CH$_4$

3.4.1 Global Tropospheric CH$_4$ Budget

In contrast to the situation for CO_2 and N_2O, the marine system plays a relatively minor role in the global tropospheric CH_4 budget, representing a small net natural contribution (Tables 3.2 and 3.3). However, in common with other global CH_4 sources, marine-derived CH_4 has proven difficult to quantify with any great certainty (Table 3.3). Although detailed CH_4 surveys in specific ocean basins have been available since the 1970s, they are comparatively limited in number and many of the early measurements were derived in the absence of reliable solubility data (Reeburgh 2007). The global marine CH_4 dataset is thus rather limited in comparison to CO_2 or N_2O. Detailed maps of the global surface ocean distribution remain to be compiled, with the recent MEMENTO initiative (Bange et al. 2009) working towards this goal (Chap. 5).

3.4.2 Formation and Removal Processes for Methane

Methanogenesis is the final stage of organic matter decomposition and is a form of anaerobic respiration carried out exclusively by single celled archaea whose growth is severely O_2-limited. The terminal electron acceptor is therefore not O_2, but carbon from low molecular weight compounds. Carbon dioxide and acetic acid (CH_3COOH) are the most familiar:

$$CO_2 + 4H_2 \rightarrow CH_4 + 2H_2O \quad (3.11)$$

$$CH_3COOH \rightarrow CH_4 + CO_2 \quad (3.12)$$

Other low molecular weight compounds acting as methanogenic substrates include formic acid (HCOOH), methanol (CH_3OH), methylamine (CH_3NH_2), dimethyl-sulphide (CH_3SCH_3) and methanethiol (CH_3SH). Unsurprisingly, anoxic coastal marine sediments (Middelburg et al. 1996) and strongly O_2-deficient waters (Naqvi et al.

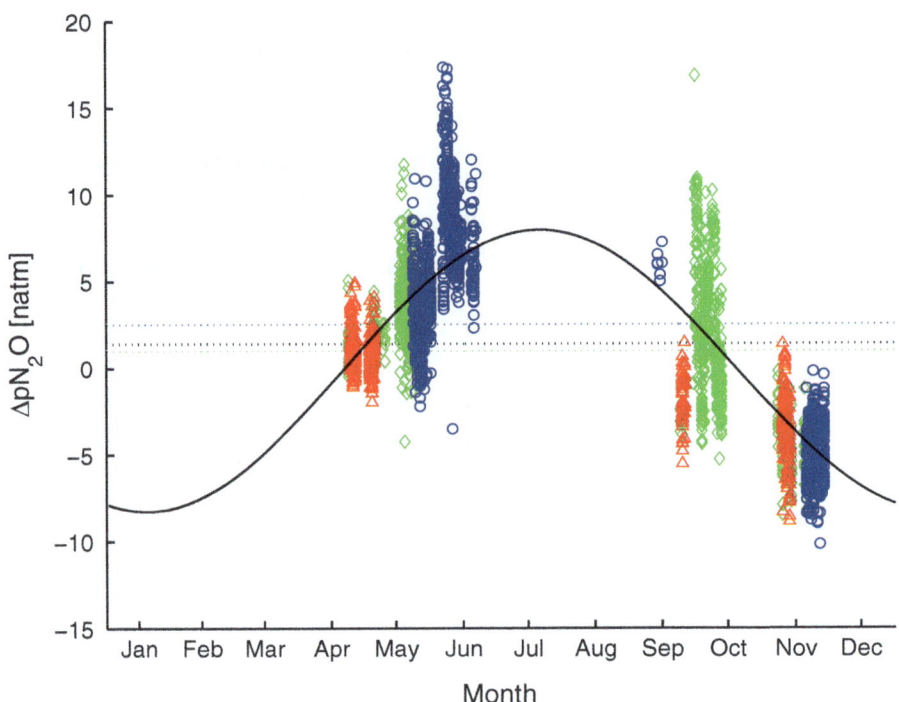

Fig. 3.13 ΔpN_2O (in natm) in the North Atlantic Ocean (19–42°N 10–66°W): Tropical (*red triangles*), western subtropical (*green diamonds*) and eastern subtropical (*blue circles*) regions. The *solid black line* denotes a fitted harmonic function. The *dotted lines* denote the respective annual mean for all data points (*black, middle line*), the western (*green, bottom line*) and the eastern (*blue, top line*) basin. (The figure is from Freing (2009))

2005) are major sites of methanogenesis. The ability of sulphate-reducing bacteria (SRB) to outcompete methanogens (Capone and Kiene 1988) means that CH_4 concentrations are generally low in near-surface sediment pore waters where sulphate reduction is active and that they are maximal below the depth where dissolved sulphate (SO_4^{2-}) becomes fully depleted (Blair and Aller 1995). This can be anywhere between several centimeters and several meters below the sediment surface. The role of SRB is an important consideration in estuaries, where water column SO_4^{2-} concentrations vary from micro- to milli-molar along the salinity gradient. In these situations the depth of SO_4^{2-} depletion increases and maximal CH_4 production generally decreases seaward. Methanogenesis rates have thus been shown to decrease by up to two orders of magnitude seaward (Abril and Borges 2004). In addition to SO_4^{2-} availability and associated SRB activity, rates of organic matter sedimentation and the availability of alternative electron acceptors also influence methanogenesis rates.

Despite inhibition of methanogenesis by both O_2 and SO_4^{2-} supply, CH_4 is typically supersaturated in the open ocean mixed layer (Sect. 3.4.3). Until recently the prevailing explanation for this so-called "marine CH_4 paradox" has been methanogenesis (Eqs. 3.11 and 3.12) within "anoxic microniches" inside zooplankton guts and suspended particles (Oremland 1979; De Angelis and Lee 1994). This notion is supported by an "oxic" methanogenic archaea isolated from coastal waters (Cynar and Yayanos 1991), the identification of methanogens in marine zooplankton guts and particles (Marty et al. 1997) and CH_4 release from sinking particles inferred from $\delta^{13}CH_4$ measurements (Sasakawa et al. 2008). Even though significant CH_4 release has been observed from mixed zooplankton-phytoplankton cultures (De Angelis and Lee 1994) and correlations of CH_4 with primary productivity indicators, such as chlorophyll *a*, have been found (Oudot et al. 2002), such correlations are weak (Upstill-Goddard et al. 1999; Holmes et al. 2000; Forster et al. 2009).

Recent investigations propose two alternative CH_4 production mechanisms, both implicating nutrient limitation in the control of mixed layer CH_4 formation. In the first hypothesis bacterioplankton successfully exploit phosphate-depleted waters, where nitrate is in excess, by deriving phosphorus from phosphonates such as methyl phosphonate (Karl et al. 2008). Methane is thus produced aerobically as a byproduct of methyl phosphonate decomposition. The second hypothesis proposes that certain microbes can catabolise dimethylsulphoniopropionate (DMSP) as a carbon source through methylotrophic methanogenesis in NO_3^- depleted waters, where phosphate (PO_4^{3-}) is plentiful (Damm et al. 2010). Based on the conversion of DMSP to hydrogen sulphide (H_2S) and CH_3SH by DMSP-utilising bacteria (Kiene et al. 2000) and a recent proposal for the intermediate formation of CH_3SH during CH_4 oxidation (Moran et al. 2008), Damm et al. (2010) proposed a thermodynamically plausible reverse reaction for the aerobic production of CH_4 from CH_3SH. The two mechanisms are entirely compatible and raise the intriguing possibility that deviations from the Redfield N:P ratio could be indirectly responsible for the marine CH_4 paradox through planktonic succession favouring species able to exploit alternative marine phosphorus and nitrogen stores.

Whatever the mechanisms responsible for upper ocean marine CH_4 production, emission of CH_4 to the troposphere is strongly moderated by aerobic and anaerobic microbial oxidation (Boetius et al. 2000). Aerobic CH_4 oxidation occurs in oxygenated water columns and oxic sediment pore waters:

$$CH_4 + 2O_2 \rightarrow CO_2 + 2H_2O \quad (3.13)$$

Anaerobic oxidation of methane (AOM) occurs both in anoxic sediment pore waters and in anoxic water columns and is believed to involve consortia of archaea and SRB:

$$CH_4 + SO_4^{2-} \rightarrow HCO_3^- + HS^- + H_2O \quad (3.14)$$

In sediment pore waters the SO_4^{2-}–CH_4 transition constrains upward CH_4 diffusion and leads to pore water CH_4 profiles having a concave-upward shape (Blair and Aller 1995).

3.4.3 Global Oceanic Distribution of Methane

Although CH_4 concentrations in the open ocean are generally rather low (a few nM), net mixed layer CH_4 production means that O_2-saturated near-surface waters are generally also supersaturated in CH_4, with typical values of 130–160 % and maxima near the base of the mixed layer (Oudot et al. 2002; Forster et al. 2009). Considerably higher mixed layer CH_4 supersaturation is, however, not uncommon. Figure 3.14 illustrates this for the upper 300 m of the water column on north–south Atlantic Ocean transects. In addition to upper ocean CH_4 production, lateral supply from continental margins (Sect. 3.4.4) has also been invoked to explain high mixed layer CH_4 concentrations (Reeburgh 2007).

Similar to N_2O, elevated CH_4 levels are also found in intermediate waters (~500–1,000 m depth) of the three major open ocean OMZs: the Eastern Tropical North and South Pacific and the Arabian Sea (Naqvi et al. 2010), and in the upwelling zones associated with these OMZs (Sect. 3.3.5). In these regions seasonal upwelling of nutrient-rich waters fuels primary productivity and enhances the downward flux of biogenic particles (Rixen et al. 1996), leading to O_2 consumption in the intermediate waters and subsequent methanogenesis.

Below the ocean mixed layer and away from the OMZ's, CH_4 concentrations progressively decrease through oxidation, such that CH_4 concentrations may approach undetectable levels in the deep ocean basins (Upstill-Goddard et al. 1999; Yoshida et al. 2011). Occasionally this deep water CH_4 signal impacts surface waters through oceanic upwelling, as is illustrated by the bottom panel of Fig. 3.14 where the effect of equatorial upwelling is evident as far north as ~15°N. Overall, on an annual basis the open ocean CH_4 budget is considered to be in steady-state with in situ production and vertical transport balancing CH_4 oxidation and emissions to the troposphere.

3.4.4 Coastal Distribution of Methane

3.4.4.1 Coastal Sediments

The total mass of CH_4 in shallow marine sediments remains unquantified but it is nevertheless thought to be substantial, with methanogenesis considered likely in at least 30 % of the global continental shelf area (Judd

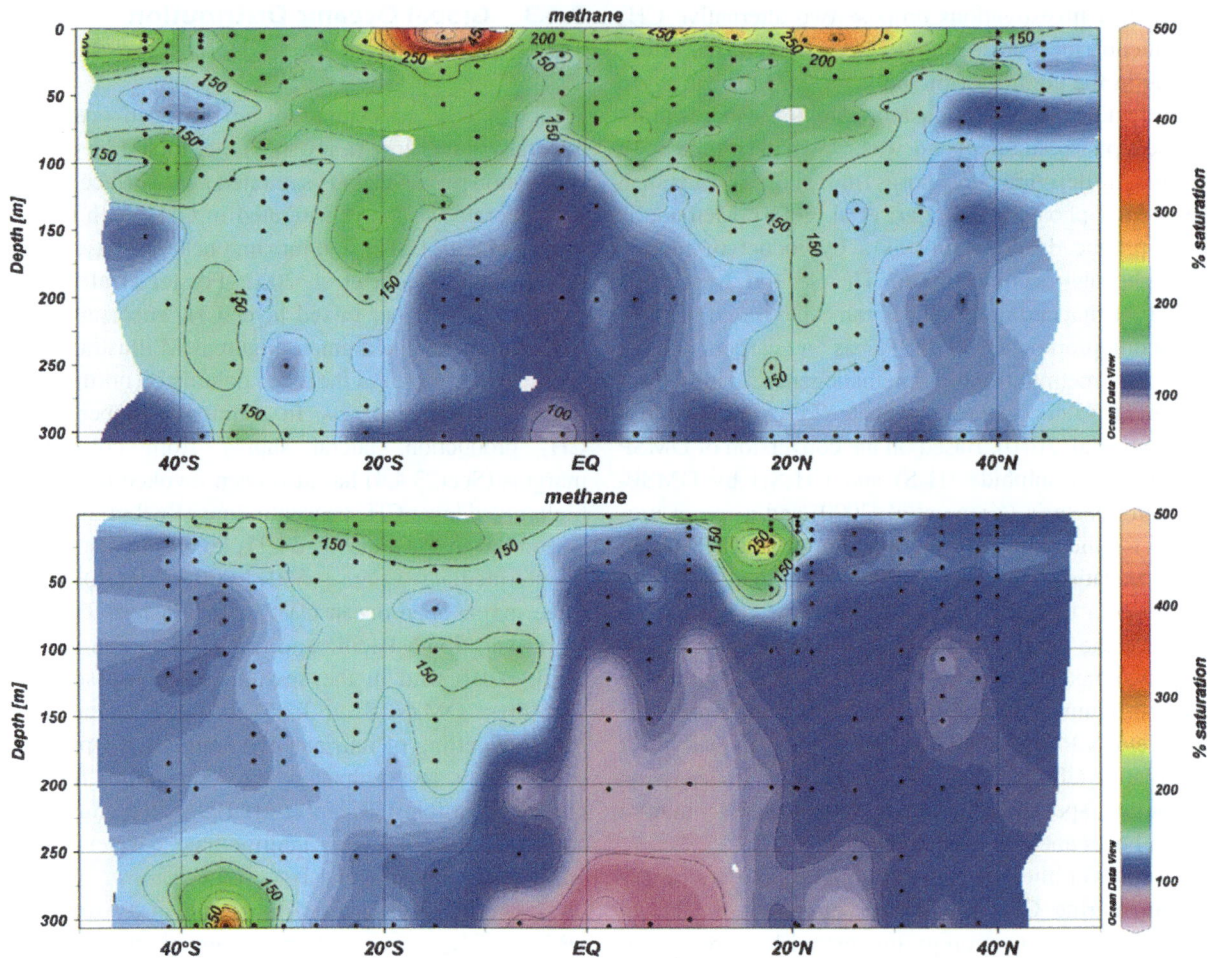

Fig. 3.14 CH_4 saturation in the upper 300 m of the Atlantic Ocean between 50°N and 52°S during 2003. *Top panel*: April–May; *bottom panel*: September–October. *Dots* represent water samples. (Reproduced from Forster et al. (2009) by permission of Elsevier)

and Hovland 2007), i.e. at least 7.4×10^6 km^2 (Laruelle et al. 2010). Although it is estimated that more than 90 % of the sediment CH_4 inventory may be consumed by AOM (Sect. 3.4.2) prior to sediment-water exchange (Dale et al. 2006), CH_4 emissions from individual sites can be very high per unit area (Middelburg et al. 1996; Abril and Iversen 2002) and can significantly impact the CH_4 signal in the overlying water. Figure 3.15 is an example from a UK estuary (Tyne) where the broad CH_4 maximum between salinities 5 and 20 may reflect inputs from anoxic intertidal mudflats in mid-estuary (Upstill-Goddard et al. 2000 and discussion below).

In some circumstances organic carbon burial may be sufficiently intense that the resulting rate of methanogenesis contributes to raising the total pore water gas concentration above the hydrostatic pressure in the sediment, with the result that gas bubbles are formed (Wever et al. 1998). It has been shown that bubbles start to form at CH_4 concentrations well below its solubility (~1 mM) and that these may contain ~40–100 % CH_4 (Chanton et al. 1989). This results in a rapid, episodic release of CH_4 enriched bubbles to the water column (ebullition) and potentially directly to the troposphere with minimal oxidation (Dimitrov 2002). CH_4 ebullition may typically exceed the diffusional sediment CH_4 flux by more than an order of magnitude (Ostrovsky 2003; Barnes et al. 2006; Nirmal Rajkumar et al. 2008). However, ebullition is notoriously difficult to quantify because spatial and temporal variability can confound attempts to accurately capture a representative sample. Perhaps

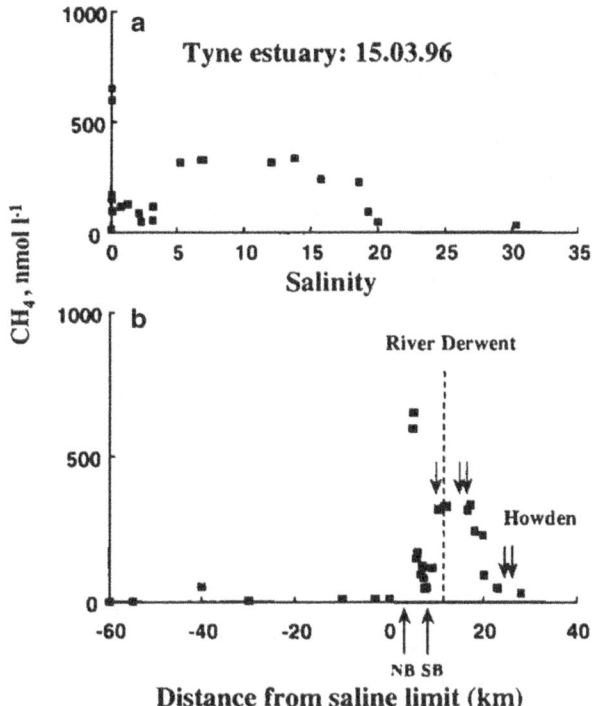

Fig. 3.15 Dissolved CH_4 in the Tyne estuary, UK. (**a**) CH_4 versus salinity; (**b**) CH_4 versus distance from the tidal limit (positive is downstream; negative is upstream). *Dotted line*: location of major tributary (Derwent); *arrows*: locations of additional freshwater discharges (Reproduced from Upstill-Goddard et al. (2000) by permission of the American Geophysical Union)

unsurprisingly, few studies have directly studied sediment CH_4 ebullition and hence the controlling processes have not been well quantified. Temperature is clearly important through solubility effects and its control of methanogenesis rates, as is water depth. In coastal waters ~3 m deep, minimal bubble dissolution was observed during migration to the air-sea surface (Martens and Klump 1980), but in deeper waters complete bubble dissolution may occur, before the bubbles reach the sea surface (Joyce and Jewell 2003). Contamination of CH_4-enriched bubbles by surfactants can also significantly reduce their rise velocities, thereby increasing the potential for dissolution (Leifer and Patro 2002). However, surfactants on the bubble surfaces will also decrease the rate of gas exchange between the bubbles and the water (e.g. Tsai and Liu 2003). Shakhova et al. (2010) reported bubbles of CH_4 entrapped in fast sea ice in the East Siberian Arctic Shelf, which they attributed to ebullition from underlying sediment. If sea ice acts to moderate the emission of CH_4 in this way, this has clear implications for Arctic CH_4 emissions as a consequence of sea ice retreat.

Vegetation strongly impacts the distribution and transport of CH_4 in coastal sediments. It has been proposed that plant-mediated CH_4 transport and CH_4 ebullition are mutually exclusive processes in tidal marshes (Van der Nat and Middelburg 1998, 2000). Vegetation impacts CH_4 concentrations in coastal sediments via the release of labile organic compounds that may stimulate methanogenesis. Vegetation also acts as a conduit for the transport of CH_4 to the troposphere and for transport of tropospheric O_2 to the rhizosphere favouring CH_4 oxidation, both transport pathways reducing CH_4 concentrations in near-surface sediments (Van der Nat and Middelburg 1998, 2000; Biswas et al. 2006). In addition, plants promote CH_4 oxidation at depth, where methanotrophs occur adjacent to or within macrophyte roots (Gerard and Chanton 1993; King 1994). Plant-mediated CH_4 transport is both passive via molecular diffusion and active via convective flow due to pressure gradients and is maximal during daylight hours in the growing season (Van der Nat et al. 1998; Van der Nat and Middelburg 2000). In two Tanzanian mangrove systems sediment-to-air CH_4 fluxes were enhanced up to fivefold in the presence of pneumatophores (above ground root systems) and it was estimated that transport via this pathway accounted for 38–64 % of the total sediment CH_4 source to air at low tide (Kristensen et al. 2008). In some situations both diffusive CH_4 exchange and ebullition can be enhanced by "tidal pumping" related to falling hydrostatic pressure (Sect. 3.3.4) (Barnes et al. 2006), with pressure changes of only a few percent having a large effect (Ostrovsky 2003). Figure 3.12 shows an inverse relation of CH_4 (and N_2O) concentrations with tidal height in a tropical mangrove creek. This indicates tidal pumping with CH_4-rich sediment pore waters seeping into the overlying creek waters at low tide, but remaining in the sediment as the hydrostatic pressure rises again. Barnes et al. (2006) suggested that tidal pumping is a major control of CH_4 and N_2O emissions from mangrove systems (Sect. 3.3.4).

Ancient microbial, thermogenic, and abiogenic CH_4 in coastal shelf sediments can become "geologically focussed", which may result in episodic CH_4 ebullition on a potentially large scale (Judd and Hovland 2007). The ephemeral nature of these "seep fluxes" is well illustrated by observations at a seep site in the outer Firth of Forth (North Sea). Dissolved CH_4 concentrations in the water column strongly increased towards the shallow seabed (~1,500 % CH_4 saturation at 90 m), but 1 year later the dissolved CH_4 concentration was only mildly supersaturated (Upstill-Goddard 2011). Shallow seeps clearly influence surface water CH_4 concentrations (Damm et al. 2005; Schmale et al. 2005). The North Sea has numerous well-documented "pock marks" (Dando et al. 1991); evidence for significant past CH_4 seepage.

3.4.4.2 Coastal Waters

Inland waters (lakes, reservoirs, streams, and rivers) can be substantial CH_4 sources to the coastal zone, although they are currently not well integrated in global greenhouse gas budgets (Table 3.3) (Bastviken et al. 2011). Most rivers studied to date are highly CH_4 supersaturated, including pristine, well-oxygenated regimes with minimal sediment cover or anthropogenic disturbance (Upstill-Goddard et al. 2000). Large CH_4 inputs from adjacent forest and/or agricultural soils have been suggested (Devol et al. 1990; Yavitt and Fahey 1991). The CH_4 concentration in river water is a complex function of catchment hydrology, vegetation cover, microbial activity, and re-aeration rates. Upstill-Goddard et al. (2000) found a general decrease in dissolved CH_4 with increasing river discharge. Their compilation of published CH_4 saturations in rivers worldwide revealed a typical range of ~1,000–40,000 %, with one extreme value for an organic-rich Amazon tributary exceeding 400,000 % (Upstill-Goddard et al. 2000). Data from 474 freshwater ecosystems point to a major CH_4 source in inland waters (Bastviken et al. 2011). Part of this contribution is almost certainly included in the source estimate for global wetlands (Table 3.3). The CH_4 source from inland waters should perhaps be formally specified as an important component in the global tropospheric CH_4 budget.

Most estuarine CH_4 data are for temperate systems. Methane saturations of up to 8,000 % have been reported for some shallow coastal embayments (Ferrón et al. 2007; Kitidis et al. 2007). A CH_4 saturation of up to 20,000 % may be typical of the mid- to upper-inner estuary (Upstill-Goddard et al. 2000; Abril and Iversen 2002; Middelburg et al. 2002). As much as 158,000 % CH_4 saturation has been reported in the Sado estuary, Portugal (Middelburg et al. 2002). The highest value recorded exceeds 3,000,000 % for the small, polluted subtropical Aydar estuary in SE India (Nirmal Rajkumar et al. 2008). This may be considered exceptional, resulting from a high organic carbon input in fresh domestic organic wastes and intense sediment methanogenesis. Indeed, methanogenesis rates were estimated to be close to maximal for the ambient temperature (Nirmal Rajkumar et al. 2008).

An important aspect of CH_4 cycling in many well-mixed macrotidal estuaries is strong tidal asymmetry with the velocity of the flood tide exceeding that of the ebb tide. This gives rise to the net transport of suspended particles upstream and retains river borne particles in a well defined turbidity maximum zone at low salinity (Uncles and Stephens 1993). High microorganism numbers associated with the suspended particles (Plummer et al. 1987) and long particle lifetimes promote enhanced biogeochemical cycling in such regions and CH_4 concentrations can significantly exceed those in the input rivers (Fig. 3.15) (Upstill-Goddard et al. 2000). While this could result from active CH_4 release from the underlying sediments during particle resuspension (as discussed above), it could also reflect in situ water column production of CH_4 through attachment of methanogenic archaea to tidally suspended particles in the turbid, O_2–poor waters, analogous to the "anoxic microniche" hypothesis in the open ocean (Barnes and Upstill-Goddard 2011). By contrast, Abril et al. (2007) found significant CH_4 oxidation in laboratory sediment suspensions, although at higher turbidity than measured in situ by Upstill-Goddard et al. (2000). The conflicting results may reflect competition between methanogenesis and oxidation, implying complexity in the relationship between CH_4 concentrations and estuarine turbidity.

Mangrove ecosystems are significant contributors to the marine source of tropospheric CH_4 (Fig. 3.12; Table 3.7). Typically one half of mangrove net primary production is retained within the system (Dittmar et al. 2006; Bouillon et al. 2008) and this carbon is buried and/or recycled, resulting in significant CH_4 production in heterotrophic sediments and in the overlying water. The mangrove CH_4 source appears rather constant throughout the year, possibly as a result of small annual temperature excursions in mangrove systems that give minimal variability in methanogenesis rates (Barnes et al. 2006; Ramesh et al. 2007). CH_4 saturations in waters surrounding mangroves are spatially and temporally variable. Up to ~30,000 % saturation was observed in mangrove creek waters (Barnes et al. 2006), but regionally ~2,000–3,000 % might be more typical (Biswas et al. 2007).

Coastal shelf seas are almost always supersaturated in CH_4. Data for coastal shelf seas tend to be restricted to the temperate northern hemisphere with hardly any data available for coastal waters along much of the Russian Arctic, South America, East and West Africa and Antarctica. Bange (2006b) provides the most extensive regional data compilation to date, for a European shelf area estimated at ~3 × 10^6 km^2 or ~12 % of the global shelf sea area (Table 3.8). The variability between sites may partly reflect seasonal variation, but it is not straightforward to assess the extent of this seasonality due to a lack of seasonal measurements at individual locations.

Marine sediments are a major CH_4 source to coastal waters, in spite of the inhibition of methanogenesis by SRB and CH_4 oxidation by O_2 and SO_4^{2-}. Indeed, substantial accumulation of dissolved CH_4 in well-oxygenated bottom waters overlying organic-rich coastal sediments is well documented (Martens and Klump 1980). Order of magnitude higher concentrations have been observed in anoxic bottom waters overlying sediments with moderate organic carbon content in the Arabian Sea (Jayakumar et al. 2001). Production of CH_4 in sediments and its supply to bottom waters is primarily linked to biological productivity in the overlying waters, with hypoxia in the water exerting a secondary effect (Bange et al. 2010a). Most regions with high productivity are associated with coastal upwellings, the extent of which is set by Laruelle et al. (2010) at ~2.3 × 10^6 km^2. While this is only ~0.6 % of the total ocean area, these regions are "hot spots" of CH_4 emissions to the troposphere, with typical surface CH_4 saturations of ~150–250 % (Naqvi et al. 2005; Kock et al. 2008). Maximal water column CH_4 accumulation occurs in sulphidic deep waters within enclosed basins, such as the Black Sea and the Cariaco Basin, but this CH_4 is largely of geological origin (Kessler et al. 2005). By contrast, over the highly productive Namibian shelf, where sulphidic bottom water is also characteristic, the large CH_4 emission from underlying organic-rich sediments is a consequence of contemporary methanogenesis (Brüchert et al. 2009). In coastal zones, where the hypoxic conditions are caused by anthropogenic activities, the CH_4 distribution is highly variable. For example, maximum CH_4 concentrations in the Gulf of Mexico (Kelley 2003) are ~15 times higher than in the Changjiang Estuary and the East China Sea (Zhang et al. 2008).

3.4.4.3 Methane Hydrates

Methane hydrate is a quasi-stable solid, resembling ice, in which CH_4 molecules are trapped within the crystalline structure of water. Hydrate stability decreases with increasing temperature and decreasing pressure (Kvenvolden 1993). Hydrate occurs extensively in buried sediments and seabed outcrops along continental margins where water depths exceed 500 m (Beauchamp 2004) and along gravitationally unstable regions of the continental slope (Fig. 3.16). The distribution and extent of stable CH_4 hydrate can be predicted from in situ temperature and pressure, such that a theoretical gas hydrate stability zone (GHSZ) can be defined. Figure 3.17 is a schematic of the GHSZs in sediments at shallow and deep marine sites. The CH_4 source may be biogenic, volcanic, hydrothermal or thermogenic. Biogenic CH_4 from sediment methanogenesis tends to dominate (Sloan 2003).

Methane hydrates can be categorised into two broad types: structural and stratigraphic. The formation of structural hydrates involves migration of CH_4 along

Table 3.7 CH_4 emissions from marine waters

	Area (10^6 km^2)	CH_4 emission (Tg C year^{-1})	Reference
Open ocean	334	0.3	Bates et al. (1996b)
Coastal upwellings	2.3	0.02–0.15	Naqvi et al. (2005), Kock et al. (2008)
Continental shelves	24.7	0.38–7.3	EPA (2010)
Estuaries	1.0	0.08–2.3	Upstill-Goddard (2011)
Mangroves	0.2	1.7	Barnes et al. (2006)
Continental margin seeps	?	7.5–36	Hornafius et al. (1999), Kvenvolden and Rogers (2005)
Total	362.2	10–48	

Table 3.8 CH_4 saturations in European shelf surface waters (excluding estuaries). *SD* standard deviation (Modified from Bange (2006b))

Region	Date	CH_4 saturation (%) (range or SD)	Reference
Barents Sea	Aug 1991	120 (115–125)	Lammers et al. (1995)
Baltic Sea	1992[a]	254 (113–395)	Bange et al. (1994)
Southern N. Sea	Nov 1980	140	Conrad and Seiler (1988)
	Aug 1993	338 (118–701)	Upstill-Goddard et al. (2000)
German Bight	Sep 1991	126 ± 8	Bange et al. (1994)
Southern Bight	Mar 1989	113 (95–130)	Scranton and McShane (1991)
Central N. Sea	May 1994	215 (120–332)	Rehder et al. (1998)
UK East coast	1995–1999	129 (112–136)	Upstill-Goddard et al. (2000)
Bay of Biscay	Nov 1980	100	Conrad and Seiler (1988)
Adriatic Sea	Aug 1996	425 (420–450)[a]	Leip (1999)
E. Ionian Sea	Jul 1993	148 ± 22	Bange et al. (1996)
N. Aegean Sea	Jul 1993	231 ± 32	Bange et al. (1996)
NW Black Sea	Jul 1995	567	Amouroux et al. (2002)
Average		222 ± 142	

[a]Includes seasonal/interannual sampling

geological faults from deep sources and subsequent crystallisation on seawater contact (O'Connor et al. 2010). This can lead to accumulation of high CH_4 concentrations in domes or underneath impermeable sediments (Archer 2007). Structural CH_4 hydrates occur at relatively shallow depths and are typically "massive", i.e. they displace sediment to generate large hydrate chunks potentially filling tens of percent of the sediment volume (Tréhua et al. 2004). However, the majority of hydrate deposits are stratigraphic. These deposits are typically dilute, accounting for only a few percent of the sediment volume and are generally located some hundreds of metres below the sea floor. Gornitz and Fung (1994) further drew a distinction between marine hydrate synthesis in "passive" and "active" margins. Local sediment accumulation constrains hydrate formation in passive margins, whereas scavenging of organics deriving from adjacent areas leads to higher hydrate abundance in active margins. Hydrate formation is also sensitive to the O_2 concentration in the overlying water: a 40 μM decrease in the deep water O_2 concentration may enhance the CH_4 inventory twofold (Buffett and Archer 2004).

Knowledge of the true extent of marine CH_4 hydrate remains incomplete. Direct observations from well logs (e.g. Sloan and Koh 2008) are limited in number. Interpretations based on sea floor organic carbon derived from sea surface chlorophyll *a* can be unsuccessful (Gornitz and Fung 1994) and indirect hydrate detection via seismic profiling may prove similarly inconclusive (Sloan 2003). The perceived CH_4 hydrate inventory has been downscaled by 3–4 orders of magnitude since the 1970s due to improvements in hydrate understanding and detection. An early, widely adopted global estimate of ~10^4 Pg C in CH_4 hydrate, including a small contribution from terrestrial permafrost (Kvenvolden 1999), was more than twice the known

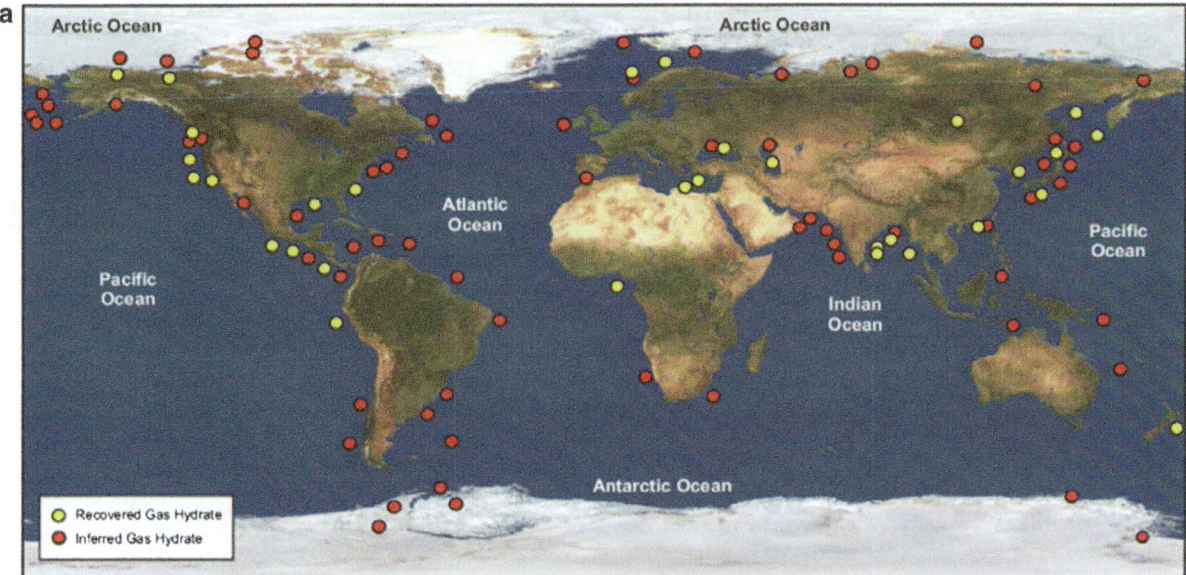
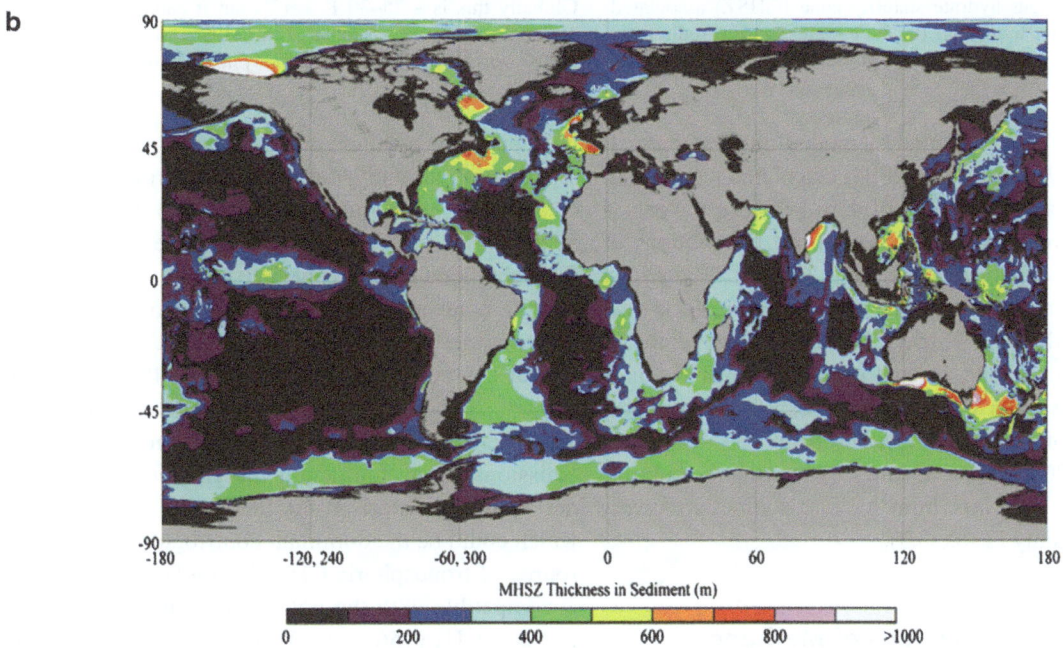

Fig. 3.16 (a) Distribution of known and inferred methane hydrate accumulations; (b) estimated thickness of the gas hydrate stability zone (GHSZ) in seafloor sediments (Reproduced from Krey et al. (2009) by permission of IOP publishing Ltd.)

fossil fuel carbon inventory. The most recent data synthesis gives a much lower estimate (170–1,000 Pg C) for marine CH_4 hydrate and a similar amount in associated free CH_4 bubbles (Archer 2007; Archer et al. 2009). O'Connor et al. (2010) set an order of magnitude uncertainty on these figures.

In excess of 250 CH_4 bubble plumes were recently observed over the GHSZ west of Spitsbergen, an area showing ~1 °C warming of bottom waters during the last 30 years (Westbrook et al. 2009). These bubble plumes were interpreted as upward migrating free gas formerly trapped below the GHSZ, implying a strong link between current warming and hydrate dissociation,

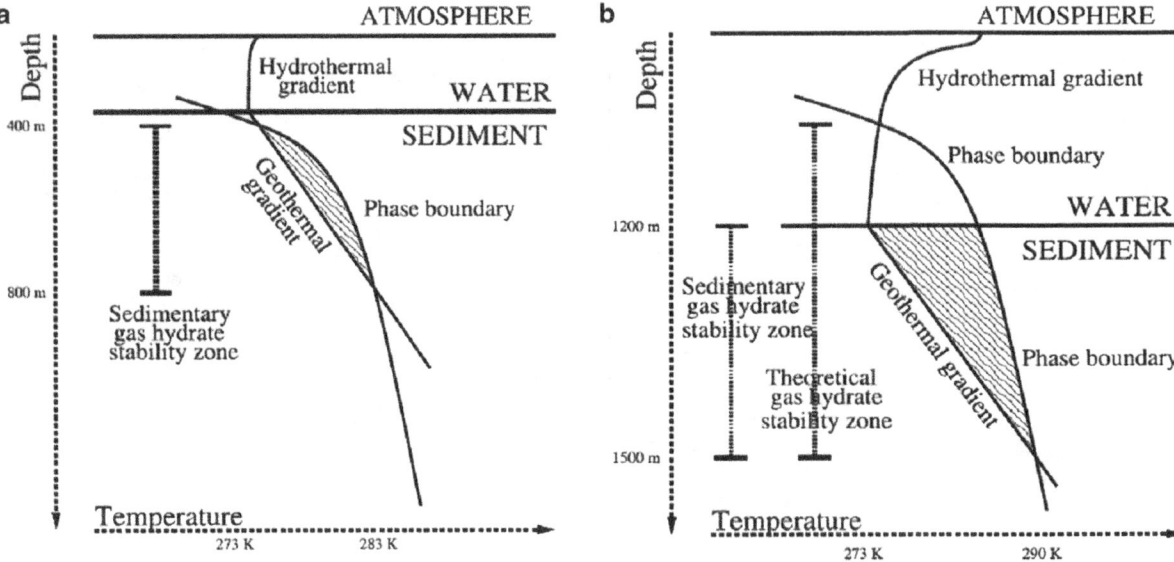

Fig. 3.17 The gas hydrate stability zone (GHSZ) associated with (**a**) shallow water and (**b**) deep water marine sediments. GHSZ volume is determined by the local geothermal gradient. Globally this is ~ 25–30 K km^{-1}, but it can show significant regional variability. (Adapted from O'Connor et al. (2010) by permission of the American Geophysical Union)

a conclusion later supported by modelling (Reagan and Moridis 2009). Lamarque (2008) calculated a potential CH$_4$ release at the sea floor of 420–1,605 Tg C year^{-1} following hydrate destabilisation from a doubling of tropospheric CO$_2$. Adjusting for 1 % CH$_4$ leakage as observed at a large field site (Mau et al. 2007), this value was reduced to 4–16 Tg C year^{-1}. However these estimates do not take account of subsequent CH$_4$ oxidation in the water column, which can easily account for 90 % of the CH$_4$ released from the sea floor (Dale et al. 2006). Consequently, estimating the current tropospheric CH$_4$ flux from hydrate sources involves large uncertainty (Table 3.3).

3.4.5 Marine Emissions of Methane

Marine CH$_4$ emissions reflect the balance of rates of formation, removal and transport. Table 3.7 emphasises the dominance of coastal over open ocean waters in the marine CH$_4$ budget, reinforcing the conclusions of an early synthesis in which more than 75 % of marine CH$_4$ emissions were ascribed to coastal waters (Bange et al. 1994). The total CH$_4$ emission from open ocean areas, experiencing O$_2$ depletion in the water column, is rather small (0.3 Tg C year^{-1}) (Bates et al. 1996b). Similarly, in coastal zones where hypoxia is anthropogenically influenced, the available information suggests that these regions are currently only minor contributors to total coastal CH$_4$ emissions (less than 0.026 Tg C year^{-1}), although this estimate does not include a contribution from ebullition (Naqvi et al. 2010).

It is noteworthy that the low end of the range of total marine CH$_4$ emissions in Table 3.7 is at the high end of the range given in the IPCC 4th Assessment Report, as shown in Table 3.3 (Denman et al. 2007). The discrepancy can be largely accounted for by CH$_4$ emissions from continental margin seeps, which are not well characterised in the IPCC synthesis, but which could be the dominant contributor to the marine source of tropospheric CH$_4$. Such a large seep source is compatible with the notion of a "missing" fraction of fossil CH$_4$ (~56 ± 11 Tg C year^{-1}) deduced from tropospheric ^{14}C data (Crutzen 1991). The uncertainty in the seep estimate is compounded by the episodic nature of the seeps and by the migration of source regions on the sea floor (Kvenvolden and Rogers 2005). There is also large uncertainty in the spectrum of bubble sizes emitted and dissolution of the gas phase into seawater (Judd and Hovland 2007). Moreover, most current estimates of seep CH$_4$ emissions are indirect and involve large extrapolations. Overall, CH$_4$ emissions in seeps have a large uncertainty.

Estimating CH_4 emissions from estuaries and mangrove sites (Table 3.7) is compounded by uncertainties over ebullition rates. The importance of this mechanism is unique to CH_4 among reactive trace gases, as it results from intense methanogenesis in sediments. Importantly, although ebullition can account for more than 90 % of CH_4 emissions at some locations (Ostrovsky 2003; Barnes et al. 2006; Nirmal Rajkumar et al. 2008), it is excluded from routine air-sea emission estimates based on gas exchange relations applied to dissolved gas gradients (Upstill-Goddard 2006). In tidal estuaries and mangrove settings some fraction of the CH_4 emission occurs directly from sediments during emersion. While some progress has been made towards estimating the global surface area of estuarine water bodies (Dürr et al. 2011), no robust estimate is currently available for the global inter-tidal area. Borges and Abril (2011) made a crude estimate of estuarine inter-tidal areas and derived a global estuarine CH_4 emission ~5 Tg C year^{-1}, notably higher than the estimate of 0.1–2.3 Tg C year^{-1} given in Upstill-Goddard (2011). Constraining marine CH_4 emissions more accurately clearly requires additional detailed studies.

3.5 Impact of Global Change

3.5.1 Future Changes in the Physics of the Oceanic Surface Layer

In order to assess the influence of global climate change on the air-sea exchange of long-lived greenhouse gases (Table 3.2), we need to separately discuss CO_2 from N_2O and CH_4, since the latter two gases have a fundamentally different global balance between the ocean and troposphere. In the case of CO_2, the ocean ultimately controls the tropospheric content of CO_2, with the CO_2 concentration in the surface layer being the immediate determinant of this ocean–atmosphere balance. This surface ocean concentration is controlled by physical, chemical, and biological processes that also create very important sources and sinks within the surface layer, causing the response of CO_2 to climate change to be complex. For N_2O and CH_4, the ocean acts as a net source of these gases to the troposphere, but their ultimate concentrations in the troposphere are on average controlled by other factors, such as their tropospheric lifetimes and terrestrial sources. Furthermore, excluding coastal regions, N_2O and CH_4 in marine waters are mostly produced away from the surface in the ocean's interior, so that the surface layer primarily acts as a conduit between the net ocean source and the net tropospheric sink.

3.5.1.1 Carbon Dioxide in the Open Ocean

For CO_2, it is furthermore of considerable help to distinguish clearly between the climate change processes acting upon the oceanic uptake of anthropogenic CO_2, and those that influence the cycling of natural carbon (Fig. 3.18). For the former, it is essentially sufficient to consider only CO_2 induced changes in the oceanic buffer capacity and in what way ocean circulation will change the net downward transport of anthropogenic CO_2. Changes in wind regimes are essentially irrelevant because air-sea exchange is not a rate limiting step for the oceanic uptake of anthropogenic CO_2 (Sarmiento et al. 1992). Changes in sea ice will locally impact the uptake of anthropogenic CO_2 significantly, but have a relatively small effect globally. Changes in temperature also have a negligible direct effect on the uptake of anthropogenic CO_2, since the buffer factor is essentially independent of temperature (Sarmiento and Gruber 2002).

Regarding the cycling of natural CO_2, changes in temperature are very important, as they directly impact CO_2 solubility in surface water. Changes in the ocean's biogeochemical loop are of fundamental importance. This loop (Sect. 3.2.2) consists of the downward (biological) component, often referred to as the biological pump, and an upward component driven by transport and mixing, which brings carbon-rich waters from the sub-surface ocean back to the surface (Gruber and Sarmiento 2002).

One of the most consistently predicted impacts of climate change on the ocean is an increase in upper ocean stratification (e.g. Bopp et al. 2002). This will largely result from continued oceanic uptake of excess heat from the troposphere that will warm the upper ocean more than the deep ocean (at least during a transient period of several hundred years). In addition, many high-latitude regions are predicted to become fresher in response to an acceleration of the hydrological cycle (Curry et al. 2003), increasing stratification there as well. In the lower latitudes, the enhanced evaporation will actually increase salinity, but this effect is much smaller than that of the higher temperature, so that the surface ocean is predicted to become

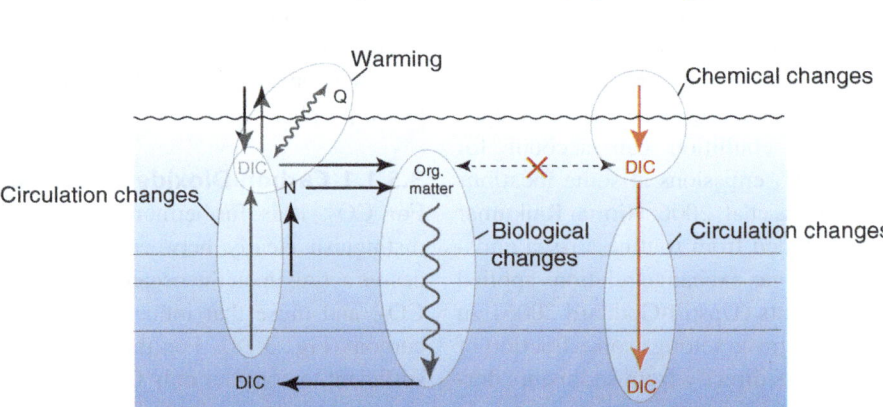

Fig. 3.18 Overview of the most important processes that affect the natural carbon cycle (*left*, *black*) and anthropogenic CO_2 uptake (*right*, *red*) in a changing world. At the center of the natural carbon cycle is the net formation of organic carbon by photosynthesis. Part of this organic matter is exported to depth, where it remineralises back to dissolved inorganic carbon (DIC). Circulation and mixing close this loop. Gas exchange through the air-sea interface connects this loop to the atmosphere. The magnitude of this exchange is governed by the balance of upward transport of DIC and downward transport of organic matter, as well as warming and cooling at the sea surface. The increase in atmospheric CO_2 has caused an additional flux across the air-sea interface, i.e. that of anthropogenic CO_2, leading to an anthropogenic increase in DIC (*red*). This DIC is then further transported to depth by mixing and transport. To first order, the anthropogenic CO_2 is not interacting with the natural carbon cycle, namely it does not affect the magnitude and pattern of the organic matter production and export (*red cross*). When looking closer, this is not entirely correct, as there is increasing evidence that the chemical changes associated with the uptake of anthropogenic CO_2 affect ocean biology. The *grey ellipses* indicate the processes that are most vulnerable to climate change, leading to a change in the net ocean–atmosphere balance of CO_2

more stably stratified nearly everywhere. Important exceptions might be coastal upwelling regions, and perhaps the Southern Ocean, where increased winds may cause enhanced upwelling, partially compensating for the increased stratification.

An increase in stratification and consequently in the surface-to-depth transport will have a strongly reducing impact on the uptake of anthropogenic CO_2. In contrast, increased stratification will produce a much more complex response of the ocean's natural carbon cycle. On the one hand, the circulation (upward) component of the biogeochemical loop will be reduced, but the net effect of this will depend on how ocean biology responds. If the net export of organic matter remains unchanged, the reduction in the upward transport will lead to a substantial increase in the rate of CO_2 uptake from the troposphere. This is because the soft-tissue part of the biological pump will continue to reduce surface CO_2 by fixing it into organic matter and exporting it to depth, while the re-supply by mixing from depth will be greatly reduced. Instead, the lost CO_2 will be replaced by uptake from the troposphere, constituting a negative feedback with regard to tropospheric CO_2. If the net export of organic matter increases, so will the uptake of CO_2 from the troposphere, but if it decreases then the net uptake of CO_2 from the troposphere will also decrease. The overall effect of expected changes in the biogeochemical loop on tropospheric CO_2 can be summarised by the efficiency of the soft-tissue pump, which describes the relative balance between the amount of (inorganic) carbon that is brought to the near-surface by mixing/transport and the amount of (organic) carbon that is exported to depth. It is currently expected that globally the efficiency of the soft-tissue pump will increase, such that the overall effect of the biogeochemical loop will be to enhance CO_2 uptake from the troposphere (Fig. 3.19). However, this conclusion critically depends on the biological response, which at present remains poorly understood.

At the same time, warming of the surface ocean will lead to a loss of CO_2 from the surface, constituting a clear positive feedback. The magnitude of this solubility-driven response is relatively well

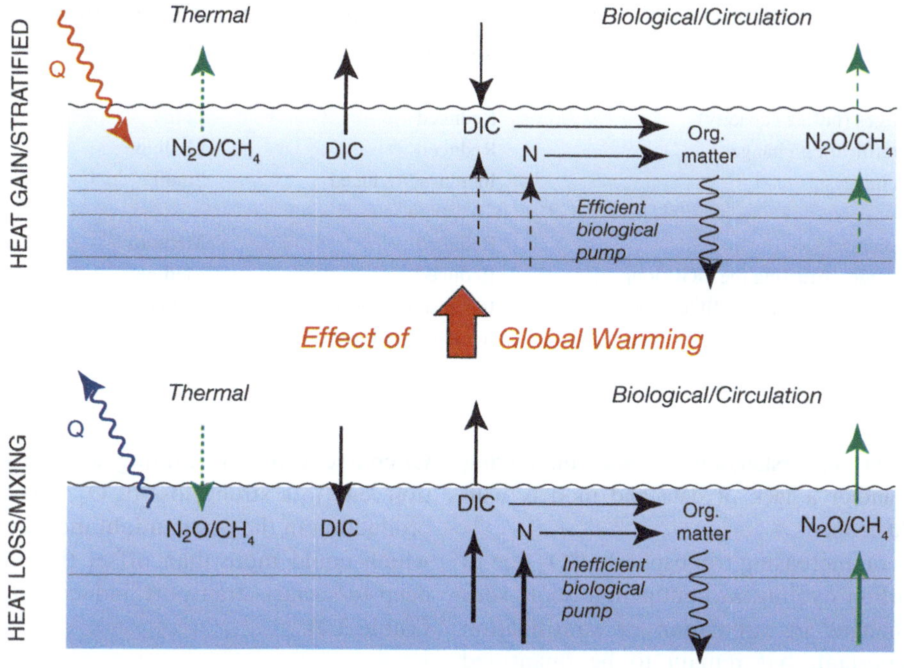

Fig. 3.19 Schematic diagram highlighting two typical situations and how they influence the exchange of natural CO_2, N_2O, and CH_4 across the air-sea interface. The *upper panel* shows a typical low-latitude situation characterised by a net heat gain from the atmosphere and stably stratified conditions. The *lower panel* represents a typical high latitude situation with a net loss of heat and weakly stratified conditions. Global warming will tend to make more regions of the global ocean behave like the *upper panel*

understood, and will largely depend on the magnitude of the ocean's future heat uptake. Changes in wind and sea ice extent will influence the net exchange of CO_2 across the air-sea interface as well. However this will be of secondary importance compared to the direct temperature and circulation effects, except in specific local regions where large changes in sea ice and/or winds might occur.

Based on an annual sea ice loss of 36,000 km^2 (mostly summer ice; Cavalieri et al. 2003), the air-sea CO_2 influx would increase by 2.0 ± 0.3 Tg C year^{-1} for the Arctic Ocean (Bates et al. 2006). Bates et al. (2006) estimated that the Arctic Ocean sink for CO_2 increased from 24 to 66 Tg year^{-1} over the past three decades due to sea ice retreat and that future sea ice melting will enhance this CO_2 uptake by 20 Tg year^{-1} by 2012. Therefore, if one considers only the physical processes through which sea ice directly impacts air-sea CO_2 exchange, the decrease of summertime sea ice extent is expected to increase CO_2 uptake in the coming decades.

Summing up, the expected changes in surface temperature and upper ocean stratification will lead to a strong reduction of the uptake of anthropogenic CO_2, and a mixed response on the natural carbon side with a loss of carbon driven by solubility, and perhaps a gain from the biogeochemical loop (Table 3.9). The net balance is likely to be a decrease in the net uptake of CO_2 from the troposphere in response to climate change, with current models suggesting a reduction in the net ocean uptake of about -16 to -33 Pg °C^{-1} warming until 2100 (Table 3.2) (Roy et al. 2011).

3.5.1.2 Carbon Dioxide in Coastal Seas

The potential feedbacks on increasing tropospheric CO_2 from changes in carbon flows in the coastal ocean could be disproportionately higher than in the open ocean. According to Borges (2011), the changes in carbon flows and related potential feedbacks in the coastal ocean could be driven by four main processes (Table 3.10): (i) changes in coastal physics (This section); (ii) changes in seawater carbonate chemistry (ocean acidification) (Sect. 3.5.2); (iii) changes in land use, waste water inputs and the use of agricultural fertilisers; (iv) changes in the hydrological cycle. These potential feedbacks remain largely unquantified

Table 3.9 Environmental influences on the future net ocean uptake of tropospheric CO_2. *BGC* biogeochemical

	Net ocean uptake	Magnitude	Understanding
Anthropogenic CO_2			
Chemical changes (buffer capacity)	Reduced	Large	High
Circulation/stratification changes	Reduced	Medium	Medium
Wind/sea ice changes	Increased/reduced	Small	Low
Natural CO_2			
Temperature changes	Reduced	Medium	Medium
Circulation changes affecting the BGC loop	Reduced	Medium	Low-medium
Biological changes affecting the BGC loop	Increased/reduced	Unknown	Very low
Wind/sea ice changes	Increased/reduced	Low	Low

due to a poor understanding of the underlying mechanisms, and/or a lack of data and models with which to evaluate them.

Feedbacks on increasing tropospheric CO_2 due to effects of carbon cycling in continental shelf seas related to changes in circulation or stratification could be important, but remain to be quantified (Table 3.10). The effect of changes of stratification on air-sea CO_2 fluxes are not straightforward to quantify since enhancement of stratification will depress primary production and export production due to the decrease of nutrient inputs, but at the same time will also decrease the vertical inputs of DIC and hence release of CO_2 to the troposphere. For example, on the Tasman shelf and in the adjacent open ocean the overall effect of enhanced stratification seems to be an enhancement of the CO_2 sink (Borges et al. 2008).

3.5.1.3 Nitrous Oxide and Methane

For N_2O and CH_4 the upper ocean acts mostly as a transport conduit from the ocean to the troposphere, where they are ultimately decomposed by chemical reactions. Changes in solubility are likely to exert a major effect on the exchange of these two gases in the open ocean, since their degree of supersaturation in the surface ocean is mostly determined by temperature and only to a smaller degree by the magnitude of the transport of supersaturated waters from below to the surface. Therefore, the increase in stratification induced by surface warming (and also by changes in salinity) is likely to lead to only a modest reduction in the ocean source strength of these two gases, provided that their rates of production remain about the same.

Of course, if ocean warming and enhanced stratification lead to a strong loss of O_2, then N_2O and CH_4 production in the ocean might increase substantially, which could more than offset the reduction in the deep-to-surface transport induced by stratification (Table 3.2).

3.5.2 Ocean Acidification

3.5.2.1 Carbon Dioxide

Models predict that by the end of the twenty-first century the pH of seawater will decrease by another 0.3–0.4 units (Caldeira and Wickett 2003). This decrease will be accompanied by a reduction in the CO_3^{2-} concentration and in the saturation state for calcite and aragonite (Sect. 3.2.2). Surface waters in the Southern Ocean and the Subarctic Pacific Ocean are expected to become undersaturated with respect to aragonite by the year 2100 (Orr et al. 2005).

Ocean acidification will affect organic carbon production and calcification, but exactly how it will do this is uncertain and is a topic of ongoing research in ocean acidification programmes around the world. For example, ocean acidification may impact the shells of pelagic calcifiers by reducing the ability of organisms to calcify (Comeau et al. 2011) and by dissolving calcareous shells (Orr et al. 2005). However, a few studies have found an increase in calcification on ocean acidification (e.g. Iglesias-Rodriguez et al. 2008). In particular, coral reefs are highly susceptible to increases in temperature and CO_2 concentration (Kleypas et al. 1999). Coral reefs globally may have ceased to grow and started to dissolve by the time the atmospheric mole fraction of

Table 3.10 Global change forcing of carbon cycling in the coastal ocean and associated feedback from increasing tropospheric CO_2 to the year 2100 (Adapted from Borges 2011)

Global change forcings	Feedback	Changes in CO_2 sources and sinks (Pg C year^{-1})	Comment
Changes in coastal physics			
Enhanced stratification	− ?	?	1
Enhanced coastal upwelling	+ ?	?	2
Impact of expanding OMZ in coastal upwelling regions	+	?	3
Enhancement of air-sea CO_2 fluxes related to Arctic sea-ice retreat	−	0.002	4
Changes in land use, waste water inputs, agricultural fertilisers and changes in the hydrological cycle			
Increase of river organic carbon delivery to the Arctic Ocean	+	?	
Increase of river nutrients delivery to the Arctic Ocean	−	?	
Global increase in river nutrient and organic matter delivery	−	0.200	5
Global increase in nutrient atmospheric deposition	−	?	6
Expanding hypoxic and anoxic zones	+	?	3
Changes in seawater carbonate chemistry (ocean acidification)			
Decrease of benthic calcification			
Coral reefs	−	0.015–0.026	
Other benthic environments	−	0.025–0.046	
Decrease of pelagic calcification			
Coccolithophorids	−	0.013–0.019	
Other pelagic calcifiers	−	?	
Dissolution of metastable $CaCO_3$ in sediment porewaters	−	0.022	7
Enhancement of primary production and export production due to increasing [CO_2]	−	0.108–0.216	8

1 Negative feedback only reported in Tasman shelf (Borges et al. 2008) assuming pCO_2 behaviour during warm years is representative of response to global warming, if extrapolated globally would produce a negative feed-back of the order of ~0.1 PgC year^{-1}.
2 Assuming exact opposite response of model output (Plattner et al. 2004) with decreasing upwelling favourable winds
3 Assuming enhanced denitrification leading to decreased primary production
4 Feedback computed for the next decade and not until 2100 (Bates et al. 2006)
5 The enhancement of primary production by nutrient inputs balances the additional CO_2 production by organic matter inputs (Mackenzie et al. 2004).
6 Not taking into account enhancement of acidification of surface waters by sulphur atmospheric deposition (Doney et al. 2007)
7 Based on Andersson et al. (2003)
8 Based on a single mesocosm experiment with mixed diatom and coccolithophorid assemblage (Riebesell et al. 2007)

CO_2 reaches 560 ppm (Silverman et al. 2009). Ocean acidification may well increase primary production by some phytoplankton species (Rost and Riebesell 2004) and may change ratios of carbon to nitrogen uptake by phytoplankton and subsequent carbon export (Riebesell et al. 2007).

Ocean acidification will also impact marine biogeochemical processes. These include changes in the availability of essential trace metals like iron, which are modified in speciation by pH change (Santana-Casiano et al. 2006). A decrease in pH may affect the solubility of some minerals (Liu and Millero 2002) and also the distribution of chemical species, favouring the free dissolved forms of metals, and exerting significant physiological, ecological and toxicological effects on organisms.

On timescales of a few 100 years, a change in the hard tissue $CaCO_3$ pump is estimated to have a small impact on global CO_2 uptake by the oceans (Table 3.2) (Denman et al. 2007), although the effect could be very substantial on glacial to interglacial timescales (e.g. Archer et al. 2000; Matsumoto et al. 2002). The effects of ocean acidification have been estimated for coastal waters (Table 3.10). Overall, ocean acidification is likely to decrease pelagic and benthic calcification in coastal waters and to result in the dissolution of $CaCO_3$

in coastal sediments. These combined effects have been estimated to increase the uptake of tropospheric CO_2 by ~0.02 Pg C year^{-1}, an amount equivalent to ~10 % of the modern-day CO_2 sink in coastal seas (Tables 3.2 and 3.10). The increase in export production upon ocean acidification could also provide a significant negative feedback to increasing tropospheric CO_2, although the conclusions are based on a single perturbation experiment (Table 3.10) (Riebesell et al. 2007).

Denman et al. (2007) concluded that the combined effects of climate change and ocean acidification on the biological carbon pump are not clear and could either increase or decrease the uptake of tropospheric CO_2. What is more clear is that ocean acidification will likely result in the dissolution of $CaCO_3$ sediments in the interior ocean on a time scale of 40 kyear, thus creating a negative feedback on the increase in tropospheric CO_2.

3.5.2.2 Nitrous Oxide and Methane

One other important consequence of ocean acidification is a shift of the NH_3/NH_4^+ equilibrium towards NH_4^+ (e.g. Bange 2008). Beman et al. (2011) recently showed that nitrification rates decreased significantly when the pH was lowered to values expected to occur in the future ocean. One explanation for the pH sensitivity of nitrification rates is that the ammonia mono-oxygenase enzyme uses NH_3 and not NH_4^+ as substrate in the first step of the nitrification sequence. On the basis of these results Beman et al. (2011) suggested that future oceanic N_2O production via nitrification might decrease by up to 44 %, although Freing et al. (2012) contend that this scenario could be an over simplification. Nitrification is part of organic matter remineralisation (i.e. oxidation of organic matter with O_2 to CO_2) and it leads to decreases in both pH and O_2 (Sect. 3.3.2). In general decreasing O_2 concentrations lead to increasing N_2O production during nitrification, such that there seems to be only a minor effect of decreasing pH on N_2O production via nitrification as part of the organic matter remineralisation process. Laboratory experiments to verify the effect of ocean acidification on N_2O production in the ocean are yet to be carried out (Table 3.2). Similarly, the potential effect of ocean acidification on oceanic CH_4 production remains to be investigated (Table 3.2).

3.5.3 Deoxygenation and Suboxia in the Open Ocean

One of the most important effects of global change on the oceans will be the deoxygenation of seawater arising from surface water warming and increased stratification of the upper ocean. These processes will lead to a decrease in O_2 solubility and its supply to subsurface waters, respectively (Keeling et al. 2010). There is compelling evidence to show that this may already be happening (Joos et al. 2003; Stramma et al. 2008; Keeling et al. 2010; Helm et al. 2011). The model-predicted decrease in the oceanic O_2 inventory ranges from 1 % to 7 % by the year 2100 (Keeling et al. 2010). The effects of such a decrease are expected to be greatest in the oceanic OMZs. For example, while a 1 °C warming of the upper ocean, which would lower the O_2 solubility by ~5 µM, may lead to an increase in the volume of hypoxic waters by 10 %, the volume of suboxic waters may increase by a factor of 3 (Deutsch et al. 2011). The ongoing expansion and intensification of the OMZs (Stramma et al. 2008) is expected to profoundly impact the biogeochemical cycling of redox-sensitive elements, especially nitrogen, and will, in conjunction with ocean acidification, adversely affect marine life (Brewer and Peltzer 2009). This may also involve modification of the oceanic source terms for climatically important gases that are sensitive to O_2 concentrations, such as N_2O and CH_4 (Sects. 3.3.2 and 3.4.2).

N_2O formation by nitrification is enhanced when O_2 concentrations are lowered (Sect. 3.3.2). Stramma et al. (2008) showed that intermediate ocean waters (300–700 m water depth) have been losing O_2 at rates ranging from 0.09 ± 0.21 µmol kg^{-1} year^{-1} in the eastern equatorial Indian Ocean to 0.34 ± 0.13 µmol kg^{-1} year^{-1} in the eastern tropical Atlantic Ocean during the last 50 years. Assuming a mean $\Delta[N_2O]/AOU$ (apparent oxygen utilisation) ratio of 10^{-4} (Walter et al. 2006), Bange et al. (2010) computed a maximum additional N_2O contribution from deoxygenation of 6 % above the mean N_2O background concentration in the intermediate waters of the tropical North Atlantic Ocean. It therefore seems reasonable to conclude that ongoing open ocean deoxygenation will have a minor effect on oceanic N_2O production and emissions (Table 3.2).

The total CH_4 emission from open ocean areas experiencing O_2 depletion in the water column is quite small (0.3 Tg C year^{-1}) (Table 3.7) (Bates et al. 1996b) and is unlikely to increase significantly due to the future expansion of open ocean OMZs (Table 3.2) (Naqvi et al. 2010).

3.5.4 Coastal Euthrophication and Hypoxia

Human activities related to the production of food and energy are causing the release of large quantities of nutrients, such as nitrogen and phosphorus, to the environment, a substantial fraction of which gets transported to coastal waters (Seitzinger et al. 2002; Smith et al. 2003). By 2100 changes in biological activity due to the increased nutrient delivery in rivers might cause a negative feedback on increasing tropospheric CO_2 similar in magnitude to the present-day CO_2 sink in coastal seas (0.2 Pg C year^{-1}) (Tables 3.2 and 3.10).

Stimulation of primary productivity and the degradation of photosynthesised organic matter due to nutrient over-enrichment (eutrophication) often results in dissolved O_2 depletion in the bottom waters of seasonally stratified shelf waters. Thus, over 400 hypoxic zones have developed in coastal areas all over the world in the last few decades (Diaz and Rosenberg 2008). Due to their severely detrimental ecological effects, such as the exclusion of higher animals, these hypoxic zones are often popularly referred to as "dead zones". Although these "dead zones" differ from the naturally-formed O_2-deficient zones that occur on the continental shelves in eastern-boundary upwelling regions, there is now evidence that the latter are also intensifying as a result of anthropogenic nutrient loading and/or changes in circulation (Naqvi et al. 2000; Chan et al. 2008).

The expansion and intensification of O_2 deficiency in coastal areas is expected to affect the future cycling of N_2O and CH_4 in these regions. Currently, anthropogenic "dead zones" make a relatively insignificant (less than 0.043 Tg N year^{-1}) contribution to the total marine source of tropospheric N_2O (Naqvi et al. 2010). Nevertheless, it is likely that the comparatively large N_2O emissions from natural hypoxic zones include an anthropogenically-enhanced component. Taken together with the sensitivity of N_2O cycling in aquatic systems to minor changes in already low O_2 concentrations, this implies that further expansion and intensification of coastal hypoxia may significantly impact the global tropospheric N_2O budget (Table 3.2).

From the limited information available on CH_4 emissions from "dead zones", it would appear that these regions do not contribute much (less than 0.026 Tg C year^{-1}) to the total marine source of tropospheric CH_4 (Naqvi et al. 2010), although it should be noted that this estimate does not include bubble ebullition. Given the lack of evidence for a primary control of bottom water hypoxia on sedimentary CH_4 production it would appear that future intensification and/or expansion of coastal hypoxia is unlikely to significantly increase total marine emissions of CH_4 to the troposphere (Table 3.2).

3.5.5 Changes in Methane Hydrates

There is substantial debate over the role of CH_4 hydrates in potential future climate change (O'Connor et al. 2010). What is clear is that the currently estimated CH_4 hydrate inventory (Sect. 3.4.4) is sufficiently large that the release of even a modest fraction to the troposphere over a 12-year period, the tropospheric lifetime of CH_4 (Table 3.1), could enhance greenhouse forcing by an amount equivalent to increasing the tropospheric CO_2 concentration by a factor larger than 10 (Archer 2007). The global CH_4 hydrate reservoir thus has the potential to promote substantial global warming (Table 3.2).

One region in which the effects of CH_4 hydrate destabilisation are likely to be most clearly manifested is the Arctic Ocean; a marine ecosystem that is highly susceptible to global change (Doney et al. 2012). Romanovski et al. (2005) modelled the extent and temporal evolution of submarine permafrost on the shelves of the Laptev Sea and East Siberian Sea and deduced that the entire Arctic shelf is underlain by relic permafrost stable enough to support CH_4 hydrate. Much of this Arctic CH_4 hydrate is comparatively shallow structural hydrate (Sect. 3.4.4) and the gas hydrate stability zone is ~200 m below the sea surface over much of the region. If dislodged from the sediment, for example during a submarine landslide (Brewer et al. 2002; Paull et al. 2003) or by erosion, large amounts of hydrate could potentially survive largely intact during ascent to the ocean surface. Contemporary hydrate melting is indeed apparent along the Siberian margin, due to rapid coastline recession that is exposing sub-sea floor deposits to overlying seawater at rates of ~10–15 km year^{-1}. The hydrate melting collapses further land into the sea, increasing the exposure of hydrates and leading to further melting, a process which is thought to have occurred continually over the past 7,500 years, resulting in ~100–500 km recession of the coastline (Hubberten and Romanovski 2001). As such these CH_4 releases are not abrupt, but rather tend to modestly increase the CH_4 background over time. For example, contemporary dissolved CH_4 saturations on the East-Siberian and Laptev Sea shelves are ~2,500 % in surface waters and in excess of 4,000 % in bottom waters, consistent with such seabed release (Shakhova and Semiletov 2007; Shakhova et al. 2007). Similarly, surface waters over the North Slope of Alaska are highly supersaturated in CH_4 (Kvenvolden 1999) and the release of hydrate-derived CH_4 has been clearly observed in the Beaufort Sea (Paull et al. 2008).

Notwithstanding such evidence for the ongoing release of hydrate-derived CH_4, Archer and Buffett (2005) and Archer (2007) argue that any significant climate induced hydrate melting response is likely to be on the time scale of millennia or longer, because the vast majority of CH_4 hydrate is of the stratigraphic type (Sect. 3.4.4) and is sufficiently insulated from the sediment surface by many hundreds of metres of overlying sediment. They also argue that any hydrate melting will occur below the gas hydrate stability zone, forming CH_4 bubbles whose fate is uncertain. Possibilities include their retention in the sediment, upward escape through the stability zone, or the initiation of submarine landslides through sediment column destabilisation. Although such a landslide would potentially cause abrupt CH_4 release, it is estimated that a landslide the size of the Storegga slide off Norway would typically release CH_4 sufficient only to affect climate on a scale comparable to a large volcanic eruption for ~10 years (Archer 2007).

Estimates of the rate at which melting hydrates are likely to increase the tropospheric CH_4 inventory over the timescale of decades (Table 3.2) are much less well constrained than changes in other CH_4 sources such as peat decomposition in thawing permafrost, fossil fuels and agriculture, although the potential rates may be comparable (Archer 2007). Major uncertainties exist over the rate and extent of CH_4 escape to the overlying water and troposphere, which is related to sediment stability and permeability and the ability of the gas hydrate stability zone to trap CH_4 bubbles (Archer 2007). On geologic timescales, due to the relative tropospheric lifetimes of CH_4 and CO_2 (Table 3.1), the largest climate impact will likely be from CO_2 deriving from CH_4 oxidation (Schmidt and Shindell 2003; Archer and Buffett 2005; Archer 2007). Following the cessation of hydrate CH_4 release, the enhanced CO_2 concentration will persist, while tropospheric CH_4 will recover relatively rapidly to a lower steady state (Schmidt and Shindell 2003). Significant oxidation of hydrate-derived CH_4 to CO_2 in the oceans would reduce the climate impact over several decades, but on timescales of millennia or more the climate impact might be significant, because of equilibration of this oceanic CO_2 with the troposphere over several hundred years. Indeed, there may well be positive climate feedback linking tropospheric CO_2, deep ocean temperature and CO_2 production from hydrate-derived CH_4 (Archer and Buffett 2005). These authors propose that in a worst case scenario, on the timescale of millennia to hundreds of millennia, the total global CH_4 hydrate source of tropospheric CO_2 could equal that from fossil fuels.

3.6 Key Uncertainties in the Air-Sea Transfer of CO_2, N_2O and CH_4

3.6.1 Outgassing of Riverine Carbon Inputs

The error bars on carbon inputs by rivers and estuarine outgassing create considerable uncertainty when attempting to convert from net contemporary fluxes to anthropogenic fluxes (Gruber et al. 2009; Takahashi et al. 2009). If the outgassing of river borne CO_2 by the open ocean has indeed been overestimated by ~0.2 Pg C year^{-1} as argued in Sect. 3.2.4, the anthropogenic CO_2 sink derived from CO_2 climatologies (e.g. Takahashi et al. 2009) would have been over-estimated by the same amount.

3.6.2 Heterogeneity in Coastal Systems

Obtaining meaningful air-sea gas transfer estimates for CO_2, CH_4 and N_2O in coastal systems is a substantial challenge and the ranges in Tables 3.4, 3.5 and 3.7 have large inherent uncertainties. These uncertainties reflect the heterogeneity and biogeochemical complexity of coastal systems and include: (i) gross scaling errors arising from the degree to which study sites are representative globally of each "compartment"; (ii) bias in the CO_2, N_2O and CH_4 values reflecting incomplete spatial and temporal data resolution; (iii) uncertainties in gas exchange rates arising from selected gas transfer relations and representative wind speeds (Chap. 2) (Upstill-Goddard 2006; Wanninkhof et al. 2009). Minimising gross scaling errors largely relies on the availability of accurate area determinations. General problems of defining the extent of oceanic upwellings are well documented (Nevison et al. 2004; Naqvi et al. 2005), as are the difficulties of defining representative estuarine areas (Barnes and Upstill-Goddard 2011). Minimising measurement bias has a clear seasonal aspect; production and consumption of CO_2, N_2O and CH_4 are biologically driven and as such have strong temperature dependence. Seasonality also affects the intensity of upwelling and river runoff, which affects nutrient and carbon supply. Unfortunately, most sampling campaigns take place during summer. There is also a regional aspect; most coastal regions outside the northern hemisphere are either undersampled or are not sampled at all.

Given the likelihood that coastal regions will play an important role in future trace gas budgets due to increased economic and population pressures, additional studies of CO_2, N_2O and CH_4 fluxes from key coastal regions will be required. Sampling campaigns should be spatially and temporally focused and ideally coordinated internationally. In particular CH_4 emissions by ebullition and from seeps and inland waters need more to be more accurately quantified.

3.6.3 Sea Ice

The role of sea ice in air-sea gas exchange remains poorly understood (Chap. 2). Until recently sea ice was regarded as a lid that effectively precluded air-sea gas exchange, as evidenced by under ice concentrations of dissolved gases (CFCs, O_2, CO_2) far from equilibrium with their tropospheric contents (Gordon et al. 1984; Weiss et al. 1992; Klatt et al. 2002; Bakker et al. 2008). However, recent evidence now points to significant gas exchange between sea ice and the troposphere (Delille et al. 2007; Geilfus et al. 2012), highlighting a need for more detailed research in this area.

3.6.4 Parameterising Air-Sea Gas Transfer

The choice of turbulence-driven air-sea gas exchange relations may introduce significant bias (Chap. 2), especially where bottom-driven turbulence is a major contributor to gas exchange in shallow systems (Upstill-Goddard 2006, 2011). In addition, wind speeds in coastal systems have short spatio-temporal variability and often the appropriate wind speed distributions required for gas exchange relations are not available. Similarly, current speeds and water depths required for non-wind speed driven gas exchange relations may also be lacking (Upstill-Goddard 2006).

3.6.5 Data Collection, Data Quality and Data Synthesis

The detailed study of interannual and decadal variations and trends in regional surface water fCO_2 and air-sea CO_2 fluxes has only recently become possible. Our future understanding of these trends and of the underlying mechanisms responsible is expected to improve due to more extensive data coverage and as longer observational records become available. Future long term data collection and data synthesis will require the development of instrumentation that is more reliable and accurate while being less labour intensive. In addition the rigorous standardisation of data collection and quality control procedures will be essential. Such technical developments will require substantial and sustained funding (Borges et al. 2010; Byrne et al. 2010; Feely et al. 2010; Gruber et al. 2010; Monteiro et al. 2010).

Encouragingly, the marine CO_2 community is already making good progress towards coordinated data collection and synthesis. Notable is the agreed use of certified reference materials for the analysis of DIC and total alkalinity and traceable calibration gases for fCO_2 analysis (DOE 1994; Dickson et al. 2007). There is also agreement on recommendations for reporting fCO_2 measurements (IOCCP 2004). The International Ocean Carbon Coordination Project (IOCCP) plays the key role in coordinating the marine carbon community and in ensuring the implementation of international agreements. By contrast, the status of marine N_2O and CH_4 research is far less mature in this regard. The N_2O and CH_4 communities have yet to progress towards discussing the adoption of either internationally agreed analytical standards or recommended analytical protocols.

Recent synthesis products for ocean carbon enable the intercomparison of model-data, the analysis of variation and trends in CO_2 air-sea fluxes, and the processes driving these. A vast and expanding global surface water fCO_2 database is now available for assessing CO_2 air-sea climatologies (Takahashi et al. 1997, 2002, 2009) and since 2011 the Surface Ocean CO_2 Atlas (SOCAT; www.socat.info) enables public access to a large CO_2 data archive for the global oceans and coastal seas (Bakker et al. 2012; Pfeil et al. 2013; Sabine et al. 2013) (Chap. 5). Other data synthesis efforts are also underway, notably MEMENTO for surface ocean N_2O and CH_4 concentrations (Chap. 5) and syntheses of carbon in the ocean interior: GLODAP (Global Ocean Data Analysis Project), CARINA (CARbon IN the Atlantic Ocean), PACIFICA (PACIFic ocean Interior Carbon) and GLODAP-2 (Key et al. 2004; Bange et al. 2009; Tanhua et al. 2010; Suzuki et al. 2013). Such data synthesis efforts, along with modelling and data-model intercomparisons (Gruber et al. 2009) are critical to improving our current understanding of the exchanges of CO_2, N_2O and CH_4 between the troposphere, coastal seas, the surface open ocean and the ocean interior.

3.7 Conclusions and Outlook

3.7.1 Carbon Dioxide

In general, the enhanced rate of change in tropospheric CO_2 observed today (Global Carbon Project 2011) will have a strong impact on future climate and environmental change. Presently it appears that neither the oceans nor terrestrial systems will absorb CO_2 as efficiently in the future as they do today. In addition, ocean acidification will become, and probably is already, an issue for ecosystems in the ocean. Both the climate and environmental effects and feedbacks make it very difficult to firmly predict future changes in ocean sources and sinks for CO_2. From time-series we can clearly identify the rate of change in CO_2 uptake and in ocean acidification. From surface water CO_2 measurements on Voluntary Observing Ships we can make CO_2 air-sea flux maps and assess temporal and spatial variation in the oceanic uptake of CO_2. Finally, repeat hydrography elucidates the storage of carbon in the interior ocean. More important will be to predict the effect of changes in processes controlling oceanic uptake and release of carbon, notably of changes in ocean circulation, the magnitude of the biological pumps and carbon storage. This will require a comprehensive measuring system for future observations and improved modelling tools. Today, there is a clear lack of the observations required to reduce the uncertainties in air-sea CO_2 exchange and to predict its future behaviour. This in turn is very important for predicting the occurrence of levels of ocean acidification that are harmful to ecosystems. The latter also requires results from biological perturbation experiments under varying CO_2 scenarios. The future vision is:

- The design of a comprehensive network of VOS, repeat hydrographic sections and time-series. Optimalisation might be obtained through modelling, statistical analysis and experience, based upon the existing network;
- The development of automated ocean stations. Cable-based systems might be applied along coastal areas, and moored systems and buoys in open ocean situations. Remotely controlled floats and other moving platforms, such as gliders, would provide additional process information;
- The development of data storage, data reduction, data synthesis, data assimilation and visualisation techniques, as well as continuation, automation and expansion of ongoing data synthesis efforts;
- The development of models that can be validated by data;

(Borges et al. 2010; Byrne et al. 2010; Feely et al. 2010; Gruber et al. 2010; Monteiro et al. 2010).

3.7.2 Nitrous Oxide and Methane

While our knowledge of the oceanic distribution, the formation pathways and the oceanic emissions of N_2O and CH_4 has increased considerably during the last four decades, we are far from having a comprehensive picture. Major questions and technical challenges remain to be solved:

- Reliable and fast, high-precision N_2O and CH_4 sensors for use at open ocean and coastal time-series stations and on ships of opportunity should be developed in order to expand the spatial and temporal coverage of oceanic N_2O and CH_4 measurements. Recently developed OA-ICOS (Off-Axis Integrated Cavity Output Spectrometer) instruments coupled to a continuously working equilibration device are a promising technology for use on VOS lines (Gülzow et al. 2011).
- We still only have a rudimentary understanding of N_2O and CH_4 cycling in coastal areas. We need to know more about seasonality in the major formation, consumption and transport pathways and the driving forces behind these. In this context, the ongoing dramatic increase in the number of coastal "dead zones" is a critical consideration because ongoing coastal eutrophication may well modify greatly, current emissions of N_2O and CH_4 from coastal areas.

Open Access This chapter is distributed under the terms of the Creative Commons Attribution Noncommercial License, which permits any noncommercial use, distribution, and reproduction in any medium, provided the original author(s) and source are credited.

References

Abril G, Borges AV (2004) Carbon dioxide and methane emissions from estuaries. In: Tremblay A, Varfalvy L, Roehm C, Garneau M (eds) Greenhouse gas emissions: fluxes and processes, hydroelectric reservoirs and natural environments. Springer, Berlin, pp 187–207

Abril G, Iversen N (2002) Methane dynamics in a shallow, non-tidal estuary (Randers Fjord, Denmark). Mar Ecol Prog Ser 230:171–181

Abril G, Nogueira E, Hetcheber H, Cabeçadas G, Lemaire E, Brogueira MJ (2002) Behaviour of organic carbon in nine contrasting European estuaries. Estuar Coast Shelf Sci 54:241–262

Abril G, Commarieu MV, Guerin F (2007) Enhanced methane oxidation in an estuarine turbidity maximum. Limnol Oceanogr 52:470–475

Amouroux D, Roberts G, Rapsomanikis S, Andreae MO (2002) Biogenic gas (CH_4, N_2O, DMS) emission to the atmosphere from near-shore and shelf waters of the north-western Black Sea. Estuar Coast Shelf Sci 54:575–587

Anderson LG, Falck E, Jones EP, Jutterström S, Swift J (2004) Enhanced uptake of atmospheric CO_2 during freezing of seawater: a field study in Storfjorden, Svalbard. J Geophys Res 109, C06004. doi:10.1029/2003JC002120

Andersson AJ, Mackenzie FT, Ver LM (2003) Solution of shallow-water carbonates: an insignificant buffer against rising atmospheric CO_2. Geology 31:513–516

Archer D (2007) Methane hydrate stability and anthropogenic climate change. Biogeosciences 4:521–544. doi:10.5194/bg-4-521-2007

Archer D, Buffett B (2005) Time-dependent response of the global ocean clathrate reservoir to climatic and anthropogenic forcing. Geochem Geophys Geosyst 6, Q03002. doi:10.1029/2004GC000854

Archer D, Kheshgi H, Maier-Reimer E (1997) Multiple timescales for neutralization of fossil fuel CO_2. Geophys Res Lett 24:405–408

Archer D, Winguth A, Lea D, Mahowald N (2000) What casued the glacial/interglacial atmospheric pCO_2 cycles? Rev Geophys 38(2):159–189

Archer D, Buffett B, Brovkin V (2009) Ocean methane hydrates as a slow tipping point in the global carbon cycle. Proc Natl Acad Sci USA 106(49):20596–20601

Aydin M, Verhulst KR, Saltzman ES, Battle MO, Montzka SA, Blake DR, Tang Q, Prather MJ (2011) Recent decreases in fossil-fuel emissions of ethane and methane derived from firn air. Nature 476:198–201. doi:10.1038/nature10352

Baggs E, Philippot L (2010) Microbial terrestrial pathways to nitrous oxide. In: Smith K (ed) Nitrous oxide and climate change. Earthscan, London, pp 36–62

Baker DF, Law RM, Gurney KR, Rayner P, Peylin P, Denning AS, Bousquet P, Bruhwiler L, Chen Y-H, Ciais P, Fung IY, Heimann M, John J, Maki T, Maksyutov S, Masarie K, Prather M, Pak B, Taguchi S, Zhu Z (2006) TransCom 3 inversion intercomparison: impact of the transport model errors on the interannual variability of regional CO_2 fluxes, 1988–2003. Global Biogeochem Cycle 20, GB1002. doi:10.1029/2004GB002439

Bakker DCE, Hoppema M, Schröder M, Geibert W, De Baar HJW (2008) A rapid transition from ice covered CO_2–rich waters to a biologically mediated CO_2 sink in the eastern Weddell Gyre. Biogeosciences 5:1373–1386. doi:10.5194/bg-5-1373-2008

Bakker DCE, Pfeil B, Olsen A, Sabine CL, Metzl N, Hankin S, Koyuk H, Kozyr A, Malczyk J, Manke A, Telszewski M (2012) Global data products help assess changes to the ocean carbon sink. Eos Trans Am Geophys Union 93(12):125–126. doi:10.1029/2012EO120001

Ballantyre AP, Alden CB, Miller JB, Tans PP, White JWC (2012) Increase in observed net carbon dioxide uptake by land and oceans during the past 50 years. Nature 488:70–73. doi:10.1038/nature11299

Bange HW (2006a) New directions: the importance of the oceanic nitrous oxide emissions. Atmos Environ 40:198–199

Bange HW (2006b) Nitrous oxide and methane in European coastal waters. Estuar Coast Shelf Sci 70:361–374

Bange HW (2008) Gaseous nitrogen compounds (NO, N_2O, N_2, NH_3) in the ocean. In: Capone DG, Bronk DA, Mulholland MR, Carpenter EJ (eds) Nitrogen in the marine environment, 2nd edn. Elsevier, Amsterdam, pp 51–94

Bange HW, Andreae MO (1999) Nitrous oxide in the deep waters of the world's oceans. Global Biogeochem Cycle 13(4):1127–1135

Bange HW, Bartell UH, Rapsomanikis S, Andreae MO (1994) Methane in the Baltic and North Seas and a reassessment of the marine emissions of methane. Global Biogeochem Cycle 8:465–480

Bange HW, Rapsomanikis S, Andreae MO (1996) The Aegean Sea as a source of atmospheric nitrous oxide and methane. Mar Chem 53:41–49

Bange HW, Bell TG, Cornejo M, Freing A, Uher G, Upstill-Goddard RC, Zhang G (2009) MEMENTO: a proposal to develop a database of marine nitrous oxide and methane measurements. Environ Chem 6:195–197

Bange HW, Bergmann K, Hansen HP, Kock A, Koppe R, Malien F, Ostrau C (2010a) Dissolved methane during hypoxic events at the Boknis Eck time series station (Eckernförde Bay, SW Baltic Sea). Biogeosciences 7:1279–1284

Bange HW, Freing A, Kock A, Löscher C (2010b) Marine pathways to nitrous oxide. In: Smith K (ed) Nitrous oxide and climate change. Earthscan, London, pp 36–62

Barnes J, Upstill-Goddard RC (2011) N_2O seasonal distribution and air-sea exchange in UK estuaries: implications for tropospheric N_2O source from European coastal waters. J Geophys Res 116, G01006. doi:10.1029/2009JG001156

Barnes J, Ramesh R, Purvaja R, Nirmal Rajkumar A, Senthil Kumar B, Krithika K, Ravichandran K, Uher G, Upstill-Goddard RC (2006) Tidal dynamics and rainfall control N_2O and CH_4 emissions from a pristine mangrove creek. Geophys Res Lett 33, L15405. doi:10.1029/2006GL026829

Bastviken D, Tranvik LJ, Downing JA, Crill PM, Enrich-Prast A (2011) Freshwater methane emissions offset the continental carbon sink. Science 331:50

Bates NR, Michaels AF, Knap AH (1996a) Seasonal and inter-annual variability of oceanic carbon dioxide species at the U.S. JGOFS Bermuda Atlantic Time-series Study (BATS) site. Deep-Sea Res Part II 43(2–3):347–383

Bates TS, Kelly KC, Johnson JE, Gammon RH (1996b) A reevaluation of the open ocean source of methane to the atmosphere. J Geophys Res 101:6953–6961

Bates NR, Samuels L, Merlivat L (2001) Biogeochemical and physical factors influencing seawater fCO_2 and air-sea CO_2 exchange on the Bermuda coral reef. Limnol Oceanogr 46(4):833–846

Bates NR, Moran SB, Hansell DA, Mathis JT (2006) An increasing CO_2 sink in the Arctic Ocean due to sea-ice loss. Geophys Res Lett 33, L23609. doi:10.1029/2006GL027028

Battin TJ, Kaplan LA, Findlay S, Hopkinson CS, Marti E, Packman AI, Newbold JD, Sabater F (2008) Biophysical controls on organic carbon fluxes in fluvial networks. Nat Geosci 1:95–100

Bauzá JF, Morrell JM, Corredor JE (2002) Biogeochemistry of nitrous oxide production in the Red Mangrove (*Rhizophora mangle*) forest sediments. Estuar Coast Shelf Sci 55:697–704

Beauchamp B (2004) Natural gas hydrates: myths, facts and issues. C R Geosci 336:751–765. doi:10.1016/j.crte.2004.04.003

Beman JM, Chow C-E, King AL, Feng Y, Fuhrman JA, Andersson A, Bates NR, Popp BN, Hutchins DA (2011) Global declines in oceanic nitrification rates as a consequence of ocean acidification. Proc Natl Acad Sci USA 108 (1):208–213. doi:101073/pnas.1011053108

Bender ML, Ho DT, Hendricks MB, Mika R, Battle MO, Tans PP, Conway TJ, Sturtevant B, Cassar N (2005) Atmospheric O_2/N_2 changes, 1993–2002: implications for the partitioning of fossil fuel CO_2 sequestration. Global Biogeochem Cycle 19, GB4017. doi:10.1029/2004GB002410

Biscaye PE, Anderson R (1994) Particle fluxes on the slope of the southern Mid-Atlantic Bight: SEEP-II. Deep-Sea Res Part II 41:459–469

Biscaye PE, Anderson R, Deck BL (1988) Fluxes of particles and constituents to the Eastern United States continental slope and rise: SEEP-I. Cont Shelf Res 8:888–904

Biswas H, Mukhopadhayay SK, De TK, Sen S, Jana TK (2006) Methane emission from the wetland rice fields in Sagar Island, NE coast of Bay of Bengal, India. Int J Agric Res 1(1):78–86

Biswas H, Mukhopadhyay SK, Sen S, Jana TK (2007) Spatial and temporal patterns of methane dynamics in the tropical mangrove dominated estuary, NE coast of Bay of Bengal, India. J Mar Syst 68:55–64

Blair NE, Aller RC (1995) Anaerobic methane oxidation on the Amazon shelf. Geochim Cosmochim Acta 59:3707–3715

Boetius A, Ravenschlag K, Schubert CJ, Rickert D, Widdel F, Gieseke A, Amann R, Jørgensen BB, Witte U, Pfannkuche O (2000) A marine microbial consortium apparently mediating anaerobic oxidation of methane. Nature 407:623–626

Bopp L, Le Quéré C, Heimann M, Manning AC (2002) Climate-induced oceanic oxygen fluxes: implications for the contemporary carbon budget. Global Biogeochem Cycle 16(2):1022. doi:10.1029/2011GB001445

Borges AV (2005) Do we have enough pieces of the jigsaw to integrate CO_2 fluxes in the Coastal Ocean? Estuaries 28(1):3–27

Borges AV (2011) Present day carbon dioxide fluxes in the coastal ocean and possible feedbacks under global change. In: Duarte PM, Santana-Casiano JM (eds) Oceans and the atmospheric carbon content. Springer, Berlin, pp 47–77

Borges AV, Abril G (2011) Carbon dioxide and methane dynamics in estuaries. In: Wolanski E, McLusky DS (eds) Treatise on estuarine and coastal science, vol 5, Biogeochemistry. Elsevier, Amsterdam, pp 119–161. doi:10.1016/B978-0-12-374711-2.00504-0

Borges AV, Gypens N (2010) Carbonate chemistry in the coastal zone responds more strongly to eutrophication than to ocean acidification. Limnol Oceanogr 55:346–353

Borges AV, Djenidi S, Lacroix G, Théate J, Delille B, Frankignoulle M (2003) Atmospheric CO_2 flux from mangrove surrounding waters. Geophys Res Lett 30(11):1558. doi:10.1029/2003GL017143

Borges AV, Delille B, Frankignoulle M (2005) Budgeting sinks and sources of CO_2 in the coastal ocean: diversity of ecosystems counts. Geophys Res Lett 32, L14601. doi:10.1029/2005GL023053

Borges AV, Schiettecatte L-S, Abril G, Delille B, Gazeau F (2006) Carbon dioxide in European coastal waters. Estuar Coast Shelf Sci 70(3):375–387

Borges AV, Tilbrook B, Metzl N, Lenton A, Delille B (2008) Inter-annual variability of the carbon dioxide oceanic sink south of Tasmania. Biogeosciences 5:141–155

Borges AV, Alin SR, Chavez FP, Vlahos P, Johnson KS, Holt JT, Balch WM, Bates N, Brainard R, Cai W-J, Chen CTA, Currie K, Dai M, Degrandpré M, Delille B, Dickson A, Evans W, Feely RA, Friederich GE, Gong G-C, Hales B, Hardman-Mountford N, Hendee J, Hernandez-Ayon JM, Hood M, Huertas E, Hydes D, Ianson D, Krasakopoulou E, Litt E, Luchetta A, Mathis J, McGillis WR, Murata A, Newton J, Ólafsson J, Omar A, Perez FF, Sabine C, Salisbury

JE, Salm R, Sarma VVSS, Schneider B, Sigler M, Thomas H, Turk D, Vandemark D, Wanninkhof R, Ward B (2010) A global sea surface carbon observing system: inorganic and organic carbon dynamics in coastal oceans. In: Hall J, Harrison DE, Stammer D (eds) Proceedings of OceanObs'09: sustained ocean observations and information for society, vol 2, Venice, Italy, 21–25 Sept 2009, ESA publication WPP-306. doi:10.5270/OceanObs09.cwp.07

Bouillon S, Middelburg JJ, Dehairs F, Borges AV, Abril G, Flindt MR, Ulomi S, Kristensen E (2008) Importance of intertidal sediment processes and pore water exchange on the water column biogeochemistry in a pristine mangrove creek (Ras Dege, Tanzania). Biogeosciences 4:311–322

Boutin J, Etcheto J, Dandonneau Y, Bakker DCE, Feely RA, Inoue HY, Ishii M, Ling RD, Nightingale PD, Metzl N, Wanninkhof R (1999) Satellite sea surface temperature: a powerful tool for interpreting in situ pCO_2 measurements in the equatorial Pacific Ocean. Tellus 51B:490–508

Brewer PG, Peltzer ET (2009) Limits to marine life. Science 324:347–348

Brewer PG, Paull C, Peltzer ET, Ussler W, Rehder G, Friederich G (2002) Measurement of the fate of gas hydrates during transit through the ocean water column. Geophys Res Lett 29:38. doi:10.1029/2002GL014727

Brix H, Gruber N, Keeling CD (2004) Interannual variability of the upper ocean carbon cycle at station ALOHA near Hawaii. Global Biogeochem Cycle 18, GB4019. doi:10.1029/2004GB002245

Broecker WS, Peng T-H (1982) Tracers in the sea. Eldigio Press, Lamont-Doherty Geological Observatory, Columbia University, Palisades

Brüchert V, Currie B, Peard KR (2009) Hydrogen sulphide and methane emissions on the central Namibian shelf. Prog Oceanogr 83:169–179

Buffett B, Archer D (2004) Global inventory of methane clathrate: sensitivity to changes in the deep ocean. Earth Planet Sci Lett 227:185–199. doi:10.1016/j.epsl.2004.09.005

Buitenhuis E, Van Bleijswijk J, Bakker DCE, Veldhuis MJW (1996) Trends in inorganic and organic carbon in a bloom of *Emiliania huxleyi* in the North Sea. Mar Ecol Prog Ser 143:271–282

Byrne RH, DeGrandpre MD, Short RT, Martz TR, Merlivat L, McNeil C, Sayles FL, Bell R, Fietzek P (2010) Sensors and systems for in situ observations of marine carbon dioxide system variables. In: Hall J, Harrison DE, Stammer D (eds) Proceedings of OceanObs'09: sustained ocean observations and information for society, vol 2, Venice, Italy, 21–25 Sept 2009, ESA publication WPP-306. doi:10.5270/OceanObs09.cwp.13

Cabello P, Roldán MD, Moreno-Vivián C (2004) Nitrate reduction and the nitrogen cycle in archaea. Microbiology 150 (11):3527–3546

Cai W-J (2011) Estuarine and coastal ocean carbon paradox: CO_2 sinks or sites of terrestrial carbon incineration? Annu Rev Mar Sci 3:123–145

Cai W-J, Dai MH, Wang YC (2006) Air-sea exchange of carbon dioxide in ocean margins: a province-based synthesis. Geophys Res Lett 33, L12603. doi:10.1029/2006GL026219

Caldeira K, Wickett ME (2003) Anthropogenic carbon and ocean pH. Nature 425:365

Cantera JJL, Stein LY (2007) Role of nitrite reductase in the ammonia-oxidizing pathway of *Nitrosomonas europaea*. Arch Microbiol 188(4):349–354

Capone DG, Kiene RP (1988) Comparison of microbial dynamics in marine and freshwater sediments: contrasts in anaerobic carbon catabolism. Limnol Oceanogr 33(4 part 2):725–749

Cavalieri DJ, Parkinson CL, Vinnikov KY (2003) 30-Year satellite record reveals contrasting Arctic and Antarctic decadal sea ice variability. Geophys Res Lett 30(18):1970. doi:10.1029/2003GL018031

Chan F, Barth JA, Lubchenco J, Kirincich A, Weeks H, Peterson WT, Menge BA (2008) Emergence of anoxia in the California current large marine ecosystem. Science 319:920–920

Chanton JP, Martens CS, Kelley CA (1989) Gas transport from methane-saturated, tidal freshwater and wetland sediments. Limnol Oceanogr 34(5):807–819

Chen CTA, Borges AV (2009) Reconciling opposing views on carbon cycling in the coastal ocean: continental shelves as sinks and near-shore ecosystems as sources of atmospheric CO_2. Deep-Sea Res Part II 56(8–10):578–590

Codispoti LA (2010) Interesting times for marine N_2O. Science 327:1339–1340

Codispoti LA, Elkins JW, Friederich GE, Packard TT, Sakamoto CM, Yoshinari T (1992) On the nitrous oxide flux from productive regions that contain low oxygen waters. In: Desai BN (ed) Oceanography of the Indian ocean. Oxford-IBH, New Delhi, pp 271–284

Codispoti LA, Flagg C, Kelly V (2005) Hydrographic conditions during the 2002 SBI process experiments. Deep-Sea Res Part II 52(24–26):3199–3226

Cole JA (1988) Assimilatory and dissimilatory reduction of nitrate to ammonia. Symp Soc Gen Microbiol 42:281–329

Comeau S, Gattuso JP, Nisumaa AM, Orr J (2011) Impact of aragonite saturation state changes on migratory pteropods. Proc Royal Soc B Biol Sci. doi:10.1098/rspb.2011.0910

Conrad R, Seiler W (1988) Methane and hydrogen in seawater (Atlantic Ocean). Deep-Sea Res 35:1903–1917

Corbière A, Metzl N, Reverdin G, Brunet C, Takahashi T (2007) Interannual and decadal variability of the oceanic carbon sink in the North Atlantic subpolar gyre. Tellus 59B:168–178

Cornejo M, Farías L, Paulmier A (2006) Temporal variability in N_2O water content and its air-sea exchange in an upwelling area off central Chile (36°S). Mar Chem 101:85–94

Cornejo M, Farías L, Gallegos M (2007) Seasonal cycle of N2O vertical distribution and air-sea fluxes over the continental shelf waters off central Chile (36°S). Prog Oceanogr 75:383–395

Crutzen PJ (1970) The influence of nitrogen oxides on the atmospheric ozone content. Q J R Meteorol Soc 96:320–325

Crutzen PJ (1991) Methane's sinks and sources. Nature 350:380–381

Curry R, Dickson R, Yashayaev I (2003) A change in the freshwater balance of the Atlantic Ocean over the past four decades. Nature 426:826–829

Cynar FJ, Yayanos AA (1991) Enrichment and characterization of a methanogenic bacterium from the oxic upper layer of the ocean. Curr Microbiol 23:89–96

Dale AW, Regnier P, Van Cappellen P (2006) Bioenergetic controls on anaerobic oxidation of methane (AOM) in coastal marine sediments: a theoretical analysis. Am J Sci 306:246–294. doi:10.2475/ajs.306.4.246

Damm E, Mackensen A, Budéus G, Faber E, Hanfland C (2005) Pathways of methane in seawater: plume spreading in an Arctic shelf environment (SW Spitsbergen). Cont Shelf Res 25:1453–1472

Damm E, Helmke E, Thoms S, Schauer U, Nöthig E, Bakker K, Kiene R (2010) Methane production in aerobic oligotrophic surface water in the central Arctic Ocean. Biogeosciences 7:1099–1108

Dando PR, Austen MC, Burke RA, Kendall MA, Kennicutt MC, Judd AG, Moore DC, Ohara SCM, Schmaljohann R, Southward AJ (1991) Ecology of a North-Sea pockmark with an active methane seep. Mar Ecol Prog Ser 70:49–63

De Angelis MA, Lee C (1994) Methane production during zooplankton grazing on marine phytoplankton. Limnol Oceanogr 39:1298–1308

Delille B, Jourdain B, Borges AV, Tison J-L, Delille D (2007) Biogas (CO_2, O_2, dimethylsulfide) dynamics in spring Antarctic fast ice. Limnol Oceanogr 52(4):1367–1379

Denman KL, Brasseur G, Chidthaisong A, Ciais P, Cox PM, Dickinson RE, Hauglustaine D, Heinze C, Holland E, Jacob D, Lohman U, Ramachandran S, Da Silva Dias PL, Wofsy SC, Zhang X (2007) Couplings between changes in the climate system and biogeochemistry. In: Solomon S, Qin D, Manning M, Chen Z, Marquis M, Averyt KB, Tignor M, Miller HL (eds) Climate change 2007: The physical science basis. contribution of working group i to the fourth assessment report of the intergovermental panel on climate change. Cambridge University Press, Cambridge/New York, pp 499–587

Deutsch C, Brix H, Ito T, Frenzel H, Thompson L (2011) Climate-forced variability of ocean hypoxia. Science 333:336–339

Devol AH, Richey JE, Forsburg BR, Martinelli LA (1990) Seasonal dynamics in methane emissions from the Amazon river floodplain. J Geophys Res 95:16417–16426

Diaz RJ, Rosenberg R (2008) Spreading dead zones and consequences for marine ecosystems. Science 321:926–929

Dickson AG, Sabine CL, Christian JR (eds) (2007) Guide to best practices for ocean CO_2 measurements, vol 3, PICES special publication. North Pacific Marine Science Organization, Sidney, BC, Canada

Dimitrov L (2002) Contribution to atmospheric methane by natural seepages on the Bulgarian continental shelf. Cont Shelf Res 22:2429–2442

Dittmar T, Hertkorn N, Kattner G, Lara RJ (2006) Mangroves, a major source of dissolved organic carbon to the oceans. Global Biogeochem Cycle 20, GB1012. doi:10.1029/2005GB002570

Dlugokencky EJ, Houweling S, Bruhwiler L, Masarie KA, Lang PM, Miller JB, Tans PP (2003) Atmospheric methane levels off: temporary pause or a new steady-state? Geophys Res Lett 30(19):1992. doi:10.1029/2003GL018126

Dlugokencky EJ, Bruhwiler L, White JWC, Emmons LK, Novelli PC, Montzka SA, Masarie KA, Lang PM, Crotwell AM, Miller JB, Gatti LV (2009) Observational constraints on recent increases in the atmospheric CH_4 burden. Geophys Res Lett 36, L18803. doi:10.1029/2009GL039780

DOE (1994) In: Dickson AG, Goyet C (eds) Handbook of methods for the analysis of the various parameters of the carbon system in sea water; version 2. ORNL/CDIAC-74, Carbon Dioxide Information Analysis Center, Oak Ridge National Laboratory, US Department of Energy, Oak Ridge, Tennessee, USA

Doney SC, Mahowald N, Lima I, Feely RA, Mackenzie FT, Lamarque J-F, Rasch PJ (2007) Impact of anthropogenic atmospheric nitrogen and sulfur deposition on ocean acidification and the inorganic carbon system. Proc Natl Acad Sci USA 104(37):14580–14585

Doney SC, Ruckelshaus M, Duffy JE, Barry JP, Chan F, English CA, Galindo HM, Grebmeier JM, Hollowed AB, Knowlton N, Polovina J, Rabalais NN, Sydeman WJ, Talley LD (2012) Climate change impacts on marine ecosystems. Annu Rev Mar Sci 4:11–37. doi:10.1146/annurev-marine-041911-111611

Dore JE, Karl DM (1996) Nitrification in the euphotic zone as a source for nitrite, nitrate, and nitrous oxide at station ALOHA. Limnol Oceanogr 41:1619–1628

Dore JE, Lukas R, Sadler DW, Karl DM (2003) Climate-driven changes to the atmospheric CO_2 sink in the subtropical North Pacific Ocean. Nature 424:754–757

Dürr HH, Laruelle GG, Van Kempen CM, Slomp CP, Meybeck M, Middelkoop H (2011) Worldwide typology of nearshore coastal systems: defining the estuarine filter of river inputs to the oceans. Estuar Coasts 34:441–458

EPA (2010) Methane and nitrous oxide emissions from natural sources. 430-R-10-001. Office of Atmospheric Programs (6207J), United States Environmental Protection Agency, Washington, DC

EPICA Community Members (2004) Eight glacial cycles from an Antarctic ice core. Nature 429:623–628

Fazzolari E, Mariotti A, Germon JC (1990a) Nitrate reduction to ammonia – a dissimilatory process in *Enterobacter-Amnigenus*. Can J Microbiol 36(11):779–785

Fazzolari E, Mariotti A, Germon JC (1990b) Dissimilatory ammonia production vs. denitrification in vitro and in inoculated agricultural soil samples. Can J Microbiol 36(11):786–793

Feely RA, Sabine CL, Lee K, Berelson W, Kleypas J, Fabry VJ, Millero FJ (2004) Impact of anthropogenic CO_2 on the $CaCO_3$ system in the oceans. Science 305:362–366

Feely RA, Takahashi T, Wanninkhof R, McPhaden MJ, Cosca CE, Sutherland SC, Carr M-E (2006) Decadal variability of the air-sea CO_2 fluxes in the equatorial Pacific Ocean. J Geophys Res 111, C08S90. doi:10.1029/2005JC003129

Feely RA, Fabry VJ, Dickson AG, Gattuso J-P, Bijma J, Riebesell U, Doney S, Turley C, Saino T, Lee K, Anthony K, Kleypas J (2010) An international observational network for ocean acidification. In: Hall J, Harrison DE, Stammer D (eds) Proceedings of OceanObs'09: sustained ocean observations and information for society, vol 2, Venice, Italy, 21–25 Sept 2009, ESA publication WPP-306. doi:10.5270/OceanObs09.cwp.29

Ferrón S, Ortega T, Gómez-Parra A, Forja JM (2007) Seasonal study of dissolved CH_4, CO_2 and N_2O in a shallow tidal system of the bay of Cádiz (SW Spain). J Mar Syst 66:244–257. doi:10.1016/j.jmarsys.2006.03.021

Flückiger J, Blunier T, Stauffer B, Chappellaz J, Spahni R, Kawamura K, Schwander J, Stocker TF, Dahl-Jensen D (2004) N_2O and CH_4 variations during the last glacial epoch: insight into global processes. Global Biogeochem Cycle 18, GB1020. doi:10.1029/2003GB002122

Forster P, Ramaswamy V, Artaxo P, Berntsen T, Betts R, Fahey DW, Haywood J, Lean J, Lowe DC, Myhre G, Nganga J, Prinn R, Rage G, Schulz M, Van Dorland R (2007) Changes in atmospheric constituents and in radiative forcing.

In: Solomon S, Qin D, Manning M, Chen Z, Marquis M, Averyt KB, Tignor M, Miller HL (eds) Climate change 2007: the physical science basis. Contribution of working group I to the fourth assessment report of the intergovermental panel on climate change. Cambridge University Press, Cambridge/New York. pp 130–234

Forster G, Upstill-Goddard RC, Gist N, Robinson C, Uher G, Woodward EMS (2009) Nitrous oxide and methane in the Atlantic Ocean between 50°N and 52°S: latitudinal distribution and sea-to-air flux. Deep-Sea Res Part II 56:964–976. doi:10.1016/j.dsr2.2008

Francis CA, Beman JM, Kuypers MM (2007) New processes and players in the nitrogen cycle: the microbial ecology of anaerobic and archaeal ammonia oxidation. ISME J 1(1):19–27

Frankignoulle M, Canon C, Gattuso J-P (1994) Marine calcification as a source of carbon dioxide: positive feedback of increasing CO_2. Limnol Oceanogr 39(2):458–462

Frankignoulle M, Gattuso J-P, Biondo R, Bourge I, Copin-Montégut G, Pichon M (1996) Carbon fluxes in coral reefs 2. Eulerian study of inorganic carbon dynamics and measurement of air-sea CO_2 exchanges. Mar Ecol Prog Ser 145:123–132

Frankignoulle M, Abril G, Borges A, Bourge I, Canon C, Delille B, Libert E, Théate MJ (1998) Carbon dioxide emission from European estuaries. Science 282:434–436. doi:10.1126/science.282.5388.434

Freing A (2009) Production and emissions of oceanic nitrous oxide. Ph.D. thesis. University of Kiel, Kiel

Freing A, Wallace DWR, Bange HW (2012) Global oceanic production of nitrous oxide (N_2O). Philos Trans R Soc B 367:1245–1255

Friedlingstein P, Houghton RA, Marland G, Hacker J, Boden TA, Conway TJ, Canadell JC, Raupach MR, Ciais P, Le Quéré C (2010) Update on CO_2 emissions. Nat Geosci 3:811–812. doi:10-1038/ngeo1022

Gattuso J-P, Pichon M, Delesalle B, Frankignoulle M (1993) Community metabolism and air-sea CO_2 fluxes in a coral reef ecosystem (Moorea, French Polynesia). Mar Ecol Prog Ser 96:259–267

Gattuso J-P, Payri CE, Pichon M, Delesalle B, Frankignoulle M (1997) Primary production, calcification, and air-sea CO_2 fluxes of a macroalgal-dominated coral reef community (Moorea, French Polynesia). J Phycol 33(5):729–738

Gattuso J-P, Frankignoulle M, Wollast R (1998) Carbon and carbonate metabolism in coastal aquatic ecosystems. Annu Rev Ecol Syst 29:405–433

Gazeau F, Smith SV, Gentili B, Frankignoulle M, Gattuso J-P (2004) The European coastal zone: characterization and first assessment of ecosystem metabolism. Estuar Coast Shelf Sci 60(4):673–694

Geibert W, Assmy P, Bakker DCE, Hanfland C, Hoppema M, Pichevin L, Schröder M, Schwarz JN, Stimac I, Usbeck U, Webb A (2010) High productivity in an ice melting hotspot at the eastern boundary of the Weddell Gyre. Global Biogeochem Cycle 24, GB3007. doi:10.1029/2009GB003657

Geilfus N-X, Carnat G, Papakyriakou T, Tison J-L, Else B, Thomas H, Shadwick E, Delille B (2012) Dynamics of pCO_2 and related air-ice CO_2 fluxes in the Arctic coastal zone (Amundsen Gulf, Beaufort Sea). J Geophys Res 117, C00G10. doi:10.1029/2011JC007117

Gerard G, Chanton J (1993) Quantification of methane oxidation in the rhizosphere of emergent aquatic macrophytes: defining upper limits. Biogeochemistry 23:79–97

Global Carbon Project (2011) Carbon budget and trends 2010. www.globalcarbonproject.org/carbonbudget. Accessed 13 Jan 2012

González-Dávila M, Santana-Casiano JM, Rueda MJ, Llinás O, González-Dávila E-F (2003) Seasonal and interannual variability of sea-surface carbon dioxide species at the European Station for Time Series in the Ocean at the Canary Islands (ESTOC) between 1996 and 2000. Global Biogeochem Cycle 17(3):1076. doi:10.1029/2002GB001993

González-Dávila M, Santana-Casiano JM, Rueda MJ, Llinás O (2010) The water column distribution of carbonate system variables at the ESTOC site from 1995 to 2004. Biogeosciences 7:3067–3081

Gordon AL, Huber BA (1990) Southern Ocean winter mixed layer. J Geophys Res 95:11655–11672

Goreau TJ, Kaplan WA, Wofsy SC, McElroy MB, Valois FW, Watson SW (1980) Production of NO_2^- and N_2O by nitrifying bacteria at reduced concentrations of oxygen. Appl Environ Microbiol 40:526–532

Gornitz V, Fung I (1994) Potential distribution of methane hydrates in the world's oceans. Global Biogeochem Cycle 8(3):335–347. doi:10.1029/94GB00766

Gruber N, Keeling CD (2001) An improved estimate of the isotopic air-sea disequilibrium of CO_2: implications for the oceanic uptake of anthropogenic CO_2. Geophys Res Lett 28 (3):555–558. doi:10.1029/2000GL011853

Gruber N, Sarmiento JL (2002) Large-scale biogeochemical/physical interactions in elemental cycles. In: Robinson AR, McCarthy JJ, Rothschild BJ (eds) The sea, vol 12. Wiley, New York, pp 337–399

Gruber N, Keeling CD, Bates NR (2002) Interannual variability in the North Atlantic Ocean carbon sink. Science 298:2374–2378

Gruber N, Gloor M, Mikaloff Fletcher SE, Doney SC, Dutkiewicz S, Follows MJ, Gerber M, Jacobson AR, Joos F, Lindsay K, Menemenlis D, Mouchet A, Müller SA, Takahashi T (2009) Oceanic sources, sinks, and transport of atmospheric CO_2. Global Biogeochem Cycle 23, GB1005. doi:10.1029/2008GB003349

Gruber N, Körtzinger A, Borges A, Claustre H, Doney SC, Feely RA, Hood M, Ishii M, Kozyr A, Monteiro P, Nojiri Y, Sabine CL, Schuster U, Wallace DWR, Wanninkhof R (2010) Plenary Paper: Toward an integrated observing system for ocean carbon and biogeochemistry at a time of change. In: Hall J, Harrison DE, Stammer D (eds) Proceedings of OceanObs'09: sustained ocean observations and information for society, vol 1, Venice, Italy, 21–25 Sept 2009, ESA publication WPP-306. doi:10.5270/OceanObs09. p 18

Gülzow W, Rehder G, Schneider B, Schneider von Deimling J, Sadkowiak B (2011) A new method for continuous measurement of methane and carbon dioxide in surface waters using off-axis integrated cavity output spectroscopy (ICOS): an example from the Baltic Sea. Limnol Oceanogr Methods 9:168–174

Gurney KR, Law RM, Denning AS, Rayner PJ, Pak BC, Baker D, Bousquet P, Bruhwiler L, Chen Y-H, Ciais P, Fung IY, Heimann M, John J, Maki T, Maksyutov S,

Peylin P, Prather M, Taguchi S (2004) Transcom 3 inversion intercomparison: model mean results for the estimation of seasonal carbon sources and sinks. Global Biogeochem Cycle 18, GB1010. doi:10.1029/2003GB002111

Gypens N, Borges AV, Lancelot C (2009) Effect of eutrophication on air-sea CO_2 fluxes in the coastal Southern North Sea: a model study of the past 50 years. Global Change Biol 15(4):1040–1056

Hall TM, Prather MJ (1993) Simulations of the trend and annual cycle in stratospheric CO_2. J Geophys Res 98:10573–10581. doi:10.1029/93JD00325

Harlay J, Borges AV, Van Der Zee C, Delille B, Godoi RHM, Schiettecatte L-S, Roevros N, Aerts K, Lapernat P-E, Rebreanu L, Groom S, Daro M-H, Van Grieken R, Chou L (2010) Biogeochemical study of a coccolithophorid bloom in the northern Bay of Biscay (NE Atlantic Ocean) in June 2004. Prog Oceanogr 80:317–336. doi:10.1016/j.pocean.2010.04.029

Harlay J, Chou L, De Bodt C, Van Oostende N, Piontek J, Suykens K, Engel A, Sabbe K, Groom S, Delille B, Borges AV (2011) Biogeochemistry and carbon mass balance of a coccolithophore bloom in the northern Bay of Biscay (June 2006). Deep-Sea Res Part I 58:111–127

Heinze C, Maier-Reimer E, Winn K (1991) Glacial pCO_2 reduction by the world ocean: experiments with the Hamburg carbon cycle model. Paleoceanography 6(4):395–430

Heip C, Goosen NK, Herman PMJ, Kromkamp J, Middelburg JJ, Soetaert K (1995) Production and consumption of biological particles in temperate tidal estuaries. Oceanogr Mar Biol: Annu Rev 33:1–149

Helm KP, Bindoff NL, Church JA (2011) Observed decreases in oxygen content of the global ocean. Geophys Res Lett 38, L23602. doi:10.1029/2011GL049513

Ho DT, Law CS, Smith MJ, Schlosser P, Harvey M, Hill P (2006) Measurements of air-sea gas exchange at high wind speeds in the Southern Ocean: implications for global parameterizations. Geophys Res Lett 33, L16611. doi:10.1029/2006GL026817

Holmes ME, Sansone FJ, Rust TM, Popp BN (2000) Methane production, consumption, and air-sea exchange in the open ocean: an evaluation based on carbon isotopic ratios. Global Biogeochem Cycle 14:1–10

Holt J, Wakelin S, Huthnance J (2008) Down-welling circulation of the northwest European continental shelf: a driving mechanism for the continental shelf carbon pump. Geophys Res Lett 36, L14602. doi:10.1029/2009GL038997

Hopkinson CSJ, Smith EM (2005) Estuarine respiration: an overview of benthic, pelagic and whole system respiration. In: Del Giorgio PA, Williams PJL (eds) Respiration in aquatic ecosystems. Oxford University Press, Oxford

Hornafius JS, Quigley DC, Luyendyk BP (1999) The world's most spectacular marine hydrocarbons seeps (Coal Oil Point, Santa Barbara Channel, California): quantification of emissions. J Geophys Res 104(C9):20703–20711

Hubberten H-W, Romanovski NN (2001) Terrestrial and offshore permafrost evolution of the Laptev Sea region during the last Pleistocene-Holocene glacial-eustatic cycle. In: Paepe R, Melnikov V (eds) Permafrost response on economic development, environmental security and natural resources. Proc-NATO-ARW, Novosibirsk, 1998. Kluwer, Dordrecht, pp 43–60

Huthnance JM, Holt JT, Wakelin SL (2009) Deep ocean exchange with west-European shelf seas. Ocean Sci 5:621–634

Iglesias-Rodriguez M, Halloran PR, Rickaby REM, Hall IR, Colmenero-Hidalgo E, Gittins JR, Green DRH, Tyrrell T, Gibbs SJ, Von Dassow P, Rehm E, Armbrust EV, Boessenkool KP (2008) Phytoplankton calcification in a High-CO_2 World. Science 320(5874):336–340. doi:10.1126/science.1154122

IOCCP (2004) Ocean surface pCO_2, data integration and database development workshop, National Institute for Environmental Studies, Tsukuba, Japan, 14–17 Jan 2004. IOCCP (International Ocean Carbon Coordination Project) report 2. www.ioccp.org

IPCC (2007) In: Solomon S, Qin D, Manning M, Chen Z, Marquis M, Averyt KB, Tignor M, Miller HL (eds) Climate change 2007: the physical science basis. Contribution of working group I to the fourth assessment report of the intergovernmental panel on climate change. Cambridge University Press, Cambridge/New York

Ishii M, Inoue HY, Midorikawa T, Saito S, Tokieda T, Sasano D, Nakadate A, Nemoto K, Metzl N, Wong CS, Feely RA (2009) Spatial variability and decadal trend of the oceanic CO_2 in the western equatorial Pacific warm/fresh water. Deep-Sea Res Part II 56(8–10):591–606

Jackson MA, Tiedje JM, Averill BA (1991) Evidence for a no-rebound mechanism for production of N_2O from nitrite by the copper-containing nitrite reductase from *Achromobacter-Cycloclastes*. FEBS Lett 29(1):41–44

Jacobson AR, Mikaloff Fletcher SE, Gruber N, Sarmiento JL, Gloor M (2007) A joint atmosphere–ocean inversion for surface fluxes of carbon dioxide: 1. Methods and global-scale fluxes. Global Biogeochem Cycle 21, GB1019. doi:10.1029/2005GB002556

Jansen E, Overpeck J, Briffa KR, Duplessy J-C, Joos F, Masson-Delmotte V, Olago D, Otto-Bliesner B, Peltier WR, Rahmstorf S, Ramesh R, Raynaud D, Rind D, Solomina O, Villalba R, Zhang D (2007) Paleoclimate. In: Solomon S, Qin D, Manning M, Chen Z, Marquis M, Averyt KB, Tignor M, Miller HL (eds) Climate change 2007: the physical science basis. Contribution of working group I to the fourth assessment report of the intergovermental panel on climate change. Cambridge University Press, Cambridge/New York, pp 434–497

Jansson BPM, Malandrin L, Johannson HE (2000) Cell cycle arrest in Archaea by the hypusination inhibitor N1-Guanyl-1,7-Diaminoheptane. J Bacteriol 182:1158–1161. doi:10.1128/JB.182.4.1158-1161

Jayakumar DA, Naqvi SWA, Narvekar PV, George MD (2001) Methane in coastal and offshore waters of the Arabian Sea. Mar Chem 74:1–13

Jin X, Gruber N, Dunne JP, Sarmiento JL, Armstrong RA (2006) Diagnosing the contribution of phytoplankton functional groups to the production and export of particulate organic carbon, $CaCO_3$, and opal from global nutrient and alkalinity distributions. Global Biogeochem Cycle 20, GB2015. doi:10.1029/2005GB002532

Joos F, Meyer R, Bruno M, Leuenberger M (1999) The variability in the carbon sinks as reconstructed for the last 1000 years. Geophys Res Lett 26:1437–1441

Joos F, Prentice IC, Sitch S, Meyer R, Hooss G, Plattner G-K, Gerber S, Hasselmann K (2001) Global warming feedbacks on terrestrial carbon uptake under the Intergovernmental Panel on Climate Change (IPCC) emission scenarios. Global Biogeochem Cycle 15(4):891–907

Joos F, Plattner G-K, Stocker TF, Körtzinger A, Wallace DWR (2003) Trends in marine dissolved oxygen: implications for ocean circulation changes and the carbon budget. Eos Trans Am Geophys Union 84(21):197. doi:10.1029/2003EO210001

Joyce J, Jewell PW (2003) Physical controls on methane ebullition from reservoirs and lakes. Environ Eng Geosci 9:167–178

Judd A, Hovland M (2007) Seabed fluid flow. Impact on geology, biology and the marine environment. Cambridge University Press, Cambridge, UK

Kai FM, Tyler SC, Randerson JT, Blake DR (2011) Reduced methane growth rate explained by decreased northern hemisphere microbial sources. Nature 476:194–197. doi:10.1038/nature10259

Kaiser M, Attrill M, Jennings S, Thomas DN, Barnes D, Brierley A, Hiddink JG, Kaartokallio H, Polunin NVC, Raffaelli D (2011) Marine ecology: processes, systems and impacts, 2nd edn. Oxford University Press, Oxford

Karl DM, Beversdorf L, Björkman KM, Church MJ, Martinez A, Delong EF (2008) Aerobic production of methane in the sea. Nat Geosci 1:473–478

Keeling RF, Körtzinger A, Gruber N (2010) Ocean deoxygenation in a warming world. Annu Rev Mar Sci 2:199–229

Kelley C (2003) Methane oxidation potential in the water column of two diverse coastal marine sites. Biogeochemistry 65:105–120

Kemp WM, Smith EM, Marvin-DiPasquale M, Boynton WR (1997) Organic carbon-balance and net ecosystem metabolism in Chesapeake Bay. Mar Ecol Prog Ser 150:229–248

Kessler JD, Reeburgh WS, Southon J, Varela R (2005) Fossil methane source dominates Cariaco Basin water column methane geochemistry. Geophys Res Lett 32, L12609. doi:10.1029/2005GL022984

Key RM, Kozyr A, Sabine CL, Lee K, Wanninkhof R, Bullister J, Feely RA, Millero F, Mordy C, Peng T-H (2004) A global ocean carbon climatology: results from GLODAP. Global Biogeochem Cycle 18, GB4031

Kiene RP, Linn LJ, Bruton JA (2000) New and important roles for DMSP in marine microbial communities. J Sea Res 43:209–224

King GM (1994) Ecophysiological characteristics of obligate methanotrophic bacteria and methane oxidation in situ. In: Murrell JC, Kelly DP (eds) Microbial growth on C1 compounds. Intercept Press, Andover, pp 303–313

Kitidis V, Tizzard L, Uher G, Judd A, Upstill-Goddard RC, Head IM, Gray ND, Taylor G, Durán R, Diez R, Iglesias J, García-Gil S (2007) The biogeochemical cycling of methane in Ria de Vigo, NW Spain: sediment processing and sea–air exchange. J Mar Syst 66:258–271

Klatt O, Roether W, Hoppema M, Bulsiewicz K, Fleischmann U, Rodehacke C, Fahrbach E, Weiss RF, Bullister JL (2002) Repeated CFC sections at the Greenwich Meridian in the Weddell Sea. J Geophys Res 107:3030. doi:10.1029/2000JC000731

Kleypas JA, Buddemeier RW, Archer D, Gattuso J-P, Langdon C, Opdyke BN (1999) Geochemical consequences of increased atmospheric carbon dioxide on coral reefs. Science 284:118–120

Kleypas JA, Feely RA, Fabry VJ, Langdon C, Sabine CL, Robins LL (2006) Impacts of ocean acidification on coral reefs and other marine calcifiers: a guide for future research. Report of a workshop, 2005. St. Petersburg, Florida. NSF, NOAA and US Geological Survey

Kock A, Gebhardt S, Bange HW (2008) Methane emissions from the upwelling area off Mauritania (NW Africa). Biogeosciences 5:1119–1125

Kock A, Schafstall J, Dengler M, Brandt P, Bange HW (2012) Sea-to-air and diapycnal nitrous oxide fluxes in the eastern tropical North Atlantic Ocean. Biogeosciences 9:957–964

Koné YJM, Abril G, Kouadio KN, Delille B, Borges AV (2009) Seasonal variability of carbon dioxide in the rivers and lagoons of Ivory Coast (West Africa). Estuar Coasts 32:246–260

Kreuzwieser J, Buchholz J, Rennenberg H (2003) Emission of methane and nitrous oxide by Australian mangrove ecosystems. Plant Biol 5:423–431

Krey V, Canadell JG, Nakicenovic N, Abe Y, Andruleit H, Archer D, Grubler A, Hamilton NTM, Johnson A, Kostov V, Lamarque J-F, Langhorne N, Nisbet EG, O'Neill B, Riahi K, Riedel M, Wang W, Yakushev V (2009) Gas hydrates: entrance to a methane age or climate threat? Environ Res Lett 4, 034007

Kristensen E, Flindt MR, Borges AV, Bouillon S (2008) Emission of CO_2 and CH_4 to the atmosphere by sediments and open waters in two Tanzanian mangrove forests. Mar Ecol Prog Ser 370:53–67

Kroeze C, Dumont E, Seitzinger SP (2005) New estimates of global emissions of N_2O from rivers and estuaries. Environ Sci 2:159–165

Kvenvolden KA (1993) Gas hydrates – geological perspective and global change. Rev Geophys 31(2):173–187. doi:10.1029/93RG00268

Kvenvolden KA (1999) Potential effects of gas hydrate on human welfare. Proc Natl Acad Sci USA 96:3420–3426

Kvenvolden KA, Rogers BW (2005) Gaia's breath – global methane exhalations. Mar Pet Geol 22(4):579–590

Lam P, Jensen MM, Lavik G, van de Vossenberg J, Schmid M, Woebken D, Gutierrez D, Amann R, Jetten MSM, Kuypers MMM (2009) Revising the nitrogen cycle in the Peruvian oxygen minimum zone. Proc Natl Acad Sci USA 106(12):4752–4757

Lamarque J-F (2008) Estimating the potential for methane clathrate instability in the 1% CO_2 IPCC AR4 simulations. Geophys Res Lett 35, L19806. doi:10.1029/2008GL035291

Lammers S, Suess E, Hovland M (1995) A large methane plume east of Bear Island (Barents Sea): implications for the marine methane cycle. Geol Rundsch 84:59–66

Laruelle GG, Dürr HH, Slomp CP, Borges AV (2010) Evaluation of sinks and sources of CO_2 in the global coastal ocean using a spatially-explicit typology of estuaries and continental shelves. Geophys Res Lett 37, L15607. doi:10.1029/2010GL043691

Le Quéré C, Rödenbeck C, Buitenhuis ET, Conway TJ, Lagenfelds R, Gomez A, Labuschagne C, Ramonet M, Nakazawa T, Metzl N, Gillett N, Heimann M (2007) Saturation of the Southern Ocean CO_2 sink due to recent climate change. Science 316:1735–1738

Le Quéré C, Takahashi T, Buitenhuis ET, Rödenbeck C, Sutherland SC (2010) Impact of climate change and variability on the global oceanic sink of CO_2. Global Biogeochem Cycles 24, GB4007

Lee K, Sabine CL, Tanhua T, Kim T-W, Feely RA, Kim H-C (2011) Roles of marginal seas in absorbing and storing fossil fuel CO_2. Energy Environ Sci 4:1133

Lefèvre N, Watson AJ, Olsen A, Ríos AF, Pérez FF, Johannessen T (2004) A decrease in the sink for atmospheric CO_2 in the North Atlantic. Geophys Res Lett 31(7), L07306

Lefèvre N, Watson AJ, Watson AR (2005) A comparison of multiple regression and neural network techniques for mapping in situ pCO_2 data. Tellus 57B:375–384

Leifer I, Patro RK (2002) The bubble mechanism for methane transport from the shallow sea bed to the surface: a review and sensitivity study. Cont Shelf Res 22:2409–2428

Leip A (1999) Nitrous oxide (N_2O) emissions from a coastal catchment in the delta of the Po river: measurements and modeling of fluxes from a Mediterranean lagoon and agricultural soils. Ph.D. thesis, University of Bayreuth, Bayreuth

Lenton A, Cordon G, Bopp L, Metzl N, Cadule P, Tagliabue A, Le Sommer J (2009) Stratospheric ozone depletion reduces ocean carbon uptake and enhances ocean acidification. Geophys Res Lett 36, L12606. doi:10.1029/2009GL038227

Liss PS, Merlivat L (1986) Air-sea exchange rates: introduction and synthesis. In: Buat-Ménard P (ed) The role of air-sea exchange in geochemical cycling. D. Reidel Publishing, Dordrecht, pp 113–127

Liu X, Millero FJ (2002) The solubility of Fe(III) in seawater. Mar Chem 77:43–54

Liu KK, Atkinson L, Quiñones RA, Talahue-McManus L (2010) Biogeochemistry of continental margins in a global context. In: Liu KK, Atkinson L, Quiñones RA, Talahue-McManus L (eds) Carbon and nutrient fluxes in continental margins. Springer, Berlin etc., pp 3–24

Löscher CR, Kock A, Könneke, M, LaRoche J, Bange HW, Schmitz RA (2012) Production of oceanic nitrous oxide by ammonia-oxidizing archaea. Biogeosciences 9:2419–2429. doi:10.5194/bg-9-2419-2012

Lovenduski NS, Gruber N, Doney SC, Lima ID (2007) Enhanced CO_2 outgassing in the Southern Ocean from a positive phase of the Southern Annular Mode. Global Biogeochem Cycle 21, GB2026. doi:10.1029/2006GB002900

Lovenduski NS, Gruber N, Doney SC (2008) Toward a mechanistic understanding of the decadal trends in the Southern Ocean carbon sink. Global Biogeochem Cycle 22, GB3016. doi:10.1029/2007GB003139

Lüthi D, Le Floch M, Bereiter B, Blunier T, Barnola J-M, Siegenthaler U, Raynaud D, Jouzel J, Fischer H, Kawamura K, Stocker TF (2008) High-resolution carbon dioxide concentration record 650,000–800,000 years before present. Nature 453:379–382. doi:10.1038/nature06949

Mackenzie FT, Lerman A, Andersson AJ (2004) Past and present of sediment and carbon biogeochemical cycling models. Biogeosciences 1(1):11–32

Manning AC, Keeling RF (2006) Global oceanic and land biotic carbon sinks from the Scripps atmospheric oxygen flask sampling network. Tellus 58B:95–116. doi:10.1111/j.1600-0889.2006.00175.x

Martens CS, Klump JV (1980) Biogeochemical cycling in an organic-rich coastal marine basin – I. Methane sediment-water exchange processes. Geochim Cosmochim Acta 44:471–490

Martens-Habbena WPM, Berube PM, Urakawa H, De la Torre J, Stahl DA (2009) Ammonia oxidation kinetics determine niche separation of nitrifying Archaea and Bacteria. Nature 461(7266):976–979

Marty D, Nival P, Yoon WD (1997) Methanoarchaea associated with sinking particles and zooplankton collected in the Northeastern tropical Atlantic. Oceanol Acta 20:863–869

Matsumoto K, Sarmiento J, Brezeinski MA (2002) Silicic acid leakage from the Southern Ocean: a possible explanation for glacial atmospheric pCO_2. Global Biogeochem Cycle 16(3):5. doi:10.1029/2001GB001442

Matthews BJH (1999) The rate of air-sea CO_2 exchange: chemical enhancement and catalysis by marine microalgae. Ph.D. thesis, University of East Anglia, Norwich

Mau S, Valentine DL, Clark JF, Reed J, Camilli R, Washburn L (2007) Dissolved methane distributions and air-sea flux in the plume of a massive seep field, Coal Oil Point, California. Geophys Res Lett 34, L22603. doi:10.1029/2007GL031344

McKinley GA, Fay AR, Takahashi T, Metzl N (2011) Convergence of atmospheric and North Atlantic carbon dioxide trends on multidecadal timescales. Nat Geosci 4:606–609. doi:10.1038/NGEO1193

Metzl N (2009) Decadal increase of oceanic carbon dioxide in Southern Indian Ocean surface waters (1991–2007). Deep-Sea Res Part II 56:607–619

Michelsen HA, Irion FW, Manney GL, Toon GC, Gunson MR (2000) Features and trends in Atmospheric Trace Molecule Spectroscopy (ATMOS) version 3 stratospheric water vapor and methane measurements. J Geophys Res 105(D18):22713–22724

Middelburg JJ, Klaver G, Nieuwenhuize J, Wielemaker A, de Haas W, Van der Nat JFWA (1996) Organic matter mineralization in intertidal sediments along an estuarine gradient. Mar Ecol Prog Ser 132:157–168

Middelburg JJ, Nieuwenhuize J, Iversen N, Høgh N, De Wilde H, Helder W, Seifert R, Christof O (2002) Methane distribution in European tidal estuaries. Biogeochemistry 59:95–119

Mikaloff Fletcher SE, Gruber N, Jacobson AR, Doney SC, Dutkiewicz S, Gerber M, Gloor M, Follows M, Joos F, Lindsay K, Menemenlis D, Mouchet A, Müller SA, Sarmiento JL (2007) Inverse estimates of the oceanic sources and sinks of natural CO_2 and the implied oceanic transport. Global Biogeochem Cycle 21, GB1010. doi:10.1029/2006GB002751

Miller LA, Papakyriakou TN, Collins RE, Deming JW, Ehn JK, Macdonald RW, Mucci A, Owens O, Raudsepp M, Sutherland N (2011) Carbon dynamics in sea ice: a winter flux time series. J Geophys Res 116(C2), C02028. doi:10.1029/2009jc006058

Monaco A, Biscay P, Soyer J, Pocklington R, Heussner S (1990) Particle fluxes and ecosystem respons on a continental margin: the 1985–1988 Mediterranean ECOMARGE Experiment. Cont Shelf Res 10(9–11):809–839. doi:10.1016/0278-4343(90)90061-P

Monteiro PMS, Schuster U, Hood M, Lenton A, Metzl N, Olsen A, Rogers K, Sabine CL, Takahashi T, Tilbrook B, Yoder J, Wanninkhof R, Watson AJ (2010) Community white paper. A global sea surface carbon observing system: Assessment of changing sea surface CO_2 and air-sea CO_2 fluxes. In: Hall J,

Harrison DE, Stammer D (eds) Proceedings of OceanObs'09: sustained ocean observations and information for society, vol 2, Venice, Italy, 21–25 Sept 2009, ESA publication WPP-306. doi:10.5270/OceanObs09.cwp.64

Montzka SA, Dlugokencky EJ, Butler JH (2011) Non-CO_2 greenhouse gases and climate change. Nature 476:43–50

Moran JJ, Beal EJ, Vrentas JM, Orphan VJ, Freeman KH, House CH (2008) Methyl sulphides as intermediates in the anaerobic oxidation of methane. Environ Microbiol 10:162–173

Morell JM, Capella J, Mercado A, Bauzá J, Corredor JE (2001) Nitrous oxide fluxes in Caribbean and tropical Atlantic waters: evidence for near surface production. Mar Chem 74:131–143

Mucci A (1983) The solubility of calcite and aragonite in seawater at various salinities, temperatures, and one atmosphere total pressure. Am J Sci 283:780–799

Mudelsee M (2001) The phase relations among atmospheric CO_2 content, temperature and global ice volume over the past 420 ka. Q Sci Rev 20:583–589

Naik H, Naqvi SWA, Suresh T, Narvekar PV (2008) Impact of a tropical cyclone on biogeochemistry of the central Arabian Sea. Global Biogeochem Cycle 22, GB3020. doi:10.1029/2007GB003028

Naqvi SWA (2008) The Indian Ocean. In: Capone DG, Carpenter EJ, Bronk DA (eds) Nitrogen in the marine environment, 2nd edn. Elsevier, Amsterdam, pp 631–681

Naqvi SWA, Jayakumar DA, Narvekar PV, Naik H, Sarma VVSS, D'Souza W, Joseph S, George MD (2000) Increased marine production of N_2O due to intensifying anoxia on the Indian continental shelf. Nature 408:346–349

Naqvi SWA, Bange HW, Gibb SW, Goyet C, Hatton AD, Upstill-Goddard RC (2005) Biogeochemical ocean–atmosphere transfers in the Arabian Sea. Prog Oceanogr 65:116–144

Naqvi SWA, Bange HW, Farías L, Monteiro PMS, Scranton MI, Zhang J (2010) Marine hypoxia/anoxia as a source of CH_4 and N_2O. Biogeosciences 7:2159–2190

Naudin JJ, Cauwet G, Chrétiennot-Dinet MJ, Deniaux B, Devenon JL, Pauc H (1997) River discharge and wind influence upon particulate transfer at the land-ocean interaction: case study of the Rhône river plume. Estuar Coast Shelf Sci 45:303–316. doi:10.1006/ecss.1996.0190

Nevison CD, Weiss RF, Erickson DJ III (1995) Global oceanic emissions of nitrous oxide. J Geophys Res 100:15809–15820

Nevison C, Lueker T, Weiss RF (2004) Quantifying the nitrous oxide source from coastal upwelling. Global Biogeochem Cycle 18, GB1018. doi:1010.1029/2003GB002110

Nightingale PD, Malin G, Law CS, Watson AJ, Liss PS, Liddicoat MI, Boutin J, Upstill-Goddard RC (2000) In-situ evaluation of air-sea gas exchange parameterisations using novel conservative and volatile tracers. Global Biogeochem Cycle 14(1):373–387

Nirmal Rajkumar A, Barnes J, Ramesh R, Purvaja R, Upstill-Goddard RC (2008) Methane and nitrous oxide fluxes in the polluted Adyar River and estuary. SE India Mar Pollut Bull 56:2043–2051

O'Connor FM, Boucher O, Gedney N, Jones CD, Folberth GA, Coppell R, Friedlingstein P, Collins WJ, Chappellaz J, Ridley J, Johnson CE (2010) Possible role of wetlands, permafrost, and methane hydrates in the methane cycle under future climate change: a review. Rev Geophys 48, RG4005. doi:10.1029/2010RG000326

Odum HT, Hoskin CM (1958) Comparative studies of the metabolism of Texas bays. Publ Inst Mar Sci Univ Tex 5:16–46

Odum HT, Wilson R (1962) Further studies on the reaeration and metabolism of Texas bays. Publ Inst Mar Sci Univ Tex 8:23–55

Ohde S, van Woesik R (1999) Carbon dioxide flux and metabolic processes of a coral reef, Okinawa. Bull Mar Sci 65:559–576

Olsen A, Brown KR, Chierici M, Johannessen T, Neill C (2008) Sea surface CO_2 fugacity in the subpolar North Atlantic. Biogeosciences 5:535–547, www.biogeosciences-5/535/2008/

Omar A, Johannessen T, Bellerby RGJ, Olsen A, Anderson LG, Kivimäe C, Omar A, Johannessen T, Bellerby RGJ, Olsen A, Anderson LG, Kivimäe C (2005) Sea-ice and brine formation in Storfjorden: implications for Arctic wintertime air-sea CO_2 flux. In: Drange H, Dokken T, Furevik T, Gerdes R, Berger W (eds) The Nordic seas: an integrated perspective, vol 158, American Geophysical Union Geophysical Monograph. American Geophysical Union, Washington, DC, pp 117–187

Oremland RS (1979) Methanogenic activity in plankton samples and fish intestines: a mechanism for in situ methanogenesis in ocean surface waters. Limnol Oceanogr 24:1136–1141

Orr JC, Fabry VJ, Aumont O, Bopp L, Doney SC, Feely RA, Gnanadesikan A, Gruber N, Ishida A, Joos F, Key RM, Lindsay K, Maier-Reimer E, Matear R, Monfray P, Mouchet A, Najjar RG, Plattner G-K, Rodgers KB, Sabine CL, Sarmiento JL, Schlitzer R, Slater RD, Totterdell IJ, Weirig M-F, Yamanaka Y, Yool A (2005) Anthropogenic ocean acidification over the twenty-first century and its impact on calcifying organisms. Nature 437:681–686

Ostrovsky I (2003) Methane bubbles in Lake Kinneret: quantification and temporal and spatial heterogeneity. Limnol Oceanogr 48:1030–1036

Oudot C, Jean-Baptiste P, Fourré E, Mormiche C, Gueve M, Ternon J-F, Le Corre P (2002) Transatlantic equatorial distribution of nitrous oxide and methane. Deep-Sea Res Part I 49(7):1175–1193

Paull CK, Brewer PG, Ussler W III, Peltzer ET, Rehder G, Clague D (2003) An experiment demonstrating that marine slumping is a mechanism to transfer methane from seafloor gas-hydrate deposits into the upper ocean and atmosphere. Geo-Mar Lett 22:198–203. doi:10.1007/s00367-002-0113-y

Paull CK, Ussler W, Holbrook S, Hill TM, Keaten R, Mienert J, Haflidaon H, Johnson JE, Winters WG, Lorenson TD (2008) Origin of pockmarks and chimney structures on the flanks of the Storegga slide, offshore Norway. Geo-Mar Lett 28:43–51. doi:10.1007/s00367-007-0088-9

Peters GP, Marland G, Le Quéré C, Boden T, Canadell JG, Raupach MR (2012) Rapid growth in CO_2 emissions after the 2008–2009 global financial crisis. Nat Clim Change 2:2–4. doi:10.1038/nclimate1332

Petit JR, Jouzel J, Raynaud D, Barkov NI, Barnola J-M, Basile I, Bender M, Chappellaz J, Davis M, Delaygue G, Delmotte M, Kotlyakov VM, Legrand M, Lipenkov VY, Lorius C, Pépin L, Ritz C, Saltzman E, Stievenard M (1999) Climate and atmospheric history of the past 420,000 years from the Vostok ice core, Antarctica. Nature 399:429–436. doi:10.1038/20859

Pfeil, B, Olsen, A, Bakker, DCE, Hankin S, Koyuk H, Kozyr A, Malczyk J, Manke A, Metzl N, Sabine CL, Akl J, Alin SR, Bates N, Bellerby RGJ, Borges A, Boutin J, Brown PJ, Cai W-J, Chavez FP, Chen A, Cosca C, Fassbender AJ, Feely RA, González-Dávila M, Goyet C, Hales B, Hardman-Mountford N, Heinze C, Hood M, Hoppema M, Hunt CW, Hydes D, Ishii M, Johannessen T, Jones SD, Key RM, Körtzinger A, Landschützer P, Lauvset SK, Lefèvre N, Lenton A, Lourantou A, Merlivat L, Midorikawa T, Mintrop L, Miyazaki C, Murata A, Nakadate A, Nakano Y, Nakaoka S, Nojiri Y, Omar AM, Padin XA, Park G-H, Paterson K, Perez FF, Pierrot D, Poisson A, Ríos AF, Salisbury J, Santana-Casiano JM, Sarma VVSS, Schlitzer R, Schneider B, Schuster U, Sieger R, Skjelvan I, Steinhoff T, Suzuki T, Takahashi T, Tedesco K, Telszewski M, Thomas H, Tilbrook B, Tjiputra J, Vandemark D, Veness T, Wanninkhof R, Watson AJ, Weiss R, Wong CS, Yoshikawa-Inoue H (2013) A uniform, quality controlled Surface Ocean CO2 Atlas (SOCAT). Earth Syst Sci Data 5:125–143. doi:10.5194/essd-5-125-2013

Philippot L (2002) Denitrifying genes in bacterial and Archaeal genomes. Biochim Biophys Acta Gene Struct Expr 1577 (3):355–376

Plattner G-K, Frenzel H, Gruber N, Leinweber A, McWilliams JC (2004) Changing winds and coastal carbon cycle: a case study for an upwelling region. The ocean in a high-CO_2 world. UNESCO, Paris

Plummer DH, Owens NJP, Herbert RA (1987) Bacteria-particle interactions in turbid estuarine environments. Cont Shelf Res 7:1429–1433. doi:10.1016/0278-4343(87)90050-1

Prytherch J, Yelland MJ, Pascal RW, Moat BI, Skjelvan I, Srokosz MA (2010) Open ocean gas transfer velocity derived from long-term direct measurements of the CO_2 flux. Geophys Res Lett 37, L23607. doi:10.1029/2010GL045597

Ramesh R, Purvaja R, Neetha V, Divia J, Barnes J, Upstill-Goddard RC (2007) CO_2 and CH_4 emissions from Indian mangroves and surrounding waters. In: Tateda Y, Upstill-Goddard RC, Goreau T, Alongi D, Nose A, Kristensen E, Wattayakorn G (eds) Greenhouse gas and carbon balances in mangrove coastal ecosytems. Gendai Tosho, Kanagawa, pp 153–164

Rangama Y, Boutin J, Etcheto J, Merlivat L, Takahashi T, Delille B, Frankignoulle M, Bakker DCE (2005) Variability of the net air-sea CO_2 flux inferred from shipboard and satellite measurements in the Southern Ocean south of Tasmania and New Zealand. J Geophys Res 110:1–17. C09005. doi: 10.1029/2004JC002619

Raven J, Caldeira K, Elderfield H, Hoegh-Guldberg O, Liss P, Riebesell U, Shepherd J, Turley C, Watson AJ (2005) Ocean acidification due to increasing atmospheric carbon dioxide, vol 12/05, Policy document. The Royal Society, London

Ravishankara AR, Daniel JS, Portmann RW (2009) Nitrous oxide (N_2O): the dominant ozone-depleting substance emitted in the 21st century. Science 326:123–125

Reagan MT, Moridis GJ (2009) Large-scale simulation of methane hydrate dissociation along the West Spitsbergen Margin. Geophys Res Lett 36, L23612. doi:10.1029/2009GL041332

Redfield AC, Ketchum BH, Richards FA (1963) The influence of organisms on the composition of seawater. In: Hill MN (ed) The sea, vol 2. Wiley Interscience, New York, pp 26–77

Reeburgh WS (2007) Oceanic methane biogeochemistry. Chem Rev 107(2):486–513

Rehder G, Keir RS, Suess E, Pohlmann T (1998) The multiple sources and patterns of methane in North Sea waters. Aquat Geochem 4:403–427

Rhee TS, Kettle AJ, Andreae MO (2009) Methane and nitrous oxide emissions from the ocean: a reassessment using basin-wide observations in the Atlantic. J Geophys Res 114, D12304. doi:10.1029/2008JD011662

Riebesell U, Schulz KG, Bellerby RGJ, Botros M, Fritsche P, Meyerhofer M, Neill C, Nondal G, Oschlies A, Wohlers J, Zollner E (2007) Enhanced biological carbon consumption in a high CO_2 ocean. Nature 450:545–548

Rigby M, Prinn RG, Fraser PJ, Simmonds PG, Langenfelds RL, Huang J, Cunnold DM, Steele LP, Krummel PB, Weiss RF, O'Doherty S, Salameh PK, Wang HJ, Harth CM, Mühle J, Porter LW (2008) Renewed growth of atmospheric methane. Geophys Res Lett 35, L22805. doi:10.1029/2008GL036037

Rixen T, Haake B, Ittekkot V, Guptha MVS, Nair RR, Schlüssel P (1996) Coupling between SW monsoon-related surface and deep ocean processes as discerned from continuous particle flux measurements and correlated satellite data. J Geophys Res 101:28569–28582

Robertson JE, Watson AJ (1992) Thermal skin effect of the surface ocean and its implications for CO_2 uptake. Nature 358:738–740

Romanovski NN, Hubberten H-W, Gavrilov AV, Eliseeva AA, Tipenkio GS (2005) Offshore permafrost and gas hydrate stability zone on the shelf of East Siberian Seas. Geo-Mar Lett 25(2–3):167–182. doi:10.1007/s00367-004-0198-6

Rost B, Riebesell U (2004) Coccolithophores and the biological pump: responses to environmental changes. In: Thierstein HR, Young JR (eds) Coccolithophores: from molecular processes to global impact. Springer, Berlin, pp 99–125

Roy T, Bopp L, Gehlen M, Schneider B, Cadule P, Fröhlicher TL, Segschneider J, Tjiputra J, Heinze C, Joos F (2011) Regional impacts of climate change and atmospheric CO_2 on future ocean carbon uptake: a multimodel linear feedback analysis. J Climate 24(9):2300–2318. doi:10.1175/2010JCLI3787.1

Rysgaard S, Bendtsen J, Delille B, Dieckmann GS, Glud R, Kennedy H, Mortensen J, Papadimitriou S, Thomas DN, Tison JL (2011) Sea ice contribution to the air–sea CO_2 exchange in the Arctic and Southern Oceans. Tellus 63B (5):1–8. doi:10.1111/j.1600-0889.2011.00571.x

Sabine CL, Feely RA, Gruber N, Key RM, Lee K, Bullister JL, Wanninkhof R, Wong CS, Wallace DWR, Tilbrook B, Millero FJ, Peng T-H, Kozyr A, Ono T, Ríos AF (2004) The oceanic sink for anthropogenic CO_2. Science 305:367–371

Sabine CL, Hankin S, Koyuk H, Bakker DCE, Pfeil B, Olsen A, Metzl N, Kozyr A, Fassbender A, Manke A, Malczyk J, Akl J, Alin SR, Bellerby RGJ, Borges A, Boutin J, Brown PJ, Cai W-J, Chavez FP, Chen A, Cosca C, Feely RA, González-Dávila M, Goyet C, Hardman-Mountford N, Heinze C, Hoppema M, Hunt CW, Hydes D, Ishii M, Johannessen T, Key RM, Körtzinger A, Landschützer P, Lauvset SK, Lefèvre N, Lenton A, Lourantou A, Merlivat L, Midorikawa T, Mintrop L, Miyazaki C, Murata A, Nakadate A, Nakano Y, Nakaoka S, Nojiri Y, Omar AM, Padin XA, Park G-H,

Paterson K, Perez FF, Pierrot D, Poisson A, Ríos AF, Salisbury J, Santana-Casiano JM, Sarma VVSS, Schlitzer R, Schneider B, Schuster U, Sieger R, Skjelvan I, Steinhoff T, Suzuki T, Takahashi T, Tedesco K, Telszewski M, Thomas H, Tilbrook B, Vandemark D, Veness T, Watson AJ, Weiss R, Wong CS, Yoshikawa-Inoue H (2013) Gridding of the Surface Ocean CO_2 Atlas (SOCAT) Gridded data products. Earth Syst Sci Data 5:145–153. doi:10.5194/essd-5-145-2013

Santana-Casiano JM, González-Dávila M (2011) pH decrease and effects on the chemistry of seawater. In: Duarte P, Santana-Casiano JM (eds) Oceans and the atmospheric carbon content. Springer, Berlin, pp 95–114

Santana-Casiano JM, González-Davila M, Millero FJ (2006) The role of Fe(II) species on the oxidation of Fe(II) in natural waters in the presence of O_2 and H_2O_2. Mar Chem 99:70–82

Santoro AE, Buchwald C, McIlvin MR, Casciotti KL (2011) Isotopic signature of N_2O produced by marine ammonia-oxidizing archaea. Science 333(6047):1282–1285. doi:10.1126/science.1208239

Sarmiento JL, Gruber N (2002) Sinks for anthropogenic carbon. Phys Today 55:30–36

Sarmiento JL, Gruber N (2006) Ocean biogeochemical dynamics. Princeton University Press, Princeton

Sarmiento JL, Sundquist ET (1992) Revised budget for the oceanic uptake of anthropogenic carbon dioxide. Nature 356:589–593

Sarmiento JL, Orr JC, Siegenthaler U (1992) A perturbation simulation of CO_2 uptake in an ocean general circulation model. J Geophys Res 97(C3):3621–3646

Sarmiento JL, Dunne J, Gnanadesikan A, Key RM, Matsumoto K, Slater R (2002) A new estimate of the CaCO3 to organic carbon export ratio. Global Biogeochem Cycle 16(4):1107. doi:10.1029/2002GB001919

Sasakawa M, Tsunogai U, Kameyama S, Nakagawa F, Nojiri Y, Tsuda A (2008) Carbon isotopic characterization for the origin of excess methane in subsurface seawater. J Geophys Res 113, C03012. doi:10.1029/2007JC004217

Schlünz B, Schneider RR (2000) Transport of terrestrial organic carbon to the oceans by rivers: re-estimating flux and burial rates. Int J Earth Sci 88:599–606

Schmale O, Greinert J, Rehder G (2005) Methane emission from high intensity marine gas-seeps in the Black Sea into the atmosphere. Geophys Res Lett 32, L07609. doi:10.1029/2004GL021138

Schmidt GA, Shindell DT (2003) Atmospheric composition, radiative forcing, and climate change as a consequence of a massive methane release from gas hydrates. Paleoceanography 18(1):1004. doi:10.1029/2002PA000757

Schuster U, Watson AJ (2007) A variable and decreasing sink for atmospheric CO_2 in the North Atlantic. J Geophys Res 112(C11), C11006

Schuster U, Watson AJ, Bates NR, Corbière A, González-Dávila M, Metzl N, Pierrot D, Santana-Casiano M (2009) Trends in North Atlantic sea-surface fCO_2 from 1990 to 2006. Deep-Sea Res Part II 56(8–10):620–629

Scranton MI, McShane K (1991) Methane fluxes in the southern North Sea: the role of European rivers. Cont Shelf Res 11:37–52

Seitzinger SP, Kroeze C (1998) Global distribution of nitrous oxide production and N inputs in freshwater and coastal marine ecosystems. Global Biogeochem Cycle 12:93–113

Seitzinger SP, Kroeze C, Bouwman AE, Caraco N, Dentener F, Styles RV (2002) Global patterns of dissolved inorganic and particulate nitrogen inputs to coastal systems. Estuaries 25:640–655

Shakhova N, Semiletov I (2007) Methane release and coastal environment in the East Siberian Arctic shelf. J Mar Syst 66:227–243

Shakhova N, Semiletov I, Salyuk AN, Belcheva N, Kosmach D (2007) Methane anomalies in the near-water atmospheric layer above the shelf of East Siberian Arctic Shelf. Trans Russ Acad Sci 415:764–768

Shakhova N, Semiletov I, Salyuk A, Yusupov V, Kosmach D, Gustafsson O (2010) Extensive methane venting to the atmosphere from sediments of the East Siberian Arctic Shelf. Science 327:1246–1250. doi:10.1126/science.1182221

Silverman J, Lazar B, Cao L, Caldeira K, Erez J (2009) Coral reefs may start dissolving when atmospheric CO_2 doubles. Geophys Res Lett 36, L05606. doi:10.1029/2008GL036282

Sloan ED (2003) Fundamental principles and applications of natural gas hydrates. Nature 426:353–363. doi:10.1038/nature02135

Sloan ED, Koh CA (2008) Clathrate hydates of natural gases, 3rd edn. CRC Press, Boca Raton

Smith SV, Hollibaugh JT (1993) Coastal metabolism and the oceanic carbon balance. Rev Geophys 31:75–89

Smith SV, Mackenzie FT (1987) The ocean as a net heterotrophic system: implications from the carbon biogeochemical cycle. Global Biogeochem Cycles 1:187–198

Smith SV, Swaney DP, Talaue-McManus L, Bartley JD, Sandhei PT, McLaughlin CJ, Dupra VC, Crossland CJ, Buddemeier RW, Maxwell BA, Wulff F (2003) Humans, hydrology, and the distribution of inorganic nutrient loading to the ocean. Bioscience 53:235–245

Sowers TA, Alley R, Jubenville J (2003) Elemental and isotopic records of atmospheric nitrous oxide covering the last 106,000 years from the GISP II and Taylor Dome ice cores. Science 301:945–948

Spahni R, Chappellaz J, Stocker TF, Loulergue L, Hausammann G, Kawamura K, Flückiger J, Schwander J, Raynaud D, Masson-Delmotte V, Jouzel J (2005) Atmospheric methane and nitrous oxide of the late Pleistocene from Antarctic ice cores. Science 310(5752):1317–1321

Stramma L, Johnson GC, Sprintall J, Mohrholz V (2008) Expanding oxygen-minimum zones in the tropical oceans. Science 320:655–658

Stramma L, Schmidtko S, Levin A, Johnson GC (2010) Ocean oxygen minima expansions and their biological impacts. Deep-Sea Res Part I 57(4):587–595

Suntharalingam P, Sarmiento JL (2000) Factors governing the oceanic nitrous oxide distribution: simulations with an ocean general circulation model. Global Biogeochem Cycle 14:429–454

Suykens K, Delille B, Chou L, De Bodt C, Harlay J, Borges AV (2010) Dissolved inorganic carbon dynamics and air-sea carbon dioxide fluxes during coccolithophore blooms in the northwest European continental margin (northern Bay of Biscay). Global Biogeochem Cycle 24, GB3022. doi:10.2029/2009GB003730

Suzuki A, Kawahata K (2004) Reef water CO_2 system and carbon production of coral reefs: topographic control of system-level perfrormance. In: Shiyomi M, Kawahata H,

Koizumi H, Tsuda A, Awaya Y (eds) Global environmental change in the ocean and on land. Terrapub, Tokyo, pp 229–248

Suzuki T, Ishii M, Aoyama M, Christian JR, Enyo K, Kawano T, Key RM, Kosugi N, Kozyr A, Miller LA, Murata A, Nakano T, Ono T, Saino T, Sasaki K, Sasano D, Takatani Y, Wakita M, Sabine CL (2013) PACIFICA Data Synthesis Project. ORNL/CDIAC-159, NDP-092. Carbon Dioxide Information Analysis Center, Oak Ridge National Laboratory, U.S. Department of Energy, Oak Ridge, Tennessee, USA. doi:10.3334/CDIAC/OTG. PACIFICA_NDP092

Sweeney C, Gloor E, Jacobson AR, Key RM, McKinley G, Sarmiento JL, Wanninkhof R (2007) Constraining global air-sea gas exchange for CO_2 with recent bomb ^{14}C measurements. Global Biogeochem Cycle 21, GB2015. doi:10.1029/2006GB002784

Takahashi T, Feely RA, Weiss RF, Wanninkhof RH, Chipman DW, Sutherland SC, Takahashi TT (1997) Global air-sea flux of CO_2: an estimate based on measurements of sea-air pCO_2 difference. Proc Natl Acad Sci USA 94:8292–8299

Takahashi T, Sutherland SC, Sweeney C, Poisson A, Metzl N, Tilbrook B, Bates N, Wanninkhof RH, Feely RA, Sabine CL, Olafsson J, Nojiri Y (2002) Global sea-air CO_2 flux based on climatological surface ocean pCO_2, and seasonal biological and temperature effects. Deep-Sea Res Part II 49:1601–1622

Takahashi T, Sutherland SC, Wanninkhof R, Sweeney C, Feely RA, Chipman DW, Hales B, Friederich G, Chavez F, Sabine C, Watson AJ, Bakker DCE, Schuster U, Metzl N, Inoue HY, Ishii M, Midorikawa T, Nojiri Y, Koertzinger A, Steinhoff T, Hoppema JMJ, Olafsson J, Arnarson TS, Tilbrook B, Johannessen T, Olsen A, Bellerby R, Wong CS, Delille B, Bates NR, de Baar HJW (2009) Climatological mean and decadal change in surface ocean pCO_2, and net sea-air CO_2 flux over the global oceans. Deep-Sea Res Part II 56:544–577. doi:10.1016/j.dsr2.2008.12.009

Takahashi T, Sutherland SC, Kozyr A (2011) Global ocean surface water partial pressure of CO_2 database: Measurements performed during 1957–2010 (Version 2010). ORNL/CDIAC-159, NDP-088(V2010). Carbon Dioxide Information Analysis Center, Oak Ridge Nat. Lab., US Department of Energy, Oak Ridge, Tenn, doi:10.3334/CDIAC/otg.ndp088 (V2010)

Tanhua T, van Heuven S, Key RM, Velo A, Olsen A, Schirnick C (2010) Quality control procedures and methods of the CARINA database. Earth Syst Sci Data 2:35–49. doi:10.5194/essd-2-35-2010

Telszewski M, Chazottes A, Schuster U, Watson AJ, Moulin C, Bakker DCE, González-Dávila M, Johannessen T, Körtzinger A, Lüger H, Olsen A, Omar A, Padin XA, Ríos A, Steinhoff T, Santana-Casiano M, Wallace DWR, Wanninkhof RH (2009) Estimating the monthly pCO_2 distribution in the North Atlantic using a self-organizing neural network. Biogeosciences 6:1405–1421, http://www.biogeosciences.net/6/1405/2009

Thomas H, Bozec Y, Elkalay K, de Baar HJW (2004) Enhanced open ocean storage of CO_2 from shelf sea pumping. Science 304(5673):1005–1008

Thomas H, Prowe F, van Heuven S, Bozec Y, de Baar HJW, Schiettecatte L-S, Suykens K, Koné M, Borges AV, Lima ID, Doney SC (2007) Rapid decline of the CO_2 buffering capacity in the North Sea and implications for the North Atlantic Ocean. Global Biogeochem Cycle 21, GB4001. doi:10.1029/2006GB002825

Thomas H, Prowe AEF, Lima I, Doney SC, Wanninkhof R, Greatbatch RJ, Schuster U, Corbière A (2008) Changes in the North Atlantic Oscillation influence CO_2 uptake in the North Atlantic over the past 2 decades. Global Biogeochem Cycle 22, GB4027. doi:10.1029/2007GB003167

Tréhua AM, Longb PE, Torresa ME, Bohrmannc G, Rackd FR, Collette TS, Goldbergf DS, Milkovg AV, Riedelh M, Schultheissi P, Bangsj NL, Barrk SR, Borowskil WS, Claypoolm GE, Delwichen ME, Dickenso GR, Graciap E, Guerinf G, Hollandq M, Johnsona JE, Leer Y-J, Lius C-S, Sut X, Teichertu B, Tomaruv H, Vannestew M, Watanabex M, Weinbergery JL (2004) Three-dimensional distribution of gas hydrate beneath southern Hydrate Ridge: constraints from ODP Leg 204. Earth Planet Sci Lett 222:845–862

Tsai W, Liu K-K (2003) An assessment of the effect of sea surface surfactant on global atmosphere–ocean CO_2 flux. J Geophys Res 108:3127. doi:10.1029/2000JC000740

Tsunogai S, Watanabe S, Sato T (1999) Is there a "continental shelf pump" for the absorption of atmospheric CO_2? Tellus 51B(3):701–712

Uncles RJ, Stephens JA (1993) The freshwater-saltwater interface and its relationship to the turbidity maximum in the Tamar estuary, United Kingdom. Estuaries 16:126–141. doi:10.2307/1352770

Upstill-Goddard RC (2006) Air-sea exchange in the coastal zone. Estuar Coast Shelf Sci 70:388–404

Upstill-Goddard RC (2011) The production of trace gases in the estuarine and coastal environment. In: Wolanski E, McLusky DS (eds) Treatise on estuarine and coastal science, vol 2, Geochemistry of estuaries and coasts. Elsevier, Amsterdam, pp 271–309

Upstill-Goddard RC, Owens NJP, Barnes J (1999) Nitrous oxide and methane during the 1994 SW monsoon in the Arabian Sea/northwestern Indian Ocean. J Geophys Res 104:30067–30084

Upstill-Goddard RC, Barnes J, Frost T, Punshon S, Owens NJP (2000) Methane in the southern North Sea: low-salinity inputs, estuarine removal, and atmospheric flux. Global Biogeochem Cycle 14:1205–1217

Van der Nat F-JWA, Middelburg JJ (1998) Seasonal variation in methane oxidation by the rhizosphere of *Phragmites australis* and *Scirpus lacustris*. Aquat Bot 61(2):95–110

Van der Nat F-JWA, Middelburg JJ (2000) Methane emission from tidal freshwater marshes. Biogeochemistry 49(2):103–121

Van der Nat F-JWA, Middelburg JJ, Van Meteren D, Wielemakers A (1998) Diel methane emissions patterns from *Scirpus lacustris* and *Phragmites australis*. Biogeochemistry 41:1–22

Van Scoy KA, Morris KP, Robertson JE, Watson AJ (1995) Thermal skin effect and the air-sea flux of carbon dioxide: a seasonal high-resolution estimate. Global Biogeochem Cycle 9(2):253–262

Volk T, Hoffert MI (1985) Ocean carbon pumps: analysis of relative strengths and efficiencies in ocean-driven atmospheric CO_2 changes. In: Sundquist E, Broecker WS (eds) The carbon cycle and atmospheric CO_2: natural variations Archean to present, vol 32, American Geophysical Union

Geophysical Monograph. American Geophysical Union, Washington, DC, pp 99–110

Wakelin SL, Holt JT, Blackford JC, Allen JI, Butenschön M, Artioli Y (2012) Modeling the carbon fluxes of the northwest European continental shelf: validation and budgets. J Geophys Res 117, C05020. doi:10.1029/2011JC007402

Walsh JJ (1988) On the nature of continental shelves. Academic Press, San Diego

Walter S, Bange HW, Breitenbach U, Wallace DWR (2006) Nitrous oxide in the North Atlantic Ocean. Biogeosciences 3:607–619

Wanninkhof RH (1992) Relationship between wind speed and gas exchange over the ocean. J Geophys Res 97(C5):7373–7382

Wanninkhof RH, McGillis WR (1999) A cubic relationship between air-sea CO_2 exchange and wind speed. Geophys Res Lett 26(13):1889–1892

Wanninkhof RH, Asher WE, Ho DT, Sweeney C, McGillis WR (2009) Advances in quantifying air-sea gas exchange and environmental forcing. Annu Rev Mar Sci 1:213–244

Watson AJ, Orr JC (2003) Carbon dioxide fluxes in the global ocean. In: Fasham MJR (ed) Ocean biogeochemistry: a JGOFS synthesis, Global change. IGBP series. Springer, Berlin, pp 123–143

Watson AJ, Schuster U, Bakker DCE, Bates N, Corbière A, González-Dávila M, Friedrich T, Hauck J, Heinze C, Johannessen T, Körtzinger A, Metzl N, Olaffson J, Oschlies A, Pfeil B, Olsen A, Oschlies A, Santano-Casiano JM, Steinhoff T, Telszewski M, Ríos A, Wallace DWR, Wanninkhof RH (2009) Tracking the variable North Atlantic sink for atmospheric CO_2. Science 326(5958):1391–1393

Wayne RP (2000) Chemistry of atmospheres, 3rd edn. Oxford University Press, Oxford, 775 pp

WDCGG (2012) World Meteorological Organization (WMO) Global Atmosphere Watch (GAW) Data. Greenhouse gases and other atmospheric gases. WMO WDCGG (World Data Centre for Greenhouses Gases) 36(4). Japan Meteorological Agency and the WMO, Tokyo, p. 100

Weiss RF (1974) Carbon dioxide in water and seawater: the solubility of a non-ideal gas. Mar Chem 2:203–205

Weiss RF, Van Woy FA, Salameh PK (1992) Surface water and atmospheric carbon dioxide and nitrous oxide observations by shipboard automated gas chromatography: Results from expeditions between 1977 and 1990. Scripps Institution of Oceanography reference 92–11, ORNL/CDIAC-59, NDP-044, Carbon Dioxide Information Analysis Center, Oak Ridge National Laboratory, US Department of Energy, Oak Ridge, Tennessee, USA

Westbrook GK, Thatcher KE, Rohling EJ, Piotrowski AM, Palike H, Osborne AH, Nisbet EG, Minshull TA, Lanoiselle M, James RH, Huhnerbach V, Green D, Fisher RE, Crocker AJ, Chabert A, Bolton C, Beszczynska-Moller A, Berndt C, Aquilina A (2009) Escape of methane gas from the seabed along the West Spitsbergen continental margin. Geophys Res Lett 36, L15608. doi:10.1029/2009GL039191

Wetzel P, Winguth A, Maier-Reimer E (2005) Sea-to-air fluxes CO_2 fluxes from 1948 to 2003: a model study. Global Biogeochem Cycle 19, GB2005. doi:10.1029/2004GB002339

Wever TF, Abegg F, Fiedler HM, Fechner G, Stender IH (1998) Shallow gas in the muddy sediments of Eckernförde Bay, Germany. Cont Shelf Res 18:1715–1739

Wittke F, Kock A, Bange HW (2010) Nitrous oxide emissions from the upwelling off Mauritania (NW Africa). Geophys Res Lett 37, L12601. doi:10.1029/2010GL042442

Wollast R (1998) Evaluation and comparison of the global carbon cycle in the coastal zone and in the open ocean. In: Brink KH, Robinson AR (eds) The global coastal ocean. Wiley, New York, pp 213–252

Wollast R, Chou L (2001) The carbon cycle at the ocean margin in the northern Gulf of Biscay. Deep-Sea Res Part II 48:3265–3293

Wong CS, Christian JR, Wong S-KE, Page J, Xie L, Johannessen S (2010) Carbon dioxide in surface sea water of the eastern North Pacific Ocean (Line P), 1973–2005. Deep-Sea Res Part I 57:687–695

Woodwell GM, Rich PH, Hall CAS (1973) Carbon in estuaries. In: Woodwell GM, Pecan EV (eds) Carbon and the biosphere. United States Atomic Energy Commission, Springfield, pp 221–240

Wuchter C, Abbas B, Coolen MJL, Herfort L, van Bleijswijk J, Timmers P, Strous M, Teira E, Herndl GJ, Middelburg JJ, Schouten S, Damste JSS (2006) Archaeal nitrification in the ocean. Proc Natl Acad Sci USA 103(33):12317–12322

Wuchter C, Abbas B, Coolen MJL, Herfort L, van Bleijswijk J, Timmers P, Strous M, Teira E, Herndl GJ, Middelburg JJ, Schouten S, Damste JSS (2007) Archaeal nitrification in the ocean (vol 103, pg 12317, 2006). Proc Natl Acad Sci USA 104(13):5704–5704

Yavitt JB, Fahey TJ (1991) Production of methane and nitrous oxide by organic soils within a northern hardwood forest ecosystem. In: Oremland RS (ed) Biogeochemistry of global change: radiatively active trace gases. Chapman and Hall, New York, pp 261–277

Yoshida O, Inoue HY, Watanabe S, Suzuki K, Noriki S (2011) Dissolved methane distribution in the South Pacific and the Southern Ocean in austral summer. J Geophys Res 116, C07008. doi:10.1029/2009JC006089

Zhang G-L, Zhang J, Liu SM, Ren J-L, Xu J, Zhang F (2008) Methane in the Changjiang (Yangtze River) Estuary and its adjacent marine area: riverine input, sediment release and atmospheric fluxes. Biogeochemistry 91:71–84

Zhang G-L, Zhang J, Liu S-M, Ren J-L, Zhao Y-C (2010) Nitrous oxide in the Changjiang (Yangtze River) estuary and its adjacent marine area: riverine input, sediment release and atmospheric fluxes. Biogeosciences 7:3505–3516

Zumft WG (1997) Cell biology and molecular basis of denitrification. Microbiol Mol Biol Rev 61:533–616

Ocean–Atmosphere Interactions of Particles

Gerrit de Leeuw, Cécile Guieu, Almuth Arneth, Nicolas Bellouin, Laurent Bopp, Philip W. Boyd, Hugo A.C. Denier van der Gon, Karine V. Desboeufs, François Dulac, M. Cristina Facchini, Brett Gantt, Baerbel Langmann, Natalie M. Mahowald, Emilio Marañón, Colin O'Dowd, Nazli Olgun, Elvira Pulido-Villena, Matteo Rinaldi, Euripides G. Stephanou, and Thibaut Wagener

Abstract

This chapter provides an overview of the current knowledge on aerosols in the marine atmosphere and the effects of aerosols on climate and on processes in the oceanic surface layer. Aerosol particles in the marine atmosphere originate predominantly from direct production at the sea surface due to the interaction between wind and waves (sea spray aerosol, or SSA) and indirect production by gas to particle conversion. These aerosols are supplemented by aerosols produced over the continents, as well as aerosols emitted by volcanoes and ship traffic, a large part of it being deposited to the ocean surface by dry and wet deposition. The SSA sources, chemical composition and ensuing physical and optical effects, are discussed. An overview is presented of continental sources and their ageing and mixing processes during transport. The current status of our knowledge on effects of marine aerosols on the Earth radiative balance, both direct by their interaction with solar radiation and indirect through their effects on cloud properties, is discussed. The deposition on the ocean surface of some key species, such as nutrients, their bioavailability and how they impact biogeochemical cycles are shown and discussed through different time and space scales approaches.

4.1 Introduction

An aerosol consists of a suspension of particles and its surrounding medium. For atmospheric aerosols the surrounding medium is the air in which the aerosol particles (or droplets) are suspended. Often the term aerosol is used to refer to only the particles or droplets. In this contribution we follow this convention and refer to aerosol particles whether they occur as particles or as droplets, i.e. chemicals in their liquid phase, or dissolved in a liquid, are also referred to as particles. Aerosol sources are numerous and can be of natural or anthropogenic origin, they can enter the atmosphere directly as particles (primary aerosol) or form in the atmosphere from their precursors in the gas phase through physical and chemical reactions (secondary aerosol formation).

Aerosols are an important constituent of the atmospheric boundary layer. Aerosol particles provide surfaces for heterogeneous chemical processes, they also act as a condensation sink for atmospheric trace

G. de Leeuw (✉)
e-mail: gerrit.leeuw@fmi.fi

C. Guieu (✉)
e-mail: guieu@obs-vlfr.fr

gases. Hygroscopic particles serve as cloud condensation nuclei. The chemical and physical properties of aerosol particles are very variable in both space and time and depend on the proximity of sources and sinks and the chemical and physical transformation during their atmospheric lifetime. Particle sizes may vary from a few nanometres to some tens of micrometres. Particles at the high end of this size range are sufficiently heavy that their atmospheric residence time is very short and hence their concentrations are negligible, although in hurricane conditions such large sea spray particles may be important in the transfer of ocean–atmosphere transfer of heat and water vapour (e.g., Andreas et al. 2008). Physical processes, in particular the vertical transport and removal of particles by dry deposition to the surface, depend on particle size. Very small particles are subject to growth by condensation and coagulation. Transport is determined by turbulence and Brownian diffusion. Very large particles having sufficient mass are subject to gravitational forces resulting in rapid sedimentation. These processes, in addition to formation by direct emission and secondary processes, chemical transformations and in-cloud processing, determine the number concentrations of aerosol particles which may vary by 10 orders of magnitude depending on size.

The particle size distribution describes the variation of the aerosol number concentration as function of particle size (radius or diameter, specified for a certain relative humidity to account for differences due to hygroscopic growth). Figure 4.1 shows a schematic representation of a hypothetical particle size distribution, presented as number size distribution (bottom) or mass size distribution (top) with five different modes (cluster, nucleation, Aitken, accumulation and coarse modes). The predominant origin of particles contributing to a certain size range is indicated and will be further discussed in Sect. 4.2. As a result of various interacting processes, the most abundant aerosol particles in the atmosphere are those with a radius of a few tenths of microns. Particles in this size range experience a minimum in the efficiency of dry deposition and they are often referred to as accumulation mode particles. The atmospheric lifetime of these particles is relatively long, on the order of a few days to a few weeks depending on the surface roughness and related deposition velocity, and their main removal mechanism is wet deposition.

Aerosols in the marine atmosphere originate from a variety of production and transformation processes. Sea spray aerosol is directly produced at the sea surface through the interactions between wind and surface waves. Ship emissions and volcanoes also contribute to primary aerosol in the marine atmosphere. Secondary aerosol formation from gases released from the sea surface also contribute significantly to marine atmosphere aerosol loading. In addition, aerosols formed over land by either primary or secondary formation processes are transported over the oceans and contribute substantially to the aerosol concentrations over most of the world's oceans. Estimates of the mass concentrations show that the largest aerosol contributions on a global scale are from sea spray aerosol and desert dust (Andreae and Rosenfeld 2008; Jickells et al. 2005). A comprehensive review of marine aerosols was published by Lewis and Schwartz (2004). An update on the status of sea spray aerosol production was published by de Leeuw et al. (2011).

Aerosol particles are important both because they affect atmospheric processes and, after deposition to the sea surface, because they affect processes in sea water. Their effects in the atmosphere are very diverse and depend on the chemical and physical properties of the aerosol particles. Aerosols have a strong impact on climate both due to scattering and absorption of incoming solar radiation (direct effect) and through their effects on cloud properties and associated cloud albedo (first indirect effect) and precipitation (second indirect effect). Optical properties of aerosol particles are determined by their size relative to the wavelength of incident light and their chemical composition (which determines their complex refractive index). The optical properties such as angular scattering (the aerosol phase function) and absorption for a given aerosol size distribution and chemical composition can be computed using a Mie code (Mie 1908). For non-spherical particles more sophisticated codes need to be applied. The scattering and absorption efficiency are near zero for very small particles and near two for very large particles, where small and large are relative to the wavelength, λ, of the incident radiation, i.e. determined by the size parameter $2\pi r/\lambda$ (r is particle in situ radius). Therefore, the particles most important for climate, as determined by the product of the particle size distribution and the scattering efficiency, are in the accumulation mode.

4.1 Introduction

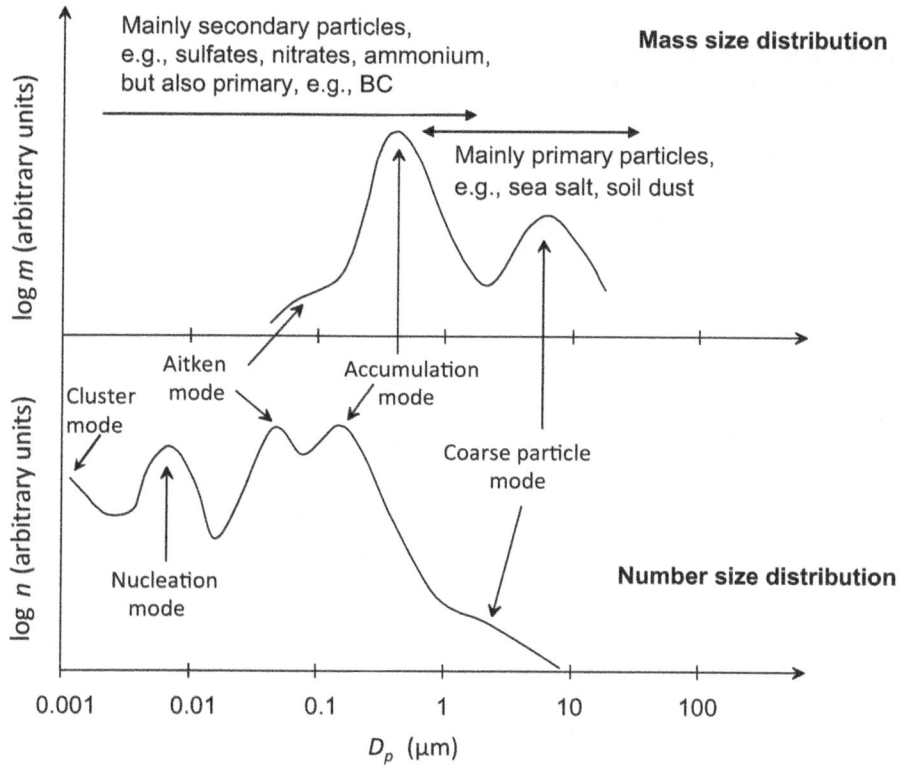

Fig. 4.1 Schematic representation of a characteristic particle size distribution showing the different modes and the origin of particles which predominantly contribute to a certain size range (see text for further explanation of sources). The *bottom figure* shows a number size distribution and the *top figure* shows the associated mass size distributions. Mass is related to number by m = $\pi D_p^3 \rho/6$, where D_p is the dry particle diameter (m) and ρ is the particle density (g m^{-3}) (Figure courtesy of Dr. Aki Virkkula, Finnish Meteorological Institute, Helsinki, Finland)

When aerosols are deposited at the ocean surface, a number of processes between dissolved forms and particles (such as dissolution, adsorption, and aggregation) take place on various timescales. Many of these processes impact surface ocean chemistry and in particular the cycles of elements of biogeochemical interest (macro and micronutrients such as nitrogen, phosphorus, iron etc.). New atmospheric nutrients entering the surface layer of the ocean may fertilise phytoplankton growth and as such may affect CO_2 draw-down and production of other gases such as DMS, that may modify global climate on long time scales. The complex interactions between atmospheric deposition, marine biogeochemistry, carbon export and sequestration in the deep ocean and their feedbacks on climate have been recognised as an important field of research over the past 20 years and a synthesis of current knowledge is presented here.

In Sect. 4.2 we discuss the occurrence of aerosol particles in the marine atmosphere through their production, transport and transformation and deposition. The discussion on production includes the direct production of sea spray aerosol, its organic enrichment and indirect production of aerosols in the marine atmosphere. Non-marine sources are discussed as well as their transport and transformation in the marine atmosphere and removal by deposition. In Sects. 4.3 and 4.4 effects of aerosols in the marine atmosphere, i.e. direct and indirect effects on climate and effects of atmospheric aerosol deposition on marine biology are discussed. Section 4.5 discusses atmospheric deposition and its biogeochemical impacts into the ocean.

4.2 Aerosol Production and Transport in the Marine Atmosphere

In this section an overview of marine sources of aerosol is presented, starting with the direct production of sea spray aerosol and the enrichment of sea spray aerosol in organic material, followed by secondary production in the marine atmosphere. Subsequently non-marine aerosol sources are discussed.

4.2.1 Sources of Aerosol in the Marine Atmosphere

Aerosols are produced directly at the sea surface from the interaction between wind and waves, or indirectly through secondary processes involving gases exchanged between ocean and atmosphere or gases produced from anthropogenic activities over the ocean (for instance, shipping producing extensive amounts of NOx). The aerosol load in the marine atmosphere is further augmented by the direct production of aerosols over land and their subsequent transport over the ocean. Most particles in the marine atmosphere are deposited to the ocean where they may affect biological processes. Particles which are produced from the ocean and transported over land may affect regional air quality (visibility and corrosion) and climate, and they may play a role in cycling from the marine to the terrestrial environments.

4.2.1.1 Sea Spray Aerosol Production

Sea spray aerosol (SSA) consists of a suspension, in air, of particles that are directly produced at the sea surface (de Leeuw et al. 2011a). Sea spray aerosols may be enriched in certain substances, i.e. their chemical composition may deviate from that of sea salt. These particles exist mainly in the liquid phase (i.e. as drops). The radii of these particles vary from around 10 nm to at least several millimeters. The atmospheric residence times of the particles vary from seconds to minutes for larger particles, for which gravitational sedimentation is the principal removal mechanism, to days for smaller particles, for which removal is primarily by precipitation. The size of an SSA particle is commonly specified by its equilibrium radius at a relative humidity (RH) of 80 %, r_{80}.

Many measurements indicate that the relative concentrations of the major solutes in sea spray particles are similar to their relative concentrations in bulk seawater, although this may not be the situation for some substances as a consequence of the formation process, or of exchange with the atmosphere subsequent to formation. SSA particles are said to be enriched or depleted in such substances, and the enrichment factor, defined as the ratio of the concentration of a substance to the concentration of one of the major constituents of bulk seawater (typically sodium) in the particle to the same ratio for bulk seawater, may be less (depletion) or greater than unity (enrichment).

The aerosol consisting of sea spray particles in the atmosphere has traditionally been termed "sea salt aerosol", but in de Leeuw et al. (2011a) it was denoted "sea spray aerosol" in recognition that the composition of the particles may differ from that of bulk seawater. One consequence of this difference is that the hygroscopic and cloud droplet activation properties of sea spray particles may differ from those calculated under the assumption that the particles are composed only of sea salt. In particular, in biologically active waters, sea spray aerosol has been observed to be enriched in organic matter (OM) and the contribution of OM to sea spray aerosol has been an important area of recent research, as discussed in Sect. 4.2.2.2.

The production of sea spray aerosol (SSA) was recently reviewed by de Leeuw et al. (2011a) who critically examined laboratory and field experimental results on sea spray production, on the enrichment in organic matter, and on the measurement and parameterisation of whitecap coverage, and placed it in the context of previous understanding which was comprehensively reviewed by Lewis and Schwartz (2004). The review by de Leeuw et al. (2011a) included material published in the peer-reviewed literature until early 2010. These authors considered 13 production flux formulations, as well as fluxes measured by Norris et al. (2008) using the eddy correlation method, and these formulations are provided in Fig. 4.1 and in the Appendix of de Leeuw et al. (2011a). Below we briefly summarise some of that material to provide background and context for an overview of more recent work.

SSA particles are formed at the sea surface mainly by breaking waves via bubble bursting and by the tearing off of wave crests at elevated wind speeds (>9 ms^{-1}). When a wave breaks, air is entrained into the water and dispersed into a cloud of bubbles (Thorpe 1992). The resulting white coloured area of the sea surface is often denoted a "whitecap" on account of enhanced, wavelength-independent scattering of visible radiation by the interfaces between water and bubbles. The fraction of the sea surface covered by white area is defined as the whitecap fraction, W.

The bubbles rise to the surface and float, where water drains off. The film cap of the bubble becomes unstable and when it opens it fragments into many small droplets, the so-called film droplets, and the water jet rising in the remaining cavity breaks up into a stream of 1–6 'jet' droplets.

The production flux of SSA particles can be specified as either the interfacial flux, i.e. the flux of those particles leaving the sea surface, or as the effective flux, which is defined as the flux of those particles produced at the sea surface that attain a given height, typically 10 m, above mean sea level. They thus remain in the atmosphere for a sufficiently long time to participate in processes such as the scattering and absorption of solar radiation, cloud formation and atmospheric chemistry. For small SSA particles (i.e. those with r_{80} smaller than about 1 μm), the effective flux can, for all practical purposes, be considered to be the same as the interfacial flux. For medium SSA particles (those with r_{80} between about 1 and 25 μm), the effective flux becomes increasingly less than the interfacial flux with increasing r_{80}. For larger SSA particles, which have short atmospheric residence times and typically do not attain heights more than a few meters above the sea surface, the effective flux is essentially zero.

The SSSF (Sea Spray aerosol Source Function) is a numerical representation of the size-dependent production flux of SSA particles:

$$f(r_{80}) = dF(r_{80})/d\log_{10}r_{80} \quad (4.1)$$

where $f(r_{80})$ denotes the number of particles in a given infinitesimal range of the common logarithm of r_{80}, $d\log_{10}r_{80}$, introduced into the atmosphere per unit area per unit time, and $F(r_{80})$ is the total number flux of particles of size less than r_{80}.

An expression for the SSSF required as input to models would represent the size-dependent production flux expressed by Eq. 4.1 as a function of the controlling ambient variables a, b, …; i.e. $f(r_{80}; a, b,…)$. The near-surface wind speed, commonly measured and expressed at a reference height of 10 m, U_{10}, is thought to be the dominant factor affecting sea spray production. Other factors that are expected to affect the SSA production flux are those affecting sea state, such as fetch (the upwind distance over the water of nearly constant wind velocity) and atmospheric stability (often parameterised by the air-sea temperature difference), which also affects vertical transport; seawater temperature and salinity; and the presence, amount, and nature of surface-active substances.

As discussed in de Leeuw et al. (2011a), the effect of water temperature on the resultant size distribution was investigated by Mårtensson et al. (2003) (at 2 °C, 5 °C, 15 °C, and 25 °C) and by Sellegri et al. (2006) (at 4 °C and 23 °C), while effects of salinity were investigated by Mårtensson et al. (2003) and Tyree et al. (2007). Nilsson et al. (2007) compared the Mårtensson et al. (2003) parameterisation with production fluxes derived from eddy covariance measurements at Mace Head (assumed water temperature of 12 °C) and the Clarke et al. (2006) parameterisation derived from profile measurements at the coast of Hawaii (water temperature ca. 25 °C). Both comparisons provided favourable results thus confirming the effect of water temperature on the SSA source flux from two independent types of measurements.

The Mårtensson et al. (2003) experimental data was used by Sofiev et al. (2011) to derive a modification of the Monahan et al. (1986) SSSF formulation which resulted in a temperature and salinity dependent SSSF. This modified SSSF was implemented in the dispersion model SILAM (Sofiev et al. 2006) and applied to compute the distribution of sea salt over the North Atlantic and Western Europe, as well as globally. The influence of sea surface temperature and salinity were evaluated using data from several campaigns, long-term in situ and satellite data (MODIS AOD).

An approach combining satellite observations, in situ data from six cruises and model results was presented by Jaeglé et al. (2011). These authors compared model results (GEOS-Chem, with the Gong (2003) formulation for the SSSF) with MODIS and AERONET AOD observations and in situ aerosol measurements. Modelled mass concentrations of coarse mode sea salt aerosol (SS) were overestimated at high wind speeds over the Southern, North Pacific and North Atlantic Oceans, but underestimated over warm tropical waters of the Central Pacific, Atlantic and Indian Oceans. The in situ observations were used to derive an empirical SS source function depending on both wind speed and SST. This resulted in a correction to the Gong (2003) source function. Using Gong (2003) with this correction, the model results for AOD agree significantly better with the MODIS and AERONET observations and provide an explanation for the high AOD observed over the tropical oceans.

In contrast, Witek et al. (2007a, b) did not find a water temperature dependence of the difference between modelled and measured mass concentrations, where the NAAPS model was compared with measurements from five open-ocean shipboard campaigns covering a range of water temperatures from less than 10 °C to about 30 °C. The Mårtensson et al. (2003) data show a size dependent effect of water temperature which crosses over at r_{80} of about 30–40 nm. One may argue that in the mass concentration the water temperature effect would cancel out, but because the mass is dominated by larger particles one would expect that the mass increases with increasing water temperature. Witek et al. included particles with aerodynamic radius at 55 % RH of up to 5 µm in their calculation of the sea salt aerosol mass. In NAAPS the sea salt dry mass emission flux is simply parameterised as a function of wind speed only (F = 1.37×10^{-13} $U_{10}^{3.41}$ [kg m^{-2} s^{-1}]) and has no SST dependence.

Hultin et al. (2011) conducted wave tank experiments using fresh Baltic Sea water with a salinity of 6–7, much lower than oceanic sea water (salinity ~33). These authors observed a clear dependence of aerosol production (r_{80} between 0.01 and 0.9 µm) on water temperature, i.e. a distinct decrease for all particle sizes with increasing water temperature accompanied by a decrease of dissolved oxygen. As discussed by Hultin et al. (2011) several authors have studied the effect of dissolved gases on the production of sea salt aerosol. For instance, Stramska et al. (1990) observed that more sea salt aerosol particles are produced in water in which dissolved oxygen is super-saturated than in water where dissolved oxygen is sub-saturated. Dissolved oxygen affects the bubble size distribution and thus also the resulting aerosol spectral flux (Lewis and Schwartz 2004). However, Hultin et al. (2011) conclude that the range of dissolved oxygen encountered during their measurements is too small to significantly affect the bubble size distributions in the size range (>2 mm) of importance for their measurements and speculate that the biological activity responsible for the decreased dissolved oxygen concentrations also alters the surface chemistry and the surfactant concentrations which in turn reduce particle production.

Norris et al. (2012) used micrometeorological measurements of SSA fluxes at the open North Atlantic to formulate a source function in terms of only U_{10}. This source function lies within the range of earlier formulations for particles with $r_{80} < 1$ µm but decreases more rapidly for larger particles.

4.2.1.2 Organic Enrichment of Particulate Organic Matter in Sea Spray Aerosol

In biologically rich seawater, accumulation of organic substances at the sea surface can result in enrichment of organic matter in sea spray particles, especially for submicron particles (Blanchard 1964; Middlebrook et al. 1998; O'Dowd et al. 2004). As far back as 1948, Woodcock (1948) showed that drops produced by bubbles bursting in areas with high concentrations of plankton could carry irritants across the air-sea interface. Blanchard (1963, 1964) extended research into enrichment of organic matter in sea spray and its subsequent transfer into the atmosphere.

Surface-active OM of biogenic origin (such as lipidic and proteinaceous material and humic substances), enriched in the oceanic surface layer and transferred to the atmosphere by bubble-bursting processes, are the most likely candidates to contribute to the observed organic fraction in marine aerosol (Gershey 1983; Mochida et al. 2002).

The observed organic aerosol characteristics are consistent with laboratory studies on aerosol generated from Atlantic sea water (Gershey 1983) that showed a peak in organic aerosol concentration, and a concomitant increase in WIOC (water insoluble organic carbon) and high-molecular-mass surface-active fractions, during periods of phytoplankton blooming. Moreover, the increasing enrichment of the aerosol organic fraction with decreasing size is consistent with thermodynamic predictions (Oppo et al. 1999) of bubble-bursting processes under conditions in which the ocean surface layer becomes concentrated with surfactant material that can be incorporated into sea spray drops in addition to inorganic salts.

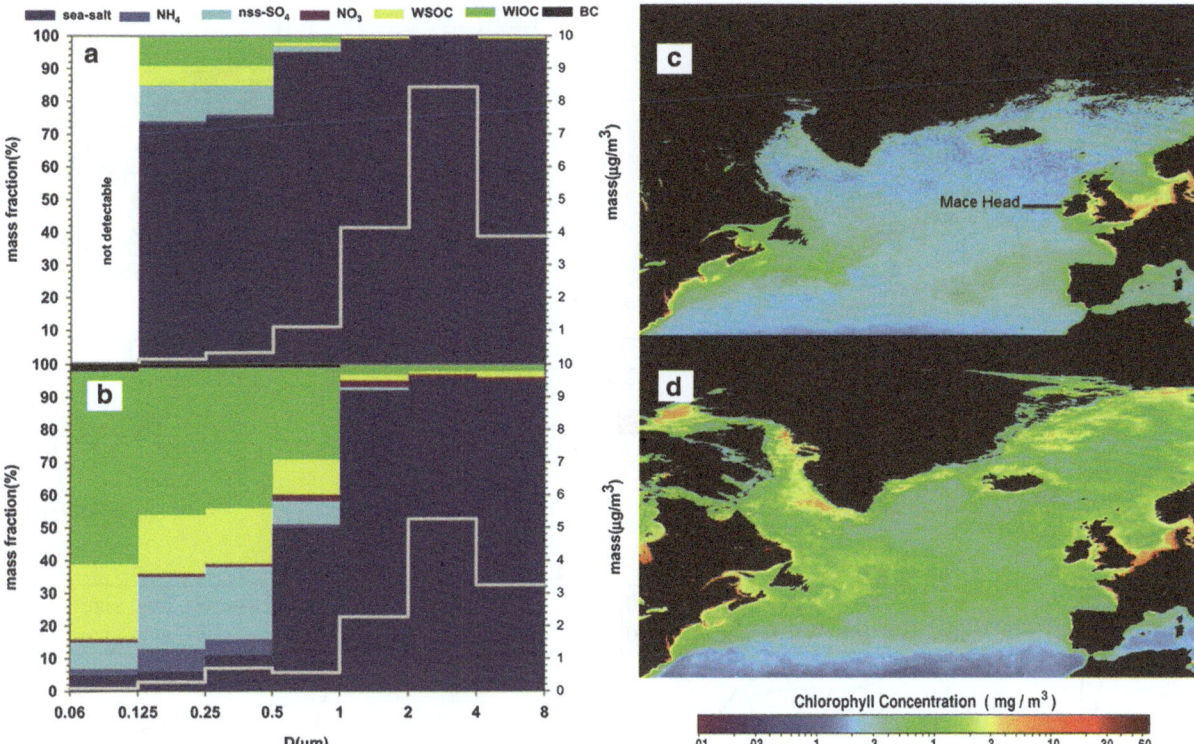

Fig. 4.2 Average mass concentration of total particulate matter (*white line, right axis*) and mass fraction (*colours, left axis*) of sea salt, NH4, nss-SO4, NO3, water-soluble organic matter (*WSOM*), water-insoluble organic matter (*WIOM*), and black carbon (*BC*) in several size ranges for North Atlantic marine aerosol sampled at Mace Head, Ireland, in clean marine air during periods of (**a**) low biological activity, November (2002) January (2003) and February (2003); and (**b**) high biological activity, March-October, (2002). Radius corresponds to relative humidity of approximately 70 %. For low biological activity mass concentrations of aerosol constituents other than sea salt were below detection limits for the size range 0.03–0.06 μm. Oceanic chlorophyll *a* concentrations over the North Atlantic for periods of (**c**) low and (**d**) high biological activity are 5 year averages (1998–2002) over the same months as for the composition measurements, based on satellite measurements of ocean colour (Courtesy of SeaWiFS Project, NASA/Goddard Space Flight Center and ORBIMAGE) (Adapted from O'Dowd et al. (2004))

In 2004, O'Dowd et al. (2004) and Cavalli et al. (2004) reported significant organic mass enrichment in submicron aerosol (Fig. 4.2) that possessed a strong seasonality following the chlorophyll *a* seasonal pattern. The organic matter comprised both water soluble and water insoluble organic matter (WSOM/WIOM). These studies, and a more extended study by Yoon et al. (2007) suggested that the WIOM was primary in origin. This suggestion was corroborated by gradient flux measurements (Ceburnis et al. 2008) at Mace Head which demonstrated that the WIOM had a gradient similar to sea salt, indicating a surface (i.e. primary) source while WSOM possessed a gradient identical to non-sea-salt (nss) sulphate, indicating transfer from gas phase to aerosol surfaces (i.e. secondary aerosol production).

These N.E. Atlantic results were corroborated by measurements in the South Atlantic at Amsterdam Island (Sciare et al. 2009). Since 2008, an Aerodyne high resolution Time of Flight aerosol mass spectrometer has been continuously deployed at Mace Head and this continuous database of real-time chemical composition has led to further elucidation of sea spray aerosol chemical properties. In particular, Ovadnevaite et al. (2011a) report the regular occurrence of significant primary organic aerosol plumes at concentrations often exceeding those reported in heavily polluted air (e.g. 4 μg m^{-3}) and extending for periods exceeding 24 h. They also reported a unique primary organic mass spectral fingerprint hitherto unreported. In that study, it was also reported that the organic aerosol was 55 % oxygenated and

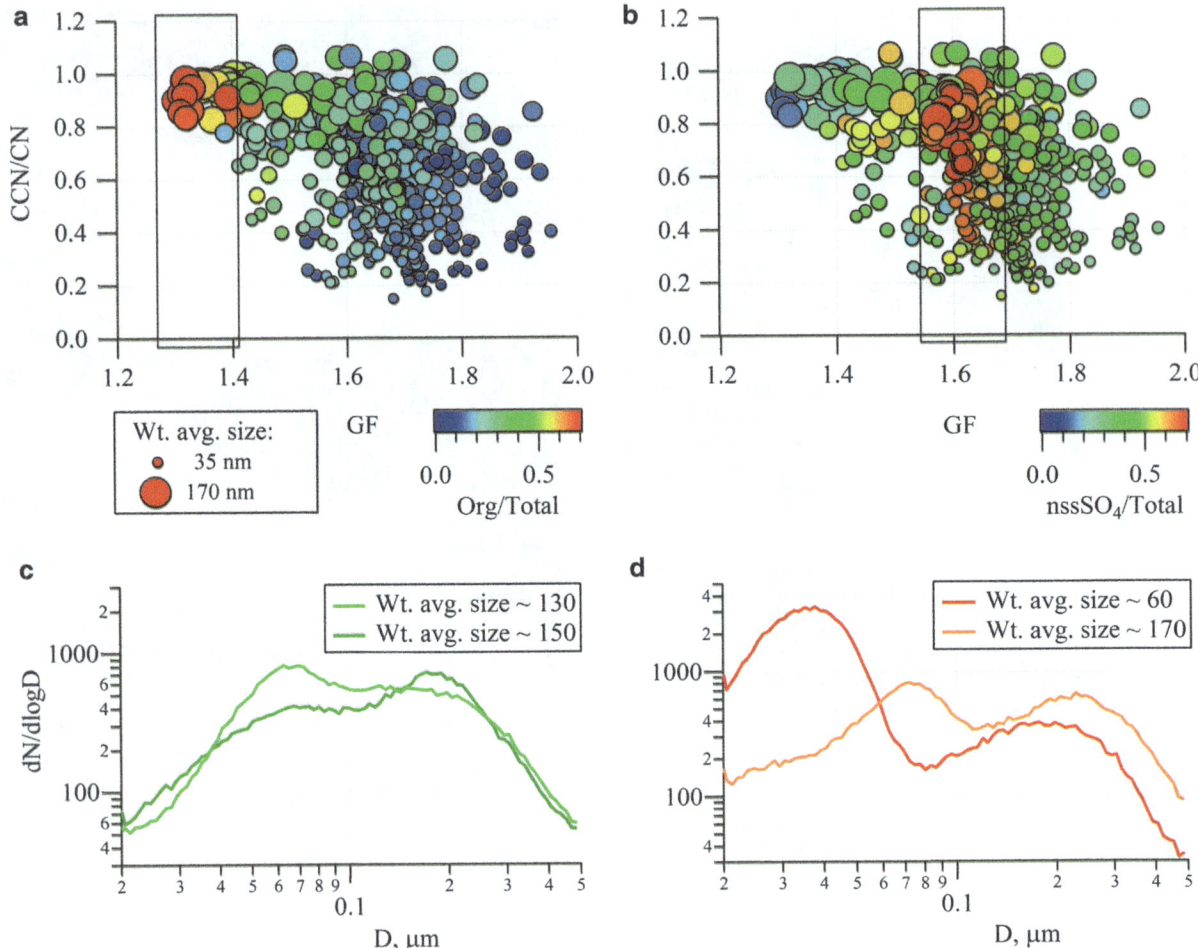

Fig. 4.3 (**a** and **b**) CCN 0.75 % activity (CCN/CN) as a function of GF (at 90 % RH), chemical composition (*colour scale*) and weighted average particle size (*size of the circle*). CN is the total particle number above 20 nm in diameter; the colour scale represents the dominance of a given chemical species. Measurement periods: 02nd – 27th May 2009, 11th – 28th August 2009 and 14th July – 12th August 2010. Note, the measurement periods cover periods much longer than individual plume events. In Figure (**a**), the *boxed region* highlights particles dominated by primary organic matter while Figure (**b**) highlights the particles dominated by sulphate. Particles to the extreme right of both figures are dominated by sea salt mass. (**c**) Two organic-dominated size distributions (on 00:00 UTC – 22:00 UTC 16th August 2009 and 13:30 UTC-16:30 UTC 05th August 2010) and their resultant weighted diameters and (**d**) the same for sulphate-dominated distributions (on 21:00 UTC – 22:00 UTC 02nd August 2010 and 06:00 UTC – 10:00 UTC 09th August 2010) (Reproduced from Ovadnevaite et al. (2011b) by permission of the American Geophysical Union)

45 % hydrocarbon – like. The oxygenated component must have a very low solubility to be consistent with the previous off-line WSOM/WIOM ratios reported. The correlation coefficient between the AMS hydrocarbon-like and oxygenated organics was 0.97, pointing to a common source, and a degree of chemical ageing of sea spray. The large enrichment of hydrocarbon-like organics suggest low water uptake, as shown in Fig. 4.3 (Ovadnevaite et al. 2011b) where hygroscopic growth factors for highly enriched organic particles have a growth factor of 1.3; however, these aerosols have almost a 100 % CCN activation efficiency. So, while water uptake is low at subsaturated conditions (negatively influencing the direct radiative effect), it can be high in supersaturated conditions (positively influencing the indirect aerosol radiative effect).

It has also been speculated that the observed properties of marine primary organic aerosols may be driven by the peculiar physico-chemical properties of marine hydrogels transferred into the atmosphere through the bubble bursting process. Evidence of the transfer of biogenic mucus-like exopolymers in sea spray aerosol have been provided, for instance, by Leck and Bigg (2005) from transmission electron microscopy analyses.

Russell et al. (2010) observed an ocean derived primary organic aerosol component in marine aerosol dominated by carbohydrate-like material, based on FTIR measurements of submicron marine aerosol over the North Atlantic and Arctic Oceans and on Positive Matrix Factorisation data elaboration. According to the authors, the primary marine signal in submicron marine aerosol is made on average for 88 % of hydroxyl groups. The apparent solubility of the carbohydrate-like components in an aqueous phase suggests that DOC provides the source of most primary organics, although the authors recognise that in bloom conditions in productive waters POC could also contribute, as shown by Facchini et al. (2008).

More recently, Decesari at al. (2011) presented the results of a multi-technique investigation of the chemical properties of marine organic aerosol collected during a cruise in the NE Atlantic Ocean that downsized the role of hydroxyl groups in favour of carboxyls and carbonyls. Moreover, the work of Decesari et al. (2011) pointed out that both primary and secondary processes contribute to the observed organic aerosol load over remote oceanic regions, with secondary products comprising both the atmospheric evolution of primary organics and the gas-to-particle conversion processes of volatile organic precursors emitted by marine biota.

Laboratory Studies

Laboratory studies have partially elucidated the properties of organic enriched sea spray. For example, Sellegri et al. (2006) investigated the impact of the artificial surfactant sodium dodecyl sulphate (SDS) on bubble-mediated spray production. In particular, they examined the physical size distribution using online spectrometers and found that the sub-micron sea salt distribution was tri-modal with modal diameters of 50, 110 and 350 nm resulting in a distribution which peaks at a (dry) diameter of 100 nm. With the addition of SDS, however, the peak diameter reduced to 50–80 nm, depending on the type of bubbling. The mode at 350 nm became more prominent when SDS was introduced and when the foam was artificially burst by blowing air over the foam, this mode dominated. The prominence of the large-diameter mode in the presence of a surfactant is consistent with the suggested increase in mean spray size when enriched in organics (O'Dowd et al. 2004; Yoon et al. 2007). In contrast, Tyree et al. (2007) investigated spray size distributions produced from natural sea water for winter and summer DOC concentrations and found little difference regardless of whether or not artificial, filtered or unfiltered sea water was used. Changes were seen, however, in number concentration as 20–40 % more spray droplets were observed for the winter sample compared to that of the summer. More recent studies by Fuentes et al. (2011) focussed on evaluating the impact of nanogel and DOC plankton exudates both in the laboratory and during research cruises. They found an increase in the production of particles smaller than 100 nm for organic carbon concentrations >175 μM. The sea spray produced contained a volume fraction of organic carbon 8–37 % which was somewhat lower than the maximum enrichment fraction observed in the field. Fuentes et al. (2011) suggest that the observed shift to larger mean sizes for enriched sea spray aerosol observed by Yoon et al. (2007) is inconsistent with their results. However, Fuentes et al. (2011) conducted experiments in natural seawater enriched with organics released by algal laboratory cultures which were subjected to 0.2 μm filtration in order to remove bacteria and avoid biodegradation of the organic matter, which may have caused the discrepancy.

Keene et al. (2007) conducted bubble bursting experiments using highly oligotrophic seawater from near the Bermuda coast and found enrichment at all sizes with the enrichment factor increasing with reducing particle size. The most detailed off-line chemical laboratory study was conducted by Facchini et al. (2008a) who produced sea spray in plankton rich North East Atlantic waters amidst a large bloom. They found that the mass fraction of organic matter approached 75 % (Fig. 4.4) for the smallest sizes (down to 0.062 μm diameter) to 20 % at sizes less than 1 μm. Supermicron particles contained less than a few percent organic mass fraction. The majority of the enriched organic matter was water insoluble organic matter (WIOM) and the mass fraction of WIOM and sea salt, as a function of size, replicated very closely the mass fraction observed in air (during the same

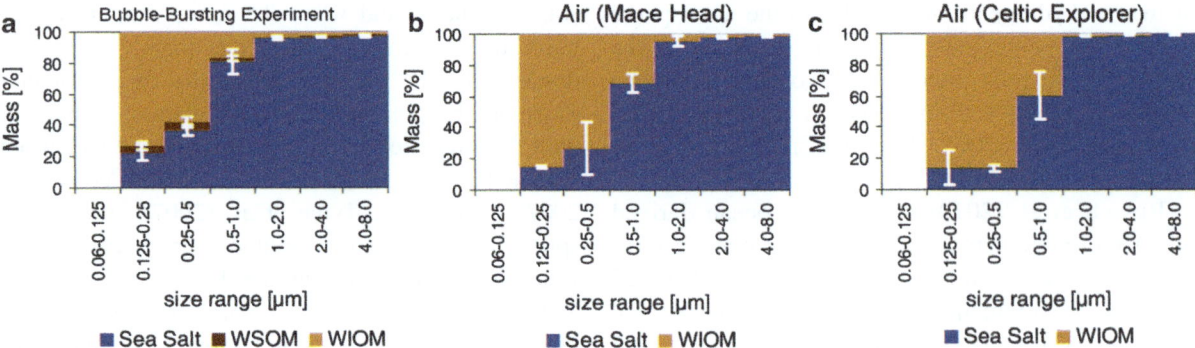

Fig. 4.4 (a) Mass fraction of sea salt, water-soluble organic matter (*WSOM*), and water-insoluble organic matter (*WIOM*) as a function of particle radius sampled at approximately 70 % RH, (a) for seawater bubble-bursting chamber experiments with fresh seawater, conducted in a shipboard laboratory in a plankton bloom over the N.E. Atlantic (May–June 2006), (b) for clean marine air at Mace Head, Ireland, May-June 2006, and (c) for clean marine air 200–300 km offshore west–northwest of Mace Head in a plankton bloom coincident in time with aforementioned samples (Adapted from Facchini et al. (2008a))

cruise) and that sampled previously at Mace Head. This comparison suggests that the vast majority of the WIOM observed in clean air samples are primary in origin. Facchini et al. (2008a) reported that the WIOM consisted of colloids and aggregates exuded by phytoplankton.

Global Distribution of Organic Enrichment

A number of international studies have expanded the measurement picture emerging from NE Atlantic waters, corroborating the findings of OM enrichment in seaspray. Namely, the following cruises: MAP (Marine Aerosol Production, e.g., Facchini et al. 2008a), OOMPH (Organics over the Ocean Modifying Particles; Zorn et al. 2008), ICEALOT (International Chemistry Experiment in the Arctic LOwer Troposphere; Russell et al. 2010; Frossard et al. 2011), and RHaMBLe (Reactive Halogens in the Marine Boundary Layer; Lee et al. 2010), all demonstrating the enrichment of OM in sea spray, albeit to different degrees. O'Dowd et al. (2008) integrated the studies of O'Dowd et al. (2004), Cavalli et al. (2004) and Yoon et al. (2007) with the eddy correlation microphysical flux measurements of Geever et al. (2005) to produce the first combined organic–inorganic sea spray source function and applied it to the REMOTE (REgional MOdel with Tracer Extension) regional climate model. This "chemical" parameterisation for organic enrichment could be applied to any sea spray physical source function and was indeed applied to global budgets by Langmann et al. (2008), Vignati et al. (2010) and Myriokefalitakis et al. (2010).

Figure 4.5 illustrates the global distribution of submicron sea salt and water insoluble organic matter using the parameterisation of Vignati et al. (2010) with the TM5 (Tracer Model 5; Krol et al. 2005) chemical transport model.

The studies by Lapina et al. (2011) and Gantt et al. (2011) are more advanced in that Lapina et al. (2011) apply a water temperature dependent source function, while Gantt et al. (2011) extended the scheme to include a wind speed dependency of organic enrichment – that is, the OM enrichment decreases with wind speed, but still the net OM increased with wind speed. Using this new scheme, Gantt et al. (2011) improved significantly the agreement between measured and predicted OM mass, as illustrated in Fig. 4.6.

Meskhidze et al. (2011) applied the Gantt et al. (2011) scheme and the Vignati et al. (2010) parameterisation, modified using Facchini et al. (2008a) to include size dependence, with the NCAR Community Atmosphere Model CAM5. The findings of Meskhidze et al. (2011) are that different mechanisms contribute to the marine OM fluxes, with a major contribution of marine organic aerosols to the submicron organic aerosol mass over the tropical and mid-latitudes, while methane sulphonate dominates at high latitudes. The Gantt et al. (2011) parameterisation yields a more accurate representation of the seasonal cycle of marine organic aerosol mass concentrations than Vignati et al. (2010).

One question, however, continues to arise given that the OM enrichment scheme is based on chlorophyll *a* concentration fields and that is: "Is

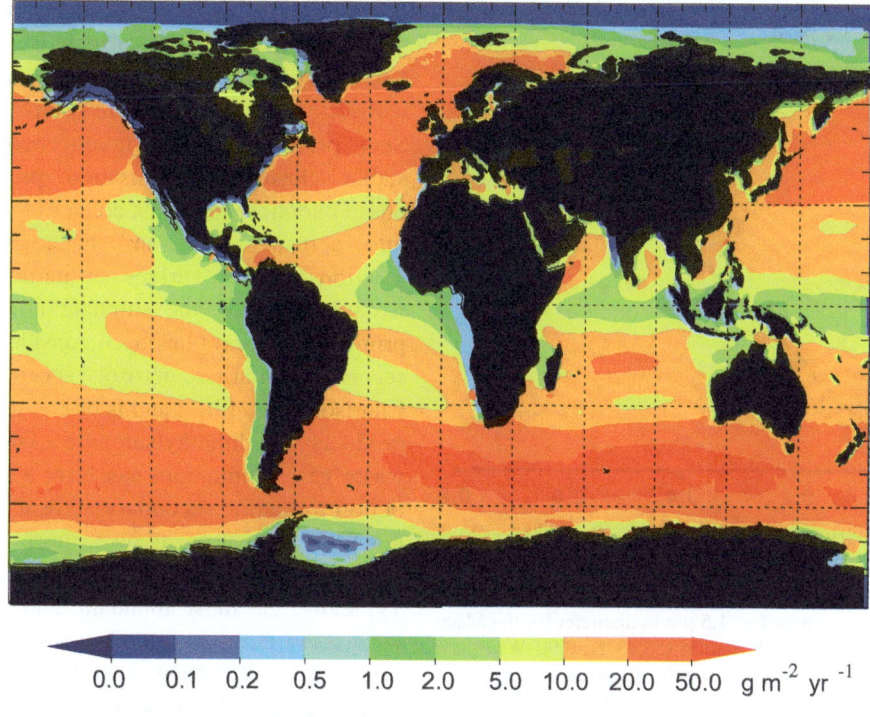

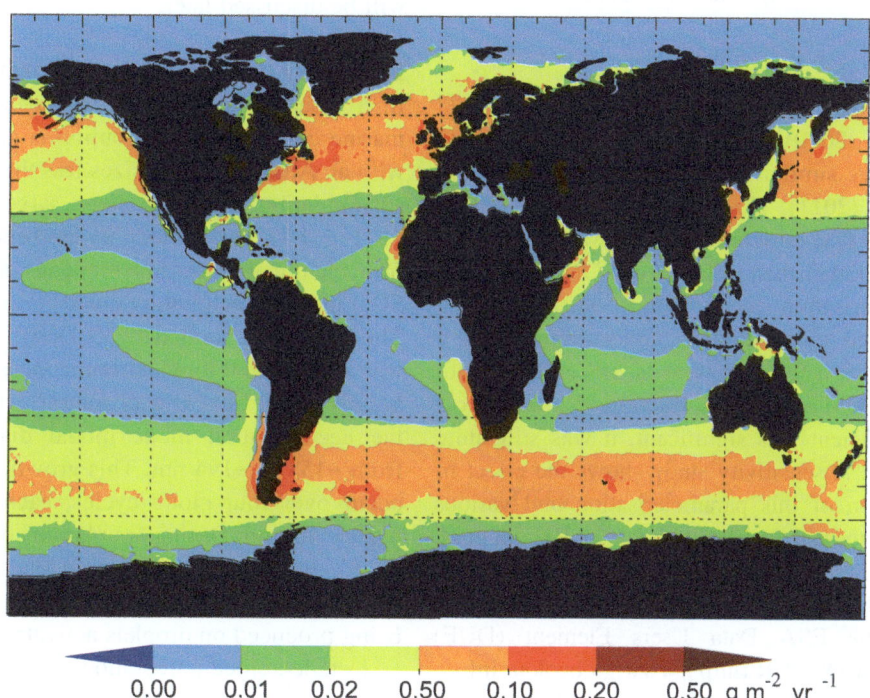

Fig. 4.5 Global distribution of mass flux of sea salt (*upper panel*) and water-insoluble organic matter WIOM (*lower panel*) in sea spray, with 0.1 μm < r_{80} < 1 μm averaged over a 1-year period in 2002–2003 using the TM5 chemical transport model as described in Vignati et al. (2010) (E. Vignati, private communication 2010)

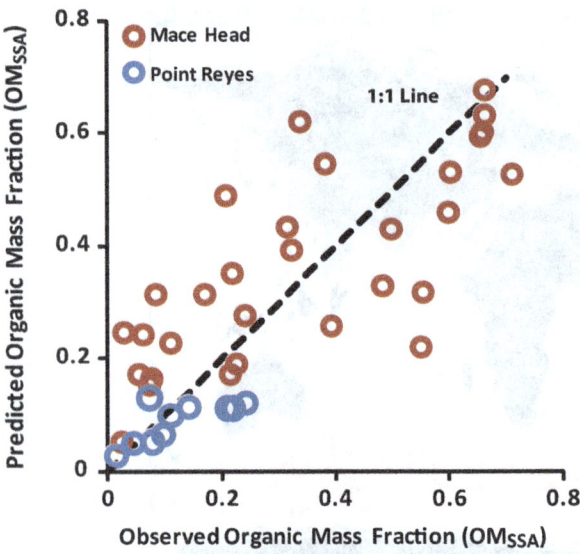

Fig. 4.6 Scatterplot of predicted versus observed organic mass fraction of sea spray aerosol < 1.5 μm in diameter for the Mace Head Atmospheric Research Station (53.33°N, 9.90°W) in *red* and of sea spray aerosol < 2.5 μm in diameter for the Point Reyes National Seashore IMPROVE site (38.12°N, 122.91°W) in *blue*, with the 1:1 line as a *black dotted line* (Adapted from Gantt et al. 2011)

4.2.1.3 Secondary Aerosol Formation in the Marine Atmospheric Boundary Layer

Secondary aerosol formation is the production of aerosol via gas-to-particle conversion processes. Such processes include homogeneous nucleation of stable clusters, condensation processes, aqueous phase chemical reactions converting dissolved gasses into aerosol mass, and heterogeneous chemical reactions on the surface of particles. Historically, nss-sulphate (i.e. the sulphate not associated with primary sea spray) has been considered the dominant secondary marine aerosol species (Shaw 1983; Charlson et al. 1987) and is one of the main oxidation products of dimethylsulphide (DMS), a plankton waste gas. Ammonium can form a significant contribution to marine secondary aerosol; however, it does not form aerosol on its own, more so, it requires the presence of an acid aerosol which it can neutralise. Typically, the most abundant acid is sulphuric acid; however, a range of organic acids, including methane sulphonic acid (MSA), also an oxidation product of DMS, can also be present in significant amounts, as will be discussed later.

Secondary Inorganic Aerosol Formation

In a cloudy marine boundary layer, the submicron marine aerosol size distribution is generally bimodal with an Aitken mode at sizes less than 100 nm and an accumulation mode at sizes larger than 100 nm (cf. Fig. 4.1). This bimodality has been shown to result from chemical processing (aqueous phase oxidation of SO_2) in non-precipitating clouds (Hoppel et al. 1986). This was corroborated by O'Dowd et al. (1999a, b) who illustrated through airborne measurements of progressive cloud cycling in marine stratocumulus that the accumulation mode modal diameter increased from ~158 to ~194 nm. This growth was observed to occur over four cloud cycles, each taking approximately 40 min. During simulations of this case study, 30 % nss-sulphate production occurred in droplets activated on sulphate nuclei, with the remainder being produced on droplets activated on sea salt.

Detailed cloud parcel modelling studies of the heterogeneous oxidation of SO_2 to aerosol sulphate were conducted by O'Dowd et al. (2000) both for activated droplets and un-activated haze particles. The simulations showed that dissolved ozone and hydrogen peroxide were the dominant oxidants, with the ozone oxidation pathway dominant on sea salt

chlorophyll *a* the best surrogate for OM enrichment?" Russell et al. (2010) suggest there was little or no relationship. A similar conclusion was reached by Lapina et al. (2011). However, Gantt et al. (2011) found that the OM enrichment was best correlated to chlorophyll *a* rather than to DOM or POM. The original parameterisation was based on correlating satellite-derived chlorophyll *a* concentrations, in a box 1,000 km × 1,000 km west of Mace Head, with the measured OM enrichment fraction. While the correlation coefficient was significant, it was still quite low (r = 0.3). A sensitivity study shows the effect of the formulation of this parameterisation and various modifications thereof (Albert et al. 2012). Rinaldi et al. (2013) reassessed the relationship between OM enrichment and chlorophyll *a* using analysis chlorophyll *a* data from the ESA Data Users Element (DUE) project GLOBCOLOUR (http://www.globcolour.info/). The reanalysis combined SeaWifs, MODIS and MERIS platforms, interpolated to reduce data loss due to clouds. The results are impressive with *r* increasing from 0.3 to 0.75–0.8. This study also reports that chlorophyll *a* is the best surrogate for OM enrichment in sea spray.

activated droplets. The production of sulphate across the aerosol size distribution is non-linear and a significant amount of sulphate production (75–90 %) occurred in sea salt based droplets. The number concentration of activated salt nuclei also significantly influences the total amount of sulphate produced. Below cloud, the amount of sulphate produced in sea salt aerosol is limited by the carbonate buffering capacity (Sievering et al. 1992). However, once activated, in-cloud production can exceed by many times the cloud-free production due to the transient buffering capacity of activated droplets (O'Dowd et al. 2000). Up to 2 $\mu g\ m^{-3}$ nss-sulphate could be produced at SO_2 concentrations of the order of 500 ppt. Organic acids and nitric acid tend to be more associated with the sea salt modes and reduces alkalinity and consequently the amount of sulphate produced.

In terms of sulphate aerosol, nss-sulphate is typically partially neutralised by ammonia to different degrees, leading to sulphuric acid, ammonium bisulphate or ammonium sulphate. In Polar Regions, nss-sulphate is typically in the form of sulphuric acid (O'Dowd et al. 1997) with the ammonium to sulphate molar ratio increasing at lower latitudes.

Secondary Organic Marine Aerosol

The second most abundant aerosol sulphur species is methane sulphonic acid (MSA), an organic acid which is also an oxidation product of DMS. It is the single most dominant secondary organic species (Facchini et al. 2008b). MSA in the aerosol phase results from condensation of gas phase MSA; however, it appears to be semi-volatile as gas phase concentrations have been observed to be inversely correlated with dew point, reflecting a sensitive equilibrium partitioning (Berresheim et al. 2002).

Over the past few years studies over marine remote regions have allowed identification of typical marine SOA components, other than MSA and DMS oxidation products. The presence of monomethylammonium (MMA^+), dimethylammonium (DMA^+) and trimethylammonium (TMA^+) salts in marine aerosol particles was reported for the first time by Gibb et al. (1999). Their presence was attributed to secondary production, suggesting the condensation of volatile alkyl amines, degassed from the sea, through acid–base reactions, in analogy with NH_4^+. This hypothesis has been strengthened, in recent years, by the observation that alkyl amines participate in SOA formation in different environments reacting with acids (Murphy et al. 2007; Angelino et al. 2001; Tan et al. 2002).

Facchini et al. (2008b) report DMA^+ and diethylammonium (DEA^+) salt concentrations ranging, together between <0.4 and 56 ng m^{-3} in submicron marine aerosol particles collected over the North Atlantic Ocean during Spring and Summer. The authors highlight the importance of alkylammonium salts in marine aerosols, observing that they are the most abundant organic species, after MSA, in submicron marine particles. Alkyl-ammonium salts represent on average 11 % of the marine SOA and 35 % of the aerosol water soluble organic nitrogen (WSON). Facchini et al. (2008b) present also considerable evidence that DMA^+ and DEA^+ are secondary aerosol components, originating from biogenic precursors emitted by the ocean. The maxima in the accumulation mode, as is the case for other well-known secondary components ($nssSO_4$, NH_4, MSA), supports the hypothesis of a gas-to-particle conversion process responsible for the accumulation of alkyl-ammonium salts in the fine aerosol fraction. Moreover, DMA^+ and DEA^+ concentrations measured at Mace Head were always higher in clean marine samples than in polluted air masses, as for MSA, therefore a natural biogenic source is very likely.

Confirming the findings of Facchini et al. (2008b), Müller et al. (2009) reported non-negligible monomethylammonium (MA^+), DMA^+ and DEA^+ in submicrometer particles at Cape Verde, during the season of enhanced oceanic biological activity. Also, Sorooshian et al. (2009) observed DEA^+ in submicron particles over the North Pacific Ocean, reporting a certain correlation with chlorophyll a sea surface concentrations, further supporting the hypothesis of a biogenic origin of marine aerosol amines.

Besides alkylammonium salts and MSA, carboxylic and di-carboxylic acids are commonly reported in marine aerosol (Kawamura and Sakaguchi 1999; Claeys et al. 2010; Crahan et al. 2004; Sorooshian et al. 2007; Aggarwal and Kawamura 2008), accounting for less than 10 % of total particulate organic carbon. Usually, a secondary origin is attributed to the detected di-carboxylic acids (Kawamura et al. 2010), of which oxalic acid is often reported as the most abundant (Kawamura et al. 1996a, b). However, oxidised organics, such as C_5–C_{10} carboxylic or di-carboxylic acids, can also be produced by the oxidative degradation of primary particles generated by sea spray and rich in fatty acids (Kawamura and Sakaguchi 1999). Recently, Rinaldi et al. (2011) presented convincing evidence that an important fraction of marine aerosol oxalic acid may derive from the oxidation, in clouds, of glyoxal.

Recent instrumental advances have allowed a deeper insight into the chemical composition of marine organic aerosols. Using liquid chromatography/negative ion electrospray ionisation mass spectrometry, Claeys et al. (2010) investigated marine organic aerosol chemical composition at Amsterdam Island (Southern Indian Ocean). They managed to characterise about 25 % of the analysed marine aerosol WSOC, which is a remarkable result. MSA (17–21 %), oxalate (5 ± 2 %), malonate (1.8 ± 0.9 %) and organosulphates (0.8 ± 1.5 %) were the major identified components. The organosulphates characterised in Claeys et al. (2010) can be considered tracers for an SOA formation process that is specific to the marine environment, that is, oxidation of marine biomass. More specifically, the organosulphates correspond to sulphate esters of C9–C13 hydroxyl carboxylic acids, which are attributed to oxidation of unsaturated fatty acid of phytoplanktonic origin.

In addition, Decesari et al. (2011) demonstrated, through an ensemble of NMR and LC-MS analyses, that marine aerosol WSOC, the fraction traditionally associated to SOA, is the combination of a less hydrophilic fraction, consisting of a distribution of C_8–C_9 alkanoic acids and diacids, and of a more oxidised, more hydrophilic fraction, where sulphate esters of C_6–C_{11} hydroxycarboxylic acids were found together with MSA and low-molecular weight amines and acids. These results highlight the complexity of the chemical composition of marine SOA and provide evidence for the coexistence in marine SOA of the products of the atmospheric oxidation of primary biogenic materials emitted within sea spray and of compounds deriving from the gas-to-particle conversion of volatile organic compounds emitted by marine biota.

Further, Zhou et al. (2008) found evidence that primary organic matter emitted within sea spray is a dominant sink for the OH radical, with its consequent degradation and the likely production of a series of low-molecular weight organic compounds. These can

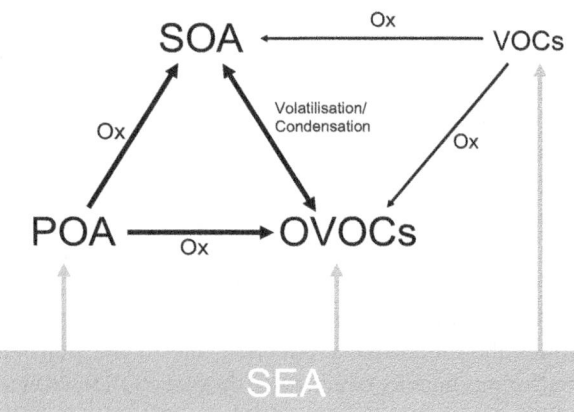

Fig. 4.7 Conceptual picture of the interactions of organic aerosol sources within the marine boundary layer

partition into the gas phase and contribute to SOA formation.

The picture emerging from these results is a complex one, in which primary and secondary aerosol sources interact to generate the observed organic aerosol burden in the MBL. These interactions are schematically shown in Fig. 4.7. Notwithstanding recent improvements, current knowledge of the chemical composition of marine SOA and on their formation mechanisms remains limited and further research is required to address the many unresolved issues.

New Particle Formation in the Marine Boundary Layer?

While the majority of the secondary marine aerosol mass is thought to be produced via heterogeneous and/or condensation processes, the number concentration of secondary marine aerosol is determined by homogeneous nucleation, or new particle formation. New particle production may occur in the marine boundary layer (Russell et al. 1994; Pandis et al. 1994) or in the free troposphere after which it can be entrained into the marine boundary layer (Raes 1995), although the relative importance of these two sources has been an issue of debate. Over the open oceans, there have only been a few recorded observations of new particle formation (Covert et al. 1992, 1996a, b; Clarke et al. 1998; Hoppel et al. 1994; Ehn et al. 2010). Observations of a significant nucleation event by Kollias et al. (2004) in the southeastern Pacific appear to be due to emissions from South America (Tomlinson et al. 2007).

4.2 Aerosol Production and Transport in the Marine Atmosphere

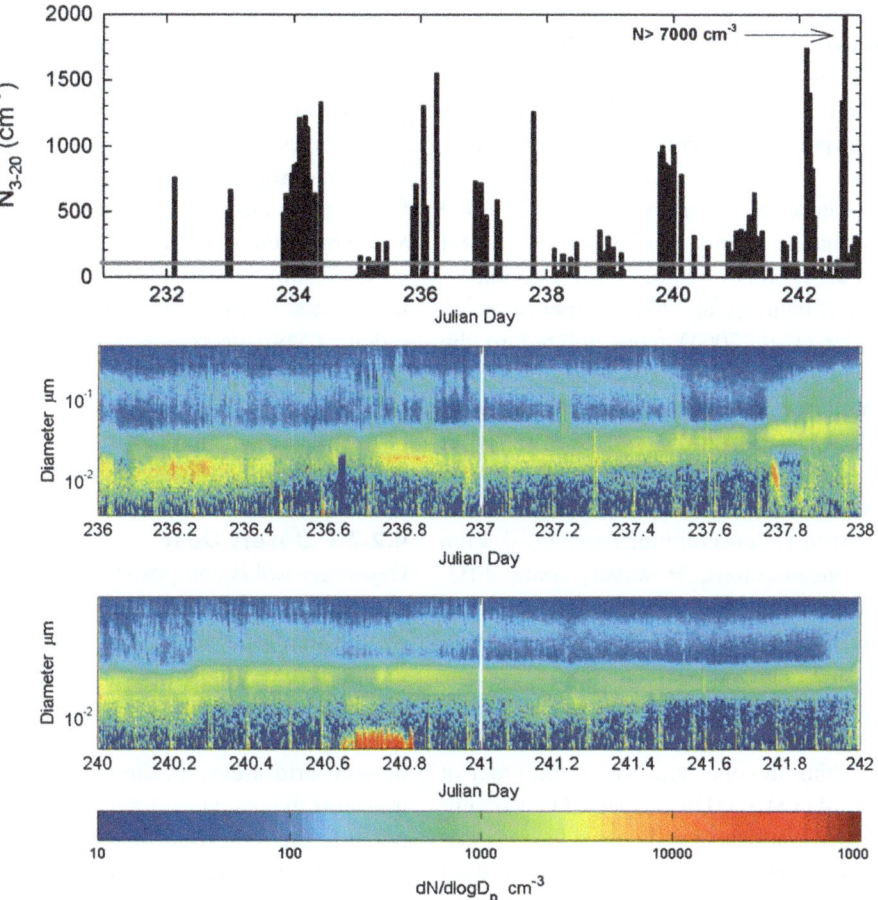

Fig. 4.8 (*Top*) N_{3-20} concentrations in clean marine air at Mace Head for a 12 day period in August 2009, illustrating clean nucleation mode event occurrence. (*Middle*) Combined nSMPS and standard SMPS-derived aerosol size distributions (3–500 nm diameter) corresponding to selected aerosol growth event on JD 236–237 and (*bottom*) on JD 240–241. A coastal event is evident and distinct from the open ocean event on JD240.6-JD240.8 (Reproduced from O'Dowd et al. (2010) by permission of the American Geophysical Union)

Examination of some of the experimental data indicates that under typical marine conditions, nucleation does not occur. Ultrafine particles observed over the Pacific by Covert et al. (1992) could be explained by entrainment from the free troposphere, while ultrafine particle concentrations observed in the tropics were considered to result from *in situ* particle nucleation relating to the natural DMS cycle (Clarke et al. 1998). However, ultra-fine particles appear to occur in polar marine air masses in the Antarctic region (O'Dowd et al. 1997) and in a very recent analysis of nucleation and Aitkin mode aerosol in North East Atlantic air sampled at Mace Head, O'Dowd et al. (2010) report the regular appearance of a recently formed nucleation mode (D ~ 10–15 nm) followed by subsequent growth to sizes of 50 nm over 24–48 h timescales (O'Dowd et al. 2010), as shown in Fig. 4.8.

A detailed aerosol nucleation and dynamics parcel modelling study, with monte carlo simulations, by Pirjola et al. (2000) evaluated the likelihood of new particle formation occurring in the marine boundary layer, taking DMS-derived sulphuric acid as the main nucleating candidate. These authors considered both binary nucleation of sulphuric acid and water and ternary nucleation of sulphuric acid, water and ammonia under a range of realistic aerosol regimes. They concluded that the occurrence of new particles in the unperturbed marine boundary could not be explained by known natural sources of sulphur species or DMS;

and the occurrence of new or ultra-fine particles could only be explained by the presence of additional condensible species, which are required to grow newly formed clusters to detectable sizes before they are scavenged by the pre-existing aerosols through coagulation processes.

Iodine oxides have also been implicated in marine new particle formation events, particularly in coastal air (O'Dowd et al. 1998, 1999a, b, 2002). Laboratory experiments (Hoffmann et al. 2001; Jimenez et al. 2003; Burkholder et al. 2004) have pointed to the rapid photolysis and subsequent oxidation by ozone of CH_2I_2, released from macro-algae. McFiggans et al. (2004), Saiz-Lopez et al. (2005) and Sellegri et al. (2005) demonstrated that I_2 was the dominant precursor with Sellegri et al. (2005) reporting a linear relationship between I_2 concentrations and 3.0–3.4 nm sized particle concentrations. However, some differences in the exact form of iodine oxides is still an issue of debate (Hoffmann et al. 2001; Jimenez et al. 2003; Saunders and Plane 2005).

Such new particle formation events are related to strong coastal emissions of halogen precursors (Mäkelä et al. 2002), resulting in concentrations of the order of 10^7 cm^{-3}. O'Dowd et al. (2002b) suggested that significantly lower concentrations of iodine oxide precursor condensible vapours (i.e. away from strong coastal sources) could provide either the nucleation mechanism for embryo formation (typically <1 nm in size) and/or the additional condensible vapours required to grow stable sulphate clusters into stable aerosol sizes of a few nanometers (>5–10 nm). If a particle grows to a size of 6 nm, it has a 100 times greater likelihood to survive coagulation loss as compared to a 1–2 nm sized particle.

While iodine oxides may be involved in open ocean nucleation and/or growth, due to the very short lifetimes of these species, it has proved very difficult to specify their role. For that matter, it has also been impossible to elucidate, from experimental field studies, what species are actually involved in the nucleation process and if additional species are required to explain growth of clusters into aerosol particles. This is because of the inability, until recently, to measure particles in the nucleation size range.

However, recent deployment of mass spectrometers has provided realtime information on condensing vapours as detected in conjunction with the appearance and growth of a nucleation mode (Dall'Osto et al. 2012). While not direct evidence of participation of either nucleation or growth of clusters into sizes larger than 3 nm, the quantification of additional condensible aerosol mass during such an event strongly suggests that the same condensible vapours are also responsible for the initial stages of growth. This study pointed to the coincidence of nitrogenated and aliphatic condensible vapours as being responsible for the observed nucleation modes and subsequent growth. In the same study it was found using quantum-chemistry calculations that nucleation of sulphuric acid, dimethylamine and subsequent condensation of MSA decreases cluster evaporation rates (and hence promotes cluster-to-aerosol formation rates).

4.2.2 Non-Marine Sources

4.2.2.1 Desert Dust

Desert aerosol is composed of mineral crustal particles suspended from surface soils by aeolian erosion. Desert and semi-arid areas are a major source of particles in the global atmosphere (e.g. Prospero et al. 2002) and large oceanic areas are regularly under the direct influence of turbid air masses transported from deserts or semi-arid areas, as shown by Fig. 4.9. These dust transport events cause long-range transfers from continent to the remote surface ocean of huge amounts of matter. It has long been recognised that mineral dust deposition to the remote surface ocean significantly influences trace element biogeochemistry (Buat-Ménard and Chesselet 1979; Graham and Duce 1979), marine productivity (Martin and Fitzwater 1988), and deep-sea sedimentation (Venkatarathnam and Ryan 1971; Loÿe-Pilot et al. 1986). More recently, it has been suggested that suspended dust particles affect the optical properties of clear surface waters (Claustre et al. 2002).

Due to the trade wind regimes, dust from the Sahara and Sahel is encountered all year long over the tropical Atlantic (Moulin et al. 1997) and as a consequence dominates the solar extinction by aerosol particles in this region on a yearly time scale (Mahowald et al. 2009). Other regions of the world ocean are subject to dust transport and deposition from more seasonal or episodic dust events. Whereas the Sahara-Sahel region dominates dust deposition in the Mediterranean, in most of the North Atlantic and even in regions of the North tropical Indian and Pacific Oceans, source regions in China dominate dust deposition in the North Pacific and Arctic. Middle East sources are most important for

Fig. 4.9 Desert dust on the move over the tropical Atlantic. This colour combination of visible and infrared Meteosat images shows a huge desert dust plume transported from Africa. The total mass of dust transported was several millions of tons (Figure courtesy of X. Schneider, CEA)

dust deposition in the Arabian Sea and Bay of Bengal. Sources in North America provide dust deposited in the North tropical Pacific and northwestern Atlantic while those in South America dominate dust deposition in the South tropical Pacific, the southern Atlantic and the Indian Ocean. Sources in South Africa contribute dust to the South tropical Atlantic and Indian Oceans and Australian sources dominate dust deposition in the southern Pacific and subtropical South Indian Ocean, (Grousset and Biscaye 2005; Mahowald 2007, Fig. 4.10).

Fine dust particles in the long-range transported aerosol size range (diameter smaller than ~20 μm) are strongly bound and not easily mobilised in arid soils, but tend to form large aggregates of fine particles (e.g. clays) or stick to larger particles (e.g. sand grains) which compose the particle size modes found in arid soils (Chatenet et al. 1996). The production of soil dust aerosols is a two-step process (Gomes et al. 1990) firstly resulting from the mobilisation of those large loose soil grains when the surface wind reaches a sufficient velocity (>5–15 m s^{-1} depending on surface characteristics; Goudie and Middleton 2006) to lift them and produce the so-called saltation of particles, bouncing downstream as commonly seen on sand beaches. Secondly saltating grains then produce small aerosol particles by a sandblasting effect when they settle out and impact the surface soil, disintegrating aggregates or producing small debris. The most productive source areas combine the presence of fine sand grains of ~60–100 μm in diameter that require the minimum threshold friction velocity for being lifted (Iversen and White 1982) and alluvial deposits of aggregated fine clay materials (Prospero et al. 2002) that have the best potential to produce fine aerosol particles (Marticorena et al. 1997).

According to Alfaro et al. (1998) the particle size distribution of aeolian dust aerosol can be approximated by a sum of three lognormal modes with respective mass-median diameters of the order of 1.5, 6.7, and 14.2 μm. (a similar particle size distribution is also produced when clay aggregates from arid soils are crushed and sieved for a long time, allowing the production of a dust aerosol model for use in experiments; Guieu et al. 2010). The relative proportions of the modes are a function of surface wind speed, with larger wind velocities producing more fine particles (Alfaro et al. 1997). Following settling of the largest particles during transport, the mass-median diameter of desert dust over ocean is generally a few microns (e.g. Arimoto et al. 1985, 1997; Dulac et al. 1989; Dubovik et al. 2002; Reid et al. 2003a). However, a number of authors report observations of "giant" sand-sized (>62.5 μm in diameter) dust particles of aeolian origin at very long distances from sources (e.g. Betzer et al. 1988, see also Goudie and Middleton 2006, p. 31). Compared to other super-micron sized aerosol particles such as sea salt aerosol particles which adsorb water, desert dust aerosol particles are characterised by a non-spherical shape, which can be explained by the laminar structure of the clay minerals that are abundant in desert dust aerosols. This irregular shape results in optical properties, i.e. the angular scattering of solar light, which complicate the inversion of dust properties from passive remote sensing data in the solar spectrum (Mishchenko et al. 1995; Dubovik et al. 2002). The irregular particles cause a relatively high rate of depolarization of the scattered light which is useful for identifying dust layers in aerosol lidar remote sensing (Sassen 2000). Another specific optical property of desert dust, responsible for their colour, is their ability to absorb the shortest (UV-blue) solar wavelengths (Moulin et al. 2001). This absorption is controlled by the presence of iron oxides (Alfaro et al. 2004).

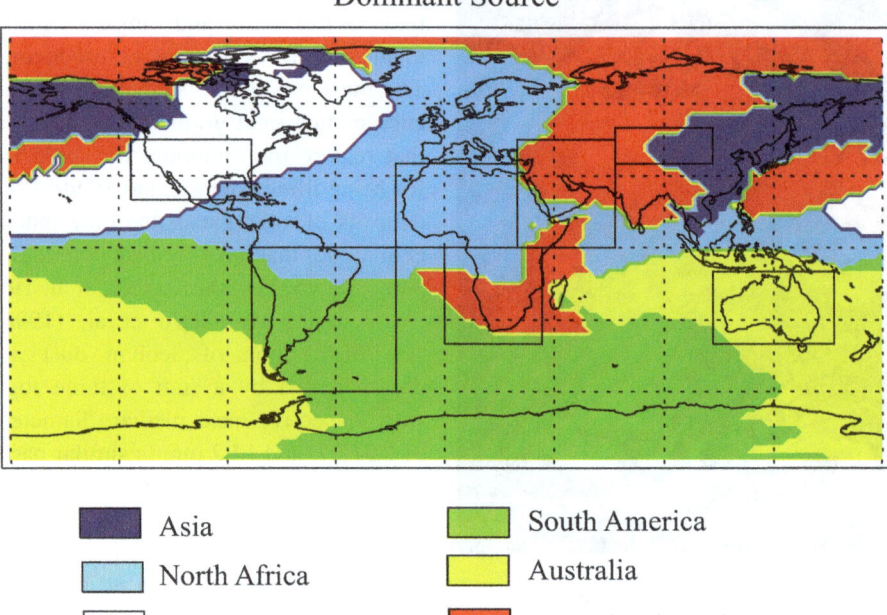

Fig. 4.10 A model view of source regions dominating the distribution of dust deposition fluxes (From Mahowald et al. 2009). Models generally reproduce data on dust provenance (Grousset and Biscaye 2005) with the exception of the Greenland region which is dominated by Asian dust (Bory et al. 2003)

It causes radiative heating of the atmospheric turbid layers (Alpert et al. 1998) and reinforces a decrease in incoming UV radiation that impacts dissolved organic matter and phytoplankton organisms in surface waters (Tedetti and Sempéré 2006).

Due to the minimum threshold surface wind speed required for aeolian erosion, and to the fact that the dust emission flux varies with the cube of the surface wind speed (Greeley and Iversen 1985), dust events are episodic and show a very high variability in strength. In terms of desert dust atmospheric concentrations or deposition at a given place, the intensity distribution at yearly or longer time scales generally shows a tail towards infrequent high values that control long-term means (Loye-Pilot and Martin 1996; Mahowald et al. 2009 and references therein on observations series).

Desert dust transport is generally characterised by maximum concentrations of aerosol particles in turbid air layers lifted above the marine atmospheric boundary layer, as was observed in the North Pacific (Kritz et al. 1990), the tropical (Carlson and Prospero 1972; Karyampudi et al. 1999; Dulac et al. 2001) and subtropical (Chazette et al. 2001) Atlantic Ocean and the Mediterranean Sea (Dulac et al. 1996; Hamonou et al. 1999). Desert dust settles from the atmosphere to the surface both through gravitational settling and wet deposition, with precipitation either within or below the clouds. Wet deposition of Saharan dust often produces highly concentrated red rains (Avila et al. 1998); mentioned as 'blood' rain in Homer's *Illiad*. Dry deposition is generally dominant only close to source regions and is controlled by the small fraction of the largest dust particles (Dulac et al. 1989; Arimoto et al. 1997). It seems that atmospheric processes such as the large-scale vertical upward movement of dust-loaded air masses counteract the gravitational settling velocity of dust particles (Dulac et al. 1992; Maring et al. 2003).

The chemical composition of desert dust aerosol particles reflects that of the average Earth surface rocks with a dominance of SiO_2 (~60 %) and Al_2O_3 (10–16 %) resulting from the dominance of quartz and clay minerals (Goudie and Middleton 2006). Either Si or Al are considered as chemical tracers of the soil dust fraction in aerosols. The relative contributions of various clay minerals or minor constituents such as Fe and Ca can generally be used to identify various dust source regions (Bergametti et al. 1989, Goudie and Middleton 2006). The high carbonate content of desert dust is principally responsible for relatively high pH values of rainfall affected by dust particles, with values sometimes larger than six (Loÿe-Pilot et al. 1986; Avila and Rodà 2002). The deposition of desert dust may be a significant source of limiting nutrients such as Fe (Jickells et al. 2005; Mahowald et al. 2009) to remote surface waters (e.g. Bergametti et al. 1992; Ridame and Guieu 2002), despite the relatively low solubility as compared to that of anthropogenic aerosols (Bonnet and Guieu 2004; Baker et al. 2006a). However, solubility is likely to be enhanced by the condensation/evaporation cycle encountered in cloud formation (Desboeufs et al. 2001). Contrary to previous hypotheses, it has been recently shown that the solubility of iron from desert dust is controlled by the iron content in clay minerals rather than by the abundance of iron oxides (Journet et al. 2008).

The interested reader is referred to Goudie and Middleton (2006) for a more comprehensive review of the literature on desert dust related questions.

4.2.2.2 Volcanic Gases, Aerosols and Ash

Volcanic emissions are important sources of atmospheric gases (e.g. Bardintzeff and McBirney 2000), aerosols and ash (e.g. Mastin et al. 2009). Volcanic gas emissions consist primarily of H_2O, followed by CO_2, SO_2, HCl, HF and other compounds. These gases and their oxidation products (in particular sulphate aerosols) may play an important role in the tropospheric and stratospheric chemistry and can impact terrestrial and oceanic ecosystems and human health. H_2O and CO_2 are important greenhouse gases, but their atmospheric concentrations are so large that volcanic eruptions have only a negligible effect on their concentrations (Robock 2000), although locally the release of CO_2 might have important environment effects.

Volcanic ash is a size class referring to fragmented fine-grained particles with diameters of submicron to less than 2 mm. Tephra is the general term for fragmented volcanic material produced during volcanic eruptions that includes ash particles (<2 mm), lapilli (2–64 mm), and bombs and blocks (>64 mm) (Fisher and Schmincke 1984; Schmincke 2004).

Volcanic emissions can be released continuously by passive degassing or diffusive (soil) degassing into the troposphere. Most of the volcanic SO_2 in the atmosphere is released from relatively less explosive continuous volcanic activity compared to episodic large scale eruptions (Andres and Kasgnoc 1998). About 99 % of volcanic SO_2 is released continuously, while only 1 % is released during sporadic eruptions (Andres and Kasgnoc 1998) (see Table 4.1 for the frequency of eruptions based on the eruption magnitude).

One of the most important climatic effects of explosive volcanic eruptions is through their emission of sulphur species to the stratosphere, mainly in the form of SO_2 which reacts with OH and H_2O to form sulfate aerosols on a timescale of weeks, producing one of the dominant radiative effects from volcanic eruptions (Robock 2000). Volcanic SO_2 release into the atmosphere on a 100 year scale (between 1900 and 2000) is estimated to be 8–11 $\times 10^{12}$ g S year^{-1}, contributing 8–11 % of the total global sulphur emissions of 100×10^{12} g S year^{-1}, which includes emissions from biomass burning, other anthropogenic sources and the marine-derived dimethylsulphide (Halmer et al. 2002).

The global annual direct radiative forcing of sulphate aerosols at the top of the atmosphere by volcanic sulphate is estimated to make up 33 % of the total sulphate forcing (Graf et al. 1997) thereby exceeding the percentage contribution of volcanic SO_2 emissions by a factor of about three. Even the silent degassing volcanoes release their emissions into higher atmospheric levels compared to most anthropogenic sulphur emissions, and therefore provide longer atmospheric lifetime of volcanic sulfur species. In particular, volcanic sulphate aerosols from plinian eruptions, like the Pinatubo June 1991 eruption, may influence solar radiation reaching the Earth surface for years, as indicated by the enhanced aerosol optical depth (AOD) after the eruption (Fig. 4.11). Similarly, the Mt. Hudson (Chile) August 1991 eruption may have contributed considerably to the Southern Hemisphere AOD. Reduced solar radiation also affects marine primary productivity (MPP). It should be also considered in satellite retrieval algorithms e.g. for surface ocean chlorophyll *a* concentration.

Volcanic ash and aerosols can be transported over long distances to remote parts of the ocean (Fig. 4.12). Upon deposition in the ocean, volcanic ash can release nutrients as well as toxic substances into the seawater (Frogner et al. 2001; Duggen et al. 2007; Jones and Gislason 2008; Hamme et al. 2010; Langmann et al.

Table 4.1 The type and frequency of volcanic activity based on Volcanic Explosivity Index (*VEI*) (Newhall and Self 1982; Simkin and Siebert 1994)

VEI	Plume height (km)	Eruptive volume (km³)	Eruption type	Frequency	Example
0	0.1	10^{-6}	Hawaiian	Continuous	Kilauea
1	0.1–1	10^{-5}	Hawaiian/Strombolian	Months	Stromboli
2	1–5	10^{-3}	Strombolian/Vulcanian	Months/year	Galeras
3	3–15	10^{-2}	Vulcanian	Year/few years	Puyehue (2011)
4	10–25	10^{-1}	Vulcanian/Plinian	Year/few years	Eyjafjallaökull (2010)
5	>25	1	Plinian	5–10 years	Pinatubo (1991)
6	>25	10	Plinian/Ultra-Plinian	1,000 years	Krakatoa (1883)
7	>25	100	Ultra-Plinian	10,000 years	Tambora (1815)
8	>25	1,000	Ultra-Plinian	100,000 years	Toba (74 ka)

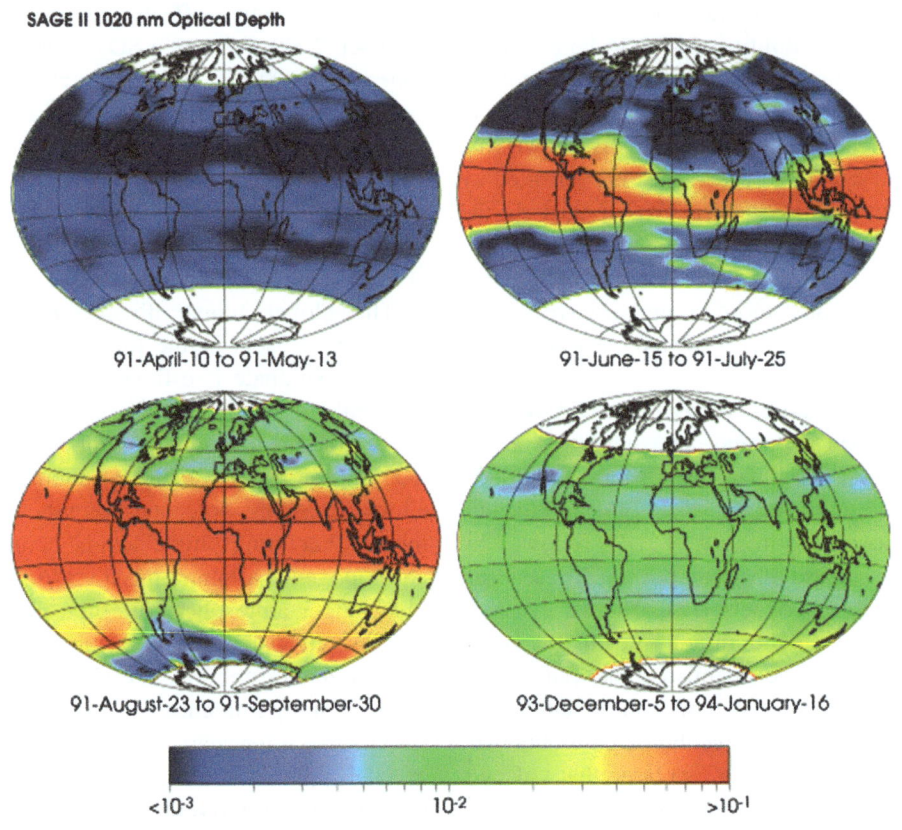

Fig. 4.11 Optical depth of stratospheric aerosol during four periods between April 1991 and January 1994 (http://www-sage2.larc.nasa.gov/Introduction.html). The eruption of Pinatubo took place in June 1991, Mt. Hudson erupted in August 1991 (Figure courtesy of NASA)

2010; Lin et al. 2011; Olgun et al. 2011). Therefore, volcanic ash may affect marine primary productivity, phytoplankton community structure, atmospheric CO_2 concentrations and can eventually (directly or indirectly) impact higher trophic levels the oceanic food-web (e.g. of zooplankton, fish). For the marine ecosystem response related to volcanic eruptions (Chap. 5: Sect. 5.2.2).

The importance of volcanic eruptions for the biogeochemistry of the surface ocean, however, has gained limited attention compared to the much better investigated effects of mineral dust. This is despite the

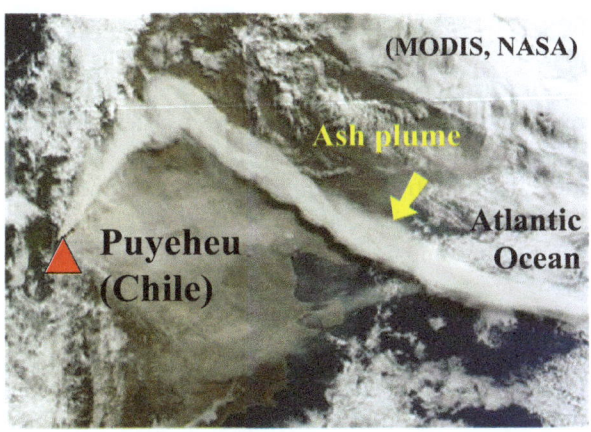

Fig. 4.12 Satellite image shows the long distance transport of volcanic eruption plumes that are ejected to high altitudes in the atmosphere as illustrated by the Puyehue eruption in Chile on 6 June 2011 with the white ash plumes reaching more than 10 km altitude and transported across Argentina towards the Atlantic Ocean (Captured by MODIS, NASA)

fact that an average of about 20 volcanoes erupt at any given time, 50–70 volcanoes erupt every year, and at least one large eruption occurs every year (e.g. Puyehue (Chile) and Grimsvötn (Iceland) in 2011, Table 4.1). Recent estimates based on marine sediment core data show that about $128\text{–}221 \times 10^{15}$ g ka^{-1} (ka = 1,000 years) of volcanic ash has been deposited into the Pacific Ocean, the largest ocean basin covering 70 % of the iron-limited ocean regions (Olgun et al. 2011). The flux of volcanic ash is of the same order of magnitude as that of mineral dust, which is around $39\text{–}519 \times 10^{15}$ g ka^{-1} (Rea 1994; Mahowald et al. 2005). On longer time-scales (e.g. during Holocene) the amount of volcanic ash deposition is comparable to that of mineral dust (Olgun et al. 2011), although marine biogeochemical impacts probably differ. Ocean regions with higher likelihood of volcanic ash deposition are shown in Chap. 5: Sect. 5.2.2.

The amount of volcanic ash and bio-available iron attached to the ash surface deposited into the ocean during large episodic volcanic eruptions may exceed the annual dust flux significantly. For example, iron input during the large eruption of Mount Hudson (Chile) between 12th and 15th August 1991 has been found to be equivalent to ~500 years of Patagonian iron dust fallout (Gaiero et al. 2003). Re-mobilization of well-preserved tephra deposits in dry regions can also impact the marine ecosystems after the eruptions (post-eruption impacts). The eruption of Mount Hudson, for example, created several volcanic ash clouds (ash storms) which would have different chemical behaviour compared to mineral dust (Wilson et al. 2011).

4.2.2.3 Global Emissions of Biogenic Volatile Organis Compounds (BVOC's) from Terrestrial Ecosystems

The term biogenic volatile organic compounds summarises a large number of compounds emitted from terrestrial biota comprising in total an estimated > 1,000 Tg C year^{-1} (Guenther et al. 1995). From an atmospheric chemistry and climate perspective, the isoprenoids (isoprene C_5H_8, and its monoterpenes and sesquiterpenes derivatives) have been the main focus of attention, reflecting the large mass emitted (isoprene), and/or fast atmospheric reactivity (isoprene, monoterpenes, sesquiterpenes) and related importance for the atmospheric burdens of O_3, SOA, OH and CH_4 (Atkinson 2000; Atkinson and Arey 2003a, b).

For BVOC emissions no regional or global scale observations exist to provide estimates of their past or present emission strength, distribution and seasonality. Global scale analyses thus have to depend on modelling studies, and these are to date only extensively published and evaluated for isoprene, and to a lesser degree, monoterpenes (Arneth et al. 2008). Current estimates of global isoprene emissions range between approximately 400 and 600 Tg C year^{-1}, while variability in monoterpene emission estimates is larger at ca. 30–130 Tg C year^{-1} (Arneth et al. 2008). The bottom-up model experiments are complemented by top-down approaches that seek to infer regional emissions of isoprene from remotely sensed formaldehyde column signals (Chance et al. 2000; Palmer et al. 2003; Barkley et al. 2008; Stavrakou et al. 2009), since formaldehyde is one of the chief isoprene oxidation products in the atmosphere. However, linking the formaldehyde retrievals directly to isoprene emissions is hampered by the need to use a chemistry transport model to account for atmospheric isoprene oxidations which are incompletely understood, adding considerable uncertainty to these types of analyses (Barkley et al. 2011, 2012).

Typically, global scale model experiments rely on algorithms that vary diurnal emissions in response to temperature and light (Guenther et al. 1995) that have been found to be the main drivers underlying emission observations in the short-term. Other attempts seek to link emissions and their variability to photosynthetic electron transport rate, reflecting the chloroplastic metabolic pathway of isoprene and monoterpene production and presence or absence of tissue storage of some BVOC, which underlies the observed light and temperature sensitivity (Niinemets et al. 1999; Arneth et al. 2007). Both approaches have to rely on specifying a leaf-level emission capacity which is defined for standard light and temperature conditions. More recently, the use of a canopy-scale emission capacity has been proposed (Guenther et al. 2006) as a model product that relies on the combination of leaf-level measurements and a canopy transfer model. Field observations demonstrate that emission capacities are not constant, but vary strongly during the year and over longer periods, for instance in response to leaf development, previous weather conditions or atmospheric CO_2 levels. In a set of recent reviews, emission capacities have been identified as the largest uncertainty in global BVOC emission models (Niinemets et al. 2010a, b, 2011). Emission capacities are species-specific and need to be set for larger plant functional units to be applicable in large-scale models. Emission algorithms are therefore either linked to global vegetation distribution classes derived from remote sensing information, or to dynamic global vegetation models. Only the latter are capable of estimating changes in emissions in future or past environments (see e.g. Arneth et al. 2008 and references therein).

Current models agree on emissions from tropical ecosystems dominating the global totals of isoprene emissions, a combination of tropical vegetation having high emission potential as well as the warm temperatures and high light conditions throughout the year. By contrast, the global monoterpene emission estimates assume also a substantial contribution from the coniferous and evergreen deciduous forests of mid- to high latitudes, even though the period of high emissions in these regions is restricted to few months of favourable weather (Arneth et al. 2011; Guenther et al. 2006; Lathière et al. 2005). While regional differences in emission seasonal patterns and overall strength for BVOC emissions appear to be large, probably much larger than for emission and uptake of CO_2, their interannual variability seems small, around 5–10 % of the mean (Arneth et al. 2011).

In the absence of regional to global observational data, model outputs from BVOC simulation experiments are severely hampered by lack of evaluation possibilities. At present, canopy-atmosphere flux measurements by micrometeorological techniques are the sole possibility to provide constraints and reality-checks on the scale of the ecosystem and with the potential to cover a period of months to years. Flux measurements from aircraft could be used to extrapolate to larger regions, although these are restricted to short-term campaigns. On canopy scale, only for isoprene are sensors sufficiently robust to allow long-term observations. But still, only one study location reported measurements covering several years (Pressley et al. 2005). Hence questions on the magnitude of interanual variability essentially remain unanswered. For other BVOC species, flux measurements rely on techniques using gas chromatography or proton transfer reaction spectrometers as sensors. A few short-term campaign studies from a very limited number of ecosystems have been published to-date (a list of example data are provided in Rinne et al. 2009, Pacifico et al. 2011, Arneth et al. 2007, Lathière et al. 2006) which is insufficient for model evaluation.

For future emission estimates large uncertainties exist with respect to how changes in climate, atmospheric CO_2, N deposition, tropospheric O_3 and human land use/land cover changes interact to alter BVOC emissions in direct and indirect ways (Arneth et al. 2010; Niinemets et al. 2010a, b). In addition to improved understanding of the processes of BVOC production and emissions in response to these various environmental changes, observation and modelling efforts clearly also have to move towards substances beyond isoprene and monoterpenes to reflect the multiple open questions on BVOC-atmosphere and climate interactions (Goldstein and Galbally 2007; Holzinger et al. 2005; Lelieveld et al. 2008).

BVOC directly or via their atmospheric oxidation products, contribute to the formation and growth of secondary organic aerosol (Kulmala et al. 2004). Similar to BVOC emissions, the total mass and number concentration of SOA particles that are formed from BVOCs are highly uncertain. Current estimates are around 12–70 Tg year^{-1}, but these numbers have also been challenged as being too small, perhaps by up to a factor of 10 (Kanakidou et al. 2004; Carslaw et al. 2010; Hallquist et al. 2009). Most modelling work to-date has estimated SOA based on monoterpene emissions only using a constant mass yield of typically 10 % of emissions. Yet, both sesquiterepenes (Bonn and Moortgat 2003) and isoprene (Claeys et al.

2004) have been identified as important precursor sources. While the SOA yield from isoprene may be low, its source strength and the gas-particle partitioning characteristics of its oxidation products are efficient to the point where it may promote SOA growth at higher altitudes and enhance the SOA formation from other sources (Claeys et al. 2004; Henze and Seinfeld 2006).

Quantification of the future direct climate impact of SOA in terms of radiative forcing have so far only considered the case of increasing BVOC emissions over the twenty-first century; these model scenarios result in a substantial cooling (up to -24 W m^{-2}; Carslaw et al. 2010) due to the increased scattering and reflection of radiation by the larger SOA mass. In addition, SOA can grow to particle size classes that act as cloud condensation nuclei, with associated indirect climate effects if cloud albedo and lifetime are affected (Lohmann and Feichter 2005). The full chain of processes from emissions, particle nucleation and subsequent growth to aerosol direct and indirect effects is only just now beginning to be included in global climate models. Initial results indicate large differences between past, present-day and future SOA number concentration and SOA radiative forcing if these processes are treated explicitly (Makkonen et al. 2012).

4.2.2.4 Anthropogenic Emissions

Emission inventories of reactive gases and aerosols are needed as input for climate and atmospheric chemistry and transport models (CTMs) to be able to assess and predict the climate impacts, air pollution concentrations or deposition of elements to ecosystems (Sect. 4.5.1). Emissions can be separated into natural emissions and anthropogenic emissions. The latter implies that the emissions are produced as a result of human activities and can be influenced by changes in technologies and/or emission reduction policies. Emissions need to be spatially distributed, e.g. on a grid, to be suitable as (deposition) model input because the location of emissions is important for their impact, especially for reactive and/or short-lived species. Examples of such gridded emissions can be found at the GEIA/ACCENT emissions portal (http://geiacenter.org).

The most important precursors of secondary anthropogenic aerosol are NOx, SOx, NH$_3$ and Non-Methane Volatile Organic Compounds (NMVOC). For a detailed review of aerosol formation we refer to Seinfeld and Pandis (2006). To quantify the impact of anthropogenic emissions on the ocean (through deposition of nutrients and/or particles) it is necessary to quantify both the primary aerosol emissions and the aerosol precursor emissions. In this section we distinguish between land-based anthropogenic emissions, biomass burning emissions and emissions from international shipping. The latter will in almost all circumstances directly influence the marine ecosystem, although a substantial part of the emissions may also deposit on land surfaces.

Anthropogenic Land-Based Emissions

Recently Lamarque et al. (2010) developed a new emissions dataset covering the 1850–2000 period, based on the combination and harmonisation of published and publicly available datasets, in support of the Intergovernmental Panel on Climate Change (IPCC) Fifth Assessment Report (AR5). This so-called ACCMIP emission data set also acts as a starting point for the emission projections up to the year 2100 given by different representative concentrations pathways (RCPs) as used in the AR5 (Van Vuuren et al. 2011). The ACCMIP output is the most recent and widely used global emission data. It is distributed on a $0.5° \times 0.5°$ degree grid and made available through ftp://ftp-ipcc.fz-juelich.de/pub/emissions/. Lamarque et al. (2010) provide emission estimates for the precursors of secondary anthropogenic aerosol and for primary carbonaceous particulate matter (BC and OC) (Table 4.2) but not for total PM10.

Uncertainty in Global Anthropogenic Emissions

Antropogenic uncertainty in emission estimates arise from uncertainty in the activity data, the fuel composition and the emission factors for all individual sources. The resulting uncertainty leads to a range of possible emissions for a given process and base year that varies strongly between regions, sectors, and pollutants. A consistent uncertainty analysis for all pollutants in Table 4.2 is a complex and laborious task, as it would have to be done for all air pollutant/source/technology/country combinations separately and has only been performed for a few species e.g. BC and OC (Bond et al. 2004)/SO$_2$ (Smith et al. 2011). To get an impression of the uncertainties Granier et al. (2011) compiled all currently available consistent global and regional emission inventories and calculated the ratio between the lowest and highest

Table 4.2 Global anthropogenic land-based emissions (Tg year^{-1},) for the year 2000 (Lamarque et al. 2010; Granier et al. 2011)

	NOx[a]	VOC	NH$_3$	SO$_2$	BC[b]	OC[b]
Anthropogenic land-based	57	130	37	93	5.0	13
Shipping	12	3	0	11	0.13	0.14
Biomass Burning	9.7	78	11	3.8	2.6	23

[a]*NOx* in Tg NO year^{-1}
[b]*BC* particulate black carbon, *OC* particulate organic carbon (Tg C year^{-1})

emissions for each species and each region for selected base years. The ratios for the global year 2000 estimates for NOx, SO$_2$, and BC were 1.17, 1.40 and 1.13, respectively (Granier et al. 2011). The ratios for the global inventories were small compared to the variation in the regional inventories which usually ranged between 1.5 and 3.0. The spread gives an impression of consensus but is not an uncertainty analysis. For example, Bond et al. (2004) estimated global fossil black carbon emissions in 1996 as 3.0 Tg C, with an uncertainty range of 2.0–7.4 Tg C, or +150 % and −30 %; this is in strong contrast with the ratio of 1.13 between highest and lowest global inventory for this pollutant. Smith et al. (2011) performed an uncertainty analysis for global and regional sulphur dioxide emissions and concluded that the overall global uncertainty is relatively small: 6–10 % over the twentieth century, but regional uncertainties ranged up to 30 %. For the SO$_2$ year 2000 and values presented in Table 4.2, the uncertainty, based on Smith et al. (2011), would be ~10 % and ~30 % for anthropogenic land-based sources and shipping, respectively. The calculated global SO$_2$ uncertainty bounds are relatively small: the low value is due to cancellation between source categories and regions. This uncertainty level would appear to be unrealistically low given that a number of previous global sulphur dioxide emissions estimates do not fall within this estimated uncertainty bound (the ratio high-low inventories was 1.40; see Granier et al. (2011) for a compilation). The reason is that additional, essentially correlated uncertainties are present that add to the uncertainty value estimated above. Examples include reporting or other biases in global data sets for energy, sulfur removal, and other driver data, methodological assumptions, and the use of common default assumptions for sources where little data exists. Sulphur emissions are less uncertain than emissions of the other air pollutants listed in Table 4.2 because emissions depend largely on sulphur contents in fuels. Uncertainty for other air pollutants is controlled more by combustion conditions and installed technologies and their uncertainty is considerably larger, as quoted above for BC. A qualitative indication is that for VOC, NH$_3$ and OC the uncertainty range will be similar to the BC ranges of Bond et al. (2004) while NOx will be in between the SO$_2$ and BC ranges.

Global Biomass Burning Emissions

Biomass burning emissions are highly variable from year to year as a result of different environmental and human factors. Schultz et al. (2008) provided a detailed literature review on continental scale estimates of biomass burning emissions and constructed a global emissions data set with monthly time resolution for the period 1960–2000. The previously discussed ACCMIP historical dataset (Lamarque et al. 2010) provides decadal monthly mean biomass burning emissions, mostly based on Schulz et al. (2008). As can be seen from Table 4.2, biomass burning emissions contribute significantly to total global emissions of aerosols and their precursors but its relevance differs by substance, from being quite modest for SO$_2$ to dominant for OC. Over the past 20 years estimates of biomass burning emissions, including their spatial location, have greatly improved due to the availability of earth observation data from satellites. The activity data detected from space (burned area or fire radiative power) includes all major grassland, savanna, and forest fires (including deforestation fires) (e.g. Van der Werf et al. 2006, 2010; Kloster et al. 2010). To estimate emissions, satellite-derived burned areas (Giglio et al. 2010) drive the fire module of a biogeochemical model that calculates fuel loads for each month and grid cell, which are then combined with emission factors (Andreae and Merlet 2001). A good example of this methodology is the Global Fire Emissions Database (GFED) which contains emissions from open fires for the 1997–2004 period (van der Werf et al. 2006). A new version of the inventory that covers the 1997–2009 period, called GFED-v3, was made available at the beginning of 2010 (van der Werf et al. 2010).

International Shipping Emissions

The ACCMIP emissions dataset (Table 4.2) provides ship emissions including international shipping, domestic shipping and fishing, but excluding military

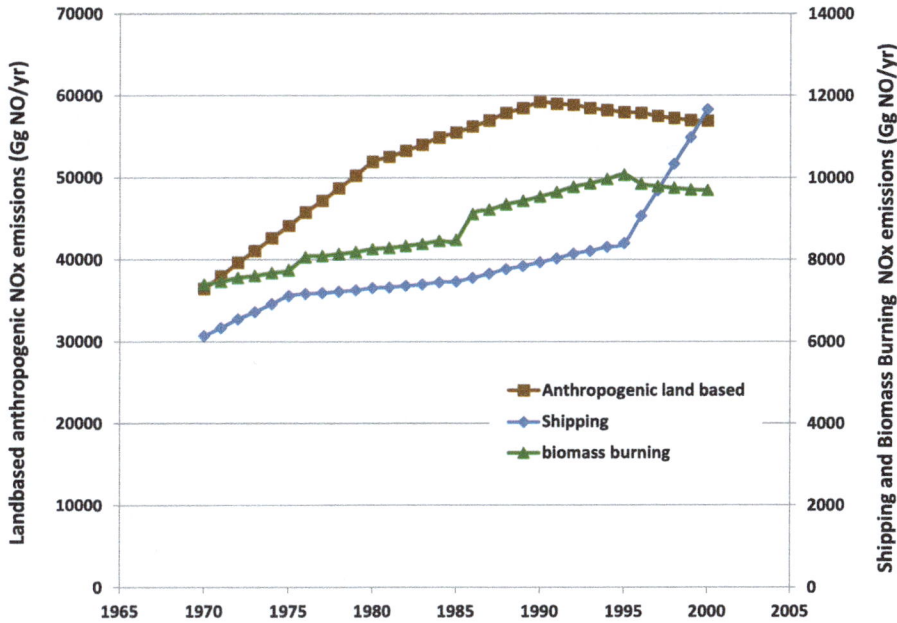

Fig. 4.13 Historic development of yearly anthropogenic NOx emissions for the years 1970–2000 (data courtesy of Schultz et al., 2008; Granier et al. 2011)

vessels, based on Eyring et al. (2009). These authors used data from the International Maritime Organization (IMO) study discussed in Buhaug et al. (2008), while non-CO_2 emission totals were derived as a mean of previous studies (Corbett and Köhler 2003; Eyring et al. 2005; Endresen et al. 2003, 2007). Ship emissions are distributed over the globe using the International Comprehensive Ocean–atmosphere Data Set (ICOADS; Wang et al. 2008), which provides changing shipping patterns on a monthly basis. Eyring et al. (2009) estimated that in 2000, the emissions released by the ocean-going registered fleet for nitrogen oxides (NOx), sulphur oxides (SOx), and particulate matter (PM) were 11.6 Tg NO, 11 Tg SO_2, and 1.4 Tg PM, respectively, within a bounded range of 6–21 Tg NO, 6–20 Tg SO_2 and 0.4–3.4 Tg PM.

Comparison and Evaluation of Different Emission Datasets

A detailed comparison and evaluation of different emission datasets of anthropogenic and biomass burning emissions for the 1980–2010 period was made by Granier et al. (2011). They identified large discrepancies between the global and regional emission data sets, showing that there is still no consensus on the best estimates for surface emissions of atmospheric compounds. Hence the data presented in Table 4.2 have a substantial uncertainty; about 10–30 % for land-based anthropogenic emissions and shipping and 50–80 % for biomass burning (Granier et al. 2011). For a full description of the various global and regional emission datasets currently available we refer to Granier et al. (2011). To understand the impact of anthropogenic emissions on marine ecosystems it is important to know not only the absolute emission in a certain base year but also the trend over time. As an example we show the historic development of NOx emissions for the period 1970–2000 in Fig. 4.13.

Clearly all emissions have increased over the past decades, with the land-based emission slowing down somewhat in recent years but shipping emissions showing a steep increase. The significant increase in the emissions from wildfires throughout the period from 1960 to 2000 due to the increasing importance of forest and peat soil burning is remarkable (Schultz et al. 2008). Annual global carbon emissions averaged at 1,660 Tg C year^{-1} during the 1960s and rose to an average of 2,560 Tg C year^{-1} during the 1990s. The most important contribution to the trend comes from enhanced deforestation in the tropical regions. The implication of Fig. 4.13 is that the input from the atmosphere to the ocean will have significantly increased over the past decades and that this is bound to have influenced the cycling of elements and nutrients as well as other processes. Trends in other substances may vary somewhat from the NOx trend depicted here but overall the pattern is consistent. For a full description of the various global and regional emission datasets currently available and emission trends for 1980–2010 we refer to Granier et al. (2011); historic emission trends since 1850 are provided by Lamarque et al. (2010).

4.2.3 Ageing and Mixing of Aerosols During Transport

Aerosol particles produced over the continents are present in remote oceanic regions as a result of long-range transport from high-emission regions. From the moment an aerosol particle forms or is emitted until its removal from the atmosphere by deposition, it has undergone various physical and chemical transformations. This aerosol ageing is due to coagulation, condensation/evaporation of semi-volatile components, adsorption/absorption of volatile components and chemical reactions. The dynamic exchange of semi-volatile substances facilitates the exchange of molecules between particles of different origin. As a consequence, ambient aerosols are often complex internal and external mixtures of both inorganic and organic components. The degree to which aerosol particles are transformed controls their chemical composition, size, shape, and surface area and hence determines the aerosol chemical evolution as well as the particles hygroscopic and optical properties.

Recent studies on aerosol ageing have been focussed on dust and organic aerosol (OA). Current understanding of secondary organic aerosol (SOA) formation is incomplete (e.g. de Gouw et al. 2005; Heald et al. 2005, O'Dowd and de Leeuw 2007) and models underpredict SOA concentrations, notably in aging plumes (Alvarado and Prinn 2009; Grieshop et al. 2009). Moreover, despite extensive work on dust evolution during transport, in particular on Asian dust, the chemical mechanisms involved in dust aging are still in question, implying large uncertainties in estimation of dust cycling and impact (Jickells et al. 2005; Su et al. 2008; Formenti et al. 2011).

Recent studies show an explosion of identified organic species in the aerosol phase, notably due to the improvement of analytical techniques. In contrast to the clear evolution of sulphate which is irreversibly oxidised and condensed, the OA presents versatile transformation highly dependent on meteorological and chemical conditions. The chemistry of these species is not well-known and there is a large need for further studies.

4.2.3.1 Chemical Ageing of Organic Aerosols

The organic composition of aerosols continues to change long after initial particle formation as a result of chemical reactions between particle constituents, reactive uptake of gaseous semivolatile molecules, direct photochemical processes inside the particle, and condensation and evaporation of water on the particle (Pöschl 2005; Robinson et al. 2007; Rinaldi et al. 2010). The additional gas-phase oxidation during transport appears to be principally responsible for most of the aging of organic aerosol particles (e.g. Lambe et al. 2009). However, other dominant ageing mechanisms such as condensational ageing (uptake of oxidised organic vapors) or homogeneous ageing (condensed-phase chemistry such as oligomerisation) has also been identified (Volkamer et al. 2007; Jimenez et al. 2009). These processes are schematically shown in Fig. 4.14.

The ageing takes place on a time scale ranging from minutes to days (e.g. Dunlea et al. 2009), and may significantly change the molecular composition of particles before their removal from the atmosphere. For example, in biomass burning plumes, the unreacted primary organic aerosol, traced by levoglucosan concentrations, represents only 17 % of the OA mass after 3.5 h (Hennigan et al. 2011), suggesting that the majority of OA is transformed via photo-oxidation. Fu et al. (2011) show that over tropical regions, atmospheric oxidation products can account for 47–59 % of the total organics. Typically, field and laboratory measurements indicate that organic material changes its carbon number and becomes increasingly oxidised, less volatile, and more hygroscopic as a result of continuous ageing in the atmosphere (Zhang et al. 2007; Capes et al. 2008; DeCarlo et al. 2008; Jimenez et al. 2009; Morgan et al. 2010, Ng et al. 2010; Kroll et al. 2011). The extent of oxidation of OA, generally estimated from Aerosol Mass Spectrometer (AMS) measurements, increases as the aerosol is exposed to atmospheric oxidants, notably OH radicals (Grieshop et al. 2009; Sage

4.2 Aerosol Production and Transport in the Marine Atmosphere

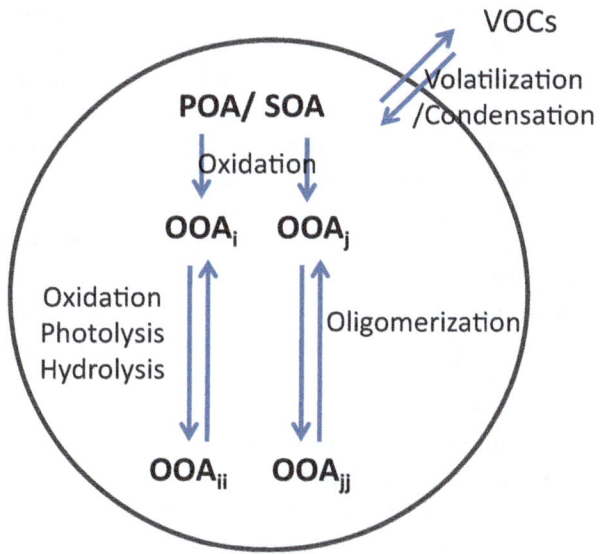

Fig. 4.14 Main pathways of OA aging. *OOA* corresponds to oxygenated organic aerosol

et al. 2008). Thus, diurnal oxidant cycling involves a significant daily variation of OA composition with more oxidized OA in the afternoon (Aiken et al. 2008; Hildebrandt et al. 2010; Adler et al. 2011).

As a result of the large number of reaction pathways, intermediates and products, aerosol particles sampled from ambient air contain thousands of chemically distinct organic compounds (e.g. Fu et al. 2011). Ageing processes in the atmosphere, resulting in further oxidation and oligo-polymerisation of the organic compounds, ultimately transform them to highly oxidised macromolecular material such as humic-like substances (HULIS) (Holmes and Petrucci 2006; Hallquist et al. 2009). Moreover, due to their semi-volatile properties and high atmospheric reactivity, organic species are often observed in internal mixing with inorganic particles. Thus, inorganic sulphate or nitrate aerosols are usually coated or aggregated with organic particles, forming organic sulphate or organic nitrate aerosols (Surratt et al. 2007; Froyd et al. 2010; Hawkins et al. 2010; Li and Shao 2010). The detailed characterisation of the huge set of OA compounds is beyond the capabilities of most analytical techniques (Kroll et al. 2011). As a consequence, several recent studies aim to identify organic species or find a metric which can be used as a tracer of organic ageing, such as the determination of molecular markers (e.g. Claeys et al. 2010), the mass spectral diagnostic (e.g. via f_{44} increase, Ng et al. 2010) or the use of the average carbon oxidation state coupled with carbon number (Kroll et al. 2011).

Despite this extreme chemical complexity, it seems that organic matter produced from pollutants in urban atmospheres, from biogenic emissions, or from terpenes exposed to photochemical reactions in smog chambers, is processed to particulate organic matter of similar oxidation state, hygroscopicity, volatility, and molecular mass (Jimenez et al. 2009). In other words, organic aerosols from very diverse origin are observed to evolve in the atmosphere to particles with similar chemical and physical properties (Jimenez et al. 2009; Ng et al. 2010).

4.2.3.2 Internal Mixing
Dust/Inorganic Species

The transformation of dust during atmospheric transport leads mainly to coating by sulphates or nitrates (Formenti et al. 2011). First observations of internal mixing have been made in Asia where desert sources are located close to highly polluted areas (e.g. Iwasaka et al. 1988; Okada et al. 1990; Yamato and Tanaka 1994; Zhou et al. 1996; Fan et al. 1996). Strong internal mixing of dust and sulphate or nitrate was also reported for the eastern Mediterranean (e.g. Falkovich et al. 2001; Sobanska et al. 2003; Putaud et al. 2004; Koçak et al. 2007; Coz et al. 2009) and over the Atlantic Ocean (Kandler et al. 2007; Dall'Osto et al. 2010) in polluted European and North African air masses. The sulphate/nitrate coating is the result of different heterogeneous chemical processes, such as the uptake of gaseous SO_2 or HNO_3 (and/or NOx) onto the particle surface and their subsequent conversion to SO_4^{2-} and NO_3^- (Dentener et al. 1996; Kim and Park 2001; Usher et al. 2002) and collision/coalescence between dust and aerosol ammonium sulphate or nitrate (Mori et al. 1998; Sullivan et al. 2007; Suzuki et al. 2010). The heterogeneous processes responsible for sulphate mixing with dust may be season-dependent with predominance of coagulation process in summer and SO_2 uptake in spring (Suzuki et al. 2010). Recent observations of strong internal mixing of dust with nitrate or sulphate in mesoscale convective clouds during the AMMA campaign (Crumeyrolle et al. 2008; Matsuki et al. 2010a) suggest in-cloud processing involving aqueous phase oxidation of NOx/SO_2, as previously assumed by Levin and Ganor (1996) and Liu et al. (2005). Carbonate minerals in dust particles, notably calcite ($CaCO_3$), have been identified as the main substrate for adsorbing sulfur and nitrogen gaseous species (Dentener et al. 1996; Krueger et al. 2004). Thus, the heterogeneous reactions involve the conversion of calcium carbonate to other calcium salts such as calcium nitrate ($Ca(NO_3)_2$), calcium sulphate (gypsum, $CaSO_4$) or their ammonium salts ((NH_4)NO_3 and (NH_4)HSO_4 or (NH_4)$_2SO_4$) as first demonstrated in laboratory experiments (Krueger et al. 2004) and then observed in recent field studies (Laskin et al. 2005; Matsuki et al. 2005, 2010b; Sullivan and Prather 2007; Shi et al. 2008; Huang et al. 2010; Suzuki et al. 2010).

The reactivity of inorganic acids with dust is determined by several factors: chemical mineralogy of dust (Sullivan et al. 2007), transport pathways, the extent to which dust is transported across polluted sources (Sullivan et al. 2007; McKendry et al. 2008) and meteorological and chemical processing. In particular, sulphate and nitrate coating on dust surfaces is favoured in the marine atmosphere where the relative humidity is high (Hanisch and Crowley 2001; Usher et al. 2002; Trochkine et al. 2003; Zhang et al. 2003a; Okada and Kai 2004; Ooki and Uematsu 2005; Matsuki et al. 2005). Dall'osto et al. (2010), who compare African dust mixing state at different distances from the emission source, demonstrate a continuous chemical evolution of dust particle composition during atmospheric transport, consistent with the relatively slow atmospheric oxidation of sulphur dioxide. Moreover, measurements on individual particles show evidence of the mineralogy-dependent formation of sulphate/nitrate on dust particles (Matsuki et al. 2005; Laskin et al. 2005; Sullivan and Prather 2007). Sulphate formation is favoured on aluminosilicate-rich particles (Laskin et al. 2005; Shi et al. 2008) while preferential nitrate formation on carbonate-rich dust is observed (Ro et al. 2005; Sullivan et al. 2007; Matsuki et al. 2010a; Fairlie et al. 2010). It was proposed that the preferential association of sulphate with Al-rich dust is partly due to the oxidation of SO_2 to H_2SO_4 catalysed by transition-metals, mainly iron, present in aluminosilica minerals (Sullivan et al. 2007; Sullivan and Prather 2007). Yet, the opposite behavior of sulphate and nitrate formation on carbonates could be explained by their difference of hygroscopicity (Sullivan et al. 2009; Formenti et al. 2011). Calcium sulphate is poorly water soluble, preventing further uptake of water and other gaseous species, and hence suppressing the transformation of sulphur dioxide to sulphate in these particles. In contrast, calcium nitrate is highly hydrophilic, enhancing uptake of water and resulting in a positive feedback, transforming all calcium in the particles to calcium nitrate. The calcium-rich spherical particles observed in Asian dust plumes in Japan and in polluted urban air masses in China (Fig. 4.15; Matsuki et al. 2005; Okada et al. 2005), as well as in the Eastern Mediterranean (Laskin et al. 2005) or in convective systems over the Sahel during the monsoon period (Matsuki et al. 2010a) provide field evidence of this feedback process. This mineralogy-dependent salt formation could imply dust source dependence.

In addition, mixing between dust and chloride has also been observed close to the Asian coast, over the North Pacific (Zhang and Iwasaka 2001; Murphy et al. 2006; Sullivan et al. 2007; Tobo et al. 2009, 2010) and over the North Atlantic (Sullivan et al. 2007). The absorption of hydrogen chloride (HCl) seems to be responsible for the chloride coating, forming calcium chloride $CaCl_2$ (Kelly and Wexler 2005; Tobo et al. 2009). This heterogeneous pathway could be predominant in the remote marine boundary layer with respect to sulphate and nitrate formation (Ma and Choi 2007; Tobo et al. 2009, 2010). The main source of gaseous HCl is volatilization from sea salt particles during heterogeneous reaction of sea salt with HNO_3 or H_2SO_4 (Tobo et al. 2009).

The mixing of dust in polluted air masses also favors mixing with other anthropogenic compounds, such as metals (Cu, As, Ni, Cd, Zn, Pb) (Sun et al. 2005; Zhang et al. 2005b; Erel et al. 2006; Huang et al. 2010; Wang et al. 2011), probably playing a role on the atmospheric deposition of nutrients to the marine biosphere (Sun et al. 2005).

Dust/Organic Species

Internal mixing between dust and organic carbon has been observed in African biomass burning plumes (Hand et al. 2010) and in polluted Asian air masses (Fig. 4.15, Leaitch et al. 2009; Geng et al. 2009; Li and Shao 2010; Stone et al. 2011). Recent studies show that ATOFMS (Aerosol Time of Flight Mass Spectrometry) in providing aerosol particles composition as a function of particle size, is a pertinent technology to identify organic species mixed with dust (Sullivan and Prather 2007; Yang et al. 2009; Dall'Osto et al. 2010). Thus, Sullivan and Prather (2007) found from their experiments during ACE-Asia that oxalic acid

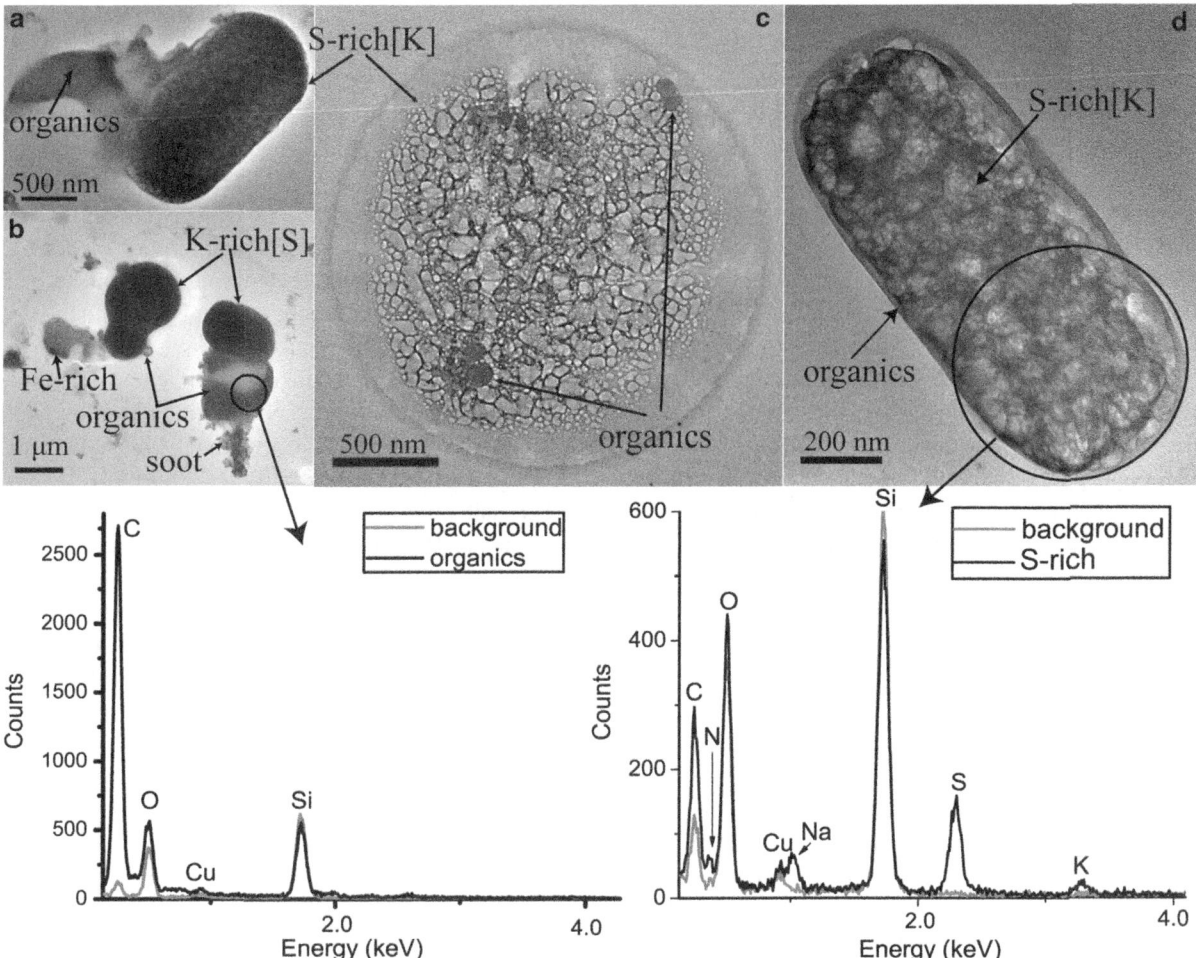

Fig. 4.15 Illustration of different mixing states between dust (*Si-rich* and *K-rich*), organic species and soot particles by Transmission Electron Microscopy (*TEM*) images and Energy-Dispersive X-ray (*EDX*) spectra. The circles show the sites of EDX measurements. The carbon peaks from EDX spectra were compared between the background (*grey spectrum* from an area without particles) and the particle (*black spectrum*). (**a**) S-rich particle with organic coating aggregated with an organic particle. (**b**) K-rich particle aggregated with an organic particle and soot, including a fine Fe-rich particle. (**c**) S-rich particle with organic inclusions. (**d**) S-rich particles with organic coating (Re-produced from Li and Shao (2010) by permission of the American Geophysical Union)

and malonic acid were predominantly internally mixed with mineral dust and aged sea salt particles. They also observed a diurnal enrichment of oxalic acid in mineral dust, indicating a probable gas-phase photochemical production of dicarboxylic acid followed by partitioning into the particle phase (Sullivan et al. 2007). Yang et al. (2009) also observed oxalate-containing dust and sea salt particles in Shangai but they associated oxalate formation to oxalic acid production by heterogeneous reaction occurring in hydrated/deliquesced aerosol. Leaitch et al. (2009) reported enrichment of dust particles by various organic species, notably formate and acetate in Asian dust collected on the North American coast, probably due to uptake of organic particle precursors by dust nearer Asian anthropogenic sources. Internal mixing between dust and carboxylic acid has also been suggested for African dust in source areas during monsoon periods by in-cloud processes in convective systems (Desboeufs et al. 2010) and in zones of transport in Israel (Falkovich et al. 2001, 2004) and over the French Alps (Aymoz et al. 2004). Due to the numerous observations of mixing between dust and carboxylic acids, which are tracers of SOA, dust is suspected to play a role in SOA formation (Mochida et al. 2007; Duvall et al. 2008; Wang et al. 2009; Stone et al. 2011). Nevertheless, there is no clear evidence of the formation of secondary organic aerosol on dust in these studies. If carboxylic acids are the most common identified organic species in mixing with dust, internal mixing of dust has been observed with MSA (Dall'Osto et al. 2010); PAH (Falkovich et al. 2004; Stone et al. 2011); pesticides (Falkovich et al. 2004); fatty acids (Mochida et al. 2007); hopane and levoglucosan (Stone et al. 2011) and organic nitrogen (Dall'Osto et al. 2010). The abundant organic matter found on dust surfaces suggests that dust is an excellent medium for long-range transport of pollution in the troposphere (Falkovich et al. 2004). The extent organic compounds coat onto dust depends on the transport pathway, reactivity of organic species, ambient concentration and ambient humidity, notably for high water-soluble species like carboxylic acids (Falkovich et al. 2004; Dall'Osto et al. 2010).

Sea Salt

Sea salt aerosols can also participate in heterogeneous reactions with nitric and sulphuric acids, leading to chloride (and also other halogens like Br) depletion (notably Cl-depletion) through HCl volatilisation and the production of halogen radicals, particularly in relatively polluted marine air (Sturges and Shaw 1993; Johansen et al. 1999; Kumar et al. 2008). The release of reactive chloride is considered to be an important intermediate in the oxidation reactions associated with the removal of light hydrocarbons and ozone in the marine atmosphere (Singh and Kasting 1988; Vogt et al. 1996). The magnitude of Cl-depletion of marine aerosols has been demonstrated to usually increase with decreasing sea salt particle size (Mouri and Okada 1993; Kerminen et al. 1998; Yao et al. 2003; Hsu et al. 2007). This reaction produces sea salt particles coated with sulphate and nitrate over Asian and Pacific areas (Matsumoto et al. 2004; Matsuki et al. 2005; Yang et al. 2009) and in the Mediterranean region (Tursic et al. 2006). In the clean atmosphere, methanesulphonate is the major species involved in chloride depletion as observed in Finland (Kerminen et al. 1998) and in the Arctic (Maskey et al. 2011). The S-containing sea salt particles are generated by reactions of the sea salt particles with MSA and/or H_2SO_4 from biogenic sources rather than anthropogenic ones. Mixing between sea salt and oxalate is also observed (Kerminen et al. 1998; Yang et al. 2009). A number of studies have reported large variations in the magnitude of Cl-depletion (ranging from few percent to 100 %) over different oceanic regions (Graedel and Keene 1995; Song and Carmichael 1999; Maxwell-Meier et al. 2004; Quinn and Bates 2005; Hsu et al. 2007). This variability has been attributed to the complex interplay of several factors, including variable wind-field, turbulence in the vicinity of the sampling site and the influence of anthropogenic components. Thus, the primary mechanisms governing differential Cl loss from sea salt particles in diverse oceanic regions are now well understood (e.g. Song and Carmichael 1999; Maxwell-Meier et al. 2004; Quinn and Bates 2005).

Sea salt mixing with dust has also been reported: more than 60 % of the total particle population collected in Japan was found to be modified by sea salt as concluded from individual particle analysis (Okada et al. 1990; Niimura et al. 1998; Zhang et al. 2003b; Ma et al. 2005) and the interaction of dust with sea salt was likely an important process in size and composition changes of dust aerosols during their long-range transport (Zhang and Iwasaka 2004, 2006). In contrast, mixtures of dust and sea salt were reported to occur only to a minor degree in the African dust plume outflow (e.g. Reid et al. 2003b; Niemi et al. 2005; Kandler et al. 2007). The mechanisms responsible for the mixing of dust particles and sea salt have not been elucidated in detail (Andreae et al. 1986; Zhang et al. 2005a; Andreae and Rosenfeld 2008), while in-cloud processing was suggested as a major route for agglomerate formation (Andreae et al. 1986; Niimura et al. 1998).

Future Directions

Since the first models of aerosol ageing (e.g. Song and Carmichael 1999), progress in the understanding of chemical multiphase processes enabled, at best, taking into account the chemical reactivity of aerosols. However, recent studies show an explosion of identified organic species in the aerosol phase, notably due to the improvement in analytical techniques. In contrast to the clear evolution of sulphate which irreversibily oxidised and condensed, the OA presents versatile transformations which are highly dependent on meteorological and chemical conditions. The chemistry of the OA species is not well-known and there is a large need for further studies.

While the convergence of physico-chemical properties with ageing is observed for OA, this is not the case for dust and sea salt particles. Thus, even while numerous studies focussed on the chemical evolution of these aerosol particles, the challenge remains to improve our understanding of the link between aerosol ageing and their properties, notably size, ability to release nutrients and hygroscopicity. This point also needs to extend to measurements, up to now largely situated in Asian regions and the Pacific, to other oceanic regions, notably regions where high climate change is expected such as the Mediterranean basin (Giorgi 2006; IPCC 2007).

4.2.4 Dust-Mediated Transport of Living Organisms and Pollutants

A major argument for studying inter-regional and inter-continental transport of dust and its impacts is the recognition of ecological and human health damage risks. Owing to advances in meteorology, analytical instrumentation, satellite technology and image interpretation, more precise information on source areas, transport patterns and depositional zones for dust can now be obtained.

As an example, fungi transported by Saharan dust have been found to damage Caribbean corals (Colarco 2003a, b). Several health studies in urban and suburban environments have shown a higher risk of mortality from high exposure to fine (particles with diameter <10 μm or PM_{10}) particulate matter (e.g. Lippmann 2007). Airborne particles emitted from geological strata also pose threats to human health and the environment worldwide due to expansion of infrastructure development to serve the increasing population. After generation, dust can be carried by wind into sensitive environments and adverse health effects of respiratory "dust" on human health are well documented (Love et al. 1997).

Two misconceptions exist: desert soils are too inhospitable to accommodate a diverse microbial community and if present, the microorganisms will not be able to withstand the physical stresses (UV, desiccation, temperature) of atmospheric transport. However, dust events have been shown to introduce a significant pulse of microorganisms (Griffin 2007; Schlesinger et al. 2006; Polymenakou et al. 2008) and other microbiological materials (e.g. cellular fragments, fungal spores) into the atmosphere (Jaenicke 2005). Dust-borne transport of microorganisms, mainly over the marine environment, should be enhanced due to tolerable humidity levels and attenuation of UV by the particle load of the various dust clouds. Polymenakou et al. (2008) examined the microbial quality of aerosols over the Eastern Mediterranean region (Island of Crete, Greece) during an African dust storm. Bacterial communities associated with aerosol particles of six different size ranges were characterised using molecular culture-independent methods (analysis of 16S rRNA genes). Spore forming bacteria such as Firmicutes were found to be present in all aerosol particles and dominated the large particle sizes. Besides the dominance of Firmicutes in dust particles, phylogenetic neighbours to human pathogens associated with the respirable particles were also detected. These pathogens have been linked to several diseases such as pneumonia, meningitis, and bacteremia or suspected to induce pathologic reactions such as endocarditis (i.e. *S. pneumoniae, S. mitis, S. gordonii, H. parainfluenzae, A. lwoffi, A. johnsonii, P. acnes*).

The amount of toxic waste stemming from obsolete pesticides in Africa is higher than previously estimated (Mandavilli 2006). In Africa alone the amounts of toxic wastes are estimated at ca. 120,000 tonnes, with more than 500,000 tonnes worldwide. An estimated 30 % of the waste is believed to be persistent organic pollutants (POPs). The POPs family includes dioxins (polychlorinated dibenzodioxins and dibenzofurans), polychlorinated biphenyls (PCBs) and several organochlorine pesticides such as dichlorodiphenyltrichloroethane (DDT) and hexachlorobenzene. It should be pointed out that the POPs composition of dust in air masses changes in relation to modifications in land use, intensity of pesticide use, and burning of synthetic materials and biomass in the dust source regions and in areas swept by dust air masses (Garrison et al. 2003). There are few studies of the transport of POPs from Africa to the Mediterranean. It has been shown that the air masses arriving in the Eastern Mediterranean from Africa contained levels of polychlorinated biphenyls (PCBs) as high (>100 pg m^{-3}) as those in the air masses coming from industrialised Western Europe (Mandalakis and Stephanou 2002). Concurrently, in another study (Garrison et al. 2006) aimed at elucidating the potential role that Saharan dust might play in the degradation of Caribbean ecosystems, a series of persistent organic pollutants (POPs), trace metals and viable microorganisms were identified in the atmosphere over dust source areas of West Africa and in the Caribbean.

Microbes and fungi present on dust can survive a transcontinental journey and stay alive for centuries (Gorbushina et al. 2007). It was shown that Saharan dust collected over the Atlantic by Charles Darwin in the nineteenth century contained members of the spore-forming bacteria *Bacillales* (Firmicutes) and fungi attached to the particles. By combining geochemical, microbiological and microscopic methods to analyse their almost 200-year-old samples, Gorbushina et al. (2007) were able to show beyond doubt that dust, which clearly originated from West Africa, transported viable microorganisms across the Atlantic Ocean.

4.3 Direct Radiative Effects (DRE)

The direct radiative effect (DRE) is the change in net downward radiative flux, measured in Wm^{-2}, due to aerosol scattering and absorption of radiation (Forster et al. 2007). The DRE exerted by marine aerosols has received less attention that the anthropogenic DRE, which is a climate forcing. However, there are good reasons for quantifying the marine DRE. Firstly, it gives insight into pre-industrial radiative effects of aerosols since the distinction between natural and anthropogenic aerosols is important in the study of the climate system. Secondly, the decrease in sea-ice extent due to climate change leads to an increase in sea salt generation (Bellouin et al. 2011). Climate change also impacts the activity of ocean biogeochemistry and the associated DMS emissions into the atmosphere (Halloran et al. 2010). These changes impact the DRE of marine aerosols and the Earth's radiative budget.

In the shortwave spectrum and at the top of the atmosphere, the sign of the DRE depends on the balance between the decrease in net flux due to aerosol scattering and the increase due to aerosol absorption. Marine aerosols lack elemental carbon and iron oxides and are therefore weakly absorbing in the shortwave spectrum (Irshad et al. 2009). Consequently, their shortwave DRE at the top of the atmosphere is negative, and more negative per unit mass than absorbing aerosols of similar sizes. The marine aerosol DRE is exerted predominantly in cloud-free conditions, except where clouds are too thin to mask it. Aerosols are more efficient at exerting a DRE when their radius is comparable to the wavelength of the radiation. The size distribution of marine aerosols, especially sea salt, covers both the sub- and super-micron ranges (Dubovik et al. 2002). Their DRE therefore covers both the shortwave and longwave spectra. In the longwave spectrum, marine DRE is positive at the top of the atmosphere because the aerosol layer is typically colder than the surface (Reddy et al. 2005). In addition, since aerosol size is a key parameter for the DRE, hygroscopic growth is important, especially at the high relative humidity experienced by marine aerosols, and is a strong function of chemical composition (Randles et al. 2004).

Four aerosol characteristics are needed to quantify the DRE: optical depth, single-scattering albedo, size distribution, and vertical profile (Yu et al. 2006). Among those parameters, aerosol optical depth is better constrained, with dedicated ground-based and satellite instruments providing routine retrievals (Holben et al. 2001; Remer et al. 2008; Kokhanovsky and de Leeuw 2009; de Leeuw et al. 2011b). Retrievals are however limited to cloud-free conditions, a restriction detrimental to the sampling of marine aerosols, especially at high latitudes of the southern hemisphere where near-surface wind speeds, cloud cover, and sea salt optical depths are large. Satellite retrievals are more accurate over dark ocean surfaces, where marine aerosols are located, than over the relatively brighter land surfaces (King et al. 1999). Satellite products also include information on the aerosol size distribution, through the Ångström coefficient or optical depth of the fine mode (Anderson et al. 2005). Ground-based sun-photometer networks provide retrievals of the single-scattering albedo (Schuster et al. 2005) and size distribution (Dubovik et al. 2002). Coastal and island sites and ship cruises are useful to characterise marine aerosols locally (Smirnov et al. 2011). The aerosol vertical profile is available from ground-based and spaceborne lidars (Winker et al. 2010).

DRE estimates from observations use retrieved aerosol properties coupled to radiative transfer calculations (e.g. Bellouin et al. 2008) or attribute a fraction of broadband radiative fluxes measured by radiometers to aerosols (e.g. Loeb and Manalo-Smith 2005). The shortwave cloud-free top-of-atmosphere total (natural and anthropogenic) DRE is estimated to be in the range -4 to $-6\ Wm^{-2}$ over global oceans and around $-5\ Wm^{-2}$ over land (Yu et al. 2006). In the

4.3 Direct Radiative Effects (DRE)

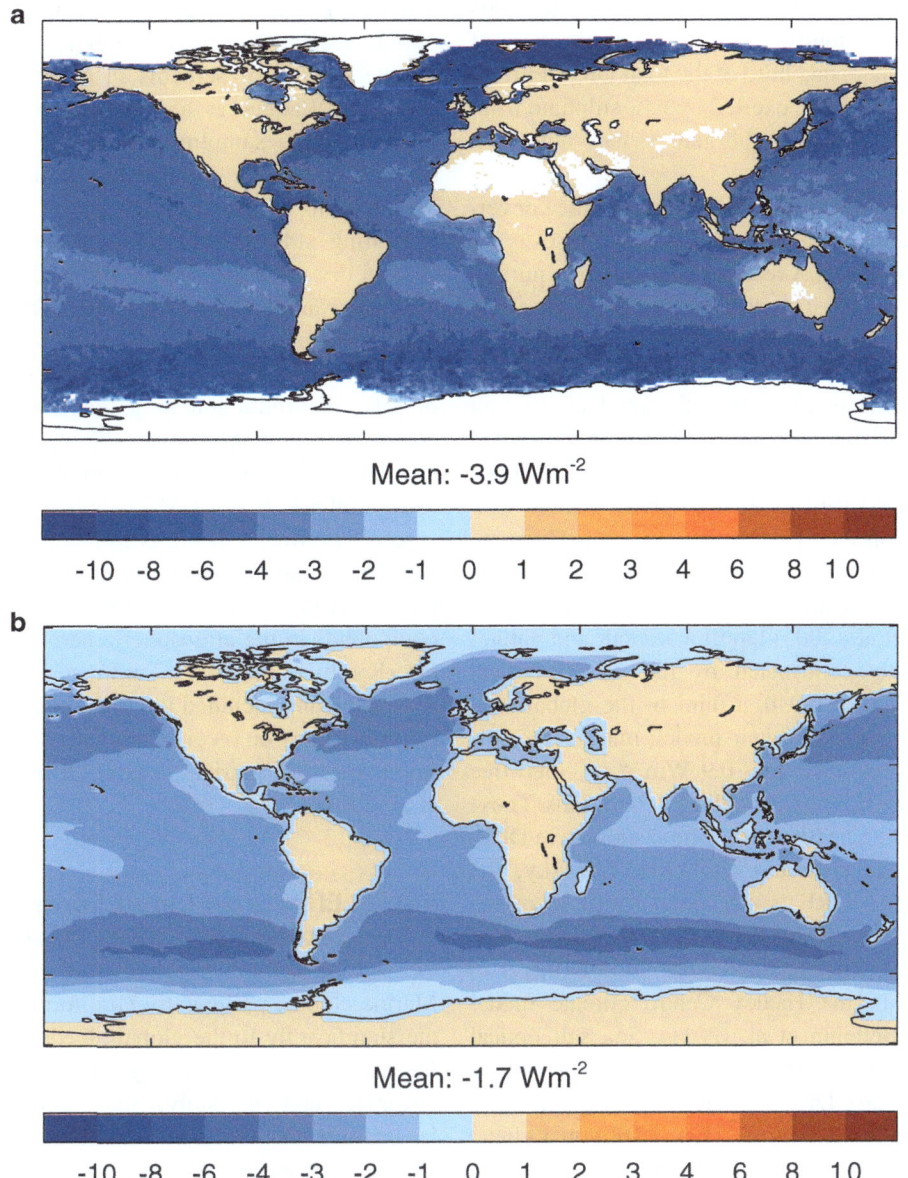

Fig. 4.16 (a) Shortwave direct radiative effect, in Wm^{-2}, of marine aerosols at the cloud-free top of atmosphere as estimated from MODIS collection five aerosol retrievals by Bellouin et al. (2008) for the year 2006. (b) Same but for sea salt aerosols modelled in the Hadley Centre climate model HadGEM2 for present-day conditions

southern ocean where marine aerosols are expected to contribute most to the total cloud-free DRE, the latter is estimated at −4 to −6 Wm^{-2} (Yu et al. 2006). Aerosol observations characterise the ambient aerosol. The contribution of different aerosol types to the total DRE is therefore difficult to quantify from observations alone. Aerosol size retrievals have been used as a useful but imperfect proxy for this task (Kaufman et al. 2005; Bellouin et al. 2008). Marine aerosols are found to be the dominant contributor to the global direct effect. Bellouin et al. (2008) estimated that the sea salt shortwave DRE for the year 2006 is −3.9 Wm^{-2} (Fig. 4.16a) at the cloud-free top of the atmosphere, half of the total estimated DRE. Zhao et al. (2011) mix observation with modelling to attribute the total DRE to individual species. They find a global-averaged sea salt shortwave DRE at the cloud-free top of the atmosphere of -2.2 ± 0.6 Wm^{-2}, one third of the total DRE. Further speciation into the DRE exerted by nss- sulphate from ocean DMS, and marine organics has not been attempted.

Modelling estimates of the DRE involve the conversion of simulated dry masses of various aerosol types into optical properties. A realistic conversion requires a good simulation of size distribution (Vignati et al. 2004), mixing state (Stier et al. 2007), and hygroscopic growth. In a comparison of global numerical aerosol modelling (Kinne et al. 2006), the median sea salt optical depth is 0.030 at 0.55 μm on a global average, 24 % of the total aerosol optical depth. Sea salt is the second largest contributor after sulphate, which itself includes an unspecified marine component. Diversity among participating models is large and the sea salt fraction of the optical depth varies between 18 % and 50 %. Reddy et al. (2005) report a sea salt and natural sulphate (mostly contributed by the ocean) optical depth at 0.55 μm of 0.040, a third of the global total aerosol optical depth. In their model, marine aerosols exert a shortwave DRE of -0.9 Wm^{-2} in cloud-free conditions and -0.5 Wm^{-2} in all-sky conditions. Those numbers represent nearly half the total shortwave DRE. In the longwave spectrum and cloud-free sky, the simulated marine DRE is +0.2 Wm^{-2}, 40 % of the total DRE in that spectrum, the remainder being exerted by mineral dust. The sea salt contribution is larger in simulations with the Hadley Centre climate model described by Bellouin et al. (2011). Sea salt optical depth at 0.55 μm is 0.052, 44 % of the total optical depth. Its shortwave DRE in cloud-free sky at the top of the atmosphere is -1.7 Wm^{-2} (Fig. 4.16b) on a global average, 40 % of the total DRE. In all-sky, the sea salt DRE is estimated at -0.8 Wm^{-2} for a total DRE of -2.2 Wm^{-2}. In cloud-free longwave, sea salt DRE is +0.4 Wm^{-2}, half of the total DRE in that spectrum. Modelled estimates are typically weaker than observational estimates, although the incomplete coverage of the Earth's surface by satellite aerosol retrievals complicates the comparison.

The results above show that observation-based and modelling estimates of the total and marine DRE remain associated with large ranges of values. Uncertainties in the aerosol optical properties, mixing, and identification of different components need to be reduced in order to improve our knowledge of the total and marine DRE. Progress is being made in aerosol remote-sensing techniques (Dubovik et al. 2011), with remote-sensing of aerosols in cloudy sky becoming possible (Waquet et al. 2009; Omar et al. 2009). Improving the sampling of remote oceanic regions through a synergy of satellite instruments would be beneficial to observations of marine aerosols. On the modelling side, the diversity in the simulations of aerosol dry mass and optical properties contributes to the large range of simulated DRE. There is a factor 5 among model simulations of sea salt dry mass (Kinne et al. 2006), and improvements in the parameterisation of sea salt production as a function of wind speed could reduce that range, although differences in simulated wind speed will certainly remain. Sea salt and sulphate from DMS oxidation are the only aerosols produced at the ocean–atmosphere interface in current atmosphere–ocean general circulation models. However, evidence for a major role of organic compounds in the atmospheric aerosol cycle is growing (Kirkby et al. 2011). Parameterisations should be developed and included in global numerical models in order to follow on recent observational evidence of the importance of marine organic emissions (O'Dowd et al. 2004).

4.4 Effects on Cloud Formation and Indirect Radiative Effects

Marine clouds, particularly stratiform clouds, contribute significantly to the global albedo for two reasons: firstly, they contribute a large fraction of global cloud coverage and secondly, they comprise reflecting layers over the dark, absorbing, ocean surface. Furthermore, given that marine clouds are typically clean clouds (i.e. low cloud droplet number concentration), they are far more susceptible to perturbations in cloud nuclei availability compared to continental clouds. For example, Slingo (1990) estimated that a 10–15 % decrease in effective radius, corresponding to increase in cloud droplet concentration of 30–45 % would be sufficient to off-set global warming increase relating to a doubling of CO_2.

Although in the early days of cloud physics, sea salt was considered the dominant marine cloud condensation species (Mason 1957), by the 1980s, sea salt was more or less dismissed as having little or no role in cloud formation (Charlson et al. 1987) on the basis that the number concentration of sea salt CCN rarely exceeded 1 cm^{-3} and was typically too large to mix up to cloud base. Despite this, Latham and Smith (1990) postulated that sea salt could be involved in a wind-speed related negative feedback system whereby with increasing global temperatures, zonal wind speeds would increase, leading to increased sea salt CCN generation, increased cloud droplet concentrations and ultimately increased cloud albedo. Their measurements were based on the existence of sea salt CCN down to sizes of ~0.2 μm radius. O'Dowd and Smith (1993) later demonstrated that sea salt nuclei extended into sizes as small as 0.05 μm radius and, under moderately high wind speeds, could account for ~70 % of the nuclei larger than this size. Furthermore, O'Dowd et al. (1993) demonstrated through combined measurement and modelling studies that there was significant competition between sea salt and nss-sulphate nuclei and that under certain conditions, the addition of a small number of sea salt nuclei under moderately low updraft velocity conditions could lead to a reduction in cloud droplet concentrations as these nuclei cloud be preferentially activated over sulphate nuclei, leading to a suppression of the peak supersaturation achieved in clouds.

Korhonen et al. (2010a), taking observations of the accelerating tropospheric westerly jet, estimated an increase in wind speed of 0.45 m s^{-1} decade^{-1} at 50–65°S since 1980 and that this wind speed increase has produced an increase in cloud condensation nuclei of 22 % on average and up to 85 % in some regions, leading to increased cloud albedo. The recognition of the importance of sea salt nuclei in marine cloud processes has led to the interesting suggestion that global warming could be ameliorated somewhat through controlled enhancement of albedo and lifetime in low level marine clouds (Latham 2002), in other words, geo-engineering, whereby artificial floating sea spray generators are deployed in regions of persistent stratocumulus clouds. While intriguing as a geo-engineering solution to global warming, Korhonen et al. (2010b), using a global model, and simulating emissions from a fleet of spray-emitting vessels in four regions of persistent stratocumulus fields found increases in cloud droplet concentration of maximum 20 %, and even a reduction was predicted in one region.

The above studies focussed on sea salt as opposed to sea spray nuclei, the latter which can comprise varying degrees of organic matter enrichment. As discussed in Sect. 4.2.1.2, O'Dowd et al. (2004) quantified the relative contributions of marine aerosol in terms of sea salt, sulphate, WSOM and WIOM during seasons of low and high biological activity over the North East Atlantic. They found that organic matter dominated the sub-micron mass fraction during the high biological activity period and suggested that a significant fraction of the organic matter was of primary, sea spray, origin. This was later corroborated through laboratory and gradient flux studies which demonstrated that the WIOM was almost exclusively produced from bubble bursting at the ocean surface. Ovadnevaite et al. (2011a) further identified, using aerosol mass spectrometry, significant primary organic aerosol plumes approaching 4 μg m^{-3} for extended periods, although the mass spectrometry revealed significant oxygenated organic matter in the plume. It should be noted that solubility is a relative definition and the detection of oxygenated organic matter in the plumes is not inconsistent with the previous identification of almost exclusively WIOM. The question is whether or not this primary organic matter which is apparently water insoluble, will have a negative or positive impact on cloud droplet concentration? In a follow-on study, Ovadnevaite et al. (2011b) found that the aerosol particles dominating these so-called primary organic plumes have a low hygroscopicity (~1.2–1.25), they have almost a 100 % activation efficiency at 0.25 % supersaturation even for Aitken mode particles.

Ovadnevaite et al. (2011b) calculated the weighted number of (organically-enriched) sea spray and nss-sulphate nuclei and compared the number concentration of these nuclei with CCN and cloud droplet concentration (Fig. 4.17) for the cloud forming on the plume. The total (combined sulphate and sea spray) calculated nuclei concentration agreed almost perfectly with the CCN concentration at 0.75 % supersaturation, while the sea spray concentration agreed very closely to the cloud droplet concentration, leading to a correlation coefficient of r = 0.76, while nss-sulphate was anti-correlated to the cloud droplet concentration. Not only were these results surprising in that what was apparently non- or very low-solubility organic spray aerosol acting as highly efficient cloud nuclei, but the number concentration and resultant cloud droplet concentration exceeded 350 cm^{-3} which can be regarded as a very high droplet concentration for a maritime stratiform cloud, and more typical of polluted or continental clouds. It was postulated that the organic

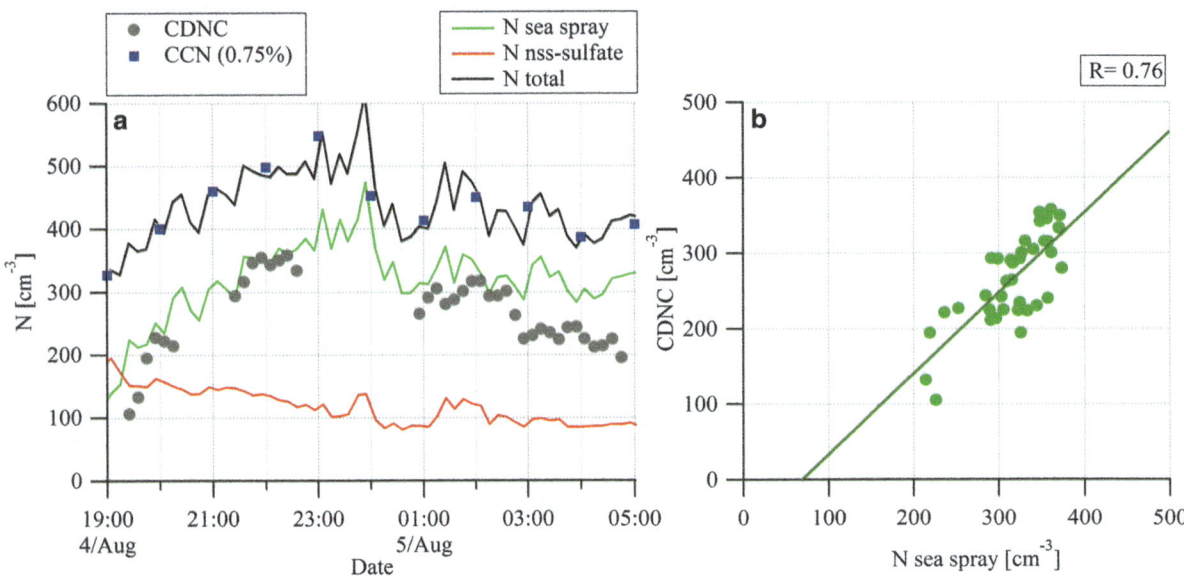

Fig. 4.17 (**a**) CDNC, measured CCN$_{0.75\%}$, calculated sea spray, sulphate and total nuclei concentration. (**b**) CDNC as a function of sea spray particle concentration (Reproduced from Ovadnevaite et al. (2011b) by permission of the American Geophysical Union)

matter in the sea spray could be a marine hydrogel resulting in such behaviour. To summarise, sea salt aerosol is likely to have an important impact on the indirect effect, and organic enrichment, in certain regions, may have an even more important impact.

4.5 Deposition of Aerosol Particles to the Ocean Surface and Impacts

Studying the key interactions between the atmosphere and the ocean is essential to understand the present functioning of biogeochemical cycles in the ocean and to predict their evolution in the future. Over the past two decades a considerable effort has been made to improve our understanding of the relevant processes involved in the delivery of bioavailable atmospheric nutrients to the surface of the ocean, their impact on marine biogeochemical cycles, biota response and carbon export. As shown in the following sections, different temporal and spatial scales have been explored from the large-scale experiments to the microcosm studies and through modelling approaches, in both high-nutrient low-chlorophyll (HNLC) and low-nutrient-low chlorophyll (LNLC) areas of the ocean. Recent progress is presented in the following section showing that we are beginning to understand the links between atmospheric deposition and global/regional biogeochemical cycles.

4.5.1 Deposition

Atmospheric deposition of aerosols can impact the biogeochemical cycles of several important nutrients in the ocean, especially iron, nitrogen and phosphorus (Martin et al. 1991) (Duce 1986; Callaghan et al. 2008; Falkowski et al. 1998; Fung et al. 2000) (Krishnamurthy et al. 2010), although deposition of other species may also be important (e.g. Nozaki 1997). Some atmospheric inputs (e.g. toxic metals or dissociation products of strong acids and bases) may actually reduce ocean productivity (Paytan et al. 2009; Doney et al. 2007). The most important constituent for ocean biogeochemistry is likely iron, with nitrogen, phosphorus, toxic metals and acidic species also important in some regions (Doney et al. 2007; Hunter et al. 2011; Krishnamurthy et al. 2010; Okin et al. 2011; Paytan et al. 2009).

4.5.1.1 Iron

Iron is a micronutrient, and required in small quantities by biota. In some regions there is insufficient iron, and thus iron limitations occurs (Martin et al. 1991). Therefore, atmospheric deposition can play a critical role in supplying new iron to the surface ocean (Fung et al. 2000). Most of the iron deposited onto the ocean surface comes from atmospheric mineral aerosols which is approximately 3.5 % iron, with small contributions from combustion sources (Luo et al. 2008); because of the dominance of mineral aerosols to the iron budget, the largest deposition of iron to the oceans occurs downwind of the large desert regions (Fig. 4.18a).

Not all forms of iron are thought to be equally bioavailable (Jickells and Spokes 2001; Mahowald et al. 2009). Most soils have relatively insoluble iron forms, while atmospheric aerosols appear to be more soluble, arguing that atmospheric processing may be occurring (Jickells and Spokes 2001) and plausible mechanisms based on acidity have been proposed (e.g. Jickells and Spokes 2001; Meskhidze et al. 2005; Zhu et al. 1997), although this can be very sensitive to the mineralogical composition of aerosol (Journet et al. 2008). Smaller particles are slightly more soluble than larger particles (e.g. Baker and Jickells 2006; Chen and Siefert 2004; Hand et al. 2004), which can largely be explained by the longer residence time of smaller particles (Hand et al. 2004). Recent studies have suggested a role for combustion sources of soluble iron (Chuang et al. 2005; Guieu et al. 2005; Sedwick et al. 2007), however estimates from observations and models still suggest that the largest deposition of soluble iron occurs downwind of the main desert dust source areas (Fig. 4.18b). Extrapolations of the limited data on soluble iron from combustion and atmospheric processing suggest that soluble iron deposition may have doubled over the last century, assuming constant mineral aerosol composition because of increased pollution (Luo et al. 2008; Mahowald et al. 2009). More recent estimates of changes in desert dust over the past century based on observations suggest there may have been almost a doubling in desert dust between 1900 and 2000 (Mahowald et al. 2010), suggesting almost a quadrupling of soluble iron inputs to the oceans over this time period.

The importance of atmospheric iron was probably initially overestimated (e.g. Ridgwell and Watson 2002), because of an underestimate of the ocean sediment sources of iron (Lam and Bishop 2008) and a lack of understanding of the role of colloids in the ocean in maintaining iron supply (Parekh et al. 2004). However, current understanding suggests that changes in iron between glacial and interglacial times can play an important, if secondary, role in facilitating the drawdown of atmospheric carbon dioxide (Kohfeld et al. 2005), and changes in soluble iron deposition over the anthropocene may have driven small (5 ppm) changes in atmospheric carbon dioxide (Mahowald et al. 2010).

4.5.1.2 Phosphorus

On longer time scales, ocean productivity is limited by phosphorus (Falkowski et al. 1998), and in some regions on shorter time scales (Mills et al. 2004; Moore et al. 2006; Wu et al. 2000). While the source of phosphorus to the ocean from rivers is thought to be much larger than from atmospheric deposition (11 Tg P year^{-1} vs. 0.6 Tg P year^{-1}) (Seitzinger et al. 2005; Mahowald et al. 2008), much of the riverine inputs may be sequestered in estuaries, and not be available to open ocean biota. Thus, atmospheric deposition of phosphorus can be important. Similar to iron, most atmospheric phosphorus is thought to be in the form of aerosols (Graham and Duce 1979), predominately mineral aerosol particles (83 %) (Mahowald et al. 2008), since crustal material is on average 700 ppm phosphorus. Thus, atmospheric phosphorus deposition is similar to iron in being largest downwind of desert regions (Fig. 4.18e). Other sources of phosphorus include primary biogenic emission of aerosol particles and aerosol emission from biomass burning, fossil and bio-fuel burning, volcanoes, as well as sea salt aerosol (Mahowald et al. 2008). Not all phosphorus is likely to be soluble in the oceanic mixed layer; soluble P or phosphate is measured to be between 7 % and 100 % of total phosphorus in aerosols (e.g. Graham and Duce 1979; Mahowald et al. 2008). It is likely that mineral aerosols are less soluble than other sources of phosphorus (e.g. Mahowald et al. 2008), but there may be atmospheric processing by acids of phosphorus to make it more soluble (Baker et al. 2006b; Nenes et al. 2011). Estimates of phosphate deposition are limited by shortage of measurements of phosphorus and phosphate, as well as limitations in the understanding of the atmospheric phosphorus cycling (e.g. Mahowald et al. 2008). Because of the large reservoir of phosphorus in the oceans, atmospheric deposition is not thought to be a dominant control on global ocean productivity (Krishnamurthy et al. 2010).

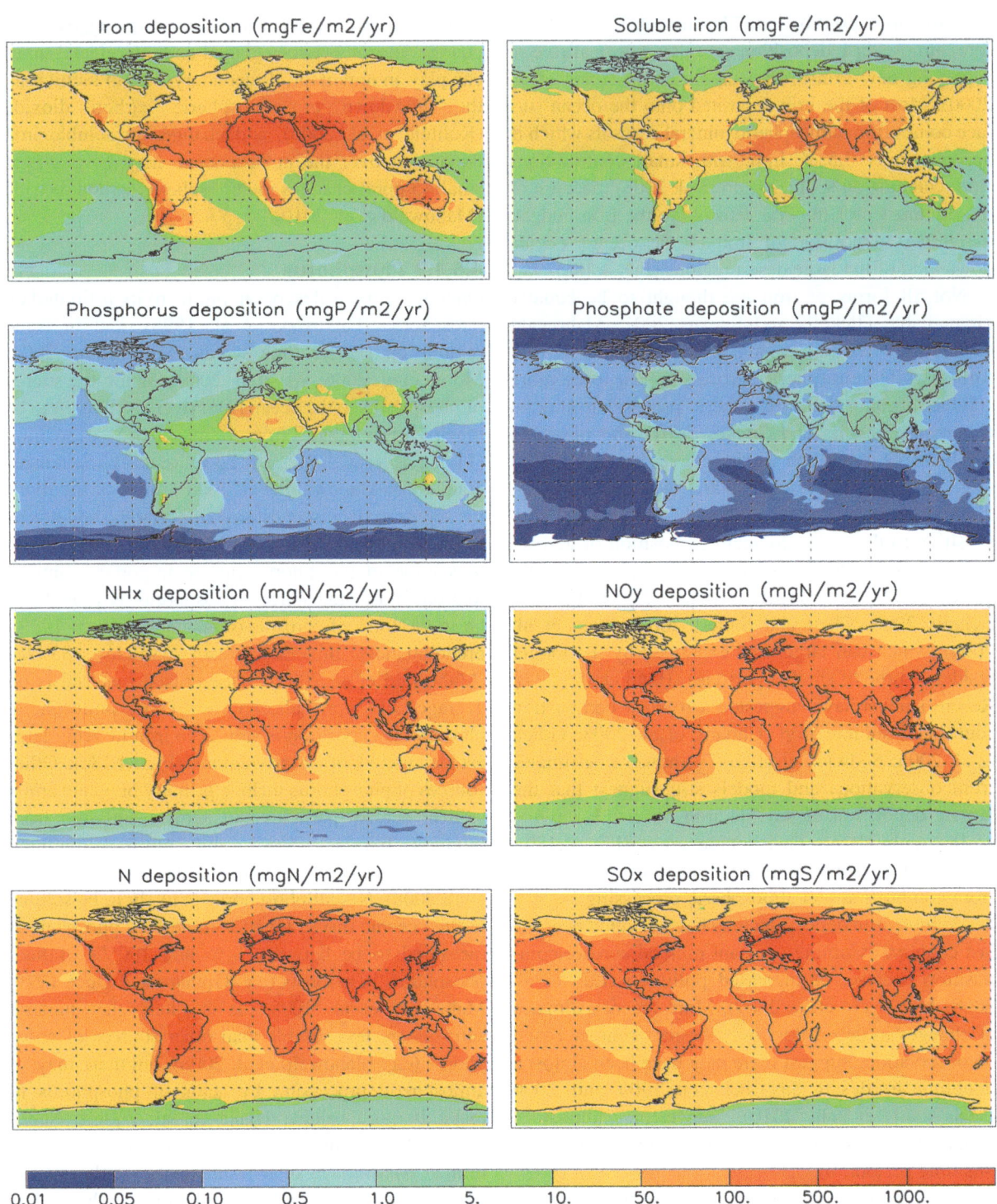

Fig. 4.18 Deposition maps of Fe (**a**) (Mahowald et al. 2005), soluble Fe (**b**) (Okin et al. 2011), P (**c**) (Okin et al. 2011) soluble P (**d**) (Okin et al. 2011) NHx (**e**) (Dentener et al. 2006), NOy (**f**) (Dentener et al. 2006), Total N (**g**) (Dentener et al. 2006), SOx (**h**) (Dentener et al. 2006)

4.5.1.3 Nitrogen

The two known sources of new nitrogen to the ocean are biological N_2 fixation and atmospheric deposition. Although molecular nitrogen gas (N_2) is only available for diazotrophs organisms, nitrogen fixation can be an important process induced by atmospheric deposition of other limiting nutrient and is discussed elsewhere (see Sect. 4.5.4.2). Nitrogen aerosol deposition comes predominately from combustion sources (NOx) and agricultural sources (NH_4). The atmospheric reactive nitrogen flux (NOx, NH_4, organic nitrogen compounds) has dramatically increased over the last century (Duce et al. 2008; Galloway et al. 2008) and most of it is bioavailable (Duce et al. 2008). While released into the atmosphere as a gas, about half of the reactive nitrogen is deposited as aerosols (Adams et al. 1999). Nitrogen aerosols tend to have a relatively short lifetime in the atmosphere (few days), which is seen in the estimated distribution of reactive nitrogen deposition (Fig. 4.18c, d). Because the pool of nitrogen in the ocean is so large, it is unlikely that anthropogenic new nitrogen is important to net ocean uptake of carbon (Krishnamurthy et al. 2010; Okin et al. 2011). However, large parts of the ocean are thought to be nitrogen limited, additional N deposition could locally lead to additional productivity, especially because the effects of increasing atmospheric nitrogen deposition are expected to continue to grow in the future (Duce et al. 2008; Krishnamurthy et al. 2010).

4.5.1.4 Deposition of Other Species

While iron, nitrogen and phosphorus deposition are thought to provide nutrients which increase ocean productivity, deposition of toxic metals (e.g. Cu) or acidic species are thought to reduce ocean productivity. While Cu is deposited into the ocean predominately by mineral aerosols, deposition is thought to be enhanced by anthropogenic activity (Paytan et al. 2009). Atmospheric acids are deposited to the oceans as acid rain in the form of sulphates and nitrates occurring in aerosols: thus nitrate aerosols both increase productivity by adding nitrogen, and potentially reduce productivity by increasing acidity (Doney et al. 2007) (Fig. 4.18g).

4.5.2 Elements of Biogeochemical Interest and Their Chemical Forms

As shown in the previous section, atmospheric particles from both natural and anthropogenic sources contain chemical elements which participate to marine biogeochemistry when deposited to the surface ocean. Among them, iron (Fe) has received particular attention, especially when associated with mineral dust which dominates the external input of this key element to the surface open ocean (Jickells et al. 2005). Dust-derived Fe has been proposed to be responsible of the glacial-interglacial differences in atmospheric CO_2 due to the strong Fe limitation of biological activity in high-nutrient low chlorophyll oceanic regions (Martin 1990). This has boosted the literature on the chemical forms of dust-derived Fe and the multiple factors controlling its solubility both in the atmosphere and in the marine environment (see Sect. 4.5.3). Other sources of atmospheric Fe such as biomass burning (Guieu et al. 2005) or volcanic ash (Sect. 4.2.2.2) can partially contribute to the marine Fe pool at a regional scale although their overall contribution remains low compared to aeolian dust (Mahowald et al. 2005).

As opposed to HNLC oceanic regions, in LNLC areas biological activity is often limited or co-limited by phosphorus and/or nitrogen. This has motivated interest in atmospheric sources of these two elements and their contribution to the marine pool, particularly in two LNLC areas: the Mediterranean Sea (e.g. Markaki et al. 2010) and the Atlantic Ocean (e.g. Baker et al. 2010). At a global scale, mineral dust appears as the major source of P followed by primary biogenic particles and combustion sources (Mahowald et al. 2008). Indeed, dust deposition events can transiently increase the concentration of dissolved inorganic P in the surface waters of the LNLC Mediterranean Sea (Pulido-Villena et al. 2010). The partitioning between dissolved and total P in atmospheric deposition varies widely between 7 % and 100 % (Migon and Sandroni 1999; Mahowald et al. 2008). As for Fe, numerous factors such as the aerosol source and physico-chemical transformation during transport control the dissolved fraction of atmospheric P deposition. Recent results highlight the importance of the atmospheric acidification of aerosols resulting from the mixing of polluted and dust-laden air masses as a source of dissolved P to the oceans (Nenes et al. 2011).

As opposed to Fe and P, N in the atmosphere is mainly present in gaseous form which implies a different behaviour. N deposition is dominated at a global scale by anthropogenic emissions which have significantly increased since the mid-1800s and for which future increases are expected (Dentener et al. 2006; Duce et al. 2008). Atmospheric fluxes of N are also very important in remote dusty regions. Modelling and laboratory studies have indicated that mineral dust particles can take up acids resulting in increased coarse mode nitrate (NO_3^-) (Usher et al. 2003 and references therein). When secondary NO_3^- is accumulated in the coarse mode, it can be removed more rapidly by dry or wet deposition (Dentener et al. 1996).

The bioavailable fraction of nutrients has been traditionally associated with the inorganic forms. This may explain the scarcity of data concerning atmospheric deposition of organic nutrients. Today, we are aware that not only heterotrophic organisms but also autotrophic phytoplankton is able to take up organic forms of either phosphorus or nitrogen. Organic N and P contained in atmospheric deposition can thus exert a biogeochemical effect on the surface ocean the magnitude of which remains poorly explored. Although organic nitrogen has been measured in atmospheric deposition (e.g. Cornell et al. 1995), it is only recently that a real interest on its contribution to biogeochemical cycles at a global scale has begun to emerge (Cornell 2011; Lesworth et al. 2010). Indeed, organic nitrogen can constitute 30 % of total N deposition and an important fraction is of anthropogenic origin (Duce et al. 2008 and references therein).

Virtually nothing is known about the significance of organic P in total P deposition. A few studies have been conducted in the Mediterranean Sea showing a contribution of organic P to total P deposition of 34 % and 38 % in the western and eastern basins respectively (Migon and Sandroni 1999; Markaki et al. 2010). These values are in good agreement with the results of Chen et al. (2007) who estimated that, on average, organic P constituted 31 % of total P deposition in the Gulf of Aqaba. These few reported values suggest that atmospheric inputs can be a significant source of organic phosphorus and highlight the need of evaluating the role of this external source in marine biogeochemistry.

Current models of the carbon cycle, either at regional or global scale, do not account for atmosphere–ocean exchanges of organic carbon, particularly, atmospheric inputs of organic carbon to the surface ocean. And yet, the few existing data indicate that the magnitude of these deposition processes are far from being negligible (Fig. 4.19). Willey et al. (2000) estimated the input of dissolved organic carbon (DOC) to the ocean to be 90 Tg C year^{-1}, equivalent to the magnitude of riverine input to the open ocean.

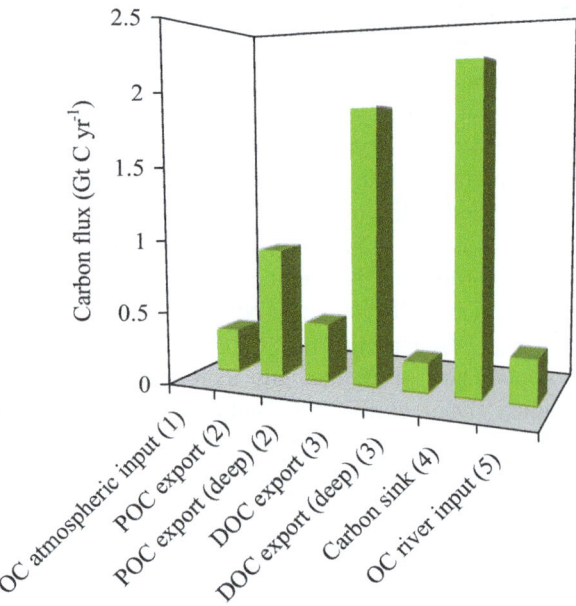

Fig. 4.19 Comparison between the only existing estimate of the global atmospheric flux of organic carbon to the ocean and selected global oceanic carbon fluxes typically considered in carbon cycle models. The aim of this figure is to represent visually the relative magnitudes, rough values are given and no error bars are included. (*1*) Global atmospheric deposition of organic carbon (Jurado et al. 2008). (*2*) Global oceanic downward flux of particulate organic carbon of which 45 % actually reaches the deep sea floor (Jahnke 1996). (*3*) Global oceanic downward flux of dissolved organic matter of which 10 % survives to depths >500 m (Hansell et al. 2009). (*4*) Current estimate of the oceanic sink of anthropogenic carbon (Denman et al. 2007). (*5*) Global riverine input of organic carbon to the ocean (Seitzinger et al. 2010). The atmospheric input of organic carbon is of the same order of magnitude as the POC and DOC deep downward flux and represents more than 10 % of the oceanic carbon sink. Moreover, it is similar to the riverine input, which has been so far the main allochthonous source of organic carbon represented in carbon cycle models

Accounting not only for rain, but also for dry deposition, Jurado et al. (2008) derived a global estimate of 245×10^{12} Tg C year^{-1}, a value largely dominated (76 %) by wet deposition of gaseous species (not including CO_2). This atmospheric source may be particularly important for the open ocean areas less affected by riverine influences.

In summary, our current knowledge of the atmospheric fluxes of nutrients to the surface ocean suffers from a disequilibrium between the organic and inorganic fractions. Further data on atmospheric deposition of organic phosphorus, nitrogen and carbon will certainly constitute a step forward towards the quantification and prediction of the impact of atmospheric deposition on marine biogeochemistry.

4.5.3 Dissolution- Scavenging Processes

In recent years, an abundant literature has been published on the effect of dust deposition on the biogeochemical functioning of the surface ocean (e.g. Landing and Paytan 2010). Even if the assumption is rough, it can be considered that microorganisms preferentially uptake elements in the dissolved form. Therefore, from a (bio)-geochemical point of view, for a given chemical element, a key issue is to assess the impact of atmospheric particle deposition on the dissolved pool of such elements in the ocean. Atmospheric deposition can transport a large amount of essential macro and micro-nutrients to the ocean (see Sect. 4.5.1). Due to the large scientific debate on the role of atmospheric dust iron deposition to the surface ocean on the climate variability at millennium time scales (Martin 1990), an extensive set of studies have been produced on the impact of dust deposition on iron stocks. Due to its importance in oligotrophic oceanic areas, an abundant literature has also been produced for phosphorus in the last two decades. However, some recent studies suggest a potential control of biological activity by other trace elements of atmospheric origin (Paytan et al. 2009; Ridame et al. 2011) in large oceanic areas. It is therefore important to emphasise that, even if most of the examples hereafter are based on iron or phosphorus chemistry, the concepts introduced have a wider importance than just for these elements.

This section will focus on the "post-depositional" processes, which include all processes that occur at the surface of aerosol particles once they have deposited at the ocean surface. The impact of atmospheric particles on elemental stocks in the ocean will be controlled by the balance between dissolution and scavenging at the surface of the atmospheric particles which settle through the water column. For a given elemental stock, dissolution will constitute a source whereas scavenging will be a sink.

A large set of literature exists on the dissolution of nutrients or trace metals from atmospheric particles, with an important emphases on iron (Baker and Croot 2010). A number of these studies focus on atmospheric processes. They have demonstrated that the solubility of different elements from atmospheric particles is controlled by the nature, origin, size and mineralogy of the particles and by the complex processes that affect particles during atmospheric transport (see Sect. 4.2.3).

Once atmospheric particles deposit in seawater, the solvent where the solubility processes occur changes drastically from atmospheric conditions, shifting from low pH value and low ionic strength in cloud droplets to high ionic strength and a slightly basic pH value for seawater. Recent studies have demonstrated that solubility of different trace elements is lower in seawater than in ultrapure deionized water (Buck et al. 2010; Chen et al. 2006). The main driving force for these differences is pH.

In the ocean, the thermodynamic solubility of ionic species can be very low. For example, the solubility of inorganic iron in oxic seawater is between 100 and 200 pM (Liu and Millero 2002). Higher concentrations of dissolved iron are possible in seawater through complexation by organic iron binding ligands. Indeed, more than 99 % of dissolved iron is in the form of organic complexes (Hunter and Boyd 2007). In this case, the seawater binding ligand concentration may control the amount of a specific element that can dissolve in seawater. For iron, this control by specific binding ligand has been demonstrated in batch dissolution experiments with dust end-member particles (Wagener et al. 2008, 2010). Moreover, the potential of free iron binding ligand to keep iron from dust particles dissolved was demonstrated in situ during a cruise in the tropical Atlantic after a Saharan dust event (Rijkenberg et al. 2008). Other elements, such as cobalt (Saito and Moffet 2001) and copper (Buck and Bruland 2005) are also found predominantly in the form of organic complexes in seawater and the same control mechanism on dissolution as for iron could occur.

The importance of the particulate concentration in seawater on the relative solubility has been discussed in earlier studies. Bonnet and Guieu (2004) have demonstrated that the relative solubility of iron decreases when the particulate concentration increases. This trend might be explained by the iron binding capacity of seawater which limits the iron that can dissolve and therefore decreases the relative amount of dissolved iron for higher particulate concentrations (Baker and Croot 2010). However, similar trends have been demonstrated for phosphorus dissolution (Ridame and Guieu 2002) whereas dissolved phosphorus concentrations are not demonstrated to be controlled by the solubility capacity of seawater.

Scavenging of dissolved elements on particles in the water column has been extensively studied on settling particles because this process is essential to explain the deep concentrations of different chemical elements. For some elements, scavenging on particles is even used to trace fluxes of biological particles in the water column: thorium-234 deficit in the water column is used to assess the export of organic carbon (Buesseler et al. 1992). Moreover, in the field of geochemistry, a rich literature exists on rare earth elements scavenging on pure mineral phases because of the importance of this geochemical process during sediment formation.

Only few studies exist on scavenging processes for atmospheric particles in the ocean. Zhuang and Duce (1993) performed a set of scavenging experiments with radiolabelled ^{59}Fe on natural aerosols in seawater and demonstrated significant scavenging on dust particles. A similar type of experiments with radiolabelled ^{33}P demonstrated scavenging of phosphorus on Saharan dust end-member particles (Ridame et al. 2003). In earlier studies, based on pure haematite particles, Honeyman and Sanchi (1991) demonstrated the scavenging of colloidal iron when freshly produced haematite particles coagulate to form larger aggregates. This points to the importance of the particle size (and certainly chemical composition) for scavenging processes.

Even fewer studies have taken into account the balance between scavenging and dissolution processes when atmospheric particles deposit to the surface ocean. In a set of batch reactor iron dissolution experiments in seawater, Bonnet and Guieu (2004) considered the importance of scavenging to suggest that the net result of dust dissolution is lower than the effective dissolution from dust particles. In a recent artificial dust seeding experiment in a large mesocosm, it was demonstrated that the net effect of dust addition was a decrease of the dissolved iron stock in the first 10 m of the water column due to scavenging on the settling dust particles (Wagener et al. 2010). The mesoscoms experiment was simulated with a 1D model which takes into account dissolution and scavenging of iron on dust particles. The equilibrium between both processes allowed definition of a critical Fe concentration above which the balance between dissolution and scavenging is in favour of scavenging, resulting in a net sink of dissolved iron (Ye et al. 2011).

One of the major challenges in future studies on abiotic processes occurring when atmospheric particles deposit at the ocean surface will be to investigate the above process in a more realistic way. This implies studying the processes involved (scavenging and dissolution) at a higher temporal and vertical resolution. So far, most of the studies on the biogeochemical impact of atmospheric deposition, have assumed that once they deposit at the ocean surface, atmospheric particles are perfectly mixed through the surface mixed layer (SML) (Fig. 4.20 – Left panel). It is considered that after a certain time, a fraction of a soluble element is dissolved (or scavenged) and will increase (or decrease) the dissolved stock of the element. Based on this postulate, and in order to asses the impact of atmospheric particles, batch reactor experiments need to be conducted where atmospheric particles are in contact with seawater at the expected concentration in the SML. The net effect of atmospheric deposition for an element would then be assessed by measuring the increase or decrease of the element after a certain amount of time. However, in order to understand the atmospheric particle deposition process in a more realistic way, two main issues (among others) need to be considered:

1. When they deposit at the surface of the ocean, particles will first encounter the sea surface micro-layer. This micro-layer is enriched in organic matter with a specific chemical composition (Frew 1997), which might greatly impact processes at the surface of atmospheric particles directly after deposition.
2. Once particles cross the micro-layer, the physico-chemical processes that occur during settling through the water column must be considered. Through the formation of aggregates (between atmospheric particles but also with organic particles), the size distribution of the particles will change when they settle through the water column. Moreover the influence of organic matter, which may play the role of a glue during these coagulation and aggregation processes, will also change the chemical nature of the particles and in particular the surface properties (Verdugo et al. 2004). These changes will influence the settling rate of the particles, but also the scavenging or dissolution processes that may occur at the particle surface.

4.5 Deposition of Aerosol Particles to the Ocean Surface and Impacts

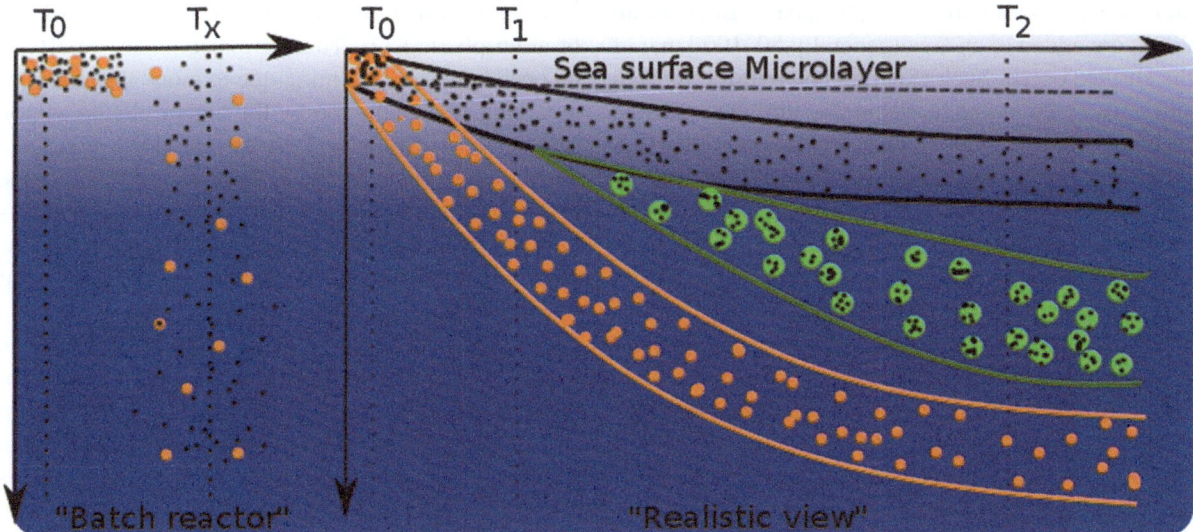

Fig. 4.20 Conceptual view of the fate of atmospheric particles after their deposition. *Left panel*: "Batch reactor" point of view with atmospheric particles perfectly mixed over the SML. *Right panel*: Hypothetical dynamics of atmospheric particles in the water column. At T_0 particles are in the surface microlayer. After T_0, bigger particles (*orange*) settle faster than smaller particles (*black*) which form aggregates with organic matter (*green*) at an intermediate depth

One challenge for future research is thus to take into account the dynamics of the particles which settle through the water column. As illustrated in Fig. 4.20 (right panel), this new approach might lead to a much more complex image of the impact of atmospheric particles in the ocean. However, to tackle these concepts, new tools able to investigate at such scales need to be developed. In particular, two major technological issues have to be solved:

1. The settling of atmospheric particles through the water column and their change in size and composition must be followed in situ with high temporal resolution. Promising results have been recently obtained in a mesocosm experiment with a combination of optical in situ measurements (Bressac et al. 2012).
2. Chemical techniques must be fast and reliable enough to investigate at these scales. Even though the recent development of new sensors for biogeochemical parameters on board new platforms (gliders, profilers,...) is promising, there is a crucial need of chemical sensors for relevant micro and macro nutrients.

In conclusion, a major perspective in assessing the impact of atmospheric deposition on the ocean is to investigate the competition between scavenging and dissolution within a dynamical process of the settling of the particles through the water column. The debate is largely open on the relevant scales that need to be investigated.

4.5.4 Atmospheric Impacts in HNLC and LNLC Areas

4.5.4.1 Experimental: Large Scale Fertilisation Experiments (Fe, P)

Direct assessment of the fate of atmospheric dust deposition and the subsequent biogeochemical impact on HNLC or LNLC waters has been extremely limited due to the episodic nature of such events and the need for researchers to 'be in the right place at the right time'. Such assessments have therefore been indirect, and mainly confined to a comparison of remotely-sensed datasets from different satellite sensors (Boyd et al. 2010). The use of such indirect approaches has in many cases led to mis-attribution of cause (aerosol deposition) and effect (altered ocean biogeochemistry via an aerosol-iron mediated phytoplankton bloom) (Boyd et al. 2010).

Due to the lack of direct information on the 'dust-nutrients-biota-biogeochemistry' linkages, we have had to rely on experimental approaches in which a known quantity of nutrients (either iron or phosphate, and in a few cases iron and phosphate, Boyd et al. 2007) are added to an area (generally 50–100 km^2) of the surface ocean. Such experiments, referred to as in situ mesoscale ocean enrichment, provide valuable insights into the wide range of processes that are influenced by an episodic nutrient addition, and as such provide a 'biogeochemical timeline' of how the nutrients released from dust deposition may have myriad effects on both surface and subsequently subsurface or atmospheric processes.

Trends from over ten of these experiments reveal convergences but also differences, which in many cases are due to regional differences between the environmental characteristics of HNLC waters in tropical to polar waters (de Baar et al. 2005). Convergent results provide confidence in compiling a biogeochemical timeline of the many processes that may be altered by aerosol deposition. In the case of iron enrichment, rapid initial changes (i.e. hours to days) include altered iron chemistry, up-regulation of cellular machinery in pelagic microbes such as photosynthetic competence (F_v/F_m), increased growth rates by all phytoplankton groups which is first manifested as an increase in stocks of cyanobacteria followed by those of nanophytoplankton (such as haptophytes). This initial increase in cyanobacterial abundances is truncated within a day or so as they are grazed down to initial levels as nano-grazers respond to higher pico-prey concentrations. A similar increase in haptophyte abundances lags the cyanobacteria, and then they are also brought under grazer control. In contrast the diatom stocks take longer to reveal an increase in abundances, and in many cases they escape grazing pressure resulting in an iron-mediated bloom (de Baar et al. 2005; Boyd et al. 2007).

This biological activity results in distinct biogeochemical signatures that produce feedbacks that are important for ocean–atmosphere interactions. For example, at some but not all experimental sites, the grazing of the haptophytes by microzooplankton may result in increased ocean DMSP inventories (Boyd et al. 2007) that consequently may in some cases result in elevated DMS concentrations in the upper ocean. However, in other studies such as SERIES in the NE subarctic Pacific no such increase in DMS was observed by the end of the bloom (Boyd et al. 2007). Concurrently there is widespread evidence of significant decreases in oceanic CO_2 concentrations due to enhanced C fixation by the blooming diatoms (de Baar et al. 2005) which may result, in time, in a drawdown of atmospheric CO_2. There are also knock-on effects on ocean physics (warming due to more absorption of incoming solar radiation by the higher phytoplankton stocks) and chemistry (uptake of added iron, and decrease of much of the inventories (with in some cases alteration of the stoichiometry of nutrient stocks – Boyd et al. 2004) of other plant nutrients such as silicate and nitrate). In some studies, the imprint of the iron-mediated blooms is eventually recorded (1–2 weeks) in the waters underlying the surface mixed layer, such varying degrees of enhancement of downward export of algal carbon from the bloom (Boyd et al. 2004; c.f. Smetacek et al. 2012), and elevated concentrations of other greenhouse gases such as N_2O and CH_4 probably due to remineralisation of the sinking particles (Boyd et al. 2007). However, in many experiments there were major uncertainties in estimating the ratio of iron added to additional carbon exported to depth from the base of the surface mixed layer.

Although fewer in situ nutrient enrichment studies have taken place in LNLC waters, relative to those in HNLC waters, they too provide evidence of the many effects of episodic nutrient enhancement, for example resulting in widespread foodweb effects such as the putative mechanisms of 'ecological tunnelling' reported for the CYCLOPS experiment in the E. Mediterranean by Thingstad et al. (2005). As there have been so few studies in LNLC waters, it is premature to comment on whether more effects are observed when an episodic pulse of nutrients is added to either HNLC or LNLC waters.

The benefits of what we have learned from these in situ mesoscale experiments for better understanding the biogeochemical effects of aerosol nutrient deposition into the ocean are twofold. Firstly, they provide us with a timeline of conspicuous changes in upper ocean properties that may help detection and attribution of the effects of episodic dust deposition events on the upper ocean using a combination of shipboard and/or satellite observations. Second, they have increased our conceptual understanding of how atmospheric processes influence those in the ocean, and how in turn those in the ocean may feedback on the lower atmosphere. However, when the magnitude, resulting elemental stoichiometry of the upper ocean, and chemistry of added nutrients of mesoscale ocean nutrient enrichment is compared with that resulting from an aerosol deposition event it is clear that the former is much greater than the latter (see Fig. 4.21). This prevents the ready extrapolation of the results from such experiments to the natural world, and suggests that some cautious downscaling – via modelling simulations – of the observed biogeochemical signatures might provide some bounds on how an aerosol deposition event might influence the upper ocean.

4.5.4.2 Experimental: Microcosms

Microcosm experiments involve the enclosure of a water sample in a container, typically 1–10 l in volume, and its incubation under simulated in situ conditions. Microcosm experiments are often used in the study of the effects of atmospheric nutrient deposition on natural plankton communities, as they represent an easy experimental approach to monitor a wide range of chemical and biological properties and their response to controlled perturbations. Microcosm experiments have the same limitations as any approach that involves in vitro confinement of water samples, including sampling bias and the difficulty of precisely simulating in situ conditions. The composition of the community can change markedly after bottle enclosure (Massana et al. 2001; Calvo-Díaz et al. 2011) and therefore most bioassay experiments have a duration that does not exceed 3–4 days (e.g. Bonnet et al. 2005; Mills et al. 2004; Moore et al. 2008), which means the recorded changes represent short-term responses to the simulated perturbation. For all these reasons, the results obtained from bioassay experiments are best interpreted in conjunction with in situ observations collected during oceanographic cruises and time-series monitoring programmes.

Published reports show a large degree of variability in the responses of surface plankton communities during microcosm fertilisation experiments intended to simulate the effects of atmospheric deposition. Part of this variability stems from the fact that the materials used to amend the water samples differ widely among studies, as they may consist of collected aerosols (Herut et al. 2005), collected rainwater (Klein et al. 1997), unprocessed desert soils (Mills et al. 2004), and atmospherically processed soils (Ternon et al. 2011). In addition to the geographical variability in the composition and solubility of natural aerosols, the relative abundance of anthropogenic particles varies widely in both space and time, and plays a crucial role in determining the biological effects of atmospheric deposition. An additional source of variability is the fact that different communities inhabiting the surface ocean have distinct community structures and experience different types and degrees of nutrient limitation (Moore et al. 2013), all of which result in diverging responses to a given perturbation. In spite of all these sources of variability, the microcosm experiments conducted during the last decade have yielded consistent patterns regarding the responses of surface plankton to atmospheric inputs in terms of biomass and abundance, community structure, and metabolic activity.

Main Results Obtained from the Microcosm Approach

Phytoplankton biomass, estimated from chlorophyll a concentration, typically increases after the addition of aerosols, although the reported increases are usually modest, e.g. around 50 % or less (Mills et al. 2004; Bonnet et al. 2005; Marañón et al. 2010). In LNLC waters, the potential of atmospheric deposition to cause phytoplankton blooms (e.g. resulting in chlorophyll a concentrations of 1 µg L^{-1} or more) is limited because, after taking into account nutrient content and solubility as well as dilution over the upper mixed layer, even large deposition events such as strong dust storms are unlikely to increase nitrate and phosphate concentrations by more than approximately 0.2 µM (Guieu et al. 2002; Ridame and Guieu 2002; Mills et al. 2004). As an example, if in a nitrogen-limited system nitrate concentration is increased by 0.2 µM and one assumes a molar C:N ratio of 6.6 for phytoplankton biomass production and a carbon to chlorophyll a ratio (g:g) of 100, the resulting maximum increase in chlorophyll a concentration will be only 0.16 µg L^{-1}. It must be taken into account, however, that LNLC waters cover vast expanses of the ocean, such as the subtropical gyres, which means that even small areal increases in biomass and productivity can have significant global impacts. The situation is different in the case of HNLC regions, where iron limits both phytoplankton standing stocks and primary production rates, and an excess of macronutrients is available. A strong dust deposition event (delivering 2 mg L^{-1} of aerosols to the upper mixed layer) can release around 2 nM Fe (Bonnet and Guieu 2004; Mills et al. 2004). If we assume that 10 % of this dissolved iron is bioavailable (Wu et al. 2001)

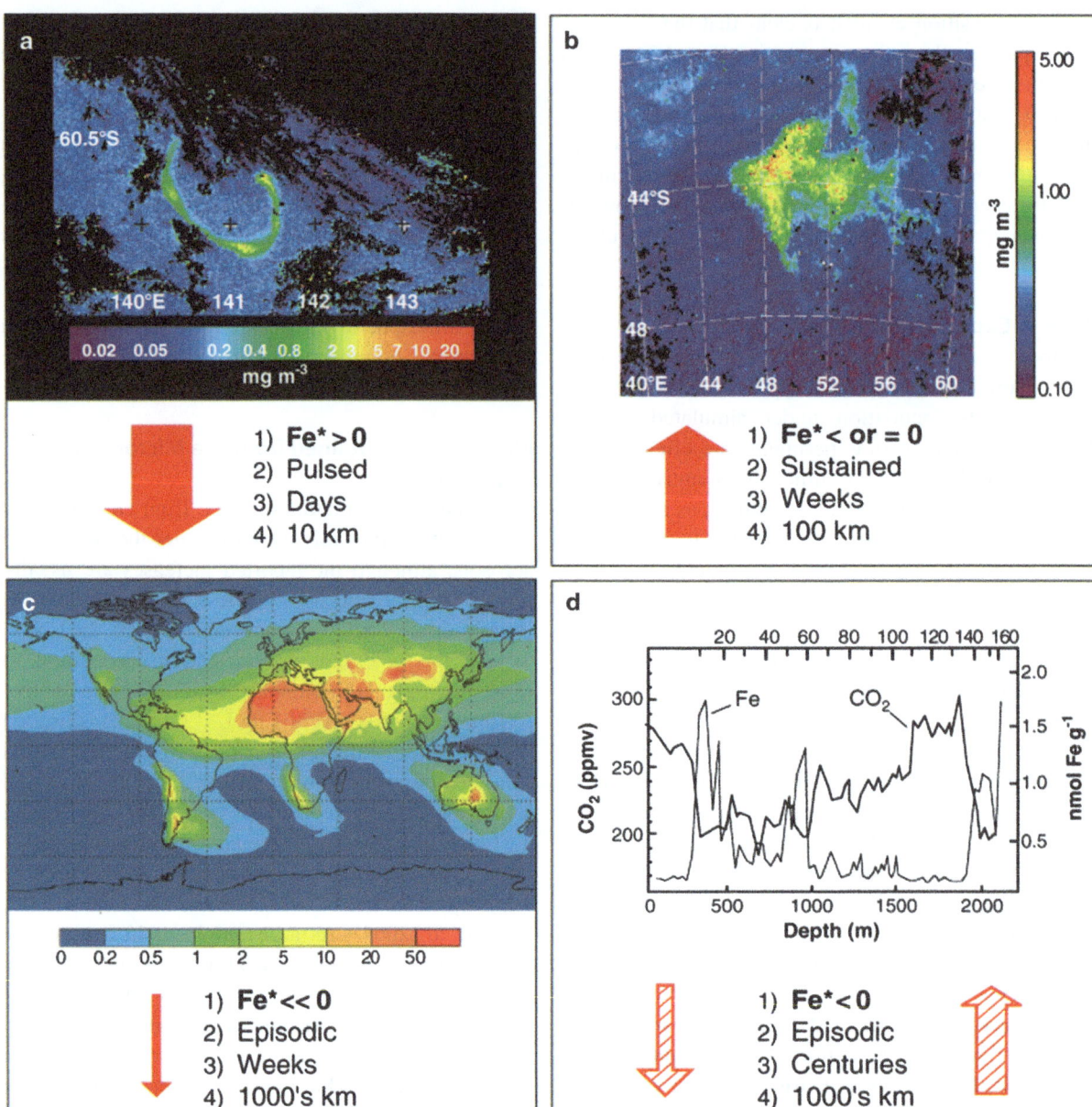

Fig. 4.21 A comparison for Southern Ocean waters of mechanisms responsible for perturbations in Fe supply. Numbers in each panel: (*1*) Fe*, the relative magnitude of Fe supply relative to macronutrient supply (Parekh et al. 2005); (*2*) the mode of Fe supply; (*3*) the time scale over which surface waters receive increased Fe supply; and (*4*) the length scales of Fe supply events. (**a**) Satellite image of a purposeful in situ Southern Ocean FeAX (Mesoscale iron addition experiments) [SOIREE (Boyd et al. 2000)]. (**b**) An FeNX (Fe natural enrichment experiments) near Crozet within the HNLC Southern Ocean, where naturally occurring blooms are evident from remote sensing (Boyd et al. 2007 and references given therein). (**c**) An atmospheric dust deposition event (dust units are g m^{-2} year^{-1}) in the modern Southern Ocean [e.g. from Patagonia (Jickells et al. 2005)]. (**d**) Fe supply to the Southern Ocean during the last glacial maxima from direct [i.e. higher dust deposition (Martin 1990; Wolff et al. 2006)] and/or indirect [i.e. upwelling of waters with higher Fe concentrations (Lefevre and Watson 1999)] sources. The magnitude of this supply is unknown; hence, Fe* is expressed as < 0. Fe* is defined as Fe* = [Fe]−{(Fe/P) algal uptake ratio × [PO4^{3-}]} (Parekh et al. 2005). If Fe* > 0, primary production is ultimately macronutrient-limited; if Fe* < 0, production is ultimately Fe-limited. The width of *red arrows* denotes the relative magnitude of changes in Fe supply; the hatched arrows in (**d**) denote uncertainties about whether Fe supply in the geological past was episodic or sustained. In (**b**) to (**d**), *downward-* and *upward-pointing arrows* represent atmospheric and oceanic (*upwelling*) supply, respectively (Reproduced from Boyd et al. 2007 by permission of American Association for the Advancement of Science)

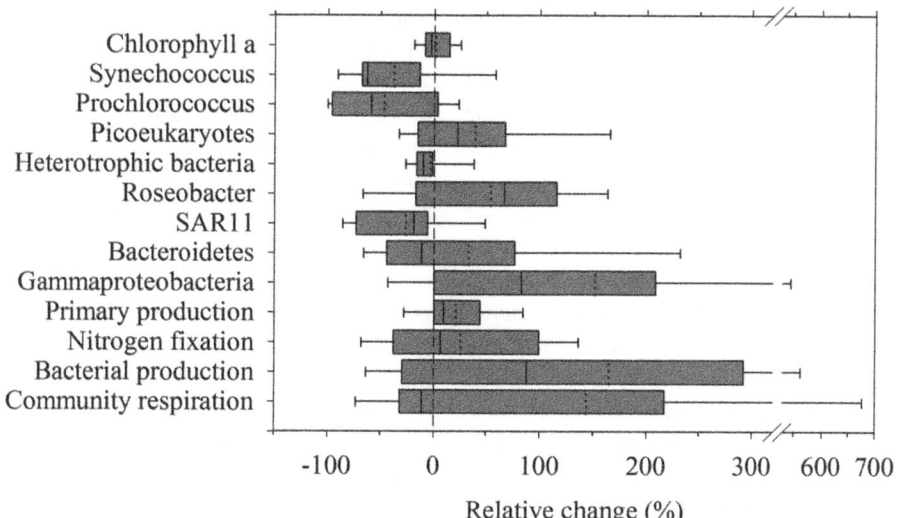

Fig. 4.22 Relative change (%) of different variables in response to dust addition during eight bioassay experiments in the tropical Atlatic Ocean. Relative change was calculated as $100 \times (D-C)/C$, where D and C are the mean value of the variable in the dust and the control treatments, respectively. *Boxes* and *bars* enclose the 25th–75th and 5th–95th percentiles, respectively, the *vertical dotted line* is the mean, and the *vertical continuous line* is the mode ($n = 8$). Variables shown are total chlorophyll *a* concentration; the abundance of *Synechococcus*, *Prochlorococcus*, picoeukaryotes and total heterotrophic bacteria; the abundance of bacteria belonging to the groups *Roseobacter*, SAR11, bacteroidetes and gammaproteobacteria; and the metabolic rates primary production, N_2 fixation, bacterial production, and community respiration (Figure modified from Marañón et al. (2010))

and take a molar C:Fe ratio of 10^5 for surface phytoplankton (Boyd et al. 2007), the resulting estimated chlorophyll *a* increase would be 2.4 µg L^{-1}. Thus, stoichiometric constraints and the different nutrient limitation conditions dictate that dust deposition events have a much larger potential for stimulation of phytoplankton biomass and production in HNLC regions than they do in LNLC regions.

Bacterial abundance and biomass have also been reported to increase after the addition of atmospheric aerosols (Herut et al. 2005; Pulido-Villena et al. 2008; Marañón et al. 2010). It is now well established that bacterial metabolism can be limited by inorganic nutrients in very oligotrophic regions (Rivkin and Anderson 1997; Mills et al. 2008) and atmospheric inputs therefore have the potential to stimulate bacterial growth. In most cases, however, the observed increases of bacterial abundance in the amended treatments, compared with the controls, are modest (<20 %). A recurrent pattern in microcosm experiments is that the response of phytoplankton and bacterioplankton to the addition of aerosols or individual nutrients is much stronger in terms of metabolic activity that in terms of abundance or biomass (see Fig. 4.22). In oligotrophic environments, a tight trophic coupling exists between picoplankton and their unicellular grazers, which means that a stimulation of production rates translates only partially, if at all, into an increase of population abundance.

In addition to considering the bulk biomass of bacterioplankton and phytoplankton, it is also important to take into account community structure and how the relative abundance of different taxonomic groups changes in response to atmospheric deposition. There is growing evidence that within both the phytoplankton and the bacterioplankton different taxonomic groups respond differently to a given input. In particular, the abundance of the cyanobacterium *Prochlorococcus* has been shown to decrease in response to the addition of Saharan soils and collected aerosols (Hill et al. 2010; Marañón et al. 2010) and also during natural deposition events (Herut et al. 2005). Adverse effects of some aerosols have also been reported for *Synechococcus* and the picoeukaryotes (Paytan et al. 2009; Marañón et al. 2010). The toxicity of some aerosols for certain groups of phytoplankton may be related to the presence of high levels of copper (Paytan et al. 2009). A negative response by some groups can be compensated by an increase in other groups, such that the bulk biomass does not change markedly. For instance, during a set of eight dust addition experiments conducted in the Atlantic Ocean it was found that both *Prochlorococcus* and *Synechococcus* decreased in the presence of dust, whereas the opposite was true for the picoeukaryotes (Marañón et al. 2010) (Fig. 4.22). Less is known about the response of bacterioplankton community structure to atmospheric deposition, but again contrasting changes in different groups have been observed. In the Atlantic Ocean, the abundance of the SAR11 lineage, which is dominant in ultraoligotrophic environments (Morris et al. 2002), tends to decrease in treatments amended with Saharan dust (Hill et al. 2010; Marañón et al. 2010), whereas the opposite is true for gammaproteobacteria, a group known to respond to increased nutrient concentrations (Horňák et al. 2006).

Metabolic rates are usually more responsive than standing stocks to aerosol additions and nutrient amendments. Phytoplankton primary production and N_2 fixation can show marked increases, by a factor of 2 or more, which are due to the fertilising effect of N, P and Fe (Mills et al. 2004; Bonnet et al. 2005; Herut et al. 2005; Blain et al. 2004; Marañón et al. 2010; Law et al. 2011; Ternon et al. 2011). Factorial nutrient addition experiments have demonstrated that N is the primary limiting nutrient for carbon fixation in LNLC waters (Moore et al. 2008), whereas P and Fe co-limit N_2 fixation in the Atlantic Ocean (Mills et al. 2008). Thus, desert dust has the potential to stimulate autotrophic productivity in the ocean through an increase in both carbon and N_2 fixation, therefore favouring CO_2 sequestration by the biological pump. However, recent evidence indicates that heterotrophic processes such as bacterial production and bacterial respiration are in fact stimulated by atmospheric deposition to a larger degree than autotrophic processes. In the NW Mediterranean Sea, it has been reported that bacterial respiration can increase by a factor of 3 in response to Saharan dust inputs (Pulido-Villena et al. 2008). Similarly, bacterial production was shown to increase by a factor of 8 during microcosms dust addition experiments in the Eastern Mediterranean Sea (Herut et al. 2005). In a set of eight dust addition experiments conducted in the Atlantic Ocean, the largest increases in metabolic rates (up to 700 %) were reported for bacterial production and community respiration (Marañón et al. 2010) (Fig. 4.22). Whether phytoplankton or bacterioplankton dominate the metabolic response to atmospheric inputs has important biogeochemical implications. If bacteria outcompete phytoplankton in using the newly supplied nutrients, and increase their production and respiration rates, the net result may be the remineralisation of dissolved organic carbon, the release of CO_2 and a weakening of the potential for biological CO_2 sequestration. This is a paradoxical outcome if we consider that atmospheric deposition is commonly regarded as a process that fertilises the open ocean. However, only a handful of studies have addressed concurrently the metabolic rates of both phytoplankton and bacteria to atmospheric inputs, and therefore more studies are needed in order to understand how competitive interactions within the planktonic community modulate the overall response in terms of the balance between CO_2 fixation and respiration. Another priority for future studies is to conduct repeated microcosm addition experiments over time and in different locations, in order to obtain general patterns that relate the initial characteristics of the community, in terms of taxonomic composition and type and degree of limitation by different nutrients, to its response to a given perturbation.

4.5.4.3 Experimental: In Situ Mesocosms

Although in situ mesocosms have been widely used to study biological responses to changes in nutrient conditions (e.g. Duarte et al. 2000) or environmental conditions such as increasing levels of atmospheric carbon dioxide on carbon uptake import or export from the upper ocean (eg Riebesell et al. 2007), impacts of actual atmospheric deposition on ecosystems was only performed recently (Guieu et al. 2010). In situ marine manipulation, although being a technical challenge, represents a significant progress compared to the approach using microcosms. Indeed, mesocosms have the potential to allow controlling physical and chemical forcing, including atmospheric deposition, and the advantage of following simultaneously the whole ecosystem and the export of matter (Fig. 4.23) the biogeochemical modelling in a 1-D model taking into account the role of atmospheric inputs on the biological carbon pump can thus be done. Such experiments remain challenging first because of the difficulty to reproduce an actual atmospheric deposition event, meaning that enough representative atmospheric material has to be produced and second because contamination has to be

4.5 Deposition of Aerosol Particles to the Ocean Surface and Impacts

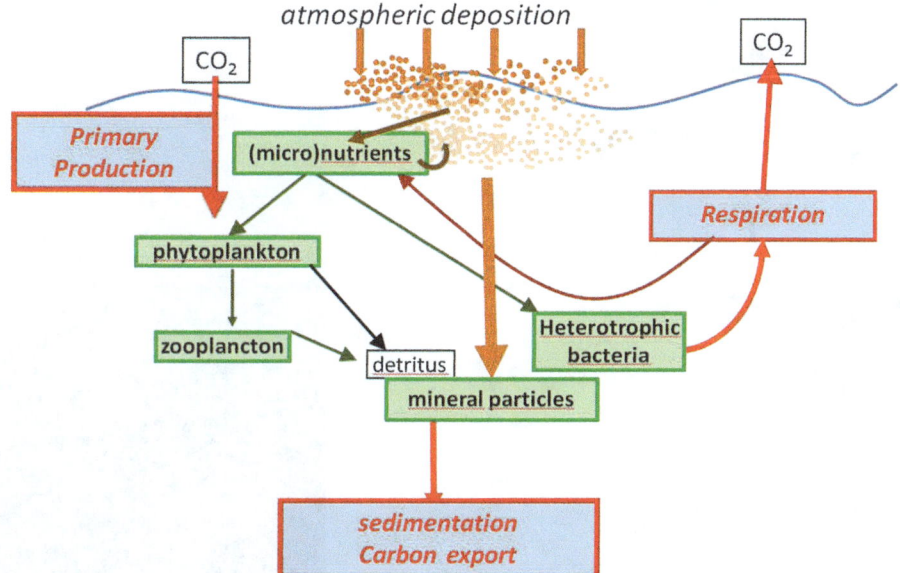

Fig. 4.23 Stocks (*in green*) and fluxes (*in blue*) that can be measured in a mesocosm experiment after the simulation of a realistic atmospheric input in a water body large enough to be representative of natural processes. As the particles are naturally sinking, those changes are closer to the 'real' processes occurring in the open ocean, compared to microcosms approaches where the particles are homogenised and not allowed to sink. Sampling in quick succession at different depths of the mesocosms allows for parameterising the complex processes involved. Sedimentation, including carbon export, can be quantified and this represents a major step forward of that methodology

avoided during the different steps of the experiments. As a matter of fact, the water tested for their response to atmospheric inputs has usually low concentrations of (micro)nutrients (typically iron and phosphorus are at nanomolar levels) and the induced changes in the biogeochemical cycle of those elements is one of the objectives of such large mesocosm experiments. In the recent experiment reported by Guieu et al. 2010, large (52 m^3) mesocosms containing no metallic part and with as little as possible induced perturbation during the sampling sequence have been used and the required conditions for biogeochemical studies in oligotrophic environments have been reproduced (Fig. 4.24). Such large clean mesocosms have been thus proven to be highly relevant tools to study in situ the response of an oligotrophic ecosystem to atmospheric deposition.

Mesocosms present the advantage of enabling studies of processes both as a function of time and depth while the atmospheric particles are sinking. According to what was pointed out in Sect. 4.5.3, this is a very important consideration, in particular regarding adsorption/desorption processes and aggregation mediated by the introduction of lithogenic particles (Bressac et al. 2012 and Sect. 4.5.5). Mesocosms are thus much more representative of the reality of the atmospheric deposition to the surface of the ocean and results obtained can be more easily extrapolated than results obtained in microcosms that rely on a fixed and homogeneously-distributed concentration of particles that can alter the dust dissolution kinetics and does not account for particle migration through the water column, precluding any estimation of C export induced by atmospheric deposition.

Those recent mesocosm experiments confirm that significant changes in the cycling of chemical elements are induced by the deposition of dust particles, and that there are strong responses of the ecosystem at different trophic levels. Pulido-Villena et al. (2010) provided a quantitative confirmation of the fertilising potential of mineral dust as a significant increase in phosphate concentration was observed soon after seeding. The dissolved inorganic phosphorus (DIP) released was completely lost after 24 h and no further increase was observed during the experiment. DIP loss might thus have been dominated by biological uptake, in agreement with the observed rapid response of both bacteria and phytoplankton. The iron cycle was also profoundly

Fig. 4.24 The different steps to reproduce a realistic atmospheric dust deposition and subsequently follow the biogeochemical changes induced near the surface of the oligotrophic ocean illustrated by experiment 2 (2010) of the DUNE project: (**a**) sampling soils in a region of South Tunisia where the Saharan aerosol is produced and is typical of inputs to the North-Western Mediterranean, (**b**) Soil collected on the three stages of the dry sieving column at the end of the fine particle production process. (**c**) Artificial ageing of the atmospheric particles in a clean room to reproduce the processes taking place during aerosols transport (**d**) a group of three mesocosms from below, (**e**) a view of the seven mesocosms during the DUNE2 campaign in June 2010 (**f**) a view above the surface of a group of three mesocosms showing the artificial seeding. (Photos a, b : F. Dulac, CEA; photo c: LISA; photos d, e, f : D. Luquet, OOV). Methodology fully described in Guieu et al. 2010

affected by the dust input to the surface seawater, as a dissolved iron (DFe) scavenging was observed rapidly after the seeding, withdrawing almost 1 nM DFe from the whole water column inside the three mesocosms amended with dust while the atmospheric particles are settling (Wagener et al. 2010). Those results were satisfactorily simulated with a one-dimensional model of the Fe biogeochemical cycle, coupled with a simple ecosystem model (Ye et al. 2011). When a second dust wet deposition was simulated, iron dissolution from the dust particles was then evidenced due to the excess Fe binding ligand concentrations produced by the enhanced biological activity (Wuttig et al. 2013). When simulating dust wet deposition at least a doubling of Chla (mainly attributed to small phytoplankton (<3 μm); Giovagnetti et al. 2013) was observed but the Chla concentrations remained very low (maximum values 0.22 μg L-1) maintaining the oligotrophic status of the tested waters (Guieu et al. 2013). Laghdass et al. (2011) investigated the effect of dust deposition on the diversity of the heterotrophic bacterial community. Combining several molecular biology approaches, the results indicate that besides a natural temporal trend observed inside and outside the mesocosms, dust deposition affected the particle-attached bacterial community. In particular, *Alteromomas macleodii* made a higher contribution to the active particle-attached bacterial community in the dust-amended than in the control mesocosms. Bacterial respiration and primary production both increased very rapidly after the seeding (by a factor of 2 and 3, respectively), showing a competition for the new nutrients between heterotrophic bacteria and phytoplankton (Guieu et al. 2012). Among phytoplankton, diazotrophs – although only responsible for few percent of the induced new production – were strongly stimulated by the atmospheric input (Ridame et al. 2013). Coupling geochemical and optical measurements have demonstrated that rapid particulate carbon export was in part due to aggregation processes between organic matter and lithogenic particles (Bressac et al. 2012). Results from mesocosm approaches indicate that the role of atmospheric deposition on oligotrophic areas cannot be seen solely as a simple fertilisation effect because (1) both autotrophs and heterotrophs are stimulated (confirming on longer time scales the findings from microcosm experiments described in Sect. 4.5.4.2) and (2) a significant export of particulate organic carbon to the deep ocean is attributed to aggregation processes (Bressac et al. 2013).

Although they represent a powerful tool to handle the complex question of the role of atmospheric deposition on ecosystem functioning, the use of such large, clean and pelagic mesocosm in the open ocean remains a difficult task for obvious logistical reasons. So far, the only two experiments conducted have been done in coastal areas chosen for being representative of waters with comparable chemical and biological characteristic to open waters. Future development could be the feasibility of identical experiments in the open ocean. Another disadvantage of the mesocosm approach is that only few treatments can be handled (during the DUNE (A DUst experiment in a low Nutrient, low chlorophyll Ecosystem) experiment, triplicates of controls and triplicates of dust deposition were the only treatments considered), contrary to microcosm approach where a lot of different treatments can easily be used.

4.5.4.4 Modelling

In the past two decades, ocean biogeochemical models (e.g. Maier-Reimer 1993) have been used to study the response of marine biogeochemistry to a variety of forcings. Many studies have focussed in particular on the response of ocean carbon storage (Sarmiento and Le Quéré 1996) and marine productivity (Bopp et al. 2001) to anthropogenic climate change. Among the various forcings employed, the impact of changes in atmospheric deposition on marine biota and marine biogeochemistry has received much less attention. One reason for this relative lack of attention is linked to the very large uncertainties in future changes of atmospheric nutrient deposition to the ocean due to climate change, as exemplified by simulated future changes in dust deposition (e.g. Mahowald and Luo 2003; Mahowald et al. 2006; Tegen et al. 2004).

Early studies have focussed on the role of changes in dust deposition on the marine carbon pump and atmospheric CO_2 during glacial times. Watson et al. (2000) used a box model of the ocean carbon cycle forced with atmospheric iron fluxes derived from the Vostok ice-core dust record (Petit et al. 1999). They showed that changes in the Southern Ocean biota caused by iron deposition could explain only a maximum of 40 ppmv drawdown of atmospheric CO_2 at the glacial maximum, half of the observed glacial-interglacial CO_2 change. Other modelling studies, however, using general circulation models and more complex biogeochemistry (including a explicit representation of the ocean iron cycle) converge towards a more moderate effect; the impact of changes in dust deposition on marine biota at glacial maximum could drive a ~15 ppm reduction in atmospheric CO_2 (e.g. Archer et al. 2000; Bopp et al. 2003).

Similar models have been used to quantify or predict the effect of changing dust deposition on marine biota and the global carbon cycle in the recent past and in the near future. Over the twentieth century, increase in dust deposition inferred from observations has been estimated to be responsible for an increase in marine productivity and an additional drawdown of carbon into the ocean (Mahowald et al. 2010). This additional drawdown could amount to 8 Pg C, more than 5 % of total anthropogenic carbon stored in the ocean over the industrial period.

While future projections of desert dust deposition over the ocean are still largely uncertain, even as regards the sign of changes (Tegen et al. 2004; Mahowald et al. 2009), several ocean modelling studies have addressed its potential impact on marine biogeochemistry. Changes in dust deposition may force changes in ocean productivity, as large as the changes in productivity forced by CO_2 increases and the resulting climate change (Mahowald et al. 2011). But there is only relatively little impact of varying aeolian Fe input on cumulative ocean CO_2 fluxes and atmospheric pCO_2 over 2000–2100 (Tagliabue et al. 2008).

The simulated sensitivity of marine biota to iron deposition is, however, largely dependent on processes that are or not explicitly included in models. Complex iron chemistry in the ocean (Tagliabue et al. 2009), adsorptive scavenging of dissolved iron and solubilisation of particulate iron (Ye et al. 2011), varying iron content of dust particles as well as varying iron solubility in seawater (Luo et al. 2006) have been investigated in individual modelling studies but an explicit representation of these processes is still lacking in most of the biogeochemical models that

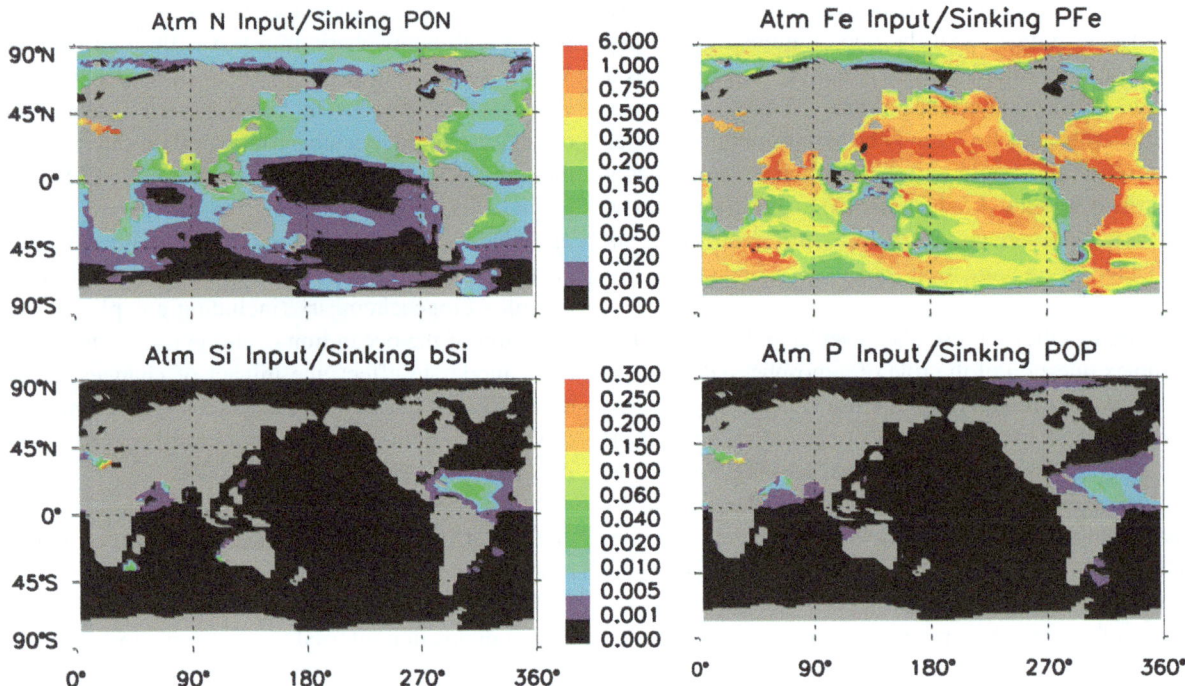

Fig. 4.25 Simulated ratios between atmospheric inputs of N, Fe, Si, and P and corresponding sinking particulate export for modern era climate conditions. Note the change in scale between *upper* and *lower panels* (Reproduced from Krishnamurthy et al. (2010) by permission of the American Geophysical Union)

do include an explicit representation of the iron cycle. In addition, Krishnamurthy et al. (2009) have shown that increasing iron deposition could not only lead to increasing marine productivity in HNLC regions, but also significantly enhance N-fixation in subtropical regions if iron limitation of N-fixation is included. Finally, iron mobilisation from sediments has been suggested to be of the same magnitude as dust deposition and its inclusion in models significantly reduces the response to changes in atmospheric iron deposition (Moore and Braucher 2008).

Apart from the effect of iron deposition, few studies have focussed on the role of atmospheric deposition of other elements on marine biota and ocean biogeochemical cycles. Using a coupled ocean general circulation/biogeochemical model, Krishnamurthy et al. (2010) have shown that atmospheric Si and P depositions have only a weak effect on marine productivity and biogeochemical cycles, as these depositions are small relative to the flux of these nutrients from below. They would contribute only to a very small fraction of export production (Fig. 4.25). Even if atmospheric nitrogen inputs have only a weak effect on marine biota and productivity at the global scale, they can contribute significantly (up to 25 %) to export production at the regional scale, especially in subtropical oligotrophic gyres (Krishnamurthy et al. 2009, 2010). The impacts of changes in N, P and Si deposition in the near future on marine biota are largely unknown.

In addition to the direct effect of nutrient deposition on the marine biota through a fertilisation effect, Doney et al. (2007) have shown that atmospheric inputs of reactive sulphur and nitrogen could potentially alter seawater alkalinity and pH. They show that inputs of strong acids such as HNO_3 and H_2SO_4 could lead to a decrease in pH. The alterations in surface water chemistry however are only a few percent of the acidification due to the oceanic uptake of anthropogenic CO_2. The impacts could be more substantial in coastal waters.

4.5.5 Particulate Matter and Carbon Export

Depending on the source of the aerosols, atmospheric deposition to the ocean is of very variable intensity, the

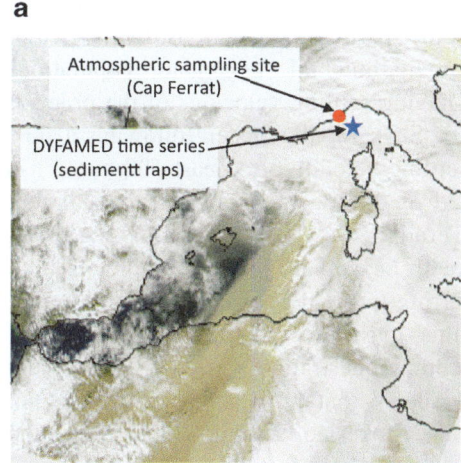
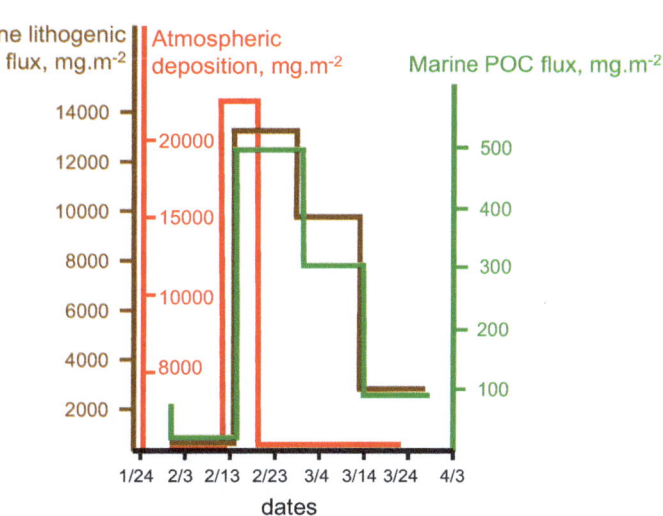

Fig. 4.26 Saharan dust event in February 2004 and its consequences for marine lithogenic material and particulate organic carbon exports in the water column. (**a**) Satellite image showing the transport of Saharan dust across the Mediterranean Sea (SeaWiFS, NASA). (**b**) Total atmospheric deposition mass flux measured at the Cap Ferrat sampling site (French Riviera) along with fluxes from sediment traps at 200 m at the nearby DYFAMED time series station (43°25′N 07°52′E; http://www.obs-vlfr.fr/sodyf/) showing the huge increase in both lithogenic and POC flux following the event (Adapted from Ternon et al. 2010)

most spectacular atmospheric fluxes being associated with lithogenic particles from the different deserts of the continents. Indeed, strong dust events can bring large amounts of particulate matter to the ocean surface in a few hours and fluxes as high as 25 g m^{-2} per event have been reported in the Mediterranean Sea (Ternon et al. 2010), see Fig. 4.26. This lithogenic material is transferred to the deep ocean and contributes significantly to sedimentation in areas such as the tropical North Pacific, the North Atlantic and the Mediterranean Sea, which are well impacted by dust deposition (Uematsu et al. 1985; Prospero et al. 2002; Loÿe-Pilot et al. 1986). But not only lithogenic export is associated with those important atmospheric depositions.

As shown in previous sections, aerosols from different sources do bring to the ocean a significant amount of elements of biogeochemical interest that feed the pool of nutrients and micronutrients in the ocean surface layer. When the body of water impacted by atmospheric deposition is depleted in those (micro) nutrient(s) brought by atmospheric deposition, (micro) nutrient(s) limited organisms such as bacteria and phytoplankton will feed on this new resource. In such conditions, atmospheric deposition has a 'fertilising' effect (see references in Sect. 4.5.4) and carbon fixation by autotrophic organisms should be followed by particulate organic carbon fluxes to the deep ocean. Although the link between addition of aerosol iron and export production was emphasised by Cassar et al. 2007, the quantification of such induced export of particulate carbon was difficult to assess in the numerous iron fertilisation experiments reported in de Baar et al. (2005) where it was only firmly proven in two of the reported experiments for which it was shown to be quite modest. Differently, Smetacek et al. (2012) concluded that following the EIFEX ocean iron fertilisation experiment a substantial portion of the induced bloom likely reached the sea floor. According to Blain et al. (2007) carbon export from such artificially induced short-term blooms are at least an order of magnitude lower than carbon export induced by natural iron fertilisation in the Southern Ocean.

On the other hand, several studies have shown the importance of lithogenic particles for the export of organic matter through an aggregation process (Hamm 2002; Passow and De la Rocha 2006; Ploug et al. 2008) and the export in this case is not related to a fertilisation effect. During 4 years of simultaneous measurements of atmospheric deposition and export of material in the water column in the NW Mediterranean Sea, Ternon et al. 2010 found a series of "lithogenic events" corresponding to both high POC and high lithogenic marine fluxes (originating mainly from either recent Saharan fallout events, or from 'old' Saharan dust 'stored' in the upper water column layer). Such "lithogenic events", are believed to result in part from organic-dust aggregation inducing a ballast effect. The most remarkable event was in February 2004 when an extreme Saharan event exported ~45 % of the total annual POC (Fig. 4.26), compared to an average of ~25 % for the bloom period, at a time of the year when the organisms are not limited by the concentrations of nutrient in the upper ocean and thus such high POC flux couldn't be the consequence of a fertilisation by the dust. Zooplankton activity and the incorporation of mineral particles into faecal pellets have also been proposed as a mechanism to explain such lithogenic events (e.g. Fowler et al. 1987; Buat-Menard et al. 1989). Ternon et al. also found another 'lithogenic event', in summer when the surface mixed layer is depleted in nutrients, where only 20 % of the POC exported could be explained by fixation of carbon induced by atmospheric fertilisation, the rest being assumed to be due to the incorporation of mineral particles acting as ballast in organic aggregates formed by the dissolved and colloidal organic matter present in the water column prior to the event.

From those few examples, we can stipulate that dust deposition can have two main effects: a POC export directly linked to the carbon fixation induced by the fertilisation via atmospheric nutrients and the POC export 'mediated' by the introduction of lithogenic particles with a ballast effect on organic matter, leading to the formation of large aggregates. As mentioned above, both effects can be combined and this was recently observed by Bressac et al. (2012) following an artificial dust addition in a large clean mesocosm: it was observed, with optical measurements, that particulate export following a strong dust deposition (10 g m^{-2}) was a nonlinear multi-step process composed of particle populations with different size distributions and composition as a function of time and depth. Interestingly, it was shown that the pattern of the size distribution and composition of particles in the mesocosm was strongly influenced by the formation of organic-mineral aggregates and that the fresh organic matter produced by the fertilisation induced by the dust deposition lead to organic-mineral aggregates with the fastest settling velocities compared to the other particles populations. Bressac et al. measured organic-mineral aggregate populations (>61.2 µm) having settling velocities of the order of 24~86 m d^{-1}. Such rapid transfer following a dust event with settling velocity of at least 100 m d^{-1} has been observed for 'lithogenic events' from sediment trap data (Ternon et al. 2010). Such fast sinking particles would be less prone to remineralisation while transferring to the deeper layers (Hedges et al. 2000) and this may have important consequences regarding carbon sequestration. Those different recent studies have shown that lithogenic particles do not sink following Stokesian settling calculations and that atmospheric particles may have a very important role in transferring rapidly particulate organic matter to the deep ocean but, so far, no quantification of this role in carbon export and budget has been performed.

The parameterisation of the complex processes involved in the sinking of both lithogenic and organic particles following an intense dust event will only be possible by realising a combination of optical, biological and chemical measurements in the upper water column when dust deposition occurs. This could be done by a combination of in situ measurements (including in situ mesocosms) and in large tanks in the laboratory (both in biotic and abiotic conditions) with approaches allowing the actual sinking of the particles. The characterisation of organic matter is very much needed in such combined approaches as organic-mineral aggregation depends mainly on the quantity and quality of the organic matter (Passow and De La Rocha 2006) meaning that the biogeochemical conditions of the water body in which the deposition occurs may be a crucial factor for the fate of particulate matter following a dust event.

4.6 Summary and Outlook

Aerosol particles in the marine atmosphere are produced either directly at the sea surface, as sea spray aerosol particles, or over land from where they can be subsequently transported over the ocean. Secondary production takes place in the atmosphere from precursor gases which are emitted either from the sea surface or from a variety of sources over land. The direct production of sea spray aerosol was recently reviewed by de Leeuw et al. (2011a) and conclusions presented there still hold. Results in the last decade show the large contribution of organic substances to SSA particles especially in locations of high biological activity, which dominates the chemical composition for particles with $r_{80} < 0.25$ μm. The consequences for hygroscopic properties of SSA particles and thus their effect on climate via light scattering and their influence on cloud properties has been recognised and is still an important topic for research. Meskhidze and co-workers (Gantt et al. 2011; Meskhidze et al. 2011) have shown the effect of wind speed on the fraction of organic matter in SSA particles in addition to the dependence on particle size (Facchini et al. 2008a). Ovadnevaite et al. (2011) show the importance of organic matter for the ability of SSA particles to act as CCN.

Production fluxes have been determined for particles with r_{80} as small as 0.01 μm, which is also important for the assessment of the role of SSA particles on climate using large-scale models. However, de Leeuw et al. (2011a) concluded that the uncertainties in the production fluxes are still very large. For particles with $r_{80} > 1$ μm the uncertainty is a multiplicative factor of 4 (Lewis and Schwartz 2004). Recent research by Jaeglé et al. (2011) confirms the large importance of the effect of SST (Mårtensson et al. 2003) on the production fluxes of coarse particles (those with $r_{80} > 1$ μm), with consequences for the effect of SSA particles on the direct radiative effect on climate.

For particles with $0.3 < r_{80} < 1$ μm and smaller all flux estimates published after Lewis and Schwartz (2004) appear to be higher than the range of values discussed by these authors. These high production fluxes would result in unrealistically large number concentrations of SSA particles in the marine atmospheric boundary layer (de Leeuw et al. 2011a) but the reason for these high production fluxes is not clear. There is no convincing reason to discard the Lewis and Schartz (2004) production flux estimates and their uncertainties, which are based on numerous observations, but the more recent formulations not only produce larger production fluxes but they also seem to converge to values which differ by about a factor of 2–3 (Clarke et al. 2006).

For SSA particles with $r_{80} < 0.3$ μm the production flux estimates rely on the whitecap method, i.e. estimates of the whitecap fraction and on the particle production flux per unit area of whitecap. The latter is mainly obtained from laboratory experiments and comparisons by de Leeuw et al. (2011a) show the large difference in the shape and magnitude of the SSA particle size spectra produced in such experiments, with a large dependence on experimental conditions (e.g. Fuentes et al. 2010). This puts into question the validity of one of the underlying assumptions of the whitecap method, i.e. that the production per unit whitecap area can be independently determined and that the total production flux is given by the product of the production per unit whitecap area and the whitecap fraction. The whitecap fraction in turn varies over a very wide range between different observations (Lewis and Schwartz 2004; Anguelova and Webster 2006) and all the most recent determinations, using novel methods, appear to result in smaller whitecap fractions than the most commonly used function by Monahan and O'Muirchaertaigh (1986) which is based on numerous previous observations. Furthermore the whitecap fraction appears not to be unambiguously determined by wind speed and depends on wind speed history (Callaghan et al. 2008).

Most of these studies use a combination of different techniques and novel technology developments. These techniques include combinations of advanced modelling using different information sources based on results from laboratory and field experiments, long term observations or intensive field campaigns at coastal locations or open ocean, using advanced techniques such as AMS, HTDMA, CCNC and shipborne eddy covariance measurements. Many of the recent studies rely on the use of earth observation (EO), i.e. satellite-based instruments, to obtain information on aerosol concentrations, forcing parameters and aerosol effects. EO is increasingly used in studies on sea spray aerosols and their effects. Nevertheless, remote sensing results also show rather large ambiguities such as in the determination of the aerosol optical depth (AOD) as determined by different remote sensing methods (Smirnov et al. 2012). The use of EO in air-sea interaction studies and effects on SSA production and their effects on climate and biogeochemical cycles has an important place in the SOLAS community.

In spite of recent progress based on combinations of modelling, EO and in situ observations, a large drawback put forward in the literature is the lack of reliable experimental data for a wide variety of conditions.

Another question posed in the literature concerns the reliability and accuracy of EO products which are further developed based on new insights and improved EO instrumentation leading to improvement of data quality and development of new products. However, satellite data alone cannot be used to assess SSA production and needs to be used in combination with other techniques such as modelling and in situ or ground-based remote sensing.

The contribution of land-based aerosol particles and precursor gases is another issue which is extremely important for the assessment of atmospheric inputs into the ocean and their biogeochemical effects (see below). These studies rely to a large extent on models to determine aerosol transport. The models in turn utilise emission data bases as input and they are often based on bottom-up emission inventories for a certain period or year and therefore represent mean emissions. Often the inventories lack detail on seasonal or diurnal variations and do not include actual episodic emission strenghts of species which are important for biogeochemical cycles such as dust, volcanic ash and biomass burning aerosol. Earth Observation (satellites) may contribute to improvement of emission estimates and several studies have recently been published on the use of EO data with advanced modelling techniques to provide this information for emissions of e.g. volcanic ash, VOCs and NO_2.

Numerous studies have been conducted to establish links between atmospheric deposition and biogeochemical cycles in the past 20 years, increasing exponentially with recent developments of new sophisticated methodologies (in particular the use of clean sampling techniques in the field and the generalisation of expertise in low levels determination of chemicals in seawater).

Studies on effects of atmospheric iron on biogeochemical processes have been an important research activity for more than 20 years because of, its recognised importance as a limiting nutrient in the large HNLC ocean. These studies no longer focus only on iron in dust but also the biogeochemical impact of other recently recognised important sources – such as volcanic ash – are now considered. Other studies have recently focussed on LNLC areas where microbes can be limited or co-limited by deposition of atmospheric inorganic nitrogen and phosphorus. In terms of biotic response to episodic atmospheric nutrient inputs, the large scale experiments conducted over the past 20 years have mainly focussed on the role of iron in HNLC areas. Although responses differ due to regional differences between the environmental characteristics of HNLC waters, such experiments have allowed simultaneous observation of the wide range of processes involved by such inputs of new nutrients, how these change the upper ocean properties and how the ocean feedbacks on the lower atmosphere. Microcosm experiments show the response of microorganisms to deposition of aerosol particles to LNLC areas. Such experiments also show consistent patterns in terms of biomass and abundance, community structure, and metabolic activity revealing that both autotrophs (including diazotrophs) and heterotrophs can be significantly stimulated. This is confirmed by mesocosm experiments where both chemical and biological processes can be studied simultaneously over longer time scales while accounting for the vertical dimension as well. In situ observations and mesocosm approaches indicate that atmospheric deposition, in particular dust, can result in carbon export through two types of processes: export directly linked to carbon fixation induced by the fertilization via atmospheric nutrients and export by carbon ballasting by the lithogenic particles. Ocean biogeochemical models to study the response of marine biota and biogeochemistry to atmospheric deposition have mainly focussed so far on dust (iron) deposition over glacial times, recent past and near future, showing moderate effects on reduction of atmospheric CO_2. The few modelling studies which recently focussed on the role of atmospheric deposition of other elements than iron have shown that atmospheric N, P and Si deposition have a weak effect on marine productivity and biogeochemical cycles contributing only to a very small fraction of export production.

All those different approaches indicate that extrapolation of the results to the 'real world' is not easy and that further studies using a combination of the different approaches are necessary to quantify the present picture and how it will change in the future. A number of research directions have been recommended, such as specific studies to understand the role of the microlayer, to establish the partitioning between organic and inorganic forms of atmospheric nutrients and their link with bioavailability, to consider the balance between scavenging/dissolution taking into account the dynamics of the particles while they settle through the water column (Law et al. 2013). Studies considering a wide range of in situ conditions should be conducted in order to understand the competition for the new nutrient resources among planktonic organisms and how the balance between CO_2 fixation and respiration impact the carbon budget. The balance between POC export linked to carbon fixation induced by the atmospheric nutrient fertilisation, and the POC export mediated by the introduction of atmospheric particles through aggregation processes, have to be better constrained. Another challenge is to correctly model the complex biogeochemical processes involved in order to quantify the impacts on biota and global carbon cycle and to predict the changes induced by the evolution of emission/deposition in the future.

Open Access This chapter is distributed under the terms of the Creative Commons Attribution Noncommercial License, which permits any noncommercial use, distribution, and reproduction in any medium, provided the original author(s) and source are credited.

References

Adams P, Seinfeld J, Koch D (1999) Global concentrations of tropospheric sulfate, nitrate and ammonium aerosol simulated in a general circulation model. J Geophys Res 104:13791–13823

Adler G, Flores JM, Riziq AA, Borrmann S, Rudich Y (2011) Chemical, physical, and optical evolution of biomass burning aerosols: a case study. Atmos Chem Phys 11:1491–1503. doi:10.5194/acp-11-1491-2011

Aggarwal SG, Kawamura K (2008) Molecular distributions and stable carbon isotopic compositions of dicarboxylic acids and related compounds in aerosols from Sapporo, Japan: implications for photochemical aging during long-range atmospheric transport. J Geophys Res 113, D14301. doi:10.1029/2007JD009365

Aiken AC, DeCarlo PF, Kroll JH et al (2008) O/C and OM/OC ratios of primary, secondary, and ambient organic aerosols with high-resolution time-of-flight aerosol mass spectrometry. Environ Sci Technol. doi:10.1021/es703009q

Albert MFMA, Schaap M, Scannell C, O'Dowd CD, de Leeuw G (2012) Uncertainties in the determination of the organic fraction of global sub-micron sea-spray emissions. Atmos Environ 57:289–300. doi:10.1016/j.atmosenv.2012.04.009

Alfaro SC, Gaudichet A, Gomes L, Maillé M (1997) Modeling the size distribution of a soil aerosol produced by sandblasting. J Geophys Res 102:11239–11249

Alfaro SC, Gaudichet A, Gomes L, Maillé M (1998) Mineral aerosol production by wind erosion: aerosol particle sizes and binding energies. Geophys Res Lett 25:991–994

Alfaro SC, Lafon S, Rajot JL, Formenti P, Gaudichet A, Maillé M (2004) Iron oxides and light absorption by pure desert dust: an experimental study. J Geophys Res 109, D08208. doi:10.1029/2003JD004374

Alpert P, Kaufman YJ, Shay-El Y, Tanré D, da Silva A, Schubert S, Joseph JH (1998) Quantification of dust-forced heating of the lower troposphere. Nature 395:367–370

Alvarado MJ, Prinn RG (2009) Formation of ozone and growth of aerosols in young smoke plumes from biomass burning: 1. Lagrangian parcel studies. J Geophys Res 114, D09306

Anderson TL, Wu Y, Chu DA, Schmid B, Redemann J, Dubovik O (2005) Testing the MODIS satellite retrieval of aerosol fine-mode fraction. J Geophys Res 110, doi:10.1029/2005JD005978

Andreae MO, Merlet P (2001) Emission of trace gases and aerosols from biomass burning. Global Biogeochem Cycles 15:955–966

Andreae MO, Rosenfeld D (2008) Aerosol-cloud-precipitation interactions. Part 1. The nature and sources of cloud-active aerosols. Earth-Sci Rev 89:13–41

Andreae MO, Charlson RJ, Bruynseels F, Storms H, Van Grieken R, Maenhaut W (1986) Internal mixture of sea salt, silicates, and excess sulfate in marine aerosols. Science 32:1620–1623

Andreas EL, Persson POG, Hare JE (2008) A bulk turbulent air-sea flux algorithm for high-wind, spray conditions. J Phys Oceanogr 38:1581–1596

Andres RJ, Kasgnoc AD (1998) A time-averaged inventory of subaerial volcanic sulphur emissions. J Geophys Res 103:25251–25261

Angelino S, Suess DT, Prather KA (2001) Formation of aerosol particles from reactions of secondary and tertiary alkylamines: characterization by aerosol time-of-flight mass spectrometry. Environ Sci Technol 35(15):3130–3138

Anguelova MD, Webster F (2006) Whitecap coverage from satellite measurements: a first step toward modeling the variability of oceanic whitecaps. J Geophys Res Oceans 111, C03017. doi:10.1029/2005JC003158

Archer D, Winguth A, Lea D, Mahowald N (2000) What caused the glacial/interglacial atmospheric pCO2 cycles? Rev Geophys 38:159–189

Arimoto R, Duce RA, Ray BJ, Unni CK (1985) Atmospheric trace element at Enewetak Atoll: 2. Transport to the ocean by wet and dry deposition. J Geophys Res 90:2391–2408

Arimoto R, Ray BJ, Lewis NF, Tomza U, Duce RA (1997) Mass particle size distribution of atmospheric dust and the dry deposition of dust to the remote ocean. J Geophys Res 102:15867–15874

Arneth A, Niinemets U, Pressley S, Back J, Hari P, Karl T, Noe S, Prentice IC, Serca D, Hickler T, Wolf A, Smith B (2007) Process-based estimates of terrestrial ecosystem isoprene emissions: incorporating the effects of a direct CO2-isoprene interaction. Atmos Chem Phys 7:31–53

Arneth A, Monson RK, Schurgers G, Niinemets U, Palmer PI (2008) Why are estimates of global isoprene emissions so similar (and why is this not so for monoterpenes)? Atmos Chem Phys 8:4605–4620

Arneth A, Sitch S, Bondeau A, Butterbach-Bahl K, Foster P, Gedney N, de Noblet-Ducoudre N, Prentice IC, Sanderson M, Thonicke K, Wania R, Zaehle S (2010) From biota to chemistry and climate: towards a comprehensive description of trace gas exchange between the biosphere and atmosphere. Biogeosciences 7:121–149

Arneth A, Schurgers G, Lathiere J, Duhl T, Beerling DJ, Hewitt CN, Martin M, Guenther A (2011) Global terrestrial isoprene emission models: sensitivity to variability in climate and vegetation. Atmos Chem Phys 11:8037–8052. doi:10.5194/acp-11-8037-2011, 2011

Atkinson R (2000) Atmospheric chemistry of VOCs and NOx. Atmos Environ 34:2063–2101

Atkinson R, Arey J (2003a) Atmospheric degradation of volatile organic compounds. Chem Rev 103:4605–4638

Atkinson R, Arey J (2003b) Gas-phase tropospheric chemistry of biogenic volatile organic compounds: a review. Atmos Environ 37(2):197–219

Avila A, Rodà F (2002) Assessing decadal changes in rainwater alkalinity at a rural Mediterranean site in the Montseny mountains (NE Spain). Atmos Environ 36:2881–2890

Avila A, Alarcón M, Queralt I (1998) The chemical composition of dust transported in red rains – its contribution to the biogeochemical cycle of a holm aok forest in Catalonia (Spain). Atmos Environ 32:179–191

Aymoz G, Jaffrezo JL, Jacob V, Colomb A, George C (2004) Evolution of organic and inorganic components of aerosol during a Saharan dust episode observed in the French Alps. Atmos Chem Phys 4:2499–2512

Baker AR, Croot PL (2010) Atmospheric and marine controls on aerosol iron solubility in seawater. Mar Chem 120:4–13

Baker A, Jickells T (2006) Mineral particle size as a control on aerosol iron solubility. Geophys Res Lett 33, doi:10.1029/2006GL026557

Baker AR, Jickells TD, Witt M, Linge KL (2006a) Trends in the solubility of iron, aluminium, manganese and phosphorus in aerosol collected over the Atlantic Ocean. Mar Chem 98:43–58

Baker A, French M, Linge K (2006b) Trends in aerosol nutrient solubility along a west–east transect of the Saharan dust plume. Geophys Res Lett 33, doi:10.1029/2005GL024764

Baker AR, Lesworth T, Adams C, Jickells TD, Ganzeveld L (2010) Estimation of atmospheric nutrient inputs to the Atlantic Ocean from 50°N to 50°S based on large-scale field sampling: fixed nitrogen and dry deposition of phosphorus. Global Biogeochem Cycles 24, GB3006. doi:10.1029/2009GB003634

Bardintzeff J-M, McBirney AR (2000) Volcanology. Second edition, Jones, Bartlett. Andres RJ, Kasgnoc AD (1998) A time-avaraged inventory of subaerial volcanic sulphur emissions. J Geophys Res 103:25252–25261

Barkley MP, Palmer PI, Kuhn U, Kesselmeier J, Chance K, Kurosu TP, Martin RV, Helmig D, Guenther A (2008) Net ecosystem fluxes of isoprene over tropical South America inferred from GOME observations of HCHO columns. J Geophys Res 113(D20), D20304. doi:10.1029/2008jd009863

Barkley MP, Palmer PI, Ganzeveld L, Arneth A, Hågberg D, Karl T, Guenther A, Paulot F, Wennberg PO, Mao J, Kurosu TP, Chance K, Muller J-F, De Smedt I, Van Roozendael M, Chen D, Wang Y, Yantosca RM (2011) Can a 'state of the art' chemistry transport model simulate Amazonian tropospehric chemistry? J Geophys Res 116, D16302. doi:10.1029/2011JD015893

Barkley MP, Kurosu TP, Chance K, De Smedt I, Van Roozendael M, Arneth A, Hagberg D, Guenther A (2012) Assessing sources of uncertainty in formaldehyde air mass factors over tropical South America: implications for top-down isoprene emission estimates. J Geophys Res 117, D13304. doi:10.1029/2011JD016827

Bellouin N, Jones A, Haywood J, Christopher S (2008) Updated estimate of aerosol direct radiative forcing from satellite observations and comparison against the Hadley Centre climate model. J Geophys Res 113, doi:10.1029/2007JD009385

Bellouin N, Rae JGL, Jones A, Johnson CE, Haywood JM, Boucher O (2011) Aerosol forcing in the Climate Model Intercomparison Project (CMIP5) simulations by HadGEM2-ES and the role of ammonium nitrate. J Geophys Res 116, D20206. doi:10.1029/2011JD016074

Bergametti G, Gomes L, Remoudaki E, Desbois M, Martin D, Buat-Ménard P (1989) Present transport and deposition patterns of African dusts to the north-western Mediterranean. In: Leinen M, Sarnthein M (eds) Paleoclimatology and paleometeorology: modern and past patterns of global atmospheric transport. Kluwer, Boston, pp 227–251

Bergametti G, Remoudaki E, Losno R, Steiner E, Chatenet B, Buat-Ménard P (1992) Source, transport and deposition of atmospheric phosphorus over the northwestern Mediterranean. J Atmos Chem 14:501–513

Berresheim H, Elste T, Tremmel HG, Allen AG, Hansson H-C, Rosman K, Dal Maso M, Mäkelä JM, Kulmala M, O'Dowd CD (2002) Gas-aerosol relationships of H2SO4, MSA, and OH: observations in the coastal marine boundary layer at Mace Head, Ireland. J Geophys Res 107(D19):8100. doi:10.1029/2000JD000229

Betzer PR, Carder KL, Duce RA, Merrill JT, Tindale NW, Uematsu M, Costello DK, Young RW, Feely RA, Breland JA, Bernstein RE, Greco AM (1988) Long-range transport of giant mineral aerosol particles. Nature 336:568–571

Blain S, Guieu C, Claustre H, Leblanc K, Moutin T, Quéguiner B, Ras J, Sarthou G (2004) Availability of iron and major nutrients for phytoplankton in the northeast Atlantic Ocean. Limnol Oceanogr 49:2095–2104

Blain S, Queguiner B, Armand L, Belviso S, Bombled B, Bopp L, Bowie A, Brunet C, Brussaard C, Carlotti F, Christaki U, Corbiere A, Durand I, Ebersbach F, Fuda J-L, Garcia N, Gerringa L, Griffiths B, Guigue C, Guillerm C, Jacquet S, Jeandel C, Laan P, Lefevre D, Lo Monaco C, Malits A, Mosseri J, Obernosterer I, Park Y-H, Picheral M, Pondaven P, Remenyi T, Sandroni V, Sarthou G, Savoye N, Scouarnec L, Souhaut M, Thuiller D, Timmermans K, Trull T, Uitz J, van Beek P, Veldhuis M, Vincent D, Viollier E, Vong L, Wagener T (2007) Effect of natural iron fertilization on carbon sequestration in the Southern Ocean. Nature 446:1070–1075

Blanchard DC (1963) The electrification of the atmosphere by particles from bubbles in the sea. Prog Oceanogr 1:73–112

Blanchard DC (1964) Sea to air transport of surface active material. Science 146:396–397

Bond TC, Streets DG, Yarber KF, Nelson SM, Woo J-H, Klimont Z (2004) A technology-based global inventory of black and organic carbon emissions from combustion. J Geophys Res 109, D14203. doi:10.1029/2003JD003697

Bonn B, Moortgat GK (2003) Sesquiterpene ozonolysis: origin of atmospheric new particle formation from biogenic hydrocarbons. Geophys Res Lett 30(11):1585. doi:10.1029/2003GL017000

Bonnet S, Guieu C (2004) Dissolution of atmospheric iron in seawater. Geophys Res Lett 31, doi:10.1029/2003GL018423

Bonnet S, Guieu C, Chiaverini J, Ras J, Stock A (2005) Effect of atmospheric nutrients on the autotrophic communities in a low nutrient, low chlorophyll system. Limnol Oceanogr 50:1810–1819

Bopp L, Monfray P, Aumont O, Dufresne J-L, Le Treut H, Madec G, Terray L, Orr JC (2001) Potential impact of climate change on marine export production. Global Biogeochem Cycles 15:81–99

Bopp L, Kohfeld KE, Le Quéré C, Aumont O (2003) Dust impact on marine biota and atmospheric CO2 during glacial periods. Paleoceanography 18:1046. doi:10.1029/2002PA000810

Bory AJ-M, Biscaye PE, Grousset FE (2003) Two distinct seasonal Asian source regions for mineral dust deposited in Greenland (North GRIP). Geophys Res Lett 30:1167. doi:10.1029/2002GL016446

Boyd PW, Watson AJ, Law CS, Abraham ER, Trull T, Murdoch R, Bakker DCE, Bowie AR, Buessler KO, Chang H, Charette MA, Croot P, Downing K, Frew RD, Gall M, Hadfield M, Hall JA, Harvey M, Jameson G, La Roche J, Liddicoat MI, Ling R, Maldonado M, McKay RM, Nodder SD, Pickmere S, Pridmore R, Rintoul S, Safi K, Sutton P, Strzepek R, Tanneberger K, Turner SM, Waite A, Zeldis J (2000) A mesoscale phytoplankton bloom in the polar Southern Ocean stimulated by iron fertilization. Nature 407:695–702

Boyd PW, Law CS, Wong CS, Nojiri Y, Tsuda A, Levasseur M, Takeda S, Rivkin R, Harrison PJ, Strzepek R, Gower J, McKay RM, Abraham E, Arychuk M, Barwell-Clarke J, Crawford W, Hale M, Harada K, Johnson K, Kiyosawa H, Kudo I, Marchetti A, Miller W, Needoba J, Nishioka J, Ogawa H, Page J, Robert M, Saito H, Sastri A, Sherry N, Soutar T, Sutherland N, Taira Y, Whitney F, Wong SE, Yoshimura T (2004) The decline and fate of an iron-induced subarctic phytoplankton bloom. Nature 428:549–553

Boyd PW, Jickells T, Law CS, Blain S et al (2007) A synthesis of mesoscale iron-enrichment experiments 1993–2005: key findings and implications for ocean biogeochemistry. Science 315:612–617

Boyd PW, Mackie DS, Hunter KA (2010) Aerosol iron deposition to the surface ocean — modes of iron supply and biological responses. Mar Chem 120:130–145. doi:10.1016/j.marchem.2009.01.008

Bressac M, Guieu C, Doxaran D, Bourrin F, Obolensky G, Grisoni J-M (2012) A mesocosm experiment coupled with optical measurements to assess the fate and sinking of atmospheric particles in clear oligotrophic waters. Geo-Mar Lett 32(2):153–164

Bressac M, Guieu C, Doxaran D, Bourrin F, Desboeufs K, Leblond N, Ridame C (2013) Quantification of the lithogenic carbon pump following a dust deposition event. Biogeoscience 10, 13639–13677. doi:10.5194/bgd-10-13639-2013, 2013

Buat-Ménard P, Chesselet R (1979) Variable influence of the atmospheric flux on the trace metal chemistry of oceanic suspended matter. Earth Planet Sci Lett 42:399–411

Buat-Ménard P, Davies PJ, Remoudaki E, Miquel J-C, Bergametti G, Lamber CE, Ezat E, Quétel CR, La Rosa J, Fowler SW (1989) Non-steady-state biological removal of atmospheric particles from Mediterranean surface waters. Nature 340:131–133

Buck KN, Bruland KW (2005) Copper speciation in San Francisco Bay: a novel approach using multiple analytical windows. Mar Chem 96:185–198

Buck CS, Landing WM, Resing JA, Measures CI (2010) The solubility and deposition of aerosol Fe and other trace elements in the North Atlantic Ocean: observations from the A16N CLIVAR/CO2 repeat hydrography section. Mar Chem 120(1–4):57–70

Buesseler KO, Bacon MP, Cochran JK, Livingston HD (1992) Carbon and nitrogen export during the JGOFS North Atlantic bloom experiment estimated from 234Th:238U disequilibria. Deep-Sea Res 39(7/8):1115–1137

Buhaug Ø, Corbett JJ, Endresen Ø, Eyring V, Faber J, Hanayama S, Lee DS, Lindstad H, Mjelde A, Palsson C, Wanquing W, Winebrake JJ, Yoshida K (2008) Updated study on greenhouse gas emissions from ships: phase I report. International Maritime Organization (IMO), London, 1 Sept 2008, p 129

Burkholder JB, Curtius J, Ravishankara AR et al (2004) Laboratory studies of the homogeneous nucleation of iodine oxides. Atmos Chem Phys 4:19–34

Callaghan A, de Leeuw G, Cohen L, O'Dowd CD (2008) Relationship of oceanic whitecap coverage to wind speed and wind history. Geophys Res Lett 35, L23609. doi:10.1029/2008GL036165

Calvo-Díaz A, Díaz-Pérez L, Suárez LA, Morán XAG, Teira E, Marañón E (2011) Decrease in the autotrophic-to-heterotrophic biomass ratio of picoplankton in oligotrophic marine waters due to bottle enclosure. Appl Environ Microbiol 77:5739–5746

Capes G, Johnson B, McFiggans G, Williams PI, Haywood J, Coe H (2008) Aging of biomass burning aerosols over West Africa Aircraft measurements of chemical composition microphysical properties and emission ratios. J Geophys Res 113, D00C15. doi:10.1029/2008JD009845

Carlson TN, Prospero JM (1972) The large-scale movement of Saharan air outbreaks over the northern Equatorial Atlantic. J Appl Meteorol 11:283–297

Carslaw KS, Boucher O, Spracklen DV, Mann GW, Rae JGL, Woodward S, Kulmala M (2010) A review of natural aerosol interactions and feedbacks within the Earth system. Atmos Chem Phys 10:1701–1737

Cassar N, Bender ML, Barnett BA, Fan S, Moxim WJ, Levy H II, Tilbrook B (2007) The southern ocean biological response to Aeolian iron deposition. Science 317:1067–1070

Cavalli F, Facchini MC, Decesari S, Mircea M, Emblico L, Fuzzi S, Ceburnis D, Yoon YJ, O'Dowd CD, Putaud J-P, Dell'Acqua A (2004) Advances in characterization of size-resolved organic matter in marine aerosol over the North Atlantic. J Geophys Res 109, D24215. doi:10.1029/2004JD005137

Ceburnis D, O'Dowd CD, Jennings SG, Facchini MC, Emblico L, Decesari S, Fuzzi S, Sakalys J (2008) Marine aerosol chemistry gradients: elucidating primary and secondary processes and fluxes. Geophys Res Lett 35, L07804. doi:10.1029/2008GL033462

Chance K, Palmer PI, Spurr RJD, Martin RV, Kurosu TP, Jacob DJ (2000) Satellite observations of formaldehyde over North America from GOME. Geophys Res Lett 27:3461–3464

Charlson RJ, Lovelock JE, Andreae MO, Warren SG (1987) Oceanic phytoplankton, atmospheric sulfur, cloud albedo and climate. Nature 326:655–661

Chatenet B, Marticorena B, Gomes L, Bergametti G (1996) Assessing the microped size distribution of desert soils erodible by wind. Sedimentology 43:901–911

Chazette P, Pelon J, Moulin C, Dulac F, Carrasco I, Guelle W, Bousquet P, Flamant P-H (2001) Lidar and satellite retrieval of dust aerosols over the Azores during SOFIA/ASTEX. Atmos Environ 35:4297–4304

Chen Y, Siefert R (2004) Sesaonal and spatial distributions and dry deposition fluxes of atmospheric total and labile iron over the tropical and subtropical North Atlantic Ocean. J Geophys Res 109, D09305. doi:09310.01029/02003JD003958

Chen Y, Street J, Paytan A (2006) Comparison between pure-water- and seawater-soluble nutrient concentrations of aerosols from the Gulf of Aqaba. Mar Chem 101:141–152

Chen Y, Mills S, Street J, Golan D, Post A, Jacobson M, Paytan A (2007) Estimates of atmospheric dry deposition and associated input of nutrients to Gulf of Aqaba seawater. J Geophys Res 112, D04309. doi:10.1029/2006JD007858

Chou C, Formenti P, Maille M, Ausset P, Helas G, Harrison M, Osborne S (2008) Size distribution, shape, and composition of mineral dust aerosols collected during the African monsoon multidisciplinary analysis special observation period 0: dust and biomass-burning experiment field campaign in Niger, January 2006. J Geophys Res 113, D00C10. doi:10.1029/2008JD009897

Chuang P, Duvall R, Shafer M, Schauer J (2005) The origin of water soluble particulate iron in the Asian atmospheric outflow. Geophys Res Lett 32, doi:10.1029/2004GL021946

Claeys M, Graham B, Vas G, Wang W, Vermeylen R, Pashynska V, Cafmeyer J, Guyon P, Andreae MO, Artaxo P, Maenhaut W (2004) Formation of secondary organic aerosols through photooxidation of isoprene. Science 303:1173–1176

Claeys M, Wang W, Vermeylen R, Kourtchev I, Chi X, Farhat Y, Surratt JD, Gómez-González Y, Sciare J, Maenhaut W (2010) Chemical characterisation of marine aerosol at Amsterdam Island during the austral summer of 2006–2007. J Aerosol Sci 41:13–22

Clarke AD, Davis D, Kapustin VN, Eisele F, Chen G, Paluch I, Lenschow D, Bandy AR, Thornton D, Moore K, Mauldin L, Tanner D, Litchy M, Carroll MA, Collings J, Albercook G (1998) Particle nucleation in the tropical boundary layer and its coupling to marine sulfur sources. Sciences 282:89–91

Clarke AD, Qwens SR, Zhou J (2006) An ultrafine sea-salt flux from breaking waves: implications for cloud condensation nuclei in the remote marine atmosphere. J Geophys Res 111, D06202. doi:10.1029/2005JD006565

Claustre H, Morel A, Hooker SB, Babin M, Antoine D, Oubelkheir K, Bricaud A, Leblanc K, Quéguiner B, Maritorena S (2002) Is desert dust making oligotrophic waters greener? Geophys Res Lett 29:1469. doi:10.1029/2001GL014056

Colarco PR, Toon OB, Holben BN (2003a) Saharan dust transport to the Caribbean during PRIDE: 1. Influence of dust sources and removal mechanisms on the timing and magnitude of downwind aerosol optical depth events from simulations of in situ and remote sensing observations. J Geophys Res D Atmos 108(19):5-1–5-20

Colarco PR, Toon OB, Reid JS, Livingston JM, Russell PB, Redemann J, Schmid B, Maring HB, Savoie D, Welton EJ, Campbell JR, Holben BN, Levy R (2003b) Saharan dust transport to the Caribbean during PRIDE: 2. Transport, vertical profiles, and deposition in simulations of in situ and remote sensing observations. J Geophys Res D Atmos 108 (19):6-1–6-16

Corbett JJ, Köhler HW (2003) Updated emissions from ocean shipping. J Geophys Res 108:4650. doi:10.1029/2003JD003751

Cornell SE (2011) Atmospheric nitrogen deposition: revisiting the importance of the organic component. Environ Pollut 159:2214–2222

Cornell SE, Rendell A, Jickells TD (1995) Atmospheric inputs of dissolved organic nitrogen to the oceans. Nature 376:243–246

Covert DS, Kapustin VN, Bates TS, Quinn PK (1992) New particle formation in the marine boundary layer. J Geophys Res 97:20581–20589

Covert DS, Wiedensohler A, Aalto P et al (1996a) Aerosol number size distributions from 3 to 500 nm diameter in the arctic marine boundary layer during summer and autumn. TELLUS ser B 48(2):197–212

Covert DS, Kapustin VN, Bates TS et al (1996b) Physical properties of marine boundary layer aerosol particles of the mid-Pacific in relation to sources and meteorological transport. J Geophys Res-Atmos 101(D3):6919–6930. doi:10.1029/95JD03068

Coz E, Gómez-Moreno FJ, Pujadas M, Casuccio GS, Lersch TL, Artíñano B (2009) Individual particle characteristics of North African dust under different long-range transport scenarios. Atmos Environ 43:1850–1863

Crahan KK, Hegg D, Covert DS, Jonsson H (2004) An exploration of aqueous oxalic acid production in the coastal marine atmosphere. Atmos Environ 38:3757–3764

Crumeyrolle S, Gomes L, Tulet P, Matsuki A, Schwarzenboeck A, Crahan K (2008) Increase of the aerosol hygroscopicity by cloud processing in a mesoscale convective system: a case study from the AMMA campaign. Atmos Chem Phys 8:6907–6924

Dall'Osto M, Harrison RM, Highwood EJ, O'Dowd C, Ceburnis D, Querol X, Achterberg EP (2010) Variation of the mixing state of Saharan dust particles with atmospheric transport. Atmos Environ 44:3135–3146

Dall-Osto M, Ceburnis D, Monahan C, Worsnop DR, Bialek J, Kulmala M, Kurtén T, Ehn M, Wenger J, Sodeau J, Healy RC, O'Dowd C (2012) Nitrogenated and aliphatic organic vapours as possible drivers for marine secondary organic aerosol growth. J Geophys Res doi:10.1029/2012JD017522

de Baar HJW, Boyd PW, Coale KH, Landry MR, Tsuda A, Assmy P, Bakker DCE, Bozec Y, Barber RT, Brzezinski

MA, Buesseler KO, Boyé M, Croot PL, Gervais F, Gorbunov MY, Harrison PJ, Hiscock WT, Laan P, Lancelot C, Law CS, Levasseur M, Marchetti A, Millero FJ, Nishioka J, Nojiri Y, van Oijen T, Riebesell U, Rijkenberg MJA, Saito H, Takeda S, Timmermans KR, Veldhuis MJW, Waite AM, Wong C-S (2005) Synthesis of iron fertilization experiments: from the iron age in the age of enlightenment. J Geoph Res 110, C09S16. doi:10.1029/2004JC002601

de Gouw JA, Middlebrook AM, Warneke C, Goldan PD, Kuster WC, Roberts JM, Fehsenfeld FC, Worsnop DR, Canagaratna MR, Pszenny AAP, Keene WC, Marchewka M, Bertman SB, Bates TS (2005) Budget of organic carbon in a polluted atmosphere: results from the New England air quality study in 2002. J Geophys Res 110, D16305. doi:10.1029/2004jd005623

de Leeuw G, Andreas EL, Anguelova MD, Fairall CW, Lewis ER, O'Dowd C, Schulz M, Schwartz SE (2011a) Production flux of sea spray aerosol. Rev Geophys 49, RG2001. doi:10.1029/2010RG000349

de Leeuw G, Kinne S, Leon JF, Pelon J, Rosenfeld D, Schaap M, Veefkind PJ, Veihelmann B, Winker DM, von Hoyningen-Huene W (2011b) Retrieval of aerosol properties. In: Burrows JP, Platt U, Borrell P (eds) The remote sensing of tropospheric composition from space. Springer, Berlin/Heidelberg, pp 359–313. doi:10.1007/978-3-642-14791-3. ISBN 978-3-642-14790-6

DeCarlo PF, Dunlea EJ, Kimmel JR, Aiken AC, Sueper D, Crounse J, Wennberg PO, Emmons L, Shinozuka Y, Clarke A, Zhou J, Tomlinson J, Collins DR, Knapp D, Weinheimer AJ, Montzka DD, Campos T, Jimenez JL (2008) Fast airborne aerosol size and chemistry measurements above Mexico City and Central Mexico during the MILAGRO campaign. Atmos Chem Phys 8:4027–4048

Decesari S, Finessi E, Rinaldi M, Paglione M, Fuzzi S, Stephanou EG, Tziaras T, Spyros A, Ceburnis D, O'Dowd CD, Dall'Osto M, Harrison RM, Allan J, Coe H, Facchini MC (2011) Primary and secondary marine organic aerosols over the North Atlantic Ocean during the MAP experiment. J Geophys Res 116, D22210. doi:10.1029/2011JD016204

Denman KL, Brasseur G, Chidthaisong A, Ciais P, Cox PM, Dickinson RE, Hauglustaine D, Heinze C, Holland E, Jacob D, Lohmann U, Ramachandran S, da Silva Dias PL, Wofsy SC, Zhang X (2007) Couplings between changes in the climate system and biogeochemistry. In: Solomon S, Qin D, Manning M, Chen Z, Marquis M, Averyt KB, Tignor M, Miller HL (eds) Climate change 2007: the physical science basis. Contribution of working group I to the fourth assessment report of the intergovernmental panel on climate change. Cambridge University Press, Cambridge, UK/New York

Dentener FJ, Carmichael GR, Zhang Y, Lelieveld J, Crutzen PJ (1996) Role of mineral aerosol as a reactive surface in the global troposphere. J Geophys Res 101:22869–22889

Dentener F, Drevet J, Lamarque JF, Bey I, Eickhout B et al (2006) Nitrogen and sulfur deposition on regional and global scales: a multimodel evaluation. Global Biogeochem Cycles 20, GB4003

Desboeufs KV, Losno R, Colin JL (2001) Factors influencing aerosol solubility during cloud processes. Atmos Environ 35:3529–3537

Desboeufs K, Journet E, Rajot JL, Chevaillier S, Triquet S, Formenti P, Zakou A (2010) Chemistry of rain events in West Africa: evidence of dust and biogenic influence in convective systems. Atmos Chem Phys 10:9283–9293

Doney S, Mahowald N, Lima I, Feeley R, Mackenzie F, Lamarque JF, Rasch P (2007) Impact of an-thropogenic atmospheric nitrogen and sulfur depositionon ocean acidification and the inorganic carbon system. PNAS 104, doi:10.1073/pnas.0702218104; 0702214580–0702214585

Duarte CM, Agustí S, Gasol JM, Vaqué D, Vazquez-Dominguez E (2000) Effect of nutrient supply on the biomass structure of planktonic communities: an experimental test on a Mediterranean coastal community. Mar Ecol Prog Ser 206:87–95

Dubovik O, Holben BN, Eck TF, Smirnov A, Kaufman YJ, King MD, Tanré D, Slutsker I (2002) Variability of absorption and optical properties of key aerosol types observed in worldwide locations. J Atmos Sci 59:590–608

Dubovik O, Herman M, Holdak A, Lapyonok T, Tanré D, Deuzé JL, Ducos F, Sinyuk A, Lopatin A (2011) Statistically optimized inversion algorithm for enhanced retrieval of aerosol properties from spectral multi-angle polarimetric satellite observations. Atmos Meas Tech 4:975–1018

Duce R (1986) The impact of atmospheric nitrogen, phosphorus and iron species on marine biological productivity. In: Buat-Menard P (ed) Geochemical cycling. D. Reidel, Norwell, pp 497–529

Duce RA et al (2008) Impacts of atmospheric nitrogen on the open ocean. Science 320:893–897

Duggen S, Croot P, Schacht U, Hoffmann L (2007) Subduction zone volcanic ash can fertilize the surface ocean and stimulate phytoplankton growth: evidence from biogeochemical experiments and satellite data. Geophys Res Lett 34, doi:10.1029/2006GL027522

Dulac F, Buat-Ménard P, Ezat U, Melki S, Bergametti G (1989) Atmospheric input of trace metals to the western Mediterranean: uncertainties in modelling dry deposition from cascade impactor data. Tellus 41B:362–378

Dulac F, Bergametti G, Losno R, Remoudaki E, Gomes L et al (1992) Dry deposition of mineral aerosol particles in the marine atmosphere: significance of the large size fraction. In: Schwartz SE, Slinn WGN (eds) Precipitation scavenging and atmosphere-surface exchange 2. Hemisphere, Washington, DC, pp 841–854

Dulac F, Moulin C, Lambert CE, Guillard F, Poitou J, Guelle W, Quétel CR, Schneider X, Ezat U (1996) Quantitative remote sensing of African dust transport to the Mediterranean. In: Guerzoni S, Chester R (eds) The impact of African dust across the Mediterranean. Kluwer, Norwell, pp 25–49

Dulac F, Chazette P, Gomes L, Chatenet B, Berger H, Vinicula Dos Santos JM (2001) A method for aerosol profiling in the lower troposphere with coupled scatter and meteorological rawindsondes and first data from the tropical Atlantic off Sahara. J Aerosol Sci 32:1069–1086

Dunlea EJ, DeCarlo PF, Aiken AC, Kimmel JR, Peltier RE, Weber RJ, Tomlinson J, Collins DR, Shinozuka Y, McNaughton CS, Howell SG, Clarke AD, Emmons LK, Apel EC, Pfister GG, van Donkelaar A, Martin RV, Millet DB, Heald CL, Jimenez JL (2009) Evolution of Asian aerosols during transpacific transport in INTEX-B. Atmos Chem Phys 9:7257–7287. doi:10.5194/acp-9-7257-2009

Duvall RM, Majestic BJ, Shafer MM, Chuang PY, Simoneit BRT, Schauer JJ (2008) The water-soluble fraction of carbon, sulfur, and crustal elements in Asian aerosols and Asian soils. Atmos Environ 42:5872–5884

Ehn M, Vuollekoski H, Petäjä T, Kerminen V-M, Vana M, Aalto P, de Leeuw G, Ceburnis D, Dupuy R, O'Dowd CD, Kulmala M (2010) Growth rates during coastal and marine new particle formation in Western Ireland. J Geophys Res. doi:10.1029/2010JD014292

Endresen Ø, Sørgard E, Sundet JK, Dalsøren SB, Isaksen ISA, Berglen TF, Gravir G (2003) Emission from international sea transportation and environmental impact. J Geophys Res 108:4560. doi:10.1029/2002JD002898

Endresen Ø, Sørgard E, Behrens HL, Brett PO, Isaksen ISA (2007) A historical reconstruction of ships' fuel consumption and emissions. J Geophys Res 112, D12301. doi:10.1029/2006JD007630

Erel Y, Dayan U, Rabi R, Rudich Y, Stein M (2006) Trans boundary transport of pollutants by atmospheric mineral dust. Environ Sci Technol 40:2996–3005

Eyring V, Köhler HW, van Aardenne J, Lauer A (2005) Emissions from international shipping: 1. The last 50 years. J Geophys Res 110, D17305

Eyring V, Isaksen ISA, Berntsen T, Collins WJ, Corbett JJ, Endresen O, Grainger RG, Moldanova J, Schlager H, Stevenson DS (2009) Transport impacts on atmosphere and climate: shipping. Atmos Environ. doi:10.1016/j.atmosenv.2009.04.059

Facchini MC, Rinaldi M, Decesari S, Carbone C, Finessi E, Mircea M, Fuzzi S, Ceburnis D, Flannigan R, Nilsson ED, de Leeuw G, Martino M, Woeltjen J, O'Dowd CD (2008a) Primary submicron marine aerosol dominated by insoluble organic colloids and aggregates. Geophys Res Lett 35, L17814. doi:10.1029/2008GL034210

Facchini MC, Decesari S, Rinaldi M, Carbone C, Finessi E, Mircea M, Fuzzi S, Moretti F, Tagliavini E, Ceburnis D, O'Dowd CD (2008b) An important source of marine secondary organic aerosol from biogenic amines. Environ Sci Technol. doi:10.1021/es8018385

Fairlie TD, Jacob DJ, Dibb JE, Alexander B, Avery MA, van Donkelaar A, Zhang L (2010) Impact of mineral dust on nitrate, sulfate, and ozone in transpacific Asian pollution plumes. Atmos Chem Phys 10:3999–4012

Falkovich AH, Ganor E, Levin Z, Formenti P, Rudich Y (2001) Chemical and mineralogical analysis of individual mineral dust particles. J Geophys Res 106:18029–18036

Falkovich AH, Schkolnik G, Ganor E, Rudich Y (2004) Adsorption of organic compounds pertinent to urban environments onto mineral dust particles. J Geophys Res 109, D02208. doi:10.1029/2003jd003919

Falkowski PG, Barber RT, Smetacek V (1998) Biogeochemical controls and feedbacks on ocean primary production. Science 281:200–206

Fan X-B, Okada K, Niimura N, Kai K, Arao K, Shi G-Y, Qin Y, Mitsuta Y (1996) Mineral particles collected in China and Japan during the same Asian dust-storm event. Atmos Environ 30:347–351

Fisher RV, Schmincke H-U (1984) Pyroclastic rocks. Springer, Berlin/Heidelberg/New York/Tokyo

Formenti P, Schütz L, Balkanski Y, Desboeufs K, Ebert M, Kandler K, Petzold A, Scheuvens D, Weinbruch S, Zhang D (2011) Recent progress in understanding physical and chemical properties of African and Asian mineral dust. Atmos Chem Phys 11:8231–8256

Forster PM, Ramaswamy V et al (2007) Changes in atmospheric constituents and in radiative forcing. In: Solomon S, Qin D, Manning M, Chen Z, Marquis M, Averyt KB, Tignor M, Miller HL (eds) Climate change 2007: the physical science basis contribution of working group I to the fourth assessment report of the intergovernmental panel on climate change. Cambridge University Press, Cambridge

Fowler SW, Buat-Ménard P, Yokoyama Y, Ballestra S, Holm E, Van Nguyen H (1987) Rapid removal of Chernobyl fallout from Mediterranean surface waters by biological activity. Nature 329:56–58

Frew NM (1997) The role of organic films in air-sea gas exchange. In: Liss PS, Duce RA (eds) The sea surface and global change. Cambridge University Press, Cambridge, pp 121–172

Frogner P, Gislason SR, Óskarsson N (2001) Fertilizing potential of volcanic ash in ocean surface water. Geology 29:487–490

Frossard AA, Shaw PM, Russell LM, Kroll JH, Canagaratna MR, Worsnop DR, Quinn PK, Bates TS (2011) Springtime Arctic haze contributions of submicron organic particles from European and Asian combustion sources. J Geophys Res 116, D05205. doi:10.1029/2010JD015178

Froyd KD, Murphy SM, Murphy DM, de Gouw JA, Eddingsaas NC, Wennberg PO (2010) Contribution of isoprene-derived organosulfates to free tropospheric aerosol mass. Proc Natl Acad Sci. doi:10.1073/pnas.1012561107

Fu PQ, Kawamura K, Miura K (2011) Molecular characterization of marine organic aerosols collected during a round-the-world cruise. J Geophys Res Atmos 116(14), D13302. doi:10.1029/2011jd015604

Fuentes E, Coe H, Green D, de Leeuw G, McFiggans G (2010) Laboratory-generated primary marine aerosol via bubble-bursting and atomization. Atmos Meas Tech 3:141–162

Fuentes E, Coe H, Green D, McFiggans G (2011) On the impacts of phytoplankton-derived organic matter on the properties of the primary marine aerosol – part 2: composition, hygroscopicity and cloud condensation activity. Atmos Chem Phys 11:2585–2602. doi:10.5194/acp-11-2585-2011

Fung I, Meyn SK, Tegen I, Doney S, John J, Bishop J (2000) Iron supply and demand in the upper ocean. Global Biogeochem Cycles 14:281–295

Gaiero DM, Probst JL, Depetris PJ, Bidart SM, Leleyter L (2003) Iron and other transition metals in Patagonian riverborne and windborne materials: geochemical control and transport to the southern South Atlantic Ocean. Geochim Cosmochim Acta 67:3603–3623

Galloway J, Townsend A, Erisman J, Bekunda M, Cai Z, Freney J, Martinelli L, Seitzinger S, Sutton M (2008) Transformation of the nitrogen cycle: recent trends, questions and potential solutions. Science 320:889–892

Gantt B, Meskhidze N, Facchini MC, Rinaldi M, Ceburnis D, O'Dowd CD (2011) Wind speed dependent size-resolved parameterization for the organic enrichment of sea spray. Atmos Chem Phys 11:1–13

Garrison VH, Shinn EA, Foreman WT, Griffin DW, Holmes CW, Kellogg CA, Majewski MS, Richardson LL, Ritchie KB, Smith GW (2003) African and Asian dust: from desert soils to coral reefs. Bioscience 53:469–479

Garrison VH, Foreman WT, Genualdi S, Griffin DW, Kellogg CA, Majewski MS, Mohammed A, Ramsubhag A, Shinn EA, Simonich SL, Smith GW (2006) Saharan dust – a carrier of persistent organic pollutants, metals and microbes to the Caribbean? Rev Biol Trop (Int J Trop Biol ISSN-0034-7744) 54(3):9–21

Geever M, O'Dowd CD, van Ekeren S, Flanagan R, Nilsson DE, de Leeuw G, Rannik Ü (2005) Sub-micron sea-spray fluxes. Geophys Res Lett. doi:10.1029/2005GL023081

Geng H, Park Y, Hwang H, Kang S, Ro CU (2009) Elevated nitrogen-containing particles observed in Asian dust aerosol samples collected at the marine boundary layer of the Bohai Sea and the Yellow Sea. Atmos Chem Phys 9:6933–6947

Gershey RM (1983) Characterization of seawater organic matter carried by bubble-generated aerosols. Limnol Oceanogr 28:309–319

Gibb SW, Mantoura RFC, Liss PS (1999) Ocean atmosphere exchange and atmospheric speciation of ammonia and methylamines in the region of the NW Arabian Sea. Global Biogeochem Cycles 13:161–178

Giglio L, Randerson JT, van der Werf GR, Kasibhatla PS, Collatz GJ, Morton DC, DeFries RS (2010) Assessing variability and long-term trends in burned area by merging multiple satellite fire products. Biogeosciences 7:1171–1186

Giorgi F (2006) Climate change hot-spots. Geophys Res Lett 33, L08707. doi:10.1029/2006gl025734

Giovagnetti V, Brunet C, Conversano F, Tramontano F, Obernosterer I, Ridame C, Guieu C (2013) Assessing the role of dust deposition on phytoplankton ecophysiology and succession in a low-nutrient low-chlorophyll ecosystem: a mesocosm experiment in the Mediterranean, Sea. Biogeosciences 10:2973–2991. doi:10.5194/bg-10-2973-2013

Goldstein AH, Galbally IE (2007) Known and unexplored organic constituents in the Earth's at-mosphere. Environ Sci Technol 41:1514–1521

Gomes L, Bergametti G, Coudé-Gaussens G, Rognon P (1990) Submicron desert dusts: a sandblasting process. J Geophys Res 95:13927–13935

Gong SL (2003) A parameterization of sea-salt aerosol source function for sub- and super-micron particles. Global Biogeochem Cycles 17:1097. doi:10.1029/2003GB002079

Gorbushina AA, Kort R, Schulte A, Lazarus D, Schnetger B, Brumsack H-J, Broughton WJ, Favet J (2007) Life in Darwin's dust: intercontinental transport and survival of microbes in the nineteenth century. Environ Microbiol 9(12):2911–2922

Goudie AS, Middleton NJ (2006) Desert dust in the global system. Springer, Berlin

Graedel TE, Keene WC (1995) Tropospheric budget of reactive chlorine. Global Biogeochem Cycles 9:47–77

Graf HF, Feichter J, Langmann B (1997) Volcanic sulfur emissions: estimates of source strength and its contribution to the global sulfate distribution. J Geophys Res-Atmos 102:10727–10738

Graham WF, Duce RA (1979) Atmospheric pathways of the phosphorus cycle. Geochimica et Cosmo-chimica Acta 43:1195–1208

Granier C, Bessagnet B, Bond T, D'Angiola A, Denier van der Gon H, Frost GJ, Heil A, Kaiser JW, Kinne S, Klimont Z et al (2011) Evolution of anthropogenic and biomass burning emissions of air pollutants at global and regional scales during the 1980–2010 period. Clim Chang. doi:10.1007/s10584-011-0154-1

Greeley R, Iversen J (1985) Wind as a geological process on Earth, Mars, Venus and Titan, vol 4, Cambridge planetary sciences series. Cambridge University Press, Cambridge, p 333

Grieshop AP, Logue JM, Donahue NM, Robinson AL (2009) Laboratory investigation of photochemical oxidation of organic aerosol from wood fires 1: measurement and simulation of organic aerosol evolution. Atmos Chem Phys 9:1263–1277

Griffin DW (2007) Atmospheric movement of microorganisms in clouds of desert dust and implications for human health. Clin Microbiol Rev 20(3):459–477. doi:10.1128/CMR.00039-06

Grousset F, Biscaye P (2005) Tracing dust sources and transport patterns using Sr, Nd and Pb isotopes. Chem Geol 222:149–167

Guenther A, Hewitt CN, Erickson D, Fall R, Geron C, Graedel T, Harley P, Klinger L, Lerdau M, McKay WA, Pierce T, Scholes B, Steinbrecher R, Tallamraju R, Taylor J, Zimmermann P (1995) A global model of natural volatile organic compound emissions. J Geophys Res 100(D5):8873–8892

Guenther A, Karl T, Harley P, Wiedinmyer C, Palmer PI, Geron C (2006) Estimates of global terrestrial isoprene emissions using MEGAN (Model of emissions of gases and Aerosols from nature). Atmos Chem Phys 6:3181–3210

Guieu C, Loye-Pilot MD, Ridame C, Thomas C (2002) Chemical characterization of the Saharan dust end-member: some biogeochemical implications for the western Mediterranean Sea. J Geophys Res Atmos 107:4258. doi:10.1029/2001JD000582

Guieu C, Bonnet S, Wagener T, Loÿe-Pilot MD (2005) Biomass burning as a source of dissolved iron to open ocean? Geophys Res Lett 32, doi:L1960810.1029/2005GL022962

Guieu C, Dulac F, Desboeufs K, Wagener T, Pulido-Villena E, Grisoni J-M, Louis F, Ridame C, Blain S, Brunet C, Bon Nguyen E, Tran S, Labiadh M, Dominici J-M (2010) Large clean mesocosms and simulated dust deposition: a new methodology to investigate responses of marine oligotrophic ecosystems to atmospheric inputs. Biogeosciences 7:2765–2784

Guieu C, Ridame C, Pulido-Villena E, Blain S, Bressac M, Desboeufs K, Dulac F, Does dust deposition change the metabolic balance of a typical oligotrophic marine environment? (in preparation)

Guieu C, Dulac F, Ridame C, Pondaven P (2013) Introduction to the project DUNE, a DUst experiment in a low Nutrient, low chlorophyll Ecosystem, Biogeosciences Discuss 10:12491–12527

Halloran PR, Bell TG, Totterdell IJ (2010) Can we trust empirical marine DMS parameterisations within projections of future climate? Biogeosciences 7:1645–1656

Hallquist M, Wenger JC, Baltensperger U, Rudich Y, Simpson D, Claeys M, Dommen J, Donahue NM, George C, Goldstein AH, Hamilton JF, Herrmann H, Hoffmann T, Iinuma Y, Jang M, Jenkin ME, Jimenez JL, Kiendler-Scharr A, Maenhaut W, McFiggans G, Mentel TF, Monod A, Prevot ASH, Seinfeld JH, Surratt JD, Szmigielski R, Wildt J (2009) The formation, properties and impact of secondary organic aerosol: current and emerging issues. Atmos Chem Phys 9(14):5155–5235

Halmer MM, Scmincke HU, Graf HF (2002) The annual volcanic gas input into the atmosphere, in particular into the stratosphere: a global data set for the past 100 years. J Volcanol Geotherm Res 115:511–528

Hamm CE (2002) Interactive aggregation and sedimentation of diatoms, and clay-sized lithogenic material. Limnol Oceanogr 47:1790–1795

Hamme RC, Webley PW, Crawford WR, Whitney FA, DeGrandpre MD, Emerson SR, Eriksen CC, Giesbrecht KE, Gower JFR, Kavanaugh MT, Peña MA, Sabine CL, Batten SD, Coogan LA, Grundle DS, Deirdre LD (2010) Volcanic ash fuels anomalous plankton bloom in subarctic northeast Pacific. Geophys Res Lett 37, L19604. doi:10.1029/2010GL044629

Hamonou E, Chazette P, Balis D, Dulac F, Schneider X, Galani E, Ancellet G, Papayannis A (1999) Characterization of the vertical structure of Saharan dust export to the Mediterranean basin. J Geophys Res 104:22257–22270

Hand J, Mahowald N, Chen Y, Siefert R, Luo C, Subramaniam A, Fung I (2004) Estimates of soluble iron from observations and a global mineral aerosol model: biogeochemical implications. J Geophys Res 109, D17205, doi:17210.11029/12004JD004574

Hand VL, Capes G, Vaughan DJ, Formenti P, Haywood JM, Coe H (2010) Evidence of internal mixing of African dust and biomass burning particles by individual particle analysis using electron beam techniques. J Geophys Res Atmos 115, D13301. doi:10.1029/2009jd012938

Hanisch F, Crowley JN (2001) The heterogeneous reactivity of gaseous nitric acid on authentic mineral dust samples, and on individual mineral and clay mineral components. Phys Chem Chem Phys 3:2474–2482

Hansell DA, Carlson CA, Repeta DJ, Schlitzer R (2009) Dissolved organic matter in the ocean: a controversy stimulates new insights. Oceanography 22:202–211

Hawkins LN, Russell LM, Covert DS, Quinn PK, Bates TS (2010) Carboxylic acids, sulfates, and organosulfates in processed continental organic aerosol over the southeast Pacific Ocean during VOCALS-REx 2008. J Geophys Res 115, D13201. doi:10.1029/2009jd013276

Heald CL, Jacob DJ, Park RJ, Russell LM, Huebert BJ, Seinfeld JH, Liao H, Weber RJ (2005) A large organic aerosol source in the free troposphere missing from current models. Geophys Res Lett 32, L18809. doi:10.1029/2005gl023831

Hedges IH, Eglinton G, Hatcher PG, Kirchman DL, Arnosti C, Derenne S, Evershed RP, Ogel-Knabner IK, de Leeuw JW, Littke R, Michaelis W, Rullkotter J (2000) The molecularly uncharacterized component of nonliving organic matter in natural environments. Organic Geochem 31:945–951

Hennigan CJ, Miracolo MA, Engelhart GJ, May AA, Presto AA, Lee T, Sullivan AP, McMeeking GR, Coe H, Wold CE, Hao WM, Gilman JB, Kuster WC, de Gouw J, Schichtel BA, Collett JL, Kreidenweis SM, Robinson AL (2011) Chemical and physical transformations of organic aerosol from the photo-oxidation of open biomass burning emissions in an environmental chamber. Atmos Chem Phys 11:7669–7686

Henze D, Seinfeld JH (2006) Global secondary organic aerosol from isoprene oxidation. Geophys Res Lett 33, L09812. doi:10.1029/2006GL025976

Herut B et al (2005) Response of East Mediterranean surface water to Saharan dust: on-board microcosm experiment and field observations. Deep-Sea Res II 52:3024–3040

Hildebrandt L, Engelhart GJ, Mohr C, Kostenidou E, Lanz VA, Bougiatioti A, DeCarlo PF, Prevot ASH, Baltensperger U, Mihalopoulos N, Donahue NM, Pandis SN (2010) Aged organic aerosol in the Eastern Mediterranean: the Finokalia aerosol measurement experiment 2008. Atmos Chem Phys 10:4167–4186. doi:10.5194/acp-10-4167-2010

Hill PG, Zubkov MV, Purdie DA (2010) Differential responses of Prochlorococcus and SAR11-dominated bacterioplankton groups to atmospheric dust inputs in the tropical Northeast Atlantic Ocean. FEMS Microbiol Lett 306:82–89

Hoffmann T, O'Dowd CD, Seinfeld JH (2001) Iodine oxide homogeneous nucleation: an explanation for coastal new particle production. Geophys Res Lett 28(10):1949–1952. doi:10.1029/2000GL012399

Holben BN, Tanre D, Smirnov A, Eck TF, Slutsker I, Abuhassan N, Newcomb WW, Schafer J, Chatenet B, Lavenue F, Kaufman YJ, Vande Castle J, Setzer A, Markham B, Clark D, Frouin R, Halthore R, Karnieli A, O'Neill NT, Pietras C, Pinker RT, Voss K, Zibordi G (2001) An emerging ground-based aerosol climatology: aerosol optical depth from AERONET. J Geophys Res 106:12067–12097

Holzinger R, Lee A, Paw U KT, Goldstein AH (2005) Observations of oxidation products above a forest imply biogenic emissions of very reactive compounds. Atmos Chem Phys 5:67–75

Holmes BJ, Petrucci GA (2006) Water-soluble oligomer formation from acid catalyzed reactions of levoglucosan in proxies of atmospheric aqueous aerosols. Environ Sci Technol 40:4983–4989

Honeyman BD, Santschi PH (1991) Coupling adsorption and particle aggregation: laboratory studies of "colloidal pumping" using 59Fe-labeled hematite. Environ Sci Technol 25:1739–1747

Hoppel WA, Frick GM, Larson RE (1986) Effect of nonprecipitating clouds on the aerosol size distribution. Geophys Res Lett 13:125–128

Hoppel WA, Frick GM, Fitzgerald J et al (1994) Marine boundary-layer measurements of new particle formation and the effects nonprecipitating clouds have on aerosol-size distribution. J Geophys Res-Atmos 99(D7):14443–14459. doi:10.1029/94JD00797

Horňák K, Jezbera J, Nedoma J, Gasol JM, Simek K (2006) Effects of resource availability and bac-terivory on leucine incorporation in different groups of freshwater bacterioplankton, assessed using microautoradiography. Aquat Microb Ecol 45:277–289

Hsu S-C, Liu SC, Kao S-J, Jeng W-L, Huang Y-T, Tseng C-M, Tsai F, Tu J-Y, Yang Y (2007) Water-soluble species in the marine aerosol from the northern South China Sea: high chloride depletion related to air pollution. J Geophys Res 112, D19304. doi:10.1029/2007jd008844

Huang K, Zhuang GS, Li JA, Wang QZ, Sun YL, Lin YF, Fu JS (2010) Mixing of Asian dust with pollution aerosol and the transformation of aerosol components during the dust storm over China in spring 2007. J Geophys Res Atmos 115, D00k13, doi:10.1029/2009jd013145

Hultin KAH, Krejci R, Pinhassi J, Gomez-Consarnau L, Mårtensson EM, Hagström Å, Nilsson ED (2011) Aerosol and bacterial emissions from Baltic seawater. Atmos Res 99:1–14

Hunter KA, Boyd PW (2007) Iron-binding ligands and their role in the ocean biogeochemistry of iron. Environ Chem 4:221–232. doi:10.1071/EN07012

Hunter K, Liss P, Surapipith V, Dentener F, Duce R, Kanakidou M, Kubilay N, Mahowald N, Okin G, Sarin M, Uematsu M, Zhu T (2011) Impacts of anthropogenic SOx, NOx and NH3 on acidification of coastal waters and shipping lanes. Geophys Res Lett 38, doi:10.1029/2011GL047720

IPCC (2007) Climate change 2007: the physical science basis. Contribution of working group I to the fourth assessment report of the intergovernmental panel on climate change. In: Solomon S, Qin D, Manning M, Chen Z, Marquis M, Averyt KB, Tignor M, Miller HL (eds). Cambridge University Press, Cambridge, UK and New York, NY, p 996

Irshad R, Grainger RG, Peters DM, McPheat RA, Smith KM, Thomas G (2009) Laboratory measurements of the optical properties of sea salt aerosol. Atmos Chem Phys 9:221–230

Iversen JB, White DR (1982) Saltation threshold on Earth, Mars and Venus. Sedimentology 29:111–119

Iwasaka Y, Yamato M, Imasu R, Ono A (1988) Transport of Asian dust (KOSA) particles; importance of weak KOSA events on the geochemical cycle of soil particles. Tellas B Chem Phys Meterol 40B:494–503

Jaeglé L, Quinn PK, Bates TS, Alexander B, Lin J-T (2011) Global distribution of sea salt aerosols: new constraints from in situ and remote sensing observations. Atmos Chem Phys 11:3137–3157

Jaenicke R (2005) Abundance of cellular material and proteins in the atmosphere. Science 308(5718):73. doi:10.1126/science.1106335

Jahnke RA (1996) The global ocean flux of particulate organic carbon: areal distribution and magnitude. Global Biogeochem Cycles 10:71–88

Jickells T, Spokes L (2001) Atmospheric iron inputs to the oceans. In: Turner DR, Hunteger K (eds) Biogeochemistry of iron in seawater. Wiley, Chichester, pp 85–121

Jickells TD, An ZS, Andersen KK, Baker AR, Bergametti G, Brooks N, Cao JJ, Boyd PW, Duce RA, Hunter KA, Kawahata H, Kubilay N, la Roche J, Liss PS, Mahowald N, Prospero JM, Ridgwell AJ, Tegen I, Torres R (2005) Global iron connections between desert dust, ocean biogeochemistry, and climate. Science 308:67–71

Jimenez JL, Bahreini R, Cocker DR III, Zhuang H, Varutbangkul V, Flagan RC, Seinfeld JH, O'Dowd CD, Hoffmann T (2003) New particle formation from photooxidation of diiodomethane (CH2 I2). J Geophys Res 108 (D10):4318. doi:10.1029/2002JD002452

Jimenez JL, Canagaratna MR, Donahue NM, Prevot ASH, Zhang Q, Kroll JH, DeCarlo PF, Allan JD, Coe H, Ng NL, Aiken AC, Docherty KS, Ulbrich IM, Grieshop AP, Robinson AL, Duplissy J, Smith JD, Wilson KR, Lanz VA, Hueglin C, Sun YL, Tian J, Laaksonen A, Raatikainen T, Rautiainen J, Vaattovaara P, Ehn M, Kulmala M, Tomlinson JM, Collins DR, Cubison MJE, Dunlea J, Huffman JA, Onasch TB, Alfarra MR, Williams PI, Bower K, Kondo Y, Schneider J, Drewnick F, Borrmann S, Weimer S, Demerjian K, Salcedo D, Cottrell L, Griffin R, Takami A, Miyoshi T, Hatakeyama S, Shimono A, Sun JY, Zhang YM, Dzepina K, Kimmel JR, Sueper D, Jayne JT, Herndon SC, Trimborn AM, Williams LR, Wood EC, Middlebrook AM, Kolb CE, Baltensperger U, Worsnop DR (2009) Evolution of organic aerosols in the atmosphere. Science 326:1525–1529

Johansen AM, Siefert RL, Hoffmann MR (1999) Chemical characterization of ambient aerosol collected during the southwest monsoon and intermonsoon seasons over the Arabian Sea: anions and cations. J Geophys Res 104:26325–26347

Jones MT, Gislason SR (2008) Rapid releases of metal salts and nutrients following the deposition of volcanic ash into aqueous environments. Geochim Cosmochim Acta 72:3661–3680

Journet E, Desboeufs KV, Caquineau S, Colin J-L (2008) Mineralogy as a critical factor of dust iron solubility. Geophys Res Lett 35, L07805. doi:10.1029/2007GL031589

Jurado E, Dachs J, Duarte CM, Simó R (2008) Atmospheric deposition of organic and black carbon to the global oceans. Atmos Environ 42:7931–7939

Kanakidou M, Seinfeld JH, Pandis SN, Barnes I, Dentener FJ, Facchini R, van Dingenen R, Ervens B, Nenes A, Nielsen CJ, Swietlicki E, Putaud JP, Balkanski Y, Fuzzi S, Horth J, Moortgat GK, Winterhalter R, Myhre CEL, Tsigaridis K, Vignati E, Stephanou EG, Wilson J (2004) Organic aerosol and global climate modelling: a review. Atmos Chem Phys 5:1053–1123

Kandler K, Benker N, Bundke U, Cuevas E, Ebert M, Knippertz P, Rodríguez S, Schütz L, Weinbruch S (2007) Chemical composition and complex refractive index of Saharan mineral dust at Izaña, Tenerife (Spain) derived by electron microscopy. Atmos Environ 41:8058–8074

Karyampudi VM, Palm SP, Reagen JA, Fang H, Grant WB, Hoff RM, Moulin C, Pierce HF, Torres O, Browell EV, Melfi SH (1999) Validation of the Saharan dust plume conceptual model using lidar, Meteosat, and ECMWF data. Bull Am Meteorol Soc 80:1045–1075

Kaufman YJ, Boucher O, Tanré D, Chin M, Remer LA, Takemura T (2005) Aerosol anthropogenic component estimated from satellite data. Geophys Res Lett 32, doi:10.1029/2005GL023125

Kawamura K, Sakaguchi F (1999) Molecular distributions of water soluble dicarboxylic acids in marine aerosols over the Pacific Ocean including tropics. J Geophys Res 104:3501–3509

Kawamura K, Kasukabe H, Barrie LA (1996a) Source and reaction pathways ofdicarboxylic acids, ketoacids and dicarbonyls in arctic aerosols at polar sunrise. Atmos Environ 30:1709–1722

Kawamura K, Semèrè R, Imai Y, Fujii Y, Hayashi M (1996b) Water soluble dicarboxylic acids and related compounds in Antarctic aerosols. J Geophys Res 101:18721–18728

Kawamura K, Kasukabe H, Barrie LA (2010) Secondary formation of water-soluble organic acids and a-dicarbonyls and their contributions to total carbon and water-soluble organic carbon: photochemical aging of organic aerosols in the Arctic spring. J Geophys Res 115, D21306. doi:10.1029/2010JD014299

Keene WC, Maring H, Maben JR, Kieber DJ, Pszenny AAP, Dahl EE, Izaguirre MA, Davis AJ, Long MS, Zhou X, Smoydzin L, Sander R (2007) Chemical and physical characteristics of nascent aerosols produced by bursting bubbles at a model air-sea interface. J Geophys Res 112, D21202. doi:10.1029/2007JD008464

Kelly JT, Wexler AS (2005) Thermodynamics of carbonates and hydrates related to heterogeneous reactions involving mineral aerosol. J Geophys Res 110, D11201. doi:10.1029/2004jd005583

Kerminen VM, Hillamo RE, Wexler AS (1998) Model Simulations on the variability of particulate MSA to non-sea-salt sulfate ratio in the marine environment. J Atmos Chem 30:345–370

Kim B-G, Park S-U (2001) Transport and evolution of a wintertime yellow sand observed in Korea. Atmos Environ 35:3191–3201

King MD, Kaufman YJ, Tanré D, Nakajima T (1999) Remote sensing of tropospheric aerosols from space: past, present, and future. Bull Am Meteorol Soc 80:2229–2259

Kinne S, Schulz M, Textor C et al (2006) An AeroCom initial assessment optical properties in aerosol component modules of global models. Atmos Chem Phys 6:1815–1834

Kirkby J, Curtius J, Almeida J et al (2011) Role of sulphuric acid, ammonia and galactic cosmic rays in atmospheric aerosol nucleation. Nature 476:429–433

Klein C, Dolan JR, Rassoulzadegan F (1997) Experimental examination of the effects of rainwater on micro-bial communities in the surface of the NW Mediterranean Sea. Mar Ecol Prog Ser 158:41–50

Kloster S, Mahowald NM, Randerson JT, Thornton PE, Hoffman FM, Levis S, Lawrence PJ, Feddema JJ, Oleson KW, Lawrence DM (2010) Fire dynamics during the 20th century simulated by the Community Land Model. Biogeosciences 7:1877–1902

Koçak M, Mihalopoulos N, Kubilay N (2007) Chemical composition of the fine and coarse fraction of aerosols in the northeastern Mediterranean. Atmos Environ 41:7351–7368

Kohfeld K, LeQuere C, Harrison S, Anderson R (2005) Role of marine biology in glacial-integlacial CO2 cycles. Science 308:74–78

Kollias P, Fairall CW, Zuidema P, Tomlinson J, Wick GA (2004) Observations of marine stratocumulus in SE Pacific during the PACS 2003 cruise. Geophys Res Lett 31, L22110. doi:10.1029/2004GL020751

Korhonen H, Carslaw KS, Forster PM, Mikkonen S, Gordon ND, Kokkola H (2010a) Aerosol climate feedback due to decadal increases in Southern Hemisphere wind speeds. Geophys Res Lett 37, L02805. doi:10.1029/2009GL041320

Korhonen H, Carslaw KS, Romakkaniemi S (2010b) Enhancement of marine cloud albedo via controlled sea spray injections: a global model study of the influence of emission rates, microphysics and transport. Atmos Chem Phys 10:4133–4143. doi:10.5194/acp-10-4133-2010

Kokhanovsky AA, de Leeuw G (eds) (2009) Satellite aerosol remote sensing over land. Springer-Praxis, Berlin, p 388. ISBN 978-3-540-69396-3

Krishnamurthy A, Moore JK, Mahowald N, Luo C, Doney SC, Lindsay K, Zender CS (2009) The impacts of increasing anthro- pogenic soluble iron and nitrogen deposition on ocean biogeochemistry. Global Biogeochem Cycles 23, GB3016. doi:10.1029/2008GB003440

Krishnamurthy A, Moore JK, Mahowald N, Luo C, Zender CS (2010) Impacts of atmospheric nutrient inputs on marine biogeochemistry. J Geophys Res 115, G01006. doi:10.1029/2009JG001115

Kritz MA, Le Roulley J-C, Danielsen EF (1990) The China Clipper—fast advective transport of radon-rich air from the Asian boundary layer to the upper troposphere near California. Tellus 42B:46–61

Krol M, Houweling S, Bregman B, van den Broek M, Segers A, van Velthoven P, Peters W, Dentener F, Bergamaschi P (2005) The two-way nested global chemistry-transport zoom model TM5: algorithm and applications. Atmos Chem Phys 5:417–432

Kroll JH, Donahue NM, Jimenez JL et al (2011) Carbon oxidation state as a metric for describing the chemistry of atmospheric organic aerosol. Nat Chem 3:133–139

Krueger BJ, Grassian VH, Cowin JP, Laskin A (2004) Heterogeneous chemistry of individual mineral dust particles from different dust source regions: the importance of particle mineralogy. Atmos Environ 38:6253–6261

Kulmala M, Suni T, Lehtinen KEJ, Dal Maso M, Boy M, Reissell A, Rannik U, Aalto P, Keronen P, Hakola H, Back JB, Hoffmann T, Vesala T, Hari P (2004) A new feedback mechanism linking forests, aerosols, and climate. Atmos Chem Phys 4:557–562

Kumar A, Sudheer AK, Sarin MM (2008) Chemical characteristics of aerosols in MABL of Bay of Bengal and Arabian Sea during spring inter-monsoon: a comparative study. Springer, Heidelberg, p 8, ALLEMAGNE

Laghdass M, Blain S, Besseling M, Catala P, Guieu C, Obernosterer I (2011) Impact of Saharan dust deposition on the bacterial diversity and activity in the NW Mediterranean Sea. Aquat Microb Ecol 62:201–213

Lam P, Bishop J (2008) The continental margin is a key sources of iron to the North Pacific Ocean. Geophys Res Lett 35, doi:10.1029/2008GL033294

Lamarque J-F, Bond TC, Eyring V, Granier C, Heil A, Klimont Z, Lee D, Liousse C, Mieville A, Owen B, Schultz MG, Shindell D, Smith SJ, Stehfest E, Van Aardenne J, Cooper OR, Kainuma M, Mahowald N, McConnell JR, Naik V, Riahi K, van Vuuren DP (2010) Historical (1850–2000) gridded anthropogenic and biomass burning emissions of reactive gases and aerosols: methodology and application. Atmos Chem Phys 10:7017–7039

Lambe AT, Miracolo MA, Hennigan CJ, Robinson AL, Donahue NM (2009) Effective rate constants and uptake coefficients for the reactions of organic molecular markers (n-Alkanes, Hopanes, and Steranes) in motor oil and diesel primary organic aerosols with hydroxyl radicals. Environ Sci Technol 43:8794–8800

Landing WM, Paytan A (2010) Aerosol chemistry and impacts on the ocean. Mar Chem 120:1–3

Langmann BC, Scannell C, O'Dowd CD (2008) Organic matter contribution to marine aerosols and cloud condensation nuclei. Atmos Environ. doi:10.1016/j.atmosenv.2008.09.002

Langmann B, Zaksek K, Hort M, Duggen S (2010) Volcanic ash as fertiliser for the surface ocean. Atmos Chem Phys 10:3891–3899

Lapina K, Heald CL, Spracklen DV, Arnold SR, Allan JD, Coe H, McFiggans G, Zorn SR, Drewnick F, Bates TS, Hawkins LN, Russell LM, Smirnov A, O'Dowd C, Hind AJ (2011) Investigating organic aerosol loading in the remote marine environment. Atmos Chem Phys 11:8847–8860

Laskin A, Wietsma TW, Krueger BJ, Grassian VH (2005) Heterogeneous chemistry of individual mineral dust particles with nitric acid: a combined CCSEM/EDX, ESEM, and ICP-MS study. J Geophys Res 110, doi:10.1029/2004jd005206

Latham J (2002) Amelioration of global warming by controlled enhancement of the albedo and longevity of low-level maritime clouds. Atmos Sci. doi:10.1006/Asle.2002.0048

Latham J, Smith MH (1990) Effect on global warming of wind-dependent aerosol generation at the ocean surface. Nature 347:372–373

Latham J (2002) Amelioration of global warming by controlled enhancement of the albedo and longevity of low-level maritime clouds. Atmos Sci. doi:10.1006/Asle.2002.0048

Lathière J, Hauglustaine DA, De Noblet-Ducoudré N (2005) Past and future changes in biogenic volatile organic compound emissions simulated with a global dynamic vegetation model. Geophys Res Lett 32, L20818. doi:10.1029/2005GL024164

Lathière J, Hauglustaine DA, Friend A, De Noblet-Ducoudré N, Viovy N, Folberth G (2006) Impact of climate variability and land use changes on global biogenic volatile organic compound emissions. Atmos Chem Phys 6:2129–2146

Law CS, Woodward EMS, Ellwood MJ, Marriner A, Bury SJ, Safic KA (2011) Response of surface nutrient inventories and nitrogen fixation to a tropical cyclone in the southwest Pacific. Limnol Oceanogr 56:1372–1385

Law CS, Brévière E, de Leeuw G, Garçon V, Guieu C, Kieber D, Kontradowitz S, Paulmier A, Quinn P, Saltzman E, Stefels J, von Glasow R (2013) Evolving research directions in Surface Ocean-Lower Atmosphere (SOLAS) Science. Environ Chem 10:1–16, http://dx.doi.org/10.1071/EN12159

Leaitch WR, Macdonald AM, Anlauf KG, Liu PSK, Toom-Sauntry D, Li SM, Liggio J, Hayden K, Wasey MA, Russell LM, Takahama S, Liu S, van Donkelaar A, Duck T, Martin RV, Zhang Q, Sun Y, McKendry I, Shantz NC, Cubison M (2009) Evidence for Asian dust effects from aerosol plume measurements during INTEX-B 2006 near Whistler, BC. Atmos Chem Phys 9:3523–3546

Lee JD, McFiggans G, Allan JD, Baker AR, Ball SM, Benton AK, Carpenter LJ, Commane R, Finley BD, Evans M, Fuentes E, Furneaux K, Goddard A, Good N, Hamilton JF, Heard DE, Herrmann H, Hollingsworth A, Hopkins JR, Ingham T, Irwin M, Jones CE, Jones RL, Keene WC, Lawler MJ, Lehmann S, Lewis AC, Long MS, Mahajan A, Methven J, Moller SJ, Müller K, Müller T, Niedermeier N, O'Doherty S, Oetjen H, Plane JMC, Pszenny AAP, Read KA, Saiz-Lopez A, Saltzman ES, Sander R, von Glasow R, Whalley L, Wiedensohler A, Young D (2010) Reactive Halogens in the Marine Boundary Layer (RHaMBLe): the tropical North Atlantic experiments. Atmos Chem Phys 10:1031–1055. doi:10.5194/acp-10-1031-2010

Leck C, Bigg EK (2005) Source and evolution of the marine aerosol—a new perspective. Geophys Res Lett 32, L19803. doi:10.1029/2005GL023651

Lefevre N, Watson AJ (1999) Modeling the geochemical cycle of iron in the oceans and its impact on atmospheric CO2 concentrations. Global Biogeochem Cycles 13:727–736

Lelieveld J, Butler TM, Crowley JN, Dillon TJ, Fischer H, Ganzeveld L, Harder H, Lawrence MG, Martinez M, Taraborrelli D, Williams J (2008) Atmospheric oxidation capacity sustained by a tropical forest. Nature 452:737–740

Lesworth T, Baker AR, Jickells T (2010) Aerosol organic nitrogen over the remote Atlantic Ocean. Atmos Environ 44:1887–1893

Levin Z, Ganor E (1996) The effect of desert particles on cloud and rain formation in the eastern Mediterranean. In: Guerzoni S, Chester R (eds) The impact of desert dust across the Mediterranean. Kluwer, Dordrecht, pp 77–86

Lewis ER, Schwartz SE (2004) Sea salt aerosol production: mechanisms, methods, measurements and models—a critical review. American Geophysical Union, Washington, DC, p 413

Li WJ, Shao LY (2010) Mixing and water-soluble characteristics of particulate organic compounds in individual urban aerosol particles. J Geophys Res Atmos 115, D02301. doi:10.1029/2009jd012575

Lin II, Hu C, Li YH, Ho TY, Fischer TP, Wong GTF, Wu J, Huang CW, Chu DA, Ko DS, Chen JP (2011) Fertilization potential of volcanic dust in the low-nutrient low-chlorophyll western North Pacific subtropical gyre: Satellite evidence and laboratory study. Global Biogeochem Cycles 25, GB1006. doi:10.1029/2009GB003758

Lippmann M (2007) Health effects of airborne particulate matter. N Engl J Med 357:2395–2397

Liu X, Millero FJ (2002) The solubility of iron in seawater. Mar Chem 77:43–54

Liu X, Zhu J, Van Espen P, Adams F, Xiao R, Dong S, Li Y (2005) Single particle characterization of spring and summer aerosols in Beijing: formation of composite sulfate of calcium and potassium. Atmos Environ 39:6909–6918

Loeb NG, Manalo-Smith N (2005) Top-of-atmosphere direct radiative effect of aerosols over global oceans from merged CERES and MODIS observations. J Climate 18:3506–3526

Lohmann U, Feichter J (2005) Global indirect aerosol effects: a review. Atmos Chem Phys 5:715–737

Love RG, Miller BG, Groat SK, Hagen S, Cowie HA, Johnston PP, Hutchison PA, Soutar CA (1997) Respiratory health effects of opencast coalmining: a cross sectional study of current workers. Occup Environ Med 54(6):416–423

Loÿe-Pilot M-D, Martin J-M (1996) Saharan dust input to the western Mediterranean: an eleven years record in Corsica. In: Guerzoni S, Chester R (eds) The impact of desert dust across the Mediterranean. Kluwer, Dordrecht, pp 191–199

Loÿe-Pilot MM, Martin J-M, Morelli J (1986) Influence of Saharan dust on the rain acidity and atmospheric input to the Mediterranean Sea. Nature 321:427–428

Luo C, Mahowald N, Meskhidze N, Chen Y, Siefert R, Baker A, Johansen A (2006) Estimation of iron solubility from observations and a global aerosol model. J Geophys Res 110, D23307, doi:10.1029/2005JD006059, http://www.agu.org/journals/jd/jd0523/2005JD006059/

Luo C, Mahowald N, Bond T, Chuang PY, Artaxo P, Siefert R, Chen Y, Schauer J (2008) Combustion iron distribution and deposition. Global Biogeochem Cycles 22, doi:10.1029/2007GB002964

Ma C-J, Tohno S, Kasahara M (2005) A case study of the size-resolved individual particles collected at a ground-based site on the west coast of Japan during an Asian dust storm event. Atmos Environ 39:739–747

Ma C-J, Choi K-C (2007) A combination of bulk and single particle analyses for Asian dust, water. Air Soil Pollut 183:3–13. doi:10.1007/s11270-006-9302-z

Mahowald NM, Luo C (2003) A less dusty future? Geophys Res Lett 30, doi:1910.1029/2003GRL017880

Mahowald N, Baker A, Bergametti G, Brooks N, Duce R, Jickells T, Kubilay N, Prospero J, Tegen I (2005) The atmospheric global dust cycle and iron inputs to the ocean. Global Biogeochem Cycles 19, GB4025, doi:4010.1029/2004GB002402

Mahowald NM, Muhs DR, Levis S, Rasch PJ, Yoshioka M, Zender CS, Luo C (2006) Change in atmospheric mineral aerosols in response to climate: last glacial period, preindustrial, modern, and doubled carbon dioxide climates. J Geophys Res 111, D10202. doi:10.1029/2005JD006653

Mahowald N (2007) Anthropocene changes in desert area: sensitivity to climate model predictions. Geophys Res Lett 34, L18817

Mahowald N, Jickells TD, Baker AR, Artaxo P, Benitez-Nelson CR, Bergametti G, Bond TC, Chen Y, Cohen DD, Herut B, Kubilay N, Losno R, Luo C, Maenhaut W, McGee KA, Okin GS, Siefert RL, Tsukuda S (2008) The global distribution of atmospheric phosphorus deposition and anthropogenic impacts. Global Biogeochem Cycles 22, doi:10.1029/2008GB003240

Mahowald N, Engelstaedter S, Luo C, Sealy A, Artaxo P, Benitez-Nelson C, Bonnet S, Chen Y, Chuang PY, Cohen DD, Dulac F, Herut B, Johansen AM, Kubilay N, Losno R, Maenhaut W, Paytan A, Prospero JM, Shank LM, Siefert RL (2009) Atmospheric Iron deposition: global distribution, variability and human perturbations. Ann Rev Mar Sci 1:245–278

Mahowald NM, Kloster S, Engelstaedter S, Moore JK, Mukhopadhyay S, McConnell JR, Albani S, Doney SC, Bhattacharya A, Curran MAJ, Flanner MG, Hoffman FM, Lawrence DM, Lindsay K, Mayewski PA, Neff J, Rothenberg D, Thomas E, Thornton PE, Zender CS (2010) Observed 20th century desert dust variability: impact on climate and biogeochemistry. Atmos Chem Phys 10:10875–10893

Mahowald NM, Lindsay K, Rothenberg D, Doney SC, Moore JK, Thornton P, Randerson JT, Jones CD (2011) Desert dust and anthropogenic aerosol interactions in the Community Climate System Model coupled-carbon-climate model. Biogeosciences 8:387–414

Maier-Reimer E (1993) Geochemical cycles in an ocean general circulation model. Preindustrial tracer distributions. Global Biogeochem Cycles 7(3):645–677. doi:10.1029/93GB01355

Mäkelä JM, Hoffmann T, Holzke C, Väkevä M, Suni T, Mattila T, Aalto PP, Tapper U, Kauppinen EI, O'Dowd CD (2002) Biogenic iodine emissions and identification of end-products in coastal ultrafine particles during nucleation bursts. J Geophys Res 107, doi:10.1029/2001JD000580

Makkonen R, Asmi A, Kerminen V-M, Boy M, Arneth A, Hari P, Kulmala M (2012) Air pollution control and decreasing new particle formation lead to strong climate warming. Atmos Chem Phys 12:1515–1524. doi:10.5194/acp-12-1515-2012

Mandalakis M, Stephanou EG (2002) Study of atmospheric PCB concentrations over the eastern Mediterranean Sea. J Geophys Res 107:4716. doi:10.1029/2001JD001566

Mandavilli A (2006) Health agency backs use of DDT against malaria. Nature 443(7109):250–251

Marañón E et al (2010) Degree of oligotrophy controls the response of microbial plankton to Saharan dust. Limnol Oceanogr 55:2339–2352

Maring H, Savoie DL, Izaguirre MA, Custals L, Reid JS (2003) Mineral dust aerosol size distribution change during atmospheric transport. J Geophys Res 108:8592. doi:10.1029/2002JD002536

Markaki Z, Loÿe-Pilot M-D, Violaki K, Mihalopoulos N (2010) Variability of atmospheric deposition of dissolved nitrogen and phosphorus in the Mediterranean and possible link to the anomalous seawater N/P ratio. Mar Chem 120:187–194

Mårtensson EM, Nilsson ED, de Leeuw G, Cohen LH, Hansson H-C (2003) Laboratory simulations and parameterization of the primary marine aerosol production. J Geophys Res 108:4297. doi:10.1029/2002JD002263

Marticorena B, Bergametti G, Aumont B, Callot Y, N'Doumé C, Legrand M (1997) Modeling the atmospheric dust cycle 2 Simulation of Saharan dust sources. J Geophys Res 102:4387–4404

Martin JH (1990) Glacial-interglacial CO2 change: the iron hypothesis. Paleoceanography 5:1–13

Martin JH, Fitzwater SE (1988) Iron deficiency limits phytoplankton growth in the North-East Pacific subarctic. Nature 331:341–343

Martin J, Gordon RM, Fitzwater SE (1991) The case for iron. Limnol Oceanogr 36:1793–1802

Maskey S, Geng H, Song YC, Hwang H, Yoon YJ, Ahn KH, Ro CU (2011) Single-particle characterization of summertime Antarctic aerosols collected at King George Island using quantitative energy-dispersive electron probe X-ray microanalysis and attenuated total reflection Fourier transform-infrared imaging techniques. Environ Sci Technol 45:6275–6282

Mason BJ (1957) The physics of clouds. Clarendon Press, Oxford, 671

Massana R, Pedrós-Alió C, Casamayor EO, Gasol JM (2001) Changes in marine bacterioplankton phylogenetic composition during incubations designed to measure biogeochemically significant parameters. Limnol Oceanogr 46:1181–1188

Mastin LG, Guffanti M, Servranckx R, Webley P, Barsotti S, Dean K, Durant A, Ewert JW, Neri A, Rose WI, Schneider D, Siebert L, Stunder B, Swanson G, Tupper A, Volentik A, Waythomas CF (2009) A multidisciplinary effort to assign realistic source parameters to models of volcanic ash-cloud transport and dispersion during eruptions. J Volcanol Geotherm Res 186:10–21

Matsuki A, Iwasaka Y, Shi G, Zhang D, Trochkine D, Yamada M, Kim Y-S, Chen B, Nagatani T, Miyazawa T, Nagatani M, Nakata H (2005) Morphological and chemical modification of mineral dust: observational insight into the heterogeneous uptake of acidic gases. Geophys Res Lett 32, L22806. doi:10.1029/2005gl024176

Matsuki A, Quennehen B, Schwarzenboeck A, Crumeyrolle S, Venzac H, Laj P, Gomes L (2010a) Temporal and vertical variations of aerosol physical and chemical properties over West Africa: AMMA aircraft campaign in summer 2006. Atmos Chem Phys 10:8437–8451

Matsuki A, Schwarzenboeck A, Venzac H, Laj P, Crumeyrolle S, Gomes L (2010b) Cloud processing of mineral dust: direct comparison of cloud residual and clear sky particles during

AMMA aircraft campaign in summer 2006. Atmos Chem Phys 10:1057–1069

Matsumoto K, Uyama Y, Hayano T, Uematsu M (2004) Transport and chemical transformation of anthropogenic and mineral aerosol in the marine boundary layer over the western North Pacific Ocean. J Geophys Res 109, D21206. doi:10.1029/2004jd004696

Maxwell-Meier K, Weber R, Song C, Orsini D, Ma Y, Carmichael GR, Streets DG, 2004 (2004) Inorganic composition of fine particles in mixed mineral dust– pollution plumes observed from airborne measurements during ACE-Asia. J Geophys Res 109, D19S07. doi:10.1029/2003jd004464

McFiggans G, Coe H, Burgess R et al (2004) Direct evidence for coastal iodine particles from Laminaria macroalgae – linkage to emissions of molecular iodine. Atmos Chem Phys 4:701–713

McKendry IG, Macdonald AM, Leaitch WR, van Donkelaar A, Zhang Q, Duck T, Martin RV (2008) Trans-Pacific dust events observed at Whistler, British Columbia during INTEX-B. Atmos Chem Phys 8:6297–6307

Meskhidze N, Chameides W, Nenes A (2005) Dust and pollution: a recipe for enhanced ocean fertiliza-tion? J Geophys Res 110, doi:10.1029/2004JD005082

Meskhidze NJX, Xu J, Gantt B, Zhang Y, Nenes A, Ghan SJ, Liu X, Easter R, Zaveri R (2011) Global distribution and climate forcing of marine organic aerosol: 1. Model improvements and evaluation. Atmos Chem Phys 11:11689–11705

Middlebrook AM, Murphy DM, Thomson DS (1998) Observation of organic material in individual particles at Cape Grim during the First Aerosol Characterization Experiment (ACE 1). J Geophys Res 103:16475–16483

Mie G (1908) Beitrge zur Optik trber Medien, speziell kolloidaler Metallsungen. Ann Phys Leipsig 25:377–445

Migon C, Sandroni V (1999) Phosphorus in rainwater: partitioning, inputs and impact on the surface coastal ocean. Limnol Oceanogr 44:1160–1165

Mills MM, Ridame C, Davey M, La Roche J, Geider RJ (2004) Iron and phosphorus co-limit nitrogen fixation in the eastern tropical North Atlantic. Nature 429:292–294

Mills MM et al (2008) Nitrogen and phosphorus co-limitation of bacterial productivity and growth in the oligotrophic subtropical North Atlantic. Limnol Oceanogr 53:824–834

Mishchenko MI, Lacis AA, Carlson BE, Travis LD (1995) Nonsphericity of dust-like tropospheric aerosols: Implications for aerosol remote sensing and climate modelling. Geophys Res Lett 22:1077–1080

Mochida M, Kitamori Y, Kawamura K, Nojiri Y, Suzuki K (2002) Fatty acids in the marine atmosphere: factors governing their concentrations and evaluation of organic films on sea salt particles. J Geophys Res 107:4325. doi:10.1029/2001JD001278

Mochida M, Umemoto N, Kawamura K, Lim HJ, Turpin BJ (2007) Bimodal size distributions of various organic acids and fatty acids in the marine atmosphere: influence of anthropogenic aerosols, Asian dusts, and sea spray off the coast of East Asia. J Geophys Res Atmos 112(13), D15209. doi:10.1029/2006jd007773

Monahan EC, O'Muircheartaigh IG (1986) Whitecaps and the passive remote sensing of the ocean surface. Int J Remote Sens 7:627–642

Monahan EC, Spiel DE, Davidson KL (1986) A model of marine aerosol generation via whitecaps and wave disruption. In: Monahan EC, MacNiochaill G (eds) Oceanic whitecaps. D. Reidel, Norwell, pp 167–193

Moore JK, Braucher O (2008) Sedimentary and mineral dust sources of dissolved iron to the world ocean. Biogeosciences 5:631–656

Moore JK, Doney S, Lindsay K, Mahowald N, Michaels A (2006) Nitrogen fixation amplifies the ocean biogeochemical response to decadal timesclae variations in mineral dust deposition. Tellus 58B:560–572

Moore CM et al (2008) Relative influence of nitrogen and phosphorus availability on phytoplankton physiology and productivity in the oligotrophic sub-tropical North Atlantic Ocean. Limnol Oceanogr 53:291–305

Moore CM, Mills MM, Arrigo KR, Berman-Frank I, Bopp L, Boyd PW, Galbraith ED, Geider RJ, Guieu C, Jaccard SL, Jickells TD, La Roche J, Lenton TM, Mahowald NM, Marañón E, Marinov I, Moore JK, Nakatsuka T, Oschlies A, Saito MA, Thingstad TF, Tsuda A, Ulloa O (2013) Processes and patterns of oceanic nutrient limitation. Nat Geosci. doi:10.1038/ngeo1765

Morgan WT, Allan JD, Bower KN, Highwood EJ, Liu D, McMeeking GR, Northway MJ, Williams PI, Krejci R, Coe H (2010) Airborne measurements of the spatial distribution of aerosol chemical composition across Europe and evolution of the organic fraction. Atmos Chem Phys 10:4065–4083

Mori I, Nishikawa M, Iwasaka Y (1998) Chemical reaction during the coagulation of ammonium sulphate and mineral particles in the atmosphere. Sci Total Environ 224:87–91

Morris RM et al (2002) SAR11 clade dominates ocean surface bacterioplankton communities. Nature 420:806–810

Moulin C, Lambert CE, Dayan U, Dulac F (1997) Control of atmospheric export of dust from North Africa by the North Atlantic Oscillation. Nature 387:691–694

Moulin C, Gordon HR, Banzon VF, Evans RH (2001) Assessment of Saharan dust absorption in the visible from SeaWiFs imagery. J Geophys Res 106:18239–18249

Mouri H, Okada K (1993) Shattering and modification of sea-salt particles in the marine atmosphere. Geophys Res Lett 20:49–52

Müller C, Iinuma Y, Karstensen J et al (2009) Seasonal variation of aliphatic amines in marine sub-micrometer particles at the Cape Verde Islands. Atmos Chem Phys 9:9587–9597

Murphy DM, Cziczo DJ, Froyd KD, Hudson PK, Matthew BM, Middlebrook AM, Peltier RE, Sullivan A, Thomson DS, Weber RJ (2006) Single-particle mass spectrometry of tropospheric aerosol particles. J Geophys Res 111, D23S32. doi:10.1029/2006jd007340

Murphy SM, Sorooshian A, Kroll JH et al (2007) Secondary aerosol formation from atmospheric reactions of aliphatic amines. Atmos Chem Phys 7:2313–2337

Myriokefalitakis S, Vignati E, Tsigaridis K, Papadimas C, Sciare J, Mihalopoulos N, Facchini MC, Rinaldi M, Dentener FJ, Ceburnis D, Hatzianastassiou N, O'Dowd CD, van Weele M, Kanakidou M (2010) Global modelling of the oceanic source of organic aerosols. Adv Meteorol 2010:939171. doi:10.1155/2010/939171

Nenes A, Krom M, Mihalopoulos N, Van Cappellen P, Shi Z, Bougiatioti A, Zarmpas P, Herubt B (2011) Atmospheric acidification of mineral aerosols: a source of bioavailable phosphorus for the oceans. Atmos Chem Phys 11:6265–6272

Newhall CG, Self S (1982) The volcanic explosivity index (VEI): an estimate of explosive magnitude for historical volcanism. J Geophys Res 87:1231–1238

Ng NL, Canagaratna MR, Zhang Q et al (2010) Organic aerosol components observed in Northern Hemispheric datasets from aerosol mass spectrometry. Atmos Chem Phys 10:4625–4641. doi:10.5194/acp-10-4625-2010

Ng NL, Canagaratna MR, Jimenez JL, Chhabra PS, Seinfeld JH, Worsnop DR (2011) Changes in organic aerosol composition with aging inferred from aerosol mass spectra. Atmos Chem Phys 11:6465–6474

Niemi JV, Tervahattu H, Virkkula A, Hillamo R, Teinilä K, Koponen IK, Kulmala M (2005) Continental impact on marine boundary layer coarse particles over the Atlantic Ocean between Europe and Antarctica. Atmos Res 75:301–321

Niimura N, Okada K, Fan X-B, Kenji K, Kimio A, Shi G-Y, Takahashi S (1998) Formation of Asian dust-storm particles mixed internally with sea salt in the atmosphere. J Meteorol Soc Japan 76:275–288

Niinemets U, Tenhunen JD, Harley PC, Steinbrecher R (1999) A model of isoprene emission based on energetic requirements for isoprene synthesis and leaf photosynthetic properties for Liquidambar and Quercus. Plant Cell Environ 22:1319–1335

Niinemets Ü, Arneth A, Kuhn U, Monson RK, Peñuelas J, Staudt M (2010a) The emission factor of volatile isoprenoids: stress, acclimation, and developmental responses. Biogeosciences 7:2203–2223

Niinemets Ü, Monson RK, Arneth A, Ciccioli P, Kesselmeier J, Kuhn U, Noe SM, Penuelas J, Staudt M (2010b) The emission factor of volatile isoprenoids: caveats, model algorithms, response shapes and scaling. Biogeosciences 7:1809–1832

Niinemets Ü, Kuhn U, Harley PC, Staudt M, Arneth A, Cescatti A, Ciccioli P, Copolovici L, Geron C, Guenther A, Kesselmeier J, Lerdau MT, Monson RK, Peñuelas J (2011) Estimations of isoprenoid emission capacity from enclosure studies: measurements, data processing, quality and standardized measurement protocols. Biogeosciences 8:2209–2246

Nilsson ED, Mårtensson EM, Van Ekeren JS, de Leeuw G, Moerman M, O'Dowd C (2007) Primary marine aerosol emissions: size resolved eddy covariance measurements with estimates of the sea salt and organic carbon fractions. Atmos Chem Phys Discuss 7:13345–13400. doi:10.5194/acpd-7-13345-2007

Norris S, Brooks I, de Leeuw G, Smith MH, Moerman M, Lingard J (2008) Eddy covariance measurements of sea spray particles over the Atlantic Ocean. Atmos Chem Phys 8:555–563

Norris SJ, Brooks IM, Hill MK, Brooks BJ, Smith MH, Sproson DAJ (2012) Eddy covariance measurements of the sea spray aerosol flux over the open ocean. J Geophys Res 117, D07210. doi:10.1029/2011JD016549

Nozaki Y (1997) A fresh look at element distribution in the North Pacific. EOS, Am Geophys Union 78(21):221

O'Dowd CD, Smith MH (1993) Physicochemical properties of aerosols over the northeast Atlantic: evidence for wind-speed-related submicron sea-salt aerosol production. J Geophys Res 98, doi:10.1029/92JD02302

O'Dowd CD, Smith MH, Jennings SG (1993) Submicron aerosol, radon and soot carbon characteristics over the northeast Atlantic. J Geophys Res 98:1123–1135

O'Dowd CD, Geever M, Hill MK, Jennings SG, Smith MH (1998) New particle formation: spatial scales and nucleation rates in the coastal environment. Geophys Res Lett 25:1661–1664

O'Dowd C, Lowe JA, Smith MH (1999a) Observations and modelling of aerosol growth in marine stratocumulus – case study. Atmos Environ 33:3053–3062

O'Dowd CD, McFiggens G, Pirjola L, Creasey DJ, Hoell C, Smith MH, Allen B, Plane JMC, Heard DE, Lee JD, Pilling MJ, Kulmala M (1999b) On the photochemical production of new particles in the coastal boundary layer. Geophys Res Lett 26:1707–1710

O'Dowd CD, Lowe JA, Clegg N, Smith MH, Clegg SL (2000) Modeling heterogeneous sulphate production in maritime stratiform clouds. J Geophys Res-Atmos 105(D6):7143–7160

O'Dowd CD, Hämeri K, Mäkelä JM, Pirjola L, Kulmala M, Jennings SG, Berresheim H, Hansson H-C, de Leeuw G, Allen AG, Hewitt CN, Jackson A, Viisanen Y, Hoffmann T (2002) A dedicated study of new particle formation and fate in the coastal environment (PARFORCE): overview of objectives and initial achievements. J Geophys Res 107, doi:10.1029/2001000555

O'Dowd CD, Facchini MC, Cavalli F, Ceburnis D, Mircea M, Decesari S, Fuzzi S, Yoon YJ, Putaud J-P (2004) Biogenically driven organic contribution to marine aerosol. Nature 431:676–680

O'Dowd CD, Langmann B, Varghese S, Scannell C, Ceburnis D, Facchini MC (2008) A combined organic–inorganic sea-spray source function. Geophys Res Letts 35, L01801. doi:10.1029/2007GL030331

O'Dowd CD, Monahan C, Dall'Osto M (2010) On the occurrence of open ocean particle production and growth events. Geophys Res Lett 37, L19805. doi:10.1029/2010GL044679

O'Dowd CD, de Leeuw G (2007) Marine aerosol production: a review of the current knowledge. Philos Trans Royal Soc A: Math Phys Eng Sci 365:1753–1774

O'Dowd CD, Davison B, Lowe JA, Smith MH, Harrison RM, Hewitt CN (1997) Biogenic sulphur emissions and inferred sulphate CCN concentrations in and around Antarctica. J Geophys Res 102:12839–12854

Okada K, Kai K (2004) Atmospheric mineral particles collected at Qira in the Taklamakan Desert, China. Atmos Environ 38:6927–6935

Okada K, Naruse H, Tanaka T, Nemoto O, Iwasaka Y, Wu P-M, Ono A, Duce RA, Uematsu M, Merrill JT, Arao K (1990) X-ray spectrometry of individual Asian dust-storm particles over the Japanese islands and the North Pacific Ocean. Atmos Environ 24:1369–1378

Okada K, Qin Y, Kai K (2005) Elemental composition and mixing properties of atmospheric mineral particles collected in Hohhot, China. Atmos Res 73:45–67

Okin G, Baker A, Tegen I, Mahowald N, Dentener F, Duce R, Galloway J, Hunter K, Kanakidou M, Kubilay N, Prospero J,

Sarin M, Surpipith V, Uematsu M, Zhu T (2011) Impacts of atmospheric nutrient deposition on marine productivity: roles of nitrogen, phosphorus and iron. Global Biogeochem Cycles 25, doi:10.1029/2010GB003858

Olgun N, Duggen S, Croot PL, Delmelle P, Dietze H, Schacht U, Óskarsson N, Siebe C, Auer A, Garbe-Schönberg D (2011) Surface ocean iron fertilization: the role of airborne volcanic ash from subduction zone and hot spot volcanoes and related iron fluxes into the Pacific Ocean. Global Biogeochem Cycles 25, GB4001. doi:10.1029/2009GB003761

Omar AH, Winker DM, Vaughan MA, Hu Y, Trepte CR, Ferrare RA, Lee KP, Hostetler CA, Kittaka C, Rogers RR, Kuehn RE, Liu Z (2009) The CALIPSO automated aerosol classification and lidar ratio aelection algorithm. J Atmos Ocean Technol 26(10):1994–2014

Ooki A, Uematsu M (2005) Chemical interactions between mineral dust particles and acid gases during Asian dust events. J Geophys Res 110, D03201. doi:10.1029/2004jd004737

Oppo C, Bellandi S, Degli Innocenti N, Stortini AM, Loglio G, Schiavuta E, Cini R (1999) Surfactant component of marine organic matter as agents for biogeochemical fractionation of pollutants transport via marine aerosol. Mar Chem 63:235–253

Ovadnevaite J, Ceburnis D, Bialek J, Monahan C, Martucci G, Rinaldi M, Facchini MC, Berresheim H, Worsnop DR, O'Dowd C (2011a) Primary marine organic aerosol: a dichotomy of low hygroscopicity and high CCN activity. Geophys Res Lett 38, L21806. doi:10.1029/2011GL048869

Ovadnevaite J, O'Dowd C, Dall'Osto M, Ceburnis D, Worsnop DR, Berresheim H (2011) Detecting high contributions of primary organic matter to marine aerosol: a case study. Geophys Res Lett 38, L02807. doi:10.1029/2010GL046083

Pacifico F, Harrison SP, Jones CD, Arneth A, Sitch S, Weedon GP, Barkley MP, Palmer PI, Serça D, Potosnak M, Fu TM, Goldstein A, Bai J, Schurgers G (2011) Evaluation of a photosynthesis-based biogenic isoprene emission scheme in JULES and simulation of isoprene emissions under modern climate conditions. Atmos Chem Phys 11:4371–4389

Palmer PI, Jacob DJ, Fiore AM, Martin RV, Chance K, Kurosu TP (2003) Mapping isoprene emissions over North America using formaldehyde column observations from space. J Geophys Res 108(D6):4180. doi:10.1029/2002JD002153

Pandis SN, Russell LM, Seinfeld JH (1994) The relationship between DMS flux and CCN concentration in remote marine regions. J Geophys Res 99:16945–16957

Parekh P, Follows MJ, Boyle E (2004) Modeling the global ocean iron cycle. Global Biogeochem Cycles 18, GB1002. doi:10.1029/2003GB002061

Parekh P, Follows MJ, Boyle EA (2005) Decoupling of iron and phosphate in the global ocean. Global Biogeochem Cycles 19, GB2020. doi:10.1029/2004GB002280

Passow U, De la Rocha C (2006) Accumulation of mineral ballast on organic aggregates. Global Biogeochem Cycles 20, GB1013. doi:10.1029/2005GB002579

Paytan A, Mackey KRM, Chen Y, Limac ID, Doneyc SC, Mahowaldd N, Labiosae R, Postf AF (2009) Toxicity of atmospheric aerosols on marine phytoplankton. Proc Natl Acad Sci 106:4601–4605

Petit JR, Jouzel J, Raynaud D, Barkov NI, Barnola J-M, Basile I, Benders M, Chappellaz J, Davis M, Delayque G, Delmotte M, Kotlyakov VM, Legrand M, Lipenkov VY, Lorius C, Pépin L, Ritz C, Saltzman E, Stievenard M (1999) Climate and atmospheric history of the past 420,000 years from the Vostok ice core, Antarctica. Nature 399:429–436

Pirjola L, O'Dowd CD, Brooks IM, Kulmala M (2000) Can new particle formation occur in the clean marine boundary layer? J Geophys Res 105:26531–26546

Ploug H, Hvitfeld Iversen M, Fischer G (2008) Ballast, sinking velocity, and apparent diffusivity within marine snow and zooplankton fecal pellets: implications for substrate turnover by attached bacteria. Limnol Oceanogr 53:1878–1886

Polymenakou PN, Mandalakis M, Stephanou EG, Tselepides A (2008) Particle size distribution of airborne microorganisms and pathogens during an intense African dust event in the Eastern Mediterranean. Environ Heal Perspect 116(3):292–296

Pöschl U (2005) Atmospheric aerosols: composition, transformation, climate and health effects. Angew Chem Int Ed 44:7520–7540. doi:10.1002/anie.200501122

Pressley S, Lamb B, Westberg H, Flaherty J, Chen J, Vogel C (2005) Long-term isoprene flux measurements above a northern hardwood forest. J Geophys Res 110, D07301. doi:10.1029/2004JD005523

Prospero JM, Ginoux P, Torres O, Nicholson SE, Gill TE (2002) Environmental characterization of global sources of atmospheric soil dust identified with the Nimbus 7 Total Ozone Mapping Spectrometer (TOMS) absorbing aerosol product. Rev Geophys 40, doi:10.1029/2000RG000095

Pulido-Villena E, Wagener T, Guieu C (2008) Bacterial response to dust pulses in the western Mediterranean: implications for carbon cycling in the oligotrophic ocean. Global Biogeochem Cycles 22, GB1020. doi:10.1029/2007GB003091

Pulido-Villena E, Rerolle V, Guieu C (2010) Transient fertilizing effect of dust in P-deficient LNLC surface ocean. Geophys Res Lett 37, L01603. doi:10.1029/2009GL041415

Putaud J-P, Van Dingenen R, Dell'Acqua A, Raes F, Matta E, Decesari S, Facchini MC, Fuzzi S (2004) Size-segregated aerosol mass closure and chemical composition in Monte Cimone (I) during MINATROC. Atmos Chem Phys 4:889–902. doi:10.5194/acp-4-889-2004

Quinn PK, Bates TS (2005) Regional aerosol properties: comparisons of boundary layer measurements from ACE 1, ACE 2, Aerosols99, INDOEX, ACE Asia, TARFOX, and NEAQS. J Geophys Res 110, D14202. doi:10.1029/2004jd004755

Raes F (1995) Entrainment of free tropospheric aerosols as a regulating mechanism for cloud condensation nuclei in the remote marine boundary layer. J Geophys Res 100:2893–2903

Randles CA, Russell LM, Ramaswamy V (2004) Hygroscopic and optical properties of organic sea salt aerosol and consequences for climate forcing. Geophys Res Lett 31, doi:10.1029/2004GL020628

Rea DK (1994) The paleoclimatic record provided by eolian deposition in the deep sea: the geologic history of wind. Rev Geophys 32:159–195

Reddy MS, Boucher O, Balkanski Y, Schulz M (2005) Aerosol optical depths and direct radiative perturbations by species and source type. Geophys Res Lett 32, doi:10.1029/2004GL021743

Reid JS, Jonsson HH, Maring HB, Smirnov A, Savoie DL, Cliff SS, Reid EA, Livingston JM, Meier MM, Dubovik O, Tsay S-C (2003a) Comparison of size and morphological measurements of coarse mode dust particles from Africa. J Geophys Res 108:8593. doi:10.1029/2002JD002485

Reid EA, Reid JS, Meier MM, Dunlap MR, Cliff SS, Broumas A, Perry K, Maring H (2003b) Characterization of African dust transported to Puerto Rico by individual particle and size segregated bulk analysis. J Geophys Res 108, doi:10.1029/2002JD002935

Remer LA, Kleidman RG, Levy RC et al (2008) Global aerosol climatology from the MODIS satellite sensors. J Geophys Res 113, doi:10.1029/2007JD009661

Ridame C, Guieu C (2002) Saharan input of phosphorus to the oligotrophic water of the open western Mediterranean. Limnol Oceanogr 47:856–869

Ridame C, Moutin T, Guieu C (2003) Does the absortion process of phosphate onto Saharan dust explain the unusual N/P ratio in the Mediterranean sea? Oceanol Acta 26:629–634

Ridame C, Le Moal M, Guieu C, Ternon E, Biegala I, L'Helguen S, Pujo-Pay M (2011) Nutrient control of N2 fixation in the oligotrophic Mediterranean Sea and the impact of Saharan dust events. Biogeosciences 8:2629–2657

Ridame C et al (2012) Strong stimulation of N_2 fixation to Saharan dust events: results from dust fertilizations in large mesocosms (in preparation)

Ridame C, Guieu C, L'Helguen S (2013) Strong stimulation of N2 fixation in oligotrophic Mediterranean Sea: results from dust addition in large in situ mesocosms. Biogeosciences Discuss 10:10581–10613

Ridgwell AJ, Watson A (2002) Feedback between aeolian dust, climate and atmospheric CO2 in glacial time. Paleoceanography 17, doi:10.1029/2001PA000729

Riebesell U, Schulz KG, Bellerby RGJ, Botros M, Fritsche P, Meyerhöfer M, Neill C, Nondal G, Oschlies A, Wohlers J, Zöllner E (2007) Enhanced biological carbon consumption in a high CO2 ocean. Nature 450:545–548

Rijkenberg MJA, Powell C, Dall'Osto M, Nielsdottir M, Patey M, Hill P, Baker AR, Jickells T, Harrison R, Achterberg E (2008) Changes in iron speciation following a Saharan dust event in the tropical North Atlantic Ocean. Mar Chem 110:56–67

Rinaldi M, Decesari S, Finessi E, Giulianelli L, Carbone C, Fuzzi S, Dowd CD, Ceburnis D, Facchini MC (2010) Primary and secondary organic marine aerosol and oceanic biological activity: recent results and new perspectives for future studies. Adv Meteorol. doi:10.1155/2010/310682

Rinaldi M, Decesari S, Carbone C, Finessi E, Fuzzi S, Ceburnis D, O'Dowd CD, Sciare J, Burrows JP, Vrekoussis M, Ervens B, Tsigaridis K, Facchini MC (2011) Evidence of a natural marine source of oxalic acid and a possible link to glyoxal. J Geophys Res 116, D16204. doi:10.1029/2011JD015659

Rinaldi M, Fuzzi S, Decesari S, Marullo S, Santoleri R, Provenzale A, von Hardenberg J, Ceburnis D, Vaishya A, O'Dowd CD, Facchini M (2013) Is chlorophyll-a the best surrogate for organic matter enrichment in submicron primary marine aerosol? J Geophys Res Atmos 118:1–10. doi:10.1002/jgrd.50417

Rinne J, Back J, Hakola H (2009) Biogenic volatile organic compound emissions from the Eurasian tai-ga: current knowledge and future directions. Boreal Environ Res 14:807–826

Rivkin RB, Anderson MR (1997) Inorganic nutrient limitation of oceanic bacterioplankton. Limnol Oceanogr 42:730–740

Ro C-U, Hwang H, Kim H, Chun Y, Van Grieken R (2005) Single-particle characterization of four Asian dust samples collected in Korea, using low-Z particle electron probe X-ray microanalysis. Environ Sci Technol 39:1409–1419

Robinson AL, Donahue NM, Shrivastava MK, Weitkamp EA, Sage AM, Grieshop AP, Lane TE, Pierce JR, Pandis SN (2007) Rethinking organic aerosols: semivolatile emissions and photochemical aging. Science 315:1259–1262

Robock A (2000) Volcanic eruptions and climate. Rev Geophys 38:191–219

Russell LM, Pandis SN, Seinfeld JH (1994) Aerosol production and growth in the marine boundary layer. J Geophys Res 9:20989–21003

Russell LM, Hawkins LN, Frossard AA, Quinn PK, Bates TS (2010) Carbohydrate-like composition of submicron atmospheric particles and their production from ocean bubble bursting. PNAS 107(15):6652–6657. doi:10.1073/pnas.0908905107

Sage AM, Weitkamp EA, Robinson AL, Donahue NM (2008) Evolving mass spectra of the oxidized component of organic aerosol: results from aerosol mass spectrometer analyses of aged diesel emissions. Atmos Chem Phys 8:1139–1152

Saito MA, Moffett JW (2001) Complexation of cobalt by natural organic ligands in the Sargasso Sea as determined by a new high-sensitivity electrochemical cobalt speciation method suitable for open ocean work. Mar Chem 75:49–68

Saiz-Lopez A, Plane JMC, McFiggans G, Williams PI, Ball SM, Bitter M, Jones RL, Hongwei C, Hoffmann T (2005) Modelling molecular iodine emissions in the coastal marine environment: the link to new particle formation. Atmos Chem Phys 5:5405–5439

Sarmiento JL, Le Quéré C (1996) Oceanic carbon dioxide uptake in a model of century-scale global warming. Science 274:1346–1350

Sassen K (2000) Lidar backscatter depolarization technique for cloud and aerosol research. In: Mishchenko ML (ed) Light scattering by nonspherical particles: theory, measurements, and geophysical applications. Academic, San Diego, pp 393–416

Saunders RW, Plane JMC (2005) Formation pathways and composition of iodine oxide ultra-fine particles. Environ Chem 2:199–303. doi:10.1071/EN05079

Schlesinger P, Mamane Y, Grishkan I (2006) Transport of microorganisms to Israel during Saharan dust events. Aerobiologia 22:259–273

Schmincke H-U (2004) Volcanism. Springer, Berlin/Heidelberg/New York, p 324. ISBN 3-540-43650-2

Schultz MG, Heil A, Hoelzemann JJ, Spessa A, Thonicke K, Goldammer J, Held AC, Pereira JM (2008) Global emissions from wildland fires from 1960 to 2000. Global Biogeochem Cycles 22, GB2002. doi:10.1029/2007GB003031

Schuster GL, Dubovik O, Holben BN, Clothiaux EE (2005) Inferring black carbon content and specific absorption from Aerosol Robotic Network (AERONET) aerosol retrievals. J Geophys Res 110, doi:10.1029/2004JD004548

Sciare J, Favez O, Oikonomou K, Sarda-Estève R, Cachier H, Kazan V (2009) Long-term observation of carbonaceous aerosols in the Austral Ocean: evidence of a marine biogenic origin. J Geophys Res 114, D15302. doi:10.1029/2009JD011998

Sedwick P, Sholkovitz E, Church T (2007) Impact of anthropogenic combustion emissions on the frac-tional solubility of aerosol iron: evidence from the Sargasso Sea. Geochem Geophys Geosyst 8, doi:10.1029/2007GC001586

Seinfeld JH, Pandis SN (2006) Atmospheric chemistry and physics: from air pollution to climate change, 2nd edn. Wiley, New York

Seitzinger S, Harrison J, Dumont E, Beusen A, Bouwman A (2005) Sources and delivery of carbon, nitrogen and phosphorus to the coastal zone: an overview of the Global Nutrient Export from Watersheds (NEWS) models and their application. Global Biogeochem Cycles 19, doi:10.1029/2005GB002606

Seitzinger SP, Mayorga E, Bouwman AF, Kroeze C, Beusen AHW, Billen G, Van Drecht G, Dumont E, Fekete BM, Garnier J, Harrison JA (2010) Global river nutrient export: a scenario analysis of past and future trends. Global Biogeochem Cycles 24, doi:10.1029/2009GB003587

Sellegri K, Yoon YJ, Jennings SG, Pirjola L, Cautenet S, O'Dowd CD (2005) Quantification of coastal new ultra-fine particles formation from in-situ and chamber measurements during the BIOFLUX campaign. Environ Chem 2:260–270

Sellegri K, O'Dowd CD, Yoon YJ, Jennings SG, de Leeuw G (2006) Surfactants and submicron sea spray generation. J Geophys Res 111, D22215. doi:10.1029/2005JD006658

Shaw G (1983) Bio-controlled thermostasis involving the sulfur cycle. Clim Chang 5:297–303

Shi Z, Zhang D, Hayashi M, Ogata H, Ji H, Fujiie W (2008) Influences of sulfate and nitrate on the hygroscopic behaviour of coarse dust particles. Atmos Environ 42:822–827

Sievering H et al (1992) Removal of sulphuer from the marine boundary layer by ozone oxidation in sea-salt aerosols. Nature 360:571–573

Simkin T, Siebert L (1994) Volcanoes of the world, 2nd edn. Geoscience Press, Tucson, p 349

Singh HB, Kasting JF (1988) Chlorine-hydrocarbon photochemistry in the marine troposphere and lower stratosphere. J Atmos Chem 7:261–285

Slingo A (1990) Sensitivity of the Earth's radiation budget to changes in low clouds. Nature 343:49–51

Smetacek V, Klaas C, Strass VH, Assmy P, Montresor M, Cisewski B, Savoye N, Webb A, d'Ovidio F, Arrieta JM, Bathmann U, Bellerby R, Mine Berg G, Croot P, Gonzalez S, Henjes J, Herndl GJ, Hoffmann LJ, Leach H, Losch M, Mills MM, Neill C, Peeken I, Röttgers R, Sachs O, Sauter E, Schmidt MM, Schwarz J, Terbrüggen A, Wolf-Gladrow D (2012) Deep carbon export from a Southern Ocean iron-fertilized diatom bloom. Nature 487:313–319

Smith SJ, van Aardenne J, Klimont Z, Andres RJ, Volke A, Delgado Arias S (2011) Anthropogenic sulfur dioxide emissions: 1850–2005. Atmos Chem Phys 11:1101–1116

Smirnov A, Holben BN, Giles DM, Slutsker I, O'Neill NT, Eck TF, Macke A, Croot P, Courcoux Y, Sakerin SM, Smyth TJ, Zielinski T, Zibordi G, Goes JI, Harvey MJ, Quinn PK, Nelson NB, Radionov VF, Duarte CM, Losno R, Sciare J, Voss KJ, Kinne S, Nalli NR, Joseph E, Krishna Moorthy K, Covert DS, Gulev SK, Milinevsky G, Larouche P, Belanger S, Horne E, Chin M, Remer LA, Kahn RA, Reid JS, Schulz M, Heald CL, Zhang J, Lapina K, Kleidman RG, Griesfeller J, Gaitley BJ, Tan Q, Diehl TL (2011) Maritime aerosol network as a component of AERONET – first results and comparison with global aerosol models and satellite retrievals. Atmos Meas Tech 4:583–597. doi:10.5194/amt-4-583-2011

Smirnov A, Sayer AM, Holben BN, Hsu NC, Sakerin SM, Macke A, Nelson NB, Courcoux Y, Smyth TJ, Croot P, Quinn PK, Sciare J, Gulev SK, Piketh S, Losno R, Kinne S, Radionov VF (2012) Effect of wind speed on aerosol optical depth over remote oceans, based on data from the maritime aerosol network. Atmos Meas Tech 5:377–388, 10.5194/amt-5-377-2012

Sobanska S, Coeur C, Maenhaut W, Adams F (2003) SEM-DEX characterization of tropospheric aerosols in the Negev Desert (Israel). J Atmos Chem 44:299–322

Sofiev M, Siljamo P, Valkama I, Ilvonen M, Kukkonen J (2006) A dispersion modelling system SILAM and its evaluation against ETEX data. Atmos Environ 40:674–685

Sofiev M, Soares J, Prank M, de Leeuw G, Kukkonen J (2011) A regional-to-global model of emission and transport of sea salt particles in the atmosphere. J Geophys Res 116, D021302. doi:10.1029/2010JD014713

Song CH, Carmichael GR (1999) The aging process of naturally emitted aerosol (sea-salt and mineral aerosol) during long range transport. Atmos Environ 33:2203–2218

Sorooshian A, Lu M-L, Brechtel FJ, Jonsson H, Feingold G, Flagan RC, Seinfeld JH (2007) On the source of organic acid aerosol layers above clouds. Environ Sci Technol 41:4647–4654

Sorooshian A, Padró LT, Nenes A et al (2009) On the link between ocean biota emissions, aerosol, and maritime clouds: airborne, ground, and satellite measurements off the coast of California. Global Biogeochem Cycles 23, GB4007

Stavrakou T, Müller JF, De Smedt I, Van Roozendael M, van der Werf GR, Giglio L, Guenther A (2009) Global emissions of non-methane hydrocarbons deduced from SCIAMACHY formaldehyde columns through 2003–2006. Atmos Chem Phys 9:3663–3679

Stier P, Seinfeld JH, Kinne S, Boucher O (2007) Aerosol absorption and radiative forcing. Atmos Chem Phys 7:5237–5261

Stone EA, Yoon SC, Schauer JJ (2011) Chemical characterization of fine and coarse particles in Gosan, Korea during springtime dust events. Aerosol Air Qual Res 11:31–43

Stramska M, Marks R, Monahan EC (1990) Bubble-mediated aerosol production as a consequence of wave breaking in supersaturated (hyperoxic) seawater. J Geophys Res 95 (C10):18281–18288

Sturges WT, Shaw GE (1993) Halogens in aerosols in Central Alaska. Atmos Environ Part A Gen Top 27:2969–2977

Su J, Jianping H, Qiang F, Minnis P, Jinming G, Jianrong B (2008) Estimation of Asian dust aerosol effect on cloud radiation forcing using Fu-Liou radiative model and CERES measurements. Atmos Chem Phys 8:2763–2771

Sullivan RC, Prather KA (2007) Investigations of the diurnal cycle and mixing state of oxalic acid in individual particles in Asian aerosol outflow. Environ Sci Technol 41:8062–8069. doi:10.1021/es071134g

Sullivan RC, Guazzotti SA, Sodeman DA, Prather KA (2007) Direct observations of the atmospheric processing of Asian mineral dust. Atmos Chem Phys 7:1213–1236

Sullivan RC, Moore MJK, Petters MD, Kreidenweis SM, Roberts GC, Prather KA (2009) Effect of chemical mixing state on the hygroscopicity and cloud nucleation properties of calcium mineral dust particles. Atmos Chem Phys 9:3303–3316

Sun Y, Zhuang G, Wang Y, Zhao X, Li J, Wang Z, An Z (2005) Chemical composition of dust storms in Beijing and implications for the mixing of mineral aerosol with pollution aerosol on the pathway. J Geophys Res 110, D24209. doi:10.1029/2005jd006054

Surratt JD, Lewandowski M, Offenberg JH, Jaoui M, Kleindienst TE, Edney EO, Seinfeld JH (2007) Effect of acidity on secondary organic aerosol formation from isoprene. Environ Sci Technol 41:5363–5369

Suzuki I, Igarashi Y, Dokiya Y, Akagi T (2010) Two extreme types of mixing of dust with urban aerosols observed in Kosa particles: 'after' mixing and 'on-the-way' mixing. Atmos Environ 44:858–866

Tagliabue A, Bopp L, Aumont O (2008) Ocean biogeochemistry exhibits contrasting responses to a large scale reduction in dust deposition. Biogeosciences 5:11–24

Tagliabue A, Bopp L, Aumont O, Arrigo K (2009) Influence of light and temperature on the marine iron cycle: from theoretical to global modeling. Global Biogeochem Cycle 23, GB2017. doi:10.1029/2008GB003214

Tan PV, Evans GJ, Tsai J et al (2002) On-line analysis of urban particulate matter focusing on elevated wintertime aerosol concentrations. Environ Sci Technol 36:3512–3518

Tedetti M, Sempéré R (2006) Penetration of ultraviolet radiation in the marine environmen. A review. Photochem Photobiol 82:389–397

Tegen I, Werner M, Harrison SP, Kohfeld KE (2004) Relative importance of climate and land use in determining present and future global soil dust emission. Geophys Res Lett 31, L05105. doi:10.1029/2003GL019216

Ternon E, Guieu C, Loÿe-Pilot M-D, Leblond N, Bosc E, Gasser B, Martin J, Miquel J-C (2010) The impact of Saharan dust on the particulate export in the water column of the North Western Mediterranean Sea. Biogeosciences 7:809–826

Ternon E, Guieu C, Ridame C, L'Helguen S, Catala P (2011) Longitudinal variability of the biogeochemical role of Mediterranean aerosols in the Mediterranean Sea. Biogeosciences 8:1067–1080

Thingstad TF, Law CS, Krom MD, Mantoura RFC, Pitta P, Psarra S, Rassoulzadegan F, Tanaka T, Wassmann P, Wexels Riser C, Zohary T (2005) Nature of phosphorus limitation in the ultraoligotrophic Eastern Mediterranean. Science 309:1068–1071. doi:10.1126/science.1112632

Thorpe SA (1992) Bubble clouds and the dynamics of the upper ocean. Q J R Meteorol Soc 118:1–22

Tobo Y, Zhang DZ, Nakata N, Yamada M, Ogata H, Hara K, Iwasaka Y (2009) Hygroscopic mineral dust particles as influenced by chlorine chemistry in the marine atmosphere. Geophys Res Lett 36, L05817. doi:10.1029/2008gl036883

Tobo Y, Zhang D, Matsuki A, Iwasaka Y (2010) Asian dust particles converted into aqueous droplets under remote marine atmospheric conditions. Proc Natl Acad Sci 107:17905–17910. doi:10.1073/pnas.1008235107

Tomlinson JM, Li R, Collins DR (2007) Physical and chemical properties of the aerosol within the southeastern Pacific marine boundary layer. J Geophys Res 112, D12211. doi:10.1029/2006JD007771

Trochkine D, Iwasaka Y, Matsuki A, Yamada M, Kim YS, Nagatani T, Zhang D, Shi GY, Shen Z (2003) Mineral aerosol particles collected in Dunhuang, China, and their comparison with chemically modified particles collected over Japan. J Geophys Res 108, doi:10.1029/2002jd003268

Tursic J, Podkrajsek B, Grgic I, Ctyroky P, Berner A, Dusek U, Hitzenberger R (2006) Chemical composition and hygroscopic properties of size-segregated aerosol particles collected at the Adriatic coast of Slovenia. Chemosphere 63:1193–1202

Tyree CA, Hellion VM, Alexandrova OA, Allen JO (2007) Foam droplets generated from natural and artificial seawaters. J Geophys Res 112, D12204. doi:10.1029/2006JD007729

Uematsu M, Duce RA, Prospero JM (1985) Deposition of atmospheric mineral particles in the North Pacific Ocean. J Atmos Chem 3:123–138

Usher CR, Al-Hosney H, Carlos-Cuellar S, Grassian VH (2002) A laboratory study of the heterogeneous uptake and oxidation of sulfur dioxide on mineral dust particles. J Geophys Res 107:4713. doi:10.1029/2002jd002051

Usher CR, Michel AE, Grassian VH (2003) Reactions on mineral dust. Chem Rev 103:4883–4939

Van der Werf GR, Randerson JT, Giglio L, Collatz GJ, Kasibhatla PS, Arellano AF (2006) Interannual variability in global biomass burning emissions from 1997 to 2004. Atmos Chem Phys 6:3423–3441

Van der Werf GR, Randerson JT, Giglio L, Collatz GJ, Mu M, Kasibhatla PS, Morton DC, DeFries RS, Jin Y, van Leeuwen TT (2010) Global fire emissions and the contribution of deforestation, savanna, forest, agricultural, and peat fires (1997–2009). Atmos Chem Phys 10:11707–11735. doi:10.5194/acp-10-11707-2010

van Vuuren DP, Edmonds J, Kainuma M, Riahi K, Thomson A, Hibbard K, Hurtt GC, Kram T, Krey V, Lamarque JF, Masui T, Meinshausen M, Nakicenovic N, Smith SJ, Rose SK (2011) The representative concentration pathways: an overview. Clim Chang 109:5–31. doi:10.1007/s10584-011-0148-z

Venkatathnam K, Ryan WBF (1971) Dispersal patterns of clay minerals in the sediments of the Eastern Mediterranean Sea. Mar Geol 11:261–282

Verdugo P, Alldredge AL, Azam F, Kirchman DL, Passow U, Santschi PH (2004) The oceanic gel phase: a bridge in the DOM-POM continuum. Mar Chem 92:67–85

Vignati E, Wilson J, Stier P (2004) M7: an efficient size resolved aerosol microphysics module for large-scale aerosol transport models. J Geophys Res 109, doi:10.1029/2003JD004485

Vignati E, Facchini MC, Rinaldi M, Scannell C, Ceburnis D, Sciare J, Kanakidou M, Myriokefalitakis S, Dentener F, O'Dowd CD (2010) Global scale emission and distribution of sea spray aerosol: sea-salt and organic enrichment. Atmos Environ 44, doi:10.1016/j.atmosenv.2009.11.013

Vogt R, Crutzen PJ, Sander R (1996) A mechanism for halogen release from sea-salt aerosol in the remote marine boundary layer. Nature 383:327–330

Volkamer R, San Martini F, Molina LT, Salcedo D, Jimenez JL, Molina MJ (2007) A missing sink for gas-phase glyoxal in Mexico city: formation of secondary organic aerosol. Geophys Res Lett 34, L19807. doi:10.1029/2007gl030752

Wagener T, Pulido-Villena E, Guieu C (2008) Dust iron dissolution in seawater: results from a one-year time-series in the Mediterranean Sea. Geophys Res Lett 35, L16601. doi:10.1029/2008GL034581

Wagener T, Guieu C, Leblond N (2010) Effects of dust deposition on iron cycle in the surface Mediterranean Sea: results from a mesocosm seeding experiment. Biogeosciences 7:3769–3781

Wang C, Corbett JJ, Firestone J (2008) Improving spatial representation of global ship emissions inventories. Environ Sci Technol 42:193–199. doi:10.1021/es0700799

Wang G, Kawamura K, Lee M (2009) Comparison of organic compositions in dust storm and normal aerosol samples collected at Gosan, Jeju Island, during spring 2005. Atmos Environ 43:219–227

Wang Q, Zhuang G, Li J, Huang K, Zhang R, Jiang Y, Lin Y, Fu JS (2011) Mixing of dust with pollution on the transport path of Asian dust – revealed from the aerosol over Yulin, the north edge of Loess Plateau. Sci Total Environ 409:573–581

Waquet F, Riédi J, Labonnote LC, Goloub P, Cairns B, Deuzé JL, Tanré D (2009) Aerosol remote sensing over clouds using the A-Train observations. J Atmos Sci 66:2468–2480

Watson AJ, Bakker DCE, Ridgwell AJ, Boyd PW, Law CS (2000) Effect of iron supply on Southern Ocean CO_2 uptake and implications for glacial atmospheric CO_2. Nature 407:730–733

Willey JD, Kieber RJ, Eyman MS, Avery GB (2000) Rainwater dissolved organic carbon: concentrations and global flux. Global Biogeochem Cycles 14:139–148

Wilson TM, Cole JW, Sreward C (2011) Ash storms: impact of wind-remobilised volcanic ash on rural communities and agriculture following the 1991 Hudson eruption, southern Patagonia, Chile. Bull Volcanol 73:223–239

Winker DM, Pelon J, Coakley JA, Ackerman SA, Charlson RJ, Colarco PR, Flamant P, Fu Q, Hoff R, Kittaka C, Kubar TL, LeTreut H, McCormick MP, Megie G, Poole L, Powell K, Trepte C, Vaughan MA, Wielicki BA (2010) The CALIPSO mission: a global 3D view of aerosols and clouds. Bull Am Meteorol Soc 91:1211–1229

Witek ML, Flatau PJ, Teixeira J, Westphal DL (2007a) Coupling an ocean wave model with a global aerosol transport model: a sea salt aerosol parameterization perspective. Geophys Res Lett 34, L14806. doi:10.1029/2007GL030106

Witek ML, Flatau PJ, Quinn PK, Westphal DL (2007b) Global sea-salt modeling: results and validation against multi-campaign shipboard measurements. J Geophys Res 112, D08215. doi:10.1029/2006JD007779

Wolff EW, Fischer H, Fundel F, Ruth U, Twarloh B, Littot GC, Mulvaney R, Röthlisberger R, de Angelis M, Boutron CF, Hansson M, Jonsell U, Hutterli MA, Lambert F, Kaufmann P, Stauffer B, Stocker TF, Steffensen JP, Bigler M, Siggaard-Andersen ML, Udisti R, Becagli S, Castellano E, Severi M, Wagenbach D, Barbante C, Gabrielli P, Gaspari V (2006) Southern Ocean sea-ice extent, productivity and iron flux over the past eight glacial cycles. Nature 440:491–496. doi:10.1038/nature04614

Woodcock AH (1948) Note concerning human respiratory irritation associated with high concentrations of plankton and mass mortality of marine organism. J Mar Res 7:56–62

Wu J, Sunda W, Boyle E, Karl D (2000) Phosphate depletion in the western North Atlantic Ocean. Science 289:759–762

Wu J, Boyle E, Sunda W, Wen LS (2001) Soluble and colloidal iron in the oligotrophic North Atlantic and North Pacific. Science 293:847–849

Wuttig K, Wagener T, Bressac M, Dammshäuser A, Streu P, Guieu C, Croot PL (2013) Impacts of dust deposition on dissolved trace metal concentrations (Mn, Al and Fe) during a mesocosm experiment. Biogeosciences 10:2583–2600. doi:10.5194/bg-10-2583-2013

Yamato M, Tanaka H (1994) Aircraft observations of aerosols in the free marine troposphere over the North Pacific Ocean: particle chemistry in relation to air mass origin. J Geophys Res 99:5353–5377

Yang F, Chen H, Wang X, Yang X, Du J, Chen J (2009) Single particle mass spectrometry of oxalic acid in ambient aerosols in Shanghai: mixing state and formation mechanism. Atmos Environ 43:3876–3882

Yao X, Fang M, Chan CK (2003) The size dependence of chloride depletion in fine and coarse sea-salt particles. Atmos Environ 37:743–751

Ye Y, Wagener T, Volker C, Guieu C, Dieter A, Wolf-Gladrow DA (2011) Dust deposition: iron source or sink? A case study. Biogeosciences 8:2107–2124

Yoon YJ, Ceburnis D, Cavalli F, Jourdan O, Putaud J-P, Facchini MC, Descari S, Fuzzi S, Sellegri K, Jennings SG, O'Dowd CD (2007) Seasonal characteristics of the physicochemical properties of North Atlantic marine atmospheric aerosols. J Geophys Res 112, D04206. doi:10.1029/2005JD007044

Yu H, Kaufman YJ, Chin M, Feingold G, Remer LA, Anderson TL, Balkanski Y, Bellouin N, Boucher O, Christopher S, DeCola P, Kahn R, Koch D, Loeb N, Reddy MS, Schulz M, Takemura T, Zhou M (2006) A review of measurement-based assessment of aerosol direct radiative effect and forcing. Atmos Chem Phys 6:613–666

Zhang D, Iwasaka Y (2001) Chlorine deposition on dust particles in marine atmosphere. Geophys Res Lett 28:3613–3616. doi:10.1029/2001gl013333

Zhang D, Iwasaka Y (2004) Size change of Asian dust particles caused by sea salt interaction: measurements in southwestern Japan. Geophys Res Lett 31, L15102. doi:10.1029/2004gl020087

Zhang DZ, Iwasaka Y (2006) Comparison of size changes of Asian dust particles caused by sea salt and sulphate. J Meteorol Soc Japan 84:939–947. doi:10.2151/jmsj.84.939

Zhang D, Iwasaka Y, Shi G, Zang J, Matsuki A, Trochkine D (2003a) Mixture state and size of Asian dust particles collected at southwestern Japan in spring 2000. J Geophys Res 108:4760. doi:10.1029/2003jd003869

Zhang D, Zang J, Shi G, Iwasaka Y, Matsuki A, Trochkine D (2003b) Mixture state of individual Asian dust particles at a coastal site of Qingdao, China. Atmos Environ 37:3895–3901

Zhang D, Iwasaka Y, Shi G (2005a) Sea salt shifts the range sizes of Asian dust. Eos Trans Am Geophys Union 86, doi:10.1029/2005EO500003

Zhang R, Arimoto R, An J, Yabuki S, Sun J (2005b) Ground observations of a strong dust storm in Beijing in March 2002. J Geophys Res 110, D18S06. doi:10.1029/2004jd004589

Zhang Q, Jimenez JL, Canagaratna MR, Allan JD, Coe H, Ulbrich I, Alfarra MR, Takami A, Middlebrook AM, Sun YL, Dzepina K, Dunlea E, Docherty K, DeCarlo PF, Salcedo D, Onasch T, Jayne JT, Miyoshi T, Shimono A, Hatakeyama S, Takegawa N, Kondo Y, Schneider J, Drewnick F, Borrmann S, Weimer S, Demerjian K, Williams P, Bower K, Bahreini R, Cottrell L, Griffin RJ, Rautiainen J, Sun JY, Zhang YM, Worsnop DR (2007) Ubiquity and dominance of oxygenated species in organic aerosols in anthropogenically-influenced Northern Hemisphere midlatitudes. Geophys Res Lett 34, L13801. doi:10.1029/2007gl029979

Zhao XJ, Zhuang GS, Wang ZF, Sun YL, Wang Y, Yuan H (2007) Variation of sources and mixing mechanism of mineral dust with pollution aerosol – revealed by the two peaks of a super dust storm in Beijing. Atmos Res 84:265–279

Zhao TXP, Loeb NG, Laszlo I, Zhou M (2011) Global component aerosol direct radiative effect at the top of atmosphere. Int J Remote Sens 32:633–655

Zhou M, Okada K, Qian F, Wu PM, Su L, Casareto BE, Shimohara T (1996) Characteristics of dust-storm particles and their long-range transport from China to Japan – case studies in April 1993. Atmos Res 40:19–31

Zhou X, Davis AJ, Kieber DJ et al (2008) Photochemical production of hydroxyl radical and hydroperoxides in water extracts of nascent marine aerosols produced by bursting bubbles from Sargasso seawater. Geophys Res Lett 35, L20803

Zhu X, Prospero J, Millero F (1997) Diel variability of soluble Fe(II) and soluble total Fe in North Africa dust in the trade winds at Barbados. J Geophys Res 102:21297–21305

Zhuang G, Duce R (1993) The adsorption of dissolved iron on marine aerosol particles in surface waters of the open ocean. Deep Sea Res 40:1413–1429

Zorn SR, Drewnick F, Schott M, Hoffmann T, Borrmann S (2008) Characterization of the South Atlantic marine boundary layer aerosol using an aerodyne aerosol mass spectrometer. Atmos Chem Phys 8:4711–4728

Perspectives and Integration in SOLAS Science

Véronique C. Garçon, Thomas G. Bell, Douglas Wallace,
Steve R. Arnold, Alex Baker, Dorothee C.E. Bakker,
Hermann W. Bange, Nicholas R. Bates, Laurent Bopp,
Jacqueline Boutin, Philip W. Boyd, Astrid Bracher, John P. Burrows,
Lucy J. Carpenter, Gerrit de Leeuw, Katja Fennel, Jordi Font,
Tobias Friedrich, Christoph S. Garbe, Nicolas Gruber, Lyatt Jaeglé,
Arancha Lana, James D. Lee, Peter S. Liss, Lisa A. Miller,
Nazli Olgun, Are Olsen, Benjamin Pfeil, Birgit Quack, Katie A. Read,
Nicolas Reul, Christian Rödenbeck, Shital S. Rohekar,
Alfonso Saiz-Lopez, Eric S. Saltzman, Oliver Schneising,
Ute Schuster, Roland Seferian, Tobias Steinhoff,
Pierre-Yves Le Traon, and Franziska Ziska

Abstract

Why a chapter on Perspectives and Integration in SOLAS Science in this book? SOLAS science by its nature deals with interactions that occur: across a wide spectrum of time and space scales, involve gases and particles, between the ocean and the atmosphere, across many disciplines including chemistry, biology, optics, physics, mathematics, computing, socio-economics and consequently interactions between many different scientists and across scientific generations. This chapter provides a guide through the remarkable diversity of cross-cutting approaches and tools in the gigantic puzzle of the SOLAS realm.

Here we overview the existing prime components of atmospheric and oceanic observing systems, with the acquisition of ocean–atmosphere observables either from *in situ* or from satellites, the rich hierarchy of models to test our knowledge of Earth System functioning, and the tremendous efforts accomplished over the last decade within the COST Action 735 and SOLAS Integration project frameworks to understand, as best we can, the current physical and biogeochemical state of the atmosphere and ocean commons. A few SOLAS integrative studies illustrate the full meaning of interactions, paving the way for even tighter connections between thematic fields. Ultimately, SOLAS research will also develop with an enhanced consideration of societal demand while preserving fundamental research coherency.

V.C. Garçon (✉)
e-mail: veronique.garcon@legos.obs-mip.fr

T.G. Bell (✉)
e-mail: tbe@pml.ac.uk

D. Wallace (✉)
e-mail: douglas.wallace@dal.ca

The exchange of energy, gases and particles across the air-sea interface is controlled by a variety of biological, chemical and physical processes that operate across broad spatial and temporal scales. These processes influence the composition, biogeochemical and chemical properties of both the oceanic and atmospheric boundary layers and ultimately shape the Earth system response to climate and environmental change, as detailed in the previous four chapters. In this crosscutting chapter we present some of the SOLAS achievements over the last decade in terms of integration, upscaling observational information from process-oriented studies and expeditionary research with key tools such as remote sensing and modelling.

Here we do not pretend to encompass the *entire* legacy of SOLAS efforts but rather offer a selective view of some of the major integrative SOLAS studies that combined available pieces of the immense jigsaw puzzle. These include, for instance, COST efforts to build up global climatologies of SOLAS relevant parameters such as dimethyl sulphide, interconnection between volcanic ash and ecosystem response in the eastern subarctic North Pacific, optimal strategy to derive basin-scale CO_2 uptake with good precision, or significant reduction of the uncertainties in sea-salt aerosol source functions. Predicting the future trajectory of Earth's climate and habitability is the main task ahead. Some possible routes for the SOLAS scientific community to reach this overarching goal conclude the chapter.

5.1 Perspectives: In Situ Observations, Remote Sensing, Modelling and Synthesis

The scope of SOLAS science depends on multidisciplinary and multi-scale approaches being applied to the complex problems and challenges within the field. Laboratory process studies and in situ lagrangian field experiments make substantial contributions to our understanding of the various biogeochemical processes and their feedbacks. Models are almost the only way to assess what are often complex problems and they rely on input from such studies. To truly represent the domain of SOLAS within Earth System models requires global-scale datasets of accurate measurements of relevant parameters. The most accurate data-based estimates of air-sea exchange processes require as much data as possible at large spatial and temporal scales in order to be able to validate and calibrate both model outputs and satellite data. Maintaining and further expanding existing global arrays of autonomous instrumented platforms, as well as oceanic and atmospheric fixed observatories, is a modern-day challenge. Producing integrated, quality and potential bias-controlled global datasets from the collection of these measurements is our ongoing responsibility. This section of Chap. 5 considers a variety of data collation and synthesis projects relevant to SOLAS science presented in the previous four chapters. The following subsections briefly introduce each project.

5.1.1 In Situ Observations

5.1.1.1 ARGO (T, S, O_2)

In November 2007, the international Argo programme reached its initial target of 3,000 profiling floats (http://www.argo.ucsd.edu/). Every 10 days these floats measure temperature and salinity throughout the global ocean, diving down to 2,000 m and delivering data both in real time for operational users and, after careful scientific quality control, for climate change research and monitoring. Argo results from an outstanding international cooperation. More than 30 countries are involved in the development and maintenance of the array. Argo is the major systematic source of data about the interior of the ocean. Argo aims to maintain a global array of in situ measurements integrated with

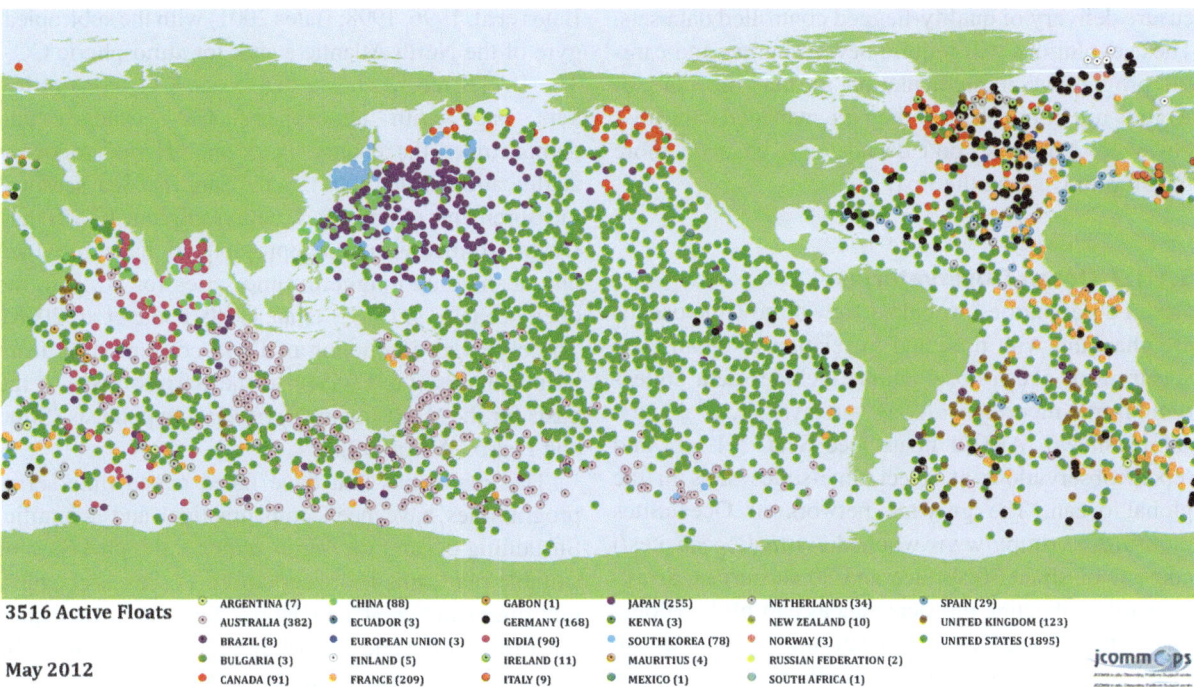

Fig. 5.1 Position of active Argo floats (Figure courtesy of JCOMMOPS, Argo Information Centre, http://argo.jcommops.org)

other elements of the climate observing system (in particular satellite observations) to:

- Detect climate variability from seasonal to decadal scales and provide long-term observations of climate change in the oceans. This includes regional and global changes in temperature and ocean heat content, salinity and freshwater content, steric height and large-scale ocean circulation.
- Provide data to constrain global and regional ocean analysis and forecasting models, to initialise seasonal and decadal forecasting ocean/atmosphere coupled models and to validate climate models.
- Provide information necessary for the calibration and validation of satellite data.

An overview of the achievements of the first decade of Argo is given in Freeland et al. (2010). Argo data (Fig. 5.1) have been used to dramatically improve estimation of heat stored by the ocean, to better understand global sea level rise, to analyse large-scale ocean circulation variations and deep convection areas. Argo has also brought remarkable advances in ocean analysis and forecasting capability. Argo data can be accessed at http://www.nodc.noaa.gov/argo/.

The last OceanObs09 conference discussed the main priorities of the international community for Argo (Roemmich et al. 2009; Freeland et al. 2010; Claustre et al. 2010). Based on the assessment that climate change research requires long-term, sustained, high quality and global observations, the leading priority and challenge for Argo must be to complete and sustain the global array. This requires deploying between 800 and 900 new floats every year. Several developments of the Argo core mission have also been proposed. Minor changes include the extension of the array into seasonal ice zones and marginal seas. Major expansions of Argo will include monitoring the deep ocean below 2,000 m and marine ecosystems. Deeper measurements are needed to constrain the deep ocean property fields for climate monitoring and long-term prediction. Recent technological advances in biogeochemical sensors will permit the acquisition of new observations of the ocean interior (e.g. Claustre et al. 2010; Adornato et al. 2010). The main parameters that are considered for initial implementation are oxygen, nitrate, chlorophyll *a* and particulate carbon. Pilot experiments have already begun, in particular for dissolved oxygen (almost 200 Argo floats are today equipped with an oxygen sensor). Potential systematic errors due to different measurement techniques or sensors need to be analysed and further corrected to ensure delivery of quality-flagged controlled data sets. These evolutions will require new resources and careful progressive implementation so that the core of Argo is not diminished. If they are managed carefully, however, Argo's second decade will be even more transformative than the first.

5.1.1.2 Ocean Observatories

Ocean observatories provide a view of how the oceans are changing with time and in relation to depth. The spectrum of ocean observations includes data collection from moorings, AUV surveys (e.g. gliders, submersibles), ARGO floats (see Sect. 5.1.1.1) and repeat observations at select time-series sites in the global ocean. The growing network of OceanSites moorings (http://www.whoi.edu/virtual/oceansites/) consists of about 30 surface and 30 subsurface arrays primarily collecting physical data as part of the Global Ocean Observing System. Ocean time-series sites are fewer in number but they allow collection of critically needed data that illustrates temporal variability on ocean–atmosphere exchange and water-column processes over seasonal to multi-decadal timescales. Four of the longest ocean time-series stations include: (1) Hydrostation S, ($32°50'$N, $64°10'$W; 1954-present) located near Bermuda in the NW Atlantic Ocean (Steinberg et al. 2001); (2) BATS (Bermuda Atlantic Time-series Study), located near Bermuda ($32°10'$N, $64°30'$W; 1988 – present) in the NW Atlantic Ocean (Steinberg et al. 2001; Bates 2007); (3) ALOHA (A Long-term Oligotrophic Habitat Assessment) or HOT site, located near Hawaii ($22°45'$N, $158°$W; 1988 – present) in the North Pacific Ocean (Dore et al. 2009); and; (4) ESTOC; European Station for Time-series in the Ocean Canary Islands ($29°10'$N, $15°,30'$W 1994-present) (González-Dávila et al. 2010).

The monthly BATS and biweekly Hydrostation 'S' programmes play a pivotal role in better understanding of the seasonality and long-term changes in ocean–atmosphere exchange of gases and particles. Both sites serve as important frameworks for larger-scale field and modelling studies in the subtropical gyre of the North Atlantic Ocean (or Sargasso Sea). Over the last 50 years, the surface ocean has warmed by ~ 0.3–0.5 °C while salinity has increased by ~ 0.15. The BATS site exhibits strong seasonality in the ocean–atmosphere exchange of gases such as oxygen (e.g. Ono et al. 2001) and carbon dioxide (CO_2 e.g. Bates et al. 1996, 1998; Bates 2001) with the subtropical gyre of the North Atlantic a sink for atmospheric CO_2. The BATS record shows that ocean CO_2 content has kept pace with atmospheric CO_2 changes and demonstrates the change in ocean pH due to ocean acidification (Bates 2007). In the Sargasso Sea, seasonal measurements of oceanic dimethyl sulphide (DMS) and DMSP (dimethylsulphoniopropionate) have provided one of the only long-term time series for DMS in the open ocean (e.g. Dacey et al. 1998; Toole et al. 2008). The observed decoupling of DMS concentration from its precursors (i.e. DMSP) in the Sargasso Sea is the basis for the '*DMS summer paradox*' hypothesis (Simó and Pedrós-Alió 1999).

The challenges that have faced ocean time-series programmes have been both practical and scientific. Sustaining ocean time-series requires the provision of suitable platforms for observation (i.e. research ships, moorings) and creates logistical demands if the continuity of funding and frequency of occupation are to be maintained. Scientific questions include reconciling time-variations within the context of the four dimensional state (space and time) of the ocean, which includes substantial mesoscale and sub-mesoscale variability (e.g. McGillicuddy et al. 1999, 2007). In the future, ocean time-series programmes will be integrated with the Global Ocean Observing System (http://www.ioc-goos.org/), with these sites acting as important nodes for new observing technologies such as gliders and AUV platforms (Dickey et al. 2009). Understanding ocean and climate relevant processes that influence the ocean–atmosphere exchange of gases and particles requires an improved synergy between (sustained) observation and hypothesis testing over a variety of scales, both spatial and temporal.

5.1.1.3 Atmospheric Observatories

The relative homogeneity of marine vs terrestrial air provides an exceptional opportunity to test aspects of surface atmospheric photochemistry. There is a paucity of long-term measurements of reactive trace gases and aerosols in clean marine environments, but the reported studies have revealed important insights into ocean–atmosphere interactions and their consequences for atmospheric composition and climate. Seasonal observations of sulphur-containing gases and aerosols at Cape Grim atmospheric observatory, Tasmania ($40.7°$S, $144.7°$E) showed summer maxima and winter minima in dimethyl sulphide (DMS), methanesulphonic acid (MSA, a unique product of DMS oxidation), non-sea-salt(nss) sulphate aerosol, and the concentration of cloud condensation nuclei (CCN) (Ayers and Gras 1991; Ayers et al. 1997), supporting proposed mechanisms for DMS oxidation to particulate sulphate. DMS emissions are believed to contribute to a significant fraction of remote marine CCN concentrations, up to 46 % in summer over the Southern Ocean between 30°S and 45°S (Korhonen et al. 2008). Measurements of sea-salt aerosol composition at Cape Grim confirmed that bromine deficits (a decrease of the bromine to sodium ratio of sea- salt aerosol compared to sea water) were linked to the availability of sulphate acidity in the aerosol (Ayers et al. 1992), as proposed by modelling studies which suggest the importance of acid catalysis in the dehalogenation process (Sander and Crutzen 1996; Vogt et al. 1996). This provided experimental evidence for the net transfer of bromine from sea salt aerosol to the gas phase, where it catalyses photochemical ozone destruction and modifies the concentrations of many important tropospheric gases (Sander and Crutzen 1996; Vogt et al. 1996; von Glasow et al. 2004).

Aspects of O_3 photochemistry have also been confirmed by long-term marine observations; at Cape Grim, diurnal cycling of hydrogen peroxide, one of the major products of HO_x radical (OH and HO_2) recombination reactions, is in opposite phase to that of O_3 (Ayers and Gras 1991), as expected in clean low-NO_x air. The observed relationships between free radical levels and the O_3 photolysis rate (jO^1D) change according to NO_x levels, indicating the critical NO concentration required to switch from O_3 destruction to O_3 production (Carpenter et al. 1997). In the northern hemisphere, a longer than 20-year record in baseline O_3 at the coastal Irish station at Mace Head (53.28°N, 9.02°W) showed that mixing ratios rose steadily during the 1980s and 1990s, probably due to increased tropospheric ozone production from methane oxidation in the presence of nitrogen oxides (NO_x), and stabilised during the 2000s (Derwent et al. 2007) with sporadic increases over the period due to boreal biomass burning events. Background O_3 can be an important contributor to the levels experienced in urban regions, which are not declining in developed regions including Europe despite decreasing precursor emissions.

The northern tropical Atlantic ocean is subject to sporadic but significant dust deposition originating in the African Sahara and Sahel regions. Dust emission has immediate impacts on humans, plus widespread influence on the radiative balance and on marine biological production and biogeochemical cycles, and is believed to have increased due to changes in land use practices (Jickells et al. 2005). A wealth of information comes from the long-term record (since 1965) of airborne desert dust measured in Barbados (13.17°N, 59.43°W). At this site, mineral dust concentrations are correlated with rainfall deficits in the sub-Saharan region (Prospero and Nees 1986; Prospero and Lamb 2003), and the net light scattering of dust exceeds that of nss-sulphate aerosol by about a factor of 4 (Li et al. 1996). There are still, however, considerable uncertainties associated with the global radiative forcing of mineral dust, given the high variability of dust loadings and limited knowledge of dust optical properties (Andreae et al. 2002). In the tropical North Atlantic, Saharan dust has been shown to stimulate nitrogen fixation, which is co-limited by iron and phosphorus (Mills et al. 2004 and references therein; Moore et al. 2009; Rijkenberg et al. 2011), and microbial species diversity (Hill et al. 2010, 2012). Recently, Okin et al. (2011) show that atmospheric deposition of iron can potentially contribute considerably to rates of marine productivity in high-nutrient-low-chlorophyll (HNLC) regions, and that iron is likely to be much more important than nitrogen in supporting net primary productivity globally.

5.1.1.4 Monitoring Reactive Trace Species in the Marine Atmosphere: Highlights from the Cape Verde Observatory

The tropics are particularly under-populated with marine reactive gas measurements, and this was a major motivation for setting up the Cape Verde Atmospheric Observatory (CVAO) in 2005. The location of the site (Fig. 5.2) about 800 km off the northwest coast of Africa in the tropical east Atlantic allows the study of clean marine air from diverse origins including North America, the Atlantic, Arctic, Europe and Africa. The station, now part of the global WMO-GAW long-term observing network, has been developed jointly by UK and German scientists at the Universities of York (UK), MPI-Jena (Germany) and IfT-Leipzig (Germany), in collaboration with the Cape Verdean meteorological service (INMG). Atmospheric measurements focus on reactive trace gases, greenhouse gases and aerosols (Fig. 5.3). The CVAO links with a sister oceanographic station, the Cape

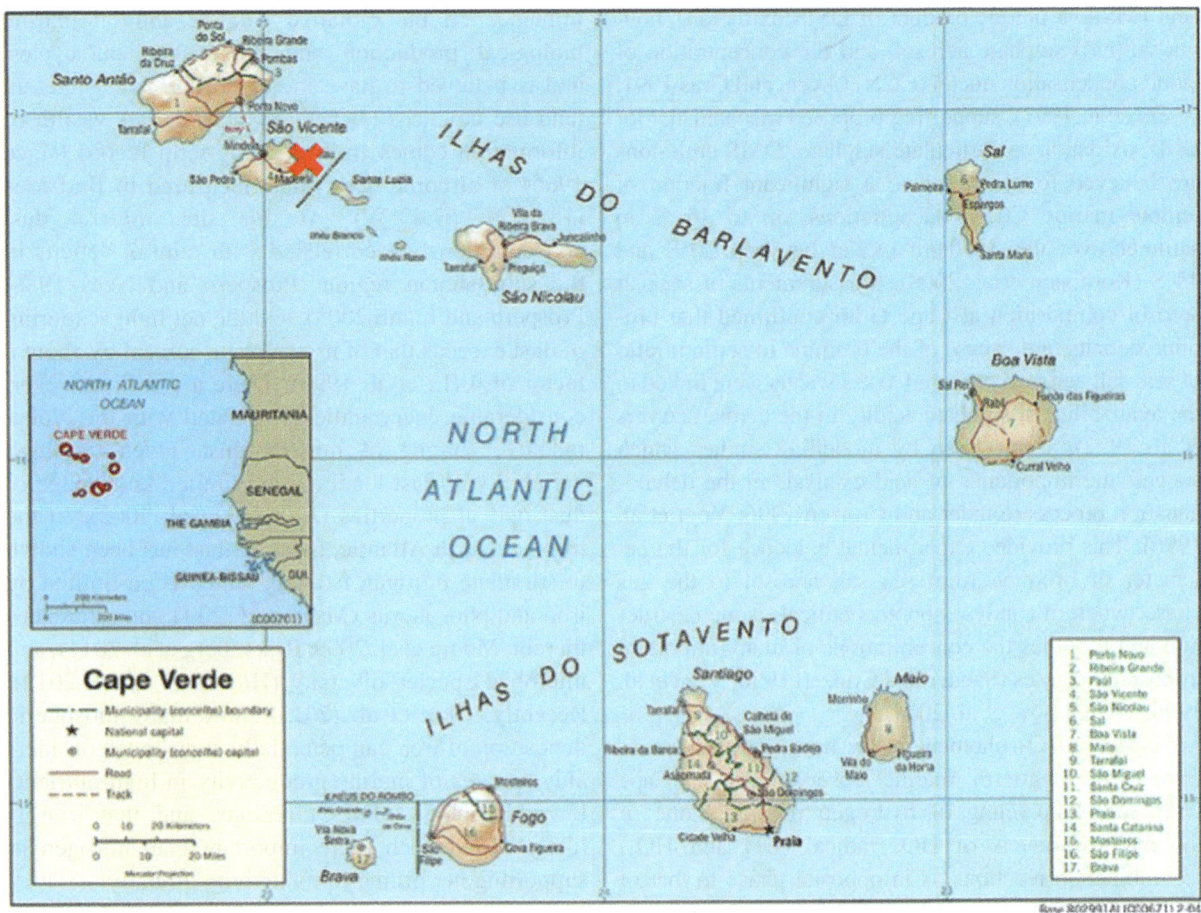

Fig. 5.2 The Cape Verde Islands and location of the CVAO, marked on São Vicente with a *red cross*. Prevailing trade winds are from the north-east (Map from http://commons.wikimedia.org/wiki/File:Cape_Verde_Map.jpg)

Verde Oceanic Observatory, located upwind of the CVAO at 17.59°N, 24.25°W. This was set-up by IFM-Geomar (Germany) in collaboration with the Cape Verde Fisheries Institute (INDP).

Similarly to Barbados, the Cape Verde archipelago is a region subject to very large depositions of dust, originating from the Sahel region, as well as from north-western Saharan sources and as far east as the Bodélé depression in Chad. In this region, dust is deposited mainly by dry deposition, peaking in the winter months when African desert dust is exported across the Atlantic within the lower troposphere. The ionic composition of the aerosol at Cape Verde is dominated by sea salt, but in Saharan dust episodes iron typically constitutes ca. 3.8 % of the total aerosol mass, with aluminium, a tracer of mineral dust, at a slightly higher concentration of ca. 7 % (Trapp et al. 2010; Carpenter et al. 2010). Total iron content reaches up to 33 μg m^{-3} in winter, and the soluble Fe content is between 0.1 % and 15.7 % (Carpenter et al. 2010). Higher solubilities are measured at lower atmospheric dust concentrations, a ubiquitous feature among aerosol solubility datasets.

Although the total aerosol mass is dominated by sea salt, in aerosol particles < 0.14 μm diameter, non-sea-salt components contribute about 80 % of the mass. These components include low-molecular-weight dicarboxylic acids (DCAs) and hydroxylated DCAs, methanesulphonic acid (MSA) and aliphatic amines. A bimodal size distribution for the DCA oxalic acid and coarse mode concentration maxima for the other DCAs are observed, as is typical for marine aerosols. The MSA concentration closely follows that of non-sea-salt-sulphate and the size distribution shows a

5.1 Perspectives: In Situ Observations, Remote Sensing, Modelling and Synthesis

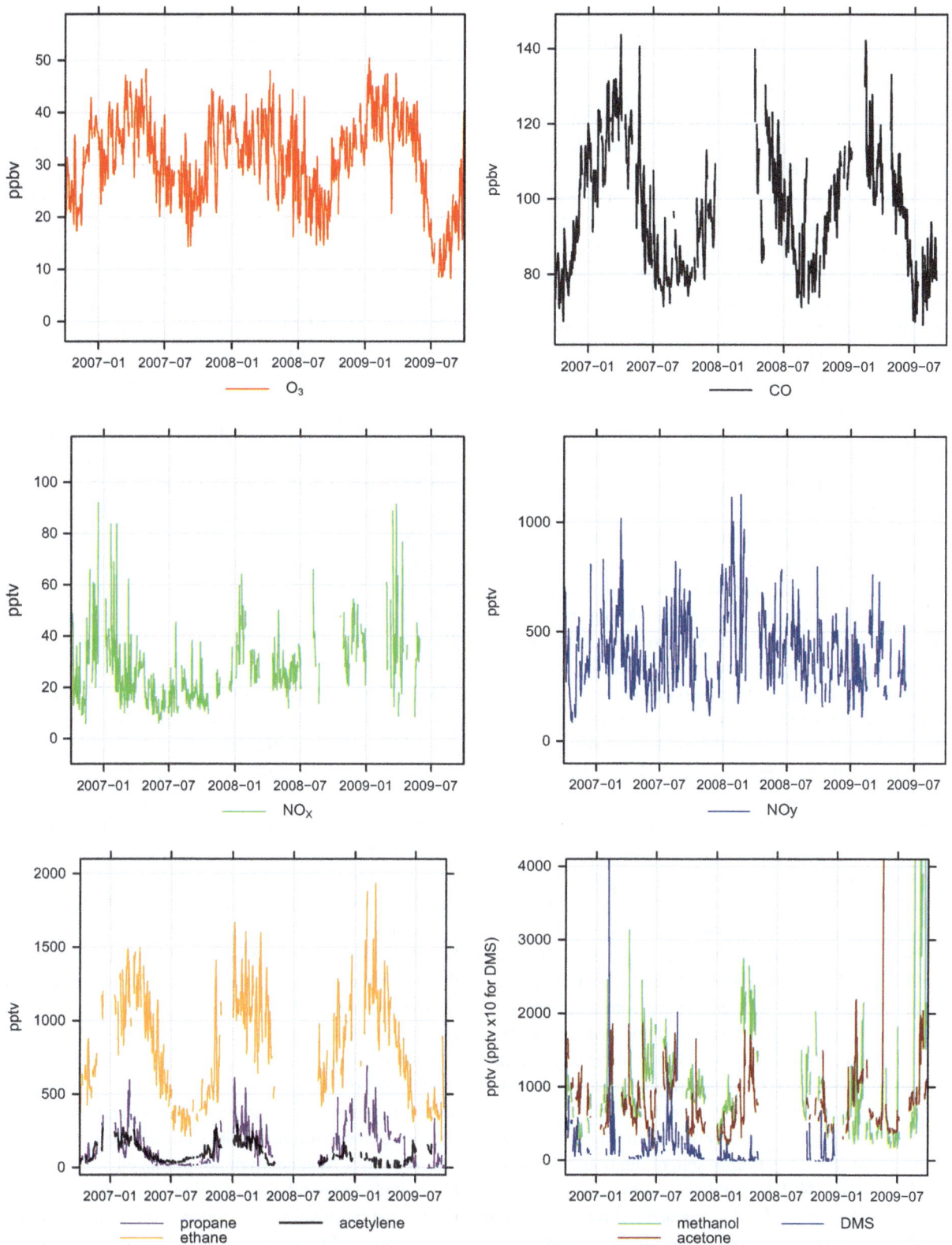

Fig. 5.3 Three-year time series (Oct 2006–Sept 2009) of daily averaged O_3, CO, NO_x, NO_y and VOC (propane, ethane, acetylene, acetone, methanol and DMS) mixing ratios measured at the Cape Verde Atmospheric Observatory (From Carpenter et al. 2010)

maximum mean concentration in the accumulation mode and in sea-salt particles. Aliphatic amines, assumed to be important in the growing process of sulphuric acid clusters, are correlated with phytoplankton activity in the subtropical North Atlantic, especially during an unexpected winter algal bloom (Müller et al. 2009).

The CVAO is one of the few global GAW stations that measures nitrogen oxides (NO_x) (as well as NO_y – both with low pptv detection limits) and VOCs (including oxygenated (O)VOCs). There is very little information on the abundance and distribution of these gases in the marine boundary layer (MBL) in part due to an inability to observe some of these compounds at the very low concentrations characteristic of this environment. Nitrogen oxides act as a catalyst for O_3 production and cycle HO_2 to OH, so are central to determining both the concentration of O_3 and CH_4. Observations show that in this region NO_x levels peak in winter (at 35–45 pptv), when air masses from Africa and Europe prevail (Lee et al. 2009a; Carpenter et al. 2010). This seasonality is attributed mainly to increased NO_x transported from the West African continent from e.g. soils, particularly after rainfall events over the Sahel region during the summer monsoon (Jaeglé et al. 2005; Stewart et al. 2008) or from anthropogenic sources, either directly as NO_x or locally produced from transported reservoir species (e.g. peroxy acetyl nitrate (PAN) or nitric acid (HNO_3)). These reservoir species may undergo decomposition within long-range plumes re-releasing NO_x, particularly for PAN as air masses descend and reach higher temperatures. Averaged NO mixing ratios at Cape Verde (daily averages are between 2 and 8 pptv) are negatively correlated with observed photochemical O_3 destruction; these observations were reproduced using a simple box model and together imply that the presence of 17–34 pptv of NO would be required to turn the tropical North Atlantic from an O_3 destroying to an O_3 producing regime (Lee et al. 2009a). Since NO_x emissions from shipping (e.g. Dalsøren et al. 2010) and African anthropogenic sources (Clarke et al. 2007) are believed to be increasing, future trends of background O_3 in this region could be of major concern for air quality and climate (Lelieveld et al. 2004).

OVOCs are generally present in higher concentrations in the lower atmosphere than non-methane hydrocarbons (NMHC) and have a comparable if not greater effect on oxidising capacity through reaction with hydroxyl radical (OH). Upon photodecomposition, they produce organic radicals that can form organic nitrate compounds such as PAN, sequestering NO_x and transporting it to remote regions of the atmosphere, thus affecting the tropospheric ozone budget and concentrations of OH (Singh et al. 1995; Tie et al. 2003). In the remote marine atmosphere, oceanic sources and sinks are expected to play a significant role in controlling OVOC concentrations, however both the magnitude and direction of OVOC fluxes are a matter of debate (Heikes et al. 2002; Carpenter et al. 2004; Williams et al. 2004; Jacob et al. 2005). Five years of acetone, methanol and acetaldehyde data from the CVAO (October 2006–September 2011) have recently been analysed using the CAM-Chem chemistry-transport model (Read et al. 2012). Observed annual mean mixing ratios of acetone, methanol and acetaldehyde were 763 ± 126 pptv, $1,029 \pm 151$ pptv and 511 ± 106 pptv, respectively. All three OVOCs show a similar cycle with maxima in spring (March) and autumn (between July and September with particularly high peaks in some years in September), lower levels in summer and generally the lowest levels in winter (Nov-Jan). The model reproduced the acetone concentrations fairly well in magnitude (annual average 670 ± 41 pptv) although underestimating the measured autumn peak, possibly due to underestimation of African biogenic sources. The modeled methanol levels (annual average 355 ± 17 pptv) were almost a factor of 3 lower than the observations, showed considerably less variability, and did not capture the pronounced peaks in spring and in summer – autumn. Possible reasons for the discrepancies include an underestimate by the model of methane concentrations and of terrestrial biogenic sources of methanol, and/or that the tropical North Atlantic is a significant net source of methanol, as suggested by recent data (Beale et al. 2013). Indeed, including estimates of the sea-air flux of methanol based upon these measurements led to an increase in the simulated levels by a factor of ~ 2.5 (new modeled annual average 879 ± 84 pptv) (Read et al. 2012). Of the three OVOCs, the most pronounced model-measured discrepancy was for acetaldehyde, with a model underestimation of over a factor of 10 (modelled annual mean mixing ratio 38 ± 7 pptv) and a predicted strong seasonal minimum in summer, contrary to the observations. Acetaldehyde is produced in the ocean through photodegradation of coloured dissolved organic matter and has a strong dependence on sunlight (Kieber and Mopper 1990). Including estimates of the sea-air flux of acetaldehyde from the measurements of Beale et al.

(2013) led to a significant increase in concentrations of acetaldehyde (annual average 139 pptv ± 46). The model thus still falls short of the observations (especially in September through to December – a period of high coastal and continental African influence) for reasons that are currently unknown.

Reactive marine-derived halogens have been proposed to exert a globally significant effect on the concentration and lifetimes of climatically active gases through gas and aerosol phases of the marine boundary layer (Vogt et al. 1996; von Glasow et al. 2002a, b). Bromine and iodine-containing reactive halogen species can influence tropospheric oxidation capacity via a number of reaction cycles including catalytic O_3 destruction, modification of NO_x and HO_x cycles with resulting effects on the lifetimes of other climatically important trace gases (Keene et al. 2009), oxidation of DMS (von Glasow et al. 2004); and oxidation of sulphur(IV) in acidified sea-salt aerosol and cloud droplets (Vogt et al. 1996). Observations of halogen oxide radicals, ozone, and supporting data at Cape Verde made in 2007 provided the first direct experimental evidence for halogen-catalysed tropospheric ozone destruction (Read et al. 2008). More recently, the presence of such halogens at only at a few pptv has been shown to constitute nearly 20 % of the instantaneous sink of HO_2 in this region (Whalley et al. 2010). Iodine monoxide (IO) radicals are believed to be produced mainly via photolysis of iodine-containing halocarbons volatilised from the ocean, yet recent data shows that the sea-air flux of these compounds is sufficient to explain only ~ 20–25 % of the levels of IO observed at Cape Verde (Jones et al. 2010; Mahajan et al. 2010). Recent research (Martino et al. 2009; Jammoul et al. 2009; Carpenter et al. 2013) suggests a role for sea surface chemistry in producing additional halogens, however the significance of such mechanisms remains an open question.

Cape Verde researchers aim to build on these first few years of measurements at the CVAO over the next decade by, for example, quantifying the nitrogen oxides budget, elucidating the nature and magnitude of oceanic iodine emissions, evaluating the influence of dust on the ocean heat budget, understanding oceanic nitrogen fixation, quantifying air-sea exchange fluxes of important gases in the west African upwelling area, and analysing long-term trends in trace gases and aerosols in the context of environmental and climate change. CVAO data can be accessed at http://badc.nerc.ac.uk/data/solas/projects/capeverde.html and http://gosic.org/gcos/GAW-data-access.htm.

5.1.1.5 Conclusions

Oceanic and atmospheric time-series sites started in the 1950s, and their contribution has been invaluable. Some have been in continuous operation without any interruption, some have stopped temporarily and some indefinitely. Since the early 2000s, the launch of the Argo programme with the target of 3,000 floats per year cruising the global ocean has brought a new perspective since it provides a unique and systematic source of information about the interior of the ocean. Only the combination of eulerian and lagrangian observatories in an integrated framework will allow a four dimensional vision of the state of the ocean. One basic key to success of these networks is the constant quest for the best procedures for quality checking, intercomparability and treatment of the data collected. Another key is for the data to be archived in a responsible manner, meaning ensuring proper software developments and addressing management challenges of really huge datasets to secure online delivery and long-term security.

5.1.2 Earth Observation Products

In 1957, the Sputnik was successfully launched and the space age initiated, heralding the evolution of Earth Observation, i.e. the scientific study of the Earth's surface and atmospheric composition from space. Since the early 1970s, satellite oceanography has made huge progress and global satellite observations are now crucial elements of the global climate observing systems (GCOS). Remotely sensed data are also basic ingredients of any oceanic and atmospheric process study. Sea surface temperature, sea level, wave height, winds, sea surface salinity, sea ice and ocean colour are ocean–atmosphere observables monitored with near global coverage on a daily to monthly basis. Satellite measurements of concentrations of trace gases and long lived greenhouse gases, when combined appropriately with atmospheric chemistry models, have over the past 30 years provided a continuously improving picture of the distribution of the surface fluxes of these gases at the air-sea interface. Satellite observations are instrumental tools of SOLAS science to address seasonal to multi-decadal time scale variability in the ocean–atmosphere exchange of gases and aerosol-borne chemicals. In situ observations, both oceanic and atmospheric, presented in Sect. 5.1.1, provide groundtruthing for the calibration algorithms necessary for deriving these oceanographic and atmospheric properties from space. Remotely sensed data provide global-scale data sets at an unprecedented spatio-temporal resolution. One major challenge ahead is to avoid any discontinuity of operating satellites for the long-term archive and to minimise calibration drift for performing proper climate studies. Earth observations represent a unique observational capability to detect changes in the ocean–atmosphere system and to better understand how planet Earth functions as a complex adaptive system. The following subsections briefly introduce each type of ocean–atmosphere observation.

5.1.2.1 Altimetry, SST, Winds, Sea State

The advent of **satellite altimetry** has given oceanographers a unique tool for studying oceanic circulation and its changes with time. From the vantage point of space, a radar altimeter is able to measure the shape of the sea surface globally and frequently. Due to three decades of international effort, satellite altimetry has benefited from a series of missions, leading to an improvement in measurement accuracy by three orders of magnitude, from tens of meters to a few centimeters (Fu and Cazenave 2001). The evolution from Seasat (1978), Geosat (1985–1989), ERS-1/2 (1991–2011), TOPEX/POSEIDON (1992–2008), Jason1 (2001 to present), OSTM/Jason2 (2008 to present), ENVISAT (2000–2012) to Cryosat-2 (2010 to present) has produced and will produce a wealth of data of progressively improving quality (http://www.aviso.oceanobs.com/, http://sealevel.jpl.nasa.gov). The Saral/Altikal mission, launched in 2012, will ensure, in association with Jason-2, the continuity of the service currently provided by the altimeters onboard Envisat and Jason-1, and will contribute to building a global ocean observing system. The Surface Water and Ocean Topography (SWOT, http://swot.jpl.nasa.gov/ mission) mission to be launched in 2020 will revolutionise our conceptual view of ocean dynamics since it will characterise mesoscale and submesoscale circulation with a 10 km space resolution or better. In order to meet the long wavelength calibration accuracy requirements, topography profile measurements will be available with an accuracy equal to or better than the Jason series of altimeters and radiometers (see

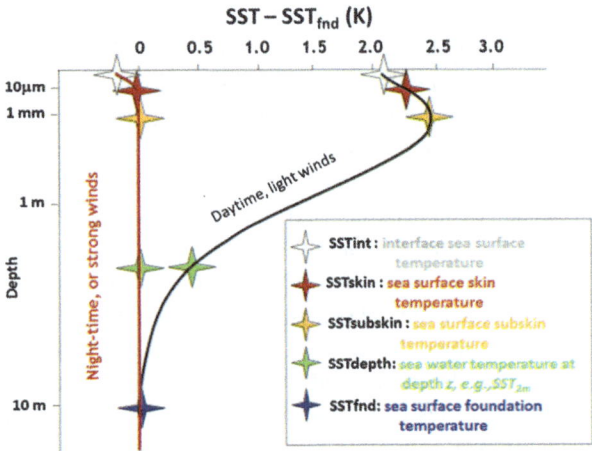

Fig. 5.4 The hypothetical vertical profiles of temperature for the upper 10 m of the ocean surface in high wind speed conditions or during the night (*red*) and for low wind speed during the day (*black*) (https://www.ghrsst.org/ghrsst-science/sst-definitions) (© American Meteorological Society. From Donlon et al. 2007. Used with permission)

SWOT Science Requirements Document, Version 1.1, 2012). Satellite altimetry observations, often assimilated by global ocean circulation and coupled numerical models, constitute the first global synoptic data sets for the study of the following topics: large scale circulation, mesoscale eddies, boundary currents, tropical circulation, large-scale variability on time scales from intraseasonal to interannual in relation to forcing mechanisms, El Niño and La Niña, planetary wave dynamics, eddy dynamics, to list a small sample of topics.

Sea Surface Temperature (SST) is a difficult parameter to define exactly because the upper ocean (~ 10 m) has a complex and variable vertical temperature structure that is related to ocean turbulence and the air-sea fluxes of heat, moisture and momentum. Figure 5.4 presents a schematic diagram that summarises the definition of SST in the upper 10 m of the ocean. The skin temperature (SSTskin) is defined as the temperature measured by an infrared radiometer typically operating at wavelengths 3.7–12 µm (chosen for consistency with the majority of infrared satellite measurements) that represents the temperature within the conductive diffusion-dominated sub-layer at a depth of ~ 10–20 µm. Merging measurements of SST made by different satellite and in situ instruments on drifting or moored buoys requires a proper framework to understand the information content and relationships between these measurements.

The latest reprocessing (Pathfinder Version 5.2) is a new reanalysis of the Advanced Very High Resolution Radiometer (AVHRR) data stream developed by the University of Miami RSMAS and the NOAA National Oceanographic Data Center. It uses an improved version of the Pathfinder algorithm and processing steps to produce twice-daily SST and related parameters dating back to 1981, at an areal resolution of approximately 4 km, the highest possible for a global AVHRR (see http://www.nodc.noaa.gov/SatelliteData/pathfinder4km/). The through-cloud capabilities of microwave radiometers (AMSR-E, TMI) provide a global daily SST map without missing data due to orbital gaps or environmental conditions precluding SST retrieval. Microwave optimally-interpolated products have been proposed at ¼° resolution, and by blending with infrared SSTs from MODIS at 0.09° resolution.

Under the Global Ocean Data Assimilation Experiment (GODAE) umbrella, the *Group for High Resolution SST* (GHRSST) aims at providing the best quality sea surface temperature data, without missing data, for applications in short, medium and decadal/climate time scales in the most cost effective and efficient manner. Each day the *GHRSST Multi-product Ensemble (GMPE)* experiment produces a median *SST* map and associated standard deviation map using *SST* analysis data collected over the previous 24 h period (i.e. yesterday). Thus, the nominal analysis time for the *GMPE* median ensemble *SST* is 12:00 for the previous day (i.e. T-1). The *GMPE* median ensemble *SST* is computed as a median average using a variety of *GHRSST L4* analysis products after their differing analysis grids have been homogenised by area averaging onto a standard ½° latitude longitude grid (https://www.ghrsst.org/data/todays-global-sst/). The median-ensemble *SST* coverage is restricted by the use of the *OSTIA* analysis land mask. The *GMPE* median ensemble *SST* is currently derived using the following inputs: the Met Office *OSTIA SST* analysis http://ghrsst-pp.metoffice.com/pages/latest_analysis/ostia.html, the *NCEP RTG_SST_HR SST* analysis http://polar.ncep.noaa.gov/sst/, the *NAVOCEANO NAVO K10 SST* observations https://www.navo.navy.mil/ops.htm, JMA *MGDSST SST* analysis http://goos.kishou.go.jp/rrtdb-cgi/jma-analysis/jmaanalysis.cgi, the RSS *MW Fusion SST* and MW + IR Fusion SST analyses mentioned above http://www.remss.com/sst/microwave_oi_sst_browse.html, the *FNMOC GHRSST-PP SST* analysis http://www.usgodae.org/cgi-bin/datalist.pl?summary=Go&dset=fnmoc_ghrsst, the MERSEA *ODYSSEA SST* analysis http://www.mersea.eu.org/Satellite/sst_validation.html, the *NOAA AVHRR OI* (Reynolds) http://www.ncdc.noaa.gov/oa/climate/research/sst/oi-daily.php, the *Meteorological Service of Canada (CMC) 1/3° SST analysis* http://www.msc-smc.ec.gc.ca/contents_e.html, and the BMRC *GAMSSA SST* analysis http://podaac.jpl.nasa.gov/dataset/ABOM-L4LRfnd-GLOB-GAMSSA_28km.

As an example central to ocean–atmosphere interactions at the heart of SOLAS science, Fig. 5.5 presents the *GMPE ensemble SST anomaly map* for January 18, 2012 showing clearly that La Niña conditions are present across the Equatorial Pacific. SSTs are at least 0.5 °C below average across much of the central and eastern equatorial Pacific ocean. A horseshoe pattern of above-average SSTs extends from the Maritime Continent into the mid-latitudes of the Pacific Ocean. The sea surface height anomalies from Jason-2 for January 20, 2012 confirm that La Niña is peaking in intensity in the equatorial Pacific (Fig. 5.6). This image is based on the average of 10 days of data centered on January 20, 2012. It depicts places where the Pacific sea surface height is higher than normal (due to warm water), and where the sea surface is lower than normal (due to cool water). Green colour indicates near-normal conditions. The La Niña episode changes global weather patterns and is associated with less moisture in the air over cooler ocean waters. This results in less rain along the coasts of North and South America and along the equator, and more rain in the far Western Pacific.

Ocean surface winds are needed to estimate momentum transfer (surface stress) and gas transfer velocity between the atmosphere and the ocean, and are instrumental for determining large-scale ocean circulation and transport. Accurate wind speeds are essential for ensuring reliable computations of air-sea heat and mass fluxes making surface winds critically important for budgeting energy, moisture, gases and particles (Fairall et al. 2010). Several reviews of space-based wind measurements and applications have been published (i.e. Liu 2002; Liu et al. 2008; Bourassa et al. 2010). The challenge is to continuously improve the present ocean wind system by means of better bias removal and calibration for low and very high wind speeds, increased temporal sampling using a constellation of instruments, finer spatial resolution and improved methods of fusing observations from multiple platforms.

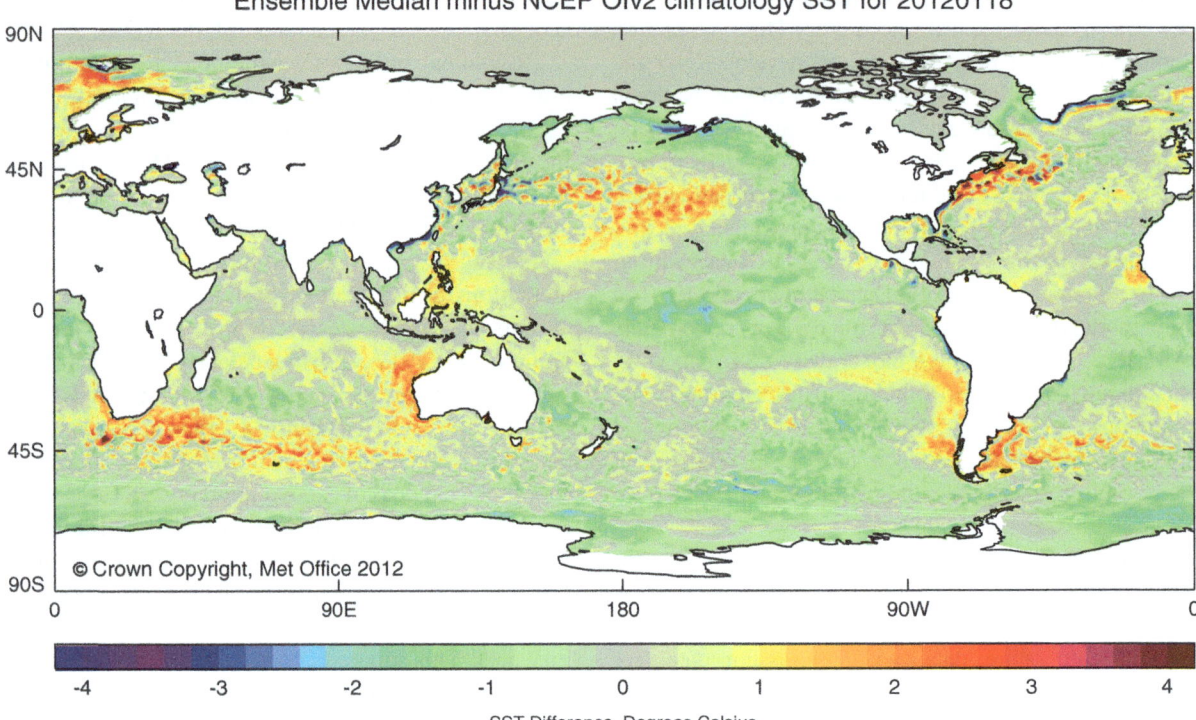

Fig. 5.5 GMPE ensemble SST anomaly map for January 18, 2012; climatology is derived from NCEP/NOAA between 1985 and 2001 (https://www.ghrsst.org/data/todays-global-sst/; Martin et al. 2012) (Figure provided by J. Roberts-Jones, Met Office UK, Crown Copyright)

Instruments that are routinely used to measure vector winds (speed and direction or two vector components) include scatterometers, passive polarimetric sensors and Synthetic Aperture Radar (SAR), and those measuring scalar winds (speed only) include passive microwave radiometers and altimeters. The SeaWinds scatterometer on the QuikSCAT satellite measures surface winds with a resolution of ~ 25 km across a swath width of ~ 1,600 km. The temporal sampling is a function of the orbit and the swath width. The main weaknesses of scatterometers are rain contamination for some rain conditions and lack of data near land (~ 15 km for QuikSCAT) (Weissman et al. 2002; Draper and Long 2004; Nie and Long 2008). Fusion of data from multiple scatterometers significantly improves the temporal coverage (Liu et al. 2008). The assimilation of scatterometer winds in Numerical Weather Prediction (NWP) models has improved the quality of forecasts of tropical cyclones (e.g. in wave forecasting and hurricane force warnings for nowcasting applications). High winds play a large role in Earth's climate, dramatically enhancing gas exchange of greenhouse and trace gases and marine aerosols. However, validation under high winds is difficult due to the scarcity of such events and their tendency to occur in high latitude regions, together with uncertainty in data from buoys and/or ships in rough seas due to wave sheltering.

The root mean square (rms) difference between remotely sensed and buoy wind speeds is generally less than 1 m s^{-1} in non-rainy regions, provided that atmospheric stability and surface currents are taken into account as satellite measurements are physically more related to wind stress than to atmospheric wind speed. The sensitivity of satellite microwave (synthetic aperture radar, scatterometer, altimeter and microwave radiometer) measurements to the **sea state** varies according to the instrument type and to its operating frequencies.

In the future, the combination of multi-frequency, multi-angular and multi-sensor measurements should improve the characterisation of the sea state and

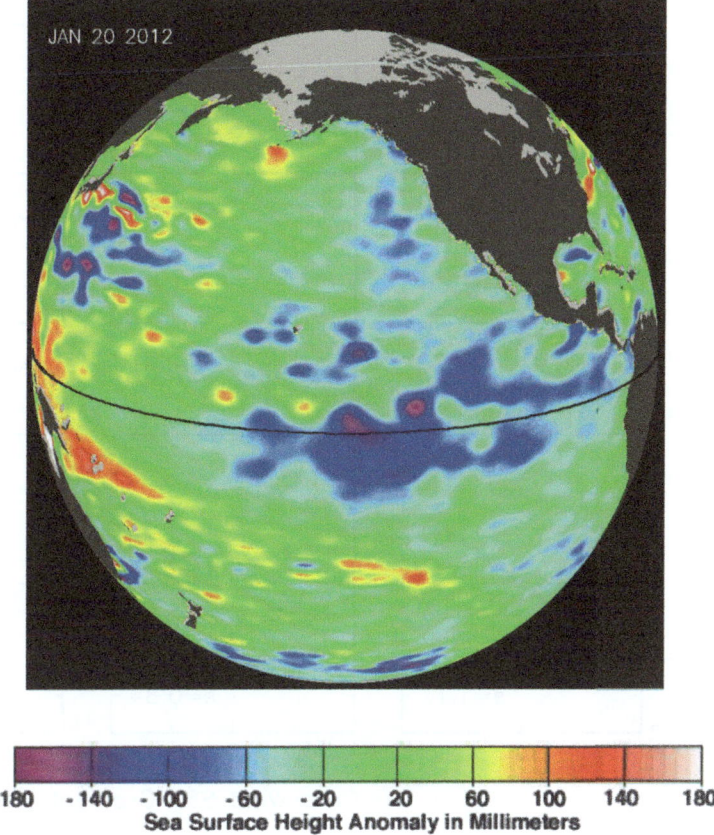

Fig. 5.6 Jason-2 Sea surface height anomalies centered on January 20th, 2012 (Figure courtesy of NASA/JPL-Caltech)

spatial resolution, especially in rainy cases and in very strong wind speed cases. Recent studies show that the retrieval of wind speed in rainy regions, with a degraded precision, is now possible using WindSat radiometric measurements (Meissner and Wentz 2009), that ocean satellite altimetry can retrieve wind speeds from the radar backscatter in gale to storm conditions (Quilfen et al. 2011) and that the new L-band radiometer measurements, much less affected by rain than at higher frequencies, allow measurements of winds in a cyclone up to 50 m s^{-1} (Reul et al. 2012b). Apart from wind speed, the combination of existing and future instruments should also help to retrieve wave parameters.

Altimeters enable the measurement of wave height and mean square slope (mss) in various wavelength ranges depending on the altimeter frequency. Until now, altimeters (e.g. Jason, ENVISAT) have operated at three frequencies (S, C, Ku bands), corresponding to wavelengths of 2.2, 5.6 and 9.3 cm. The future Saral/Altika altimeter operating in the Ka band (wavelength of 8 mm) will complement existing measurements. First retrievals of mean square slope using Global Navigation Satellite System-Reflectometry (GNSS-R) are encouraging (e.g. Clarizia et al. 2009).

Among the new planned missions, the French-Chinese CFOSAT mission (see http://smsc.cnes.fr/CFOSAT/index.htm) will, in addition to wind speed, provide the directional wave spectrum.

Scatterometer observations together with microwave observations of SST have facilitated the collection of data that couple air-sea processes at scales smaller than the regional, typically at the mesoscale. Indeed Chelton et al. (2004) and Small et al. (2008) have identified an intimate link between modification of the dynamics of the atmospheric boundary layer by the SST and the feedback of this modification on the ocean through wind surface stress and heat flux. This link has been observed between sharp SST gradients and surface winds – the so called 'Chelton effect', where winds tend to accelerate over warm and decelerate over cold waters in the frontal zone, resulting in a

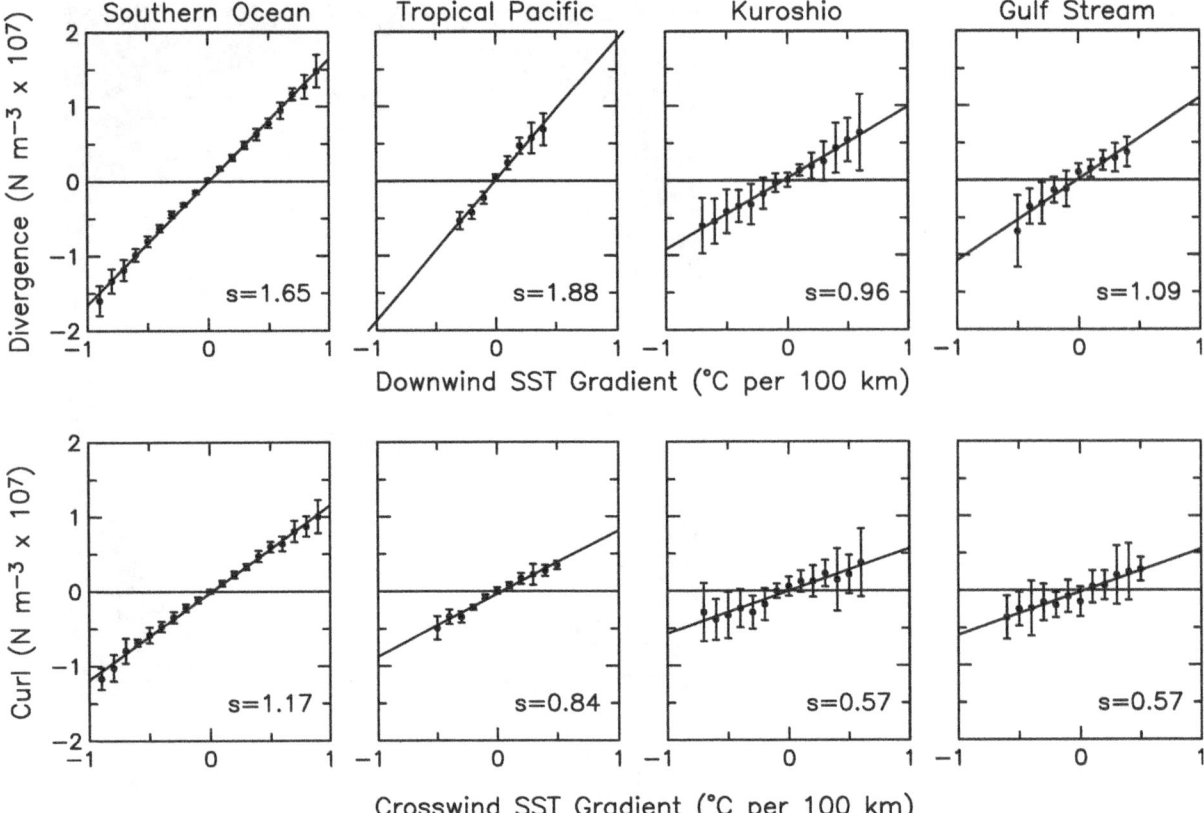

Fig. 5.7 Frontal-scale SST effects on wind stress divergence and curl. Shown are binned scatter plots of spatial high-pass filtered fields of the wind stress divergence as a function of the downwind SST gradient (*top row*) and the wind stress curl as a function of the crosswind SST gradient (*bottom row*) for four geographical regions: the Southern Ocean, the eastern tropical Pacific, the Kuroshio Extension and the Gulf Stream. The points in each panel are the means within each bin computed from 12 overlapping 6-week averages, and the error bars are ± 1 standard deviation over the 12 samples in each bin. The wind stress divergence and curl are multiplied by 10^7 and the units are Nm^{-3} (From Chelton et al. 2004)

quasi-linear relationship between the curl (divergence) of the wind and the SST gradient according to a perpendicular (parallel) direction to the wind (Fig. 5.7). The remarkable 10-year QuikSCAT data record has given some insight into the nature of this variability and the dynamic and thermodynamic impacts of this atmosphere–ocean coupling on ocean circulation and atmospheric weather patterns. However, the intricacy of interaction between the atmospheric and oceanic boundary layers through the 'Chelton effect' has yet to be merged with the still unresolved biogeochemical impact of mesoscale eddies (Siegel et al. 2011).

5.1.2.2 Sea Surface Salinity

While sea surface temperature, sea level, sea ice and sea state are relatively well monitored as an intrinsic part of the global climate observing system (GCOS 2009), until 2009 **sea surface salinity** (SSS) was not measured from space. Salinity is recognised as an essential climate variable (GCOS 2010) and satellite SSS is expected to be highly complementary to existing in situ salinity measurements (Lagerloef et al. 2010).

The feasibility of measuring SSS from space was first demonstrated in the frame of the Skylab mission launched in 1973. However, at L-band frequencies around 1.4 GHz, the sensitivity of radiometric measurements to salinity is low and the radiometric resolution of the instruments remained an obstacle to the development of new satellite missions until the 1990s. Since then, the development of new technologies (Lagerloef et al. 1995) has contributed to two satellite missions accepted by space agencies: the Soil Moisture and Ocean Salinity (SMOS) mission of the European Space Agency and the Aquarius/SAC-D

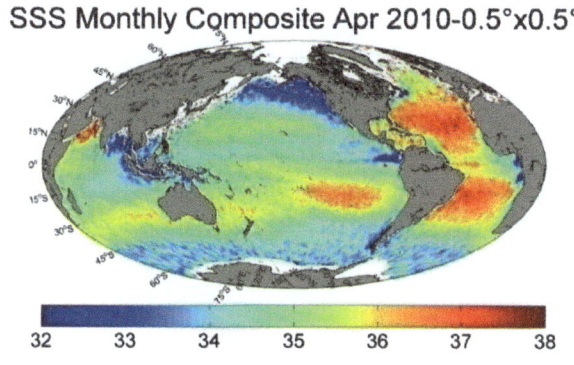
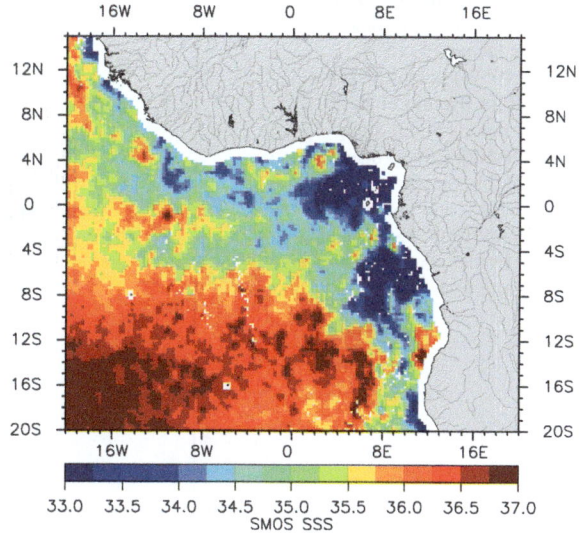

Fig. 5.8 SMOS SSS maps derived at (CATDS-CEC-OS), after applying a thorough filtering of the outliers. (**a**) global map in April 2010; (**b**) in the Gulf of Guinea from 21 to 30 April 2010 (See http://www.catds.fr/ for CATDS processing activities; N. Reul and J. Tenerelly, SMOS Level 3 SSS Research products -Algorithm Theoretical Breadboard Document, 2011, available on http://www.ifremer.fr/naiad/salinityremotesensing.ifremer.fr/CATDS_CECOS_SMOS_Level3Products_ATBD.pdf)

mission of the NASA/CONAE agencies. SMOS uses a new antenna concept (synthetic aperture) for spaceborne radiometry applications and was launched in November 2009; Aquarius uses a large size real aperture antenna and was launched in June 2011. The goal of these two missions is to achieve SSS accuracy of ~ 0.2 or better when averaged over 150–200 km and a monthly timescale.

An overview of the first retrievals of SMOS SSS is given in Font et al. (2012) and is detailed in papers submitted to the IEEE TGRS SMOS special issue (May 2012). Recently, numerous improvements have been made; below we present examples obtained from recently reprocessed data (Fig. 5.8).

Data for the entire year 2010 has been reprocessed at the Centre Aval de Traitement des Données SMOS (CATDS-CEC-OS). In order to remove outliers linked to radio-frequency interferences (RFI), consistency checks based on yearly SMOS data have been performed before retrieving SMOS SSS. When comparing monthly 1° SMOS SSS with in situ SSS data from ships, ARGO and moorings at a global scale, the error standard deviation is 0.6 globally and 0.4 in the tropics (Reul et al. 2012a).

SSS derived in July 2010 with Version v5 of the ESA processors (v5 will be used for the whole mission reprocessing until the end of 2011), led to a precision of 0.2 for SSS averaged over 10 days and 100 km in the subtropical Atlantic. In the rainy tropical Pacific Ocean between 5°N and 5°S, the SMOS-ARGO SSS scattering is greater due to the SSS vertical gradient: SMOS exhibits a mean freshening of 0.1 in the surface water with respect to 5 m depth (Fig. 5.9 Boutin et al. 2012).

The new SMOS data processing demonstrates the capability of retrieving SSS from satellites with a precision of about 0.3 or even better in warm areas. The combination of experience from the SMOS and Aquarius missions is expected to improve this precision.

New satellite SSS will be a key tool for studying air-sea interactions, e.g. the spread of fresh river water into the open ocean, rain surface freshening and the detection of fronts (see BEC processings on http://www.smos-bec.icm.csic.es/).

5.1.2.3 Marine Carbon Observations from Satellite Data: Ocean Color/PIC/POC

Phytoplankton are the basis of marine food webs and contribute up to about 50 % of global primary production. Phytoplankton play a role in the budgets of both organic trace gases and aerosols as a marine source: for example, dimethylsulphide production by

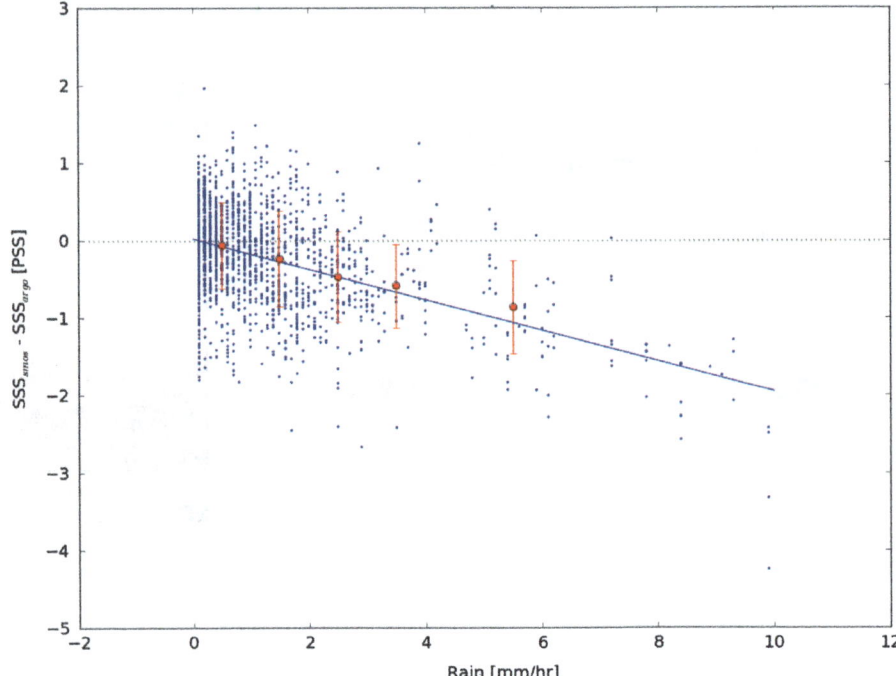

Fig. 5.9 SMOS SSS at 1 cm depth minus ARGO SSS at 5 m depth in the tropical Pacific Ocean (5–15°N; 180–110°W) versus SSMI Rain Rate (Boutin et al. 2012) (More details about along-track SMOS ESA processing is available on www.argans.co.uk/smos/ and about LOCEAN/IPSL SMOS Cal/Val activities on www.locean-ipsl.upmc.fr/smos)

oceanic phytoplankton, leading to the formation of sulphate aerosol and cloud condensation nuclei formation/growth in the marine atmosphere has been studied extensively (e.g. Liss et al. 1997). A link between oceanic chlorophyll *a* content (chl *a*), the indicator of phytoplankton biomass, and cloud droplet numbers over the Southern Ocean (Plass-Dülmer et al. 1995) has been observed, as well as enhanced organic mass in marine aerosols during periods of enhanced ocean biological activity (Singh et al. 2003; O'Dowd et al. 2004). Bromoform observations in the tropical eastern Atlantic Ocean (Quack et al. 2004, 2007) have revealed a pronounced subsurface maximum at the depth of the subsurface chl *a* maximum, suggesting a phytoplanktonic source of bromoform. Ocean-emitted volatile organic compounds also appear to be related to phytoplankton activity (e.g. Gantt et al. 2009; Yassaa et al. 2008). An improved quantification of the dependence of atmospheric composition on marine biological activity is important for studying the ocean–atmosphere interactions of gases and particles and understanding the Earth's climate system and its response to anthropogenic influence. In order to study the dynamics of phytoplankton distribution over longer timescales, optical remote sensing of ocean phytoplankton (ocean colour) provides data on phytoplankton distribution and related parameters with near global coverage on a daily to monthly time resolution. Using relationships derived from in situ oceanic and atmospheric data at or just above the sea surface, ocean colour data products can then be used to infer emission rates of trace gases on longer time scales with reasonable temporal and spatial resolution (down to 1 km), e.g. Arnold et al. (2009, 2010).

Ocean colour has been focused since the 1980s on the detection of chl *a*, due to its strong absorption properties. Merged chl *a* and reflectance satellite data products are available from the SeaWiFS, MODIS and MERIS sensors (1997 until present), through NASA and ESA efforts to produce essential climate variables (Maritorena et al. 2010). However, chl *a* concentration changes with species composition and physiological state and cannot be converted directly into carbon biomass, which is the currency used in ocean carbon models. Therefore, recent advances in ocean colour have focused on the quantification of carbon pools, such as particulate organic carbon (POC), particulate inorganic carbon (PIC) and dissolved organic carbon (DOC), as well as on the assessment of different phytoplankton functional types (PFTs). Different PFTs have distinct impacts on the marine food web and biogeochemical cycling, e.g. variable relationships of different PFTs to isoprene production have been observed in laboratory experiments (Bonsang et al. 2010). Ocean colour satellite observation is restricted to the near-surface layer, which varies from meters to about 60 m thick depending on the presence of optically-significant water constituents and the wavelength considered (Smith and Baker 1978). Products derived from ocean colour satellite data are integrated over the first penetration depth.

POC algorithms are based on empirically-derived relationships of POC to either inherent optical properties (particle backscattering or attenuation coefficients), which are related to reflectances, measured by remote sensing at several wavelengths (Stramski 1999, Loisel et al. 2001, 2002; Gardner et al. 2006), or to the blue-to-green reflectance ratio (Stramski et al. 2008). The latter algorithm seems less sensitive to regional variability, but is closely related to chl a as it uses the same input parameters (the blue to green reflectance ratio). For biogeochemical studies, satellite near-surface POC data are insufficient because they correspond to the first attenuation layer only and deep chl a maxima often exist. Moreover, biogenic detrital particles, heterotrophic bacteria and viruses also contribute to POC in variable proportions throughout the entire water column. Stramski et al. (2008) provided quantitative estimates of the POC reservoir in three oceanic layers: the attenuation, the mixed-layer (MLD) and the 200 m layer depth. In oligotrophic waters, this approach may underestimate the POC reservoir where high POC accompanies deep chl a maxima, because it assumes that POC is uniform throughout the MLD and equals the near-surface POC concentration. This work has been improved by the empirical algorithm of Duforêt-Gaurier et al. (2010) who derive integrated euphotic zone POC from the entire SeaWiFS satellite data set.

The PIC ocean colour product represents biogenic particles composed of calcium carbonate which is produced by several phytoplankton groups, mainly coccolithophores. These create massive blooms in the ocean, and their PIC, being white, strongly reflects light, which imparts a turquoise-blue-white colour to the ocean. The blooms are easily observed in the pseudo-true-colour images from satellites and can be monitored using ocean colour (Brown and Yoder 1994). Algorithms have been elaborated to quantitatively retrieve PIC at regional and global scales (Gordon et al. 2001; Balch et al. 2005) and the most recent have been used to process the whole SeaWiFS and MODIS data set. An important constraint of these algorithms is that, at typical non-bloom concentrations, the PIC scattering represents only a few percent of the total scattering. Thus, to maximise the signal to noise ratio, satellite pixels must be aggregated in space and time, in order to define accurate mean concentrations. Currently, more verification with PIC field measurements is underway to optimise the use of PIC ocean-colour data in models (Balch et al. 2011). PIC is also produced by certain zooplanktonic organisms but these particles are too large to be detected by ocean colour (Balch et al. 1996).

Different bio-optical and ecological methods have been established that use ocean colour data to identify and differentiate between PFTs or phytoplankton size classes (PSCs) in the surface ocean. These can be summarised into four main types: spectral-response methods which are based on differences in the shape of the light reflectance/absorption spectrum for different PFTs/PSCs (Sathyendranath et al. 2004; Alvain et al. 2005, 2008; Ciotti and Bricaud 2006; Bracher et al. 2009; Sadeghi et al. 2011; Brewin et al. 2010a; Devred et al. 2011), methods which use information on the magnitude of chlorophyll a biomass or light absorption to distinguish between PFTs or PSCs (Devred et al. 2006; Uitz et al. 2006; Hirata et al. 2008; Brewin et al. 2010b; Hirata et al. 2011; Mouw and Yoder 2010), methods that retrieve the particle size distribution from satellite-derived backscattering signal and derive PSCs (Kostadinov et al. 2010), and ecological-based approaches which use information on environmental factors, such as temperature and wind stress, to supplement the bio-optical data for investigating PFTs (Raitsos et al. 2008). All methods derive dominant phytoplankton groups, while Uitz et al. (2006), Bracher et al. (2009, improved by Sadeghi et al. 2011) and Hirata et al. (2011) also give chl a for the different PFTs. Nearly all the PFT methods mentioned use information from the multispectral ocean colour sensors SeaWiFS, MERIS or MODIS and are based on the parameterisation of a large global or regional in situ data set in order to yield PFTs from satellite chl a or normalised water leaving radiances. Unexpected changes in the relationships between these parameters resulting from a regional or temporal sampling bias leads to a bias in the

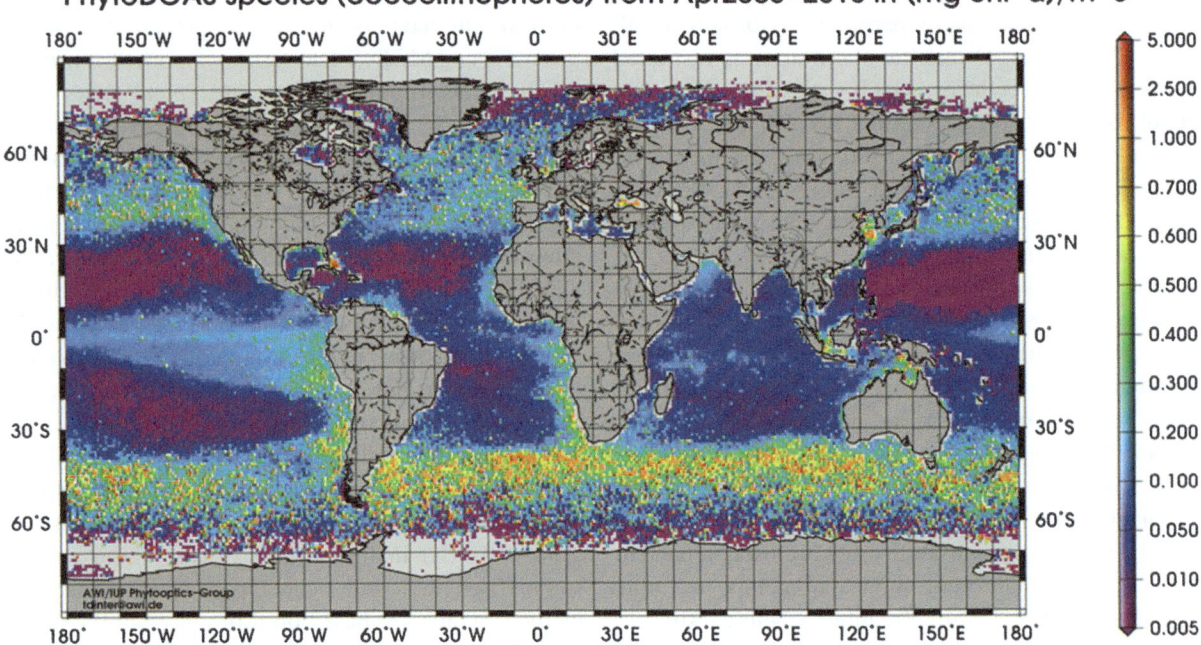

Fig. 5.10 Global climatology for April of coccolithophore biomass (given as chlorophyll *a* concentration) derived from the average of all SCIAMACHY data retrieved via PhytoDOAS multitarget-fitting (according to Sadeghi et al. 2011, 2012) (Source: T. Dinter, A. Bracher, Phytooptics, AWI-IUP)

detection of PFTs. In contrast, the PhytoDOAS method of Bracher et al. (2009, improved by Sadeghi et al. 2011) exploits the whole spectrum by using hyperspectral data of the satellite sensor SCIAMACHY and discriminates different PFTs by their characteristic absorption. Diatoms, cyanobacteria, dinoflagellates and coccolithophores are quantified (example in Fig. 5.10) without assuming empirical relationships as in the case of other PFT methods. Recent PFT algorithm intercomparison studies show the robustness of the abundance-based approaches at detecting dominant PSCs (Brewin et al. 2011). A new PFT algorithm intercomparison has been initiated where the quantitative assessment of PFT distributions will be compared. PFT monthly resolved products are available for the whole SeaWiFS or SCIAMACHY missions, using the methods of Alvain et al. (2008) and Hirata et al. (2011) covering 1998–2009 for the first data sets and the PhytoDOAS method covering 2002–2011.

5.1.2.4 Sea Ice

Methods for studying sea ice, including Earth observation products, have recently been thoroughly reviewed by Eicken et al. (2009), which includes a comprehensive reference list. Here we provide a brief summary of the approaches most relevant to the science of surface ocean-lower atmosphere exchanges.

In addition to satellite-borne remote sensing tools (e.g. Massom 2009), automated in situ systems (see Perovich 2009), mainly mounted on buoys, are proving to be a critical component of the global sea-ice observing network. In contrast to open water, sea ice is a very complex and variable surface, complicating calibration of remote sensing signals, and information from multiple sources is generally required to resolve measurements. Therefore, both multiple satellite sensors and widely distributed automatic measurement stations (also called ice-based observatories) remain critical to interpreting Earth observation data on sea ice (see Massom 2009 for a detailed discussion).

Sea ice distribution and motion are fundamental parameters needed for research on air-sea exchange in polar oceans. Passive microwave data (e.g. from the SSM/I series satellites) may be the most valuable tool available for tracking sea ice. Not only does microwave radiation provide information in the dark and through clouds, but in the Arctic, where summer brine flushing is common, it can also distinguish between first- and multi-year ice. While the relatively low resolution of passive microwave data precludes their use for studying landfast ice, synthetic aperture radar (SAR; aboard, for example, the ERS series and RADARSAT satellites) provides very high resolution data, to as good as 1 m, which is adequate not only for fast ice, but also for detailed ice edge information. Visible and infrared sensors (e.g. MODIS and AVHRR) also provide data on ice distribution, although only far-infrared sensors are useful during winter. Deriving ice motion from time series of satellite images carries a high uncertainty, and data from drifting ice buoy arrays are a vital component of accurate ice motion estimates.

Beyond the simple presence or absence of sea ice, ice thickness and the rates of freezing or melting strongly impact air-ice-ocean exchanges. To date, ice mass balance buoys (e.g. Metocean) and moored sonar have been the most accurate tools available to determine sea ice thickness over distributed areas. Kwok (2010) reviewed satellite remote sensing of sea-ice thickness and concluded that at least in the Arctic, ice thickness from radar and lidar altimetry is maturing and its shortcomings are relatively well understood. In contrast, Southern Ocean ice cover, in which flooding and snow-ice formation cause substantial density variations and where there are fewer observations to help address processing deficiencies, sea-ice thickness retrieval has not achieved the same level of accuracy as in the Arctic. The CryoSat-2-satellite, launched in 2010, was designed to use SAR to measure ice thickness very precisely, and appears to be meeting expectations.

While visible-IR sensors can provide surface skin temperatures, sea ice is often covered by a snow layer of variable thickness, and the temperature at the ice-snow interface is an important parameter in determining ice-atmosphere exchanges. Passive microwave has shown potential for providing information on both the snow thickness and snow-ice interface temperatures, but these applications are still in development. The impurity content of the snow (i.e. its salinity) can be derived from visible-IR imagery, and the salinity of sea ice substantially influences its radar transparency, providing a potential tool for remotely sensing sea ice bulk salinity that has not yet been fully developed. Frost flowers, which form on the surface of new ice under very cold and still conditions, appear to play an important role in transferring sea-ice salts, organohalides, and other organic matter to the atmosphere and can be identified with synthetic aperture radar or with a combination of active and passive microwave sensors (Kaleschke and Heygster 2004).

Regardless of which sensor data are used, classification of ice types, including leads or polynias, from remote sensing involves refined analytical tools (e.g. Soh et al. 2004; Qin and Clausi 2010). Most are formulated in Bayesian frameworks, which require substantial ground truth data to train the analytical classifier. To help meet this need, Clausi et al. (2010) developed a tool for generating high-resolution (pixel-based) maps for SAR images. They provide estimates of ice concentrations, types, and floe sizes derived from manually classified ice charts of low spatial resolution, such as those produced by the Canadian Ice Service. Similarly, Röhrs et al. (2012) used manual observation of visible satellite images to test and validate an algorithm for identifying leads in passive microwave imagery. Note that although these approaches provide a lot of data for training classification algorithms, the data are not truly ground-truthed, which requires instruments at the surface.

The autonomous O-buoys developed by the Ocean–Atmosphere-Sea Ice-Snow (OASIS) programme during IPY (Fig. 5.11) have successfully measured both ozone and CO_2 over sea ice for at least 3 months during the winter-spring transition (Knepp et al. 2010). As these buoys are further developed and more are deployed, they will provide valuable ground-truthing data to supplement those from satellite-borne absorption spectrometers (see Sects. 5.1.2.5 and 5.1.2.6) measuring aerosol and trace gas emissions from sea ice.

As yet, surface waters below the ice are inaccessible to satellites except through openings in the ice, such as leads and polynias. Therefore, the only available tools that potentially could provide operational information on ice-sea water exchanges are ice buoys, such as the Autonomous Ocean Flux Buoys (AOFB), and ice-tethered profilers (Krishfield et al. 2008). Gliders also have the potential to provide surface-ocean information at higher temporal and spatial resolution than ship-borne measurements, but their use at shallow depths under sea ice can be complicated by ice keels.

As noted before, interpreting remotely sensed data on sea ice requires synthesising information from

Fig. 5.11 O-buoy OB-5 shortly after deployment on August 5, 2011, at 78° 0.4′ N, 139° 55.5′ W in the Beaufort Sea (Photo: John W. Halfacre)

numerous tools, including different satellite sensors and buoy arrays. Recent initiatives to coordinate observations in ice-covered seas should substantially improve the utility of Earth observation products for sea ice research. Notably, the International Polar Year programme on integrated Arctic Ocean Observing Systems initiated basin-wide research coordination in the Arctic Ocean, including deployments of ice buoys and ice-tethered profilers (e.g. Dickson 2009; Perovich et al. 2012, IAOOS: Ice-Atmosphere- Arctic Ocean Observing System see http://www.iaoos-equipex.upmc.fr/en/index.html). A similar initiative in the Southern Ocean (SOOS, the Southern Ocean Observing System) is also now underway (Rintoul et al. 2012).

5.1.2.5 Aerosols

Properties of aerosols in the marine atmosphere are extensively described in Chap. 4 of this book (see also Sects. 5.1.1.3 and 5.1.1.4), including the use of satellite remote sensing for the determination of the organic mass fraction in sea spray aerosol. In this section we further elaborate on the use of satellite remote sensing for the determination of aerosol properties, which started some three decades ago with the retrieval of the aerosol optical depth (AOD, often also called aerosol optical thickness or AOT) over the ocean. The retrieval of aerosol properties over the oceans can be achieved due to the relatively small reflection of solar radiation by the ocean surface, as compared to the reflection by aerosol particles, at wavelengths in the visible and near infrared part of the electromagnetic spectrum. Over land the surface reflection at these wavelengths is much larger and over bright surfaces this overwhelms the aerosol signal.

A brief description of the history of aerosol observations from space was presented by Lee et al. (2009b) and Kokhanovsky and de Leeuw (2009), including an overview of instruments used for this purpose (see also de Leeuw et al. 2011a). Currently, AOD observations are available from several instruments, including operational products such as MODIS, which provides AOD with a validated accuracy over the ocean of $\pm (0.03 + 0.05\text{AOD})$ (Remer et al. 2008). Other examples of instruments providing aerosol products are: MISR, PARASOL, AATSR, MERIS, OMI, SCIAMACHY, GOME-2, MSG and CALIOP. These products include a variety of parameters, in addition to AOD at one or more wavelengths, such as fine mode fraction, absorption aerosol index (AAI) and an indication of aerosol chemical composition. Products, accuracy and spatial and temporal coverage depend on the instrument characteristics. The results are validated by comparison with independent ground-based sun photometer data.

Results from a comparison of MISR and MODIS aerosol products by Kahn et al. (2009) show good correlations between the AOD products (correlation coefficient 0.9 over ocean and 0.7 over land) and the Ångström exponent (correlation coefficient 0.67 over ocean when MISR AOD values > 0.2 are considered). Kahn et al. emphasise the necessity for proper interpretation of the satellite products. In particular data-quality statements should be followed to ensure proper interpretation and use of the satellite aerosol products. Other intercomparisons of satellite products also reveal significant discrepancies between AOD (order of 0.1) from different instruments, even over the ocean (Myhre et al. 2004, 2005).

The AOD is important for the measurement-based assessment of aerosol effects on climate and chemical processes in the atmosphere. Satellite data also show the spatial distribution of aerosols over the ocean which reveals, e.g. transport patterns of dust, biomass burning aerosols and anthropogenic pollution, all of which play a role in the atmospheric input of nutrients into the ocean and their biogeochemical effects (see Chap. 4 for a description of current knowledge on this subject).

Measurements of AOD over the ocean clearly show the effect of wind speed on SSA production, as reported by e.g. Mulcahy et al. (2008); O'Dowd et al. (2010), both using local sun photometer and wind speed measurements at Mace Head, Glantz et al. (2009) using SeaWifs data and ECMWF wind speeds, Lehahn et al. (2010) using MODIS AOD and QuickSCAT, AMSR-E and SSM/I data, Huang et al. (2010) using AATSR AOD and ECMWF wind speed data, Kiliyanpilakkil and Meskhidze (2011) using CALIOP and AMSR-E data and Smirnov et al. (2011) using data from the Marine Aerosol Network (MAN, Smirnov et al. 2011). Smirnov et al. (2012) compare various AOD versus wind speed relationships showing large differences in the AOD at the same wind speed. Most of these relationships show a similar change in AOD over the wind speed range of $0-10$ ms^{-1}, with the exception of the relations of Mulcahy et al. (2008) and O'Dowd et al. (2010). The latter relations show exponential dependence of AOD on wind speed whereas Smirnov et al. (2012) found a linear dependence.

AOD has been used by several authors to evaluate or improve their model results. Sofiev et al. (2011) used MODIS AOD data to evaluate their model results over the ocean. Jaeglé et al. (2011) used AOD observations to include the effect of SST on the production flux of coarse mode sea-salt aerosol in their model through a correction to the sea salt aerosol source function (see Chap. 4 and Sect. 5.2.4). Lapina et al. (2011) used MODIS and MAN AOD data for comparison with GEOS-Chem model results that use the Jaeglé et al. (2011) correction for coarse mode sea-salt aerosol production fluxes as well as an emission scheme for organic matter (see Chap. 4). These authors found that the model AOD is lower than the mean MODIS AOD value but agrees well with MAN AOD for the studied regions. Lapina et al. (2011) argue that this may be partially explained by uncertainties in the satellite retrieval and that uncertainties in the marine OM emission scheme cannot account for the AOD estimate. They conclude that only a sea spray aerosol emission parameterisation resulting in a very different spatial distribution of sea salt could resolve this discrepancy, which may suggest that either some additional marine source of aerosol has not been accounted for or that observations used in the study are insufficient to close the marine aerosol budget.

Sea spray aerosol is principally produced from waves breaking under the action of the wind. The area of the ocean covered with whitecaps is expressed in the whitecap fraction. The retrieval of whitecap fraction using satellite data was explored by Anguelova and Webster (2006). A review of this subject can be found in Lewis and Schwartz (2004), see de Leeuw et al. (2011b) for the current status in this area and a comparison of different methods.

5.1.2.6 Satellite Measurements of Trace Gases Over the Oceans

Carbon dioxide (CO_2) and methane (CH_4,) are the two most important greenhouse gases (GHG) being modified directly by anthropogenic activity, primarily fossil fuel combustion, biomass burning and land use change. In 1957, during the International Geophysical Year, IGY, accurate measurements of the mixing ratio of CO_2 at the Mauna Loa Observatory, led by C.D. Keeling, were initiated. These revealed the growth of CO_2, attributed to fossil fuel combustion, and the annual biogeochemical seasonal cycling of CO_2. Measurements were extended to other sites and to include CH_4 and other relevant gases, resulting in a global but sparse network. However global measurements at high spatial resolution are needed to identify and assess the local and regional response of CO_2 and CH_4 surface fluxes in a warming world and for the verification of national inventories of GHG.

During the 1980s, the retrieval of the total dry atmospheric columns of CO_2 and CH_4 from space was proposed as part of the SCIAMACHY (SCanning Imaging Absorption spectroMeter for Atmospheric CartograpHY, Burrows et al. 1995 and Bovensmann et al. 1999) mission. This is achieved by the remote

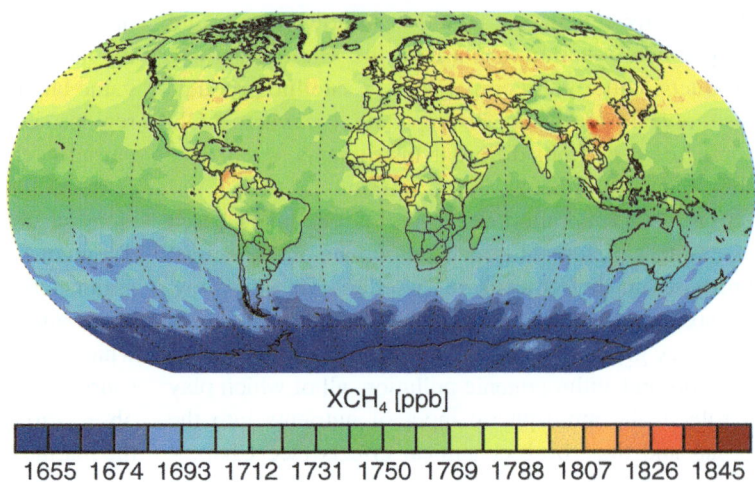

Fig. 5.12 The average global distribution of the dry column of methane retrieved from SCIAMACHY, showing the source regions such as natural wetlands and rice paddies and the hemispheric gradient

sounding in the short wave infrared spectral region and the retrieved CO_2/CH_4 data products, provided they have been adequately sampled with sufficient accuracy and are coupled with models, constrain local and regional surface fluxes. SCIAMACHY was selected in 1989, as a national contribution by Germany to the ESA ENVISAT, with The Netherlands and Belgium joining the funding consortium in Phase A and B, respectively. ENVISAT flies in a sun-synchronous orbit in descending node, having an equator crossing time of 10.00 a.m. and was launched on the 28th of February 2002. SCIAMACHY measures contiguously in eight channels scattering, reflecting and transmitting solar electromagnetic radiation upwelling from the earth's atmosphere between 214 and 2,380 nm at a channel dependent spectral resolution between 0.2 and 1.4 nm. Measurements are made alternately in limb and nadir viewing geometry and for solar and lunar occultation. Mathematical inversion of the nadir measurements of the absorptions of CO_2, CH_4 around 1.6 μ and molecular oxygen, O_2 around 0.76 μ, yields the total dry columns of CO_2 and CH_4.

The map of the average dry total column of methane is shown in Fig. 5.12. The source regions in the northern hemisphere, such as natural wetlands, rice paddies and anthropogenic regions are readily identified along with the hemispheric gradients. In Fig. 5.13 the dry column of CO_2 and CH_4 are plotted against time and sin(latitude) to show the latitudinal distribution of increase from 2003 to 2012. Combining these measurements appropriately with atmospheric models yields and constrains the surface fluxes of CH_4 and CO_2. There is now a growing body of literature combining these data and an accurate network of measurements with which to assess our understanding of surface flux distributions. SCIAMACHY has demonstrated the feasibility, but higher spatial resolution measurements and improved sampling are needed to unambiguously measure point sources and sinks of CO_2 and CH_4.

The GOSAT (Greenhouse gases Observing SATellite) was launched on 23rd of January 2009. The instrument TANSO-FTS (Thermal And Near infrared Sensor for carbon Observations - Fourier Transform Spectrometer) has now made over four years of measurements in space (Hamazaki et al. 2004). The measurements have higher spectral resolution and contain potentially more information than those of SCIAMACHY but have poorer sampling. The OCO (Orbiting Carbon Observatory) was selected by NASA but the OCO-1 launch vehicle failed in February 2009 (Crisp et al. 2004). OCO aims to make high spatial resolution measurements of the dry column of CO_2 and OCO-2 is now planned for launch in 2014. Based on the success of SCIAMACHY, the MaMap instrument was developed in 2006 to demonstrate that high spatially resolved measurements, e.g. 50 m, and high signal to noise of the total dry mole fraction of CO_2 and CH_4 are feasible from aircraft (Gerilowski et al. 2011). The results of MaMap have been used to retrieve surface fluxes of CO_2 and CH_4 from point sources. This demonstration was an essential prerequisite for the development of the CarbonSat and CarbonSat Constellation concepts. A single CarbonSat was proposed for the ESA earth explorer opportunity mission 8, and selected in November 2010 by ESA for Phase A, B1 studies,

Fig. 5.13 Plots of the dry mole fraction of CO_2 and CH_4 versus sin (latitude) from 2003 to 2012 retrieved from SCIAMACHY. These show the latitudinal increase of CO_2 and the changes in CH_4. The rate of increase accelerates in 2008 and is not yet unambiguously explained (Figures courtesy of O. Schneising, M. Buchwitz and J. P. Burrows University of Bremen)

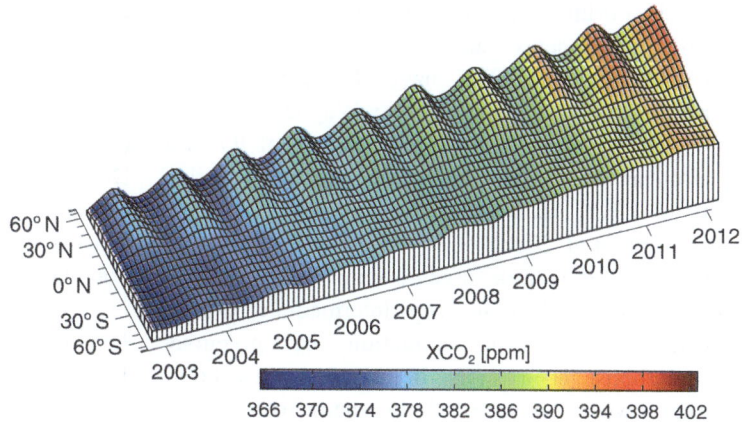

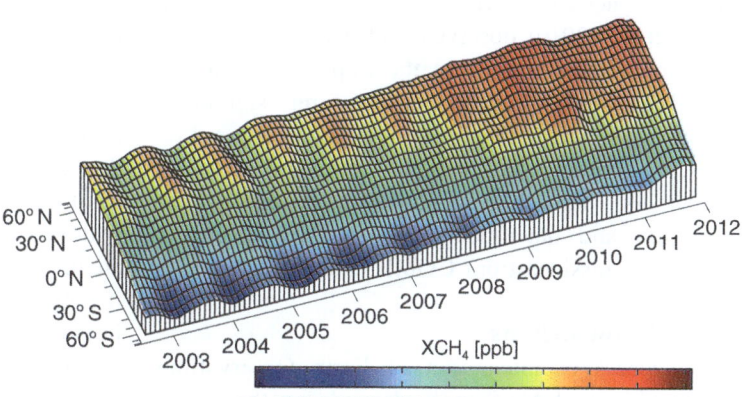

yielding 5 day global coverage. A constellation yields the daily measurements at high spatial resolution measurements required for the verification of CO_2 and CH_4 emissions in the post Kyoto era and the observation of the response of the ecosystem in a warming world.

The background concentration of most trace gases in the lower marine atmosphere is difficult to quantify from space, mainly due to low atmospheric concentrations and the low reflectance of the ocean surface. This is particularly true for ozone as its concentration in the lower troposphere is considerably lower than that throughout the rest of the ozone column. However, emissions from volcanic eruptions (e.g. SO_2), shipping routes (NO_2) and long-range transport plumes can be monitored over the oceans.

International shipping routes are a significant source of pollution in the marine boundary layer. Emissions of traces gases such as NO_2 have been measured from different satellite instruments (e.g. Richter et al. 2004; Beirle et al. 2004). For instance, Franke et al. (2009) compared modelled and satellite-observed NO_2 for the shipping lane between India and Indonesia using GOME, SCIAMACHY, OMI and GOME-2 data, finding indications of an upward trend in shipping emissions over recent years.

The background concentration of SO_2 is difficult to measure by instruments onboard satellites, although volcanic eruptions and their emission plumes can be monitored over several days. The first observation of a volcanic eruption from satellite SO_2 measurements was made using data from the Total Ozone Mapping Spectrometer (TOMS) during the El Chichón eruption in 1982 by Krueger (1983). Since then, several authors have reported satellite SO_2 observations from volcanic ash plumes across the world (e.g. Heue et al. 2010; Nowlan et al. 2011).

Reactive halogen species, such as BrO and IO, lead to ozone destruction in the lower atmosphere, especially during the polar spring. Measurements from satellite instruments such as GOME, GOME-2, SCIAMACHY and OMI have delivered detailed maps of BrO (e.g. Chance 1998; Wagner and Platt 1998; Richter et al. 1998). The detection of IO over Antarctica has also been demonstrated using SCIAMACHY data (Saiz-Lopez et al. 2007; Schönhardt et al. 2008).

Oxygenated volatile organic compounds such as HCHO and $(CHO)_2$ are key intermediate species produced during the oxidation of precursor hydrocarbons. Their short lifetime of a few hours in the lower troposphere links them to emission sources and makes them useful tracers of photochemical activity. Vrekoussis et al. (2009) and Lerot et al. (2010) observed high values of glyoxal over the oceans, mainly in the tropics close to the upwelling areas and regions having significant amounts of phytoplankton, implying oceanic biogenic activity as a possible source of glyoxal precursors. Marbach et al. (2009) have reported the detection and quantification of HCHO linked to shipping emissions from GOME data.

5.1.2.7 Conclusions

The success of integration of Earth Observations products in SOLAS Science depends on the efforts of the international atmospheric and marine science communities. They need to maintain a continuous series of sensors on satellite missions, without interruption, with a constant aim to optimise the quality of the retrieved observables. This required quality implies continuous and rigorous calibration with in situ measurements in the atmosphere and in the ocean including coastal regions. The European Space Agency has launched in 2011 a Support to Science Element Project on SOLAS Science entitled Oceanflux (http://due.esrin.esa.int/stse/projects.php) comprising three themes: Upwellings, Sea spray aerosols and Greenhouse Gases to foster the use of EO data in addressing these SOLAS science questions. Ultimately, it is up to SOLAS scientists together with Earth Observation experts to imagine and develop novel sensors, products, algorithms and methodologies to provide the long duration records of all relevant parameters of the ocean–atmosphere coupled system. Together with models, Earth Observation products constitute the key sentinels for predicting the future trajectory of Earth's climate.

5.1.3 Modelling

Modeling has emerged as a critical and successful method to answer scientific questions, to test hypotheses and to make predictions (Gruber and Doney 2008). Models range from conceptual (essentially ideas about the functioning of a system or process) to complex realistic models that push the boundaries of computational capabilities. We restrict our discussion here to mathematical models and their numerical implementations unless explicitly stated. But in all cases, models are designed for a particular purpose and therefore their transferability is often strongly limited.

Modeling has become the third pillar of the scientific method. In particular, the development of models can be thought of as an iterative process in that observations and experiments stimulate the formulation of conceptual models and hypotheses, which are then translated into mathematical models. The models can then produce predictions, which when confronted with new observations permit the developer to evaluate the models and to either corroborate their underlying hypotheses or to reject them. This leads to an iterative process of modification and improvement. The predictions do not necessarily have to lie in the future (*forecast* mode), since models can also make "predictions" for the past, i.e. when the models are run in *hindcast* mode. Finally, models can also be run in *assimilation* mode, where the model's parameters or its initial or boundary conditions are modified in order to optimally fit a given set of observations.

Physical models of oceanic and atmospheric circulation are used routinely and several mature coupled climate modelling systems exist (e.g. Gent et al. 2012; Roeckner et al. 2006; Collins et al. 2006). Ocean biogeochemical models, obtained by coupling ocean circulation models with mathematical representations of biogeochemical processes, are less mature partly because the quantitative understanding of biogeochemical processes is relatively patchy and mostly empirically-based. This is aggravated by the relative lack of critical biological and chemical observations. As a result, ocean biogeochemical/ecological models differ widely in terms of their complexity (e.g. number of functional plankton groups/state variables) and process parameterisations used (see e.g. Le Quéré et al. 2005). Arguably this is less due to differences in scientific objectives and more to the fact that an optimal compromise between realism and feasibility has not yet emerged (Anderson 2005). Also, biogeochemical predictions depend on many parameterisations and parameter choices that are not well constrained – a direct reflection of the relative paucity of biogeochemical observations and experiments.

Ocean and atmospheric models are often run independently of each other, where atmospheric model output is used as boundary information for ocean models (e.g. atmospheric temperature, humidity and carbon dioxide concentration) and vice versa. An interactive coupling between ocean and atmosphere is needed when feedbacks are of interest that lead to changes in the oceanic or atmospheric mean state, e.g. in climate models and Earth System models such as those used in the IPCC assessments. Global models are constantly being pushed to finer spatial resolutions. However, for reasons of computational efficiency and convenience models are often run at regional and local scales (as three-dimensional or one-dimensional, vertical models). Regional models allow for higher spatial resolution and more complex representation of biogeochemical processes than global models. They can thus be targeted to better resolve, for example, the scales of open ocean process studies or processes in coastal regions and on continental shelves (e.g. Gruber et al. 2011). These regional models require boundary information either from climatological observations or larger scale models.

Model simulation results are always approximations of reality. They come with uncertainties resulting from inadequacies in process resolution and parameterisations, numerical approximations and imperfections in initial and boundary information. Some of these uncertainties are not well understood or quantified. Data assimilation and inverse modeling, which encompass a variety of statistically based techniques for blending observations and models, are a way to reduce uncertainty in model simulations. Data assimilation has been used routinely for many years in numerical weather prediction for short-term forecasting and, more recently, for ocean models that are run operationally (in forecast mode) or for improving state estimates in hindcasts (e.g. Wunsch and Heimbach 2007; Brasseur et al. 2009).

For model development and validation in general and data assimilation and inverse modeling in particular, the availability of high-quality observations is essential. For example, the Argo array has led to remarkable improvements in our ability to characterise the physical state of the ocean (see Sect. 5.1.1.1). Expansion of this initiative aimed at including chemical and bio-optical measurements will likely lead to tremendous improvements in our ability to characterise the biogeochemical state of the ocean (Johnson et al. 2009; Brasseur et al. 2009).

Perhaps the most pressing modeling challenge is to provide estimates of how the mean physical and biogeochemical state of ocean and atmosphere and variations around this mean will change in the future on time scales of a century or longer. Such simulations are referred to as projections. An obvious difficulty is that these future states are outside the envelope of historical observations against which models can be validated. One strategy is to run climate models for periods in the geological past, where some information is available from palaeo-oceanographic proxies, although this solution is imperfect in that no direct and well-understood analogue to the future exists and that palaeo-proxies provide only an incomplete characterisation of the coupled ocean–atmosphere system. In IPCC assessments, attempts are made to address uncertainty in future projections by using multi-model ensembles. Experience has shown that the means of such ensembles frequently perform better than any single model (e.g. Tebaldi and Knutti 2007).

In summary, modeling is an integral part of SOLAS science. Different modeling approaches and techniques are used for different purposes depending on the scientific objective. Key examples are given in the following sections.

5.1.3.1 Global Perspective, Prognostic IPCC and Hindcast

Over the past two centuries, the ocean has taken up about 30 % of total anthropogenic CO_2 emissions, which include the emissions from the burning of fossil fuel and from land use change (Sabine et al. 2004). Although this uptake came at the cost of ocean acidification, it helped considerably to mitigate the accumulation of this anthropogenic CO_2 in the atmosphere. It is thus of great importance to determine whether the ocean will continue to provide this service to mankind, or whether feedbacks between global climate change and the ocean carbon cycle will reduce the uptake of CO_2 from the atmosphere (Sarmiento et al. 1998; Joos et al. 1999; Gruber et al. 2004). Increases in ocean stratification and marine oxygen levels can also lead to the enhanced production of marine N_2O, which would further accelerate global warming (see Chap. 3).

Global prognostic models are the only means to provide answers to such questions. In order to assess all aspects of such climate-ocean-biogeochemical feedbacks, coupled Earth system models need to be employed. In such models, all components of the system, i.e. atmosphere, ocean, land surface and sea-ice are fully prognostic, and include not only descriptions of how energy, momentum, and water are cycled between these reservoirs, but also how carbon and other important biogeochemical constituents, such as nitrogen, are transported and transformed. In the last 10 years, several such models have been developed, and they have formed an important contribution to the 4th assessment report of IPCC (Denman et al. 2007). Such models are typically forced with a prescribed set of CO_2 emission scenarios, and the model itself then determines what fraction of the emitted CO_2 stays in the atmosphere, and what amount of warming corresponds to the resulting increase in this greenhouse gas (Friedlingstein et al. 2006).

Roy et al. (2011) recently conducted an intercomparison of four such Earth system models and investigated how the net ocean CO_2 uptake was altered in response to increases in atmospheric CO_2 and global warming. Specifically, a linear sensitivity analysis was performed, where they represented the net oceanic CO_2 uptake as a sum of a CO_2-driven part, and of a temperature-driven part, i.e.,

$$\text{Int}\,(F_{as}^{net})\,dt = \gamma^* \Delta T + \beta^* \Delta CO_2$$

where F_{as}^{net} is the net ocean CO_2 uptake, ΔT is the change in global mean temperature and ΔCO_2 is the change in atmospheric CO_2, and where γ is the temperature sensitivity and β that for CO_2 (Friedlingstein et al. 2006). Figure 5.14 reveals that all values of β are negative, i.e. that the oceanic CO_2 uptake increases as atmospheric CO_2 increases. Small differences in the responses are related to differences in the buffer factor and differences in the age structure of the water that resides at the surface (Gruber et al. 2009; Roy et al. 2011). Regions where waters upwell that have not been in contact with the atmosphere for several decades and more have a large relative deficit with regard to anthropogenic CO_2 and hence have a high tendency to take it up from the atmosphere. Figure 5.14 also shows that most regions have a positive γ, i.e. that climate change decreases the uptake of atmospheric CO_2. An important exception are the high-latitude oceans, especially the Southern Ocean and the Arctic, where global warming tends to increase the uptake. These differential patterns are a result of the complex interactions occurring between ocean physics (primarily warming and vertical stratification) and ocean biology working primarily on the natural carbon cycle. Although these results represent multi-model means, the robustness of the details in these results is not yet well established. Nevertheless, the general tendency is clear. Climate change will tend to make the ocean carbon sink weaker (Gruber et al. 2004; Denman et al. 2007).

5.1.3.2 Regional Perspectives from High-Resolution Modeling

A disproportionate fraction of the air-sea fluxes of climatically relevant gases (e.g. CO_2, N_2O) is thought to occur in coastal and continental shelf regions even though they cover only about 7 % of the global ocean surface area. However, processes on continental shelves are not well described by global and basin-wide models, primarily because these models do not resolve the smaller scales relevant for shelf and coastal processes. Instead high-resolution regional models are nested within larger scale, state-of-the-art operational models such as MERCATOR (Bahurel et al. 2006) and HYCOM (Chassignet et al. 2007) and are used to quantify air-sea fluxes in these regions and to improve our understanding of the underlying mechanisms. For example, drivers of air-sea CO_2 fluxes for the wide, passive-margin shelves of the western North Atlantic were studied by Fennel et al. (2008), Fennel and Wilkin (2009) and Previdi et al. (2009) and for the semi-enclosed North Sea, a marginal sea in the eastern North Atlantic, by Prowe et al. (2009) and Kühn et al. (2010). Lachkar and Gruber (2013) recently investigated the air-sea CO_2 fluxes in the Canary and California Current Systems.

On the western North Atlantic shelves, where exchange between shelf and open ocean waters is restricted by a pronounced shelf-break front, coupling between biogeochemical processes in sediments and in the overlying water column was found to be highly relevant for air-sea fluxes of CO_2 (snapshots of the air-sea gradient in partial pressure are shown in Fig. 5.15). For example, sediment denitrification (the anaerobic remineralisation of organic matter which produces

Fig. 5.14 Separation of the regional response of integrated air-sea CO_2 fluxes into a CO_2 (**a**) and a temperature (**b**) driven component. Data are for the period from 2010 until 2100 and represent the multi-model mean (From Roy et al. (2011) © American Meteorological Society. Used with permission)

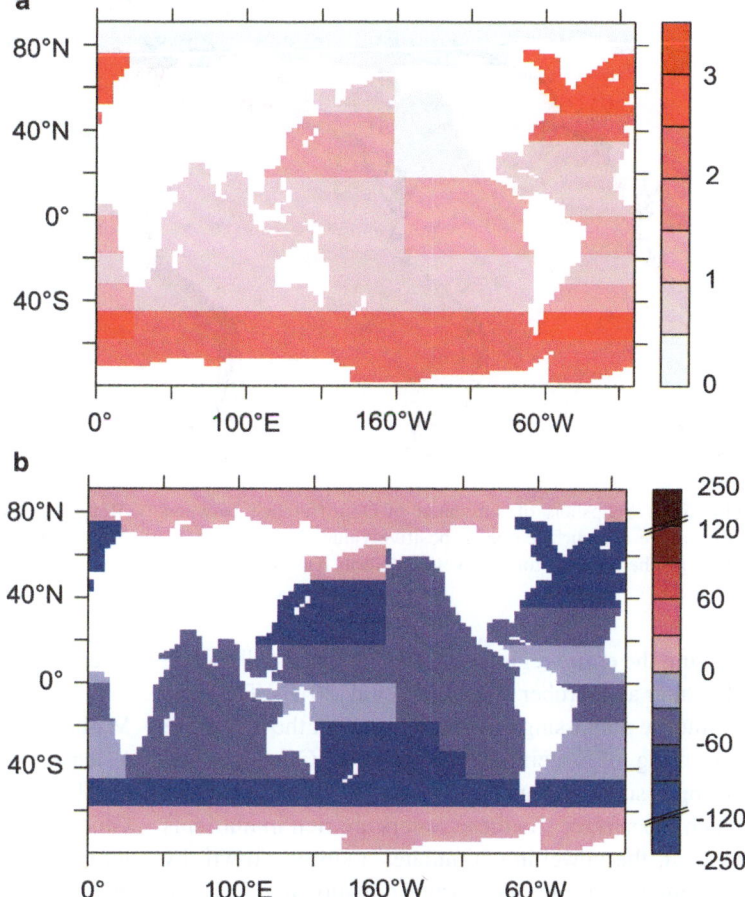

nitrogen gas, N_2) leads to decreases in primary production and increases in alkalinity, both of which alter air-sea fluxes of CO_2 (Fennel et al. 2008). Furthermore, the restricted exchange of shelf water with the open ocean prevents an efficient export of shelf-generated organic carbon to the deep ocean (Fennel and Wilkin 2009). Previdi et al. (2009) investigated interannual variations in air-sea fluxes of CO_2 and found differences in the Middle Atlantic Bight to be driven mostly by changes in wind stress while differences in the Gulf of Maine were due to changes in sea surface temperature and new production.

In the North Sea, which is seasonally stratified in its northern part and year-round tidally mixed in its southern part, pronounced spatial differences exist in terms of air-sea fluxes of CO_2. In the stratified northern part there appears to be net uptake of CO_2 from the atmosphere driven in large part by biological processes including the overflow production of semi-labile dissolved organic matter in summer. In contrast, the tidally mixed southern part is a weak source of CO_2 to the atmosphere (Prowe et al. 2009). Interannual variability of air-sea CO_2 fluxes in the North Sea appears to be driven by variability in atmospheric forcing and river inputs (Kühn et al. 2010).

Continental shelf regions are heterogeneous with respect to air-sea fluxes and regional differences result from a diversity of characteristics and mechanisms. This makes it hard to scale up from individual regions to global estimates. Since coastal regions are of most direct relevance for human activities and most directly subjected to many human perturbations (e.g. riverine and atmospheric inputs of nitrogen), regional modeling studies will continue to play an important role.

In eastern boundary upwelling regions, the net air-sea CO_2 exchange is governed by a zone of intense outgassing in the nearshore region, and a region of marginal outgassing to actual uptake further offshore,

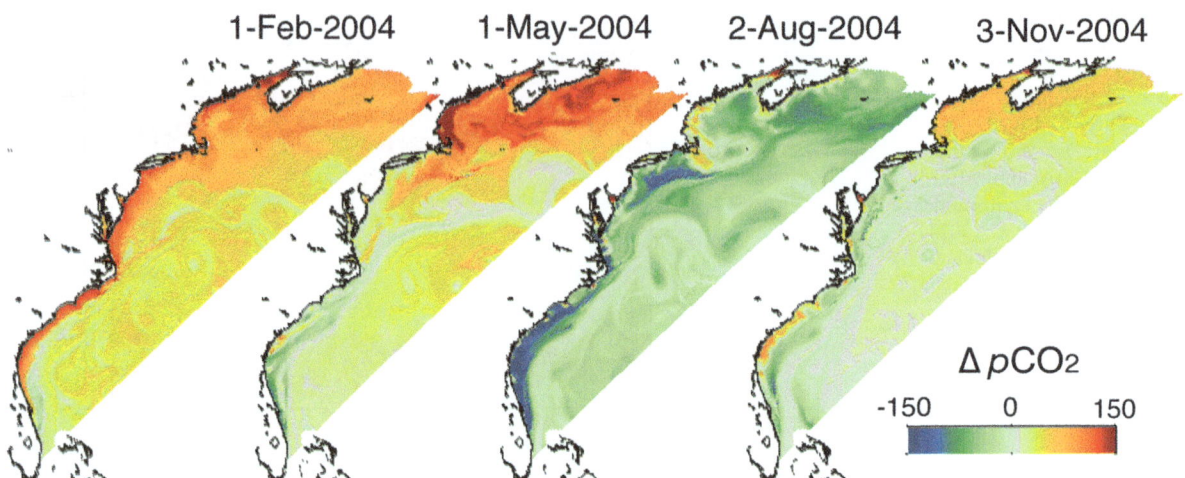

Fig. 5.15 Snapshots of simulated pCO_2 difference between atmosphere and surface ocean (positive values indicate uptake of atmospheric CO_2) for the western North American shelves (Reproduced from Fennel and Wilkin 2009 by permission of the American Geophysical Union)

making the entire region a small sink to a small source (Lachkar and Gruber 2013; Turi et al. 2013). The zone of intense outgassing is a consequence of the nearshore upwelling of waters rich in respired CO_2, creating strongly supersaturated conditions when these waters reach the surface. As these waters are rich in nutrients as well, the upwelling stimulates a strong growth by phytoplankton, creating a strong drawdown of DIC and consequently also of CO_2. Overall there appears to be a balance between upwelling-driven outgassing and biologically driven uptake, with preformed concentrations also playing a role (Hales et al. 2005).

5.1.3.3 Inverse Modelling

The aim of inverse modelling is to estimate the air-sea exchange fluxes of CO_2 (or other trace gases) on the basis of suitable data sets. This requires some model to quantitatively link the measured quantity and the CO_2 fluxes. Then, state variables or parameters of the model can be fit to the data by 'inverse methods'.

A prototypical example uses measurements of tracer abundance within the ocean or the atmosphere, respectively. This 'transport inversion' method is based on the fact that the spatial and temporal patterns of air-sea exchange, being transported and mixed away, lead to spatial gradients and temporal changes in the oceanic/atmospheric tracer field. Air-sea fluxes can thus be estimated from the condition that their corresponding tracer field, as simulated by a numerical or empirical model of oceanic/atmospheric transport, matches as closely as possible the tracer observations. Mathematically, the match is most often quantified by a 'least squares' cost function minimisation.

The oceanic transport inversion was introduced by Gloor et al. (2001) and Gruber et al. (2001). Based on inorganic carbon observations from throughout the ocean, the first set of CO_2 fluxes was estimated by Gloor et al. (2003). As inorganic carbon is not only changed by air-sea exchange but also by marine photosynthesis and remineralisation, nutrient data are needed to remove these biological influences from the data (C* method, based on Redfield ratios between biological carbon and nutrient changes). Moreover, the portion of carbon recently injected into the ocean following the anthropogenic CO_2 rise can be split off from the data (e.g. using the $\Delta C*$ technique, Gruber et al. 1996), allowing separate estimates of natural and anthropogenic air-sea CO_2 exchange.

As oceanic transport proceeds on long time scales, the ocean inversion can estimate the mean spatial pattern of air-sea exchange, but not its seasonal or interannual variability. Results are also affected by errors in the modelled transport, though this can partially be taken into account by considering ensembles of transport models (Mikaloff Fletcher et al. 2006, 2007; Gruber et al. 2009).

The atmospheric transport inversion was pioneered by Bolin and Keeling (1963), and later formalised by

Newsam and Enting (1988). Presently, more than 100 atmospheric measurement stations have been operated by many institutions, providing weekly or even hourly CO_2 mixing ratio time series, some of which extend for several decades. These data have been used to estimate the spatial patterns of CO_2 exchange (e.g. Gurney et al. 2002) or their seasonal and interannual variability (e.g. Bousquet et al. 2000; Rödenbeck et al. 2003; Baker et al. 2006, and many others).

However, the atmospheric CO_2 mixing ratios do not only reflect ocean–atmosphere exchanges, but also terrestrial and anthropogenic fluxes. Due to the diffusive nature of the atmospheric flow, the CO_2 data from the relatively few locations do not provide enough information to fully separate land and ocean fluxes. The dominance of the terrestrial signals in the atmospheric records thus largely obstructs the estimation of ocean–atmosphere exchange. Several studies therefore rely on Bayesian prior estimates of ocean fluxes, most often based on the flux climatology of Takahashi et al. (2009) calculated from measurements of CO_2 partial pressure (pCO_2) and a gas exchange parameterisation. In some cases, Bayesian priors are derived from results of ocean transport inversions; Jacobson et al. (2007) formalised this into a joint ocean–atmosphere inversion. In all these cases, however, air-sea fluxes are mainly constrained by the oceanic data from most of the globe. An exception is the Southern Ocean, where multi-decadal trends in the ocean–atmosphere CO_2 exchange have tentatively been detected from atmospheric data (Le Quéré et al. 2007).

Besides oceanic or atmospheric transport inversions, recent studies use measurements of CO_2 partial pressure (pCO_2) in an inverse context. Valsala and Maksyutov (2010) employed a tracer transport and biogeochemical model, and inversely adjusted the inorganic carbon concentration to match the pCO_2 observations. Ongoing studies involve neural networks, using inverse methods to 'learn' the relationships between pCO_2 and oceanic state variables available from ocean reanalysis projects; these relationships can then be applied to calculate the pCO_2 field at any location and time. Other ongoing studies employ simple diagnostic models of the biogeochemistry in the oceanic mixed layer, estimating ocean-internal carbon sources and sinks.

A common challenge of all these pCO_2-based studies is the need to parameterise air-sea gas exchange, involving substantial uncertainties in its formulation as well as in the driving wind fields (see Chap. 2). This problem may be solved by combining the pCO_2 approach with transport inversions, which do not depend on gas exchange parameterisations.

Further information on air-sea CO_2 fluxes may be obtained from measurements of other tracer species that share source and sink processes. For example, biological processes do not only involve uptake or release of carbon, but also of oxygen. Measurements of the small changes in the atmospheric oxygen content have been used in an atmospheric transport inversion (Rödenbeck et al. 2008) to infer interannual variations in the biogeochemistry of the tropical oceans. Also remote-sensing data offer potential to constrain air-sea CO_2 fluxes. On the one hand, measurements of the atmospheric CO_2 column mixing ratios (see Sect. 5.1.2.6) with much higher spatial density (albeit lower precision) than the surface stations, may allow atmospheric transport inversions to better separate land and ocean fluxes. Satellite observations of ocean surface properties, such as ocean colour (see Sect. 5.1.2.3) reflecting chlorophyll *a* content, can also constrain inverse models of ocean productivity as part of the CO_2 exchange.

5.1.3.4 Conclusions

The scope and ambition of modeling is rapidly expanding. For example, exponentially increasing computational power now permits scientists to simulate global and regional models at ever increasing resolution and over longer periods. The complexity of the models and of their coupling is rapidly increasing as well. While these two developments provide many fascinating new opportunities, they also come with certain risks. Our level of understanding of the results of high-resolution complex models tends to develop less rapidly, creating increasing gaps between our ability to model a system and our ability to fully decipher why a particular model produces a particular result (e.g. Anderson 2005). A second risk is that the observations and experiments that challenge the models are also not increasing as rapidly as the model development is being pushed forward. This requires sustained efforts, also in the interests of the modeling community, to maintain and expand oceanographic observations.

5.1.4 SOLAS/COST Data Synthesis Efforts

The very nature of data collection at sea leads to relatively small-scale cruises of necessarily limited spatial/temporal scope. At an international level, the elevated resource requirements and organisational difficulties associated with large-scale research campaigns further impede coordinated data collection efforts. Considering this, it is unsurprising that oceanic data coverage is fragmentary. As set out within the *Memorandum of Understanding* for COST Action 735 and the aims of SOLAS Project Integration, a major Earth System Science challenge is to translate the findings of these research campaigns into large scale datasets and climatologies that help improve our understanding of global concentrations and air/sea fluxes. Data paucity calls for centralising all concentration measurements of relevant SOLAS parameters into one common database to secure their longevity. The tasks seemed dantean because of two major issues, central to the usability of a database, to be solved as a pre-requisite. One is intercomparing measurements originating from different cruises/instruments/scientists, and another is establishing rigorous quality control and flagging of the data. To get global concentration fields, methods to inter/extrapolate in space and time are used introducing uncertainties in the resulting concentrations. Including the uncertainty in gas exchange velocity parameterisation will yield substantial uncertainties in air/sea flux calculations. Any progress in the future in this velocity parameterisation will allow recomputation of fluxes from the existing global concentrations data sets.

This section presents some specific initiatives carried out within the COST Action 735 and SOLAS Project Integration framework trying to highlight in each case the difficulties encountered, the solutions found and the recommendations proposed.

5.1.4.1 MEMENTO (MarinE MethanE and NiTrous Oxide) Database

The assessments of radiative forcing from long-lived greenhouse gases such as nitrous oxide (N_2O) and methane (CH_4) depend on an accurate synthesis of the global distribution and magnitudes of N_2O and CH_4 sources and sinks (see Chap. 3 of this book). Atmospheric dry mole fractions of N_2O and CH_4 have been routinely available since the late 1970s and they benefit from a highly coordinated global monitoring network (see e.g. Prinn et al. 2000). In stark contrast, although measurements of marine N_2O and CH_4 date back over almost four decades, they lack the temporal continuity and spatial coverage of their atmospheric counterparts. This is because almost all of them relate to single cruises or at best coordinated cruise programmes of rather limited scope. In large part this reflects the high costs and organisational difficulties of mounting large coordinated oceanographic expeditions, especially at the international level. Not surprisingly, oceanic data coverage remains fragmentary.

Seasonal and interannual variability, spatial heterogeneity in coastal areas and gradients between coastal and open ocean areas all impact the quality of marine emission estimates for N_2O and CH_4, compounded by the limited overall data coverage (Bange et al. 2009). A cost effective way of using all existing N_2O and CH_4 measurements, despite the data limitations, is to establish a global database to improve the value of marine emissions estimates. To this end the MEMENTO (MarinE MethanE and NiTrous Oxide) database has been launched as a joint initiative between SOLAS and COST Action 735. MEMENTO's major aims are to:

- Collect available N_2O and CH_4 data (i.e. underway and depth profile data) from the global ocean including coastal areas. To date 129,012 N_2O and 21,003 CH_4 measurements have been collated;
- Archive the data in a database with open access for the scientific community;
- Compute global fields of dissolved N_2O/CH_4 concentrations as well as air-sea fluxes in both the open and coastal ocean, and;
- Publish the database and the derived flux data.

Once all existing datasets have been incorporated into MEMENTO it will rapidly become a valuable tool for identifying regions of the world ocean that should be targeted in future work to improve the quality of the emissions estimates. The locations of the data archived so far in MEMENTO are shown in Fig. 5.16. Further information on MEMENTO can be obtained from https://memento.ifm-geomar.de/.

5.1.4.2 HalOcAt (Halocarbons in the Ocean and Atmosphere)

Compilation of existing air and seawater measurements of halogenated hydrocarbons into the project HalOcAt (https://halocat.geomar.de/) was

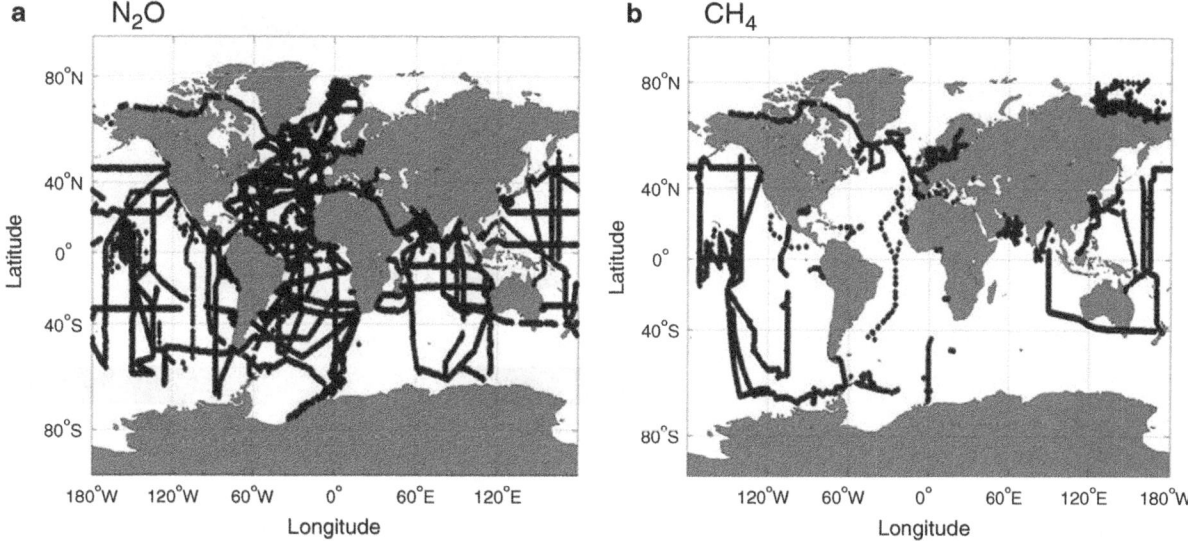

Fig. 5.16 Locations of N_2O (**a**) and CH_4 (**b**) measurements archived in MEMENTO (Version as of October 2012)

initiated through SOLAS/COST Action 735 in May 2009 and is still ongoing. Global oceanic and atmospheric halocarbon data with an emphasis on short-lived brominated and iodinated trace gases from the surface ocean and lower atmosphere have been collated in order to obtain concentration fields and air-sea fluxes. At the time of writing, the database contains 191 contributions, comprising roughly 55,400 oceanic and 476,000 atmospheric data points from depth profiles in the surface ocean and a range of heights in the lower atmosphere. Although predominantly bromoform measurements, these data represent 19 different halocarbon compounds and were collected between 1989 and 2011 during research cruises, aircraft missions and coastal studies from all over the globe. The database stems from active submission of data, literature review and publicly available data.

Quality checks as well as precision and error estimates were often missing from the contributions, coinciding with a large range in reported concentrations. Based on the current available scientific knowledge it is unfortunately not possible to retrospectively identify which data is 'correct' or 'incorrect'. Especially for oceanic data, recalibration is highly unlikely and the database is too small to perform a statistical evaluation of the data quality or to take advantage of utilising datasets that physically crossed paths for inter-calibration. There is some hope that it might be possible to check the consistency of atmospheric data from time series stations, and the community is already assessing inter-calibrations for some compounds (Jones et al. 2011). This information will be incorporated into the HalOcAt data analysis.

In spite of the challenges described above and the paucity of data on a global scale it has been possible to construct meaningful global concentration fields on a $1° \times 1°$ grid. Data interpolation, statistical regression, analysis of published distributions and information about coastal and biological sources including relation to physical and biogeochemical constraints has enabled the calculation of plausible compound distributions (Ziska et al. 2013). From these data, global climatological air-sea fluxes have been calculated for certain halocarbons such as $CHBr_3$ (see Fig. 5.17) and these fall within the published ranges of top down and bottom up emission approaches (Montzka and Reimann 2011). This novel data set now facilitates the use of more realistic data based emission fields to drive atmospheric chemistry models and enables the community to revise current emission scenarios.

Future work on and with the database will focus on its further enlargement, quality control and publication of the data products (database and climatologies of concentrations and ocean/atmosphere exchange). Although simple correlations between oceanic halocarbon concentrations and biological, physical and chemical parameters have not yielded noteworthy

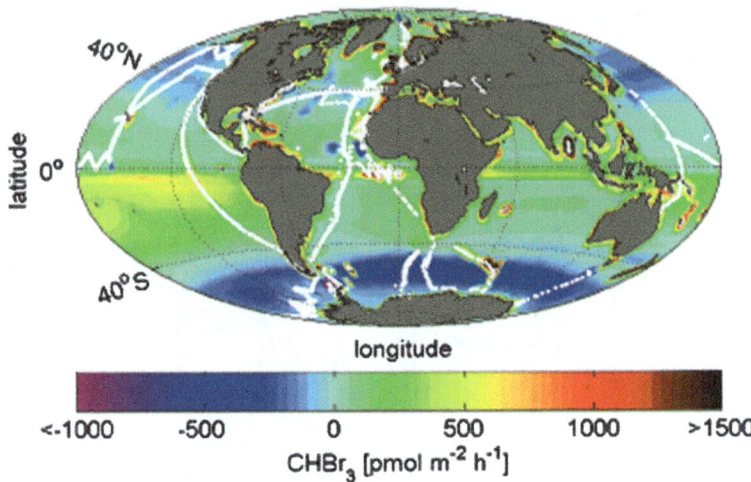

Fig. 5.17 Global air-sea flux climatology of bromoform in pmol m^{-2} h^{-1} including surface oceanic bromoform data points (*white points*) extracted from the HalOcAt database

results to date, the construction and evaluation of proxies and parameterisations for concentrations and fluxes will continue to be explored.

5.1.4.3 DMS-GO (DMS in the Global Ocean)

In 1999 Jamie Kettle and Meinrat Andreae (Kettle et al. 1999) initiated a collation of available seawater DMS measurements into a global database and subsequently created a climatology of surface ocean concentrations based on the data available. However, this important contribution to the literature (over 285 citations to date) was limited by the number of data points (15,617) of surface seawater DMS concentration and the large areas of the global ocean that lacked data, requiring crude estimates to be made. In the years since, many more DMS measurements have been made and in 2010 the DMS-GO (DMS in the Global Ocean) team initiated a call to update the database and climatology. In April 2010 the DMS database had grown to approximately 47,000 data points (Fig. 5.18), and these were used to construct an updated surface ocean DMS concentration climatology.

The construction of the climatology involved data mapping, extrapolation and interpolation onto biogeochemical provinces, smoothing, and data re-assimilation (see Lana et al. 2011 for details). The first step was the transformation of DMS concentration data into a mean value using a delineation of 1° × 1° grid squares. The second step was also based exclusively on the DMS measurements – a background DMS field was created using biogeochemical ocean provinces (Longhurst 2007), which have been defined using spatial similarities in the chemistry, physics and biology of the world ocean. The background field was created for each province for each month using an average DMS concentration. If there was a temporal paucity of data, the gaps were filled by applying an interpolation to the time series of the province, or in some cases by substituting with a scaled pattern of another province with similar biogeochemical characteristics. After these two steps, the construction of the climatology was based on the combination of 1° × 1° pixels and the background field. The new DMS climatology along with an estimate of the uncertainty (upper and lower concentration bounds) is available online for free download on the SOLAS-BODC server (http://www.bodc.ac.uk/solas_integration/implementation_products/group1/dms/).

The global sea-to-air DMS flux estimate (between 17.6 and 34.4 Tg S year^{-1}) using the new climatology has improved understanding of ocean–atmosphere DMS emissions in the global sulphur cycle and suggests an approximate 17 % increase in flux c.f. the previous climatology, mainly due to the inclusion of data from new areas of the ocean (e.g. the Indian Ocean). The updated distribution of global DMS measurements and the new DMS climatology will be useful for the validation of future ocean biogeochemical models. In addition, the continuous climatological surface ocean DMS field will likely be used as an input variable for atmospheric chemistry models. However, the updated DMS database has also highlighted areas of the global ocean that require more measurements and this is powerful information in the context of recent developments in automatic and semi-automatic DMS analysis systems. As new

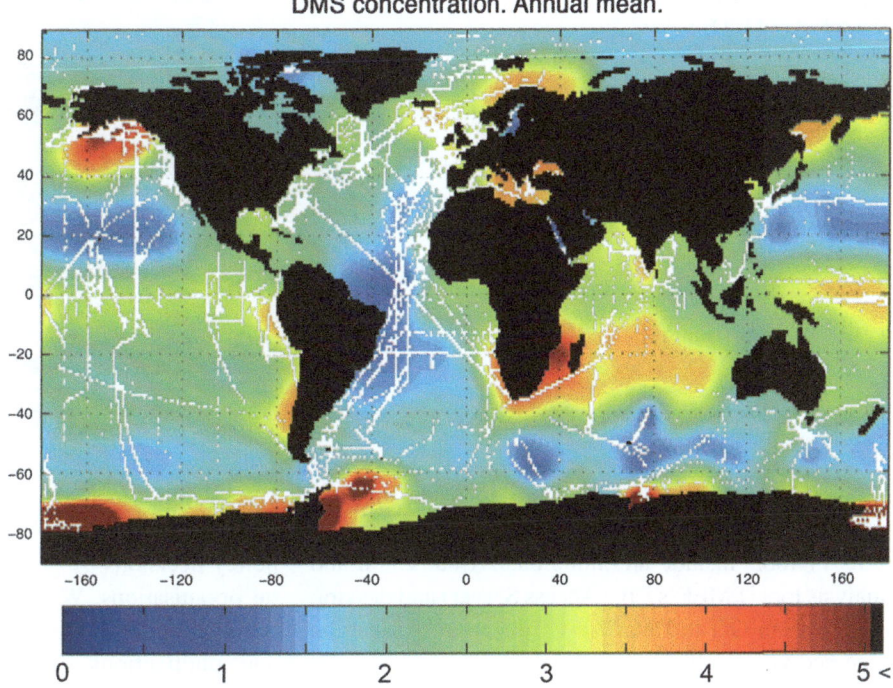

Fig. 5.18 Locations of global surface ocean DMS measurements (approx. 47,000 as of April 2010) for all months of the year (*white*) plotted over the mean climatological DMS concentration (in nM) from the same data (Data is available for download at http://saga.pmel.noaa.gov/dms/)

techniques become more widely-used the challenge will be ensuring data inter-comparability as the database swells in size (see Bell et al. 2011).

5.1.4.4 The Surface Ocean CO$_2$ ATlas (SOCAT)

The Surface Ocean CO$_2$ ATlas (SOCAT) is a global synthesis of surface ocean carbon dioxide (CO$_2$) measurements collected on research vessels, voluntary observing ships and moored as well as drifting platforms (http://www.socat.info/). The first public release of SOCAT (version 1.5) took place at UNESCO in September 2011. Version 1.5 consists of 6.3 million quality controlled, uniform format and recalculated surface water fCO$_2$ (fugacity of CO$_2$) data from 1851 voyages in the global ocean between 1968 and 2007 (Fig. 5.19).

SOCAT was initiated at the Surface Ocean CO$_2$ Variability and Vulnerabilities (SOCOVV) workshop in 2007 (IOCCP 2007). At that time surface ocean CO$_2$ data were archived in a wide range of formats and at numerous sites around the world, each with its own rules for access, and documentation of the data was frequently poor. This made it virtually impossible to generate comprehensive data synthesis products for large scale or long-term studies. To alleviate this situation the international ocean carbon community decided to initiate the Surface Ocean CO$_2$ Atlas as a community driven effort to assemble, harmonise, quality control and document the surface ocean CO$_2$ data into one open access database. While the outlook of the SOCAT products in version 1.5 has closely followed this original vision, as documented in the 2007 meeting report, the community underestimated the amount of work involved in the first release of SOCAT. It took 4 years of hard work by marine carbon scientists around the world to assemble and quality control the first version of SOCAT.

Two SOCAT products are available:
1. A global data set of recalculated surface water fCO$_2$ values in a uniform format, which has undergone 2nd level quality control;
2. A global, gridded product of monthly mean surface water fCO$_2$, with no temporal or spatial interpolation.

The above SOCAT products can be accessed at the Carbon Dioxide Information Analysis Center (CDIAC, http://cdiac.ornl.gov/oceans/) on a global and basin wide level. In addition to the concatenated data, products are all recalculated and available at the ICSU World Data Centre PANGAEA – Data Publisher for Earth & Environmental Science (http://www.pangaea.de/) as individual cruise files. Those files

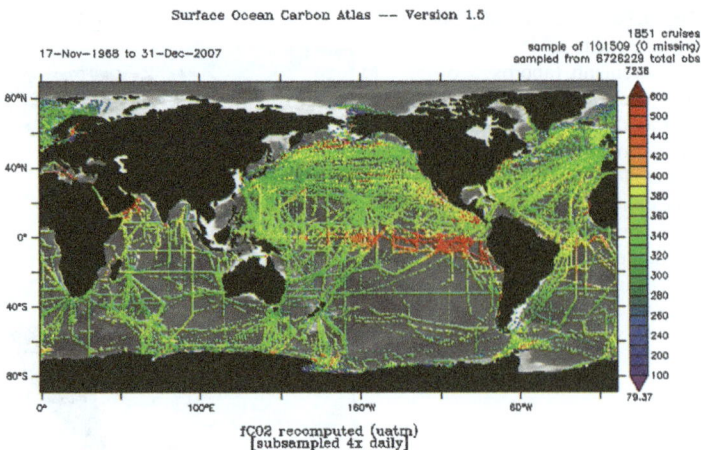

Fig. 5.19 Spatial distribution of recalculated fCO$_2$ measurements of SOCAT Version 1.5 (Courtesy of NOAA/PMEL), http://www.socat.info/

are in a uniform format and give access to the detailed metadata, input data and recalculated fCO$_2$ values. SOCAT tools include an online data visualisation and analysis tool (PMEL's Live Access Server) and desktop tools like an Ocean Data View Collection. All can be accessed via http://www.socat.info/.

The methods in SOCAT are fully documented (Pfeil et al. 2013; Sabine et al. 2013). The products and individual cruises are citable through Digital Object Identifiers (DOI-s). Cruises in SOCAT have an Expocode (a code containing Cruise ID, Year, Month and Day of Cruise), a DOI and detailed information on the data, so-called metadata (e.g. investigator name, vessel, methods, calibrations). Every data point in SOCAT has a link to its cruise file as archived at PANGAEA via the DOI string in the data file.

Preparations for the second version of SOCAT are underway. Regular releases of SOCAT are planned, e.g. every 1–2 years from the 3rd release onwards. Data can be submitted to CDIAC (http://cdiac.ornl.gov/oceans/submit.html) for inclusion in future SOCAT releases. Prompt data and metadata submission as well as citation of SOCAT in publications are deemed essential for the future existence of SOCAT. Automation of data and metadata submission and quality control as well as inclusion of additional variables (e.g. atmospheric CO$_2$ and calculated fCO$_2$ from discrete measurements) is currently being discussed.

SOCAT meets the needs of the global carbon community by making high quality surface water fCO$_2$ data available for addressing major scientific questions in the field of global change. The large number of visitors to the SOCAT website (> 1,000 hits/month) demonstrates the intense interest in these carbon synthesis products. It is anticipated that SOCAT will be used in high profile scientific analyses informing policy decisions by governments and intergovernmental organisations. We kindly ask colleagues to inform SOCAT (submit@socat.info) of recommendations for and publications that use SOCAT data products. Updates on SOCAT will be posted on http://www.socat.info/.

5.1.4.5 Aerosol and Rainwater Chemistry Database

The development of the aerosol and rainwater chemistry database is motivated by the desire to provide a repository for datasets collected from ships at sea because there had previously been no facility for collecting such data, while databases for land-based measurements are already well established (e.g. the World Data Centre for Aerosols; http://wdca.jrc.it/data/parameters/data_chem.html). The database has a deliberately broad focus in terms of the chemical species accepted, with nutrients, trace metals and organics being primary targets, but other data equally welcome. So, for instance, iron data may be a key interest for the database, but submitters of iron data are encouraged to also supply any other chemical data associated with those measurements. This ancillary data may be of use to the iron community as well as in other, unrelated research fields.

Data has been collated at BODC since 2007 and a data portal was added to the site in 2011 (http://www.bodc.ac.uk/solas_integration/implementation_products/group1/aerosol_rain/). At the time of writing the database has ~ 1,300 aerosol and ~ 80 rainfall data points, most directly downloadable from the website

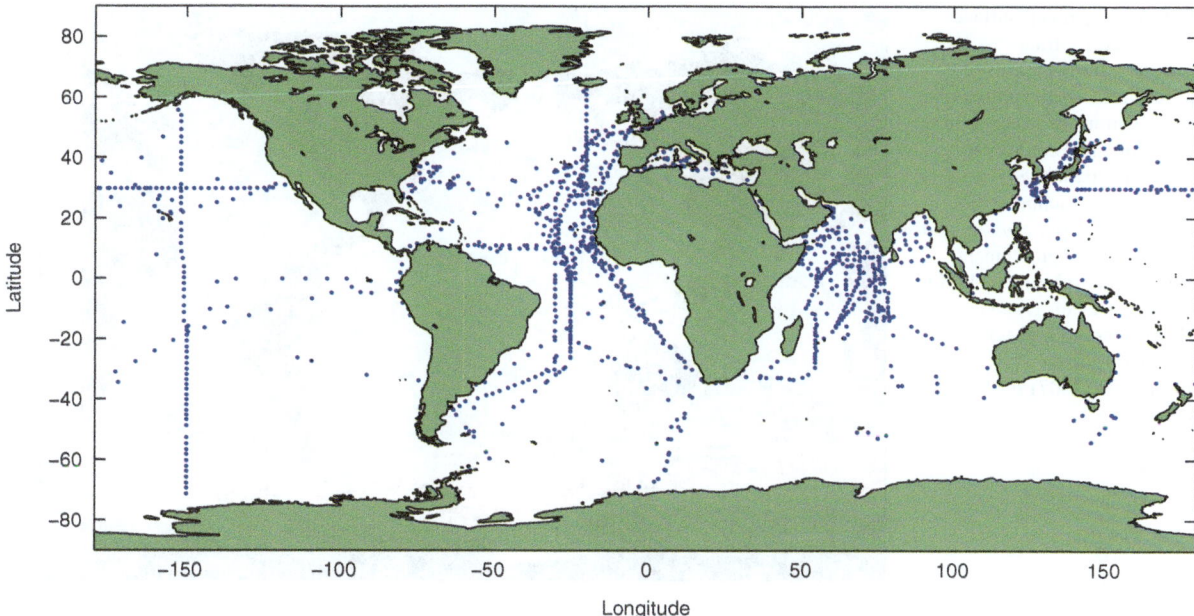

Fig. 5.20 Distribution of aerosol sample locations contained in the database as of November 2011

(Fig. 5.20). The site also contains links to other on-line databases holding related data and seven datasets of aerosol chemistry obtained from island sites which are not housed in any other publicly available database.

Baker et al.'s study (2010) is a good example of the potential benefits attainable from large ship-based datasets. Aerosol and rain data obtained during 12 cruises was used to estimate atmospheric nitrogen inputs to the Atlantic Ocean. This was achieved partly through a process of classifying the aerosol samples in their database according to which source regions they had recently passed over, and marrying these classifications, and their chemical characteristics, to an air mass climatology for the Atlantic basin. The applicability of this approach is strongly determined by the availability of data, and its spatial and seasonal distribution, as well as the seasonal variation in atmospheric source strength. For example, a subtly different approach has been required for recent attempts to estimate atmospheric iron inputs to the Atlantic (Powell et al. in preparation).

To date, the COST Action 735 database contains only chemical and very limited metadata such as sample positions, but no other potentially valuable resources such as air mass back trajectories. All of the studies discussed above used data obtained from only one research group. Similar efforts using data from multiple sources will encounter problems associated with lack of inter-comparability, particularly in 'historical' datasets. The marine aerosol community is only just starting to address these problems through an inter-comparison exercise led by the GEOTRACES programme, with active participation from within SOLAS (see http://www.geotraces.org/ for more information).

5.1.4.6 A Data Compilation of Iron Addition Experiments

In the last 3 years datasets from ten mesoscale iron enrichment experiments have been brought together through a SCOR working group (WG 131) entitled 'The Legacy of mesoscale ocean enrichment experiments'. This data collation has then been transformed into a relational database (the *Iron Synthesis Database*) by the database management team (Cyndy Chandler & Steve Gegg at BCO-DMO) based at Woods Hole Oceanographic Institution (WHOI) in the USA (http://bcodmo.org/data). The database has been structured using project and data directories, metadata, the status of data rescue for each project, and tools that are available to discover and download data of interest (see Fig. 5.21).

Datasets from the following experiments reside at the BCO-DMO site: IronEX I, IronEX II,

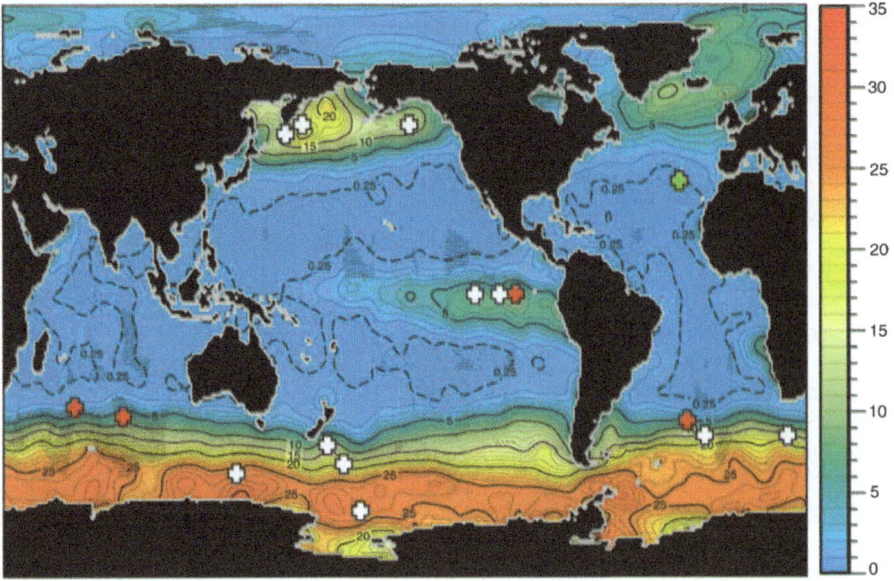

Fig. 5.21 Annual surface mixed-layer nitrate concentrations in units of μmol liter^{-1} with approximate site locations of iron addition experiments. Fe addition experiments: *white crosses*, Fe natural enrichment experiments: *red crosses*, and joint Fe and P enrichment study of the subtropical LNLC Atlantic Ocean: *green cross* (Figure reproduced with kind permission from *Science*. See Boyd et al (2007) for full details)

SOIREE, SAGE, SEEDS, SEEDS II, SOFEX North, SOFEX South and SERIES. The dataset from the Eisenex Southern Ocean study is held at the German PANGEA database (http://www.pangaea.de/). No comprehensive datasets are presently available from the EifEX Southern Ocean experiment, although some papers have been published (e.g. Hoffmann et al. 2006). The datasets within the relational database consist of physical (e.g. mixed layer depth), chemical (e.g. dissolved iron concentrations), biological (e.g. net primary production rates), optical (e.g. incident irradiance) and in some cases meteorological (e.g. wind speed) parameters.

The ten experiments all straddle High Nitrate Low Chlorophyll (HNLC) regions from polar to tropical HNLC waters and hence provide a robust test for modeling studies and comprehensive details for synthetic studies. For example, datasets concerning the production of DMSP and its subsequent transformation to DMS are available for multiple experiments including SOIREE and SERIES. Such datasets can be readily related to a wide range of environmental properties such as mixed layer depth, incident irradiance, and microbial rate processes. The interplay of these factors has been proposed as important in setting DMS concentrations following iron enrichment (Le Clainche et al. 2006). Data are also available concerning changes in the concentration of other biogenic gases following iron enrichment, and in some cases their efflux and fate.

The main challenge in setting up this relational database was obtaining datasets from some of the first in situ experiments – such as IronEX I and II – and from some of the more recent experiments. In the case of the early studies from almost two decades ago, some of the data was on very old laptops. Alternatively, for the most recent experiments, manuscripts are still being written and published and at this point there was an understandable reluctance to contribute to the public domain.

So what is the future for large-scale Fe addition experiments and what else would we really wish to better understand? Some scientists have called for larger experiments (100 km length scale) to overcome some of the artefacts such as dilution with surrounding HNLC waters, but these bring enormous logistical challenges with them. In the conventional 10 km length scale experiments conducted so far, it takes one ship 24 h to add the iron and SF$_6$ (tracer) to the 100 km^2 area of ocean in a manner that results in a coherent patch of tracer/iron. So how many ships would be required to enrich 10,000 km^2? Using planes to 'top-dress' the ocean would be equally problematic (Boyd 2008). The most promising research areas for the future are the study of naturally high iron regions where phytoplankton blooms occur, such as around the Crozet Islands in the Southern Ocean, and conducting medium scale in situ enrichments, or using mesocosms, in the generally quiescent oligotrophic waters of the lower-latitude ocean to study the environmental controls on different groups of diazotrophs.

5.1.4.7 Conclusions

The initiatives discussed in this section represent the spectrum of integration activity within SOLAS. MEMENTO and the aerosol and rainwater chemistry databases are still in relative infancy, just beginning to collate data. In contrast, the relatively mature DMS database has recently been substantially updated, tripling the number of data points and re-estimating the climatological DMS flux to the atmosphere (Lana et al. 2011). Ultimately, the SOCAT project is leading the field, with a lot of data points collated from the global ocean and well established procedures for treating the data and archiving it in a responsible manner. Its progress and willingness to continue pushing forward the boundaries of data collation and synthesis are admirable. Meanwhile, the synthesis of data and information from the various large-scale iron addition experiments has been invaluable.

Past experience suggests that such data collation exercises often prove their worth many times over, informing the scientific community in ways that were previously unforeseen. For example, the original DMS database attempted to find correlations between in situ concentrations and other ancillary parameters such as chlorophyll *a* (Kettle et al. 1999). Despite a relative lack of success at the time, understanding of the reduced sulphur cycle has benefitted enormously from this database – much of the work carried out *after* the initial data collation period has arguably proven to be much more fruitful. Without attempting to predict the future, a similar scenario appears to exist with the HalOcAt database, with initial examinations of the data suggesting no obvious correlating factors; only time will tell…

What challenges remain for the integration of SOLAS data? Success depends in part on the scope of the project. Based on the projects outlined within this chapter, data collation initiatives must overcome the following issues common to many of the databases:

Community engagement. The number of scientists willing to be a part of any initiative and who actively engage with the process can have a significant impact on the scale of data collation and the degree of community involvement. Engaging the community to deliver to data bases is an extremely time consuming effort which requires strong motivation and determination.

Support for the project. This can take the form of a website, advice on data management and/or project personnel, often data managers. Existing projects that have had resources for a data manager have benefitted enormously from this. Data management should be an intrinsic part of any project.

Approaches to data paucity. Whether extrapolating to biogeochemical provinces (e.g. DMS-GO), performing deep ocean cross over checks (e.g. SOCAT) or using aerosol chemical composition data to characterise different air masses within an Atlantic-scale dataset, these techniques are very important for 'filling' data gaps and estimating global fluxes. The techniques need to be adequately described stating clearly the assumptions being made and resulting uncertainties.

Data intercomparability. Arguably, this is one of *the* major challenges facing many datasets within SOLAS science. Collating a large-scale dataset is almost useless if the data are subsequently shown to be incomparable. An obvious way forward is to carry out regularly, at the international level, measurement intercalibration exercises. Some work has begun on the halocarbons (e.g. Butler et al. 2010; Jones et al. 2011) and for DMS (Bell et al. 2011), while GEOTRACES has plans to address this important issue for trace metal aerosol and rainwater chemical composition. SOCAT in particular has spent a lot of time on such issues, and it was noteworthy that in their 2007 meeting report the community recognised that they had probably underestimated the amount of work involved for the first release of SOCAT.

Data legacy. Managing large data sets is a challenging task. It takes dedicated efforts to set up efficient online processed data delivery (e.g. metadata added and formatted in an internationally recognised format), to mount them on an open access portal, to provide archiving and data enhancement (e.g. post processing, climatology enhancement), and to provide a data portal for archived data. International and national data centres should be the natural repositories to ensure long-term security of our community efforts.

The success of SOLAS Integration and the production of global databases depends on a concerted effort from the international atmospheric and marine science communities to not only collect data through extensive field campaigns, but also to engage with and support projects such as those outlined above.

5.2 Examples of SOLAS Integrative Studies

This section of the chapter offers a selection of SOLAS integrative studies. Integration implies a synergistic use of cross-cutting tools and eclectic data in both the atmosphere and ocean to address specific SOLAS science questions. These studies all contribute to an improved understanding of biogeochemical cycling, the establishment of present-day climatologies or the closure of global budgets.

5.2.1 DMS Ocean Climatology and DMS Marine Modelling

5.2.1.1 Global Climatologies Based on Observations

The need for a global climatology of DMS sea-surface concentrations was identified more than a decade ago. The first global database of DMS measurements was put together by Kettle et al. (1999) and Kettle and Andreae (2000). At the time, the database compiled of the order of 15,000 data points, unevenly distributed in time and space. Since then, the original database has been growing extensively and has now reached more than 45,000 data points (see: http://saga.pmel.noaa.gov/dms/). This on-going extension has been made possible thanks to the scientific community feeding the database with new measurements, but also thanks to the DMS-GO initiative (see Sect. 5.1.4.3) and SOLAS integration (Surface Ocean Lower Atmosphere Study, http://www.bodc.ac.uk/solas_integration/).

As discussed in Sect. 5.1.4.3, Lana et al. (2011) have used this updated database to obtain a new global climatology of sea surface DMS concentrations following a modified interpolation approach to the one used by Kettle et al. (1999) (Fig. 5.22). They obtained a global annual sea-to-air DMS flux, estimated at 28.1 Tg S year^{-1} (17.6–34.4), which represents a global emission increase of 17 % with respect to previous calculations. Regionally, annually-averaged concentrations show rather homogeneous values, most of them between 1 and 5 nM.

The characteristics of the seasonal cycle, already shown by Kettle et al. (1999), are mostly confirmed: maximum concentrations (up to 15–20 nM) are obtained in summer at high latitudes, in phase with chlorophyll a. Between 40°S and 40°N, however, DMS and chlorophyll a do not seem to be in phase, with high summer DMS concentrations associated with low chlorophyll a levels, a part of the DMS cycle which has been referred to as the "summer paradox".

5.2.1.2 Diagnostic Approaches: Based on Empirical Correlations

In addition to the climatology described above, several other approaches have been proposed to estimate DMS concentrations (Anderson et al. 2001; Simó and Dachs 2002; Belviso et al. 2004a). These diagnostic approaches are based on empirical relationships between DMS sea-surface concentrations and some other variables (e.g. SST, chl a, MLD), derived at a local scale, and then extrapolated to the global scale. Anderson et al. (2001) generated monthly global DMS fields using a relationship between DMS and the product of chlorophyll a, light and nutrient concentration. Simó and Dachs (2002) proposed a double-equation algorithm in which chlorophyll a and mixed layer depth are used to derive DMS sea-surface concentration. Belviso et al. (2004a) subsequently proposed a non-linear parameterisation to relate DMS concentration to chlorophyll a and an index of the community structure of marine phytoplankton.

These approaches offer some additional advantages compared to the use of DMS climatologies solely derived from in situ DMS measurements. They make use of additional information (ocean physics and biogeochemistry) in regions where no DMS measurements are available. They can be used to study the variability in time (e.g. interannual variability) of DMS sea-surface concentrations and emissions if information on ocean physics and biogeochemistry are available. These approaches have been compared in Belviso et al. (2004b) and their use in Earth System Models in the context of anthropogenic climate change has been described in Halloran et al. (2010).

5.2.1.3 Prognostic Modelling: From 1D to 3D

In addition to the diagnostic approaches described above, a number of prognostic DMS models have been developed in the last decade. Contrary to diagnostic approaches, these models include an explicit

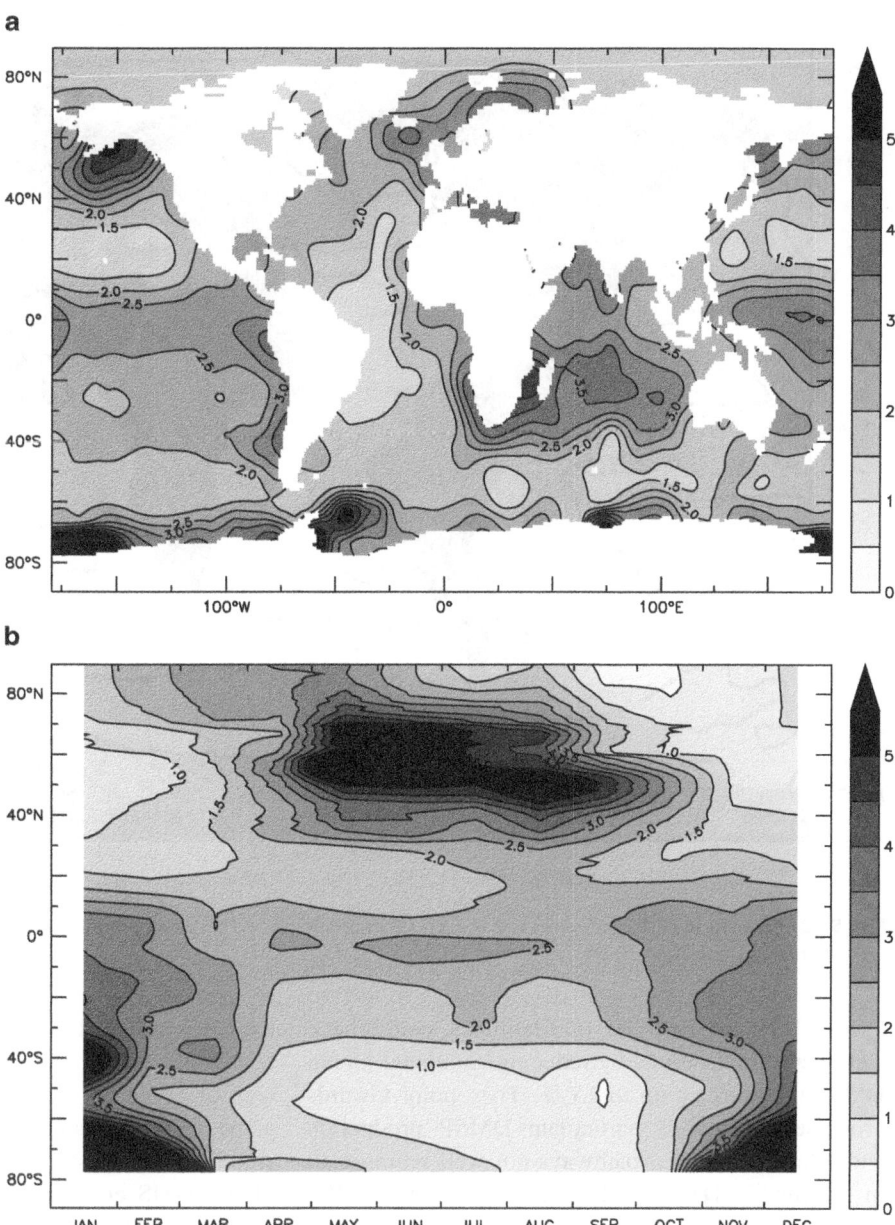

Fig. 5.22 (a) Annually-averaged DMS concentrations (nM), and (b) latitude-time (Hovmoller) diagram of DMS concentrations (nM) (Modified from Lana et al. (2011))

description of processes leading to DMS production and emission.

These models are coupled to marine biogeochemical models that represent plankton ecosystem dynamics. They have been applied at specific one-dimensional vertical column sites (1-D), but also at the global scale coupled to Ocean General Circulation Models (3-D OGCM).

Le Clainche et al. (2010) conducted a model inter-comparison exercise, CODiM, which stands for Comparison of Ocean Dimethylsulfide Models, in the frame of the SOLAS science plan and implementation strategy. They compared nine process-based models, both local one-dimensional (1-D) or global three-dimensional (3-D). From that comparison, their major point is the divergence among models related to their (in)ability to reproduce the summer peak in DMS concentrations usually observed at low- to mid- latitudes. This deficiency in simulating the summer mismatch between chl

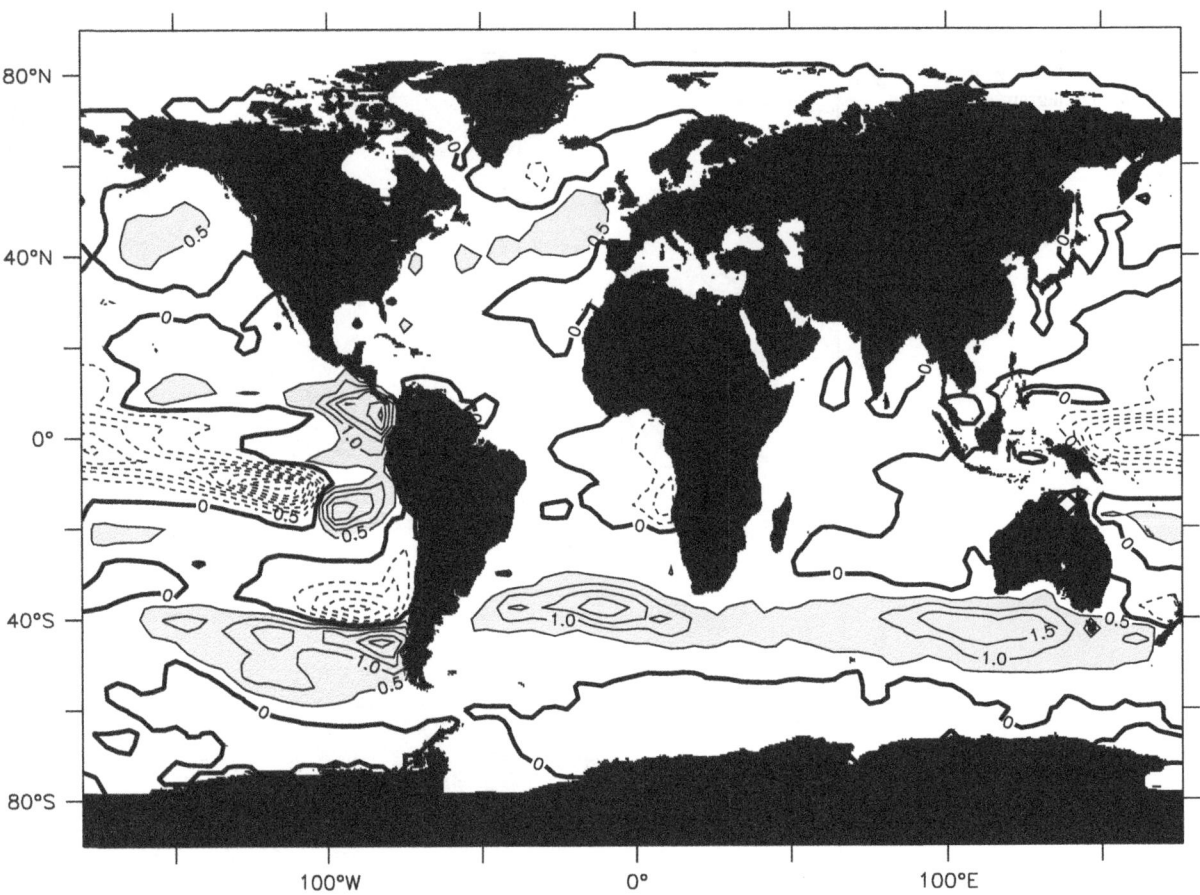

Fig. 5.23 Changes in DMS emissions at $2 \times CO_2$ (in micromol $m^{-2}\,d^{-1}$) as simulated by a coupled climate – ocean DMS model (Adapted from Bopp et al. 2003)

a and DMS at low- to mid-latitudes could have significant implications by reducing simulated global DMS emissions by up to 15 %. They point towards processes related to particulate DMSP production and release as critical pathways not well represented in prognostic DMS models, e.g. increased DMSP synthesis in phytoplankton, increased DMS release from phytoplankton under stress conditions, light-dependent DMSP to DMS conversion by bacteria.

5.2.1.4 Examples of Applications
Climate Change
Several diagnostic (Bopp et al. 2003; Gabric et al. 2004) and prognostic (Kloster et al. 2007) DMS models have been coupled to climate models to predict future DMS emissions and/or sea-surface concentrations. Due to the large variety of approaches used to model DMS emissions, there is still no consensus among the different models, even for the sign of evolution of DMS emissions with anthropogenic climate change: in summary, models tend to simulate either a slight increase or a slight decrease in global DMS emissions in a warming world. Gabric et al. (2004) for example predict an increase of DMS emissions of 14 % for a tripling of pre-industrial atmospheric CO_2. Models agree however in projecting large spatial heterogeneities in future DMS emissions. Bopp et al. (2003) for example predict an increase in the sea-to-air DMS flux of 3 % for doubled atmospheric CO_2 (Fig. 5.23), but with large spatial heterogeneities (up to + 30 % in the Southern mid-latitudes).

Iron Fertilisation
Large increases (of up to 400 %) in DMS concentrations have been documented in some

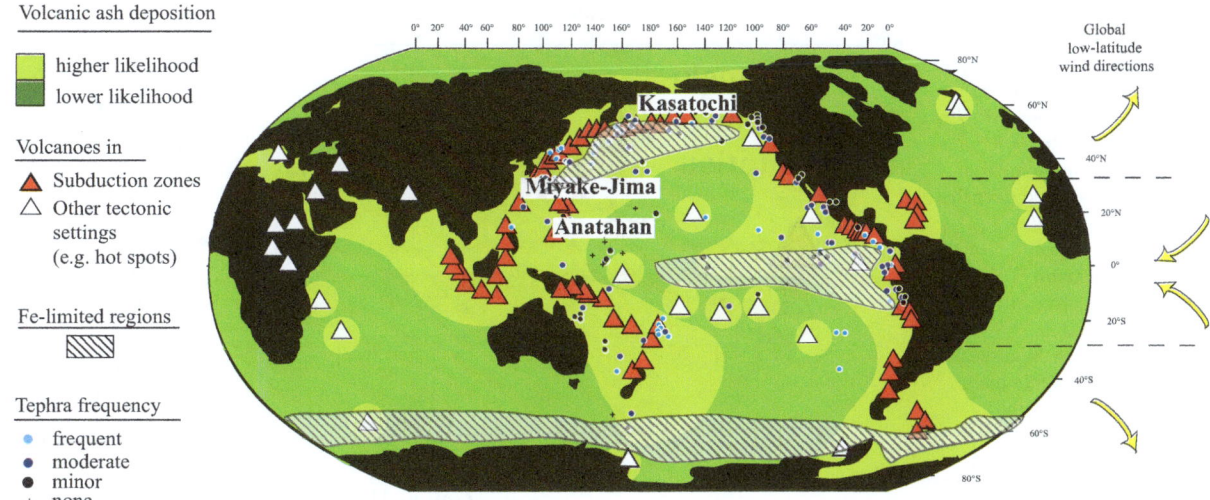

Fig. 5.24 World-map illustrating the distribution of subaerial (Holocene-) active volcanoes indicated as *triangles* (highlighted are the volcanoes cited in the text). Oceanic regions with higher likelihood of volcanic ash deposition are shown as *yellow* regions (Olgun et al. 2011), based on marine sediment core data, the frequency of tephra layers found in the ocean sediments (*coloured circles*) and global wind directions. Volcanic ash deposition can affect marine primary production in much of the Fe-limited areas that are found in the Subarctic Pacific, the Eastern Equatorial Pacific and the Southern Ocean (shown as areas with *oblique lines*)

Southern Ocean artificial iron addition experiments (Turner et al. 2004). These results have been used to suggest that iron fertilisation of the Southern Ocean could act to increase DMS fluxes from the ocean and hence amplify the potential cooling due to increased uptake of carbon (Wingenter et al 2007). Bopp et al. (2008) used a global biogeochemical model including a DMS process-based scheme. They showed that whereas patchy iron fertilisation may stimulate a short-term increase in DMS due to the initial increase in productivity, a long-term and large-scale iron fertilisation could indeed lead to reduced DMS sea-surface concentrations and emissions because of an increase in bacterial consumption rates of DMS.

5.2.2 North Pacific Volcanic Ash and Ecosystem Response

The Pacific Ocean, the largest of the ocean basins, is encircled by a ring of numerous explosive volcanoes (the *Pacific Ring of Fire*) and hosts several hot spot volcanic islands (e.g. Hawaii) (Fig. 5.24). The North Pacific is one of the most active volcanic belts in the Pacific, surrounded by more than 150 active volcanoes in Kamchatka, the Aleutians and mainland Alaska (Fig. 5.24), which create at least ten explosive volcanic eruptions each year (http://www.volcano.si.edu/). Recent geochemical and bio-incubation experiments have shown that volcanic ash rapidly releases sufficient amounts of nutrients into seawater that are potentially bio-available for phytoplankton production and growth (Frogner et al. 2001; Duggen et al. 2007; Jones and Gislason 2008; Hamme et al. 2010; Langmann et al. 2010; Lin et al. 2011; Olgun et al. 2011). Atmospheric impacts of other volcanic products such as the sulphate aerosols are discussed in Sect. 4.2.3.5 of Chap. 4.

The North Pacific, especially the subarctic region, is a high-nutrient, low-chlorophyll (HNLC) ocean region where phytoplankton growth is known to be limited by iron (Martin and Fitzwater 1988; Boyd et al. 1996; Boyd and Harrison 1999) and sporadically by silicate (Wong and Matear 1999; Whitney et al. 2005). Episodic increases of Asian mineral dust input have been observed to increase primary production in the North Pacific (Young et al. 1991). The causal connection between aeolian iron input and diatom production over longer time-scales (e.g. Quaternary) was also indicated by sediment core studies (McDonald and Pedersen 1999). In the North Pacific, volcanic ash is one of the major iron sources, due to the high flux of ash

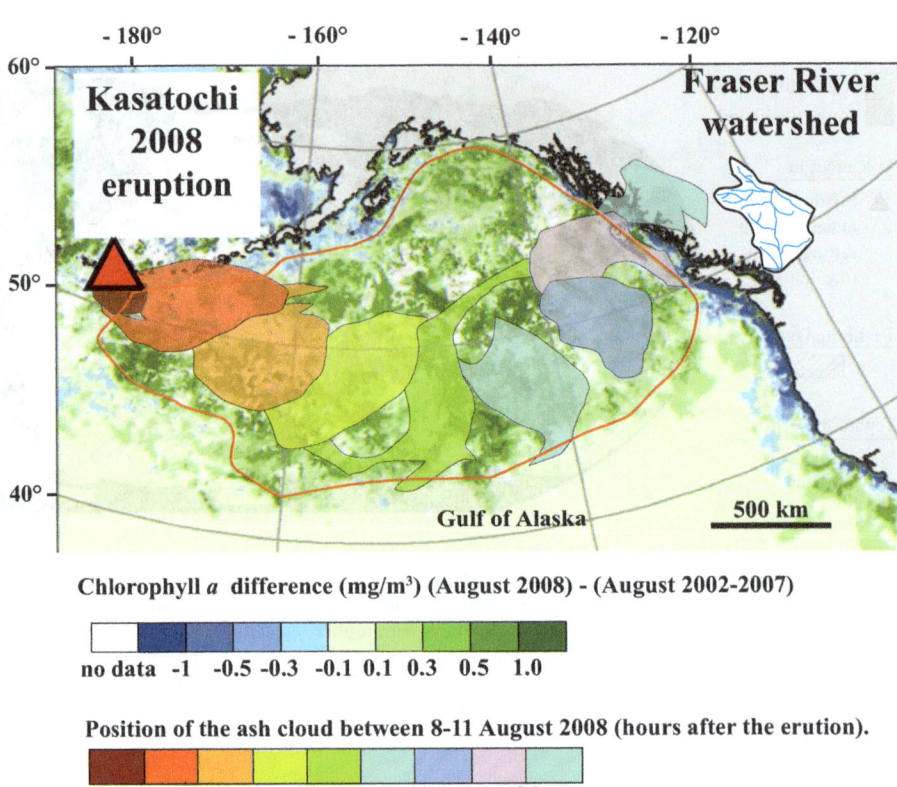

Fig. 5.25 MODIS-Aqua image showing the diatom bloom that has been related to the volcanic ash-fall during the Kasatochi eruption on 7th and 8th of August 2008, with the volcanic ash plume that has been transported over the Gulf of Alaska (*coloured areas* show the position of the plume) (Langmann et al. 2010). The relative increases in chlorophyll *a* concentrations in August 2008 are based on monthly mean August 2008 minus the monthly mean in the years between August 2002 and 2007 (Langmann et al. 2010). Also shown is the location of the Fraser River watershed area related to the discussion of increased salmon populations that are likely to have been positively affected by ocean fertilisation by the Kasatochi eruption during summer 2008

released by active volcanism in the region (Olgun et al. 2011). Deposition of volcanic ash during explosive eruptions can therefore impact phytoplankton and marine foodwebs in the North Pacific by releasing especially iron and other nutrients into seawater.

The first evidence of volcanic enhancement of marine primary production was related to the eruption of Miyake-Jima volcano (Japan, Fig. 5.24) in 2000 (Uematsu et al. 2004). The powerful Miyake-Jima 2000 eruption spread out a volcanic plume that formed ammonium-sulphate aerosols, more than 200 km away in the oligotrophic western North Pacific, resulting in an increase in chlorophyll *a* levels (Uematsu et al. 2004). Similarly, the eruption of Anahatan Volcano (in the Mariana Islands, Fig. 5.24) in 2003, produced a bloom-like patch in the western North Pacific a week after the eruption, evidenced by the MODIS images (Lin et al. 2011).

Volcanic ash fall during a recent eruption of Kasatochi volcano (a remote Aleution island, Fig. 5.25) in August 2008, has also been speculated to generate a massive diatom bloom in the Fe-limited eastern subarctic North Pacific (Fig. 5.25). The unusual phytoplankton bloom in the Gulf of Alaska started a few days after the eruption (Hamme et al. 2010; Langmann et al. 2010). Geochemical experiments also confirmed that iron released from the Kasatochi volcanic ash is sufficient to iron-fertilise the surface North Pacific (Olgun et al. 2013). The bloom area in the eastern subarctic North Pacific (Fig. 5.25) was dominated by large diatoms, and a notably high abundance of large copepods (Hamme et al. 2010), providing a good-quality food-source for the young salmon in the ocean. Lindenthal (2011) provided another source of evidence for the fertilisation of the NE Pacific Ocean by volcanic ash fertilisation from the eruption of the Kasatochi volcano. Indeed, by using an ocean biogeochemical model study and this iron source, he found a good agreement between model outputs and measured chlorophyll *a*, nutrient concentrations, pH and surface ocean pCO_2.

Volcanic eruptions are likely to have impacted the marine foodweb in the North Pacific many times in Earth's recent history (Duggen et al. 2010). The 1980 eruption of Mount St. Helens, for example, was suggested to have fertilised rivers and lakes and led to an increase in populations of golden algae (chrysophytes) and diatoms (Smith and White 1985). Notably, the Kasatochi 2008 eruption has gained further public attention because of its potential impact on fish populations (Jones 2010; Parsons and Whitney 2012; Olgun et al. 2013).

In late summer 2010, record numbers of sockeye salmon (estimated 35 million fish) returned back to Fraser River (Fig. 5.25) 2 years after the large Kasatochi eruption (Jones 2010). It has been speculated that the Kasatochi eruption provided rich food conditions (zooplankton) by enhancing marine primary production (Hamme et al. 2010; Langmann et al. 2010), and increased the marine survival of sockeye during their most critical marine life stage – the first months after they migrate into the ocean (July-October) (Parsons and Whitney 2012; Olgun et al. 2013).

Similarly, one of the largest salmon runs in 1958 followed the large eruption of Bensiammny volcano in Kamchatka. The eruption in 1912 of the Katmai volcano (Alaska) was also suggested to have fertilised the neighbouring lakes by input of phosphorus (Eicher and Rousefell 1957). High ash-loads during Katmai eruption, however, caused initial mortality of pre-smolts due to gill damage (Eicher and Rousefell 1957; Duggen et al. 2010). Despite the smaller number of spawners after the Katmai eruption, the numbers that returned in the following 4 years were as large as the pre-eruption, indicating a rapid recovery and favourable survival conditions in the ocean (Eicher and Rousefell 1957).

Further surveys of the Kasatochi 2008 eruption suggest that the local wildlife was also (directly or indirectly) affected by the eruption (del Moral 2010; Drew et al. 2010; Williams et al. 2010). The largest direct impact was probably the mortality of about 20,000–40,000 young birds (Williams et al. 2010). Most of the seabirds were displaced and the terrestrial birds did not return in 2009 (Williams et al. 2010). Marine mammals (e.g. seals) probably suffered less, but there was a clear disruption or loss of breeding habitat on the land due to meter-scale ash layers on the island (Drew et al. 2010).

In summary, the long-term response (months to years) of the ecology after major eruptions may provide new insights into how marine ecosystems are affected. Recent observations indicate that ocean fertilisation by volcanic eruptions may affect marine biomass within the ash-fall and neighbouring areas.

5.2.3 CO_2 in the North Atlantic

The North Atlantic Ocean is a major sink for atmospheric carbon dioxide (CO_2), and is therefore significant in slowing the global increase of atmospheric CO_2 caused by human activity. Until 1994, the North Atlantic stored approximately 23 % of anthropogenic carbon whilst covering only about 15 % of the world's ocean surface area (Sabine et al. 2004). Due to its significance, the North Atlantic uptake of CO_2 is continuing to be studied intensively, in order to determine the long-term trends of the uptake and its variability.

In the North Atlantic, the high-latitudes are a net sink of atmospheric CO_2 whilst the low-latitudes are a net annual source (Fig. 5.26). The mean seasonal cycle is small near the equator, a winter sink and summer source in the Subtropics, and a small winter source and strong summer sink in the subpolar region.

The air-sea flux of CO_2 is commonly estimated by the difference of CO_2 partial pressure (pCO_2) between the ocean and atmosphere combined with the transfer velocity k, which can be derived from wind speed by various parameterisations e.g. (Nightingale et al. 2000; Wanninkhof and McGillis 1999; Wanninkhof 1992, discussed in detail in Chaps. 2 and 3). The atmospheric pCO_2 is calculated from the molar fraction of CO_2 in the atmosphere (e.g. from Globalview (2011)) and atmospheric pressure. In contrast to the well-mixed atmosphere, surface seawater pCO_2 concentration is subject to considerable variability in space and time, mainly driven by temperature and biological activity. Automated measurements of pCO_2 and related parameters onboard Voluntary Observing Ships (VOS) have provided a valuable monitoring network in the North Atlantic for more than a decade. The pCO_2 of seawater is commonly determined by equilibrating a volume of air with seawater and measuring the CO_2 concentration in the gas phase (Pierrot et al. 2009). Alternatively, membrane-based systems are used, where CO_2 equilibrates (by diffusion) through a membrane and is measured within the sensor using an infra-red detector (Hales et al. 2004). An international effort to rigorously quality control such measurements has led to the creation of a global sea surface pCO_2 data set, the Surface Ocean CO_2 Atlas, SOCAT (see Sect. 5.1.4.4 of this chapter, http://www.socat.info/ and Pfeil et al. 2013). Additionally, pCO_2 can be calculated from dissolved

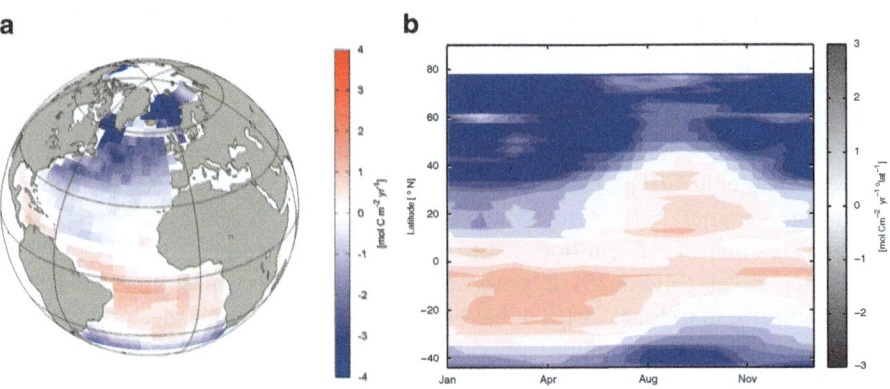

Fig. 5.26 (a) Geographical distribution of the mean air-sea CO_2 flux and (b) the zonal mean seasonal cycle of the flux in the North Atlantic (Adapted from Schuster et al. 2012)

inorganic carbon (DIC), total alkalinity (TA), temperature, salinity, and sea-surface pressure. This opens up the possibility of computing or estimating pCO_2 from parameters that can be monitored on a basin- or near basin-scale. Sea surface temperature (SST) is remotely-sensed by satellites (see Sect. 5.1.2.1) and measured by floats. Sea surface salinity (SSS) is monitored by floats and satellite data will be available soon through the Aquarius (Le Vine et al. 2007, http://aquarius.nasa.gov/) and SMOS projects (see Sect. 5.1.2.2). DIC and TA cannot yet be remotely-sensed. But for any individual ocean basin, TA can be estimated from SSS using a non-linear fit (Eden and Oschlies 2006). In addition, the SeaWiFS project provides satellite-derived estimates of the chlorophyll a concentration (see Sect. 5.1.2.3).

The sea surface pCO_2 and the air-sea CO_2 flux vary over time. A synthesis study by Watson et al. (2009) presented annual CO_2 flux estimates for the North Atlantic for the years 2002–2007. It demonstrated that the CO_2 uptake is subject to large interannual variability. Bates (2007) reported high observed variability of CO_2 fluxes due to increased wind speed over a 20 year period (1984–2005) at the Bermuda Atlantic Time Series (BATS) site. The occurrence of hurricanes in this region accounts for up to 29 % of the variability of summertime CO_2 fluxes (Bates 2007) and the increased frequency of hurricanes can potentially impact by approximately 16 % the interannual variability of CO_2 fluxes in the North Atlantic Ocean. In this context the North Atlantic Oscillation (NAO) is also a driver of variable wind speeds as it drives large scale climate variability over the North Atlantic (Hurrell 1995) and increasing values of NAO index correspond to increased CO_2 fluxes in the northern part of the North Atlantic (Olsen et al. 2003). However, Schuster et al. (2009) and Thomas et al. (2008) reported a link between the variability of NAO and CO_2 fluxes for the rest of the North Atlantic, although the heterogeneity makes the nature of the linkage difficult to identify. Modelling studies (Le Quéré et al. 2003; McKinley et al. 2004; Friedrich et al. 2006) show that SST is the primary control of North Atlantic subtropical surface pCO_2.

On a basin-scale, however, CO_2 fluxes simulated in the above model studies are not significantly correlated with the NAO index. Chemical buffering of changes in the CO_2 concentrations by the large pool of carbonate and bicarbonate ions (Broecker and Peng 1974) has been suggested as one reason for this decoupling.

Seawater pCO_2 itself is driven by sea surface temperature (SST), mixing (advective and convective) and biology (production/respiration). In the tropical North Atlantic NA, the seasonal pCO_2 cycle is strongly coupled to the SST variability. These oligotrophic waters do not support strong biological productivity due to the lack of nutrients (Longhurst 2007). The subtropical regions demonstrate low biological productivity and the seasonal cycle of pCO_2 is dominated by temperature changes. The annual amplitude of approximately 40 μatm peaks during the summer months, which corresponds to the temperature peak (Takahashi et al. 2009; Telszewski et al. 2009) but is reduced to a certain extent by low primary production. The pCO_2 in the subpolar NA is driven by high temperature variability (ΔSST $\sim$ 8 °C) and high biological productivity. Here the biological productivity during spring and summer has a strong counteracting effect on the temperature driven pCO_2 increase during this period (Körtzinger et al. 2008; Lüger et al. 2004).

Steinhoff et al. (2010) have shown that the interannual variability of convective mixing of the surface layer may influence the wintertime budget of carbon and nutrients which directly impacts the observed seawater pCO_2 and thus the ΔpCO_2. Furthermore, advective mixing in the northwest Atlantic has a strong influence. The Labrador Current transports nutrient rich, colder, and fresher water with a different CO_2 signature into the North Atlantic Drift region. On a decadal timescale, increased SST and thus increased primary production may explain the observed long-term changes of seawater pCO_2 (Corbière et al. 2007; Takahashi et al. 2009).

The North Atlantic air-sea CO_2 flux is, however, not constant over time. Based on observations and sinusoidal curve fitting, Schuster et al. (2009) found that between 45°N and 65°N, pCO_2 in the sea surface increased faster than that in the atmosphere between 1990 and 2006 (3 and 1.8 µatm year^{-1}, respectively), resulting in a decreasing sink from 0.20 to 0.09 Pg C year^{-1}, whilst winter-time observations in the Subpolar North Atlantic showed that surface $pCO2$ in winter increased even faster at 3.7 µatm year^{-1} between 1993 and 2003 (Corbière et al. 2007).

Several problems arise when using related parameters to gain an insight into basin-wide surface pCO_2. First of all, a direct calculation is not possible with the data currently available. One needs to estimate surface pCO_2 from e.g. remote sensing data using empirical fitting functions. Thus, errors in the initial observations and data gaps will impair the accuracy of the pCO_2 estimate or result in missing data. The search for a fitting function can be challenging. For example, Watson et al. (1991) and Lefèvre and Taylor (2002) reported robust linear relationships between SST and pCO_2 in the subpolar and subtropical North Atlantic, respectively. The regression coefficients in both studies were similar in magnitude, yet of opposite sign. The study by Watson et al. (1991) also documented a promising covariation of chl a and pCO_2. On the other hand, Lüger et al. (2004) found no significant correlation between chl a and pCO_2 based on measurements covering an entire seasonal cycle in the midlatitude North Atlantic.

The pioneering study by Lefèvre et al. (2005) was the first to derive maps of North Atlantic in situ pCO_2. Using mainly VOS data of the years 1995–1997, multiple regression techniques and self-organising maps (Kohonen 1982), monthly mean surface pCO_2 fields were constructed from temperature and position in time and space on a 1° × 1° grid between 50°N and 70°N. The authors determined the Subpolar Gyre uptake to be 0.13–0.15 Pg C year^{-1} based on their pCO_2 estimates and NCEP monthly wind speed. For the year 2000, the climatological mean North Atlantic sink was approximately 0.44 Pg C year^{-1} (equator to 80°N, Takahashi et al. 2009), whilst an ocean inversion model showed the mean uptake to be 0.41 Pg C year^{-1} (equator to 90°N, Gruber et al. 2009). A recent study of the air-sea CO_2 fluxes in the Atlantic gives a best estimate of the long-term mean flux for the North Atlantic of 0.46 Pg C year^{-1} for the time period of 1990–2009 (Schuster et al. 2012); this study is based on observations, atmospheric inversions, ocean inversions, and ocean biogeochemical models. The availability of satellite-derived chl a and mixed layer depth (MLD) allowed for an improvement of regression techniques as demonstrated by Jamet et al. (2007). The strongly enhanced VOS coverage achieved by the CAVASSOO and CARBOOCEAN projects and their US-American partners enabled Telszewski et al. (2009) to provide seasonal pCO_2 estimates for the years 2004–2006 in the North Atlantic between 10°N and 70°N.

The accuracy of pCO_2 mapping is a major source of uncertainty to this day. Common techniques such as a validation against the data used for deriving the mapping function or against an independent data set (e.g. additional VOS data) were used in the studies mentioned above. To what degree, however, errors derived from individual VOS were representative of the basin-wide mapping error remains unclear. The study by Friedrich and Oschlies (2009a) addressed this challenge through combining observational and modelling approaches. The authors simulated the procedure of VOS monitoring and remote sensing in an eddy-resolving biogeochemical ocean model. Subsequently, the modelled pCO_2 output was used as a ground truth to analyse the accuracy of the pCO_2 maps generated from simulated SST and chlorophyll a fields. One major finding of this study was that the conventional validation techniques underestimate the basin-scale mapping error significantly, raising questions about the reliability of error estimates in earlier studies. However, it was also shown that despite large mapping errors in some regions, the basin-wide CO_2 uptake can be determined with promising precision and that the accuracy can be improved when ARGO float data are used (Friedrich and Oschlies 2009b).

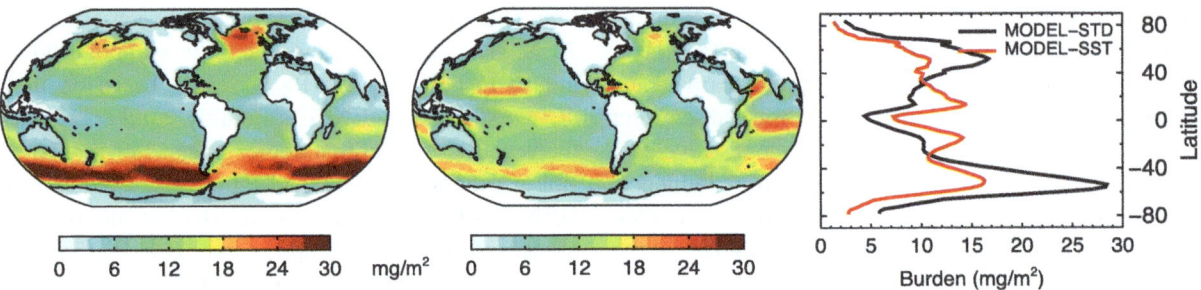

Fig. 5.27 Global burden of SSA (in mg m^{-2}) calculated with the GEOS-Chem CTM for the year 2008. *Left panel*: SSA burden calculated using the Gong (2003) source function (*MODEL-STD*). *Central panel*: SSA burden with the empirically derived source function including a sea surface temperature dependence (*MODEL-SST*). *Right panel*: Zonal average of SSA burden over the oceans (Adapted from Jaeglé et al. (2011))

In summary, the network of Voluntary Observing Ships is an efficient way to monitor the evolution of surface pCO_2 in the North Atlantic and elsewhere. Provided that sufficient coverage is maintained, VOS measurements can be combined with data from satellites and ARGO floats to generate basin-scale maps of surface pCO_2 and to estimate air-sea carbon fluxes with good precision. Future instrumentation such as floats that measure pCO_2 will be a significant asset and will help increase the accuracy of such estimates.

5.2.4 Global Distribution of Sea Salt Aerosols

The global distribution of sea salt aerosols (SSA) is generally estimated using chemical transport models (CTMs) or General Circulation Models (GCMs). Emissions are calculated by integrating a size-dependent source function of SSA at each time step over several size bins. Simultaneously, transport and depositional loss of SSA are also calculated. Models then solve the continuity equation for mass conservation in each size bin.

The resulting global distribution of SSA is highly dependent not only on the assumed sea-salt source function (see reviews by Lewis and Schwartz 2004; O'Dowd and de Leeuw 2007; de Leeuw et al. 2011b) but also on the upper size range of particles, and meteorological fields, as demonstrated in the model intercomparison study of Textor et al. (2006). These authors found a large inter-model range of 3–18 Tg for the global SSA burdens calculated in 16 different CTMs.

Calculated SSA concentrations can differ by factors of 2–3 when different source functions are used in the same CTM (e.g. Guelle et al. 2001; Pierce and Adams 2006). Most source functions are a function of whitecap coverage, with the frequently used 10 m wind speed dependence of $u_{10}^{3.41}$ (Mohanan and O'Muircheartaigh 1980). Thus, small biases in calculated wind speeds are amplified in the resulting SSA emissions. Even when SSA emissions are specified, differences in transport and deposition can result in a factor of 2 variation in predicted global SSA burdens for different models (Liu et al. 2007; Textor et al. 2007).

Despite these significant differences in the absolute concentrations of SSA, models generally display similar features in the global distribution of SSA following the spatial and seasonal distribution of surface wind speed: year-round very high SSA concentrations over the Southern Ocean, winter maximum over the Northern mid latitude storm tracks, and a minimum in the tropics and subtropics (Fig. 5.27, left panel).

This view has been challenged by three independent studies using satellite and in situ observations as top-down constraints on the global distribution of SSA. Haywood et al. (1999) examined the clear-sky solar irradiance measured by the CERES satellite and found a strong signature of SSA over the oceans away from pollution-influenced regions. The geographical pattern of SSA reflectance displayed maxima over both the Southern Ocean and the tropical oceans in the 10–20° latitude band. Satellite measurements of brightness temperature have been used to infer a first estimate of global whitecap coverage (Anguelova and Webster 2006). The resulting whitecap coverage showed a fairly uniform distribution with latitude, contrary to the expectation of strong enhancements over high latitude regions with high wind speeds and very low concentrations in the tropics.

Jaeglé et al. (2011) analysed cruise observations of supermicron SSA mass concentrations (1–10 μm diameter) from six Pacific Marine Environmental Laboratory (PMEL) cruises (Quinn and Bates 2005) with a CTM. They inferred that in addition to the well-known wind speed dependence, SSA emissions appear to depend on sea surface temperature. This is consistent with laboratory experiments reporting increased production of SSA with increasing water temperature for coarse SSA particles (Bowyer 1984, 1990; Woolf et al. 1987; Mårtensson et al. 2003). The temperature dependence leads to a decrease in the predicted SSA burden over the cold high latitude oceans, but enhanced SSA burden over the warm tropical oceans, especially in the trade wind regions. Implementing an empirical temperature- and wind speed-dependent SSA source function in the CTM leads to a more uniform global distribution of SSA (Fig. 5.27 central and right panels) and also improves agreement with in situ SSA observations and remotely sensing aerosol optical depth observations (Jaeglé et al. 2011).

These recent studies illustrate that one way to reduce the very high uncertainties associated with SSA source functions determined from laboratory and field measurements is via integrative studies that combine these source functions in CTMs and GCMs and use top-down constraints based on in situ and satellite observations. Compilations of aerosol observations such as the SOLAS aerosol database (Sect. 5.1.4.5), with a wide distribution over most of the ocean basins are extremely useful in this endeavour.

5.3 Perspectives for the Future

In this chapter we have summarised some of the major integrative SOLAS activities carried over the past decade. While there is a remarkable diversity in the scope, goals, approach, and tools associated with these efforts, they all fall into one or more of the following broad categories:

1. *Understanding biogeochemical cycles*: Compilation of geospatial data sets for the purpose of understanding biogeochemical cycles and discovering the physical and chemical controls on biogeochemistry. Such data sets permit hypothesis testing and provide a common basis for comparison of global models.
2. *Establishing current climatologies*: Establishing a baseline describing the current physical and biogeochemical state of the oceans and atmosphere. For most parameters we are far from an adequate description of the modern Earth, and as a result, we are ill prepared to detect changes.
3. *Closing global budgets:* Time series measurements of the composition of the atmosphere, such as the Mauna Loa CO_2 record, provide a unique integrated view of the global accumulation of greenhouse gases and other pollutants. Detecting change in the oceanic inventory of anthropogenic emissions is an equally critical step towards closing global budgets, but is considerably more logistically challenging.

The SOLAS integrative studies are examples of activities that could contribute to an Earth Observing System. Such a system spans a wide range of activities including process-scale laboratory and expeditionary research, development of new observational capabilities, and operational data streams from satellites, buoys, drifters and models. To be successful, an Earth Observing system will require contributions from many nations and a workforce of highly skilled Earth Scientists and engineers who operate as a global community. One of the most important and exciting aspects of the integrated SOLAS science described here is the extent to which it reflects the work of young scientists from many nations, and the development of a common vision for future priorities in global environmental science.

With this in mind, the International SOLAS Steering Committee identified five unresolved issues of significance to the global climate system that required improved international cooperation and networking (Law et al. 2013). These novel cross-cutting areas, or Mid Term Strategy Initiatives (http://www.solas-int.org/about/mid-term-strategy.html), i.e. Upwellings and associated Oxygen Minimum Zones, Sea ice biogeochemistry, Marine Aerosols, Atmospheric nutrient inputs to the oceans, and Ship Emissions, will benefit from enhanced community engagement to deliver major advances.

Looking forward, societal needs for large-scale integrated data and models are increasing. The human biogeochemical footprint on the planet is now so large that the future quality and sustainability of environmental resources will be determined by societal choices rather than natural variability. At the same time, it is critical to understand how natural Earth systems will respond to this forcing. Large scale Earth Observations and Earth System models will be essential tools underpinning societal decision-making. The future challenges include:

1. *Discovery and detection:* The pace of environmental change is increasing rapidly as a result of increasing industrialisation, population growth, and technological development. The ability to detect large-scale changes in the environment at an early stage is essential in order to allow society adequate time to avoid or mitigate the deterioration of environmental resources and services.
2. *Attribution:* When environmental changes occur, it is critical to understand the underlying causes and the extent to which they can be attributed to natural or anthropogenic causes.
3. *Prediction and uncertainties:* Prediction of future environmental change is an essential tool for the development of effective environmental policy. Almost every aspect of human society and its infrastructure is sensitive to climate and other environmental conditions. Science-based models are the principle tool used to assess the likely outcomes of current trends, the effectiveness of proposed policy options and/or geoengineering strategies, and the environmental impacts of new technologies. Quantitative assessment of uncertainties inherent in predictions of future environmental change is one of the most important products of Earth System Science models. These uncertainties contribute greatly to risk assessments of policy options.

There is enormous potential benefit to society in understanding the environmental consequences of societal trends and policies. These benefits include long term planning to take advantage of new opportunities (e.g. the melting of sea ice and the opening of the Arctic Ocean to exploration and navigation), or to avoid potential costs (i.e. minimising sea level sea level rise by mitigating greenhouse gas emissions). For the SOLAS realm, providing the types of information needed for decision-making will require improved process-level understanding of biogeochemistry, and much better observational capability for remote regions of the atmosphere and oceans. Significant investment will be required in order to maintain existing observing systems, develop and deploy new sensors for in situ and remote observations, and improve infrastructure for the archiving and distribution of data for both research and operational products.

The importance of the SOLAS realm to the future trajectory of Earth's climate and habitability is very clear. Our ability to manage and improve the quality of both natural and human systems will ultimately depend on our understanding of these interactions. The coastal zone is heavily populated and most people are well aware of the impact that coastal water quality can have on their lives and local economy. Although the open oceans cover most of Earth's surface, they are largely uninhabited, and there is a tendency for the average person to see them as remote, unchanging, and disconnected from their daily lives. The reality is far from that, and societal decision-making must take into account the myriad interactions that link us with the surface ocean and lower atmosphere. The challenges are to identify and understand these linkages, to inform the public about them, and to develop tools for integrating scientific knowledge into societal decision-making. The research summarised in this volume demonstrates the remarkable progress of SOLAS science over the past decade, and shows the capacity of the SOLAS scientific community to meet the broader challenges ahead.

Open Access This chapter is distributed under the terms of the Creative Commons Attribution Noncommercial License, which permits any noncommercial use, distribution, and reproduction in any medium, provided the original author(s) and source are credited.

References

Adornato L, Cardenas-Valencia A, Kaltenbacher E, Byrne RH, Daly K, Larkin K, Hartman S, Mowlem M, Prien RD, Garçon VC (2010) In situ nutrient sensors for ocean observing systems. In: Hall J, Harrison DE, Stammer D (eds) Proceedings of the OceanObs'09: sustained ocean observations and information for society conference, vol 2. ESA Publication, Venice, 21–25 Sept 2009, WPP-306

Alvain S, Moulin C, Danndonneau Y, Breon FM (2005) Remote sensing of phytoplankton groups on case 1 waters from global SeaWiFS imagery. Deep-Sea Res I 52:1989–2004

Alvain S, Moulin C, Danndonneau Y, Loisel H (2008) Seasonal distribution and succession of dominant phytoplankton groups in the global ocean: a satellite view. Glob Biogeochem Cycles 22, GB3001

Anderson TR (2005) Plankton functional type modelling: running before we can walk? J Plankton Res 27. doi:10.1093/plankt/fbi076

Anderson TR, Spall SA, Yool A, Cipollini P, Challenor PG, Fasham MJR (2001) Global fields of sea surface dimethylsulfide predicted from chlorophyll, nutrients and light. J Mar Syst 30(1–2):1–20. doi:10.1016/S0924-7963 (01)00028-8

Andreae TW, Andreae MO, Ichoku C, Maenhaut W, Cafmeyer J, Karnieli A, Orlovsky L (2002) Light scattering by dust and anthropogenic aerosol at a remote site in the Negev desert, Israel. J Geophys Res 107(D2):4008. doi:10.1029/2001JD9 00252

Anguelova MD, Webster F (2006) Whitecap coverage from satellite measurements: a first step toward modeling the variability of oceanic whitecaps. J Geophys Res 111: C03017. doi:10.1029/2005JC003158

Arnold SR, Spracklen DV, Williams J, Yassaa N, Sciare J, Bonsang B, Gros V, Peeken I, Lewis AC, Alvain S, Moulin C (2009) Evaluation of the global oceanic isoprene source and its impacts on marine organic carbon aerosol. Atmos Chem Phys 9(4):1253–1262. http://www.atmos-chem-phys.net/9/1253/2009/acp-9-1253-2009.html

Arnold SR, Spracklen DV, Gebhardt S, Custer T, Williams J, Peeken I, Alvain S (2010) Relationships between atmospheric organic compounds and air-mass exposure to marine biology. Environ Chem 7(3):232–241. doi:10.1071/EN09144

Ayers GP, Gras JL (1991) Seasonal relationship between cloud condensation nuclei and aerosol methanesulphonate in marine air. Nature 353:334–835

Ayers GP, Penkett SA, Gillett R, Bandy B, Galbally IE, Meyer CP, Elsworth CM, Bentley ST, Forgan BW (1992) Evidence for photochemical control of ozone concentrations in unpolluted marine air. Nature 360:446–449. doi:10.1038/360446a0

Ayers GP, Cainey JM, Gillett RW, Ivey JP (1997) Atmospheric sulphur and cloud condensation nuclei in marine air in the southern hemisphere. Phil Trans R Soc Lond B 352:203–211

Bahurel P, MERCATOR Project Team (2006) Chapter 14: MERCATOR ocean global to regional ocean monitoring and forecasting. In: Chassignet EP, Verron J (eds) Ocean weather forecasting. Springer, New York, pp 381–395

Baker DF, Bousquet P, Bruhwiler L, Chen YH, Ciais P, Denning AS, Fung IY, Gurney KR, Heimann M, John J, Law RM, Maki T, Maksyutov S, Masarie K, Pak BC, Peylin P, Prather M, Rayner PJ, Taguchi S, Zhu ZX (2006) TransCom 3 inversion intercomparison: impact of transport model errors on the interannual variability of regional CO_2 fluxes, 1988–2003. Glob Biogeochem Cycles 20:GB1002. doi:10.1029/2004GB002439

Baker AR, Lesworth T, Adams C, Jickells TD, Ganzeveld L (2010) Estimation of atmospheric nutrient inputs to the Atlantic Ocean from 50°N to 50°S based on large-scale field sampling: fixed nitrogen and dry deposition of phosphorus. Glob Biogeochem Cycles 24:GB3006. doi:10.1029/2009GB003634

Balch WM, Kilpatrick K, Holligan PM, Harbour D, Fernandez E (1996) The 1991 coccolithophore bloom in the central north Atlantic. II. Relating optics to coccolith concentration. Limnol Oceanogr 41:1684–1696

Balch WM, Gordon HR, Bowler BC, Drapeau DT, Booth ES (2005) Calcium carbonate measurements in the surface global ocean based on moderate-resolution imaging spectroradiometer data. J Geophys Res 110:C07001. doi:10.1029/2004JC002560

Balch WM, Drapeau DT, Bowler BC, Lyczskowski E, Booth ES, Alley D (2011) The contribution of coccolithophores to the optical and inorganic carbon budgets during the Southern Ocean Gas Exchange Experiment: new evidence in support of the "Great Calcite Belt" hypothesis. J Geophys Res 116: C00F06. doi:10.1029/2011JC006941

Bange HW, Bell TG, Cornejo M, Freing A, Uher G, Upstill-Goddard RC, Zhang G (2009) MEMENTO: a proposal to develop a database of marine nitrous oxide and methane measurements. Environ Chem 6:195–197. doi:10.1071/EN09033

Bates NR (2001) Interannual variability of oceanic CO_2 and biogeochemical properties in the Western North Atlantic subtropical gyre. Deep-Sea Res II 48:1507–1528. doi:10.1016/S0967-0645(00)00151-X

Bates NR (2007) Interannual variability of the oceanic CO_2 sink in the subtropical gyre of the North Atlantic Ocean over the last two decades. J Geophys Res 112(C9):C09013. doi:10.1029/2006JC003759

Bates NR, Michaels AF, Knap AH (1996) Seasonal and interannual variability of the oceanic carbon dioxide system at the U.S JGOFS Bermuda Atlantic Time series site. Deep-Sea Res II 43(2–3):347–383. doi:10.1016/0967-0645 (95)00093-3

Bates NR, Knap AH, Michaels AF (1998) Contribution of hurricanes to local and global estimates of air-sea exchange of CO_2. Nature 395:58–61. doi:10.1038/25703

Beale R, Dixon JL, Arnold SR, Liss PS, Nightingale PD (2013) Methanol, acetaldehyde and acetone in the surface waters of the Atlantic Ocean, (accepted)

Beirle S, Platt U, von Glasow R, Wenig M, Wagner T (2004) Estimate of nitrogen oxide emissions from shipping by satellite remote sensing. Geophys Res Lett 31:L18102. doi:10.1029/2004GL020312

Bell TG, Malin G, Lee GA, Stefels J, Archer S, Steinke M, Matrai P (2011) Global oceanic DMS data intercomparability. Biogeochemistry. doi:10.1007/s10533-011-9662-3

Belviso S, Moulin C, Bopp L, Stefels J (2004a) Assessment of a global climatology of oceanic dimethylsulfide (DMS) concentrations based on SeaWiFS imagery (1998–2001). Can J Fish Aquat Sci 61:804–816. doi:10.1139/F04-001

Belviso S, Bopp L, Moulin C, Orr JC, Anderson TR, Aumont O, Chu S, Elliott S, Maltrud ME, Simó R (2004b) Comparison of global climatological maps of sea surface dimethyl sulphide. Glob Biogeochem Cycles 18:GB3013. doi:10.1029/2003GB002193

Bolin B, Keeling CD (1963) Large-scale atmospheric mixing as deduced from the seasonal and meridional variations of carbon dioxide. J Geophys Res 68:3899–3920

Bonsang B, Gros V, Peeken I, Yassaa N, Bluhm K, Zöllner E, Sarda-Esteve R, Williams J (2010) Isoprene emission from phytoplankton monocultures: the relationship with chlorophyll-a, cell volume and carbon content. Environ Chem 7(6):554–563. doi:10.1071/EN09156

Bopp L, Aumont O, Belviso S, Monfray P (2003) Potential impact of climate change on marine dimethyl sulfide emissions. Tellus B 55(1):11–22

Bopp L, Aumont O, Belviso S, Blain S (2008) Modeling the effect of iron fertilization on dimethylsulfide emissions in the Southern Ocean. Deep Sea Res II 55(5–7):901–912

Bourassa, M, Stoffelen A, Bonekamp H, Chang P, Chelton D, Courtney J, Edson R, Figa J, He Y, Hersbach H, Hilburn K, Jelenak Z, Kelly K, Knabb R, Lee T, Lindstrom E, Liu W, Long D, Perrie W, Portabella M, Powell M, Rodriguez E, Smith D, Swail V, Wentz F (2010) Remotely sensed winds and wind stresses for marine forecasting and ocean modeling. In: Hall J, Harrison DE, Stammer D (eds) Proceedings of OceanObs'09: sustained ocean observations and information for society, vol 2. Venice, 21–25 Sept 2009

Bousquet P, Peylin P, Ciais P, Le Quéré C, Friedlingstein P, Tans P (2000) Regional changes in carbon dioxide fluxes of land and oceans since 1980. Science 290:1342–1346

Boutin J, Martin N, Reverdin G, Yin X, Gaillard F (2012) Sea surface freshening inferred from SMOS and ARGO salinity: impact of rain. Ocean Sci Discuss 9:3331–3357. doi:10.5194/osd-9-3331-2012

Bovensmann H, Burrows JP, Buchwitz M, Frerick J, Noël S, Rozanov VV, Chance KV, Goede APH (1999) SCIAMACHY- mission objectives and measurement modes, conference on global measurement systems for atmospheric composition, May 1997, Toronto. J Atmos Sci 56(2):127–150

Bowyer PA (1984) Aerosol production in the whitecap simulation tank as a function of water temperature (Appendix E). In: Monahan ED, Spillane MC, Bowyer PA, Higgins MR, Stabeno PJ (eds) Whitecap and the marine atmosphere, vol 7, Report. University College, Galway, pp 95–103

Bowyer PA, Woolf DK, Monahan EC (1990) Temperature dependence of the charge and aerosol production associated with a breaking wave in a whitecap simulations tank. J Geophys Res 95:5313–5319

Boyd P (2008) Implications of large-scale iron fertilization of the oceans. Mar Ecol Prog Ser 364:213–218. doi:10.3354/meps07541

Boyd P, Harrison PJ (1999) Phytoplankton dynamics in the NE subarctic Pacific. Deep-Sea Res II 46:2405–2432

Boyd PW, Muggli DL, Varela DE, Goldblatt RH, Chretien R, Orians KJ, Harrison PJ (1996) In vitro iron enrichment experiments in the NE subarctic Pacific. Mar Ecol Prog Ser 136:179–193

Boyd PW, Jickells T, Law CS, Blain S, Boyle EA, Buesseler KO, Coale KH, Cullen JJ, de Baar HJW, Follows M, Harvey M, Lancelot C, Levasseur M, Owens NPJ, Pollard R, Rivkin RB, Sarmiento J, Schoemann V, Smetacek V, Takeda S, Tsuda A, Turner S, Watson AJ (2007) Mesoscale iron enrichment experiments 1993–2005: synthesis and future directions. Science 315(5812):612–617

Bracher A, Vountas M, Dinter T, Burrows JP, Röttgers R, Peeken I (2009) Quantitative observation of cyanobacteria and diatoms from space using PhytoDOAS on SCIAMACHY data. Biogeosciences 6:751–764

Brasseur P, Gruber N, Barciela R, Brander K, Doron M, El Moussaoui A, Hobday AJ, Huret M, Kremeur A-S, Lehodey P, Matear R, Moulin C, Murtugudde R, Senina I, Svendsen E (2009) Integrating biogeochemistry and ecology into ocean data assimilation systems. Oceanography 22(3):206–215

Brewin RJW, Lavender SJ, Hardman-Mountford NJ, Hirata T (2010a) A spectral response approach for detecting dominant phytoplankton size class from satellite remote sensing. Acta Oceanol Sin 29:14–32

Brewin RJW, Sathyendranath S, Hirata T, Lavender SJ, Barciela R, Hardman-Mountford NJ (2010b) A three-component model of phytoplankton size class for the Atlantic Ocean. Ecol Model 221:1472–1483

Brewin RJW, Hardman-Mountford NJ, Lavender SJ, Raitsos DE, Hirata T, Uitz J, Devred E, Bricaud A, Ciotti A, Gentili B (2011) An intercomparison of bio-optical techniques for detecting phytoplankton size class from satellite remote sensing. Remote Sens Environ 115:325–339

Broecker WS, Peng TH (1974) Gas exchange rates between air and sea. Tellus 26:21–35

Brown CW, Yoder JA (1994) Coccolithophorid blooms in the global ocean. J Geophys Res 99:7467–7482

Burrows J, Hölzle PE, Goede APH, Visser H, Fricke W (1995) SCIAMACHY – scanning imaging absorption spectrometer for atmospheric chartography. Acta Astronaut 35(7):445–451

Butler JH, Bell TG, Hall BD, Quack B, Carpenter LJ, Williams J (2010) Technical note: ensuring consistent, global measurements of very short-lived halocarbon gases in the ocean and atmosphere. Atmos Chem Phys 10:10327–10330

Carpenter LJ, Monks PS, Galbally IE, Meyer CP, Bandy BJ, Penkett SA (1997) A study of peroxy radicals and ozone photochemistry at coastal sites in the northern and southern hemispheres. J Geophys Res 102:417–427

Carpenter LJ, Lewis AC, Hopkins JR, Read KA, Longley ID, Gallagher MW (2004) Uptake of methanol to the North Atlantic Ocean surface. Glob Biogeochem Cycles 18:GB4027

Carpenter LJ, Fleming Z, Read KA, Lee JD, Moller SJ, Hopkins JR, Purvis RM, Lewis AC, Müller K, Heinold B, Herrmann H, Fomba KW, van Pinxteren D, Müller C, Tegen I, Wiedensohler A, Müller T, Niedermeier N, Achterberg EP, Patey MD, Kozlova EA, Heimann M, Heard DE, Plane JMC, Mahajan A, Oetjen H, Ingham T, Stone D, Whalley LK, Evans MJ, Pilling MJ, Leigh RJ, Monks PS, Karunaharan A, Vaughan S, Arnold SR, Tschitter J, Pohler D, Friess U, Holla R, Mendes LM, Lopez H, Faria B, Manning AJ, Wallace DWR (2010) Seasonal characteristics of tropical marine boundary layer air measured at the Cape Verde Atmospheric Observatory. Atmos Environ 67(2–3):87–140. doi:10.1007/s10874-011-9206-1

Carpenter LJ, MacDonald SM, Shaw MD, Kumar R, Saunders RW, Parthipan R, Wilson J, Plane JMC (2013) Atmospheric iodine levels influenced by sea surface emissions of inorganic iodine. Nat Geosci 6:108–111. doi:10.1038/ngeo1687

Chance K (1998) Analysis of BrO measurements from the Global Ozone Monitoring Experiment. Geophys Res Lett 25(17):3335–3338. doi:10.1029/98GL52359

Chassignet EP, Hurlburt HE, Smedstad OM, Halliwell GR, Hogan PJ, Wallcraft AJ, Baraille R, Bleck R (2007) The HYCOM (HYbrid Coordinate Ocean Model) data assimilative system. J Mar Syst 65(1–4):60–83

Chelton DB, Schlax MG, Freilich MH, Milliff RF (2004) Satellite measurements reveal persistent small-scale features in ocean winds. Science 303:978–983

Ciotti AM, Bricaud A (2006) Retrievals of a size parameter for phytoplankton and spectral light absorption by coloured detrital matter from water-leaving radiances at SeaWiFS channels in a continental shelf off Brazil. Limnol Oceanogr Methods 4:237–253

Clarizia MP, Gommenginger CP, Gleason ST, Srokosz MA, Galdi C, Di Bisceglie M (2009) Analysis of GNSS-R delay-Doppler maps from the UK-DMC satellite over the ocean. Geophys Res Lett 36:L02608. doi:10.1029/2008GL036292

Clarke L, Edmonds J, Jacoby H, Pitcher H, Reilly J, Richels R (2007) Scenarios of greenhouse gas emissions and atmospheric concentrations, sub-report 2.1A of synthesis and assessment product 2.1 by the US Climate Change Science Program and the Subcommittee on Global Change Research. Department of Energy, Office of Biological und Environmental Research, Washington, DC, p 154

Clausi DA, Qin AK, Chowdhury MS, Yu P, Maillard P (2010) MAGIC: MAp-guided ice classification system. Can J Remote Sens 36(S1):S13–S25

Claustre H, Antoine D, Boehme L, Boss E, D'Ortenzio F, Fanton D'Andon O, Guinet C, Gruber N, Handegard NO, Hood M, Johnson K, Körtzinger A, Lampitt R, Le Traon P-Y, Lequéré C, Lewis M, Perry MJ, Platt T, Roemmich D, Sathyendranath S, Testor P, Send U, Yoder J (2010) Guidelines towards an integrated ocean observation system for ecosystems and biogeochemical cycles. In: Hall J, Harrison DE, Stammer D (eds) Proceedings of the OceanObs'09: sustained ocean observations and information for society conference, vol 1. ESA Publication, Venice, 21–25 Sept 2009, WPP-306

Collins WD et al (2006) The community climate system model version 3 (CCSM3). J Climate 19:2122–2143. doi:10.1175/JCLI3761.1

Corbière A, Metzl N, Reverdin G, Brunet C, Takahashi T (2007) Interannual and decadal variability of the oceanic carbon sink in the North Atlantic subpolar gyre. Tellus B 59(2):168–178. doi:10.1111/j.1600-0889.2006.00232

Crisp D, Atlas RM, Breon FM, Brown LR, Burrows JP, Ciais P, Connor BJ, Doney SC, Fung IY, Jacob DJ, Miller CE, O'Brien D, Pawson S, Randerson JT, Rayner P, Salawitch RJ, Sander SP, Sen B, Stephens GL, Tans PP, Toon GC, Wennberg PO, Wofsy SC, Yung YL, Kuang ZM, Chudasama B, Sprague G, Weiss B, Pollock R, Kenyon D, Schroll S (2004) The orbiting carbon observatory (OCO) mission. In: Burrows JP, Thompson AM (eds) Trace constituents in the troposphere and lower stratosphere. Adv Space Res 34(4):700–709

Dacey JW, Howse FA, Michaels AF, Wakeham SG (1998) Temporal variability of dimethylsulfide and dimethylsulfoniopropionate in the Sargasso Sea. Deep-Sea Res I 45:2085–2099. doi:10.1016/S0967-0637(98)00048-X

Dalsøren SB, Eide MS, Myhre G, Endresen Ø, Isaksen ISA, Fuglestved JS (2010) Impacts of the large increase in international ship traffic 2000–2007 on tropospheric ozone and methane. Environ Sci Technol 44:2482–2489

de Leeuw G, Kinne S, Leon JF, Pelon J, Rosenfeld D, Schaap M, Veefkind PJ, Veihelmann B, Winker DM, von Hoyningen-Huene W (2011a) Retrieval of aerosol properties. In: Burrows JP, Platt U, Borrell P (eds) The remote sensing of tropospheric composition from space, 536 pp. Springer, Berlin/Heidelberg, pp 359–313. doi:10.1007/978-3-642-14791-3. ISBN 978-3-642-14790-6

de Leeuw G, Andreas EL, Anguelova MD, Fairall CW, Lewis ER, O'Dowd C, Schulz M, Schwartz SE (2011b) Production flux of sea spray aerosol. Rev Geophys 49:RG2001. doi:10.1029/2010RG000349

del Moral R (2010) The importance of long-term studies of ecosystem reassembly after the eruption of the Kasatochi Island Volcano. Arct Antarct Alp Res 42:335–341

Denman KL, Brasseur G, Chidthaisong A, Ciais P, Cox PM, Dickinson RE, Hauglustaine D, Heinze C, Holland E, Jacob D, Lohmann U, Ramachandran S, da Silva Dias PL, Wofsy SC, Zhang X (2007) Couplings between changes in the climate system and biogeochemistry. In: Solomon S, Qin D, Manning M, Chen Z, Marquis M, Averyt KB, Tignor M, Miller HL (eds) Climate change 2007: the physical science basis. Contribution of working group I to the fourth assessment report of the intergovernmental panel on climate change. Cambridge University Press, Cambridge/New York, pp 499–587

Derwent RG, Simmonds PG, Manning AJ, Spain TG (2007) Trends over a 20-year period from 1987 to 2007 in surface ozone at the atmospheric research station, Mace Head, Ireland. Atmos Environ 41:9091–9098

Devred E, Sathyendranath S, Stuart V, Maas H, Ulloa O, Platt T (2006) A two-component model of phytoplankton absorption in the open ocean: Theory and applications. J Geophys Res 111:C03011. doi:03010.01029/02005JC002880

Devred E, Sathyendranath S, Stuart S, Platt T (2011) Absorption-derived phytoplankton cell size: application to satellite ocean-colour data in the Northwest Atlantic. Remote Sens Environ 115:2255–2266

Dickey T, Bates N, Byrne R, Chang G, Chavez F, Feely R, Hanson A, Karl D, Manov D, Moore C, Sabine C, Wanninkhof R (2009) The NOPP O-SCOPE and MOSEAN projects: advanced sensing for ocean observing systems. Oceanography 22(2):168–181

Dickson RR (2009) The integrated Arctic Ocean Observing System (iAOOS) in 2008, Report of the Arctic Ocean Sciences Board

Donlon C et al (2007) The Global Ocean Data Assimilation Experiment high-resolution sea surface temperature pilot project. Bull Am Meteorol Soc 88:1197–1213. doi:http://dx.doi.org/10.1175/BAMS-88-8-1197

Dore JE, Lukas R, Sadler DW, Church MJ, Karl DM (2009) Physical and biogeochemical modulation of ocean acidification in the central North Pacific. Proc Natl Acad Sci 106:12235–12240. doi:10.1073/pnas.0906044106

Draper DW, Long DG (2004) Evaluating the effect of rain on SeaWinds data. IEEE Trans Geosci Remote Sens 42:1411–1423

Drew GS, Dragoo GS, Renner M, Piatt JF (2010) At-sea observations of marine birds and their habitats before and after the 2008 eruption of Kasatochi Volcano, Alaska. Arct Antarct Alp Res 42:325–334

Duforêt-Gaurier L, Loisel H, Dessailly D, Nordkvist K, Alvain S (2010) Estimates of particulate organic carbon over the euphotic depth from in situ measurements. Application to satellite data over the global ocean. Deep-Sea Research I 57:351–367. doi:10.1016/j.dsr.2009.12.007

Duggen S, Croot P, Schacht U, Hoffmann L (2007) Subduction zone volcanic ash can fertilize the surface ocean and stimulate phytoplankton growth: evidence from biogeochemical experiments and satellite data. Geophys Res Lett 34: L01612

Duggen S, Olgun N, Croot P, Hoffmann L, Dietze H, Teschner C (2010) The role of airborne volcanic ash for the surface ocean biogeochemical iron-cycle: a review. Biogeosciences 7:827–844. doi:10.5194/bg-7-827-2010

Eden C, Oschlies A (2006) Adiabatic reduction of circulation-related CO_2 air-sea flux biases in a North Atlantic carbon-cycle model. Glob Biogeochem Cycles 20:GB2008. doi:10.1029/2005GB002521

Eicher GJ, Rousefell GA (1957) Effects of lake fertilization by volcanic activity on abundance of salmon. Adv Sci Limnol Oceanogr 2:70–76

Eicken H, Gradinger R, Salganek M, Shirasawa K, Perovich D, Leppäranta M (eds) (2009) Field techniques for sea ice research. University of Alaska Press, Fairbanks, p 566

Fairall C, Barnier B, Berry B, Bourassa F, Bradley F, Clayon C, de Leeuw G, Drennan W, Gille S, Gulev S, Kent E, McGillis W, Ryabinin V, Smith S, Weller R, Yelland M, Zhang H-M (2010) Observations to quantify air-sea fluxes and their role in climate variability and predictability. In: Hall J, Harrison DE, Stammer D (eds) Proceedings of OceanObs'09: sustained ocean observations and information for society, vol 2. Venice, 21–25 Sept 2009

Fennel K, Wilkin J (2009) Quantifying biological carbon export for the northwest North Atlantic continental shelves. Geophys Res Lett 36:L18605. doi:10.1029/2009GL039818

Fennel K, Wilkin J, Previdi M, Najjar R (2008) Denitrification effects on air-sea CO_2 flux in the coastal ocean: simulations for the Northwest North Atlantic. Geophys Res Lett 35: L24608. doi:10.1029/2008GL036147

Font J, Boutin J, Reul N, Spurgeon P, Ballabrera-Poy J, Chuprin A, Gabarró C, Gourrion J, Guimbard S, Hénocq C, Lavender S, Martin N, Martínez J, McCulloch M, Meirold-Mautner I, Mugérin C, Petitcolin F, Portabella M, Sabia R, Talone M, Tenerelli J, Turiel A, Vergely JL, Waldteufel P, Yin X, Zine S, Delwart S (2012) SMOS first data analysis for sea surface salinity determination. Int J Remote Sens 34:9–10. doi:10.1080/01431161.2012.716541

Franke K, Richter A, Bovensmann H, Eyring V, Jöckel P, Hoor P, Burrows JP (2009) Ship emitted NO_2 in the Indian Ocean: comparison of model results with satellite data. Atmos Chem Phys 9:7289–7301

Freeland HJ, Roemmich D, Garzoli SL, Le Traon PY, Ravichandran M, Riser S, Thierry V, Wijffels S, Belbéoch M, Gould J, Grant F, Ignazewski M, King B, Klein B, Mork KA, Owens B, Pouliquen S, Sterl A, Suga T, Suk MS, Sutton P, Troisi A, Vélez-Belchi PJ, Xu J (2010) Argo – a decade of progress in proceedings of the OceanObs'09. In: Monahan ED, Spillane MC, Bowyer PA, Higgins MR, Stabeno PJ (eds)/Hall J, Harrison DE, Stammer D (eds) Sustained ocean observations and information for society conference, vol 2. ESA Publication, Venice, 21–25 Sept 2009, WPP-306

Friedlingstein P et al (2006) Climate–carbon cycle feedback analysis: results from the C4MIP model intercomparison. J Clim 19:3337–3353

Friedrich T, Oschlies A (2009a) Basin-scale pCO_2 maps estimated from ARGO float data – a model study. J Geophys Res 114:C10012. doi:10.1029/2009JC005322

Friedrich T, Oschlies A (2009b) Neural network-based estimates of North Atlantic surface pCO_2 from satellite data: a methodological study. J Geophys Res 114:C03020. doi:10.1029/2007JC004646

Friedrich T, Oschlies A, Eden C (2006) Role of wind stress and heat fluxes in the interannual-to-decadal variability of air-sea CO_2 and O_2 fluxes in the North Atlantic. Geophys Res Lett 33:L21S04. doi:10.1029/2006GL026538

Frogner P, Gislason SR, Óskarsson N (2001) Fertilizing potential of volcanic ash in ocean surface water. Geology 29:487–490

Fu LL, Cazenave A (2001) Satellite altimetry and earth sciences, a handbook of techniques and applications, vol 69, International geophysics series. Academic, London

Gabric AJ, Simó R, Cropp RA, Hirst AC, Dachs J (2004) Modeling estimates of the global emission of dimethyl-sulfide under enhanced greenhouse conditions. Glob Biogeochem Cycles 18:GB3016. doi:10.1029/2004GB 002337

Gantt B, Meskhidze N, Kamykowski D (2009) A new physically-based quantification of marine isoprene and primary organic aerosol emissions. Atmos Chem Phys 9:4915. doi:10.5194/ACP-9-49152009

Gardner WD, Mishonov AV, Richardson MJ (2006) Global POC concentrations from in-situ and satellite data. Deep-Sea Res II 53(5–7):718–740. doi:10.1016/j.dsr2.2006.01.029

GCOS (2009) Progress report on the implementation of the global observing system for climate in support of the UNFCCC 2004–2008, GCOS-129 (WMO-TD/No. 1489, GOOS-173, GTOS-70). http://gosic.org/ios/GCOS-main-page.htm

GCOS (2010) Implementation plan for the global observing system for climate in support of the UNFCCC (2010 Update) GCOS-138, (GOOS-184, GTOS-76, WMO-TD/No. 1523). http://gosic.org/ios/GCOS-main-page.htm

Gent PR, Danabasoglu G, Donner LJ, Holland MM, Hunke EC, Jayne SR, Lawrence DM, Neale RB, Rasch PJ, Vertenstein M, Worley PH, Yang Z-L, Zhang M (2012) The community climate system model version 4. J Clim 24(19):4973–4991

Gerilowski K, Tretner A, Krings T, Buchwitz M, Bertagnolio PP, Belemezov F, Erzinger J, Burrows JP, Bovensmann H (2011) MAMAP – a new spectrometer system for column-averaged methane and carbon dioxide observations from aircraft: instrument description and performance analysis. Atmos Meas Tech 4(2):215–243. doi:10.5194/amt-4-215-2011

Glantz P, Nilsson EN, Hoyningen-Huene W (2009) Estimating a relationship between aerosol optical thickness and surface wind speed over the ocean. Atmos Res 92:58–68

GlobalView-CO_2 (2011) Cooperative atmospheric data integration project – carbon dioxide. CD-ROM, NOAA ESRL, Boulder, ftp.Cmdl.Noaa.Gov, path: Ccg/co2/globalview

Gloor M, Gruber N, Hughes TMC, Sarmiento JL (2001) Estimating net air-sea fluxes from ocean bulk data: methodology and application to the heat cycle. Glob Biogechem Cycles 15:767–782. doi:10.1029/2000GB001301

Gloor M, Gruber N, Sarmiento J, Sabine CL, Feely RA, Rödenbeck C (2003) A first estimate of present and preindustrial air-sea CO_2 flux patterns based on ocean interior carbon measurements and models. Geophys Res Lett 30:1010. doi:10.1029/2002GL015594

Gong SL (2003) A parameterization of sea-salt aerosol source function for sub- and super-micron particles. Glob Biogeochem Cycles 17(4):1097. doi:10.1029/2003GB002079

González-Dávila M, Santana-Casiano JM, Rueda MJ, Llinás O (2010) The water column distribution of carbonate system variables at the ESTOC site from 1995 to 2004. Biogeosciences 7:3067–3081. doi:10.5194/bg-7-3067-2010

Gordon HR, Boynton GC, Balch WM, Groom SB, Harbour DS, Smyth TJ (2001) Retrieval of coccolithophore calcite concentration from SeaWiFS imagery. Geophys Res Lett 28(8):1587–1590

Gruber N, Doney SC (2008) Ocean biogeochemical and ecological modeling. Encycl Ocean Sci 89–104

Gruber N, Sarmiento JL, Stocker TF (1996) An improved method for detecting anthropogenic CO_2 in the oceans. Glob Biogeochem Cycles 10(4):809–837

Gruber N, Gloor M, Fan S-M, Sarmiento JL (2001) Air-sea flux of oxygen estimated from bulk data: implications for the marine and atmospheric oxygen cycle. Glob Biogeochem Cycles 15(4):783–803

Gruber N, Friedlingstein P, Field CB, Valentini R, Heimann M, Richey JE, Romero-Lankao P, Schulze D, Chen C-TA (2004) The vulnerability of the carbon cycle in the 21st century: an assessment of carbon-climate-human interactions. In: Field CB, Raupach MR (eds) The global carbon cycle: integrating humans, climate, and the natural world. Island Press, Washington, DC, pp 45–76

Gruber N, Gloor M, Mikaloff Fletcher SE, Doney SC, Dutkiewicz S, Follows M, Gerber M, Jacobson AR, Joos F, Lindsay K, Menemenlis D, Mouchet A, Mueller SA, Sarmiento JL, Takahashi T (2009) Oceanic sources, sinks, and transport of atmospheric CO_2. Glob Biogeochem Cycles 23:GB1005. doi:10.1029/2008GB003349

Gruber N, Lachkar Z, Frenzel H, Marchesiello P, Munnich M, McWilliams JC, Nagai T, Plattner G-K (2011) Eddy-induced reduction of biological production in eastern boundary upwelling systems. Nat Geosci 4(11):787–792

Guelle W, Schulz M, Balkanski Y (2001) Influence of the source formulation on modeling the atmospheric global distribution of sea salt aerosol. J Geophys Res 106:27509–27524

Gurney K, Law RM, Denning AS, Rayner PJ, Baker D, Bousquet P, Bruhwiler L, Chen YH, Ciais P, Fan S, Fung IY, Gloor M, Heimann M, Higuchi K, John J, Maki T, Maksyutov S, Masarie K, Peylin P, Prather M, Pak BC, Randerson J, Sarmiento J, Taguchi S, Takahashi T, Yuen CW (2002) Towards robust regional estimates of CO_2 sources and sinks using atmospheric transport models. Nature 415:626–630

Hales B, Chipman D, Takahashi T (2004) High-frequency measurement of partial pressure and total concentration of carbon dioxide in seawater using microporous hydrophobic membrane contactors. Limnol Oceanogr Methods 2(2):356–364

Hales B, Takahashi T, Bandstra L (2005) Atmospheric CO_2 uptake by a coastal upwelling system. Glob Biogeochem Cycles 19:1–11. doi:10.1029/2004GB002295

Halloran PR, Bell TG, Totterdell IJ (2010) Can we trust empirical marine DMS parameterisations within projections of future climate? Biogeosciences 7:1645–1656. doi:10.5194/bg-7-1645-2010

Hamazaki T, Kuze A, Kondo K (2004) Sensor system for greenhouse gas observing satellite (GOSAT). In: Strojnik M (ed) Infrared spaceborne remote sensing XII, proceedings of the society of photo-optical instrumentation engineers (SPIE) 5543:275–282. doi:10.1117/12.560589

Hamme RC, Webley PW, Crawford WR, Whitney FA, DeGrandpre MD, Emerson SR, Eriksen CC, Giesbrecht KE, Gower JFR, Kavanaugh MT, Angelica PM, Sabine CL, Batten SD, Coogan LA, Grundle DS, Lockwood D (2010) Volcanic ash fuels anomalous plankton bloom in subarctic northeast Pacific. Geophys Res Lett 37. doi:10.1029/2010GL044629

Haywood J, Ramaswamy V, Soden B (1999) Tropospheric aerosol climate forcing in clear sky satellite observation over the oceans. Science 283:1299–1303

Heikes BG, Chang WN, Pilson MEQ, Swift E, Singh HB, Guenther A, Jacob DJ, Field BD, Fall R, Riemer D, Brand L (2002) Atmospheric methanol budget and ocean implication. Glob Biogeochem Cycles 16:1133

Heue KP, Brenninkmeijer CAM, Wagner T, Mies K, Dix B, Frieß U, Martinsson BG, Slemr F, van Velthoven PFJ (2010) Observations of the 2008 Kasatochi volcanic SO_2 plume by CARIBIC aircraft DOAS and the GOME-2 satellite. Atmos Chem Phys 10:4699–4713

Hill PG, Zubkov MV, Purdie DA (2010) Differential responses of Prochlorococcus and SAR11-dominated bacterioplankton groups to atmospheric dust inputs in the tropical Northeast Atlantic Ocean. FEMS Microbiol Lett 306(1):82–89. doi:10.1111/j.1574-6968.2010.01940.x

Hill PG, Haywood JL, Holland RJ, Purdie DA, Fuchs BM, Zubkov MV (2012) Internal and external influences on near-surface microbial community structure in the vicinity of the Cape Verde Islands. Microb Ecol 63(1):139–148. doi:10.1007/s00248-011-9952-2

Hirata T, Aiken J, Hardman-Mountford NJ, Smyth TJ, Barlow RG (2008) An absorption model to derive phytoplankton size classes from satellite ocean colour. Remote Sens Environ 112:3153–3159

Hirata T, Hardman-Mountford NJ, Brewin RJW, Aiken J, Barlow R, Suzuki K, Isada T, Howell E, Hashioka T, Noguchi-Aita M, Yamanaka Y (2011) Synoptic relationships between surface Chlorophyll-a and diagnostic pigments specific to phytoplankton functional types. Biogeosciences 8:311–327

Hoffmann LJ, Peeken I, Lochte K, Assmy P, Veldhuis M (2006) Different reactions of Southern Ocean phytoplankton size classes to iron fertilization. Limnol Oceanogr 51:1217–1229. doi:10.4319/lo.2006.51.3.1217

Huang H, Thomas GE, Grainger RG (2010) Relationship between wind speed and aerosol optical depth over remote ocean. Atmos Chem Phys 10:5943–5950. doi:10.5194/acp-10-5943-2010

Hurrell JW (1995) Decadal trends in the North Atlantic oscillation: regional temperatures and precipitation. Science 269(5224):676–679

IOCCP (2007) Surface ocean CO_2 variability and vulnerabilities workshop. IOCCP Report 7. UNESCO, Paris, 11–14 Apr 2007. http://www.ioccp.org/

Jacob DJ, Field BD, Li QB, Blake DR, de Gouw J, Warneke C, Hansel A, Wisthaler A, Singh HB, Guenther A (2005) Global budget of methanol: constraints from atmospheric observations. J Geophys Res 110:D08303

Jacobson AR, Mikaloff Fletcher SE, Gruber N, Sarmiento JL, Gloor M (2007) A joint atmosphere–ocean inversion for surface fluxes of carbon dioxide, I: methods and global-scale fluxes. Glob Biogeochem Cycles 21. doi:10.1029/2005GB002556

Jaeglé L, Steinberger L, Martin RV, Chance K (2005) Global partitioning of NOx sources using satellite observations: relative roles of fossil fuel combustion, biomass burning and soil emissions. Faraday Discuss 130:407–423

Jaeglé L, Quinn PK, Bates TS, Alexander B, Lin J-T (2011) Global distribution of sea salt aerosols: new constraints from in situ and remote sensing observations. Atmos Chem Phys 11:3137–3157. doi:10.5194/acp-11-3137-2011

Jamet C, Moulin C, Lefèvre N (2007) Estimation of the oceanic pCO2 in the North Atlantic from VOS lines in-situ measurements: parameters needed to generate seasonally mean maps. Ann Geophys 25:2247–2257

Jammoul A, Dumas S, D'Anna B, George C (2009) Photoinduced oxidation of sea salt halides by aromatic ketones: a source of halogenated radicals. Atmos Chem Phys 9(13):4229–4237

Jickells TS, An ZA, Baker AR, Bergametti G, Brooks N, Boyd PW, Duce RA, Hunter KA, Junji C, Kawahata H, Kubilay N, Anderson KK, la Roche J, Liss PS, Mahowald N, Prospero JM, Ridgwell AJ, Tegan I, Torres R (2005) Global iron connections between desert dust, ocean biogeochemistry and climate. Science 308:67–71

Johnson KS, Berelson WM, Boss ES, Chase Z, Claustre H, Emerson SR, Gruber N, Körtzinger A, Perry MJ, Riser SC (2009) Observing biogeochemical cycles at global scales with profiling floats and gliders, prospects for a global array. Oceanography 22(3):216–225

Jones N (2010) Sparks fly over theory that volcano caused salmon boom. Nat News. doi:10.1038/news.2010.572

Jones MT, Gislason SR (2008) Rapid releases of metal salts and nutrients following the deposition of volcanic ash into aqueous environments. Geochim Cosmochim Acta 72:3661–3680

Jones CE, Hornsby KE, Sommariva R, Dunk RM, von Glasow R, McFiggans G, Carpenter LJ (2010) Quantifying the contribution of marine organic gases to atmospheric iodine. Geophys Res Lett 37:L18804. doi:10.1029/2010GL043990

Jones CE, Andrews SJ, Carpenter LJ et al (2011) Results from the first national UK inter-laboratory calibration for very short-lived halocarbons. Atmos Meas Tech 4:865–874. doi:10.5194/amt-4-865-2011

Joos F, Plattner G-K, Stocker TF, Marchal O, Schmittner A (1999) Global warming and marine carbon cycle feedbacks on future atmospheric CO_2. Science 284:464–467

Kahn RA, Nelson DL, Garay MJ, Levy RC, Bull MA, Diner DJ, Martonchik MJ, Paradise SR, Hansen EG, Remer LA (2009) MISR aerosol product attributes and statistical comparisons with MODIS. IEEE Trans Geosci Remote Sens 47:4095–4114

Kaleschke L, Heygster G (2004) Towards multisensor microwave remote sensing of frost flowers on sea ice. Ann Glaciol 39:219–222

Keene WC, Long MS, Pszenny AAP, Sander R, Maben JR, Wall AJ, O'Halloran TL, Kerkweg A, Fischer EV, Schrems O (2009) Latitudinal variation in the multiphase chemical processing of inorganic halogens and related species over the eastern North and South Atlantic Oceans. Atmos Chem Phys 9:7361–7385

Kettle AJ, Andreae MO (2000) Flux of dimethylsulfide from the oceans: a comparison of updated data seas and flux models. J Geophys Res 105(D22):26793–26808

Kettle AJ, Andreae MO, Amouroux D et al (1999) A global database of sea surface dimethylsulfide (DMS) measurements and a procedure to predict sea surface DMS as a function of latitude, longitude, and month. Glob Biogeochem Cycles 13(2):399–444. doi:10.1029/1999GB900004

Kieber RJ, Mopper K (1990) Determination of picomolar concentrations of carbonyl-compounds in natural-waters, including seawater, by liquid-chromatography. Environ Sci Technol 24:1477–1481

Kiliyanpilakkil VP, Meskhidze N (2011) Deriving the effect of wind speed on clean marine aerosol optical properties using the A-Train satellites. Atmos Chem Phys 11:11401–11413. doi:10.5194/acp-11-11401-2011

Kloster S, Six KD, Feichter J, Maier-Reimer E, Roeckner E, Wetzel P, Stier P, Esch M (2007) Response of dimethyl-sulfide (DMS) in the ocean and atmosphere to global warming. J Geophys Res 112:G03005. doi:10.1029/2006JG000224

Knepp TN, Bottenheim J, Carlsen M, Carlson D, Donohoue D, Friederich G, Matrai PA, Netcheva S, Perovich DK, Santini R, Shepson PB, Simpson W, Stehle R, Valentic T, Williams C, Wyss PJ (2010) Development of an autonomous sea ice tethered buoy for the study of ocean–atmosphere-sea ice-snow pack interactions: the O-buoy. Atmos Meas Tech 3:249–261

Kohonen T (1982) Self-organized formation of topologically correct feature maps. Biol Cybern 43:59–69

Kokhanovsky AA, de Leeuw G (eds) (2009) Satellite aerosol remote sensing over land. Springer-Praxis, Berlin, p 388. ISBN 978-3-540-69396-3

Korhonen H, Carslaw KS, Spracklen DV, Mann GW, Woodhouse MT (2008) Influence of oceanic dimethyl sulfide emissions on cloud condensation nuclei concentrations and seasonality over the remote Southern Hemisphere oceans: a global model study. J Geophys Res 113:D15204

Körtzinger A, Send U, Lampitt RS, Hartman S, Wallace DWR, Karstensen J, Villagarcia MG, Llinás O, DeGrandpre MD (2008) The seasonal pCO_2 cycle at 49°N/16.5°W in the northeastern Atlantic Ocean and what it tells us about biological productivity. J Geophys Res 113:C04020

Kostadinov TS, Siegel DA, Maritorena S (2010) Global variability of phytoplankton functional types from space: assessment via the particle size distribution. Biogeosciences Discuss 7:4295–4340. doi:10.5194/bg-7-3239-2010

Krishfield R, Toole J, Proshutinsky A, Timmermans M-L (2008) Automated ice-tethered profilers for seawater observations under pack ice in all seasons. J Atmos Ocean Technol 25(11):2091–2105

Krueger AJ (1983) Sighting of El Chichón sulfur dioxide clouds with the Nimbus 7 total ozone mapping spectrometer. Science 220(4604):1377–1379

Kühn W, Pätsch J, Thomas H, Borges AV, Schiettecatte L-S, Bozec Y, Prowe AEF (2010) Nitrogen and carbon cycling in the North Sea and exchange with the North Atlantic – a model study, Part II: carbon budget and fluxes. Cont Shelf Res 30:1701–1716. doi:10.1016/j.csr.2010.07.001

Kwok R (2010) Satellite remote sensing of sea-ice thickness and kinematics: a review. J Glaciol 56(200):1129–1140

Lachkar Z, Gruber N (2013) Response of biological production and air-sea CO_2 fluxes to upwelling intensification in the California and Canary Current Systems. J Mar Syst 109–110:149–160. doi:10.1016/j.jmarsys.2012.04.003

Lagerloef G, Swift C, LeVine D (1995) Sea surface salinity: the next remote sensing challenge. Oceanography 8:44–50

Lagerloef G, Boutin J, Chao Y, Delcroix T, Font J, Niiler P, Reul N, Riser S, Schmitt R, Stammer D, Wentz F (2010) Resolving the global surface salinity field and variations by blending satellite and in situ observations. In: Hall J, Harrison DE, Stammer D (eds) Proceedings of OceanObs'09: sustained ocean observations and information for society, vol 2. ESA Publication, Venice, 21–25 Sept 2009, WPP-306. doi:10.5270/OceanObs09.cwp.51

Lana A, Bell TG, Simó R, Vallina SM, Ballabrera-Poy J, Kettle AJ, Dachs J, Bopp L, Saltzman ES, Stefels J, Johnson JE, Liss PS (2011) An updated climatology of surface dimethlysulfide concentrations and emission fluxes in the global ocean. Glob Biogeochem Cycles 25(1):GB1004. doi:10.1029/2010GB003850

Langmann B, Zaksek K, Hort M, Duggen S (2010) Volcanic ash as fertiliser for the surface ocean. Atmos Chem Phys 10:3891–3899

Lapina K, Heald CL, Spracklen DV, Arnold SR, Allan JD, Coe H, McFiggans G, Zorn SR, Drewnick F, Bates TS, Hawkins LN, Russell LM, Smirnov A, O'Dowd C, Hind AJ (2011) Investigating organic aerosol loading in the remote marine environment. Atmos Chem Phys 11:8847–8860. doi:10.5194/acp-11-8847-2011

Law CS, Brévière E, de Leeuw G, Guieu C, Garçon VC, Kieber DJ, Kontradowitz S, Paulmier A, Quinn PK, Saltzman E, Stefels J, von Glasow R (2013) Evolving research directions in Surface Ocean Lower Atmosphere (SOLAS) science. Environ Chem 10:1–16, http://dx.doi.org/10.1071/EN12159

Le Clainche Y, Levasseur M, Vezina A et al (2006) Modelling analysis of the effect of iron enrichment on DMS dynamics in the NE Pacific (SERIES experiment). J Geophys Res 111. doi:10.1029/2005JC002947

Le Clainche Y et al (2010) A first appraisal of prognostic ocean DMS models and prospects for their use in climate models. Glob Biogeochem Cycles 24:GB3021. doi:10.1029/2009GB003721

Le Quéré C, Aumont O, Monfray P, Orr J (2003) Propagation of climatic events on ocean stratification, marine biology and CO_2: case studies over the 1979–1999 period. J Geophys Res 108. doi:10.1029/2001JC000920

Le Quéré CL, Harrison SP, Prentice C, Buitenhuis ET, Aumont O, Bopp L, Claustre H, Cotrim Da Cunha L, Geider R, Giraud X, Klaas C, Kohfeld KE, Legendre L, Manizza M, Platt T, Rivkin RB, Sathyendranath S, Uitz J, Watson AJ, Wolf-Gladrow D (2005) Ecosystem dynamics based on plankton functional types for global ocean biogeochemistry models. Glob Change Biol 11:2016–2040. doi:10.1111/j.1365-2486.2005.1004.x

Le Quéré C, Rödenbeck C, Buitenhuis ET, Conway TJ, Langenfelds R, Gomez A, Labuschagne C, Ramonet M, Nakazawa T, Metzl N, Gillett N, Heimann M (2007) Saturation of the Southern Ocean CO_2 sink due to recent climate change. Science 316:1735–1738

Le Vine DML, Lagerloef GSE, Colomb FR, Yueh SH, Member S, Pellerano FA, Member S (2007) Aquarius: an instrument to monitor sea surface salinity from space IEEET. Geosci Remote 45(7):2040–2050

Lee JD, Moller SJ, Read KA, Lewis AC, Mendes L, Carpenter LJ (2009a) Year-round measurements of nitrogen oxides and ozone in the tropical North Atlantic marine boundary layer. J Geophys Res 114:D21302. doi:10.1029/2009JD011878

Lee KH, Li Z, Kim YJ, Kokhanovsky AA (2009b) Atmospheric aerosol monitoring from satellite observations: a history of three decades. In: Kim YJ, Platt U, Gu MB, Iwahashi H (eds) Atmospheric and biological environmental monitoring. Springer, Berlin

Lefèvre N, Taylor A (2002) Estimating pCO_2 from sea surface temperatures in the Atlantic gyres. Deep-Sea Res I 49(3):539–554

Lefèvre N, Watson AJ, Watson AR (2005) A comparison of multiple regression and neural network techniques for mapping in situ pCO_2 data. Tellus 57B:375–384

Lehahn Y, Koren I, Boss E, Ben-Ami Y, Altaratz O (2010) Estimating the maritime component of aerosol optical depth and its dependency on surface wind speed using satellite data. Atmos Chem Phys 10:6711–6720. doi:10.5194/acp-10-6711-2010

Lelieveld J, van Aardenne J, Fischer H, de Reus M, Williams J, Winkler P (2004) Increasing ozone over the Atlantic Ocean. Science 304:1483–1487

Lerot C, Stavrakou T, De Smedt I, Müller J-F, Van Roozendael M (2010) Glyoxal vertical columns from GOME-2 backscattered light measurements and comparisons with a global model. Atmos Chem Phys 10:12059–12072

Lewis ER, Schwartz SE (2004) Sea salt aerosol production: mechanisms, methods, measurements, and models: a critical review. American Geophysical Union, Washington, DC

Li X, Maring H, Savoie D, Voss K, Prospero JM (1996) Dominance of mineral dust in aerosol light-scattering in the North Atlantic trade winds. Nature 380(6573):416–419. doi:10.1038/380416a0

Lin II, Hu C, Li YH, Ho TY, Fischer TP, Wong GTF, Wu J, Huang CW, Chu DA, Ko DS, Chen JP (2011) Fertilization potential of volcanic dust in the low-nutrient low-chlorophyll western North Pacific subtropical gyre: satellite evidence and laboratory study. Glob Biogeochem Cycles 25:GB1006

Lindenthal A (2011) Phytoplankton growth in the NE Pacific. Diploma thesis, Institute of Geophysics, University of Hamburg, Germany (written in German)

Liss PS, Hatton AD, Malin G, Nightingale PD, Turner SM (1997) Marine sulphur emissions. Philos Trans R Soc Lond B Biol Sci 352:159. doi:10.1098/RSTB.1997.0011

Liu WT (2002) Progress in scatterometer application. J Oceanogr 58:121–136

Liu X, Penner JE, Das B, Bergmann D, Rodriguez JM, Strahan S, Wang M, Feng Y (2007) Uncertainties in global aerosol simulations, assessment using three meteorological data sets. J Geophys Res 112:D11212. doi:10.1029/2006JD008216

Liu WT, Tang W, Xie X, Navalgund R, Xu K (2008) Power density of ocean surface wind-stress from international scatterometer tandem missions. Int J Remote Sens 29:6109–6116

Loisel H, Bosc E, Stramski D, Oubelkheir K, Deschamps PY (2001) Seasonal variability of the backscattering coefficient in the Mediterranean Sea based on Satellite SeaWiFS imagery. J Geophys Res Lett 28(22):4203–4206

Loisel H, Nicolas JM, Deschamps PY, Frouin R (2002) Seasonal and inter- annual variability of particulate organic matter in the global ocean. Geophys Res Lett 29(49):2196. doi:10.1029/2002GL015948

Longhurst AR (2007) Ecological geography of the sea, 2nd edn. Academic, Burlington/Boston

Lüger H, Wallace DWR, Körtzinger A, Nojiri Y (2004) The pCO_2 varability in the midlatitude North Atlantic Ocean during a full annual cycle. Glob Biogeochem Cycles 18: GB3023. doi:10.1029/2003GB002200

Mahajan AS, Plane JMC, Oetjen H, Mendes L, Saunders RW, Saiz-Lopez A, Jones CE, Carpenter LJ, McFiggans GB (2010) Measurement and modelling of tropospheric reactive halogen species over the tropical Atlantic Ocean. Atmos Chem Phys 10:4611–4624

Marbach T, Beirle S, Platt U, Hoor P, Wittrock F, Richter A, Vrekoussis M, Grzegorski M, Burrows JP, Wagner T (2009) Satellite measurements of formaldehyde linked to shipping emissions. Atmos Chem Phys 9:8223–8234

Maritorena S, d'Andon OHF, Mangin A, Siegel DA (2010) Merged satellite ocean color data products using a bio-optical model: characteristics, benefits and issues. Remote Sens Environ 114(8):1791–1804

Mårtensson EM, Nilsson ED, de Leeuw G, Cohen LH, Hansson H-C (2003) Laboratory simulations and parameterizations of the primary marine aerosol productions. J Geophys Res 108:4297

Martin JH, Fitzwater SE (1988) Iron deficiency limits phytoplankton growth in the north-east Pacific subarctic. Nature 331:341–343

Martin M, Dash P, Ignatov A, Banzon V, Beggs H, Brasnett B, Cayula J-F, Cummings J, Donlon C, Gentemann C, Grumbine R, Ishizaki S, Maturi E, Reynolds RW, Roberts-Jones J (2012) Group for High Resolution Sea Surface temperature (GHRSST) analysis fields inter-comparisons. Part 1: a GHRSST multi-product ensemble (GMPE). Deep-Sea Res II. http://dx.doi.org/10.1016/j.dsr2.2012.04.013

Martino M, Mills GP, Woeltjen J, Liss PS (2009) A new source of volatile organoiodine compounds in surface seawater. Geophys Res Lett 36:L01609. doi:10.1029/2008GL036334

Massom RA (2009) Principal uses of remote sensing in sea ice field research. In: Eicken H, Gradinger R, Salganek M, Shirasawa K, Perovich D, Leppäranta M (eds) Field techniques for sea ice research. University of Alaska Press, Fairbanks, pp 405–466

McDonald D, Pedersen TF (1999) Multiple late Quaternary episodes of exceptional diatom production in the Gulf of Alaska. Deep-Sea Res II 46:2993–3017

McGillicuddy DJ, Johnson RJ, Siegel DA, Michaels AF, Bates NR, Knap AH (1999) Mesoscale variations of biogeochemical properties in the Sargasso Sea. J Geophys Res 104:13381–13394

McGillicuddy DJ, Anderson LA, Bates NR, Bibby T, Buesseler KO, Carlson CA, Davis CS, Ewart C, Falkowski PG, Goldthwait SA, Hansell DA, Jenkins WJ, Johnson R, Kosnyrev VK, Ledwell JR, Li QP, Siegel DA, Steinberg DK (2007) Eddy/wind interactions stimulate extraordinary mid-ocean plankton blooms. Science 316(5827):1016–1021. doi:10.1126/science.1136256

McKinley GA, Follows MJ, Marshall J (2004) Mechanisms of air-sea CO_2 flux variability in the equatorial Pacific and the North Atlantic. Glob Biogeochem Cycles 18:GB2011. doi:10.1029/2003GB002179

Meissner T, Wentz FJ (2009) Wind-vector retrievals under rain with passive satellite microwave radiometers. IEEE Trans Geosci Remote Sens 47(9):3065–3083. doi:10.1109/TGRS.2009.2027012

Mikaloff Fletcher SE, Gruber N, Jacobson AR, Doney SC, Dutkiewicz S, Gerber M, Gloor M, Follows M, Joos F, Lindsay K, Menemenlis D, Mouchet A, Müller SA, Sarmiento JL (2007) Inverse estimates of the oceanic sources and sinks of natural CO_2 and the implied oceanic transport. Glob Biogeochem Cycles 21:GB1010. doi:10.1029/2006GB002751

Mikaloff-Fletcher SE, Gruber N, Jacobson AR, Doney SC, Dutkiewicz S, Gerber M, Follows M, Joos F, Lindsay K, Menemenlis D, Mouchet A, Mueller SA, Sarmiento JL (2006) Inverse estimates of anthropogenic CO_2 uptake, transport, and storage by the ocean. Glob Biogeochem Cycles 20:GB2002. doi:10.1029/2005GB002530

Mills MM, Ridame C, Davey M, La Roche J, Geider RJ (2004) Iron and phosphorous co-limit nitrogen fixation in the eastern tropical North Atlantic. Nature 429:292–294

Mohanan EC, O'Muircheartaigh IG (1980) Optimal power-law description of oceanic whitecap coverage dependence on wind speed. J Phys Ocean 10:2094–2099

Montzka SA, Reimann S (2011) Ozone depleting substances (ODS's) and related chemicals, Chapter 1 in: scientific assessment of ozone depletion: 2010. Global Ozone Research and Monitoring Project, Report No. 52, 373 pp

Moore CM, Mills MM, Achterberg EP, Geider RJ, La Roche J, Lucas MI, McDonagh EL, Pan X, Poulton AJ, Rijkenberg MJA, Suggett DJ, Ussher SJ, Woodward MJ (2009) Large-scale distribution of Atlantic nitrogen fixation controlled by iron availability. Nat Geosci. doi:10.1038/ngeo667

Mouw CB, Yoder JA (2010) Optical determination of phytoplankton size composition from global SeaWiFS imagery. J Geophys Res 115:C12018. doi:10.1029/2010JC006337

Mulcahy JP, O'Dowd CD, Jennings SG, Ceburnis D (2008) Significant enhancement of aerosol optical depth in marine air under high wind conditions. Geophys Res Lett 35: L16810. doi:10.1029/2008GL034303

Müller C, Iinuma Y, Karstensen J, van Pinxteren D, Lehmann S, Gnauk T, Herrmann H (2009) Seasonal variation of aliphatic amines in marine sub-micrometer particles at the Cape Verde islands. Atmos Chem Phys 9(24):9587–9597

Myhre G, Stordal F, Johnsrud M, Ignatov A, Mishchenko MI, Geogdzhayev IV, Tanré D, Deuzé JL, Goloub P, Nakajima T, Higurashi A, Torres O, Holben BN (2004) Intercomparison of satellite retrieved aerosol optical depth over ocean. J Atmos Sci 61:499–513

Myhre G, Stordal F, Johnsrud M, Diner DJ, Geogdzhayev IV, Haywood JM, Holben BN, Holzer-Popp T, Ignatov A, Kahn RA, Kaufman YJ, Loeb N, Martonchik JV, Mishchenko MI, Nalli NR, Remer LA, Schroedter-

Homscheidt M, Tanré D, Torres O, Wang M (2005) Intercomparison of satellite retrieved aerosol optical depth over ocean during the period September 1997 to December 2000. Atmos Chem Phys 5:1697–1719

Newsam GN, Enting IG (1988) Inverse problems in atmospheric constituent studies: I. Determination of surface sources under a diffusive transport approximation. Inverse Probl 4:1037–1054

Nie C, Long DG (2008) A C-band scatterometer simultaneous wind/rain retrieval method. IEEE Trans Geosci Remote Sens 46:3618–3632

Nightingale PD, Malin G, Law CS, Watson AJ, Liss PS, Liddicoat MI, Boutin J, Upstill-Goddard RC (2000) In situ evaluation of air-sea gas exchange parameterizations using novel conservative and volatile tracers. Glob Biogeochem Cycles 14(1):373–387

Nowlan CR, Liu X, Chance K, Cai Z, Kurosu TP, Lee C, Martin RV (2011) Retrievals of sulfur dioxide from the Global Ozone Monitoring Experiment 2 (GOME-2) using an optimal estimation approach: algorithm and initial validation. J Geophys Res 116:D18301. doi:10.1029/2011JD015808

O'Dowd CD, de Leeuw G (2007) Marine aerosol production: a review of the current knowledge. Philos Trans R Soc A 365:1753–1774. doi:10.1098/rsta.2007.2043

O'Dowd CD, Facchini MC, Cavalli F, Ceburnis D, Mircea M, Decesari S, Fuzzi S, Yoon YL, Putaud JP (2004) Biogenically driven organic contribution to marine aerosol. Nature 431:676. doi:10.1038/NATURE02959

O'Dowd C, Scannell C, Mulcahy J, Jennings SG (2010) Wind speed influences on marine aerosol optical depth. Adv Meteorol 2010:830846. doi:10.1155/2010/830846

Okin GS, Baker AR, Tegen I, Mahowald NM, Dentener FJ, Duce RA, Galloway JN, Hunter K, Kanakidou M, Kubilay N, Prospero JM, Sarin M, Surapipith V, Uematsu M, Zhu T (2011) Impacts of atmospheric nutrient deposition on marine productivity: roles of nitrogen, phosphorus, and iron. Glob Biogeochem Cycles 25:GB2022. doi:10.1029/2010GB003858

Olgun N, Duggen S, Croot P, Dietze H, Schacht U, Oskarsson N, Siebe C, Auer A (2011) Surface ocean iron fertilization: the role of subduction zone and hotspot volcanic ash and fluxes into the Pacific Ocean. Glob Biogeochem Cycles 25. doi:10.1029/2009GB003761

Olgun N, Duggen S, Langmann B, Hort M, Waythomas CF, Hoffmann L, Croot P (2013) Geochemical evidence for oceanic iron fertilization from the Kasatochi volcanic eruption and evaluation of the potential impacts on sockeye salmon population. Mar Ecol Prog Ser (in press)

Olsen A, Bellerby RGJ, Johannesseen T, Omar AM, Skjelvan I (2003) Interannual variability in the wintertime air-sea flux of carbon dioxide in the northern North Atlantic, 1981–2001. Deep-Sea Res I 50:1323–1338

Ono S, Ennyu A, Najjar RG, Bates NR et al (2001) Shallow remineralisation in the Sargasso Sea estimated from seasonal variations in oxygen, dissolved inorganic carbon and nitrate. Deep-Sea Res II 48(8–9):1567–1582. doi:10.1016/S0967-0645(00)00154-5

Parsons TR, Whitney FA (2012) Did volcanic ash from Mt. Kasatochi in 2008 contribute to a phenomenal increase in Fraser River sockeye salmon (Oncorhynchus nerka) in 2010? Fish Oceanogr 21:374–377. doi:10.1111/j.1365-2419.2012.00630.x

Perovich D (2009) Automatic measurement stations. In: Eicken H, Gradinger R, Salganek M, Shirasawa K, Perovich D, Lepparanta M (eds) Field techniques for sea ice research. University of Alaska Press, Fairbanks, pp 383–394

Perovich DK, Eicken H, Meier W (2012) International coordination to improve studies of changes in Arctic sea ice cover. Eos 93(12):128

Pfeil B, Olsen A, Bakker DCE, Hankin S, Koyuk H, Kozyr A, Malczyk J, Manke A, Metzl N, Sabine CL, Akl J, Alin SR, Bates N, Bellerby RGJ, Borges A, Boutin J, Brown PJ, Cai W-J, Chavez FP, Chen A, Cosca C, Fassbender AJ, Feely RA, González-Dávila M, Goyet C, Hales B, Hardman-Mountford N, Heinze C, Hood M, Hoppema M, Hunt CW, Hydes D, Ishii M, Johannessen T, Jones SD, Key RM, Körtzinger A, Landschützer P, Lauvset SK, Lefèvre N, Lenton A, Lourantou A, Merlivat L, Midorikawa T, Mintrop L, Miyazaki C, Murata A, Nakadate A, Nakano Y, Nakaoka S, Nojiri Y, Omar AM, Padin XA, Park G-H, Paterson K, Perez FF, Pierrot D, Poisson A, Ríos AF, Salisbury J, Santana-Casiano JM, Sarma VVSS, Schlitzer R, Schneider B, Schuster U, Sieger R, Skjelvan I, Steinhoff T, Suzuki T, Takahashi T, Tedesco K, Telszewski M, Thomas H, Tilbrook B, Tjiputra J, Vandemark D, Veness T, Wanninkhof R, Watson AJ, Weiss R, Wong CS, Yoshikawa-Inoue H (2013) A uniform, quality controlled Surface Ocean CO_2 Atlas (SOCAT). Earth Syst Sci Data 5:125–143. doi:10.5194/essd-5-125-2013

Pierce JR, Adams PJ (2006) Global evaluation of CCN formation by direct emission of sea salt and growth of ultrafine sea salt. J Geophys Res 111:D06203. doi:10.1029/2005JD006186

Pierrot D, Neill C, Sullivan K, Castle R, Wanninkhof R, Lüger H, Johannessen T, Olsen A, Feely RA, Cosca CE (2009) Recommendations for autonomous underway pCO_2 measuring systems and data reduction routines. Deep-Sea Res II 56:512–522. doi:10.1016/j.dsr2.2008.12.005

Plass-Dülmer C, Koppmann R, Ratte M, Rudolph J (1995) Light nonmethane hydrocarbons in seawater. Glob Biogeochem Cycles 9:79. doi:10.1029/94GB02416

Powell CF, Baker AR, Jickells TD, Bange HW (in preparation) Estimation of the atmospheric flux of iron, nitrogen and phosphate to the eastern tropical North Atlantic

Previdi M, Fennel K, Wilkin J, Haidvogel DB (2009) Interannual variability in atmospheric CO_2 uptake on the Northeast US Continental Shelf. J Geophys Res 114:G04003. doi:10.1029/2008JG000881

Prinn RG, Weiss RF, Fraser PJ et al (2000) A history of chemically and radiatively important gases in air deduced from ALE/GAGE/AGAGE. J Geophys Res 105:17751–17792. doi:10.1029/2000JD900141

Prospero JM, Lamb PJ (2003) African droughts and dust transport to the Caribbean: climate change implications. Science 302:1024–1027

Prospero JM, Nees RT (1986) Impact of the North African drought and El Nino on mineral dust in the Barbados trade winds. Nature 320(6064):735–738. doi:10.1038/320735a0

Prowe AEF, Thomas H, Pätsch J, Kühn W, Bozec Y, Schiettecatte L-S, Borges AV, de Baar HJW (2009) Mechanisms controlling the air-sea CO_2 flux in the North Sea. Cont Shelf Res 29:1801–1808. doi:10.1016/j.csr.2009.06.003

Qin AK, Clausi DA (2010) Multivariate image segmentation using semantic region growing with adaptive edge penalty. IEEE Trans Image Proc 19:2157–2170

Quack B, Atlas E, Petrick G, Schauffler S, Wallace D (2004) Oceanic bromoform sources for the tropical atmosphere. Geophys Res Lett. doi:10.1029/2004GL020597

Quack B, Peeken I, Petrick G, Nachtigall K (2007) Oceanic distribution and sources of bromoform and dibromomethane in the Mauritanian upwelling. J Geophys Res 112:C10006. doi:10.1029/2006JC003803

Quilfen Y, Vandemark D, Chapron B, Feng H, Sienkiewicz J (2011) Estimating gale to hurricane force winds using the satellite altimeter. J Atmos Ocean Technol 28:453–458. doi:10.1175/JTECH-D-10-05000.1

Quinn PK, Bates TS (2005) Regional aerosol properties: comparisons of boundary layer measurements from ACE 1, ACE 2, Aerosols99, INDOEX, ACE Asia, TARFOX, and NEAQS. J Geophys Res 110(D14):D14202

Raitsos DE, Lavender SJ, Maravelias CD, Haralambous J, Richardson AJ, Reid PC (2008) Identifying four phytoplankton functional types from space: an ecological approach. Limnol Oceanogr 53(2):605–613

Read KA, Mahajan AS, Carpenter LJ, Evans MJ, Faria BE, Heard DE, Hopkins JR, Lee JD, Moller SJ, Lewis AC, Mendes L, McQuaid JB, Oetjen H, Saiz-Lopez A, Pilling MJ, Plane JMC (2008) Extensive halogen-mediated ozone destruction over the tropical Atlantic Ocean. Nature 453:1232–1235. doi:10.1038/nature07035

Read KA, Carpenter LJ, Arnold SR, Beale R, Nightingale PD, Hopkins JR, Lewis AC, Mendes L, Fleming ZL (2012) Time-series and atmospheric budgets of acetone, methanol and acetaldehyde in remote marine air at the Cape Verde Atmospheric Observatory. Environ Sci Technol 46(20):11028–11039. doi:10.1021/es302082p

Remer LA et al (2008) Global aerosol climatology from the MODIS satellite sensors. J Geophys Res 113:D14S07. doi:10.1029/2007JD009661

Reul N, Tenerelli J, Boutin J, Chapron B, Paul F, Brion E, Gaillard F, Archer O (2012a) Overview of the first SMOS sea surface salinity products. Part I: quality assessment for the second half of 2010. IEEE Trans Geosci Remote Sens 99:1–12, http://dx.doi.org/10.1109/TGRS.2012.2188408

Reul N, Tenerelli J, Chapron B, Vandemark D, Quilfen Y, Kerr Y (2012b) SMOS satellite L-band radiometer: a new capability for ocean surface remote sensing in hurricanes. J Geophys Res 117:C02006. doi:10.1029/2011JC007474

Richter A, Wittrock F, Eisinger M, Burrows JP (1998) GOME observations of tropospheric BrO in Northern Hemispheric spring and summer 1997. Geophys Res Lett 25:2683–2686

Richter A, Eyring V, Burrows JP, Bovensmann H, Lauer A, Sierk B, Crutzen PJ (2004) Satellite measurements of NO_2 from international shipping emissions. Geophys Res Lett 31:L23110. doi:10.1029/2004GL020822

Rijkenberg MJA, Langlois RJ, Mills MM, Patey MD, Hill PG, Nielsdottir MC, Compton TJ, La Roche J, Achterberg EP (2011) Environmental forcings of nitrogen fixation in the eastern (sub-) tropical North Atlantic Ocean. PlosOne 6(12):e28989

Rintoul SR, Sparrow M, Meredith MP, Wadley V, Speer K, Hofmann E, Summerhayes C, Urban E, Bellerby R (eds) (2012) The Southern Ocean observing system: initial science and implementation strategy. SOOS International Project Office, Hobart. ISBN 978-0-948277-27-6

Rödenbeck C, Houweling S, Gloor M, Heimann M (2003) CO_2 Flux History 1982–2001 Inferred from atmospheric data using a global inversion of atmospheric transport. Atmos Chem Phys 3:1919–1964

Rödenbeck C, Le Quéré C, Heimann M, Keeling R (2008) Interannual variability in oceanic biogeochemical processes inferred by inversion of atmospheric O_2/N_2 and CO_2 data. Tellus 60B:685

Roeckner E et al (2006) Sensitivity of simulated climate to horizontal and vertical resolution in the ECHAM5 atmosphere model. J Clim 19:3771–3791. doi:10.1175/JCLI3824.1

Roemmich D et al (2009) Argo: the challenge of continuing 10 years of progress. Oceanography 2(30):46–55. http://www.knmi.nl/publications/fulltexts/roemmich_et_al.oceanography_godae_09.pdf. Accessed 25 Jan 2012

Röhrs J, Kaleschke L, Bröhan D, Siligam PK (2012) An algorithm to detect sea ice leads by using AMSR-E passive microwave imagery. The Cryosphere 6(2):343–352

Roy T, Bopp L, Gehlen M, Schneider B, Cadule P, Frölicher TL, Segschneider J, Tjiputra J, Heinze C, Joos F (2011) Regional impacts of climate change and atmospheric CO_2 on future ocean carbon uptake: a multi-model linear feedback analysis. J Clim 24:2300–2318

Sabine CL, Feely RA, Gruber N, Key RM, Lee K, Bullister JL, Wanninkhof R, Wong CS, Wallace DWR, Tilbrook B, Millero FJ, Peng T-H, Kozyr A, Ono T, Ríos AF (2004) The oceanic sink for anthropogenic CO_2. Science 305:367–371

Sabine CL, Hankin S, Koyuk H, Bakker DCE, Pfeil B, Olsen A, Metzl N, Kozyr A, Fassbender A, Manke A, Malczyk J, Akl J, Alin SR, Bellerby RGJ, Borges A, Boutin J, Brown PJ, Cai W-J, Chavez FP, Chen A, Cosca C, Feely RA, González-Dávila M, Goyet C, Hardman-Mountford N, Heinze C, Hoppema M, Hunt CW, Hydes D, Ishii M, Johannessen T, Key RM, Körtzinger A, Landschützer P, Lauvset SK, Lefèvre N, Lenton A, Lourantou A, Merlivat L, Midorikawa T, Mintrop L, Miyazaki C, Murata A, Nakadate A, Nakano Y, Nakaoka S, Nojiri Y, Omar AM, Padin XA, Park G-H, Paterson K, Perez FF, Pierrot D, Poisson A, Ríos AF, Salisbury J, Santana Casiano JM, Sarma VVSS, Schlitzer R, Schneider B, Schuster U, Sieger R, Skjelvan I, Steinhoff T, Suzuki T, Takahashi T, Tedesco K, Telszewski M, Thomas H, Tilbrook B, Vandemark D, Veness T, Watson AJ, Weiss R, Wong CS, Yoshikawa-Inoue H (2013) Gridding of the Surface Ocean CO_2 Atlas (SOCAT) gridded data products. Earth Syst Sci Data 5:145–153. doi:10.5194/essd-5-145-2013

Sadeghi A, Dinter T, Vountas M, Taylor B, Peeken I, Bracher A (2011) Improvements to the PhytoDOAS method for the identification of major Phytoplankton groups using high spectrally resolved satellite data. Ocean Sci Discuss 8:2271–2311. doi:10.5194/osd-8-2271-2011

Sadeghi A, Dinter T, Vountas M, Taylor B, Altenburg-Soppa M, Bracher A (2012) Remote sensing of coccolithophore blooms in selected oceanic regions using the PhytoDOAS method applied to hyper-spectral satellite data. Biogeosciences 9:2127–2143. doi:10.5194/bg-9-2127-2012

Saiz-lopez A, Chance K, Liu X, Kurosu TP, Sander S (2007) First observations of iodine oxide from space. Geophys Res Lett 34:L12812. doi:10.1029/2007GL030111

Sander R, Crutzen PJ (1996) Model study indicating halogen activation and ozone destruction in polluted air masses transported to the sea. J Geophys Res 101:9121–9138

Sarmiento JL, Hughes TMC, Stouffer RJ, Manabe S (1998) Simulated response of the ocean carbon cycle to anthropogenic climate warming. Nature 393:245–249

Sathyendranath S, Watts L, Devred E, Platt T, Caverhill C, Maass H (2004) Discrimination of diatoms from other phytoplankton using ocean-colour data. Mar Ecol Prog Ser 272:59–68

Schönhardt A, Richter A, Wittrock F, Kirk H, Oetjen H, Roscoe HK, Burrows JP (2008) Observations of iodine monoxide columns from satellite. Atmos Chem Phys 8:637–653

Schuster U et al (2009) Trends in North Atlantic sea-surface fCO_2 from 1990 to 2006. Deep-Sea Res II 56(8–10):620–629. doi:10.1016/j.dsr2.2008.12.011

Schuster U, McKinley GA, Bates N, Chevallier F, Doney SC, Fay AR, González-Dávila M, Gruber N, Jones S, Krijnen J, Landschützer P, Lefèvre N, Manizza M, Mathis J, Metzl N, Olsen A, Rios AF, Rödenbeck C, Santana-Casiano JM, Takahashi T, Wanninkhof R, Watson AJ (2012) Atlantic and Arctic sea-air CO_2 Fluxes, 1990–2009. Biogeosciences Discuss 9:10669–10724

Siegel DA, Peterson P, McGillicuddy DJ, Maritorena S, Nelson NB (2011) Bio-optical footprints created by mesoscale eddies in the Sargasso Sea. Geophys Res Lett 38:L13608. doi:10.1029/2011GL047660

Simó R, Dachs J (2002) Global ocean emission of dimethylsulfide predicted from biogeophysical data. Glob Biogeochem Cycles 16(4):1018. doi:10.1029/2001GB001829

Simó R, Pedrós-Alio C (1999) Role of vertical mixing in controlling the oceanic production of dimethyl sulphide. Nature 402:396–399

Singh HB, Kanakidou M, Crutzen PJ, Jacob DJ (1995) High concentrations and photochemical fate of oxygenated hydrocarbons in the global troposphere. Nature 378:50–54

Singh HB, Tabazadeh A, Evans MJ, Field BD, Jacob DJ, Sachse G, Crawford JH, Shetter R, Brune WH (2003) Oxygenated volatile organic chemicals in the oceans: interferences and implications based on atmospheric observations and air–sea flux exchange models. Geophys Res Lett 30:1862. doi:10.1029/2003GL017933

Small R, deSzoeke SP, Xie SP, O'Neill LO, Seo H, Song Q, Cornillon P, Spall M, Minobe S (2008) Air-sea interaction over ocean fronts and eddies. Dyn Atmos Ocean 45:274–319

Smirnov A, Holben BN, Giles DM, Slutsker I, O'Neill NT, Eck TF, Macke A, Croot P, Courcoux Y, Sakerin SM, Smyth TJ, Zielinski T, Zibordi G, Goes JI, Harvey MJ, Quinn PK, Nelson NB, Radionov VF, Duarte CM, Losno R, Sciare J, Voss KJ, Kinne S, Nalli NR, Joseph E, Krishna Moorthy K, Covert DS, Gulev SK, Milinevsky G, Larouche P, Belanger S, Horne E, Chin M, Remer LA, Kahn RA, Reid JS, Schulz M, Heald CL, Zhang J, Lapina K, Kleidman RG, Griesfeller J, Gaitley BJ, Tan Q, Diehl TL (2011) Maritime aerosol network as a component of AERONET – first results and comparison with global aerosol models and satellite retrievals. Atmos Meas Tech 4:583–597. doi:10.5194/amt-4-583-2011

Smirnov A, Sayer AM, Holben BN, Hsu NC, Sakerin SM, Macke A, Nelson NB, Courcoux Y, Smyth TJ, Croot P, Quinn PK, Sciare J, Gulev SK, Piketh S, Losno R, Kinne S, Radionov VF (2012) Effect of wind speed on aerosol optical depth over remote oceans, based on data from the Maritime Aerosol Network. Atmos Meas Tech 5:377–388. doi:10.5194/amt-5-377-2012

Smith RC, Baker KS (1978) The bio-optical state of ocean waters and remote sensing. Limnol Oceanogr 23:247–259

Smith MA, White M (1985) Observations on lakes near Mount St. Helens: phytoplankton. J Arch Hydrobiol 104:345–363

Sofiev M, Soares J, Prank M, de Leeuw G, Kukkonen J (2011) A regional-to-global model of emission and transport of sea salt particles in the atmosphere. J Geophys Res 116:D21302. doi:10.1029/2010JD014713

Soh L-K, Tsatsoulis C, Gineris D, Bertoia C (2004) ARKTOS: an intelligent system for SAR sea ice image classification. IEEE Trans Geosci Remote Sens 42:229–248

Steinberg DK, Carlson CA, Bates NR, Johnson RJ, Michaels AF, Knap AH (2001) Overview of the US JGOFS Bermuda Atlantic Time-series Study (BATS): a decade-scale look at ocean biology and biogeochemistry. Deep-Sea Res II 48(8–9):1405–1447. doi:10.1016/S0967-0645(00)00148-X

Steinhoff T, Friedrich T, Hartman SE, Oschlies A, Wallace DWR, Körtzinger A (2010) Estimating mixed layer nitrate in the North Atlantic Ocean. Biogeosciences 7:795–807

Stewart DJ, Taylor CM, Reeves CE, McQuaid JB (2008) Biogenic nitrogen oxide emissions from soils: impact on NOx and ozone over west Africa during AMMA (African Monsoon Multidisciplinary Analysis): observational study. Atmos Chem Phys 8:2285–2297. doi:10.5194/acp-8-2285-2008

Stramski D (1999) Refractive index of planktonic cells as a measure of cellular carbon and chlorophyll a content. Deep-Sea Res I 46:335–351

Stramski D, Reynolds RA, Babin M, Kaczmarek S, Lewis MR, Röttgers R, Sciandra A, Stramska M, Twardowski MS, Claustre H (2008) Relationship between the surface concentration of particulate organic carbon and optical properties in the eastern South Pacific and eastern Atlantic Oceans. Biogeosciences 5:171–201

Takahashi T et al (2009) Climatological mean and decadal change in surface ocean pCO_2, and net sea-air CO_2 flux over the global oceans. Deep-Sea Res II 56(8–10):554–577. doi:10.1016/j.dsr2.2008.12.009

Tebaldi C, Knutti R (2007) The use of the multi-model ensemble in probabilistic climate projections. Phil Trans R Soc A 365:2053–2075

Telszewski M et al (2009) Estimating the monthly pCO_2 distribution in the North Atlantic using a self-organizing neural network. Biogeosciences 6:1405–1421

Textor C et al (2006) Analysis and quantification of the diversities of aerosol life cycles within AeroCom. Atmos Chem Phys 6:1777–1813

Textor C et al (2007) The effect of harmonized emissions on aerosol properties in global models – an AeroCom experiment. Atmos Chem Phys 7:4489–4501

Thomas H, Prowe F, Lima ID, Doney SC, Wanninkhof R, Greatbatch RJ, Schuster U, Corbière A (2008) Changes in the North Atlantic Oscillation influence CO_2 uptake in the North Atlantic over the past 2 decades. Glob Biogeochem Cycles 22(4):1–13. doi:10.1029/2007GB003167

Tie X, Guenther A, Holland E (2003) Biogenic methanol and its impacts on tropospheric oxidants. Geophys Res Lett 30:1881

Toole DA, Siegel DA, Doney SC (2008) A light-driven, one-dimensional dimethylsulfide biogeochemical cycling model for the Sargasso Sea. J Geophys Res 113:G02009. doi:10.1029/2007JG000426

Trapp JM, Millero FJ, Prospero JM (2010) Temporal variability of the elemental composition of African dust measured in trade wind aerosols at Barbados and Miami. Mar Chem 20:71–82

Turi G, Lachkar Z, Gruber N (2013) Spatiotemporal variability of air-sea CO_2 fluxes and pCO_2 in the California current system: an eddy-resolving modeling study. J Geophys Res (in press)

Turner SM, Harvey MJ, Law CS, Nightingale PD, Liss PS (2004) Iron-induced changes in oceanic sulfur biogeochemistry. Geophys Res Lett 31. doi:10.1029/2004GL020296

Uematsu M, Toratani M, Narita Y, Senga Y, Kimoto T (2004) Enhancement of primary productivity in the western North Pacific caused by the eruption of the Miyake-jima volcano. Geophys Res Lett 31:1–4

Uitz J, Claustre H, Morel A, Hooker SB (2006) Vertical distribution of phytoplankton communities in open ocean: an assessment based on surface chlorophyll. J Geophys Res 111:CO8005. doi:10.1029/2005JC003207

Valsala V, Maksyutov S (2010) Simulation and assimilation of global ocean pCO2 and air–sea CO2 fluxes using ship observations of surface ocean pCO2 in a simplified biogeochemical offline model. Tellus 62B:821–840. doi:10.1111/j.1600-0889.2010.00495.x

Vogt R, Crutzen PJ, Sander R (1996) A mechanism for halogen release from sea-salt aerosol in the remote marine boundary layer. Nature 383:327–330

von Glasow R, Sander R, Bott A, Crutzen PJ (2002a) Modeling halogen chemistry in the marine boundary layer – 1. Cloud-free MBL. J Geophys Res 107:4341. doi:10.1029/2001JD000942

von Glasow R, Sander R, Bott A, Crutzen PJ (2002b) Modeling halogen chemistry in the marine boundary layer – 2. Interactions with sulfur and the cloud-covered MBL. J Geophys Res 107:4323

von Glasow R, von Kuhlmann R, Lawrence MG, Platt U, Crutzen PJ (2004) Impact of reactive bromine chemistry in the troposphere. Atmos Chem Phys 4:2481–2497

Vrekoussis M, Wittrock F, Richter A, Burrows JP (2009) Temporal and spatial variability of glyoxal as observed from space. Atmos Chem Phys 9:4485–4504

Wagner T, Platt U (1998) Satellite mapping of enhanced BrO concentrations in the troposphere. Nature 395:486–490

Wanninkhof R (1992) Relationship between wind speed and gas exchange over the ocean. J Geophys Res 97(C5):7373–7382

Wanninkhof R, McGillis WR (1999) A cubic relationship between air-sea CO_2 gas exchange and wind speed. Geophys Res Lett 26:1889–1892

Watson AJ, Robinson C, Robinson JE, Williams PJB, Fasham MJR (1991) Spatial variability in the sink for atmospheric carbon dioxide in the North Atlantic. Nature 350:50–53

Watson AJ et al (2009) Tracking the variable North Atlantic sink for atmospheric CO_2. Science 326(5958):1391–1393. doi:10.1126/science.1177394

Weissman DE, Bourassa MA, Tongue J (2002) Effects of rain rate and wind magnitude on Sea Winds scatterometer wind speed errors. J Atmos Ocean Technol 19:738–746

Whalley LK, Furneaux KL, Goddard A, Lee JD, Mahajan A, Oetjen H, Read KA, Kaaden N, Carpenter LJ, Lewis AC, Plane JMC, Saltzman ES, Wiedensohler A, Heard DE (2010) The chemistry of OH and HO_2 radicals in the boundary layer over the tropical Atlantic Ocean. Atmos Chem Phys 10(4):1555–1576

Whitney FA, Crawford DW, Yoshimura T (2005) The uptake and export of Si and N in HNLC waters of the NE Pacific. Deep-Sea Res II 52:1055–1067

Williams JE, Holzinger R, Gros V, Xu X, Atlas E, Wallace DWR (2004) Measurements of organic species in air and seawater from the tropical Atlantic. Geophys Res Lett 31:L23S06

Williams JC, Drummond BA, Buxton RT (2010) Initial effects of the August 2008 volcanic eruption on breeding birds and marine mammals at Kasatochi Island, Alaska. Arct Antarct Alp Res 42:306–314

Wingenter OW, Elliott SM, Blake DR (2007) New directions: enhancing the natural sulfur cycle to slow global warming. Atmos Environ. doi:10.1016/j.atmosenv.2007.07.021

Wong CS, Matear RJ (1999) Sporadic silicate limitation of phytoplankton productivity in the subarctic NE Pacific. Deep-Sea Res II 46:2539–2555

Woolf DK, Bowyer PA, Monahan EC (1987) Discriminating between the film drops and jet drops produced by a simulated whitecap. J Geophys Res 92:5142–5150

Wunsch C, Heimbach P (2007) Practical global oceanic state estimation. Physica D 230:197–208. doi:10.1016/j.physd.2006.09.040

Yassaa N, Peeken I, Zöllner E, Bluhm K, Arnold S, Spracklen D, Williams J (2008) Evidence for marine production of monoterpenes. Environ Chem 5:391–401. doi:10.1071/EN08047

Young RW, Carder KL, Betzer PR, Costello DK, Duce RA, DiTuio GR, Tindale NW, Laws EA, Uematsu M, Merril JT, Feely RA (1991) Atmospheric iron inputs and primary productivity: phytoplankton responses in the North Pacific. Glob Biogeochem Cycles 5:119–134

Ziska F, Quack B, Abrahamsson K, Archer SD, Atlas E, Bell T, Butler JH, Carpenter LJ, Jones CE, Harris NRP, Hepach H, Heumann KG, Hughes C, Kuss J, Krüger K, Liss P, Moore RM, Orlikowska A, Raimund S, Reeves CE, Reifenhaeuser W, Robinson AD, Schall C, Tanhua T, Tegtmeier S, Turner S, Wang L, Wallace D, Williams J, Yamamoto H, Yvon-Lewis S, Yokouchi Y (2013) Global sea-to-air flux climatology for bromoform, dibromomethane and methyl iodide. Atmos. Chem. Phys. Discuss., 13:5601–5648, 2013 www.atmos-chem-phys-discuss.net/13/5601/2013/ doi:10.5194/acpd-13-5601-2013

Index

A
Absorption aerosol index (AAI), 266
Acetaldehyde, 26–29, 254, 255
Acetone, 26–29, 253, 254
Acetylene, 253
Acidification, 250
Activation, activated, 174, 178, 182, 205
Advanced Very High Resolution Radiometer (AVHRR), 257, 265
Advective mixing, 290
Aerobic, 18, 37
Aerosol(s), 250–252, 255, 258, 261, 262, 266–267, 270, 280–281, 283, 287, 288, 292–293
  concentration, 176, 225
  mass, 176, 177, 180, 205
  production, 174–202
  size distribution, 172, 182, 183, 185, 202
  sources, 171, 174, 184, 209
Aerosol (sea salt aerosol), 9, 10, 22, 23
Aerosol mass spectrometer (AMS), 196
Aerosol optical depth (AOD), 175, 190, 204, 225, 266, 267
Aerosol optical thickness (AOT), 266
Aerosol production, 61, 62, 64, 79, 80
Ageing, 178, 196–201, 220
Aggregation processes (aggregates), 180, 187, 197, 199, 212, 213, 219–221, 223, 224, 227
Aircraft, 27, 39
Air mass climatology, 281
Air quality, 174
Air-sea flux
  of CO_2, 272, 273, 290
  of heat, 257
Air-sea gas exchange, 115, 134, 155
Alcohols, 26, 29, 30
Aldehydes, 26, 27, 35
Aliphatic amines, 252
Alkane, 31
Alkene, 31
Alkyl amines, 37
Alkyl nitrates, 31–33
Alkyl peroxy radical (RO_2), 27, 33
A Long-term Oligotrophic Habitat Assessment (ALOHA), 250
Altimeter, 90, 99–101
Altimetry, 256–260, 265
Amines (R_xNH_y), 35–37
Ammonia (NH_3), 25, 35–37, 183
Ammonium (NH_4), 20, 35, 36, 182, 197
Ammonium monooxygenase (AMO), 20

AMS. *See* Aerosol mass spectrometer (AMS)
Anaerobic, 18, 37
Anammox, 37
Ångström exponent/coefficient, 202, 266
Anoxic, 18
Anthropogenic, 113, 115–119, 121, 124–130, 135, 142, 146–150, 153, 154
Anthropogenic aerosol, 189, 193, 202
Anthropogenic carbon, 289
AOD. *See* Aerosol optical depth (AOD)
Apparent quantum yield (AQY), 29
Aquarius/SAC-D mission, 260–261
Aqueous phase, 28, 32, 33, 35
ARGO, 248–250, 255, 261, 262, 271, 291, 292
Artificial surfactant, 68
Asian mineral dust, 287
Atlantic Ocean, 176, 179, 180, 183, 187, 188, 191, 197, 202, 209, 218
Atmospheric aerosol, 171, 204, 206, 217
Atmospheric deposition, 251
Atmospheric nitrogen inputs, 281
Atmospheric nutrient inputs, 293
Atmospheric observatories, 250–251
Atmospheric pCO_2, 221
Atmospheric pressure chemical ionization tandem mass spectrometry, 21
Atmospheric processes, 172, 188, 214
Atmospheric stability, 71, 82
Atmospheric stratification, 65, 71, 76
Atomic iodine (I), 21
Attenuation coefficients, 263
Autonomous Ocean Flux Buoys (AOFB), 265
Autonomous underwater vehicle (AUV), 250
Autotroph, 221, 226
AVHRR. *See* Advanced Very High Resolution Radiometer (AVHRR)

B
Bacteria, bacterioplankton, 179, 201, 217–220, 223
Bacterial abundance, 217
Bacterial production, 218
Bacterial respiration, 218
Base, 35
Batch, 211, 212
BATS. *See* Bermuda Atlantic Time-series Study (BATS)
BC. *See* Black carbon (BC)

Bermuda Atlantic Time-series Study (BATS), 250, 290
Bioassay experiments, 215, 217
Bioavailable, 206, 207, 209, 210, 215
Biogenic detrital particles, 263
Biogenic volatile organic compounds (BVOC), 191–192
Biogeochemical sensors, 249
Biogeochemistry, 125, 127, 129, 173, 186, 202, 206, 209–211, 213, 221
Biological activity, 176, 177, 183, 205, 209, 211, 214, 225
Biological carbon pump, 214
Biological production, 120, 134, 143
Biomass, 201, 217, 226
    burning, 189, 193, 194, 198, 209, 210, 226
    burning aerosols, 267
Biota, 179, 184, 191, 206, 207, 221, 222, 226
Black carbon (BC), 177, 194
Bloom, 15, 38, 179, 180, 213–216, 223, 224
Bodélé depression, 252
Boundary currents, 256
Brine, 95, 96
Brine flushing, 264–265
Bromine (Br), 10, 13, 16–18, 21–24, 251, 255
(bromo-)Chloroperoxidase, 18
Bromodichloromethane (CH_2BrCl_2), 17, 18
Bromoform ($CHBr_3$), 17–19, 262, 277
Bromoiodomethane (CH_2BrI), 20, 22
Bubble, 174–176, 179, 180, 205
    generation, 62
    scavenging, 62, 68
Bubble-mediated transfer, 62–64, 66, 85, 88, 91, 94, 97, 101
Budgets, 2, 7, 11, 14, 26–29, 34, 38
Buoyancy, 66, 67, 76, 90, 91
Butane, 31
Butyraldehyde, 29
BVOC. See Biogenic volatile organic compounds (BVOC)

C
Calcium carbonate ($CaCO_3$), 117, 120, 121, 126, 151, 152, 263
Calibration algorithms, 256
Calibration drift, 256
CAM-Chem chemistry-transport model, 254
Canary and California Current System, 272
Cape Grim, 251
Cape Verde Atmospheric Observatory (CVAO), 251–255
Carbon, 173, 179, 183, 194–199, 202, 206, 209, 210, 212, 214, 215, 218–224, 226
Carbonate and bicarbonate ions, 290
Carbon cycle, 116, 118, 129, 148
Carbon dioxide (CO_2), 113–157
Carbon Dioxide Information Analysis Center (CDIAC), 279, 280
Carbon disulphide (CS_2), 6, 7
Carbon monoxide (CO), 33, 38–39
CarbonSat and CarbonSat Constellation concepts, 268
Carbon tetrachloride (CCl_4), 14
Carbonyl, 29
Carbonyl sulfide (COS), 6, 7
CARBOOCEAN, 291
Carboxylic acids, 26, 27
Catalytic O_3 destruction, 260
CAVASSOO, 296
Cavity ring down, 21
CCN. See Cloud condensation nuclei (CCN)
CDIAC. See Carbon Dioxide Information Analysis Center (CDIAC)
CDOM. See Coloured dissolved organic matter (CDOM)
CERES satellite, 292
Chapman reactions, 26
Chelton effect, 259, 260
Chemical enhancement, 34
Chemical transport models (CTMs), 292, 293
Chemistry, 173, 175, 176, 180, 189, 191, 193, 196, 200, 210, 211, 213, 214, 221
chl-a maximum, 262, 263
Chlorine, 13–14, 21–23
Chloroform ($CHCl_3$), 14, 16, 19
Chloroiodomethane (CH_2ClI), 15, 20, 21
Chloromethane (CH_3Cl), 15
Chloroperoxidase, 16, 18
Chlorophyll, 262–264, 275, 282, 284, 288, 291
Chlorophyll (maximum), 17
Chlorophyll-a, 177, 180, 182, 183, 215, 217, 249, 264, 288, 290
CLAW, 2, 6, 11–13, 39
Cleavage, 3
Climate, 172, 191–193, 201, 202, 204, 211, 221, 222, 225
    change, 113–157
    feedback, 10, 11
    model, 180, 193, 203
    observing systems, 249
    variability, 249, 290
Climatology, 79, 97
Cloud, 2, 9–13, 24, 27, 28, 31, 35
    albedo, 172, 193, 205
    cycling, 182
    formation, 175, 204–206
Cloud condensation nuclei (CCN), 12, 13, 24–27, 39, 178, 204, 205, 225, 251
    formation/growth, 262
Cloud droplets, 255, 262
C* method, 274
Coagulation, 172, 186, 196, 197, 212
CO_2 air-sea climatologies, 156
COARE algorithm, 87, 90
Coarse mode, 172, 175, 210
Coastal ocean, 6, 17, 149, 151
Coastal sediments, 141–143, 152
Coccolithophores, 263, 264
CO_2 emission scenarios, 276
Coherent structure, 56, 58, 60, 61, 66, 74
Coloured dissolved organic matter (CDOM), 6, 15, 30, 37–38, 254
Community structure, 190, 215, 217, 226
Concentration, 114, 118–121, 123, 124, 130, 131, 134–144, 147, 148, 150–156
Concentration gradient, 28–29
Condensible vapour, 186
Conditions, 56, 59, 60, 62, 67, 68, 70, 71, 74, 76–79, 84–88, 93, 96
    neutral conditions, 76
Continental shelves, 126–129, 135, 137, 139–140, 144, 150, 153
Convection
    water-side convection, 66

Convective mixing, 291
Convective regions, 16, 26, 27
CO_2 partial pressure (pCO_2), 274, 275, 289–292
Copper (Cu), 198, 209, 211, 217
CO_2 uptake, 271, 272, 290, 291
Cryosat-2, 256, 265
CTMs. See Chemical transport models (CTMs)
CVAO. See Cape Verde Atmospheric Observatory (CVAO)
Cyanobacteria, 18, 19, 31, 37, 38, 214, 364

D
Dark production, 38
Dead zones, 153, 157
Degradation rate, 6, 17
Dehalogenation, 251
Demethylation, 3
Denitrification, 130–131, 134, 135, 151, 272
Density stratification, 67
Deposition, 172, 173, 186–193, 196, 198, 206–224
  deposition velocity, 11, 34, 35, 78–80, 92
  direct deposition, 78, 80
  dry deposition, 62, 74, 78–79
  particle deposition, 78
  wet deposition, 79–80
Depth profiles, 32
Desert dust, 172, 186–189, 207, 218, 221
Detrital aggregates, 19, 20
Diatoms, 3, 18, 31, 38, 264, 287, 288
Diazotrophic bacteria, 37
Dibromomethane, 15, 17
Dibromochloromethane (CH_2Br_2Cl), 17, 18
Dibromomethane (CH_2Br_2), 17–19
DIC. See Dissolved inorganic carbon (DIC)
Dichloromethane (CH_2Cl_2), 14–16
Differential optical absorption spectroscopy (DOAS), 21
Di-iodomethane (CH_2I_2), 15, 21, 22, 24
Dimethylammonium (($CH2$)$2NH2$), 35
Dimethyl selenide, 7
Dimethylsulphide (DMS), 2–13, 24, 28, 36, 39, 40, 173, 182, 183, 185, 194, 202, 204, 214
  climatology, 278
  summer paradox, 250
Dimethylsulphoniopropionate (DMSP), 2–7, 40, 250, 282, 286
Dimethylsulphoxide (DMSO), 2, 3, 9, 11
Dinoflagellates, 3, 264
Direct flux, 14, 28, 36, 39
Direct radiative effect (DRE), 178, 202–204, 225
Dispersion, 175
Dissipation rate, 58, 65, 73, 93
Dissociation constant (pK_b), 35
Dissolution, 173, 211–213, 219, 226
Dissolved form, 173, 211
Dissolved inorganic carbon (DIC), 119–122, 124, 125, 129, 148, 150, 155, 274, 290
Dissolved organic carbon (DOC), 179, 210, 262
Dissolved organic matter (DOM), 182, 210
Diurnal variation, 14, 18
Diurnal warm layer, 85
DMS. See Dimethylsulphide (DMS)
DMS climatology, 3–5, 9

DMSO. See Dimethylsulphoxide (DMSO)
DMSP. See Dimethylsulphoniopropionate (DMSP)
DMSP lyase, 2
DOC. See Dissolved organic carbon (DOC)
DOM. See Dissolved organic matter (DOM)
DRE. See Direct radiative effect (DRE)
Dry deposition, 172, 188, 210
Dry mole fraction, 113, 114, 116–118
Dust deposition, 251

E
Earth observation (EO), 194, 225, 255–270, 294
Earth observing systems, 294
Earth system models, 271, 272, 284, 294
Earth system science, 2, 40
Eastern boundary upwelling regions, 273
Ebullition, 140–142, 146, 147, 153, 155
Ecosystem, 189, 191–193, 195, 202, 218–221
  functioning, 221
  response, 287–289
Eddy correlation, 28
Eddy dynamics, 256
Effective flux, 175
El Chichón eruption, 269
Electromagnetic spectrum, 266
El Niño, 256
Emission, 113–115, 117, 119, 127, 130, 134–137, 139–144, 146–147, 152, 153, 155–157, 172, 176, 184, 186, 188–196, 198, 202, 204, 205, 207, 209, 226
Emission inventory, 193, 226
Enhancement, 56, 58, 67, 69–70, 91, 94, 101
Enrichment, 173, 174, 176–179, 199, 205, 206, 214, 216
Environmental Satellite (ENVISAT), 256, 259, 268
ENVISAT. See Environmental Satellite (ENVISAT)
EO. See Earth observation (EO)
Estuaries, 128–130, 134, 135, 137–141, 143, 144, 147
Ethane, 31, 253
Ethanol, 26, 27, 29, 30
Ethyl nitrate ($EtONO_2$), 31
Euphotic zone, 31, 38
European Iron Fertilization Experiment (EifEX), 282
European Remote Sensing (ERS) satellites, ERS-1 and -2, 256, 265
European Space Agency, 260–261, 270
European Station for Time-series in the Ocean Canary Islands (ESTOC), 250
Evasion, 67, 84
Export of matter, 218
External mixing, 196

F
fCO_2. See Surface water (fCO_2)
Fertilisation, 213–215, 221–224, 227
Fine mode, 202
Fine particles, 24
Flow structures, 81
Flux, 2, 5–11, 13, 16, 17, 19–21, 28, 29, 32, 34–39
Flux gradient, 34
Footprint, 82, 86
Forecasting capability, 249

Forecast mode, 270, 271
Formaldehyde (HCHO), 27, 29, 270
Formic acid, 26, 27
Free troposphere, 34
Freons, 15
Friction velocity, 56, 58, 60, 61, 65, 66, 70, 71, 76, 93, 101
Frost flowers, 265
Fugacity, 118, 119

G

Gases
    poorly soluble gases, 62
    soluble gases, 56, 62, 63, 72, 74, 90, 92, 94, 97
Gas exchange parametrisation, 275
Gas phase, 9, 10, 12, 21–25, 28, 31, 32, 35, 36
Gas transfer velocity, 59, 64, 65, 73, 78, 80, 81, 83, 85, 87, 88, 91, 93, 97, 98, 119, 257
GCOS. *See* global climate observing systems (GCOS)
Geo-engineering, 205
Geosat, 256
GEOTRACES, 281, 283
GFED. *See* Global fire emissions database (GFED)
GHRSST Multi-product Ensemble (GMPE), 257, 258
Gliders, 250, 265
Global annual sea-to-air DMS flux, 285
Global array, 248
Global budgets, 113
Global burden of SSA, 292
Global change, 147–154
Global climate observing systems (GCOS), 255, 260
Global fire emissions database (GFED), 194
Global Ocean Data Assimilation Experiment (GODAE), 262
Global Ocean Observing System, 255
Global prognostic models, 276
Global SSA burdens, 297
Global whitecap coverage, 297
Glyoxal, 274
GMPE. *See* GHRSST Multi-product Ensemble (GMPE)
Gravitational settling, 193
Grazing, grazers, 218, 219, 221
Greenhouse gas, 34, 38
Greenhouse gases Observing SATellite (GOSAT), 274
Groundtruthing, 261

H

Halocarbons, 14, 19
Halocarbons in the Ocean and Atmosphere (HalOcAt), 282–283, 288
HalOcAt. *See* Halocarbons in the Ocean and Atmosphere (HalOcAt)
Halogen, 14, 15, 18, 19, 21–26, 40
Halogenated hydrocarbons, 282
Haloperoxidase, 16, 21
Haptophytes, 3
Hard tissue pump, 123, 124
HCN. *See* Hydrogen cyanide (HCN)
Heat flux, 60, 63, 68, 69, 71, 73, 75, 76, 79, 84, 93
Heterogeneous, 9, 10, 22

Heterogeneous processes, 189, 202
Heterotrophic bacteria, 268
Heterotrophic organisms, 214
High-nutrient low-chlorophyll (HNLC), 211, 214, 218–227, 230, 231, 286
Hindcast mode, 275
HNLC. *See* High-nutrient low-chlorophyll (HNLC)
HOBr, 9, 22
HOCl, 9, 11, 17
HOI, 21–24, 26
Hot spot volcanic islands, 291
HO_x, 21, 22, 27
HO_x radical, 256
H_2SO_4, 9, 24, 25
Humic substances, 30
Hurricanes, 263, 295
HYbrid Coordinate Ocean Model (HYCOM), 278
Hydration, 72
Hydrazine (N_2H_4), 37
Hydrogen (H_2), 27, 37–38
Hydrogen cyanide (HCN), 33–34
Hydrogenolysis, 19
Hydrogen peroxide (H_2O_2), 18, 21, 256
Hydrogen sulphide (H_2S), 7, 9
Hydrolysis, 6, 7, 17, 19, 20, 33
Hydroperoxyl (HO_2), 256, 259, 260
Hydrostation S, 255
Hydroxide (OH), 256, 259
Hygroscopic, 176, 177, 179, 182, 200, 201, 207, 229
Hypobromite (BrO), 274
Hypoxic zones, 155, 157

I

Ice-atmosphere exchanges, 270
Ice cores, 118–121
Indirect radiative effect, 209–211
Inductively coupled plasma mass spectrometry, 22
Infrared imaging, 76
Infrared radiometer, 262
Inorganic nutrients, 221
In situ mesocosms, 223–225, 229
In-situ observations, 253–288
Integrated Arctic Ocean Observing System (IAOOS), 271
Interannual variability, 279, 281, 290, 295
Interfacial flux, 179, 180
Internal mixing, 201–205
Inventory, 86, 90, 92, 96, 99, 102–103
Inverse modelling, 279–280
I_2O_5, 25, 26
IO_3, 26
Iodine, 14, 19–26, 190, 260
Iodine dioxide (OIO), 22–25
Iodine monoxide (IO, iodine oxide), 22
Iodine monoxide (IO) radicals, 260
Iodobutanes (C_4H_9I), 20
Iodocarbon, 19–21, 24
Iodoethane (C_2H_5I), 20, 22
Iodoform (CHI_3), 21
1-and 2-iodopropane (C_3H_7I), 20, 22
$IONO_2$, 22–24

Index

Iron (Fe), 256, 257, 285–287, 291–293
   binding capacity, 216
   binding ligands, 216
Iron addition experiments, 287, 291
Iron fertilisation, 291
Isoprene, 16, 31, 32, 40, 268
Isopropyl nitrate, 32

K
Kelp, 21
Ketones, 26, 28, 35

L
Labrador current, 295
Langmuir circulation, 67, 69
Langmuir number, 69
La Niña, 261, 262
Laser speckle velocimetry (LSV), 83
Layer
   amospheric boundary layer (ABL), 73, 74, 77, 89
   atmospheric surface layer, 73
   deposition layer, 81, 82
   interfacial turbulent boundary-layer, 77
   mass boundary layer (MBL), 60, 72, 74, 75, 79, 80
   mixed layer, 68, 69, 77, 86, 87, 89
   molecular diffusion layer, 73
   sea surface microlayer, 65, 71
   turbulent boundary layer, 77
   turbulent shear layer, 63
Lidar altimetry, 270
Limitations, 86–87, 211, 213, 214, 219, 221, 223, 226
Lithogenic, 223, 224, 227–229, 231
LNLC. *See* Low-nutrient-low chlorophyll (LNLC)
Low-molecular-weight dicarboxylic acids (DCAs), 259
Low-nutrient-low chlorophyll (LNLC), 211, 214, 218–227, 231

M
Mace head, 180–182, 184, 185, 187–190, 256, 272
(Macro)algae, 14, 17, 18, 24, 38
Macrophyte, 17
Mangrove ecosystems, 137, 141, 147
Marine aerosols, 259, 263, 267, 272, 285, 298
Marine atmosphere, 177–206, 229
Marine boundary layer (MBL), 9–11, 14, 21–26, 31, 33–35, 185–191, 259, 274
MarinE MethanE and NiTrous Oxide (MEMENTO) database, 281, 287
Marine productivity, 191, 226, 231
Mass, 177, 178, 180–185, 187–189, 191, 193, 194, 196, 197, 201–204, 206, 207, 209, 214, 228
Mass median diameter, 192
Mauna Loa CO_2 record, 298
Mauna Loa Observatory, 272
MBL. *See* Marine boundary layer (MBL)
Mean square slope (mss), 264
Mean-square wave slope, 60, 62, 72
Mechanisms, 6, 12, 17, 20, 21, 32, 33, 35, 37
Mediterranean Sea, 193, 214, 223, 227, 228

MEdium Resolution Imaging Spectrometer (MERIS), 267, 269, 271
Medium scale in situ enrichments, 287
MEMENTO. *See* MarinE MethanE and NiTrous Oxide (MEMENTO) database
MERCATOR, 278
MERIS. *See* MEdium Resolution Imaging Spectrometer (MERIS)
Mesocosm, 217, 218, 223–225, 228, 229, 231
Mesoscale eddies, 261
Mesoscale ocean enrichment, 218
   experiments, 286
Metabolic activity/rates, 221–223, 231
Methane (CH_4), 2, 20, 28, 31, 37, 117–161, 256, 259, 272–274, 281
Methane hydrate, 120, 148–150, 158–159
Methane monooxygenase (MMO), 20
Methanesulphonic acid (MSA), 9, 11, 13, 256, 259
Methanethiol (CH_3SH, methylmercaptan), 7
Methanogenesis, 142–148, 151
Methanol, 26–31, 258, 259
Methanotrophic/nitrifying bacteria, 20
Methylating agents, 20
Methyl bromide (CH_3Br), 15, 17–18
Methyl cyanide (acetonitrile, CH_3CN), 33–34
Methyl ethyl ketone, 28
Methyl halide, 20
Methyl iodide (CH_3I, iodomethane), 15, 19, 20, 22
Methyl nitrate ($MeONO_2$), 32
Methyltransferase, 20
Microbes, 206, 218, 231
Microcosm, 211, 219–225, 231
Microlayer, 14, 26, 35, 65, 71, 72, 217, 231
Micrometeorology (micrometeorological), 181, 197
Microorganism, 206, 215
Microscale wave breaking, 59–63, 67, 83
Mid Term Strategy Initiatives, 293
Mixed layer, 213, 217, 219, 221
Mixed-layer depth (MLD), 268
Mixing ratios, 256, 258–260, 272, 279, 280
Model(s), 3–5, 9–13, 17–20, 22, 24–27, 32, 35, 39, 180, 185, 186, 192, 196–199, 201, 205, 207–211, 215, 217, 223–226, 229–231
   aerosol transport model, 82
   surface penetration model, 76, 84
   surface renewal model, 75, 76, 84
Moderate Resolution Imaging Spectroradiometer (MODIS), 262, 267–272, 292, 293
Molecular iodine (I_2), 22, 25
Momentum, 63, 69, 70, 73, 74, 76, 78–80, 83, 84, 93, 94
Monin-Obukhov length, 73
Monin-Obukhov Similarity Theory (MOST), 73
Monoterpene, 28, 32, 196, 197
MSA. *See* Methanesulphonic acid (MSA)
Multispectral ocean colour sensors, 269

N
National Centers for Environmental Prediction (NCEP), 262, 263, 296
Natural aerosol, 216, 221
Naval Oceanographic Office (NAVOCEANO), 262

NCEP. *See* National Centers for Environmental Prediction (NCEP)
Neutral stability, 81
New particle formation, 189–191
NH_x, 36
Nitrate, 255, 259, 287
Nitric acid (HNO3), 187, 259
Nitric oxide (NO), 21, 22, 24, 33, 35
Nitrification, 135, 136, 138, 141, 156, 157
Nitrification/denitrification, 33
Nitrogen dioxide (NO_2), 274
Nitrogen (N_2) fixation, 213, 222, 226, 256, 260
Nitrous oxide (N_2O), 117–161, 276, 277, 281
Nitryl chloride ($ClNO_2$), 21
NMHC. *See* Non-methane hydrocarbons (NMHC)
NMVOC. *See* Non-methane volatile organic compounds (NMVOC)
NO_3, 9, 10, 23, 24, 30, 31
Non-methane hydrocarbons (NMHC), 26–34, 40, 259
Non-methane volatile organic compounds (NMVOC), 198
Non-sea-salt-(nss) sulphate aerosol, 256
Normalised water leaving radiances, 269
North Atlantic, 3, 7, 8, 10, 20, 24, 29, 31, 32, 36
North Atlantic Oscillation (NAO), 295
NO_x, 10, 21–25, 28, 31
n-propyl nitrate, 32
Nucleation, 177, 185, 189–191, 197
Nucleophilic substation, 20
Numerical Weather Prediction (NWP) models, 263
Nutrients, 178, 193, 194, 198, 200, 204, 205, 211, 213–216, 218, 219, 221–228, 230, 231

O

Ocean acidification, 115, 117, 121, 130, 149–152, 156
Ocean-atmosphere exchanges, 147, 148
Ocean-atmosphere interaction, 171–227
Ocean-atmosphere interface (air-sea interface), 1
Ocean-Atmosphere-Sea Ice-Snow (OASIS), 265
Ocean Carbon-cycle Model Intercomparison Project, 97
Ocean colour, 177, 255, 262, 263, 275
Oceanflux, 270
Oceanic fertilisation by volcanic eruptions, 289
Oceanic observatories, 248
Oceanic transport inversion, 274
Ocean inversion model, 291
Ocean surface winds, 257
Ocean turbulence, 256
OH, 9–11, 14, 21–23, 26–28, 31, 34, 35, 38
OIO. *See* Iodine dioxide (OIO)
Oligotrophic, 30, 38, 179, 211, 217, 219–222
Oligotrophic environments, 217, 219
Oligotrophic waters, 263, 282–283
OMI, 266, 269
OMZ. *See* Oxygen minimum zone (OMZ)
Open ocean, 115, 117, 121–126, 131, 135–139, 143, 144, 146–150, 152–157
O_3 photochemistry, 251
O_3 photolysis rate, 251
Optical measurements, 220–221, 224
Orbiting Carbon Observatory (OCO), 268

Organic carbon, 179, 194, 198, 210, 212, 218, 221, 223
Organic carbon pump, 120
Organic matter (OM), 174, 176–182, 184, 197, 200, 205–206, 210, 212, 220–221, 223, 224
Organic nutrients, 210
Organic radicals, 254
Organohalides, 265
OSTIA, 257
OVOCs. *See* Oxygenated volatile organic compounds (OVOCs)
Oxidation, 2, 7, 9–13, 18, 20, 24, 27, 29, 30, 35–38
Oxidative stress, 20
Oxidising capacity, 254
Oxygen, 249, 268, 272, 275
Oxygenated volatile organic compounds (OVOCs), 26–31, 254, 270
Oxygen minimum zone (OMZ), 131, 132, 137, 139, 151–153, 293
Ozone (O_3), 9, 17, 20–27, 31
Ozone destruction, 251, 255, 269–270

P

Palaeo-proxies, 271
Parameterisation, 56, 62, 64, 66, 70, 73, 75, 79, 87–95, 97–99
Parasitic capillary wave, 56, 74
Partial pressure, 118, 119
  difference, 56, 85, 98
Particle
  backscattering, 263
  burst, 24
  concentration, 185, 186, 206
  fluxes, 78, 79
  settling, 212, 213
  size distribution, 172, 187, 263
Particle image velocimetry (PIV), 59, 80
Particles (IOP), 24
Particle tracking velocimetry (PTV), 80
Particulate carbon, 249
Particulate inorganic carbon (PIC), 261–264
Particulate organic carbon (POC), 179, 210, 224, 227, 261–264
Particulate organic matter (POM), 182, 197, 224
Partitioning, 183, 193, 199, 209, 226
Passive polarimetric sensors, 258
Pathfinder algorithm, 257
pCO_2. *See* CO_2 partial pressure (pCO_2)
pCO_2 mapping, 291
Péclet (Pe) number, 57
Perchloroethylene (PCE tetrachloroethylene), 14
Peroxyacetylnitrate (PAN), 27, 254
PFTs. *See* Phytoplankton functional types (PFTs)
pH, 117, 121, 122, 130, 150, 151, 189, 211, 222, 250, 288
Phaeocystis, 15, 17
Phosphorus, 173, 206–212, 219, 226, 251, 289
Photobleaching, 30
Photochemical/photochemistry, 7, 14, 19, 29, 30, 32, 33, 36–38
Photochemistry, 251
Photodecomposition, 254
PhytoDOAS, 264
Phytoplankton, 2, 3, 6, 14, 15, 17–20, 31–33, 36, 38, 120, 138, 151, 173, 176, 180, 188, 190, 210, 213–220, 223, 261, 263, 270, 274, 284, 287

Phytoplankton functional types (PFTs), 263
Phytoplankton size classes (PSCs), 263, 264
PIC. *See* Particulate inorganic carbon (PIC)
Picoplankton, 217
PIV. *See* Particle image velocimetry (PIV)
Planetary wave dynamics, 256
Plankton, 176, 179, 182, 215, 218
PM10, 193, 201
POC. *See* Particulate organic carbon (POC)
Polybrominated methanes, 17–19
Polynias, 265
Precipitation, 62, 78–80, 83, 96
Precursor hydrocarbons, 270
Primary production, 17, 71, 96, 216–218, 220, 261, 287, 291
Prochlorococcus, 19, 217
Production, 172–202, 204, 215–223
Production flux, 174, 175, 225
Production rate, 19, 27, 31
Profile, 58, 61, 66, 70–72, 74, 75, 80, 84–86, 93
Profiling floats, 248
Propanal (propional), 27, 29
Propane, 31
Propanol, 29, 30
Propene, 31
PSCs. *See* Phytoplankton size classes (PSCs)
Pyruvate, 29

Q
Quality-flagged controlled data sets, 250

R
Radar altimeter, 256, 265
Radar backscatter, 259
Radiative balance, 251
Radiative budget, 202
Radiative forcing, 24
Radical recombination, 19
Radio-frequency interferences (RFI), 261
Radiometers, 256–258
Rain, 56, 67–68, 78, 79, 93, 101
Rainwater chemistry, 280–281, 283
Reactive bromine, 17
Reactive marine-derived halogens, 255
Reactive trace gases, 250, 251
Reactive trace species, 251–255
Relative humidity (RH), 172, 174, 176–180, 198, 202
Remineralisation, 214, 218, 224
Removal, 172–174, 194–196, 200
Resonance fluorescence, 21
RH. *See* Relative humidity (RH)
Rivers, 115, 125, 129, 130, 134, 142, 151, 153–155
Roughness length, 71, 82

S
S-adenosyl-L-methionine (SAM), 20
Sahel region, 251, 252, 254
Salinity, 175, 176, 248, 250, 255, 260–261, 265, 290
SAR. *See* Synthetic aperture radar (SAR)
Satellite, 21, 27, 33, 39, 175, 177, 182, 191, 194, 201, 202, 204, 213–216, 223, 225
  measurement, 99
  oceanography, 255
  retrieval, 190, 202
  wind speeds, 90, 99, 101
Saturation (state, super, under), 16–18, 30, 32, 37
Saturation anomaly, 32
SCanning Imaging Absorption spectroMeter for Atmospheric CartograpHY (SCIAMACHY), 264, 266–270
Scatterometer, 90, 99–101, 258, 259
Scavenging, 211–213, 220, 221, 226
Schmidt number *(Sc)*
  dependence, 63, 84, 100
  exponent, 72, 78, 93
SCIAMACHY. *See* SCanning Imaging Absorption spectroMeter for Atmospheric CartograpHY (SCIAMACHY)
SDS. *See* Sodium dodecyl sulphate (SDS)
Sea ice, 95–96, 121, 125–126, 141, 147–151, 155
  bulk salinity, 265
  distribution and motion, 264
  transfer estimates, 96
Sea-ice fraction, 97
Sea salt, 174–183, 200, 204–207
  aerosol production, 61–62
  source function, 292
Sea salt aerosols (SSA), 175, 176, 200, 203, 206, 207
  emissions, 293
  mass concentrations, 293
  reflectance, 292
  source functions, 293
Seasat, 256
Seasonal(ity, cycles), 3, 14, 16–18, 32, 33, 35, 39, 180
Seasonal ice zones, 249
Sea spray, 174, 176, 178–180, 184, 205, 206
  particles, 62
Sea spray aerosol (SSA), 172–182, 224–226
Sea spray aerosol source function (SSSF), 175
Sea spray source function (SSSF), 62, 64
Sea state, 256–260
Sea surface salinity (SSS), 255, 260–261, 290
Sea surface temperature (SST), 14–17, 34, 176, 224, 256–260, 267, 273, 284, 290–292
SeaWiFS, 177, 182, 223, 262–264, 267, 290
Secondary organic aerosol (SOA), 37, 183, 184, 191, 193, 196, 199–200
Secondary production, 174, 183
Sediment core studies, 287
Self-organising maps, 291
Sensor
  closed path sensors, 83
  open path sensor, 83, 95
Sesquiterpene, 191
Shear
  air-side, 67
  wind shear, 76
Ship emissions, 172, 194–195, 293

Short-lived brominated and iodinated trace gases, 277
Short lived trace gases, 1–40
Silicate (Si), 188, 214, 222, 226
Simulation
    direct numerical simulation (DNS), 74–76
    large-eddy simulation (LES), 74–76
    large-wave simulation (LWS), 75
Single scattering albedo, 202
Sink, 6, 17, 19, 27, 28, 30, 32–39
Sinking particles, 214, 224
Skin temperature (SSTskin), 256, 265
Slick, 68, 99
SMOS. See Soil Moisture and Ocean Salinity (SMOS)
SOA. See Secondary organic aerosol (SOA)
SOCAT. See Surface Ocean CO_2 ATlas (SOCAT)
Sodium dodecyl sulphate (SDS), 179
Soil Moisture and Ocean Salinity (SMOS), 261, 262, 290
Solar radiation dose (SRD), 2, 12
Solubility, 56, 63, 64, 71, 76–77, 90–91, 95, 100, 101, 178, 189, 205, 209, 212, 215, 221
Solubility pump, 120
Soluble organic iodide (SOI), 26
Source, 2, 3, 6–9, 11, 13–23, 26–40
SRD. See Solar radiation dose (SRD)
SSA. See Sea salt aerosols (SSA); Sea spray aerosol (SSA)
SSM/I series satellites, 264
SSS. See Sea surface salinity (SSS)
SSS scattering, 261
SST. See Sea surface temperature (SST)
SSTskin. See Skin temperature (SSTskin)
Stoichiometry, 214
Stoke's drift, 65, 66
Stratification, 65, 67, 70, 71, 76, 85, 93, 124, 147, 149, 150
    atmospheric stratification, 65, 70, 71, 76
Stratosphere, 13, 17, 33
Stratosphere-troposphere exchange, 26
Stress
    kinematic stress, 76
    surface stress, 76, 99
Submersibles, 250
Sub-mesoscale variability, 250
Subpolar gyre, 291
Subtropical gyre, 250
Sulphate aerosols, 251, 262, 287
Sulphate/sulfate, 7, 9, 12, 36, 177, 178, 182–184, 186, 189, 190, 196–200, 204, 205, 209
Sulphur cycle, 4, 9, 11, 12
Sulphur dioxide (SO_2), 9–11, 269
Sulphuric acid clusters, 254
Summer paradox (or DMS summer paradox), 2, 5, 250, 284
Sun synchronous orbit, 268
Supersaturation, 118, 133, 139, 150
Surface
    divergence, 57, 58, 60
    renewal, 57, 59, 68, 74, 81
    roughness, 59, 82, 87, 90, 92
    steepness, 74
Surface Ocean CO_2 Atlas (SOCAT), 279–280, 283, 289
Surface ocean pCO_2, 288

Surface water (fCO_2) (fugacity of CO_2), 117–126, 129, 137, 156, 279, 280
Surface Water and Ocean Topography (SWOT), 256
Surfactants, 8, 34, 59, 65, 68–69, 82, 88, 90, 93
    soluble surfactants, 68
Synechococcus, 217, 218
Synthetic aperture radar (SAR), 258, 265

T
TCE. See Trichloroethylene (TCE)
Technique
    disjunct eddy covariance (DEC), 82, 83
    dual tracer technique, 68, 84–86, 94, 99
    eddy-covariance (EC), 63, 81–83, 85–88, 92–94, 101
    flow-measurement technique, 80
    mass balance technique, 80, 84, 85, 88, 89, 101
    micrometeorological technique, 62, 81–83, 85
    radon deficit, 84
    small-scale technique, 80–87
    thermographic technique, 81, 85
Ternary nucleation, 37
Terrestrial, 174, 189, 191–193
Tert-butyl alcohol, 29
Thermal structure, 58
Thermography, 85
Tidal cycles, 24
Tidal pumping, 135, 142
TOPEX/Poseidon, 256
Total Ozone Mapping Spectrometer (TOMS), 269
Trace gas cycling, 1
Trace gas exchange, 21
Transfer resistance, 21, 77, 78
Transfer velocity, 56, 57, 59, 62–65, 69–74, 76–81, 84, 87–92, 94, 97–99, 101, 257, 289
Transport, 9, 14, 18, 26, 33–36, 172, 174–202, 209–211, 220, 223, 226
Transport model, 180, 181, 191, 193
    eddy renewal model, 73
    surface penetration model, 73, 81
    surface renewal model, 73, 81
    thin film model, 72
Trichloroethylene (TCE), 14–16
Trichodesmium, 37
Tropical circulation, 256
Troposphere, 13, 17, 21, 22, 26–28, 31, 34, 35, 38, 39, 116–119, 121, 129, 130, 134, 137, 139–143, 147–150, 153, 154, 156
Tropospheric gases, 251
Tropospheric ozone, 251, 254, 255
Turbulence
    coherent turbulence, 60, 61
    Kolmogoroff scale turbulence, 74
    near-surface turbulence, 57, 58, 60, 67
    shear induced turbulence, 60, 65, 92
    surface water turbulence, 93
Turbulent kinetic energy, 60, 65, 87
Turbulent kinetic energy dissipation, 58
Turbulent transport, 72
Turnover, 3, 36

U

Ultrafine particles, 24, 25
Upper ocean dynamics, 66
Uptake, 2, 12, 15, 22, 23, 25, 30, 34–36, 178, 192, 196–199, 209, 211, 214, 216, 218
Upwelling, 7, 17, 30, 32, 38
UV radiation, 2, 3, 12, 30, 33, 38

V

Velocity parameterisation, 276
Volcanic ash, 190, 191, 209, 226, 287–289
Volcanic ash fertilisation, 288
Voluntary Observing Ships (VOS), 289, 291
VOS. *See* Voluntary Observing Ships (VOS)

W

Water insoluble organic carbon (WIOC), 176
Water insoluble organic matter (WIOM), 177, 180, 181, 205
Water soluble organic carbon (WSOC), 183
Water soluble organic matter (WSOM), 177, 178, 180, 205
Water temperature, 175, 176, 180
Wave
  capillary wave, 56, 60, 87
  gravity wave, 59, 60, 71, 74–76, 90
  mean square slope, 90, 99, 101
  wave age, 66
  wind-generated surface wave, 77
  wind-generated wave, 65–66, 74
Wave breaking
  large wave breaking, 61
  microscale wave breaking, 56–60, 65, 81
Wave field, 59, 60, 65, 72, 94
Wave height, 255, 259
Wet deposition, 172, 188, 211
Whitecap, 56, 62–64, 88, 91, 95, 174, 225, 267
  whitecap coverage, 63, 64, 292
  whitecap fraction, 175, 225, 267
Wind speed, 175, 176, 180, 187, 188, 202, 204, 205, 225
Wind stress divergence, 260
Wind surface stress, 259
WIOC. *See* Water insoluble organic carbon (WIOC)
WIOM. *See* Water insoluble organic matter (WIOM)
WSOC. *See* Water soluble organic carbon (WSOC)
WSOM. *See* Water soluble organic matter (WSOM)

Z

Zooplankton, 190, 224

The manufacturer's authorised representative in the EU is Springer Nature Customer Service Centre GmbH, Europaplatz 3, 69115 Heidelberg, Germany. If you have any concerns regarding our products, please contact ProductSafety@springernature.com

Printed and bound by CPI Group (UK) Ltd, Croydon, CR0 4YY

26/03/2026

02078998-0001